Numerical Analysis:

Historical Developments in the 20th Century

Numerical Analysis:

Historical Developments in the 20th Century

Edited by

C. Brezinski
Univ. des Sciences et Techn. de Lille
Lab. d'Analyse Numérique et d'Optimisation
UFR IEEA – M3, B.P. 36
Villeneuve d'Ascq Cedex 59655
France

L. Wuytack
Univ. Instelling Antwerpen
Departement Wiskunde en Informatica
Universiteitsplein 1
Antwerpen-Wilrijk B-2610
Belgium

2001
ELSEVIER
Amsterdam - London - New York - Oxford - Paris - Shannon - Tokyo

ELSEVIER SCIENCE B.V.
Sara Burgerhartstraat 25
P.O. Box 211, 1000 AE Amsterdam, The Netherlands

First edition 2001

Library of Congress Cataloging in Publication Data
A catalog record from the Library of Congress has been applied for.

ISBN: 0-444-50617-9

Transferred to Digital Printing 2006

Contents

Numerical analysis in the twentieth century

C. Brezinski[1] and L. Wuytack[2]

[1] *Laboratoire d'Analyse Numérique et d'Optimisation, Université des Sciences et Technologies de Lille, 59655 – Villeneuve d'Ascq cedex, France;*
[2] *Dept. Wiskunde, Universiteit Antwerpen, Universiteitsplein 1, 2610 – Wilrijk, Belgium*

Numerical analysis can be defined as that branch of mathematics interested in *constructive methods*. By constructive, we mean methods which allow to construct effectively, that is to obtain numerically, the solution of mathematical problems. A discussion of possible definitions is given in [249]. It is difficult to find who coined this appellation but a symposium with the title *Problems for the Numerical Analysis of the Future* was held at UCLA, July 29–31, 1948. It was published by the *National Bureau of Standards* as volume 15 of the *Applied Mathematics Series*. However, this was not the first conference on this subject, since a previous one with the title *Conference on Advanced Computation Techniques* (84 participants) was held at MIT, October 29–31, 1945 [9]. It seems that the first book with this expression in its title was by Douglas Rayner Hartree (1897–1958), a mathematical physicist from Cambridge, England, in 1952 [128]. He took part in the symposium at UCLA where he gave a talk on *Some unsolved problems in numerical analysis*. Of course, there were books on numerical methods a long time before Hartree's. Let us mention, for example, those of J. Vieille [262] dating from 1852 and where some methods for solving nonlinear equations are described and the more complete one by Robert Fernand Bernard, Viscount de Montessus de Ballore (1870–1937) and Robert d'Adhémar [190] in 1911. There is also a chapter on numerical and graphical methods for quadrature and differential equations by C. Runge and F.A. Willers in the *Encyklopädie der Mathematischen Wissenschaften* [219]. Also, in 1798, Joseph Louis Lagrange (1736–1813) published a paper with the words *analyse numérique* in its title (but with the meaning of "number theory") [168].

Anyway, numerical analysis is concerned with the design and the study of *algorithms*, which can be defined as a set of rules allowing to perform numerical computations. It is well-known that this word stems from the name of the arabic mathematician Mohammed Ibn Musa Abu Djefar Al-Khwarizmi (about 780–850), whose book on algebra began, in its latin translation, by *Algoritmi dixit...*

E-mail addresses: Claude.Brezinski@univ-lille1.fr (C. Brezinski), wuytack@uia.ua.ac.be (L. Wuytack).

All papers contained in this volume (except this one) are reprinted from the seven volume project of the *Journal of Computational and Applied Mathematics* for the year 2000. The aim of this project was to present the historical development of numerical analysis during the last century and to review the current research in selected domains of numerical analysis. The volumes are

Vol. I: Approximation Theory
Luc Wuytack, Jet Wimp, eds.
Vol. II: Interpolation and Extrapolation
Claude Brezinski, ed.
Vol. III: Linear Algebra,
Apostolos Hadjidimos, Henk van der Vorst, Paul Van Dooren, eds.
Vol. IV: Optimization and Nonlinear Equations
Layne T. Watson, Michael Bartholomew-Biggs, John Ford, eds.
Vol. V: Quadrature and Orthogonal Polynomials
Walter Gautschi, Francisco Marcellan, Lothar Reichel, eds.
Vol. VI: Ordinary Differential Equations and Integral Equations
Christopher Baker, Giovanni Monegato, John Pryce, Guido Vanden Berghe, eds.
Vol. VII: Partial Differential Equations
David Sloan, Stefan Van de Walle, Endre Suli, eds.

Of course, it is impossible, in this volume, to cover all aspects of numerical analysis and some quite important topics have been omitted. However, we do hope that the historical and survey papers herein will give a good idea of the development of numerical analysis during the twentieth century and will also serve as a basis for new research. Predictions for the next fifty years are given in [251].

The scenery

Mathematical analysis was developed throughout many centuries by those mathematicians who needed to solve real-world problems. The first problems were connected with the measurement of fields (as the computation of the square root) and weights (as ascending continued fractions). Then, more difficult problems were related to astronomy and, for example, the differential and integral calculus were developed for that purpose. Soon, mathematicians realized that the real-world problems they had to solve were far too complicated to be handled by their analytical methods. So, they invented special types of methods for solving these problems *numerically*. Thus, numerical analysis was first the apanage of scientists (mostly astronomers at the beginning) and applied mathematicians. This is why some numerical methods bear the names of such astronomers as Newton, Euler, Gauss, Jacobi, Lagrange, Adams, etc., and other applied scientists as Cholesky who was a cartographer in the French army and was killed during World War I. More recently, the development of some methods for the numerical integration of differential equations was pushed by problems in ballistics. One should, at this stage, make a distinction between numerical analysis whose purpose is the study of algorithms and their properties, and *applied mathematics* which is concerned in transforming a physical problem into a mathematical one, and is also often called *modeling*. Both domains could be gathered under the name of *scientific computing*. It was John von Neumann (1903–1957) who first recognized the importance of computing in science, and computers had and still have a great impact on the development of numerical analysis [186].

The work of a numerical analyst has to go through different stages. First, starting from the physical problem, he has to put it into equations. The various interesting variables (the unknowns of the problem) have to be identified and the equations they satisfied have to be written down. When these equations are too complicated, they must be simplified by neglecting the terms whose influence is small compared to that of the other terms. In many situations, these equations are ordinary or partial differential equations, or integral equations. This phase, most of the time accomplished with the help of the specialists of the problem, is called *modelisation*. Then comes the *mathematical analysis* of these equations. It is necessary to study their theoretical properties such as the conditions under which there is existence and uniqueness of the solution. If several solutions are possible, one has to characterize the solution corresponding to the physical reality. The solution of these equations are continuous functions of some parameters such as time, pressure, temperature, and so on. In practice, it will only be possible to compute the solution at discrete values of the parameters. Thus, the continuous problem is replaced by an approximate discrete one. This is the phase of *discretization*. If this problem is still too complicated, it has to be simplified, for example by linearizing the equations. This approximate problem has then to be solved by a method of *numerical analysis*. This method has to be turned into an *algorithm* to be used on a computer. The data have to be stored in the computer (which could be an important and difficult problem) and the algorithm has to be programmed in the most efficient way, taking into consideration the conditioning of the mathematical problem and the propagation of rounding errors due to the finite precision of computers (that is the numerical stability of the algorithm). If an iterative method is used, a test for stopping the iterations has to be implemented. In any case, the accuracy of the solution given by the computer has to be controlled. The computer program must be *portable*, which means that it could be run on various computers without modifications, and *robust*, that is giving the same results. Finally, the numerical results have to be analyzed and compared with the reality.

Numerical analysis covers a wide range of domains, ranging from arithmetic, probability and statistics to pure mathematics. For example, the understanding of the propagation of rounding errors in numerical computations deals with computer arithmetic, and some approaches to their study require a knowledge of probability and use statistical evaluation. The design of algorithms suitable for parallel computation is obviously closely related to computers' architecture. On the other hand, some parts of numerical analysis heavily rely on the mathematics introduced by Bourbaki (although some members of the group were not very much in favor of applied mathematics) and, in particular, on functional analysis (see [59,175,151]). This is the case of approximation theory and of the numerical methods for the solution of partial differential equations (see the early work of Jacques-Louis Lions [176]). Some other domains of numerical analysis are connected to complex function theory, or linear algebra, or geometry. Complexity theory has deep roots in pure mathematics also.

Of course, the impetus to numerical analysis was given by the development of modern computers just before and during World War II and the introduction of high-level programming languages. Alan Mathison Turing (1912–1954) (see [140,253]) and John von Neumann (1903–1957) were pioneers in this domain. Numerical analysis methods are to be programmed on computers and, vice versa, the analysis of the numerical results can lead to new ideas, new algorithms or even new theoretical results. This is exactly what happened with the discovery of fractals and chaotic behavior. Thus, there is a constant back and forth movement between the computer and the mathematician. Thus, numerical analysis can also be defined as the mixture of mathematical analysis and numerical computation. In fact, as stated by Peter Wynn [277], *numerical analysis is very much an experimental science*.

For an early history of numerical analysis see [116] and [47], where some more recent contributions are also analyzed. Apart from those contained in this volume, other historical papers dealing with recent topics in numerical analysis can be found in [194]. The reference [135] is devoted to the work conducted at the *Institute for Numerical Analysis* of the *National Bureau of Standards* during the period 1947 to 1956 where, in particular, Lanczos methods and the conjugate gradient algorithm were developed. The history of continued fractions and Padé approximants is presented in [31]. For the history of interpolation, see [32].

The actors

Let us try to trace the history of the first courses in numerical analysis in each country and to give the names of the mathematicians who were pioneers in the teaching of numerical analysis and, thus, were very influential in its development and spreading. Obviously, it is a difficult (and perilous) task to establish such a list and we would like to apologize in advance for errors and omissions. We also would like to thank all colleagues who sent us informations (their names are mentioned in parenthesis). Much other information was found on the web.

Argentina

In Argentina at the *School of Sciences, Universidad de Buenos Aires*, there was a course on the numerical solution of ordinary differential equations in 1962. It was taught by Pedro Zadunaisky; Victor Peyrera was his assistant. The textbook was P. Henrici's book on the numerical solution of ordinary differential equations [131]. Earlier, there must have been courses by Manuel Sadosky, who had written a basic book on numerical methods in Spanish. A course on numerical linear algebra by Alexander M. Ostrowski (1893–1986) in 1959 and one on numerical methods by Lothar Collatz (1910–1990) were also given around the same time.
(*Contribution by Victor Peyrera*)

Austria

Edmund Hlawka gave a two-year lecture course on "Algebra" at the Universität Wien and a numerical pendant on solving algebraic and transcendental equations at the Technische Hochschule Wien (half a year) in 1946–1948.
Since approximately 1955, there had existed a 2–year curriculum "Computational Techniques" (in German, "Rechentechnik") at the Technical University of Vienna which taught a mixture of early computer science and numerical analysis. It could either be taken as a supplement to one of the regular courses of study, or as a stand-alone study. After satisfying all requirements, the students got a document qualifying them as "Certified Computational Technicians" (in German, "Gepruefter Rechentechniker"). This must have been one of the earliest official curricula in electronic computing worldwide. There had been an IBM 650 at the University since 1958, the first "real" computer in Austria. Before that, the courses were based on punched card machinery. In 1964, an IBM 7040 arrived which remained the most powerful computer in Austria for a good number of years. This pioneering introduction of mechanical computing into academic teaching was due to Rudolf Inzinger (1907–1980), Professor in Mathematics. He was also very modern in other ways: he solicited computational projects from industry against payment which was used for the enhancement of computing facilities. This was quite unusual in Academia around 1960. The title for that chair at the Technical University of Vienna was "Computational Techniques" and it was changed to "Numerical Mathematics" upon the request of Hans J. Stetter when he took the position in 1965.
From 1959 to 1961 (?), Walter Knoedel gave a lecture entitled "Numerische Behandlung von algebraischen und transzendenten Gleichungen" each winter term and "Numerische Behandlung von linearen Gleichungssystemen und algebraischen Eigenwertaufgaben" each summer term at the

"Technische Hochschule Wien". In the winter term of 1958, he gave a lecture on "Numerische Verfahren der Algebra".
At the University of Innsbruck, it was Gerhard Wanner (now Professor in Geneva, Switzerland) who gave the first lectures in numerical analysis during the winter term 1969–1970 (it is worthwhile to mention that one of his students at that time was Ernst Hairer).
The first course in numerical mathematics at the Technical University of Graz were given by Helmut Florian during the winter semester of 1964–1965. The title of the course was "Einführung in die Numerische Mathematik".
(*Contributions by Walter Knoedel, Alexander Ostermann, Karl Perktold and Hans J. Stetter*)

Belgium

A course in numerical analysis was created at the *Université Catholique de Louvain* in October 1960. It was taught by Jean Meinguet and contained the classical topics: approximation, interpolation, integration and numerical solution of equations. A second course, on the numerical solution of partial differential equations by finite differences and finite elements methods, began in 1971 and it was also taught by Meinguet until his retirement in 1995. In 1956, he was also one of the first Belgians to use computers for the solution of scientific and technical problems.
At Leuven University, the first courses on numerical analysis which were taught under that name must have been given by Professor Ludo Buyst at the end of the 1960's and early 1970's. He was teaching mainly at the Engineering Faculty, but also attracted students from mathematics. Of course, computer programming and numerical methods went hand in hand and a (not obligatory) "Practicum" in numerical analysis was also given where the students first learned how to do calculations with the slide rule. The contents of the more advanced lecture notes were iterative methods for nonlinear equations. A separate course was on (polynomial) interpolation, which treated mainly finite differences and some ideas on numerical differentiation and integration. In a course on linear algebra, numerical methods for linear systems of equations (Gauss, etc.) were studied. Some of these topics were hidden in the computer programming course. An extensive course existed soon after that on approximation (orthogonal polynomials, least squares approximation, etc.) This one contained many more theorems than the others, which were mainly computational algorithms. The numerical methods for differential equations came somewhat later (first handwritten notes around 1970). In the early 1970's there were two courses for engineering students: 1st year: error analysis, interpolation, differentiation, integration, 2nd year: iterative methods nonlinear equations, eigenvalues, optimization.
Hugo Van de Vel was the professor giving numerical courses (solution of partial differential equations) especially for mathematics students. This must have been shortly after L. Buyst started. Hugo's courses were more of a theoretical kind: more convergence theorems than practical algorithms. Also, astronomy (more or less part of mathematics) was a group that did numerical computations (Professor Bertiau and later P. Smeyers). FORTRAN was taught in the exercises of the astronomy course.
(*Contributions by Jean Meinguet and Ronald Cools*)

Brazil

The first courses in numerical analysis were initially taught at the Engineering Schools under the classification "Numerical Calculus", due to its practical importance for engineers. Of course, at the time, there was not such strong emphasis on convergence of the methods as is required in Numerical Analysis proper, but this does not mean that these questions were ignored. "Numerical Calculus" was started as a regular course approximately at the same time in two places:

1. At the Polytechinc School of the University of Sao Paulo, in Sao Paulo, under Professor Monteiro de Camargo,
2. At the School of Engineering (founded in 1897) of the Federal University of Rio Grande do Sul, in its capital city Porto Alegre, in March 1956, under Professor Manuel da Silva Neto. Its full name was "Calculo Numerico, Grafico e Mecanico" and the program followed closely the classical book of Willers [272] translated from German. The following year, the book by Mario George Salvadori (born in 1907) and Melvin L. Baron (1927–1997) [226] was adopted for use with differences in the numerical solution of ordinary and partial differential equations. There was also a French book, with a title like "Cours de Nomographie", edited by Dunod, about 80 pages long, that was also used as a complement reference for graphical methods. The program covered the basic topics of numerical analysis. Slowly, graphical methods were set aside, with, perhaps, only one lecture devoted to it. Pedro Nowosad, originally an engineer, started his career at the School of Engineering in Porto Alegre as an Assistant to Professor Silva Neto in 1957. In 1965, he got his PhD in Mathematics from the Courant Insititute at NYU, encouraged by Professor Silva Neto, who always had a very keen notion of the importance of mathematics for engineering.
Courses in numerical analysis were also given at the State University at Campinas – UNICAMP in 1969. The Professors were Odelar Leite Linhares (who got his doctorate in Zürich) and Ivam de Queiroz Barros (who, at the time, taught both at the State University in São Paulo – USP, as well as at UNICAMP).
(Contributions by João Frederico C.A. Meyer, A.C. Moura and Pedro Nowosad)

Bulgaria
The first course in numerical analysis was given in the academic year 1959–1960 at Sofia University, Faculty of Physics and Mathematics. The lecturer was Blagovest Hristov Sendov, assistant in the Chair of Algebra. At that time, in Bulgaria, there were only mechanical calculators and one analog computer. Nevertheless, students had exercices on all basic numerical methods. Starting from 1961, Sendov was engaged in the construction of the first original Bulgarian computer which was operational in 1963.
(Contribution by Blagovest H. Sendov)

Cuba
In Cuba, numerical analysis was included for the first time in the curriculum in mathematics at the University of Havana in 1963. The teacher was an invited professor of the Carolina University of Prague named Stanislav Malon, who stayed three years and initiated these studies in the branch of applied mathematics. The contents of Numerical Analysis I were : introduction to the method of numerical analysis, solution of nonlinear equations, interpolation and numerical linear algebra. The textbook used that of Berezin and Zhidkov [17].
(Contribution by Maria Victoria Mederos)

Czechoslovakia
The situation between the wars in Czechoslovakia was such that there were two universities (Prague, Brno) in the Czech part, one (Bratislava) in Slovakia.
There were also Technical "Universities" in Prague and (maybe later) in Brno.
Numerical mathematics was taught at the Czech Technical University in Prague; there were Professors V. Láska and V. Hruška who even published a book in Czech: "Theory and Practice of Numerical

Computing", around 1935. There also was a group of mathematical methods at the Škoda-Works in Pilsen, lead by M. Hampl.
During the war 1939–1945, the Czech universities (all) were closed by the German authorities. Some professors held (illegally and privately) seminars in their homes, others were persecuted.
After the war, the universities were reopened. Mathematics was taught at faculties of natural sciences of Charles University in Prague and Masaryk University in Brno. Emphasis was on pure mathematics: algebra, topology, analysis, geometry.
The idea to supplement research in applications led to the creation of the Academy of Sciences which would not only be a body of academicians (member of the Academy) but also a number of Institutes.
Thus, in 1950, the "Central Institutes" of Mathematics, of Physics, etc., were founded; these became institutes of the Czechoslovak Academy of Sciences in 1952.
In mathematics, starting in July 1950, around 8–10 postdocs became "scientific aspirants" (that was a Russian system): they included (later well known) V. Pták, M. Fiedler, J. Mařík, O. Vejvoda, M. Zlámal, I. Babuška, J. Hájek, F. Šik, J. Kurzweil, Z. Nádeník, etc.
Some lectures were in linear algebra, numerical mathematics (Vl. Knichal), statistics, computer science (A. Svoboda, who came from the US to Czechoslovakia to introduce and support research in computer science; he left Czechoslovakia after 1958 or so).
In numerical mathematics, some of the aspirants became interested (Babuška, Fiedler, Pták, Mařík) and started research in this area.
As teaching of numerical mathematics at Charles University is concerned, it was started in September 1956. Miroslav Fiedler held a lecture (for students in their third year) in the winter semester 1956–1957 on numerical algebra, O. Vejvoda continued in the spring semester 1957 with numerical methods in differential equations.
After a few years, J. Mařík took over lecturing in numerical mathematics; he also published some materials for students.
Around 1960, the faculty of natural sciences was split, and the Faculty of Mathematics and Physics was created. A part of the faculty was then the Department of Numerical Mathematics, first led (externally) by M. Hampl; later, Ivo Marek became the head.
See also [10].
(*Contribution by Miroslav Fiedler and the help of Zdenek Strakos*)

Denmark

It seems that the first course in numerical analysis was taught in 1961 at the Technical University of Denmark by Peter Naur, at that time an Associate Professor. Naur, who was originally an astronomer, later became Professor of Computer Science at Copenhagen University.
(*Contribution by Per Christian Hansen*)

Finland

Ernst Leonard Lindelöf (1870–1946) was professor at Helsinki University and his lectures appeared as a series of five volumes [174]. A characteristic feature is his interest in numerical computations. In the first volume, he started from Lagrange interpolation and then moved to Taylor's formula [87]. Evert Johannes Nyström (1895–1960), a former student of Lindelöf and a professor of applied mathematics at the Helsinki University of Technology, also gave a course with the title "Applied mathematics" in the spring of 1948. It contained interpolation, least squares approximation, numerical integration, nomography, mechanical harmonic analyzator, and graphical integration. Later, this type of course

was given under different types of title, but mostly under "Numerical Methods" if translated word by word. Pentti Laasonen (1914–2000) also gave courses with this title at Helsinki Technical University in the 1960's.
It seems that the first thesis on a numerical analysis subject was defended at the University of Helsinki in 1914 by Johan Helo (until 1906 Helenius; 1889–1966) [130] under the supervision of Ernst Lindelöf. He studied the convergence of Newton's method for the solution of an equation and continued to work on the subject during the seven following years. Then, he studied law and started a political career after 1928 [87].
(*With contributions by Timo Eirola and Olavi Nevanlinna*)

France
A course in numerical analysis, with the title "Applied Analysis", was created at the Institut Polytechnique de Grenoble by Jean Kuntzmann (1912–1992) in 1947. The topics covered were finite differences, differential equations, zeros of functions, systems of linear equations, etc. It was the first course of this type taught in a French university and Grenoble was the only center for numerical analysis in France for some time. Kuntzmann published mimeographed notes for students as early as 1950 [162]. Practical work on computers was conducted by Jean Laborde (1912–1997) [164]. Kuntzmann created the Laboratoire de Calcul in 1951. He also became one of the founder and the first editor-in-chief of the French journal of numerical analysis *Chiffres* (now M2AN) [163]. Noël Gastinel (1925–1984) came to Grenoble in 1957 and Pierre-Jean Laurent in 1958. Gastinel was very influential in the development of numerical analysis in France [76].
At the University of Toulouse, courses in numerical analysis were taught to graduate students in mathematics and engineering by Émile Durand, a physicist who used an IBM 650 computer around 1957 for the solution of his own problems. Another course was given by Laudet around the end of the 1950's.
The first chair of numerical analysis was created at the Faculté des Sciences of the Université de Paris in 1959 for René de Possel (1905–1974), one of the first member of the Bourbaki group. He was succeeded by Jacques-Louis Lions in 1966. Lions was the first in France to be interested in problems derived from industry, by their modelisation, their mathematical treatment by functional analysis tools, and by their algorithmic solution. The school he founded does not need to be presented.
Another influential person for the development of numerical analysis in France was Henri Mineur (1899–1954), the first head of the Astrophysics Laboratory in 1939, who wrote a book on numerical analysis in 1952 [188]. Louis Pierre Couffignal (1902–1966) must also be mentioned. He was the director of the Laboratoire de Calcul Numérique at the Institut Blaise Pascal in Paris from 1946 to 1957. This laboratory was the successor to the analogue computing laboratory created by Joseph J. Pérès (1890–1962) and Lucien C. Malavard as early as 1932 [192]. Let us mention that Malavard was involved in the development of the Turbosail wind-propulsion system with Jacques-Yves Cousteau (1910–1997) in 1982. Maurice Parodi (1907–1992) was another pioneer in numerical analysis with his works on the computation of eigenvalues.
(*With contributions by Françoise Chatelin and Pierre-Jean Laurent*)

Germany
The first chair of applied mathematics was taken by Carl David Tolmé Runge (1856–1927) in Göttingen in 1904. His Habilitation (1883) was on the numerical solution of algebraic equations. Friedrich Adolf Willers (1883–1959), later in Dresden, had his first position 1928 in Freiberg, Sachsen.

He was a student of Runge (Ph.D. in 1908, Göttingen). He most likely gave courses in numerical analysis.
The first woman involved in the teaching of numerical analysis was probably Hilda Geiringer-Pollaczek (1893–1973); see [212]. At that time, she was "Assistentin" of Richard Martin Edler von Mises (1883–1953) (and his wife from 1943) at Berlin University (now called Humboldt Universität) and conducted the "Numerisches Praktikum" there in the years before the emigration in 1933. The students used mechanical desk calculators. One of the students who took this course was Lothar Collatz in 1934. Collatz had his first position in Karlsruhe from 1935 to 1943. The first lecture he gave was in summer 1937 in Karlsruhe with the title "Statik und Festigkeitslehre für Architekten". In the same year, he also gave a course in "Nomographie". In the summer 1938, he gave a course on "Numerische Methoden (Angewandte Mathematik B)".
Special courses on numerical analysis were given in 1955 by Friedrich Ludwig Bauer when he was lecturer in mathematics at the Technische Hochschule München. Then, he went to Mayence in 1958 where he gave regular introductory courses in numerical analysis as part of the curriculum, before moving to Munich in 1962.
One of the central figures in numerical mathematics and computers at that time in Germany was Alwin Walters (1898–1967) at the Technische Hochschule Darmstadt. The names of Bertram (Saarbruecken), Unger (Bonn), Lehmann (Dresden) and Romberg (Heidelberg) have also to be mentioned. Robert Sauer, who, together with Klaus Samelson (1918–1980), may be regarded as an early founder of computer science in Munich, gave a course in numerical analysis at Munich University of Technology in 1961. There have also been courses at the University of Frankfurt, given by people from the "Wetterdienst" at Offenbach (the German agency for weather reports and investigations).
(*Contributions by Friedrich L. Bauer, Martin Brokate, Rita Meyer-Spasche, Gerhard Opfer and Hans Joseph Pesch*)

Great Britain

Great Britain was certainly the country where the teaching of numerical analysis was the most developed and where courses were given at an earlier time.
Around the beginning of this century, Edmund Taylor Whittaker (1873–1956) created a Computation Laboratory at the University of Edinburgh, where various branches of numerical analysis were taught that had previously been taught systematically at no British university. His book [268] with George Robinson, a Canadian mathematician then on his staff, remained for 40 years a major text on numerical analysis. In the Preface, it is written

> *The present volume represents courses of lectures given at different times during the years 1913–1923 by Professor Whittaker to undergraduate and graduate students in the Mathematical Laboratory of the University of Edinburgh, and may be regarded as a manual of teaching and practice of the Laboratory... The manuscript of the lectures has been prepared for publication by Mr. Robinson, who has performed the whole of the work of numerical verification and has contributed additional examples.*

The book covers interpolation difference formulae, determinants and linear equations, the numerical solution of algebraic and transcendental equations, numerical integration and summation, normal frequency distributions, the method of least squares, practical Fourier analysis, the smoothing of data, correlation, the search for periodicities, the numerical solution of differential equations and some further problems.
Whittaker was the first in Scotland to teach a course in numerical analysis, and probably he was the first in the whole of the UK.

Another leading figure in numerical analysis in Edinburgh was Alexander Craig Aitken (1895–1967). Born and educated in New Zealand, he came to Edinburgh in 1923 for his Ph.D. under Whittaker and he stayed there for the remainder of his life. In 1925, he was appointed to a lectureship in Actuarial Mathematics. In his lessons, one of the topics was interpolation and numerical integration [3, p. 83]. In 1936, he became Reader in Statistics. When Whittaker retired in 1946, Aitken took his Chair, a position he held until his own retirement in 1965 [269]. Aitken taught a course entitled "Mathematical Laboratory" at least from 1952, but it certainly goes back further.
In its lectures at the London School of Economics in 1926 and 1927, just after he returned to England (see the Subsection on the USA below), Leslie John Comrie (1893–1950) stressed the importance of machines [274]. In [53], he wrote

> *In many cases they* [relay and electronic computers] *will enable numerical mathematical analysis, with its wide range of choice in parameters, to replace experiments with costly models.*

Of course, as in the USA, the war effort was essential in the development of numerical analysis [225].
A numerical analysis course was given by John Todd at King's College in London in 1946. In particular, he was teaching the Cholesky method (see below).
The first stored-program electronic digital computer, called *the baby*, started at the University of Manchester on June 21, 1948.
When he arrived at Oxford University in 1958, Leslie Fox (1918–1992) began to lecture on numerical linear algebra to the first year students [101].
There was a Senior Honours course in numerical analysis at the University in St Andrews in 1953–1954. It was given by A.R. Mitchell who moved to Dundee in 1967.
A numerical analysis course was given in Aberdeen in 1958–1959 by Fred W. Ponting. There was a lot of use of mechanical desk machines, and the book by D.R. Hartree followed [128]. Ponting said that it was the first time a course in numerical analysis had been given in Aberdeen. In those days, there were only the four ancient Scottish Universities (St Andrews, Glasgow, Aberdeen and Edinburgh) plus Queen's College Dundee (a college of St Andrews, which became the University of Dundee in 1966), and the Glasgow institution which became Strathclyde University.
(*With contributions by Mary Croarken, Roger Fletcher, Philip Heywood, George Phillips, Garry J. Tee, John Todd and Alistair Watson*)

Greece
It seems that the Mathematics Department of Patras University offered the very first courses in numerical analysis. It was in 1967–1968, the name of the course was "Numerical Analysis and Special Functions", and it was given by the Chair of (Theoretical) Mechanics, Konstantinos Goudas. The first professor in numerical analysis was appointed only in 1974–1975 and he was Kosmas Iordanidis.
The Mathematics Department of the University of Athens offered its first numerical analysis courses in 1970–1971. The Chair responsible was that of Applied Mathematics (with contents numerical analysis and computer programming) and the Professor selected in 1971 was Nikolaos Apostolatos.
The Mathematics Department of the Ioannina University was the third to follow in 1971–1972. Apostolos Hadjidimos was appointed in 1972 to the Chair of numerical analysis. It was the very first chair in a Greek University with that name.
The Mathematics Department of the University of Thessaloniki gave its first numerical analysis courses also in 1971–1972. The professor who gave the course was a Professor of Mathematical

Analysis. The Chair of numerical analysis was created in the late 70's and the professor was Elias Houstis.

Then, it was the Technical University of Athens. It must have given the first courses in numerical analysis by the mid 70's, when Alexis Bacopoulos was elected around that time. The Chair of numerical analysis belonged to one of the Engineering Schools (most probably to the Civil Engineering one).

The University of Crete, which was created in the late 70's, appointed its first Professor in Numerical Analysis in 1982. He was Vassilios Dougalis.

In 1981, the Ministry of Education passed a law for the Greek Universities which replaced that law used since 1932 (for the old one, Konstantinos Caratheodory (1873–1950) was the man responsible and the system of Chairs was very close to the French–German one). Chairs were abolished by the new law and the American system in the ranking of the academic staff was introduced. Consequently, it was much easier to appoint academic staff at lower ranks who could teach, say, numerical analysis.

Nowadays numerical analysis or similar courses are offered in all Greek Universities and not only by and in the Mathematics Departments.

(*Contributions by Apostolos Hadjidimos and George Miminis*)

Hungary

Nobody knows exactly by whom, and when, the first course in numerical analysis was given in Hungary. In different persons' opinion, it was most probably given by the great Hungarian Professor of Geometry, György Hajós (1912–1972). He taught a course called numerical analysis after the second world war around 1947–1948 at the University of Budapest which is now called Budapest University of Technology and Economy (Budapesti Mûszaki és Gazdaságtudományi Egyetem). At that time, its name was something like Hungarian Royal Joseph Palatine University of Technology and Economy (Magyar Királyi József nádor Mûszaki és Gazdaságtudományi Egyetem). "Nádor", which can be translated by "palatine", was the highest rank in the old feudal government.

(*Contribution by Maria Vicsek*)

Italy

The first course of numerical analysis seems to have been taught at the Regia Scuola di Ingegneria di Pisa by Gino Cassinis (1885–1964) as early as 1925. He published a book on the subject [45].

A course with the title "Calcoli Numerici e Grafici" was also given by Mauro Picone (1885–1977) in 1932 at the "Scuola di Scienze Statistiche e Attuariali" in Rome, a school due to the initiative of Guido Castelnuovo (1865–1952) and founded by Francesco Paolo Cantelli (1875–1966). This course was not among those offered to the students of the university since mathematicians like Gaetano Fichera (1922–1996) said that it had not the necessary characteristics for inclusion in the list of optional courses for a Masters degree in Mathematics [196]. In fact, Picone's concern in applied mathematics arose much earlier from problems encountered during World War I, and it can be seen even in his pure mathematics research just after his "Laurea". Picone was interested in theoretically justified and efficient mathematical methods which could counterbalance the modest computational instruments at his disposal. Even if the expression "numerical analysis" was not used, he defined a domain of research between classical analysis and numerical computing [7,123]. Picone was also the founder of the "Istituto Nazionale per le Applicazioni del Calcolo" in Naples in 1927 (transferred to Rome in 1937). It was the first institute in the world devoted to numerical mathematics and Picone was

a world pioneer in establishing such an institute. Picone's assistant from 1940 to 1948 was Aldo Ghizzetti (1908–1992) [94].
(Contributions by Giampietro Allasia, Michiel Bertsch, Andrea Celli, Luigi Gatteschi, Enrico Magenes, Donatella Marini and Alfio Quarteroni and the help of Michela Redivo Zaglia)

Japan
In Japan, numerical analysis was treated in practical mathematics courses and, hence, there were no courses on numerical analysis in the Universities before 1970.
Keiichi Hayashi (1879–1957) was Professor of the Faculty of Engineering of Kyushu Imperial University (Now, Kyushu University). He took up his position around 1910 and studied numerical analysis. He published five mathematical tables in German.
The first course of numerical analysis was given in 1970 by Ichizo Ninomiya (born in 1921) at the Faculty of Engineering of Nagoya University.
(Contribution by Naoki Osada)

Mexico
The first course of numerical analysis was given by Pablo Barrera Sanchez at the National University of Mexico (UNAM) in the early 70's.
(Contribution by Pablo Barrera Sanchez)

New Zealand
The first course, with the title "Statistical and Numerical Mathematics", was taught in 1963 by John C. Butcher at the University of Canterbury. The part devoted to numerical analysis dealt mainly with the numerical solution of ordinary differential equations [215, p. 144].
(Contribution by Garry J. Tee)

Norway
It seems that it was Werner Romberg (born 1909) who introduced numerical analysis in Norway. In the autumn of 1938, he came to the University of Oslo where, in different positions, he stayed until the end of 1948 (except for a period in Uppsala, Sweden, during World War II). During these years, he worked with Professor Hylleraas and, for a short period, with Professor Holtsmark. In 1949, Romberg went to the Norwegian Institute of Technology in Trondheim as an Associate Professor of Physics. In 1960, he became Professor in Applied Mathematics. In 1968, he returned to the University of Heidelberg in Germany as a Professor in "Numerische Methoden der Naturwissenschaften" [126].

Poland
The first course in numerical analysis was given by Stefan Paszkowski in 1962 at the University of Wrocław. He was, maybe, preceded by one or two years by Andrzej Kielbasinski at Warsaw University.
(Contribution by Stefan Paszkowski)

Portugal
The first course on numerical analysis at the University of Porto was given by Jaime Rios de Sousa for students in sciences and engineering.
(Contribution by Filomena Dias d'Almeida)

Spain
Some topics in numerical analysis were taught by Sixto Rios, a statistician in Madrid and by Ernesto Gardeñes in Barcelona around 1970. The first chair of numerical analysis was taken by Mariano Gasca in Bilbao in 1972. In 1973, a course of numerical analysis became obligatory for all students in mathematics.
(Contribution by Mariano Gasca)

Sweden
Numerical analysis as an academic subject in Sweden has three roots:
1) Courses at the Royal Institute of Technology and Chalmers Institute of Technology within the general subject of applied mathematics. At the Royal Institute, they started in the thirties, on the initiative of Professor Liljeström (originally a physicist), extended by Professor C.G. Esseen, better known for his contributions to probability theory. He gave well planned courses in numerical methods, which contained linear algebraic systems, single nonlinear equations, interpolation and numerical integration, Runge–Kutta's and Simpson's rule for ordinary differential equations and nomography. During 1957, the preparations started for an associate professorship in Applied Mathematics, in particular Numerical Analysis. Germund Dahlquist got this position in 1959. It was vacant in 1958, and a physicist named N. Åslund, who has a good hand with computing, acted protem and developed some courses, also containing an introduction to digital and analog computers, their logic construction and the elements of programming (Åslund later became Professor of Physics in Stockholm). Practical programming, including the running of programs written by students, started in 1959. During some of the years from 1954–1958, Dahlquist organized courses in programming for the Swedish electronic computer BESK, and organized the work on a program library for basic numerical methods, etc., but there were no regular courses on numerical methods.
During the sixties, activities grew also at other universities and institutes of technology. In 1963, full professorships were created in Stockholm (Dahlquist, jointly for the Royal Institute of Technology and the University), Gothenburg (Heinz Otto Kreiss at Chalmers), and in 1965 in Lund (Carl-Erik Fröberg). A few years later, Kreiss moved to Uppsala, and at the end of the decade Owe Axelsson and Åke Björck became Professors in Gothenburg and Linköping, respectively. From 1959, Björck and Dahlquist worked together in Stockholm in teaching (elementary courses for 500 students and a second course for 30 students) and research in numerical analysis. The first Swedish version of their textbook was printed in 1969 (the English version of 1974 is much larger [64]).
2) Independently, C.–E. Fröberg started, around 1956, courses in numerical analysis at the Department of Theoretical Physics, at Lund university, and published in 1959 a first mimeographed version of a textbook that he later considerably extended. A Swedish version (415 pages) was published in 1962, and at least two considerably modernized versions were published in English in the seventies and eighties [105]. Fröberg, who got his Ph.D. in theoretical physics, became Associate Professor of Numerical Analysis in 1956 and full professor in 1965. Around 1960, Fröberg's courses covered more material than the courses of Björck and Dahlquist in Stockholm, but they had a much larger number of students.
3) Carl Harald Cramér (1893–1985) became Professor in Actuarial Mathematics and Mathematical Statistics in Stockholm in 1929. When Dahlquist studied the subject in 1942–1943, he had to read parts of Whittaker and Robinson [268] and a Danish version of Steffensen's book on Interpolation. There was no teaching in numerical methods, but he remembers that, in the exam, one of the problems was to derive a 5th order accurate Newton–Cotes formula.

(*Contribution by Germund Dahlquist*)

Switzerland

A course with the title *Calcul Approximatif* was given as early as 1877–1879 at the University of Geneva by Charles Galopin-Schaub (deceased in 1901), then a Privat Docent. He also wrote a small book on the subject [107]. His course contains a first part on the theory of approximate computations, a second part on various approximation methods (such as series and continued fractions), a third part dealing with least squares, a fourth part on various numerical methods for the solution of transcendental equations and a fifth part with applications to analysis and mechanics.

The Department of Applied Mathematics (Institut für Angewandte Mathematik) of the Eidgenössische Technische Hochschule Zürich (ETHZ) was founded in January 1948 by Eduard Stiefel (1909–1978) with the help of Heinz Rutishauser (1918–1970) and Ambros Paul Speiser. Stiefel immediately began to lecture on numerical analysis and a Z4 computer was rented to Konrad Zuse (1910–1995) in 1949 [285]. At that time, also with Peter Henrici (1923–1987) who got his Ph.D. in mathematics in 1952 under the direction of Stiefel, ETHZ was a renowned international center for numerical analysis [124].

A course in numerical analysis was taught at the University of Basel in 1950 by Eduard Batschelet (1914–1979). At the University of Lausanne and then at École Polytechnique Fédérale de Lausanne, Charles Blanc (born in 1910), a specialist of operations research, also oriented the teaching of mathematics toward applied mathematics and numerical analysis. In 1958, he participated in a meeting at UNESCO where the French word "informatique" was coined.

(*Contributions by Olivier Besson, Jean Meinguet and Walter Gautschi*)

The Netherlands

At Delft University of Technology (TUD), on page 108 of the Program for the Academic Year 1947–1948, the following course is listed:

Prof. Dr. S.C. van Veen: Applied Analysis (Toegepaste Analyse)

A. Interpolation methods. Mechanical quadrature. Graphical differentiation and integration. Numerical solution of algebraic and transcendental equations. Theory and application of planimeters and harmonic analyzers.

B. Approximate and graphical solution of ordinary differential equations of first and second order, with boundary conditions. Determination of eigenvalues and eigenfunctions. Method of Ritz–Galerkin. Approximation formulas for elliptic integrals.

Although not called numerical analysis, it was really the subject of the course. Professor van Veen changed the name of the course to Numerical Analysis in the academic program for for 1950–1951 (page 138), without changing the contents. It was an elective for students in the third and fourth year of various engineering disciplines. Professor van Veen was a highly regarded analyst in The Netherlands. He taught for many years at TUD. He retired around 1965.

In 1956, the course was taken over by Prof. Dr. R. Timman, who changed the contents (academic program 1956–1957, page 198):

Prof. Dr. R. Timman: Numerical Analysis

A. Interpolation methods, numerical integration and differentiation. Solution methods for algebraic and transcendental equations with one unknown. Integration methods for ordinary differential equations (2 hours in the first semester).

B. Solution methods for linear systems of equations, iteration methods, numerical determination

of eigenvalues and eigenvectors of matrices. Solution methods for partial differential equations, relaxation methods and methods of characteristics (2 hours in the second semester).
Timman taught various other courses in applied mathematics. The first professor who devoted himself solely to numerical analysis was Prof. Dr. E. van Spiegel, who arrived in 1958. Pieter Wesseling succeeded him in 1977.
The first course of numerical analysis in Amsterdam was given by Prof. Dr. Adriaan van Wijngaarden.
He was born in 1916 and, for most of his career, had his main position at the Mathematical Centre (now CWI), Amsterdam, from 1947 until he retired in 1981.
He was appointed as part-time Professor at the University of Amsterdam in 1952 and presented his inaugural address on 27 October 1952 entitled "Rekenen en vertalen" (i.e., Computing and Translating).
He started his course on numerical analysis at the University of Amsterdam (presumably) in September or October 1952. The name of his course (presumably) was "Numerieke Wiskunde" (i.e. numerical analysis). He gave his course every year (weekly on Friday mornings). He put much emphasis on algorithms, programming, and later on programming languages, in particular Algol 60 and Algol 68, and design of languages.
Students assignments had the form of solving a certain problem by writing a program and then to show the program and the results in an oral examination.
Adriaan van Wijngaarden retired from the University of Amsterdam in 1983 and died in 1987 at the age of 70.
(*Contributions by Dirk Dekker and Pieter Wesseling*)

USA

Numerical analysis was not much developed in the USA before World War II and there were very few courses on this topics.
Leslie John Comrie was born in 1893 at Pukekohe in New Zealand and died in London in 1950. He studied in this Department, and graduated M.A. (Honours in Chemistry) in 1916. Throughout the second quarter of this century, Comrie was acknowledged as the leader in scientific computation. In 1923–1924, he was Assistant Professor of Mathematics and Astronomy at Swarthmore College, Philadelphia, and in 1924–1925 Assistant Professor of Astronomy at Northwestern University, Chicago. At both posts, Comrie introduced courses on computation into the standard degree curriculum [180]. He returned to England in 1925 [58].
James B. Scarborough gave a course of Engineering Mathematics in 1925 at the U.S. Naval Academy in Annapolis, Maryland, and published a book on the subject [227].
Those two courses must have been amongst the earliest courses in numerical analysis to be given in the USA.
Research in numerical analysis began with the war effort [185,205,206,217], in particular at the National Bureau of Standards [244,245]. Part of the impetus was due to the emigré mathematicians [83,19]. The emphasis was put on the computation of tables and the journal *Mathematical Tables and other Aids to Computation* (MTAC) was created in January 1943 by Raymond Clare Archibald (1875–1957) for this purpose [143]. It is very instructive to look at the first volumes of this journal to see the shift from the construction of tables to numerical analysis. Another important topic was the computation of special functions which culminated with the publication of the Handbook [1] under the supervision of Milton Abramowitz (1915–

1958) and Irene A. Stegun (born in 1919) (a new version of it is under construction, see http://dlmf.nist.gov/).

After the war, the subject of numerical analysis was taken up by universities (see, for example, the impact of war on the development of operations research [191]). As explained by Mina Spiegel Rees (1902–1997) [207]

> *Before World War II, there had been relatively little emphasis in American universities on "applied mathematics" which was strongly represented at a number of German centers, particularly Goettingen, and at other continental universities as well as on British campuses. After the war, with some of the world's most distinguished emigré mathematicians on our campuses, it was possible to contemplate a strong development in these fields in the U.S. The ONR* [Office of Naval Research] *seized the opportunity to support emerging groups like that at New York University under Richard Courant* [(1888-1972)]*whose work at Goettingen had made his name magic through the world. There were also individuals, such as* [Gabor] *Szegö* [(1895–1985)] *and Pólya at Stanford, about whom strong groups could be expected to grow....*
>
> *A number of very able young people were trained at the Institute* [for Numerical Analysis at UCLA], *and large number of older mathematicians learned the new approaches that computers required. When the Institute was terminated as a government agency, many of these people went to new positions across the country and spread the word about the computer revolution to many universities that had not yet been involved.*

A course in numerical analysis was given by Kaiser S. Kunz at Harvard around 1945–1946. He also wrote a book [165]. A similar course was given by William Edmund Milne (1890–1971) around 1949 at the Oregon State College. His book [187] contains an annotated bibliography on early texts on numerical mathematics. In a paper of 1946, it is mentioned that courses on practical mathematics were taught at Brown and New York Universities [88]. Gene Howard Golub, in 1953, followed a course on computing given by John Purcell (Jack) Nash at the University of Illinois. Then, he became assistant in the Digital Computing Laboratory there; see [267]. A seminar on numerical analysis was organized by Wolfgang Richard Wasow (1909–1993) in the 1957–1958 academic year at the University of Wisconsin in Madison. The theme was a study of iterative methods for matrix problems.

(*With contributions by Mary Croarken, Gerald Hedstrom, David R. Kincaid, Garry J. Tee and David Young*)

USSR

In 1935, a Department of Approximate Methods of Analysis was created at the Steklov Mathematical Institute of the Russian Academy of Sciences in Moscow. Its first head was A.M. Zhuravskii.

It seems that the first course in numerical analysis was given in 1948 at the University of Leningrad (now again St Petersburg) by Leonid Vitalyevich Kantorovich (1912–1986). At that time, he was interested in showing the broad possibilities for using the ideas of functional analysis in numerical mathematics.

A chair of computational mathematics was created at the University of Leningrad in 1951 and Vladimir Ivanovich Krylov (born in 1902) was appointed. Then, the position was taken up by Mark Konstantinovich Gavurin (1911–1992). The book by Vera Nikolaevna Faddeeva (1906–1983) [89] was very influential and Dmitrii Konstantinovich Faddeev (1907–1989) was also an important contributor to numerical linear algebra [159].

In 1948, a Department of Numerical Mathematics at a Research Institute in Mechanics of the Moscow State University was founded by Lazar Arionovich Lyusternik. The names of Andrei Nikolaevich Tikhonov (1906–1993) and Aleksandr Andreevich Samarskii must also be mentioned.

In the early 1950's, Sergei Lvovich Sobolev (1908–1989) began to work on computational mathematics. At the same time, the M1 computer was developed in the Laboratory of Electrosystems

at the Institute of Energy of the USSR Academy of Science under the direction of I.S. Brouk. In 1952, Sobolev became head of the first department of computational mathematics in the Soviet Union (at Moscow State University).
More recently, Guriĭ Ivanovich Marchuk has been very influential in the development of numerical analysis in the USSR as the head if the Institute of Numerical Mathematics of the Russian Academy of Sciences.
See also [10].
(*With contributions by Vera Kublanovskaya and Maria Vicsek*)

Venezuela
In Venezuela, the first course in numerical analysis was taught by Victor Pereyra in the winter semester of 1967. It was at the School of Physics and Mathematics, Faculty of Sciences, Universidad Central de Venezuela, Caracas.
(*Contribution by Godela Scherer*)

Yugoslavia
The first course of numerical analysis at the University of Ljubljana (Slovenia) was taught by France Krizanic in the academic year 1961–1962. Slovenia was until 1990 part of Yugoslavia. From then on, the course was taken over by Zvonimir Bohte (lecturer at that time) who taught the subject for the next 35 years.
(*Contribution by Zvonimir Bohte*)

Landmarks

Recently, a list of the ten top algorithms was published [82] and a list of the classic papers in numerical analysis was given by L.N. Trefethen [250]. In numerical analysis, not only algorithms are important but some concepts as well. This is the reason why we prefer to speak about landmarks and give a (non limitative) list of those we consider as the most important ones.

1901: Runge–Kutta methods

The history of methods for the numerical integration of ordinary differential equations does not began with the twentieth century. It seems that Newton was the first to solve differential equations by a numerical process and every numerical analyst knows, for example, the Euler method which goes back to 1768. There were other methods, mostly obtained by astronomers for solving celestial mechanical problems (the return of Halley's comet, for example). Their history can be found in [116] and [111].

Here, we will only discuss the case of Runge–Kutta methods. Runge published a paper in 1895 [218], where he gave two second-order two-stage methods, one based on the midpoint rule (a generalization of the Simpson method for computing definite integrals) and the other one on the trapezoidal rule. In 1900, Karl Heun (1859–1929) [136] derived a restricted class of Runge–Kutta methods (among them, those of Runge) and determined their coefficients. The general class was introduced by Martin Wilhelm Kutta (1867–1944) in 1901 [166]. He wrote down the system of nonlinear equations that must be satisfied by the coefficients to attain a specific order. In particular, he obtained the well-known fourth-order four-stage formula and a fifth-order six-stage one.

An extensive historical perspective on the numerical integration of ordinary differential equations can be found in [247]. For the developments of numerical methods for ordinary differential equations in the twentieth century, consult [44]. Applications to Volterra integral equations are reviewed in [11] and retarded differential equations discussed in [12].

1903: Cholesky decomposition

André Louis Cholesky (1875–1918) was a French military engineer. He was occupied in cartography where angles and lengths have to fulfill constraints expressing that the sum of the angles in a triangle have a value greater that π depending of the sphericity of the Earth, and that the lengths remain the same independently of the order in which the measures are taken. This procedure leads to a linear system with more unknowns than equations which has to be solved by a least squares procedure. Cholesky never published his work. So, it is difficult to date exactly his method but seems to have been found when he was in Tunisia between 1902 and 1905. It was first presented in a paper by Commandant Benoit [16] in 1924. The Cholesky method was taught by John Todd in his numerical analysis course at King's College in London as early as 1946 [245]. In [242], it is said:

In 1946 one of us (J.T.) offered a course at King's college, London (KCL) on Numerical Mathematics. While we had had some war time experience in numerical mathematics, including characteristic values of matrices, we had had little to do with the solution of systems of linear equations. In order to see how this topic should be presented, we made a survey of Math. Rev. (at that time easy!) and found a review (MR 7 (1944), 488), of a paper by Henry Jensen [146], written by E. Bodewig. Jensen stated "Cholesky's method seems to possess all advantages". So it was decided to follow Cholesky and, since the method was clearly explained, we did not try to find the original paper.

Leslie Fox, then in the newly formed Mathematics Division of the (British) National Physical Laboratory (NPL), audited the course and apparently found the Cholesky Method attractive for he took it back to NPL, where he and his colleagues studied it deeply. From these papers the Cholesky (or sometimes Choleski) Method made its way into the tool boxes of numerical linear algebraists via the textbooks of the 1950s.

In the B.A. Honours and B.Sc Special Examinations in "Mathematics, Advanced Subjects – Numerical Methods" for the internal students of King's College in 1947, Todd gave an exercise on the numerical application of Cholesky's method to a 4×4 Hilbert matrix.

The Cholesky method can also be found in papers by Paul Summer Dwyer (born 1901), the first one being [85] where it was called the "square root method" and related to a method due to Tadeusz Banachiewicz (1882–1954) [13].

The first analysis of Cholesky's method was written by Leslie Fox, Harry Douglas Huskey and James H. Wilkinson. Its numerical stability was simultaneously studied by Alan Turing [252].

As noticed in [178], Otto Toeplitz (1881–1940) proved in 1907 that an Hermitian matrix can be factorized into a product LL^* where L is lower triangular [246].

A biography of Cholesky can be found in [34] and an analysis of his method is given in [47].

1926: Aitken Δ^2 process

Many methods used in numerical analysis are iterative. Quite often, the convergence is slow and, so, the effectiveness of the method is reduced. Of course, it is sometimes possible to modify the construction of the sequence and to obtain a faster convergence. However, such a procedure highly depends on the process by which the sequence is constructed. Another way for accelerating the convergence is to transform the sequence (S_n) into another one, (T_n), which can, under some assumptions, converge faster to the same limit.

One of the most popular acceleration method is certainly Aitken Δ^2 process defined by

$$T_n = S_n - \frac{(S_{n+1} - S_n)^2}{S_{n+2} - 2S_{n+1} + S_n}, \quad n = 0, 1, \ldots$$

The method was given by Alexander Craig Aitken (1895–1967) in 1926 [2] to accelerate the convergence of the Bernoulli method for computation of the dominant zero of a polynomial. Aitken pointed out that the same method was obtained by Hans Eduard von Naegelsbach (1838– ?) in 1876 in his study of the Furstenau method for the solution of nonlinear equations [193]. The process was also given by James Clerk Maxwell (1831–1879) in his book of 1873 [181]. However, neither Naegelsbach nor Maxwell used it for the purpose of convergence acceleration. In fact, the Aitken process was known to Takakazu Seki (1642?–1708), often considered the greatest Japanese mathematician. In his book *Katsuyō Sanpō*, vol. IV, he used this process to compute the value of π, the length of a chord and the volume of a sphere. This book was written around 1680 but only published in 1712 by his disciple Murahide Araki. Parts of it can be found in [138]. For more details, see [35]. A biography and autobiography of Aitken can be found in [3]. He was also a great mental calculator as explained in [236, pp. 266–275].

What makes the Aitken process so popular is that it accelerates the convergence of all linearly converging sequences, that is sequences such that $\exists a \neq 1$

$$\lim_{n\to\infty} \frac{S_{n+1} - S}{S_n - S} = a.$$

The Aitken process was the starting point for the study of other powerful convergence acceleration methods. The first generalization was the sequence transformation due to Daniel Shanks (1917–1996) [233] and the ε–algorithm of Peter Wynn [276] (which us allows to implement it in an easy way). The Shanks transformation is related to the Padé approximation, another important topic in applied mathematics [31,33]. For the history of convergence acceleration methods, see [35].

Convergence acceleration is based on the idea of extrapolation and is related to the knowledge of an asymptotic expansion of the error [264]. Some acceleration devices also lead to new methods outside the domain of convergence acceleration since they are connected to fixed-point iterations (the Steffensen method comes out from the Aitken process, for example) and other topics.

The theoretical and practical study of these methods now form an independent chapter in numerical analysis; see, for example, [273,75,36].

1934: Rémès algorithms

It can be said that numerical methods for the best approximation of functions (that is involving the minimization of a norm) started in 1934.

Let $f, \phi_1, \ldots, \phi_n$ be functions in $C[a,b]$, the set of continuous functions on $[a,b]$. The Chebyshev approximation consists in finding

$$\phi(x) = \sum_{i=1}^{n} a_i \phi_i(x)$$

which minimizes

$$\|f - \phi\|_\infty = \max_{x\in[a,b]} |f(x) - \phi(x)|.$$

For the case $\phi_i(x) = x^{i-1}$, it was proved in 1905 by Émile Félix Édouard Borel (1871–1956) that $a_1, \ldots, a_n$ is a solution if and only if there exist $n+1$ points in $[a,b]$ where the norm is attained with alternating signs

[28]. Moreover, this solution is unique. This result was extended to functions forming a Chebyshev system (which means that every linear combination of $\phi_i, \dots, \phi_n$ has at most $n-1$ zeros in $[a,b]$) by John Wesley Young (1879–1932) in 1907 [282]. In 1918, Alfréd Haar (1885–1933) proved that a necessary and sufficient condition for the uniqueness of the solution is that the ϕ_i's form a Chebyshev system [125]. An algorithm for the solution of this problem was given by George Pólya (1887–1985) in 1913 [203], but a counterexample to its convergence was found by Jean Descloux in 1963 [78].

In 1934, Evgeny Yakovlevich Remez (1896–1975) (or Rémès as he himself spells his name in French) proposed two algorithms for the numerical solution of this problem [209,210]. Its first algorithm applies to the general Chebyshev best approximation problem. However, its convergence can be extremely slow. The second algorithm exploits the alternation property and solves a sequence of discrete problems. It was proved by László Veidinger in 1960 [261] that it converges quadratically if f and the ϕ_i's are twice continuously differentiable.

Nowadays, approximation theory forms an important branch of numerical analysis. A detailed account of its history is given in [266]; see also [41,32].

1946: The Monte Carlo method

The idea of the Monte Carlo method came to the mind of Stanislaw Marcin Ulam (1909–1984) when he was playing solitaire after an illness. Instead of making complicated combinatorial computations about his chances, he had the idea of playing one hundred times and counting the number of successful plays [86,256]. The method was developed by him, John von Neumann and Nicholas Metropolis (1915–1999) [184]. It consists in simulating a physical or mathematical system by a sampling operation whose expectation or variance gives the behavior of the system. For example, the solution of a partial differential equation which is too difficult to compute by a more classical method (for example, the neutron transport equation or Schrödinger equation) can be obtained by a stochastic process whose probability function asymptotically satisfies the equation. Of course, Monte Carlo methods need the construction of random number generators.

1946: Splines

Spline interpolation consists of piecewise polynomial interpolation. Thus, a spline function is a piecewise polynomial whose values and the values of some of its first derivatives coincide at the interpolation points. Although the subject had been studied as early as 1906 by George David Birkhoff (1884-1944) [23], its development really starts in 1946 with two papers by Isaac Jacob Schoenberg (1903-1990) [228,229]. In [230], he reported the history of his discovery (see also [32]) and said

> *A spline is a simple mechanical device for drawing smooth curves. It is a slender flexible bar made of wood or some other elastic material. The spline is placed on the sheet of graph paper and held in place at various points by means of certain heavy objects (called "dogs" or "rats") such as to take the shape of the curve we wish to draw.*

Spline functions have many applications in computer-aided design. The story began when the French automobile company *Citroën* asked Paul de Casteljau in 1958 to develop a system for helping engineers to design cars. Casteljau had the idea of using Bernstein polynomials for this purpose but, due to secrecy, he was not allowed to publish his results. However, the secret was not total since, in the early sixties, Pierre Bézier (1910-1999), an engineer at *Renault*, heard that Citroën was developing such a project and he initiated a new approach to curves and surfaces approximation. It consists of piecewise polynomials controlled by polygonal lines. The system UNISURF used by Renault was described in several publications and this is why such curves and surfaces are now known as *Bézier curves* and *surfaces*. It was R. Forrest who proved in 1972, while translating Bézier's book [18], that both approaches were identical, since Bézier curves and surfaces can be formulated in terms of Bernstein polynomials [99].

Parametric splines, that is splines with certain differentiability constraints, were introduced in Computer Aided Geometric Design (CAGD) by J. Ferguson in 1963 who was working at Boeing Co [93]. These curves were studied simultaneously by Carl de Boor and W. Gordon at General Motors for the purpose of interpolating data. B-splines were introduced by de Boor in 1972 [71]. They are connected to Bézier curves.

The case of two variables began to be treated in the sixties by a tensor product approach. In 1959, Garrett Birkhoff (1911–1996) began his consulting work with General Motors Research. The problem was the representation of automobile surfaces in order to exploit the new numerically controlled milling machines for the cutting of dies needed for the stamping of outer and inner panels. It was necessary to determine the free parameters in a flexible mathematical model so as to fit closely to measurements taken from the finished physical model of the car. Birkhoff quickly recommended use of cubic splines for the representation of smooth curves. He was familiar with their use through his contact with the David Taylor Model Basin and also through the work done at Boeing. Together with Henry Garabedian, he developed a bivariate generalization of cubic spline interpolation for interpolating a C^1 surface to a given rectangular mesh of cubic splines [22]. This was a pioneer work in splines which led to many further developments [259,281]. De Casteljau constructed a surface from a net of control points [73, p.18] and Ferguson gave a method for interpolating vector-valued surface data by composite bicubic surfaces [93]. Bézier surfaces were introduced in the late sixties and they still remain, together with B-splines surfaces, one of the most popular schemes [26]. The triangular polynomial patches of de Casteljau [72, p.10] are widely used in the finite elements method.

Radial basis functions, which are now very much in flavor, can be considered as a generalization of splines to several dimensions [204].

Many historical notes on splines can be found in [231]; see also [32]. On the history of multivariate polynomial interpolation, see [108].

1947: The simplex algorithm

A *linear programming problem* consists in finding $x \in \mathbb{R}^n$ which minimizes (c,x) and, simultaneously, satisfies the constraints $x \geq 0$ and $Ax = b$, where $A \in \mathbb{R}^{m\times n}$, $b \in \mathbb{R}^m$ and $c \in \mathbb{R}^n$ are given. Such problems arise in planning activities in economics.

The constraints define a convex polyhedral set and the linear programming problem can be solved by a descent procedure along its outside edges. This was the approach suggested independently by Jean Baptiste Joseph Fourier (1768–1830) in 1824 [100], Charles de la Vallée Poussin (1866–1962) in 1911 [74] and Frank Lauren Hitchcock (1875–1957) in 1941 [139]. The *simplex algorithm* for solving linear programming problems was proposed by George Bernard Dantzig in 1947 [66]. It follows the same idea but looking at the problem in the geometry of the columns of the matrix A rather than its rows.

Linear programming was also independently studied by L.V. Kantorovich. In his Nobel prize presentation speech in economics in 1975 he said:

see: www.nobel.se/economics/laureates/1975/kantorovich-autobio-html

The thirties was also important for me as I began my first economics. The very starting point was rather accidental. In 1938, as Professor of the University, I acted as a consultant for the Laboratory of the Plywood Trust in a very special extreme problem. Economically, it was a problem of distributing some initial raw materials in order to maximize equipment productivity under certain restrictions. Mathematically, it was a problem of maximizing a linear function on a convex polytope. The well-known general recommendation of calculus to compare the function values in the polytope vertices lost its force since the vertices number was enormous even in very simple problems.

But this accidental problem turned out to be very typical. I found many different economic problems with the same mathematical form: work distribution for equipment, the best use of sowing area, rational material cutting, use of complex resources, distribution of transport flows. This was reason enough to find an efficient method of solving the problem. The method was found under influence of ideas of functional analysis as I named the "method of resolving multipliers".

In 1939, the Leningrad University Press printed my booklet called The Mathematical Method of Production Planning and Organization which was devoted to the formulation of the basic economic problems, their mathematical form, a sketch of the solution method, and the first discussion of its economic sense. In essence, it contained the main ideas of the theories and algorithms of linear programming. The work remained unknown for many years to Western scholars. Later, Tjalling Koopmans, George Dantzing, et al, found these results and, moreover, in their own way. But their contributions remained unknown to me until the middle of the 50s.

I recognized the broad horizons offered by this work at an early stage. It could be carried forward in three directions:

1) The further development of methods of solving these extremal problems and their generalization; their application to separate classes of problems;

2) A mathematical generalization of these problems such as, non-linear problems, problems in functional spaces, the application of these methods to extremal problems of mathematics, mechanics and technical sciences;

3) The spreading of the method of description and analysis from separate economic problems to general economic systems with their application to planning problems on the level of an industry, a region, the whole national economy as well as the analysis of the structure of economic indices....

Some activity took place in the first two directions (the results were published partly immediately, partly after the war), but the third one lured me most. I hope that the reasons were clarified enough in my Nobel lecture.

The studies were interrupted by the war. During the war, I worked as Professor of the Higher School for Naval Engineers. But even then I found time to continue my deliberations in the realm of economics. It was then that I wrote the first version of my book. Having returned to Leningrad in 1944, I worked at the University and at the Mathematical Institute of the USSR Academy of Sciences, heading the Department of Approximate Methods. At that time, I became interested in computation problems, with some results in the automation of programming and in computer construction.

My economics studies progressed as well. I particularly wish to mention the work done in 1948–1950 at the Leningrad Carriage-Building Works by geometrist V. A. Zalgaller under my guidance. Here the optimal use of steel sheets was calculated by linear programming methods and saved material. Our book of 1951 summarized our experience and gave a systematic explanation of our algorithms including the combination of linear programming with the idea of dynamic programming (independently of R. Bellman).

On the history of the simplex method, see [67]. See also [173] and [5, pp. 61–79].

More recently, a new algorithm for linear programming was proposed by Narendra Karmarkar [153]. It was claimed to be faster than the simplex algorithm but, however, it was received with scepticism by some researchers because of lack of evidence for this property [96, p. 185].

1950: Nonlinear programming

The birth of nonlinear programming is dated from 1950 with the publication of a paper by Harold W. Kuhn and Albert W. Tucker (1905–1995) [160]. They gave a necessary and sufficient condition for the existence of an optimal solution to the problem of minimizing $f(x)$ subject to the constraints $g_i(x) \leq 0$ for $i = 0, \ldots, m$ and $x \in X \subseteq \mathbb{R}^n$. This paper attracted much attention while a similar result obtained by William Karush in his Master's Thesis in 1939 [154] under the supervision of Lawrence M. Graves at the University of Chicago and by Fritz John (1910–1995) in 1948 [147] were almost totally ignored (John's paper was even rejected).

For a complete discussion, see [155] and [156].

1950: Successive overrelaxation method

After having discretized, by finite differences, a linear partial differential equation (usually of elliptic type), one has to solve a system of linear equations. For a discretization of fifty points in both direction, a system of 2500 equations is obtained. Around 1945, this was much too large for applying a direct method and *relaxation methods* (such as those of Liebmann, Richardson, Jacobi, Gauss–Seidel and Southwell), were used. However, their convergence was slow, as Leslie Fox showed in his D. Phil. Thesis in 1942 under Sir Richard Vynne Southwell (1888-1970) [101]. When David Young asked Garrett Birkhoff for a thesis subject, he was offered to work on this topic. He made two key observations: for systems with a *consistently ordered* matrix, the eigenvalues of the Gauss–Seidel method are the squares of those of the Jacobi method and, instead of relaxing (that is modifying the ith component of the approximate solution for annihilating the ith component of the residual), it is better to introduce a parameter and *overrelax*. For some type of matrices, Young found the expression for the optimal value of the overrelaxation parameter (that is the value minimizing the spectral radius of the iteration matrix). It is connected to the square of the spectral radius of the Jacobi iteration matrix. The method, also studied simultaneously by Stanley P. Frankel [104], was called *successive overrelaxation* (SOR). Young defended his thesis in 1950 [279]. On the history of SOR, see [280].

Following this early work, a whole theory was built and other methods were found and studied. The method was successfully applied to the solution of many partial differential equations, which certainly had quite a great impact on the future development of numerical analysis. On these methods, see the book of Richard Stephen Varga, whose first edition goes back to 1962 and is certainly one of the most frequently quoted books of the numerical analysis literature [260].

1951: Continuation methods

For solving a system of nonlinear equations, one needs tools for localizing the area of the expected solution and for constructing acceptable data for starting an effective iteration process. Numerical continuation methods include a variety of topics such as embedding methods, homotopy methods, parameter variation methods, incremental methods, piecewise linear methods. The basic concept of the so-called incremental methods goes back to E. Lahaye in 1934 [169]. The use of continuation by differentiation started in about 1952 by Dimitrii Fedorovich Davidenko (born in 1922) [69]. Several variants of continuation methods were among others investigated in the 1970's by Herbert Bishop Keller (born in 1923), Eugene L. Allgower and Werner Carl (born in 1927) Rheinboldt (see [211] for references). Software packages (e.g. HOMPACK, MANPACK, ALCON, BIFPACK) exist that are based on continuation methods. For a survey we refer to the paper by Rheinboldt [211]. See [6] for an extensive study.

1952: Lanczos method

A *projection method* for solving the $n \times n$ system $Ax = b$ consists in constructing a sequence of iterates $x_k \in \mathbb{R}^n$ such that

$$\begin{aligned} x_k - x_0 \quad & \in \mathcal{K}_k, \\ r_k = b - Ax_k \quad & \perp \mathcal{L}_k, \end{aligned}$$

where $\mathcal{K}_k$ and $\mathcal{L}_k$ are two subspaces of dimension k of $\mathbb{R}^n$. When $\mathcal{K}_k = K_k(A, r_0) = \text{span}\,(r_0, Ar_0, \ldots, A^{k-1}r_0)$, the method is called a *Krylov subspace* method. If, in addition, $\mathcal{L}_k = K_k(A^T, y)$ where y is an (almost) arbitrary vector, we find the method obtained by Cornelius Lanczos (1893–1974) in 1952 [171]. It was Aleksei

Nikolaevich Krylov (1863–1945) who first used this type of subspace in 1931 to construct the characteristic polynomial of a matrix [157]. Lanczos encountered the mathematical problem when he was doing research for the Boeing Aircraft Company. Around this time, he wrote to Albert Einstein (1879–1955) [112]

The reason I am interested very much in dealing with methods of approximation is not the practical applicability of the solution but rather the fact that a very "economical" solution is possible only if it is very "adequate" too. To get a solution in a few steps means nearly always that one has found a method which is consistent with the intrinsic nature of the problem.

Einstein replied

Your remark concerning the significance of the adapted approximation methods is very enlightening, and I am convinced that it represents not merely a practical method but also is a promising mathematical viewpoint.

This was really a prophetic view!

In an interview given in 1974 [152], only some time before he died, Lanczos was asked

What would you say has been the most important, the most fundamental and essential aspect of your sixty years of work?

He answered

I believe my most important contribution was in the fields of mathematics, to be precise, in numerical analysis – my discovery of a method now known as the Lánczos method. It is very little used today, because there are now a number of other methods, but it was particularly interesting in that the analysis of the matrix could be carried out, that is, all the eigenvectors could be obtained by a simple procedure.

The main advantage of projection methods is that they only require products of vectors either by A and A^T, and the computation of scalar products. So, they are cheap and easy to implement.

When the matrix A is symmetric positive definite, the vector $x = A^{-1}b$ is the unique vector minimizing the strictly convex quadratic functional

$$J(u) = \frac{1}{2}(u, Au) - (u, b).$$

This minimum can be reached by a descent method such as the *conjugate gradient* algorithm proposed by Magnus Rudolph Hestenes (1906–1991) and Eduard Ludwig Stiefel (1909–1978) in 1952 [134]. Hestenes was working at the *Institute for Numerical Analysis* (INA) at UCLA. In June and early July 1951, he derived the conjugate gradient algorithm. When, in August, Stiefel arrived at INA from Switzerland for a congress, the librarian gave him the routine written by Hestenes. Shortly after, Stiefel came to Hestenes office and said about the paper "this is my talk!". Indeed, Stiefel had invented the same algorithm from a different point of view. So, they decided to write a joint paper on the algorithm and its properties [134]. In the meantime, Lanczos, an Hungarian born physicist (he worked with Einstein) and mathematician, was a member of INA. He soon observed that the conjugate gradient algorithm could be derived from his biorthogonalization algorithm for computing eigenvalues of matrices [170]. Then, he derived from it his *method of minimized iterations* for solving a system of linear equations [171]. Let us also mention that the Lanczos method is also connected to a generalization of the Aitken Δ^2 process called the topological ε–algorithm [30, pp. 184–189] and, so, to Padé approximants.

The historical development of gradient methods has been described in several publications; see [133], the preface of [132], and [135, pp. 48–65]. An annotated bibliography with historical comments can be found in [119]. For more recent developments, see [224].

Krylov subspace methods have a finite termination property, that is $\exists k \leq n$ (the dimension of the system) so that $r_k = 0$ and $x_k = x$. This is, in particular, true for the Lanczos method and the conjugate gradient algorithm. In fact, this property played against these methods for quite a long time since, indeed, if n is very large, it is not suitable to be forced to perform so many iterations to obtain the solution of the problem. However, around 1960, John K. Reid [208] observed that, in many cases, the conjugate gradient algorithm gives a very good approximation of x in a few iterations. This remark was the rebirth of the conjugate gradient algorithm. In his 1952 paper, Lanczos treated the case of any matrix while the conjugate gradient algorithm was restricted to symmetric positive definite ones. This algorithm was extended to the general case and put into an algorithmic form by Roger Fletcher in 1974 [95]. It is the *biconjugate gradient* algorithm (BiCG) which allows us to implement the Lanczos method. Other algorithms for this implementation exist. They can be derived by purely linear algebra techniques or obtained from the theory of *formal orthogonal polynomials* as reviewed in [38]. Then, the BiCG gave rise to a vast literature and was widely used. It can suffer from drawbacks as the loss of orthogonality (due to the finite precision of the computations), division by zero (breakdowns) or by a numbers close to zero (near-breakdown). Such problems could be tackled by various techniques, some of them based again on formal orthogonal polynomials; see, for example, [37].

More recently, transpose-free variants of Lanczos methods were defined: the CGS in 1989 by Peter Sonneveld [237] and the Bi-CGSTAB in 1992 by Henk Van der Vorst [257]. The GMRES obtained by Yousef Saad and Martin H. Schultz in 1986 [223] is also a Krylov subspace method. Together with Bi-CGSTAB, it is the most used method for solving linear systems.

Let us mention that a topic which has been much developed these last decades is that of orthogonal polynomials. It has important applications in interpolation, best approximation, Padé approximation, quadrature and cubature, wavelets, linear algebra, etc. A review of their applications and the numerical problems posed by their computation in given in [110]. For cubature and quadrature, see [57,172].

1953: Interior point methods

In mathematical programming, interior-point methods had a huge impact on theory, practice and computation, since the development of the simplex method. Various interior-point methods were considered in the 1950s and were investigated extensively in the 1960s. It was, however, a paper by N. Karmarkar in 1984 that made them of interest for many researchers [153]. Working with the primal problem in linear programming was easier to implement or had better complexity bounds. A new area was inaugurated by N. Megiddo in 1987 [182] giving a framework for primal–dual algorithms. It allowed extensions to convex programming and linear complementarity. Also, new algorithms with interesting properties could be obtained, e.g., a predictor–corrector algorithm by S. Mehrotra in 1989 [183]. More information on interior-point methods can be found in the survey paper by Florian A. Potra and S.J. Wright in [275].

1955: Romberg method

It is well known that, for a sufficiently smooth function f, the asymptotic expansion of the error of the trapezoidal rule for computing an approximate value of

$$I = \int_a^b f(x)\,dx$$

is given by the Euler–Maclaurin formula. In 1955, Werner Romberg gives a procedure for eliminating the successive terms in this asymptotic expansion and, thus, improving the accuracy of the trapezoidal rule [216].

It is based on the extrapolation process due to Lewis Fry Richardson (1881–1953) [213,214], itself deriving from the Neville–Aitken scheme (1932) for computing the interpolation polynomial. For more details about the history of this process, see [84] and [35, pp. 451–453].

This method can now be found in every textbook on numerical analysis without any reference to Romberg's original paper, a proof of true fame! Besides its usefulness, the Romberg method also showed to non-specialists that convergence acceleration methods can be quite powerful. Thus, it had an important impact on the development of this domain of numerical analysis. The E–algorithm, which is the most general convergence acceleration procedure presently known, is an extension of the Romberg method; see [35]. A French translation of Romberg's original paper can be found in [46, pp. 501–503] and an English one in [47, pp. 451–453].

1956: The finite element method

It can be quite difficult to assign a year to the birth of an idea. This is particularly true for the *finite element method* for solving partial differential equations (PDEs). Until the mid fifties and the beginning of the sixties, PDEs were discretized by finite differences on rectangular grids. When the domain is not a rectangle some problems for following its shape arise and it will be easier to use triangles instead of rectangles. Finite elements are based on triangles. However, the finite element method, which uses triangles, is more than that and contains several ingredients. It is based on a triangularization of the domain, on a basis of functions with finite support, on the assemblage of the triangles together, and on a variational formulation of the problem in the spirit of the Rayleigh–Ritz method. The paper by M.J. Turner et al. [255] of 1956 is usually considered as the beginning of finite elements. It deals with a local approximation and uses an assembly strategy. However, the results were derived without the use of a variational formulation. Finite elements really start with the paper by Ray William Clough, Jr. (born in 1920) in 1960 [51] on linear plane elasticity problems. From 1960 to 1965, a variety of finite elements were proposed and studied, mostly in the engineering community. Spline functions, which form an essential part of the finite element method, were developed around the same time. The mathematical theory was slower to emerge from this intense activity. The first papers in this direction are those of Richard S. Varga [258] and Garrett Birkhoff (1911–1996) et al. [21] in 1966, see [4, pp.1–15]. Two-dimensional problems began to be studied in 1968 and the paper by M. Zlamal [284] attracted the attention of many mathematicians who started working on the topic. One could mention, in particular, a convergence result obtained by Philippe G. Ciarlet in 1968 [49]. By 1972, the finite element method had become a major area in numerical analysis and applied mathematics, see [50].

The history of finite elements is given in [198,199,283]. See also [243].

1957: The fortran optimizing compiler

The first high-level programming language was FORTRAN (FORmula TRANslation) which was developed between 1953 and 1956 by John Backus and his team. The first compiler was written in 1957 for the IBM 704 computer and it contained 25,000 instructions. The immediate success of FORTRAN was due to the fact that it was very close to mathematics and it certainly helped the popularization of numerical analysis methods in the natural sciences community. Nowadays, computer scientists find that, although several versions with many improvements followed, it remains a language with many weaknesses. However, there are so many FORTRAN codes everywhere in the world that nobody will even try to rewrite them in a better language, and it will certainly still remain in use for a long time.

On Backus, see [234, pp. 5–20].

1957: Error analysis

When making numerical computations, there is an important problem which was soon realized by pioneers. It concerns the finite precision of floating-point computations. In a computer, each arithmetical operation is only performed with a finite number of digits (as by hand where, for example, only a certain number of 3 are kept when dividing 1 by 3). These *rounding errors* can accumulate in a catastrophic way and the result obtained could have nothing to do with the true one. In that case, the algorithm suffers from *numerical instability*. Upper bounds on these errors can be obtained. Such inequalities were first derived by James Hardy Wilkinson (1919–1986) [270] (see [200]). It is also possible to study these errors by a *backward error analysis*, an approach which consists in finding which perturbation on the data would have given the same results in exact arithmetic. This approach was implicitly contained in the papers of von Neumann and Herman H. Goldstine in 1947 [263] and Turing [252] in 1948. It was explicit in a report by J. Wallace Givens, Jr. (1910–1993) in 1954 [113], but it was mainly developed by Wilkinson. His book on the subject [271] was a classic for many years (see [137] for a more recent reference).

However, this approach is a pessimistic one since, hopefully, rounding errors can compensate. A more recent approach, due to Michel La Porte and Jean Vignes [167], is a statistical one. It is based on stochastic arithmetic and it provides a more realistic knowledge of rounding errors. The software CADNA allows us to implement this approach [48]. Of course, in practice, it is not enough to evaluate rounding errors and it is better to control or even to try to avoid them. For this purpose, several ways have been explored: correction of the arithmetic [202], p–adic arithmetic [121,122], multiprecision arithmetic, and interval arithmetic with guaranteed and sharp inclusion regions for the exact result [161].

There is another notion related to the finite precision of the computer: it is the concept of *condition number* which measures the sensitivity of the exact result of a mathematical problem to variations in the data. This concept should not be confused with the numerical stability of an algorithm for solving the problem. Although, this concept was clear for a long time, the term was first used by Turing [252].

1959: The QR factorization

Triangularization of a matrix by unitary similarity transformations seems to have been used for the first time in the book by Herbert Westren Turnbull (1885–1961) and Aitken in 1932 [254] in the proof of a result by Issai Schur (1874–1941) [232]. These transformations also appear in the book of William Feller (1906–1970) and George Elmer Forsythe (1917–1972) [92] in 1951. However, the systematic use of orthogonal transformations to reduce matrices to a simpler form was initiated by Givens in 1958 [114]. Reflectors were first used as an efficient tool for generating orthogonal matrices by Alston Scott Householder (1904–1993) in 1958 [141,142].

In 1961–1962, J.F.G. Francis [103] and Vera N. Kublanovskaya [158] independently turned Householder's factorization into a method for computing the eigenvalues of a matrix. It is the *QR*–algorithm which is in the style of the *LR*–algorithm obtained in 1955 by Heinz Rutishauser (1918–1970) [222]. See [120] for historical comments.

Whatever the form of an eigenvalue problem, the *QR* algorithm is likely to be useful. For example, for generalized eigenvalue problems $Ax = \lambda Bx$, the *QZ* algorithm [189] is a variant of *QR*. Another variant is used to calculate the singular value decomposition (SVD) of a matrix (see below) and *QR* can also be helpful for large eigenvalue problems . The application to the linear least squares problem is due to Gene Howard Golub in 1965 [117]. An extensive treatment is given in [24]. The use of *QR*–methods to solve least squares problems is mentioned in [197]. *QR*–based codes are used in the software packages EISPACK and LINPACK. A parallel version of *QR* can be found in ScaLAPACK.

For technical and historical details, see [239] and the paper of D.S. Watkins [265].

1960: Multigrid methods

The idea of the multigrid was introduced and analyzed by H. Brakhage [29] and R.P. Fedorenko [91], but only got greater attention in the 1970's. The central principle is based on the interplay of smoothing and coarse grid correction. These techniques can be applied to a variety of partial differential equations from Poisson equations to Navier–Stokes equations, from two-dimensional square domains to three-dimensional unstructured airfoil grids. Several forms of multigrid methods appear in the literature. We refer to the paper by K. Stueben [241] in this volume for a review of algebraic multigrid. The multigrid method has proved itself as a powerful tool for fast and efficient computations.

1961: Fractals and chaos

Until the middle of the 19th century, mathematicians were thinking that a continuous function had a derivative at almost all points. In 1872, Karl Theodor Wilhelm Weierstrass (1815–1897) defined a continuous function having no derivative at any point. In 1883, Georg Ferdinand Ludwig Philipp Cantor (1845–1918) gave the example of a segment of length, one to which an infinite number of segments of total length was subtracted and so that a non-denumerable set of points remained. The mathematicians of this time were forced to admit these examples but thought that they showed the limits of classical analysis and considered them, as Jules Henri Poincaré (1854–1912) said, only as monsters for the museum of horrors. Let us recall that, when correcting an error in the work for which he obtained the Prize offered by Oskar II, the King of Sweden and Norway, in 1889, Poincaré discovered the chaotic behavior of the solution of some differential equations when the initial conditions are changed.

A curve has dimension 1 since one parameter is sufficient to locate each of its points (the distance from a point of the curve chosen as the origin). A surface has dimension 2 since two parameters are needed for defining each of its points, and so on. In 1890, Giuseppe Peano (1858–1932) constructed a curve (of dimension 1) passing through every point in a square (of dimension 2). Thus, the definition of the dimension from the number of parameters required was unsatisfactory. In 1906, Niels Fabian Helge von Koch (1870–1924) gave a closed curve (known as the *snowflake*) with infinite length enclosing a finite area. Thus, the question arises to assign a dimension to such objects. If you have a rule, you need two of them to obtain a rule with a double length. If you have a square, you need four of them to have a square with a side of double length, and so on. More generally, for multiplying by a, the side of an object of dimension d, you need $c = a^d$ similar objects where d is the dimension. Thus $d = \log c/\log a$. This number d is called the *fractal dimension* and, if it is not an integer, the object is called a *fractal*. Applying this definition to the snowflake leads to $d = 1.2618\ldots$

Lewis Fry Richardson (the same responsible for the Richardson extrapolation method) remarked that if you measure the length L of a coast, or that of a boarder between two countries, the result depends on the length a of the instrument chosen for the measurement. Indeed, with a smaller rule you can measure smaller details. He found by experiments that $L(a) = a^{1-d}$ with $d = 1.25$ for the coast of Britany and $d = 1.15$ for the boarder between Portugal and Spain. Applying the same technique to the snowflake gives again $d = 1.2618\ldots$

The notion of fractal dimension was the subject of quite intensive research when physicists found that many objects, such as clouds, have a fractal dimension. In fact, fractals are everywhere in nature [14].

Another problem is the dynamics of the fixed point iterations $x_{k+1} = F(x_k)$, $k = 0, 1, \ldots$ with x_0 given. Such iterations model the evolution of a system from the time t to the time $t + dt$. Thus, to predict its evolution over a long time, many iterations are needed. In practice, the law F is only known approximately and only stable phenomena (that is those not very sensitive to small changes in F) are of interest. These iterations can exhibit a quite chaotic behavior. Let us consider the simple iterations $x_{k+1} = x_k^2 + c = F_c(x_k)$ in the complex plane. F_c has two fixed points $(1 \pm \sqrt{1-4c})/2$. If $c \neq 1/4$, one of them is attractive and the other one repulsive. According to the choice of x_0, the sequence (x_n) can converge or not. Let $\mathcal{A}_c(\infty)$ be the set of $x_0 \in \mathbb{C}$ such that $|x_k|$ tends to infinity with k and let J_c be the frontier of $\mathcal{A}_c(\infty)$. J_c is called the *Julia set* of F_c. Obviously, J_c depends on the value of the complex number c. The first research about such topics goes back to the beginning of the twentieth century with the works of Gaston Maurice Julia (1893–1978) in 1918 [149], Pierre Joseph Louis Fatou (1878–1929) in 1919–1920 [90], and others. However, these pioneers were not able to see the full complexity (and beauty) of the problem which was only realized with the first numerical computations; see [201]. In fact, to the surprise of the first experimenters, J_c can change very much with c and can be very complicated, in fact it can be a fractal. Let $\mathcal{K}_c$ be the complement of $\mathcal{A}_c(\infty)$ in the complex plane. It was proved that either $\mathcal{K}_c$ is a connected set, or it consists of an infinite number of distinct points. It is a connected set if and only if it contains the origin. The *Mandelbrot set* $\mathcal{M}$ of F_c is the set of values of c so that $\mathcal{K}_c$ is connected. The set $\mathcal{M}$ can be a fractal. Many of these notions were introduced and studied, from 1961, by Benoît Mandelbrot who wrote many books on the subject (see, for example, [179]).

On chaos and fractals, see [115]. An interview of Mandelbrot can be found in [4, pp. 206–225].

1963: A-stability

The first paper on stiff differential equations is due to C.F. Curtiss and J.O. Hirschfelder in 1952 [60]. The concept of A-stability, a very fundamental one for the numerical integration of stiff ordinary differential equations, was introduced by Germund Dahlquist around 1963 [62]. He showed that the order of an A-stable linear explicit multistep method cannot exceed two. It was also proved that explicit Runge–Kutta methods cannot be A-stable. In fact, as explained in [63], the whole story began in 1951 when Dahlquist and Rutishauser met at a GAMM meeting at Freiburg im Breisgau and discovered they found quite similar results on the numerical (in)stability of the leapfrog method [61,221].

Two ways of research emerge from these results: the search for requirements less limitative than A-stability, and the study of implicit methods; see [111]. The concept of A-stability is a major one. It has generated many researches and an enormous literature on the subject is available; see, for example, [127,144].

1964: Symbolic computation

Symbolic computing, also known as *computer algebra*, consists in the manipulation of mathematical formulae as done by hand with a pencil and a paper. Although the roots of computer algebra go back to the 1950's, the first system, named, ALPAK, was developed in 1964 at Bell Laboratories. It was followed by ALTRAN in 1968, REDUCE in 1968, and MACSYMA in 1970, which was the most influential of these packages. MAPLE was conceived in November 1980 and soon became very popular. Nowadays, symbolic computing is often coupled with numerical algorithms for better efficiency. Computer algebra systems are based on deep mathematics such as Gröbner bases, introduced in 1965 by Bruno Buchberger [40] and integration in finite terms, due to Joseph Liouville (1809–1882) [177, pp. 351–422].

Let $f_n = f(nT/N)$ be the values of a given function f at equidistant points. The discrete Fourier transform of f is the function $\widehat{f}$ whose values at the points $\omega_k = k/T$ are given by $\widehat{f}(\omega_k) = TF_k$ with

$$F_k = \frac{1}{N}\sum_{i=0}^{N-1} f_i \omega_N^{-ik}$$

with $\omega_N = e^{i2\pi/N}$.

The computation of $F_0, \ldots, F_{N-1}$ by these formulae requires $(N-1)^2$ complex multiplications and $N(N-1)$ complex additions if the quantities ω_N^j are known. This is too much for a real time application.

It was remarked by John Wilder Tukey (1915–2000) in 1963 that if N has the form $N = n_1 n_2$, then the series can be expressed as an n_1–term series, each term being itself a subseries with n_2 terms. This trick, which is the basic idea for the so-called *fast Fourier transform* (FFT), reduces the number of operations from N^2 to $N(n_1 + n_2)$. Moreover, in some cases, the process can be iterated and the number of operations still reduced to $N \log N$. Indeed, if $n_1 = 2$ and if we set $n_2 = m$, we have

$$F_k = \frac{1}{2}(F_k^e + \omega_N^{-k} F_k^o)$$

with the even and odd parts respectively given by

$$\begin{aligned} F_k^e &= \frac{1}{m}\sum_{i=0}^{m-1} f_{2i}\omega_{N/2}^{-ik}, \\ F_k^o &= \frac{1}{m}\sum_{i=0}^{m-1} f_{2i+1}\omega_{N/2}^{-ik}. \end{aligned}$$

Thus,

$$F_{k+m} = \frac{1}{2}(F_k^e - \omega_N^{-k} F_k^o)$$

and the procedure can be repeated if m is even.

Soon after, James William Cooley wrote a three-dimensional FFT program. The main difficulty was to save storage by overwriting data using a clever indexing scheme. Because of the many applications of the technique in industry, the possibility of a patent was discussed, but it was decided to put the FFT into the public domain. Cooley asked Tukey to be a coauthor of the paper which appeared in 1965 [56].

The idea of the FFT did not emerge from nothing. Early contributors are mentioned in [54] (see also the references quoted there) and, among them, Carl Friedrich Gauss (1777–1855) [109], H. König and C. Runge in 1924 [220], and G.C. Danielson and C. Lanczos in 1942 [65] (see [55]. In particular, Gauss discovered the FFT in 1805, at the age of 28, two years before Fourier's paper [129].

The FFT has very many applications in the physical sciences, in particular in signal processing, and many computations would not have been possible without it, even with the fastest present computers.

1965: The singular value decomposition

Let $A \in \mathbb{R}^{n\times m}$ be a rectangular matrix of rank r. There exist real numbers $\sigma_1 \geq \cdots \geq \sigma_r > 0$ and two orthogonal matrices $V = [v_1, \ldots, v_m] \in \mathbb{R}^{m\times m}$ and $U = [u_1, \ldots, u_n] \in \mathbb{R}^{n\times n}$ such that

$$\begin{array}{ll} Av_i = \sigma_i u_i,\ i = 1, \ldots, r & A^T u_i = \sigma_i v_i,\ i = 1, \ldots, r \\ Av_i = 0, \quad i = r+1, \ldots, m & A^T u_i = 0, \quad i = r+1, \ldots, n. \end{array}$$

Thus, the vectors $v_1, \ldots, v_m$ are the eigenvectors of A^TA and $u_1, \ldots, u_n$ are those of AA^T corresponding to the nonzero eigenvalues $\sigma_1^2, \ldots, \sigma_r^2$ of A^TA and AA^T.

Let Σ be the $n \times m$ matrix

$$\Sigma = \begin{pmatrix} \hat{\Sigma} & 0 \\ 0 & 0 \end{pmatrix}$$

with $\hat{\Sigma}$ =diag$(\sigma_1, \ldots, \sigma_r)$. Thus

$$A = U\Sigma V^T = \sum_{i=1}^{r} \sigma_i u_i v_i^T.$$

This formula is called the *singular value decomposition* (SVD) of A and the numbers σ_i are called its *singular values*. It is a very valuable tool in numerical linear algebra and its discovery goes back to Eugenio Beltrami (1835–1900) [15] in 1873 and Marie Ennemond Camille Jordan (1838–1922) [148] in 1874. See [238], for the complete early history.

From the numerical point of view, an algorithm for computing the singular value decomposition is due to Gene Howard Golub and William Kahan in 1965 [118]; it is based on QR factorization.

1965: Quasi-Newton methods

Let us consider the system of nonlinear equations $f(x) = 0$ where $f : \mathbb{R}^n \longmapsto \mathbb{R}^n$. The Newton method consists of the iterations

$$x_{k+1} = x_k - [f'(x_k)]^{-1} f(x_k), \quad k = 0, 1, \ldots$$

where $f'(x_k)$ is the Jacobian matrix of f at the point x_k. Thus, the Newton method requires the analytical expressions of n^2 partial derivatives, sometimes an impossible task. In quasi-Newton methods, $f'(x_k)$ (or its inverse) is replaced by an approximation and the iterations take one of the forms

$$\begin{aligned} x_{k+1} &= x_k - H_k^{-1} f(x_k), \\ x_{k+1} &= x_k - C_k f(x_k). \end{aligned}$$

These methods were introduced by William Davidon [70] at the end of the sixties for the minimization of a convex nonlinear functional (thus for the solution of an unconstrained optimization problem). If the functional is quadratic, this algorithm, when applied with exact minimization, is equivalent to the conjugate gradient algorithm as showed by Roger Fletcher and Mike J.D. Powell [97]. It is known as the Davidon–Fletcher–Powell algorithm (DFP). Soon after, another algorithm was proposed by Fletcher and Colin M. Reeves [98]. These methods gained full recognition after the methods proposed by Charles G. Broyden in 1965 [39] and the condition obtained by John E. Dennis and Jorge J. Moré in 1974 for their superlinear convergence [77]. In these methods, the approximation of the Jacobian (or of its inverse) is updated at each iteration by a rank-one or a rank-two modification. See [42] for some historical account.

There exist many methods of this type. They are very much in favor in optimization; see, for example, [96]. Their historical development is analyzed in [278]. Truncated Newton methods are reviewed in [195].

1969: Fast matrix manipulations

Let us consider the product of two $n \times n$ matrices $A = BC$. The elements of A are given by the classical formula

$$a_{ij} = \sum_{k=1}^{n} b_{ik} c_{kj}, \quad i, j = 1, \ldots, n.$$

Thus, the computation of A requires n^3 multiplications and $n^2(n-1)$ additions. This is not optimal and another formula was proposed by Volker Strassen in 1969 [240]. It requires $(7/8)n^3$ multiplications and $(7/8)n^3 + (11/4)n^2$ additions. If $n = 2^k$, the algorithm can be used repeatedly and the total number of arithmetical operations comes down to $4.7n^{2.807}$. This procedure can be used in direct methods for the solution of systems of linear equations and the number of operations goes from $\mathcal{O}(n^3)$ to $\mathcal{O}(n^{2.807})$ or even less.

This branch of numerical analysis is called *complexity*. It is an important challenge to find algorithms for reducing the number of arithmetical operations (such as in the FFT), to demonstrate that they have a lowest bound that cannot be passed and to find the corresponding algorithm if it exists. On this topic, see [248].

1976: High-performance computing and software packages

As stated in the introduction of [80], the first commercially available high-performance computer was the CRAY 1, delivered at Los Alamos in 1976. It was a vector machine. The first commercially viable parallel computer was the ALLIANT in 1985 but more highly parallel machines only appeared on the market in 1988. These architectures had, and still have, a quite important impact on the search and development of numerical algorithms, in particular in numerical linear algebra.

The increasing availability of advanced-architecture computers has had a significant effect on scientific computing, including algorithm research and software development especially in numerical linear algebra. The EISPACK [235], and LINPACK [79] software libraries were designed for supercomputers used in the 1970's and early 1980's. MATLAB was elaborated as a teaching tool during the period 1977–1984 by Cleve B. Moler at the University of New Mexico. The development of LAPACK [8] in the late 1980's was intended to make the EISPACK and LINPACK libraries run on shared memory, vector supercomputers. In 1991 began the development of ScaLAPACK [25] for distributed memory concurrent computers.

For more details on the interaction between algorithms and software, see [81] in this volume. See also [27].

1982: Wavelets

Fourier analysis has proved to be very useful in many branches of applied sciences. However, an expansion into a Fourier series is non local, which means that, for a given value of the variable, it needs the summation of an infinite number of terms. This drawback disappears if the basis functions have compact support. However, it is difficult to construct such a basis. The drawback partly disappears if the basis functions are rapidly (that is exponentially) decreasing outside a compact interval. In that case, a small number of terms will be sufficient for obtaining $f(x)$ with high precision. The aim of wavelets is to construct such a basis. There is another drawback which wavelets remedy: the Fourier transform hides information on time. Indeed, it gives the frequencies contained in the signal but does not say anything about the time when these frequencies are emitted. However, information on time is not lost but only hidden in the phases. Thus, a sum of sine and cosine functions can be amplified or cancelled out if they are simultaneous in time or not. Moreover, it is not possible to compute the phases from the Fourier coefficients with sufficient accuracy. To remedy such drawbacks, windows were introduced into Fourier analysis by Dennis Gabor (1900–1979) in 1946 [106] (Nobel prize winner in physics in 1971 for the discovery of holography). They consist in decomposing the signal into frequencies for various intervals. The size of the window does not change but is successively filled up by oscillations with various frequencies. However, if the window is too narrow, only picks and discontinuities are localized and the analysis becomes blind to low frequencies whose period is too big for the window. On the other hand, if the window is large, it is not possible to be precise in time when a discontinuity or a pick arises because information is too much diluted. Wavelets do not have these drawbacks and are satisfactory both in frequency and time. Instead of keeping the size of the window fixed and to vary the frequencies, the key idea consists of the converse: to keep the frequencies fixed and to change the size of the window. Of course, this idea was in the air for some time and it was, in fact, used independently in various branches of applied sciences. However, the starting point can be attributed to Jean Morlet, a French geophysicist, in 1982. From then, the theory and applications of wavelets developed very rapidly and an enormous literature on the subject is available. Among the numerous references, one can mention [68].

For an account of the subject and its historical developments, see [43] and [145]. A review is given in [52]. Fourier series and wavelets are presented, together with their history, in [150].

Acknowledgements: Thanks are due to all contributors who sent us information about the courses in numerical analysis in their country. Without their help, it would not have been possible to write the Section on this topic. Special thanks are also due to Philip J. Davis, Walter Gautschi and Frank W.J. Olver for their comments on an earlier version. We acknowledge the improvements offered by some of the participants to a meeting held in Luminy in October 2000. We also would like to thank all authors who accepted the challenge of writing the papers contained in the 7 volumes of J. Comput. Appl. Math. and the staff of North-Holland, in particular Keith Jones, for their support and efficiency during this project.

References

[1] M. Abramowitz, I.A. Stegun, eds., Handbook of Mathematical Functions with Formulas, Graphs and Mathematical Tables, National Bureau of Standards, U.S. Government Printing Office, Washington, 1964.
[2] A.C. Aitken, On Bernoulli's numerical solution of algebraic equations, Proc. R. Soc. Edinb., 46 (1926) 289–305.
[3] A.C. Aitken, To Catch the Spirit, University of Otago Press, Dunedin, 1995.
[4] D.J. Albers, G.L. Alexanderson, eds., Mathematical People. Profiles and Interviews, Birkhäuser, Boston, 1985.
[5] D.J. Albers, G.L. Alexanderson, C. Reid, eds., More Mathematical People. Contemporary Conversations, Harcourt Brace Jovanovich, Boston, 1990.
[6] E.L. Allgower, K. Georg, Numerical Continuation Methods. An Introduction, Springer-Verlag, Berlin, 1990.
[7] L. Amerio, Mauro Picone, Rend., Sci. Mat. Appl., A, 112 (1978) 2–8.
[8] E. Anderson, Z. Bai, C. Bischof, J. Demmel, J.J. Dongarra, J. Du Croz, A. Greenbaum, S. Hammarling, A. McKenney, S. Ostrouchov, D. Sorensen, LAPACK User's Guide, 2nd ed., SIAM, Philadelphia, 1995.
[9] R.C. Archibald, Conference on Advanced Computation Techniques, MTAC, 2 (1946–1947) 65–68.
[10] I. Babuška, The development of computational mathematics in Czechoslovakia and the USSR, in: A History of Scientific Computing, S.G. Nash ed., Addison-Wesley, Reading, 1990, pp. 314–328.
[11] C.T.H. Baker, A perspective on the numerical treatment of Volterra equations, J. Comput. Appl. Math., 125 (2000) 217–249 and this volume.
[12] C.T.H. Baker, Retarded differential equations, J. Comput. Appl. Math., 125 (2000) 309–335 and this volume.

[13] T. Banachiewicz, An outline of the Cracovian algorithm of the method of least squares, Astr. Jour., 50 (1942) 38–41.
[14] M. Barnsley, Fractals Everywhere, Academic Press, San Diego, 1988.
[15] E. Beltrami, Sulle funzioni bilineari, Giorn. di Mat., 11 (1873) 98–106.
[16] Cdt. Benoit, Note sur une méthode de résolution des équations normales provenant de l'application de la méthode des moindres carrés à un système d'équations linéaires en nombre inférieur à celui des inconnues, (Procédé du Commandant Cholesky), Bulletin Géodésique, 2 (1924) 67–77.
[17] I.S. Berezin, N.P. Zhidkov, Computing Methods, Pergamon Press, Oxford, 1965.
[18] P. Bézier, Numerical Control: Mathematics and Applications, Wiley, New York, 1972.
[19] G. Birkhoff, Richard von Mises' years at Harvard, Z. Angew. Math. Mech., 63 (1983) 283–284.
[20] G. Birkhoff, Fluid dynamics, reactor computations, and surface representation, in: A History of Scientific Computing, S.G. Nash ed., ACM Press, New York, 1990, pp. 63–87.
[21] G. Birkhoff, C. de Boor, M.H. Schultz, B. Wendroff, Rayleigh–Ritz approximation by piecewise cubic polynomials, SIAM J. Numer. Anal., 3 (1966) 188–203.
[22] G. Birkhoff, H.L. Garabedian, Smooth surface interpolation, J. Math. Phys., 39 (1960) 258–268.
[23] G.D. Birkhoff, General mean value and remainder theorems with applications to mechanical differentiation and quadrature, Trans. Amer. Math. Soc., 7 (1906) 107–136.
[24] Å. Björck, Numerical Methods for Least Squares Problems, SIAM, Philadelphia, 1996.
[25] L.S. Blackford, J. Choi, A. Cleary, E. D'Azevedo, J. Demmel, I. Dhillon, J. Dongarra, S. Hammarling, G. Henry, A. Petitet, K. Stanley, D. Walker, R. C. Whaley, ScaLAPACK User's Guide, SIAM, Philadelphia, 1997.
[26] W. Böhm, G. Farin, J. Kahmann, A survey of curve and surface methods in CAGD, Comput. Aided Geom. Des., 1 (1984) 1–60.
[27] R.F. Boisvert, Mathematical software: past, present, and future, Math. Comput. Simulation, 54 (2000) 227–241.
[28] É. Borel, Leçons sur les Fonctions de Variables Réelles, Gauthier-Villars, Paris, 1905.
[29] H. Brakhage, Über die numerische Behandlung von Integralgleichungen nach der Quadraturformelmethode, Numer. Math., 2, 1960, 183–196.
[30] C. Brezinski, Padé-Type Approximation and General Orthogonal Polynomials, Birkhäuser Verlag, Basel, 1980.
[31] C. Brezinski, History of Continued Fractions and Padé Approximants, Springer-Verlag, Berlin, 1991.
[32] C. Brezinski, Historical perspective on interpolation, approximation and quadrature, in: Handbook of Numerical Analysis, vol. III, P.G. Ciarlet and J.L. Lions eds., North-Holland, Amsterdam, 1994, pp. 3–46.
[33] C. Brezinski, Extrapolation algorithms and Padé approximations: a historical survey, Appl. Numer. Math., 20 (1996) 299–318.
[34] C. Brezinski, André Louis Cholesky, in: Numerical Analysis, A Numerical Analysis Conference in Honour of Jean Meinguet, Bull. Soc. Math. Belg., 1996, pp. 45–50.
[35] C. Brezinski, Convergence acceleration during the 20th century, J. Comput. Appl. Math., 122 (2000) 1–21 and this volume.
[36] C. Brezinski, M. Redivo Zaglia, Extrapolation Methods. Theory and Practice, North-Holland, Amsterdam, 1991.
[37] C. Brezinski, M. Redivo Zaglia, H. Sadok, The matrix and polynomial approaches to Lanczos-type algorithms, J. Comput. Appl. Math., 123 (2000) 241–260.
[38] C. Brezinski, M. Redivo Zaglia, H. Sadok, A review of formal orthogonality in Lanczos-based methods, J. Comput. Appl. Math., to appear.
[39] C.G. Broyden, A class of methods for solving nonlinear simultaneous equations, Math. Comput., 19 (1965) 577–593.
[40] B. Buchberger, Ein Algorithmus zum Auffinden der Basiselemente des Restklassenringes nach einem Nulldimensionalen Polynomideal, Doctoral Dissertation, Mathematische Institut, Universität Innsbruck, Austria, 1965.
[41] R.C. Buck, Survey of recent Russian literature on approximation, in: On Numerical Approximation, R.E. Langer ed., The University of Wisconsin Press, Madison, 1959, pp. 341–359.
[42] M.D. Buhmann, R. Fletcher, M.J.D. Powell's contributions to numerical mathematics, in: Approximation Theory and Optimization. Tributes to M.J.D. Powell, M.D. Buhmann and A. Iserles eds., Cambridge University Press, Cambridge, 1997, pp. 1–30.
[43] B. Burke Hubbard, Ondes et Ondelettes. La Saga d'un Outil Mathématique, Pour la Science, Diffusion Belin, Paris, 1995.
[44] J.C. Butcher, Numerical methods for ordinary differential equations in the twentieth century, J. Comput. Appl. Math., 125 (2000) 1–29 and this volume.
[45] G. Cassinis, Calcoli Numerici Grafici e Meccanici, Tipografia Editrice Mariotti–Pacini, Pisa, 1928.
[46] J.L. Chabert et al., Histoire d'Algorithmes. Du Caillou à la Puce, Belin, Paris, 1994.
[47] J.L. Chabert et al., A History of Algorithms. From the Pebble to the Microchip, Springer-Verlag, Berlin, 1999.
[48] J.M. Chesneaux, Contribution to computer arithmetic and self-validating numerical methods, in Study of the Computing Accuracy by Using Probabilistic Approach, C. Ulrich ed., IMACS Ann. Comput. Appl. Math., Vol. 7, 1990, pp. 19–30.
[49] P.G. Ciarlet, An $\mathcal{O}(h^2)$ method for a non-smooth boundary-value problem, Aequationes Math., 2 (1968) 39–49.
[50] P.G. Ciarlet, The Finite Element Method for Elliptic Problems, North-Holland, Amsterdam, 1978.
[51] R.W. Clough, The finite element method in plane stress analysis, in: Proceedings of the Second ASCE Conference on Electronic Computation, 1960.
[52] A. Cohen, Wavelets methods in numerical analysis, in: Handbook of Numerical Analysis, vol. VII, P.G. Ciarlet and J.L. Lions eds., North-Holland, Amsterdam, 2000, pp. 417–711.

[53] L.J. Comrie, The application of commercial calculating machines to scientific computing, MTAC, 2 (1946–1947) 149–159.
[54] J.W. Cooley, How the FFT gained acceptance, in: A History of Scientific Computing, S.G. Nash ed., ACM Press, New York, 1990, pp. 133–140.
[55] J.W. Cooley, Lanczos and the FFT: a discovery before its time, in: Proceedings of the Cornelius Lanczos International Centenary Conference, J.D. Brown, M.T. Chu, D.C. Ellison, R.J. Plemmons eds., SIAM, Philadelphia, 1994, pp. 3–9.
[56] J.W. Cooley, J.W. Tukey, An algorithm for the machine calculation of complex Fourier series, Math. Comput. 19 (1965) 297–301.
[57] R. Cools, I. P. Mysovskikh, H. J. Schmid, Cubature formulae and orthogonal polynomials, J. Comput. Appl. Math., 127 (2001) 121–152 and this volume.
[58] M. Croarken, L.J. Comrie: a forgotten figure in the history of numerical computation, Mathematics Today, 36 (2000) 114–118.
[59] C.W. Cryer, Numerical Functional Analysis, Clarendon Press, Oxford, 1982.
[60] C.F. Curtiss, J.O. Hirschfelder, Integration of stiff equations, Proc. U.S. Nat. Acad. Sci., 38 (1952) 235–243.
[61] G. Dahlquist, Fehlerabschätzungen bei Differenzenmethoden zur numerische Integration gewöhnlicher Differentialgleichungen, Z. Angew. Math. Mech., 31 (1951) 239–240.
[62] G. Dahlquist, A special stability problem for linear multistep methods, BIT, 3 (1963) 27–43.
[63] G. Dahlquist, 33 years of numerical instability, part I, BIT, 25 (1985) 188–204.
[64] G. Dahlquist, Å. Björck, Numerical Mathods, Prentice-Hall, Englewood Cliffs, 1974.
[65] G.C. Danielson, C. Lanczos, Some improvements in practical Fourier analysis and their application to X-ray scattering from liquids, J. Franklin Inst., 233 (1942) 365–380; 435–452.
[66] G.B. Dantzig, Linear programming, in: Problems for the Numerical Analysis of the Future, Proceedings of the Symposium on Modern Calculating Machinery and Numerical Methods, UCLA, July 29–31, 1948, Appl. Math. vol. 15, National Bureau of Standards, 1951, pp. 18–21.
[67] G.B. Dantzig, Origins of the simplex method, in: A History of Scientific Computing, S.G. Nash ed., ACM Press, New York, 1990, pp. 141–151.
[68] I. Daubechies, Ten Lectures on Wavelets, SIAM, Philadelphia, 1992.
[69] D. Davidenko, On a new method of the numerical solution of systems of nonlinear equations (in Russian), Dokl. Akad. Nauk USSR, 88 (1953) 601–602.
[70] W. Davidon, Variable metric methods for minimization, A.E.C. Res. and Develop. Rept. ANL-5990, Argonne National Laboratory, Argonne, Illinois, 1959.
[71] C. de Boor, On calculating B-splines, J. Approx. Theory, 6 (1972) 50–62.
[72] P. de Casteljau, Outillage, Méthodes de Calcul, Automobiles André Citröen S.A., Paris, 1959.
[73] P. de Casteljau, Courbes et Surfaces à Pôles, Automobiles André Citröen S.A., Paris, 1963.
[74] C. de la Vallée Poussin, Sur la méthode de l'approximation minimum, Ann. Soc. Sci. Bruxelles, sér. I, 35 (1910–1911) 1–16.
[75] J.P. Delahaye, Sequence Transformations, Springer-Verlag, Berlin, 1988.
[76] J. Della Dora, F. Robert, Noël Gastinel and computational mathematics, Linear Algebra Appl., 78 (1986) 237–247.
[77] J.E. Dennis, J.J. Moré, A characterization of superlinear convergence and its applications to quasi-Newton methods, Math. Comput., 28 (1974) 549–560.
[78] J. Descloux, Approximations in L^p and Chebyshev approximations, SIAM J., 11 (1963) 1017–1026.
[79] J.J. Dongarra, J.R. Bunch, C.B. Moler, G.W. Stewart, Linpack User's Guide, SIAM, Philadelphia, 1979.
[80] J.J. Dongarra, I.S. Duff, D.C. Sorensen, H.A. Van der Vorst, Numerical Linear Algebra for High-Performance Computers, SIAM, Philadelphia, 1998.
[81] J.J. Dongarra, V. Eijkhout, Numerical linear algebra algorithms and software, J. Comput. Appl. Math., 123 (2000) 489–514 and this volume.
[82] J.J. Dongarra, F. Sullivan, eds., Top ten algorithms, IEEE/AIP Comput. Sci. Tech., Jan.–Feb. 2000.
[83] A. Dresden, The migration of mathematicians, Am. Math. Mon., 49 (1942) 415–429.
[84] J. Dutka, Richardson extrapolation and Romberg integration, Hist. Math., 11 (1984) 3–21.
[85] P.S. Dwyer, A matrix presentation of least squares and correlation theory with matrix justification of improved method of solution, Ann. Math. Stat., 15 (1944) 82–89.
[86] R. Eckhardt, Stan Ulam, John von Neumann, and the Monte Carlo method, in N.G. Cooper ed., *From Cardinals to Chaos. Reflections on the Life and Legacy of Stanislaw Ulam*, Cambridge University Press, Cambridge, 1989, pp. 131–143.
[87] G. Elfving, The History of Mathematics in Finland, 1828–1918, Frenckell, Helsinki, 1981.
[88] A. Erdélyi, J. Todd, Advanced instruction in practical mathematics, Nature, 158 (1946) 690–692.
[89] V.N. Faddeeva, Computational Methods of Linear Algebra, State Technical Press, Moscow–Leningrad, 1950.
[90] P. Fatou, Sur les équations fonctionnelles, Bull. Soc. Math. France, 47 (1919) 161–271; 48 (1920) 33–94; 208–314.
[91] R.P. Fedorenko, A relaxation method for solving elliptic difference equations, USSR Comput. Math. Phys., 1961, 1092–1096.
[92] W. Feller, G.E. Forsythe, New matrix transformations for obtaining characteristic vectors, Q. Appl. Math., 8 (1951) 325–331.
[93] J.C. Ferguson, Multivariate curve interpolation, J.A.C.M., 2 (1964) 221–228.
[94] G. Fichera, Ricordo di Aldo Ghizzetti, Rend. Mat. Appl., VII Ser., 14 (1994) 9–36.
[95] R. Fletcher, Conjugate gradient methods for indefinite systems, in: Numerical Analysis, G.A. Watson ed., LNM vol. 506, Springer-Verlag, Berlin, 1976, pp. 73–89.

[96] R. Fletcher, Practical Methods of Optimization, Second Edition, Wiley, New York, 1991.
[97] R. Fletcher, M.J.D. Powell, A rapidly convergent descent method for minimization, Comput. J., 6 (1963) 163–168.
[98] R. Fletcher, C.M. Reeves, Function minimization by conjugate gradients, Computer J., 7 (1964) 149–154.
[99] A.R. Forrest, Interactive interpolation and approximation by Bézier polynomials, Comput. J., 15 (1972) 71–79.
[100] J.B.J. Fourier, Solution d'une question particulière du calcul des inégalités, Nouveau Bull. Sci. Soc. Philom. Paris, (1826) 99–100.
[101] L. Fox, Early numerical analysis in the United Kingdom, in: A History of Scientific Computing, S.G. Nash ed., Addison-Wesley, Reading, 1990, pp. 280–300.
[102] L. Fox, H.D. Huskey, J.H. Wilkinson, Notes on the solution of algebraic linear simultaneous equations, Quart. J. Mech. Appl. Math., 1 (1948) 149–173.
[103] J.G.F. Francis, The *QR* transformation: a unitary analogue to the *LR* transformation, Computer J., 4 (1961–1962) 265–271; 332–345.
[104] S.P. Frankel, Convergence rates of iterative treatments of partial differential equations, Math. Tables Aids Comput., 4 (1950) 65–75.
[105] C.-E. Fröberg, Introduction to Numerical Analysis, Addison-Wesley, Reading, 1965.
[106] D. Gabor, Theory of communication, J. Inst. Elect. Engng., 93 (1946) 429–457.
[107] C. Galopin-Schaub, Théorie des Approximations Numériques. Notions de Calcul Approximatif, H. Georg, Genève, 1884.
[108] M. Gasca, T. Sauer, On the history of multivariate polynomial interpolation, J. Comput. Appl. Math., 122 (2000) 23–35 and this volume.
[109] C.F. Gauss, Nachlass: Theoria interpolationis methodo nova tractata, in: Carl Friedrich Gauss, Werke, Band 3, Koniglichen Gesellschaft der Wissenschaften, Göttingen, 1866, pp. 265–303.
[110] W. Gautschi, Orthogonal polynomials: applications and computation, in: Acta Numerica 1996, A. Iserles ed., Cambridge University Press, Cambridge, 1996, pp. 45–119.
[111] C.W. Gear, R.D. Skeel, The development of ODE methods: a symbiosis between hardware and numerical analysis, in: A History of Scientific Computing, S.G. Nash ed., ACM Press, New York, 1990, pp. 88–105.
[112] B. Gellai, Cornelius Lanczos, a biographical essay, in: Proceedings of the Cornelius Lanczos International Centenary Conference, J.D. Brown, M.T. Chu, D.C. Ellison, R.J. Plemmons eds., SIAM, Philadelphia, 1994, pp. xxi–xlviii.
[113] W. Givens, Numerical computation of the charasteristic values of a real symmetric matrix, Oak Ridge National Laboratory, ORNL-1574, 1954.
[114] W. Givens, Computation of plane unitary rotations transforming a general matrix to triangular form, SIAM J. Appl. Math., 6 (1958) 26–50.
[115] J. Gleick, Chaos: Making a New Science, Viking, New York, 1987.
[116] H.H. Goldstine, A History of Numerical Analysis from the 16th through the 19th Century, Springer-Verlag, New York, 1977.
[117] G.H. Golub, Numerical methods for solving least squares problems, Numer. Math., 7 (1965) 206–216.
[118] G.H. Golub, W. Kahan, Calculating the singular values and pseudo-inverse of a matrix, SIAM J. Numer. Anal., Ser. B, 2 (1965) 205–224.
[119] G.H. Golub, D.P. O'Leary, Some history if the conjugate gradient and Lanczos methods, SIAM Rev., 31 (1989) 50–100.
[120] G. Golub, H. Van der Vorst, Eigenvalue computation in the 20th century, J. Comput. Appl. Math., 123 (2000) 35–65 and this volume.
[121] R.T. Gregory, E.V. Kryshnamurthy, Methods and Applications of Error-Free Computation, Springer-Verlag, New York, 1984.
[122] R.T. Gregory, E.V. Kryshnamurthy, Error-Free Computation: Why it is Needed and Methods for Doing it, Springer-Verlag, New York, 1985.
[123] A. Guerraggi, L'analisi, in: La Matematica Italiana dopo l'Unita, S. Di Sieno, A. Guerragio, P. Nastasi eds., Marcos y Marcos, Milano, 1998, pp. 1–158.
[124] M.H. Gutknecht, The pioneer days of scientific computing in Switzerland, in: A History of Scientific Computing, S.G. Nash ed., Addison-Wesley, Reading, 1990, pp. 301–313.
[125] A. Haar, Die Minkowskische Geometrie und die Annaherung an stetige Funktionen, Math. Ann., 78 (1918) 294–311.
[126] T. Håvie ed., Romberg Seminar on Quadrature/Interpolation/Extrapolation and Rational Approximations (Padé-Continued Fractions, Det Kong. Norske Vid. Selsk. Skr., 2 (1989).
[127] E. Hairer, G. Wanner, Solving Ordinary Differential Equations II. Stiff and Differential-Algebraic Problems, Springer-Verlag, Berlin, 1991.
[128] D.R. Hartree, Numerical Analysis, Oxford University Press, Oxford, 1952.
[129] M.T. Heideman, D.H. Johnson, C.S. Burrus, Gauss and the history of the Fast Fourier Transform, Arch. Hist. Exact Sci., 34 (1985) 265–277.
[130] J. Helo, Über die Newtonsche Näherungsmethode und eine Verallgemeinerung derselben, Thesis, University of Helsinki, 1914.
[131] P. Henrici, Discrete Variable Methods in Ordinary Differential Equations, Wiley, New York, 1962.
[132] M.R. Hestenes, Conjugate Direction Methods in Optimization, Springer-Verlag, New York, 1980.
[133] M.R. Hestenes, Conjugacy and gradients, in: A History of Scientific Computing, S.G. Nash ed., ACM Press, New York, 1990, pp. 167–179.

[134] M.R. Hestenes, E. Stiefel, Methods of conjugate gradients for solving linear systems, J. Res. Natl. Bur. Stand., 49 (1952) 409–436.
[135] M.R. Hestenes, J. Todd, Mathematicians Learning to Use Computers, Special Publication 730, National Institute of Standards and Technology, U.S. Government Printing Office, Washington, 1991.
[136] K. Heun, Neue Methode zur approximativen Integration der Differentialgleichungen einer unabhängigen Veränderlichen, Zeit. Math. Phys., 45 (1900) 23–38.
[137] N.J. Higham, Accuracy and Stability of Numerical Algorithms, SIAM, Philadelphia, 1996.
[138] A. Hirayama, K. Shimodaira, H. Hirose, Takakazu Seki's Collected Works Edited with Explanations, Osaka Kyoiku Tosho, Osaka, 1974.
[139] F.L. Hitchcock, The distribution of a product from several sources to numerous localities, J. Math. Phys., 20 (1941) 224–230.
[140] A. Hodges, Alan Turing: The Enigma of Intelligence, Counterpoint, London, 1983.
[141] A.S. Householder, A class of methods for inverting matrices, J. Soc. Ind. Appl. Math., 6 (1958) 189–195.
[142] A.S. Householder, Unitary triangularization of a nonsymmetric matrix, J. Assoc. Comput. Mach., 5 (1958) 339–342.
[143] E. Isaacson, The origin of *Mathematics of Computation* and some personal recollections, in: A History of Scientific Computing, S.G. Nash ed., ACM Press, New York, 1990, pp. 211–216.
[144] A. Iserles, S.P. Nørsett, Order Stars, Chapman & Hall, London, 1991.
[145] S. Jaffard, Décomposition en ondelettes, in: Development of Mathematics, 1950–2000, J.-P. Pier ed., Birkhäuser, Basel, 2000, pp. 609–634.
[146] H. Jensen, An attempt at a systematic classification of some methods for the solution of normal equations, Geodaetisk Institut Kobenhan, Meddelelse no. 18, 1944, 45 pp.
[147] F. John, Extremum problems with inequalities as subsidiary conditions, in: Studies and Essays Presented to R. Courant on his 60th Birthday, January 8, 1948, Interscience, New York, 1948, pp. 187–204.
[148] C. Jordan, Mémoire sur les formes bilinéaires, J. Math. Pures Appl., Sér. 2, 19 (1874) 35–54.
[149] G. Julia, Mémoire sur l'itération des fonctions rationnelles, J. Math. Pures Appl., 4 (1918) 47–245.
[150] J.-P. Kahane, P.-G. Lemarié-Rieusset, Fourier Series and Wavelets, Gordon and Breach, Reading, 1996.
[151] L.V. Kantorovich, G.P. Akilov, Functional Analysis, 2nd ed., Pergamon Press, Oxford, 1982.
[152] I. Kardos, Scientists Face to Face, Corvina Kidó, Budapest, 1978.
[153] N. Karmarkar, A new polynomial-time algorithm for linear programming, Combinatoria, 4 (1984) 373–395.
[154] W. Karush, Minima of Functions of Several Variables with Inequalities as Side Conditions, Master's Thesis, University of Chicago, 1939.
[155] T.N. Kjeldsen, The emergence of nonlinear programming: interactions between practical mathematics and mathematics proper, Math. Intell., 22 (2000) 50–54.
[156] T.N. Kjeldsen, A contextualized historical analysis of the Kuhn-Tucker theorem in nonlinear programming: the impact of World War II, Historia Math., 27 (2000) 331–361.
[157] A.N. Krylov, On the numerical solution of equations whose solution determines the frequency of small vibrations of meterial systems (In Russian), Izv. Akad. Nauk. SSSR Otd. Mat. Estest., 1 (1931) 491–539.
[158] V.N. Kublanovskaya, On some algorithms for the solution of the complete eigenvalue problem, USSR Comp. Math. Phys., 3 (1961) 637–657.
[159] V.N. Kublanovskaya, The contribution of the Leningrad mathematicians to the development of numerical linear algebra in the period 1950–1986, in: A History of Scientific Computing, S.G. Nash ed., Addison-Wesley, Reading, 1990, pp. 329–338.
[160] H.W. Kuhn, A.W. Tucker, Nonlinear programming, in: Proceedings of the Second Berkeley Symposium on Mathematical Statistics and Probability, J. Neyman ed., Berkeley, 1950, pp. 481–492.
[161] U.W. Kulisch, W.L. Miranker, Computer Arithmetic in Theory and Practice, Academic press, New York, 1981.
[162] J. Kuntzmann, Cours d'Analyse Appliquée, Tomes I et II, C.D.U., Paris, 1950–1951.
[163] J. Kuntzmann, Le calcul à Grenoble avant-hier, in: Computers and Computing, P. Chenin, C. di Crescenzo, F. Robert, eds., Masson, Paris, and Wiley, Chichester, 1986.
[164] J. Kuntzmann, J. Laborde, Travaux Pratiques de Calcul, C.D.U., Paris, 1952.
[165] K.S. Kunz, Numerical Analysis, McGraw-Hill, New York, 1957.
[166] M.W. Kutta, Beitrag zur näherungsweisen Integration oder Differentialgleichungen, Zeit. Math. Phys., 46 (1901) 435–453.
[167] M. La Porte, J. Vignes, Étude statistique des erreurs dans l'arithmétique des ordinateurs. Application au contrôle des résultats d'algorithmes numériques, Numer. Math., 23 (1974) 63–72.
[168] J.L. Lagrange, Éssai d'analyse numérique sur la transformation des fractions, J. Éc. Polytechnique, 5 (1798) 93–114..
[169] E. Lahaye, Une méthode de résolution d'une catégorie d'équations transcendentes, C.R. Acad. Sci. Paris, 198 (1934) 1840–1842.
[170] C. Lanczos, An iteration method for the solution of the eigenvalue problem of linear differential and integral operators, J. Res. Natl. Bur. Stand., 45 (1950) 255–282.
[171] C. Lanczos, Solution of systems of linear equations by minimized iterations, J. Res. Natl. Bur. Stand., 49 (1952) 33–53.
[172] D.P. Laurie, Computation of Gauss-type quadrature formulas, J. Comput. Appl. Math., 127 (2001) 201–217 and this volume.
[173] J.K. Lenstra, A.H.G. Rinnooy Kan, A. Schrijver eds., History of Mathematical Programming, North-Holland, Amsterdam, 1991.
[174] E. Lindelöf, Differentiali–ja Integralilasku ja sen Sovellutukset, vol. 1, Mercator, Helsinki, 1928.

[175] P. Linz, Theoretical Numerical Analysis, Wiley, New York, 1979.
[176] J.L. Lions, Problèmes aux limites en théorie des distributions, Acta Math., 94 (1955) 13–153.
[177] J. Lützen, Joseph Liouville, 1809–1882: Master of Pure and Applied Mathematics, Springer-Verlag, New York, 1990.
[178] C.C. MacDuffee, The Theory of Matrices, Springer-Verlag, Berlin, 1933.
[179] B.B. Mandelbrot, Fractals: Form, Chance, and Dimension, W.H. Freeman and Co., San Francisco, 1977.
[180] H.S.W. Massey, Leslie John Comrie, Obituary Notices of Fellows of the Royal Society, 8 (1952–1953) 96–107.
[181] J.C. Maxwell, A Treatise on Electricity and Magnetism, Oxford University Press, Oxford, 1873.
[182] N. Megiddo, Pathways to the optimal set in linear programming, in: Progress in Mathematical Programming: Interior-Point and Related Methods, N. Megiddo ed., Springer-Verlag, New York, 1989, pp. 131–158.
[183] S. Mehrotra, On the implementation of a primal–dual interior point method, SIAM J. Optim., 2 (1992) 575–601.
[184] N. Metropolis, The beginning of the Monte Carlo method, in N.G. Cooper ed., *From Cardinals to Chaos. Reflections on the Life and Legacy of Stanislaw Ulam*, Cambridge University Press, Cambridge, 1989, pp. 125–130.
[185] N. Metropolis, The Los Alamos experience, 1943–1954, in: A History of Scientific Computing, S.G. Nash ed., Addison-Wesley, Reading, 1990, pp. 237–250.
[186] N. Metropolis, J. Howlett, G.-C. Rota eds., A History of Computing in the Twentieth Century. A Collection of Essays, Academic Press, New York, 1980.
[187] W.E. Milne, Numerical Calculus. Approximations, Interpolation, Finite Differences, Numerical Integration, and Curve Fitting, Princeton University Press, Princeton, 1949.
[188] H. Mineur, Techniques de Calcul Numérique, à l'Usage des Mathématiciens, Astronomes, Physiciens et Ingénieurs, Béranger, Paris, 1952.
[189] C.B. Moler, G.W. Stewart, An algorithm for generalized matrix eigenvalue problems, SIAM J. Numer. Anal., 10 (1973) 241–256.
[190] R. de Montessus, R. D'Adhémar, Calcul Numérique, Gaston Doin et Cie, Paris, 1911.
[191] P.M. Morse, Mathematical problems in operations research, Bull. Am. Math. Soc., 54 (1948) 602–621.
[192] P.-E. Mounier-Kuhn, L'Institut Blaise-Pascal du CNRS (1946–1969), in: De Bourbaki à la Machine à Lire. Journée d'Hommage à René de Possel, 1905–1974, Institut Blaise Pascal, unpublished.
[193] H. Naegelsbach, Studien zu Fürstenau's neuer Methode der Darstellung und Berechnung der Wurzeln algebraischer Gleichungen durch Determinanten der Coefficienten, Arch. Math. Phys., 59 (1876) 147–192; 61 (1877) 19–85.
[194] S.G. Nash ed., A History of Scientific Computing, ACM Press, New York, 1990.
[195] S. G. Nash, A survey of truncated-Newton methods, J. Comput. Appl. Math., 124 (2000) 45–59 and this volume.
[196] P. Nastasi, Il Quadro Istituzionale, in: La Matematica Italiana dopo l'Unita, S. Di Sieno, A. Guerragio, P. Nastasi eds., Marcos y Marcos, Milano, 1998, pp. 817–943.
[197] Y. Nievergelt, A tutorial history of least squares with applications to astronomy and geodesy, J. Comput. Appl. Math., 121 (2000) 37–72 and this volume.
[198] J.T. Oden, Historical comments on finite elements, in: A History of Scientific Computing, S.G. Nash ed., ACM Press, New York, 1990, pp. 152–166.
[199] J.T. Oden, Finite elements: an introduction, in: Handbook of Numerical Analysis, vol. II, P.G. Ciarlet and J.L. Lions eds., North-Holland, Amsterdam, 1991, pp. 3–15.
[200] B.N. Parlett, The contribution of J.H. Wilkinson to numerical analysis, in: A History of Scientific Computing, S.G. Nash ed., ACM Press, New York, 1990, pp. 17–30.
[201] H.O. Peitgen, H. Jürgens, D. Saupe, Chaos and Fractals. New Frontiers of Science, Springer-Verlag, New York, 1992.
[202] M. Pichat, Correction d'une somme en arithmétique à virgule flottante, Numer. Math., 19 (1972) 400–406.
[203] G. Pólya, Sur un algorithme toujours convergent pour obtenir les polynômes de meilleure approximation de Tchebysheff pour une fonction continue quelconque, C.R. Acad. Sci. Paris, 157 (1913) 840–843.
[204] M.J.D. Powell, The theory of radial basis function approximation in 1990, in: Advances in Numerical Analysis. Volume II: Wavelets, Subdivision Algorithms, and Radial Basis Functions, W. Light ed., Oxford University Press, Oxford, 1992, pp. 105–210.
[205] G.B. Price, Adjustments in mathematics to the impact of war, Am. Math. Mon., 50 (1943) 31–34.
[206] M. Rees, The mathematical sciences and World War II, Am. Math. Mon., 87 (1980) 607–621.
[207] M. Rees, Early years of the mathematics program at ONR, in: Mathematics. People, Problems, Results, vol. III, D.M. Campbell and J.C. Higgins eds., Wadsworth, Belmont, 1984, vol. III, pp. 201–206.
[208] J.K. Reid, On the method of conjugate gradients for the solution of large sparse systems of linear equations, in: Large Sparse Sets of Linear Equations, J.K. Reid ed., Academic Press, New York, 1971, pp. 231–254.
[209] E.Y. Rémès, Sur un procédé convergent d'approximations successives pour déterminer les polynômes d'approximation, C.R. Acad. Sci. Paris, 198 (1934) 2063–2065.
[210] E.Y. Rémès, Sur le calcul effectif des polynômes d'approximation de Tchebichef, C.R. Acad. Sci. Paris, 199 (1934) 337–340.
[211] W.C. Rheinboldt, Numerical continuation methods: a perspective, J. Comput. Appl. Math., 124 (2000) 229–244.
[212] J.L. Richards, Hilda Geiringer von Mises (1893–1973) in: Women of Mathematics, L.S. Grinstein and P.J. Campbell, eds., Greenwood Press, Westport, 1987, pp. 41–46.

[213] L.F. Richardson, The approximate arithmetical solution by finite difference of physical problems involving differential equations, with an application to the stress in a masonry dam, Philos. Trans. Roy. Soc. London, ser. A, 210 (1910) 307–357.
[214] L.F. Richardson, The deferred approach to the limit. I: Single lattice, Philos. Trans. Roy. Soc. London, ser. A, 226 (1927) 299–349.
[215] H.S. Roberts ed., A History of Statistics in New Zealand, New Zealand Statistical Association, Wellington, 2000.
[216] W. Romberg, Vereinfachte numerische Integration, Kgl. Norske Vid. Selsk. Forsk., 28 (1955) 30–36.
[217] J.B. Rosser, Mathematics and mathematicians in World War II, Notices Am. Math. Soc., 29 (1982) 509–515.
[218] C.D.T. Runge, Über die numerische Auflösung von Differentialgleichungen, Math. Ann., 46 (1895) 167–178.
[219] C. Runge, F.A. Willers, Numerische und graphische Quadratur und Integration gewöhnlicher und partieller Differentialgleichungen, Encyk. d. Math. Wissenschaften, II 3, Heft 2, B.G. Teubner, Leipzig.
[220] C. Runge, H. König, Vorlesungen über Numerisches Rechnen, Grundlehren Math. Wiss., Band XI,, Julius Springer, Berlin, 1924.
[221] H. Rutishauser, Über die Instabilität von Methoden zur Integration gewöhnlicher Differentialgleichungen, Z. Angew. Math. Mech., 3 (1952) 65–74.
[222] H. Rutishauser, Une méthode pour la détermination des valeurs propres d'une matrice, C.R. Acad. Sci. Paris, 240 (1955) 34–36.
[223] Y. Saad, M.H. Schultz, GMRES: A generalized minimal residual algorithm for solving nonsymmetric linear systems, SIAM J. Sci. Statist. Comput., 7 (1986) 856–869.
[224] Y. Saad, H. Van der Vorst, Iterative solution of linear systems in the 20th century, J. Comput. Appl. Math., 123 (2000) 1–33 and this volume.
[225] D.H. Sadler, J. Todd, Mathematics in government service and industry, Nature, 157 (1946) 571–573.
[226] M.G. Salvadori, M.L. Baron, Numerical Methods in Engineering, Prentice Hall, Englewood Cliffs, 1952.
[227] J.B. Scarborough, Numerical Mathematical Analysis, The Johns Hopkins Press, Baltimore, 1930.
[228] I.J. Schoenberg, Contributions to the problem of approximation of equidistant data by analytic functions. Part A: On the problem of smoothing or graduation. A first class of analytic approximation formulae, Quart. Appl. Math., 4 (1946) 45–99.
[229] I.J. Schoenberg, Contributions to the problem of approximation of equidistant data by analytic functions. Part B: On the problem of osculatory interpolation. A second class of analytic approximation formulae, Quart. Appl. Math., 4 (1946) 112–141.
[230] I.J. Schoenberg, A brief account of my life and work, in: I.J. Schoenberg Selected Papers, C. de Boor ed., Birkhäuser, Basel, 1988, vol. 1, pp. 1–10.
[231] L.L. Schumaker, Spline Functions: Basic Theory, Wiley, New York, 1981.
[232] I. Schur, Über die charakterischen Würzeln einer lineare Sunstitution mit einer Anwendung auf die Theorie des Integralgleichungen, Math. Ann., 66 (1909) 448–510.
[233] D. Shanks, Non linear transformations of divergent and slowly convergent sequences, J. Math. Phys., 34 (1955) 1–42.
[234] D. Shasha, C. Lazere, Out of Their Minds. The Lives and Discovery of 15 Great Computer Scientists, Springer-Verlag, New York, 1995.
[235] B.T. Smith, J.M. Boyle, J.J. Dongarra, B.S. Garbow, Y. Ikebe, V.C. Klema, C.B. Moler, Matrix Eigensystem Routines-EISPACK Guide (second edition), Springer, Berlin, 1976.
[236] S.B. Smith, The Great Mental Calculators, Columbia University Press, New York, 1983.
[237] P. Sonneveld, CGS, a fast Lanczos-type solver for nonsymmetric linear systems, SIAM J. Sci. Stat. Comp., 10 (1989) 36–52.
[238] G.W. Stewart, On the early history of the singular value decomposition, SIAM Rev., 35 (1993) 551–566.
[239] G.W. Stewart, Matrix Algorithms. Volume I: Basic Decompositions, SIAM, Philadelphia, 1998.
[240] V. Strassen, Gaussian elimination is not optimal, Numer. Math., 13 (1969) 354–356.
[241] K. Stueben, A review of algebraic multigrid, J. Comput. Appl. Math., 128 (2001) 281–309 and this volume.
[242] O. Taussky, J. Todd, Cholesky, Toeplitz and the triangular factorization of symmetric matrices, unpublished (due to the courtesy of J. Todd).
[243] V. Thomée, From finite differences to finite elements (a short history of numerical analysis of PDEs), J. Comput. Appl. Math., 128 (2001) 1–54 and this volume.
[244] J. Todd, Numerical analysis at the National Bureau of Standards, SIAM Rev., 17 (1975) 361–370.
[245] J. Todd, The prehistory and early history of computation at the U.S. National Bureau of Standards, in: A History of Scientific Computing, S.G. Nash ed., Addison-Wesley, Reading, 1990, pp. 251–268.
[246] O. Toeplitz, Die Jacobische Transformation der quadratischen Formen von unendlich vielen Veränderlichen, Nachr. Akad. Wiss. Göttingen, Math.-Phys. Kl., II (1907) 101–110.
[247] D. Tournès, L'Intégration Approchée des Équations Différentielles Ordinaires (1671–1914), Thèse, Université de Paris VII – Denis Diderot, 1996, Presses Universitaires du Septentrion, Villeneuve d'Ascq, 1999.
[248] J.F. Traub, H. Woźniakowski, A General Theory of Optimal Algorithms, Academic Press, New York, 1980.
[249] L.N. Trefethen, The definition of numerical analysis, SIAM News, nov. 1992.
[250] L.N. Trefethen, Classic papers in numerical analysis, NA-Digest, vol. 93, issue 19, 9 May 1993.
[251] L.N. Trefethen, Predictions for scientific computing fifty years from now, Mathematics Today, 36 (2000) 53–57.
[252] A.M. Turing, Rounding-off errors in matrix processes, Quart. J. Mech. Appl. Math., 1 (1948) 287–308.
[253] S. Turing, Alan M. Turing, W. Heffer & Sons, Cambridge, 1959.

[254] H.W. Turnbull, A.C. Aitken, An Introduction to the Theory of Canonical Matrices, Blackie and Son, London, 1932.
[255] M.J. Turner, R.W. Clough, H.C. Martin, L.J. Topp, Stiffness and deflection analysis of complex structures, J. Aeronaut. Sci., 23 (1956) 805–823.
[256] S.M. Ulam, Adventures of a Mathematician, Charles Scribner's Sons, New York, 1976.
[257] H.A. Van der Vorst, Bi-CGSTAB: a fast and smoothly converging variant of Bi-CG for the solution of nonsymmetric linear systems, SIAM J. Sci. Statis. Comput., 13 (1992) 631–644.
[258] R.S. Varga, Hermite interpolation-type Ritz methods for two-point boundary value problems, in: Numerical Solution of Partial Differential Equations, J.H. Bramble ed., Academic Press, New York, 1966, pp. 365–373.
[259] R.S. Varga, In Memoriam, Garrett Birkhoff, J. Approx. Theory, 95 (1998) 1–4.
[260] R.S. Varga, Matrix Iterative Analysis, 2nd ed., Springer-Verlag, Berlin, 2000.
[261] L. Veidinger, On the numerical determination of the best approximation in the Chebyshev sense, Numer. Math., 2 (1960) 99–105.
[262] J. Vieille, Théorie Générale des Approximations Numériques suivie d'une Application à la Résolution des Équations Numériques, Mallet–Bachelier, Paris, 1852.
[263] J. Von Neumann, H.H. Goldstine, Numerical inverting of matrices of high order, Bull. Amer. Math. Soc., 53 (1947) 1021–1099.
[264] G. Walz, Asymptotics and Extrapolation, Akademie Verlag, Berlin, 1996.
[265] D.S. Watkins, QR-like algorithms for eigenvalue problems, J. Comput. Appl. Math., 123 (2000) 67–83.
[266] G.A. Watson, Approximation in normed linear spaces, J. Comput. Appl. Math., 121 (2000) 1–36 and this volume.
[267] D.J. Wheeler, Programmed computing at the Universities of Cambridge and Illinois in the early fifties, in: A History of Scientific Computing, S.G. Nash ed., Addison-Wesley, Reading, 1990, pp. 269–279.
[268] E.T. Whittaker, G. Robinson, The Calculus of Observations. A Treatise on Numerical Mathematics, Blackie & Son, London, 1924.
[269] J.M. Whittaker, Alexander Craig Aitken, Biographical Memoirs of Fellows of the Royal Society, 14 (1968) 1–14.
[270] J.H. Wilkinson, Error analysis of floating-point computation, Numer. Math., 2 (1960) 319–340.
[271] J.H. Wilkinson, Rounding Errors in Algebraic Processes, Prentice Hall, Englewood Cliffs, 1963.
[272] F.A. Willers, Practical Analysis: Graphical and Numerical Methods, Dover, New York, 1948.
[273] J. Wimp, Sequence Transformations and their Applications, Academic Press, New York, 1981.
[274] J.R. Womersley, Scientific computing in Great Britain, MTAC, 2 (1946–1947) 110–117.
[275] S.J. Wright, F.A. Potra, Interior point methods, J. Comput. Appl. Math. 124 (2000) 281–302.
[276] P. Wynn, On a device for computing the $e_m(S_n)$ transformation, MTAC, 10 (1956) 91–96.
[277] P. Wynn, On some recent developments in the theory and application of continued fractions, SIAM J. Numer. Anal., ser. B, 1 (1964) 177–197.
[278] T. Yamamoto, Historical developments in convergence analysis for Newton's and Newton-like methods, J. Comput. Appl. Math., 124 (2000) 1–23 and this volume.
[279] D.M. Young, Iterative Methods for Solving Partial Differential Equations of Elliptic Type, Doctoral Thesis, Harvard University, 1950.
[280] D.M. Young, A historical review of iterative methods, in: A History of Scientific Computing, S.G. Nash ed., ACM Press, New York, 1990, pp. 180–194.
[281] D.M. Young, Garrett Birkhoff and applied mathematics, Notices Am. Math. Soc., 44 (1997) 1446–1450.
[282] J.W. Young, General theory of approximation by functions involving a given number of arbitrary parameters, Trans. Am. Math. Soc., 8 (1907) 331–344.
[283] O.C. Zienkiewicz, Origins, milestones and directions of the finite element method – A personal view, in: Handbook of Numerical Analysis, vol. IV, P.G. Ciarlet and J.L. Lions eds., North-Holland, Amsterdam, 1996, pp. 3–67.
[284] M. Zlamal, On the finite element method, Numer. Math., 12 (1968) 394–409.
[285] K. Zuse, The Computer – My Life, Springer-Verlag, Berlin, 1993.

Approximation in normed linear spaces

G.A. Watson

Department of Mathematics, University of Dundee, Dundee DD1 4HN, Scotland, UK

Abstract

A historical account is given of the development of methods for solving approximation problems set in normed linear spaces. Approximation of both real functions and real data is considered, with particular reference to L_p (or l_p) and Chebyshev norms. As well as coverage of methods for the usual linear problems, an account is given of the development of methods for approximation by functions which are nonlinear in the free parameters, and special attention is paid to some particular nonlinear approximating families.

1. Introduction

The purpose of this paper is to give a historical account of the development of numerical methods for a range of problems in best approximation, that is problems which involve the minimization of a norm. A treatment is given of approximation of both real functions and data. For the approximation of functions, the emphasis is on the use of the Chebyshev norm, while for data approximation, we consider a wider range of criteria, including the other l_p norms, $1 \leqslant p < \infty$. As well as the usual linear problems, a general account is given of nonlinear best approximation, and we also consider some special cases. Only a passing mention is made of least-squares problems, as that is considered elsewhere. The focus is also entirely on the approximation of real quantities, and so best approximation of complex quantities is not covered. A partial justification of this is that dealing with problems in generality as complex ones would introduce additional complication not entirely justified by the additional algorithmic initiatives.

Since we are concerned here with historical development, technical details are not included for their own sake. The intention is, where appropriate, to be descriptive, rather to give a technically rigorous and detailed account of methods. However, it seemed necessary at times for the sake of comprehensibility, and in order to fully appreciate algorithmic developments, to include a reasonable amount of technical detail.

E-mail address: gawatson@maths.dundee.ac.uk (G.A. Watson).

Obviously a major factor in the development of methods has been the advent of powerful computing facilities, as this has opened up opportunities to tackle a wide range of practical problems. Whereas at one time, the main consideration may have been elegance and simplicity, with attention perhaps focussed on a set of problems satisfying “classical” assumptions, those considerations now usually have to take second place to the treatment of problems which are seen to be of practical importance, for which algorithms have to be robust and efficient.

The paper is effectively divided into two parts, the first (Section 2) being concerned with approximation by linear families, and the second (Section 3) being concerned with approximation by nonlinear families. These sections themselves further subdivide into two parts, where we consider separately approximation of data and of functions, and these are dealt with in that order within the two sections, with a further breakdown in what seems to be a reasonably natural way to take account of important special cases.

For the approximation of functions, we are primarily concerned with univariate functions on an interval $[a,b]$, because that is where most effort has been concentrated. However, some relevant comments are made on the extent to which multivariate functions may also be treated, with a few references made to this.

2. Linear approximation

The approximation of a given function defined on an interval by a linear combination of given functions is the most fundamental problem in approximation theory. The functions involved are usually continuous, and this can be thought of as a continuous infinite dimensional approximation problem. If the functions are replaced by vectors in $\mathbb{R}^m$, then we have a class of finite dimensional or discrete problems, many of which have their origins in data fitting. That solutions to linear best approximation problems always exist is a result which goes back at least to Riesz in 1918 [174]. We will consider the finite dimensional problem first, and begin by making some general remarks, before looking at special cases.

2.1. Linear approximation in $\mathbb{R}^m$

Let $A \in \mathbb{R}^{m\times n}$ where $m \geqslant n$, and let $\boldsymbol{b} \in \mathbb{R}^m$. Then the statement of a linear best approximation problem in $\mathbb{R}^m$ can be given as

$$\text{find } \boldsymbol{x} \in \mathbb{R}^n \text{ to minimize } \|\boldsymbol{r}\|, \qquad (1)$$

where

$$\boldsymbol{r} = A\boldsymbol{x} - \boldsymbol{b},$$

and $\|.\|$ is a given norm on $\mathbb{R}^m$. The dependence of $\boldsymbol{r}$ on $\boldsymbol{x}$ will generally be suppressed, unless confusion is possible.

This particular problem has attracted enormous interest. It will be assumed throughout that $\operatorname{rank}(A) = n$, and there is no $\boldsymbol{x}$ such that $\boldsymbol{r} = 0$. These are not essential, neither in theory nor in practice; however, they are conditions that are normally satisfied in practice, and their assumption considerably simplifies the presentation. If the norm is a differentiable function of $\boldsymbol{x}$, then we can easily characterize a minimum by zero derivative conditions: these are necessary, and, exploiting convexity, also sufficient. The best known example is when the norm is the least-squares norm, when zero derivative conditions just give the usual normal equations

$$A^{\mathrm{T}}A\boldsymbol{x} = A^{\mathrm{T}}\boldsymbol{b}.$$

The method of least squares is considered in detail elsewhere. But in a data fitting context, other l_p norms, particularly those for values of p satisfying $1 \leqslant p < 2$ are also important. The reason for this is that it is common for the usual conditions justifying the use of the l_2 norm not to hold, for example there may be wild points or gross errors in the data, and these other norms give reduced weight to these wild points. This is considered in Sections 2.2 and 2.3. Of great interest also has been the use of the Chebyshev norm; this is perhaps of less value in a data fitting context, but problems arise for example in continuous function approximation when the region of approximation is discretized. The problem is rich in structure and the theory is a beautiful one; we consider this case in Section 2.4.

We will restrict attention here to the problem (1), although there are many modifications of that problem which are relevant in a data fitting context. Most modifications have only been given serious treatment comparatively recently, and so they are of lesser interest from a historical point of view.

2.2. Linear l_1 approximation in $\mathbb{R}^m$

Consider now the problem (1) with the l_1 norm

$$\|\boldsymbol{r}\|_1 = \sum_{i=1}^{m} |r_i|. \tag{2}$$

This problem has a long history: its statement goes back well into the mid eighteenth century, and predates the introduction of least squares. Certainly, it was used in work of Laplace in 1786, in solving the overdetermined system of linear equations determining planetary orbits [110]. The first systematic methods for solving this problem seem due to Edgeworth [61]; in 1887 he gave a method based on tabulation, and in 1888 a method for the case when $n = 2$ which was essentially graphical and conceptual, but based on calculating descent directions. In 1930, Rhodes [167], motivated by the problem of fitting a parabola to data, tried Edgeworth's later method but found it "cumbrous". He gave a method where each iteration was calculated by solving 2 interpolation conditions for 2 of the parameters, and minimizing with respect to the remaining parameter. A proof that this kind of approach can give a solution was established by Singleton in 1940 [182]. A detailed historical account is given by Farebrother in a 1987 paper [63], covering the period 1793 to 1930. [1]

The first modern systematic study of this problem appears to be by Motzkin and Walsh [131,132] in the late 1950s, and characterization results are given in the 1964 book by Rice [172]. A convenient form of these may be deduced from these results or as a simple consequence of applying to this special case known results in abstract approximation theory: we will not attempt to go down that historical route, since it is something of a diversion from the main theme. However, it is the case

[1] The 1999 book by Farebrother [64] is also relevant.

that a vector $\boldsymbol{x} \in \mathbb{R}^n$ solves the l_1 problem if and only if there exists a vector $\boldsymbol{v} \in \mathbb{R}^m$ satisfying

$$A^T \boldsymbol{v} = 0,$$

where $\|\boldsymbol{v}\|_\infty \leqslant 1$, and $v_i = \text{sign}(r_i)$ whenever $r_i \neq 0$. The first simple (direct) proof of this was probably given by Watson [199] in 1980. A formal treatment of the important result that when A has rank n, a solution will be such that n components of r_i are zero, was given by Motzkin and Walsh [131] in 1955. In the context of the l_1 problem, any point characterized in this way can be defined to be a *vertex*. The interpolation result (in special cases) appears to have been known to Gauss, and to have been used in early methods: for example, the methods of Rhodes and Singleton are essentially vertex to vertex descent methods.

The results of Motzkin and Walsh were arrived at by direct consideration of the problem. However, its relationship with a linear programming problem was recognized around the same time, [2] and linear programming theory provides a parallel route to the same properties. Around 1947, Dantzig did his pioneering work on the simplex method of linear programming, and over the next few years, duality theory was developed, largely by von Neumann, Gale, Kuhn and Tucker. The significance of these developments for numerical methods for the l_1 (and the l_∞) problem cannot be overemphasized.

The first representation of the l_1 problem as a tractable linear programming problem seems due to Charnes et al. [35] in 1955. The key observation is that if extra variables $\boldsymbol{u}$ and $\boldsymbol{v} \in \mathbb{R}^m$ are introduced, then the problem can be posed as

$$\text{minimize} \sum_{i=1}^{m} (u_i + v_i) \text{ subject to} \tag{3}$$

$$[I : -I : A] \begin{bmatrix} \boldsymbol{u} \\ \boldsymbol{v} \\ \boldsymbol{x} \end{bmatrix} = \boldsymbol{b}$$

$$\boldsymbol{u} \geqslant 0, \quad \boldsymbol{v} \geqslant 0.$$

Since in the simplex method, no columns of I and $-I$ can simultaneously be basic, then

$$u_i v_i = 0, \quad i = 1, \ldots, m.$$

It follows that $u_i + v_i = |u_i - v_i|$ for all i and the equivalence of the simplex method applied to this problem with the minimization of (2) can readily be established.

Another version of the primal can be stated:

$$\text{minimize } \boldsymbol{e}^T \boldsymbol{s} \text{ subject to}$$

$$-\boldsymbol{s} \leqslant A\boldsymbol{x} - \boldsymbol{b} \leqslant \boldsymbol{s}.$$

This goes back at least to the 1964 Russian edition of the book by Zuhovitskii and Avdeyeva [211]. However, this form of the problem does not seem to have attracted as much attention as (3). The zero residuals will result in a form of degeneracy.

---

[2] Farebrother [63] in his 1987 paper interprets the work of Edgeworth in this context, and states that "..it must be conceded that Edgeworth had developed a fully operational, if somewhat complex, linear programming procedure for the L_1 estimation problem in 1888".

Fisher [66] in 1961 gave some publicity to (3) for the benefit of the statistical community, and this form was also used by Barrodale and Young [13] in 1966, who provided an Algol implementation and numerical results. The fact that the components of $\boldsymbol{x}$ may be non-negative is not a major problem in this context: for example, they can each be replaced by the difference of two non-negative variables. It was also noted that no first phase simplex calculation is required because an initial basic feasible solution can readily be obtained: if $b_i < 0$ then $\boldsymbol{e}_i$ can be present in the initial basis, if $b_i > 0$ then $-\boldsymbol{e}_i$ can be, with either used if $b_i = 0$.

The linear programming connection is sometimes wrongly credited to Wagner [192] in 1959, who posed the problem as a bounded variable or interval programming problem. In fact the form of the problem considered by Wagner [192] can be interpreted as the dual of (3). This can be written as

$$\text{maximize } \boldsymbol{b}^{\mathrm{T}}\boldsymbol{v} \text{ subject to} \tag{4}$$

$$A^{\mathrm{T}}\boldsymbol{v} = \boldsymbol{0}$$

$$-\boldsymbol{e} \leqslant \boldsymbol{v} \leqslant \boldsymbol{e},$$

where $\boldsymbol{e}$ is a vector with every component equal to 1. Attention was re-focussed on (4) by Robers and Ben-Israel [175] in 1969, and Robers and Robers [176] in 1973, who argued the advantages of that approach, which included computational efficiency: the problem with the primal appeared to be the large number of extra variables required. However, an improved version of the primal linear programming method was given by Davies [53] in 1967 and Barrodale and Roberts [10] in 1970, where a special pivot column selection rule was employed, and in 1973, both Spyropoulos et al. [183] and Barrodale and Roberts [11] gave efficient implementations of the simplex method applied to the primal which fully exploited the structure. The Barrodale and Roberts method achieved efficiency by taking multiple pivot steps, exploiting the fact that descent can continue beyond the usual point when feasibility is lost, because feasibility can readily be recovered by swapping certain variables into and out of the basis. Further efficiency was achieved by imposing certain restrictions on the choice of variables to enter and leave the basis. A Fortran programme and numerical results were provided, together with favourable comparisons with some other primal and dual methods [12].

In 1975, Abdelmalik [2] developed a special purpose method for the dual, using the dual simplex method, and his method seemed comparable with that of Barrodale and Roberts [10]. This turned out not really to be surprising, because, as pointed out by Armstrong and Godfrey [6] in 1979, the application of the dual simplex method to the dual is equivalent to applying the primal simplex method to the primal. So apart from implementation aspects, the methods were the same.

A basic feasible solution to (3) in which all columns of A are present in the basis can readily be shown to correspond to a vertex as defined above. Therefore, once the columns of A are present in the basis, the simplex method is a vertex to vertex descent method. There are many other variants of these linear programming methods, but away from a linear programming context, *direct* descent methods were being considered. For given $\boldsymbol{x}$, let

$$Z = \{i : r_i = 0\}.$$

Then since for full rank problems the solution occurs at a point $\boldsymbol{x}$ with Z containing n indices (a vertex), we want to systematically descend to such a point. Perhaps the first modern direct descent methods were given by Usow [189] in 1967, and Claerbout and Muir [43] in 1973. A natural way to implement descent methods is by first finding a vertex, and then descending through a sequence of vertices. Thus there are two types of step depending on whether at the current point, Z contains (a) fewer than n indices (b) exactly n indices. (The possibility that Z contains more than n indices corresponds to a degenerate situation, and although there are ways round it, will for our purposes be ignored.) Then in case (a) movement as far as possible is made in the direction $\boldsymbol{d}$ in such a way that the number of indices in Z at the new point is increased, and in case (b) movement as far as possible is made in the direction $\boldsymbol{d}$ in such a way that the number of indices in Z is maintained. Effective methods of this type, therefore, have this strategy in common, and are distinguished by the way the descent direction is calculated. There are mainly two approaches, (i) *reduced gradient methods*, where the "active constraints" are used to express certain variables in terms of others, the objective function is expressed in terms of the latter group, and its gradient is obtained in terms of those, and (ii) *projected gradient methods*, where the gradient is obtained by projecting the gradient of the objective function onto the orthogonal complement of the span of the active constraints.

Bartels et al. [18] in 1978 gave a projected gradient method, and reduced gradient methods were given by Osborne [147,148] in 1985 and 1987. Both projected and reduced gradient methods were analyzed in detail by Osborne [147] in 1985, and he pointed out that although reduced gradient methods seem more suitable for implementation using a tableau format, with updating, in fact such organization is available for implementing both methods. On relationships with linear programming methods, he showed that there is an exact equivalence between the possible options available in implementing the simplex method and those available in the direct application of the reduced gradient method. Thus these algorithms are equivalent: only the implementational details are different. The usual simplex step corresponds to a particular option in the reduced gradient method, based on an unnormalized steepest edge test for determining the variable to leave the basis. A different way of choosing this variable (a normalized steepest edge test, which is scale invariant) was used by Bloomfield and Stieger [26] in 1983, and their evidence showed that this can lead to improvement.

Nearly all the good methods considered to the end of the 1980s were vertex to vertex methods, which exploit the polyhedral nature of the function to be minimized, and (in the absence of degeneracy) they are finite. There has been recent interest in interior point methods for linear programming problems, stimulated by the results of Karmarker [102] in 1984. In conjunction with a formal connection with classical barrier methods for constrained optimization problems, this has resulted in renewed interest in linear programming, and there has of course been an impact on special cases such as the l_1 problem.

The use of interior point methods for l_1 problems goes back at least as far as work of Meketon [127] in 1987, and methods have been given since then by Ruzinsky and Olsen [178] in 1989, Zhang in 1993 [209] and Duarte and Vanderbei [56] in 1994. Portnoy and Koenker [157] in 1997 make a case for the superiority of interior point methods over simplex-based methods for large problems. Based on comparisons of l_1 problems having n up to 16 and m from 1000 to 200 000, they conclude that there is "a compelling general case for the superiority of interior point methods over traditional simplex methods for large linear programming problems". Their algorithm of choice for the l_1 problem is based on a primal–dual log barrier method due to Mehrotra [123] in 1992, and includes a statistical preprocessing approach which estimates whether a residual is zero or not. The opposition is represented by a variant of the Barrodale and Roberts method.

Meantime, two other types of smoothing method were being developed for the l_1 problem.[3] The first of these is typified by an algorithm of Coleman and Li [46] in 1992, which is based on affine scaling: while not strictly speaking an interior point method, it is nevertheless in the spirit of such methods. Here, an attempt is made to satisfy the characterization conditions by an iterative descent method which has the following characteristics: (a) it generates a sequence of points which are such that Z is empty, so that derivatives exist, (b) it is globally convergent, (c) it ultimately takes damped Newton steps (damped to satisfy (a)), but with sufficiently accurate approximations to the full Newton step to permit quadratic convergence (under nondegeneracy conditions). Careful implementation of the method can avoid difficulties with near-zero components of $\boldsymbol{r}$ and the approach seems promising for large problems as it is insensitive to problem size. Some comparisons show that it is superior to Meketon's interior point method for problems with n up to 200, m up to 1000.

A second approach to smoothing the l_1 problem was developed by Madsen and Nielsen [116] in 1993. It is based on the use of the Huber M-estimator, defined by

$$\psi_\gamma \equiv \psi_\gamma(\boldsymbol{r}) = \sum_{i=1}^{m} \rho(r_i), \tag{5}$$

where

$$\rho(t) = \begin{cases} t^2/2, & |t| \leqslant \gamma, \\ \gamma(|t| - \gamma/2), & |t| > \gamma, \end{cases} \tag{6}$$

and γ is a scale factor or tuning constant. The function (5) is convex and once continuously differentiable, but has discontinuous second derivatives at points where $|r_i| = \gamma$. The mathematical structure of the Huber M-estimator seems first to have been considered in detail by Clark [44] in 1985. Clearly if γ is chosen large enough, then ψ_γ is just the least-squares function; in addition if γ tends to zero, then limit points of the set of solutions may be shown to minimize the l_1 norm. It is the latter property which concerns us here.

It has been suggested by Madsen and Nielsen [116] in 1993 and also by Li and Swetits [113] in 1998 that the preferred method for solving the l_1 problem is via a sequence of Huber problems for a sequence of scale values $\gamma \to 0$. This algorithmic development has lead to increased interest in the relationship between the Huber M-estimator and the l_1 problem; for example there is recent work of Madsen et al. [117] in 1994, and Li and Swetits [113] in 1998. The method of Madsen and Nielsen generates Huber solutions for a sequence of values of γ, tending to zero. The solutions are obtained by solving least-square problems, exploiting structure so that new solutions can be obtained using updating often in $O(n^2)$ operations. A key feature is that it is not necessary to let γ reach zero; once a sufficiently small value is identified, then the l_1 solution may be obtained by solving an $n \times n$ linear system. Madsen and Neilsen give some comparisons (for randomly generated problems, and with m mostly set to $2n$ for m up to 1620) with the method of Barrodale and Roberts [10] and claim superiority.

An important issue as far as the implementation of simplex type methods is the efficiency of the line search. The Barrodale and Roberts [10] method incorporates the equivalent of a comparison sort, and this leaves room for considerable improvement. Bloomfield and Stieger [26] considered

[3] The observation that a best approximation can always be computed as the limit of a sequence of l_p approximations as $p \to 1$ is due to Fischer [65] in 1983 (an algorithm based on this was in fact given by Abdelmalik [1] in 1971), although this is not a very practical approach.

this aspect in their 1983 book, and suggested using a fast median method. An alternative based on the use of the secant algorithm was considered (in a related context) by George and Osborne [71] in 1990, and again by Osborne [147] in 1985. Numerical experiments were reported by Osborne and Watson [154] in 1996, where the secant-based method was seen to be as good as fast median methods on randomly generated problems, and to perform considerably better on problems with systematic data. Comparisons of other types of method with simplex methods really need to take this into account before definitive conclusions can be drawn.

2.3. Linear l_p approximation in $\mathbb{R}^m$, $1 < p < \infty$, $p \neq 2$

For given $\boldsymbol{x} \in \mathbb{R}^n$, let $D_{|r|}$ be defined by

$$D_{|r|} = \text{diag}\{|r_1|, \dots, |r_m|\}.$$

Then $\boldsymbol{x}$ minimizes

$$||\boldsymbol{r}||_p^p = \sum_{i=1}^{m} |r_i|^p$$

with $1 < p < \infty$ if and only if derivatives with respect to $\boldsymbol{x}$ are zero, that is if

$$A^{\mathrm{T}} D_{|r|}^{p-1} \theta = 0, \tag{7}$$

where $\theta_i = \text{sign}(r_i)$, $i = 1, \dots, m$. This is a nonlinear system of equations for $\boldsymbol{x}$.

This criterion (for p even) was mentioned by Gauss as a generalization of his least-squares criterion. Apart from this special case, the more general l_p problem only seems to have attracted relatively recent computational attention. The range $1 < p < 2$ is of particular interest computationally because there is potentially reduced smoothness: problems with $p \geqslant 2$ are twice differentiable, but problems with $1 < p < 2$ may be only once differentiable. If $p \geqslant 2$ or if $1 < p < 2$ and no component of $\boldsymbol{r}$ is zero then twice differentiability is guaranteed and so (7) can be written as

$$A^{\mathrm{T}} D \boldsymbol{r} = 0, \tag{8}$$

where

$$D = D_{|r|}^{p-2},$$

and this is a particularly convenient form with which to work. It represents a generalized system of normal equations, effectively a least-squares problem apart from the "weighting" matrix D. Fixing $\boldsymbol{x}$ to an approximate value in D and solving this weighted system for a new approximation gives an example of the technique known as *iteratively reweighted least squares* or IRLS, which seems to have been introduced by Beaton and Tukey [20] in 1974. Since good software for (weighted) least-squares problems was then available, this seemed an attractive idea, additionally so since there are some apparently good theoretical properties: this simple iteration process will converge locally if p is close to 2, and if zero components of $\boldsymbol{r}$ are avoided, it is globally convergent (from any initial approximation) for $1 < p < 2$. The last result seems first to have been given by Dutter [60] in 1975. However, convergence can be slow, particularly as p nears 1 (it is linear with convergence constant $|p - 2|$, as shown by Wolfe [206] in 1979), and there are potential numerical difficulties for reasons which will be clear from the previous section. The matrix D (which may not exist) can be replaced by approximations (even by the unit matrix), and this gives rise to variants of the IRLS technique, but again convergence can be very slow.

Most recent algorithms for solving (8) are based on Newton's method, and many variants were proposed in the 1970s. It is interesting that the Newton step is just $1/(p-1)$ times the IRLS step (as measured by the difference between successive approximations), as pointed out by Watson [196] in 1977, and this gave an explanation of some success obtained by Merle and Späth [128] in 1974 in using a damped IRLS procedure with step length $(p-1)$. Thus apart from differences due to the line search, IRLS and Newton's method with line searches *are essentially the same method.* It is easily seen that the region of convergence of Newton's method is proportional to $|(p-1)/(p-2)|$, so good line search procedures are needed even with the basic method, certainly far from $p=2$. However, for $p>2$, Newton's method with line search is usually perfectly satisfactory.

Since from a practical point of views the interesting cases are those when $1<p<2$, different strategies have been proposed for getting round the difficulties arising from zero (or near zero) components of $\boldsymbol{r}$. These included the substitution of small nonzero values, solving a slightly perturbed problem, or identifying and so removing these components from the set. However, not just zero components but *nearly zero* components are potentially troublesome. There is some evidence, however, that these phenomena are not *by themselves* a major problem, but only if they are accompanied by p being close to 1. The main difficulty appears to be due the fact that as p approaches 1, we are coming closer to a discontinuous problem, effectively to a constrained problem. It seems necessary to recognize this in a satisfactory algorithm, and consider some of the elements of the l_1 problem in devising an approach which will deal in a satisfactory with small values of p. This is the philosophy in a recent method due to Li [114] in 1993, which is essentially equivalent to the method for the l_1 problem of Coleman and Li [46] referred to in the previous section. Numerical results show that the new method is clearly superior to IRLS (with the same line search) for values of p close to 1, with the gap between the two methods widening as p approaches 1. There is little difference for values of $p \geqslant 1.5$ or so. As with the l_1 case, the number of iterations appears to be independent of the problem size.

2.4. Linear Chebyshev approximation in $\mathbb{R}^m$

The use of the criterion now known as the Chebyshev norm

$$\|\boldsymbol{r}\|_\infty = \max_i |r_i|, \tag{9}$$

seems to go back to Laplace in 1786, who gave a solution procedure for $n=2$. Cauchy in 1814 and Fourier in 1824 gave descent methods. A detailed historical account is given by Farebrother [63] in his 1987 paper, covering the period 1793 to 1824. The function space analogue was studied first by Chebyshev[4] from the 1850s, arising from an analysis of a steam engine linking, and both continuous and discrete problems now carry his name.

For any $\boldsymbol{x} \in \mathbb{R}^n$, let

$$\bar{I}(\boldsymbol{x}) = \{i\colon |r_i(\boldsymbol{x})| = \|\boldsymbol{r}\|_\infty\}.$$

[4] The number of variants in the western literature which have been used for Chebyshev is legendary, but most people now seemed to have settled on this one. Butzer and Jongmans [33] in 1999 gave a detailed account of Chebyshev's life and work.

Then $\boldsymbol{x}$ is a solution if and only if there exists a subset $I \subset \bar{I}$ containing at most $n+1$ indices, and a nontrivial vector $\lambda \in \mathbb{R}^m$ such that

$$\lambda_i = 0, \quad i \notin I,$$

$$A^{\mathrm{T}}\lambda = 0,$$

$$\lambda_i \operatorname{sign}(r_i) \geqslant 0, \quad i \in I.$$

This is an example of a "zero in the convex hull" type of characterization result, and is the discrete analogue of Kirchburger's 1903 result for the continuous problem [106]. A simple consequence is that for the full rank problem, there always exists a solution with $\bar{I}$ containing $n+1$ indices. Such a point can be thought of as a *vertex*.

An early method for the minimization of (9) was the Polya algorithm [156], which solves a sequence of l_p problems with $p \to \infty$: the assumption here is that the l_p problems are relatively easy to solve, being differentiable for large finite p. This method was given (in fact for continuous functions) in 1913, and convergence is readily established if the Chebyshev solution is unique. A proof of convergence to a particular Chebyshev approximation called the strict Chebyshev approximation (in the event of nonuniqueness of the Chebyshev solution) was given by Descloux [54] in 1963. Fletcher et al. [69] in 1974 used an extrapolation technique to accelerate convergence of the Polya algorithm, and in the same year Boggs [27] used a technique based on deriving a differential equation describing the l_p solution as a function of p. An algorithm due to Lawson [111] in 1961 was based on the solution of a sequence of weighted least-squares problems, but like the Polya algorithm, it can be very slowly convergent. Indeed none of these methods has been regarded as giving a particularly practical approach.

A fundamental assumption which was identified as important at an early stage was the *Haar condition*, that every $n \times n$ submatrix of A is nonsingular. This is sufficient for uniqueness of $\boldsymbol{x}$ minimizing (9) (and also necessary in the case when $m = n+1$). This "classical" assumption played a major role in the minimization of (9) until the 1960s. It goes back to Haar [79] in 1918.

Before proceeding, it is helpful to point out an important property which is satisfied at a minimum of (9). The result, due to de la Vallée Poussin [190] in 1911, tells us that if J runs through all subsets of $n+1$ indices from $\{1, \ldots, m\}$, then

$$\min_{x} \max_{i} |r_i| = \max_{J} \left\{ \min_{x} \max_{i \in J} |r_i| \right\}. \tag{10}$$

Therefore, if we can identify a set J where the maximum on the right-hand side is attained, solving a Chebyshev problem *on that subset* (and this is relatively easy) will give a solution to the original problem. For any J such that the corresponding problem matrix is full rank, the solution on J will occur at a vertex. Therefore, if A has full rank, so that the problem has a solution at a vertex, then it is sufficient to investigate all the vertices in a systematic way.

An exchange algorithm for finding an extremal subset or optimal vertex was given by Stiefel [184] in 1959. It assumed that A satisfied the Haar condition, and worked with a sequence $J_1, J_2, \ldots$ of subsets of $n+1$ components of $\boldsymbol{r}$. The key aspect of the method was that J_{k+1} differed from J_k by one index, and a rule was given for exchanging one of the indices in J_k for another index outside it to give J_{k+1} in such a way that

$$h_{k+1} > h_k,$$

where

$$r_i(\boldsymbol{x}) = \theta_i h_k, \quad i \in J_k, \tag{11}$$

with $|\theta_i| = 1$, $i \in J_k$. Thus, we have an example of a vertex-to-vertex *ascent method.* Because there are only a finite number of selections of $n + 1$ indices from m the method must terminate, when from (10) it follows that a solution has been obtained. If the Haar condition is not satisfied, then strict inequality may hold for successive h_k values and the theory of the method is compromised.

Because the function of $\boldsymbol{x}$ given by (9) is piecewise linear, the Chebyshev problem may be posed as a linear programming problem, and as in the l_1 case, properties of the problem are again available through this route. Let $h = \max_i |r_i|$. Then the problem may be stated

minimize h subject to

$$-h \leqslant r_i \leqslant h, \quad i = 1, \ldots, m.$$

In terms of the variables h and $\boldsymbol{x}$, this may be restated

$$\text{minimize } z = \boldsymbol{e}_{n+1}^{\mathrm{T}} \begin{bmatrix} \boldsymbol{x} \\ h \end{bmatrix} \text{ subject to}$$

$$\begin{bmatrix} A & \boldsymbol{e} \\ -A & \boldsymbol{e} \end{bmatrix} \begin{bmatrix} \boldsymbol{x} \\ h \end{bmatrix} \geqslant \begin{bmatrix} \boldsymbol{b} \\ -\boldsymbol{b} \end{bmatrix}.$$

One of the first to consider the linear programming formulation of the problem was Zuhovickii [211,212] in a series of papers originating in the Russian literature in the early 1950s. The above form is not particularly suitable for the application of standard techniques such as the simplex method because $2m$ slack variables are required, the basis matrices will be $2m \times 2m$, and although h is nonnegative, this is not true in general of the components of $\boldsymbol{x}$.

All of these difficulties are overcome by turning to the dual problem, which is

$$\text{maximize } z = [\boldsymbol{b}^{\mathrm{T}}, \ -\boldsymbol{b}^{\mathrm{T}}]\boldsymbol{w} \text{ subject to}$$

$$[A^{\mathrm{T}} \quad -A^{\mathrm{T}}]\boldsymbol{w} = 0,$$

$$[\boldsymbol{e}^{\mathrm{T}} \quad \boldsymbol{e}^{\mathrm{T}}]\boldsymbol{w} \leqslant 1,$$

$$\boldsymbol{w} \geqslant 0.$$

Only one slack variable is required (to make the inequality an equality), the basis matrices are only $(n + 1) \times (n + 1)$, and all the variables are nonnegative. The advantage in using the dual seems to have been first pointed out by Kelley in [105] 1958 in an application to curve fitting. Standard linear programming theory tells us that if a variable is dual basic, then the corresponding primal constraint holds with equality. Thus a basic feasible solution corresponds precisely to a solution to a set of equations having the form (11). It would appear therefore that there is a precise equivalence between a step of the simplex method applied to the dual, and a step of the Stiefel exchange method. This result was known to Stiefel [185] in 1960, who gave an indication of it by considering a small problem and using a geometric argument. He also (unnecessarily) eliminated the unconstrained variables from the primal before proceeding to the dual.

The precise equivalence was first worked out in detail by Osborne and Watson [149] in 1967, although Bittner [24] in 1961 examined how linear programming could be used to relax the Haar condition. In order for the usual simplex method to be applied to the dual, the Haar condition is not required, only the nonsingularity of successive basis matrices: for this it is only necessary for A to have full rank n. The point here is that the simplex method does not permit nonsingular basis matrices. Note however that if the Haar condition does not hold, degeneracy may cause cycling in the simplex algorithm, although this can always be resolved.

A programme implementing the standard simplex method for the problem considered here seems first to have been given by Barrodale and Young [13] in 1966, who gave an Algol programme along with numerical results, and Bartels and Golub [19] gave a version in 1968 which used a numerically stable factorization procedure. In 1975, Barrodale and Phillips [8] used the special structure present in the dual formulation to greatly reduce the number of iterations required: conditions were imposed on variables entering and leaving the basis, and the fact exploited that some variables could easily be exchanged for others. The usual simplex rules were modified to permit ascent through a number of vertices, beyond the one which would usually be reached in a simplex step, by exploiting the fact that feasibility could easily be regained by such exchanges. Modifications of this basic technique to allow more than one index to be exchanged at each step were given by Hopper and Powell [90] in 1977 and by Armstrong and Kung [7] in 1979.

The Stiefel exchange method and variants which solve the dual formulation of the problem are examples of ascent methods, whose justification is based on (10). However, it is possible to solve the problem by a *descent* process. The primal linear programming problem is an example of a descent method, and although its direct solution is not recommended for the reasons already given, it is nevertheless possible to implement satisfactory descent methods.

As for the l_1 problem, good direct descent methods might be expected to follow the common strategy of having (in the absence of degeneracy) basically two types of step depending on whether the current point x is such that $\bar{I}$ contains (a) fewer than $n+1$ indices, or (b) exactly $n+1$ indices. In a manner precisely analogous to that considered for the l_1 problem, a strategy can be developed which ultimately gives a vertex-to-vertex descent process. Methods of reduced gradient type were given by Cheney and Goldstein [37] in 1959 and Cline [45] in 1976. A projected gradient method was given by Bartels et al. [16] in 1978. It appeared to be the case that such methods did not seriously compete with ascent methods. However, improvements in descent methods were considered by Bartels et al. [17] in 1989: they argued that the good performance of the Barrodale and Phillips method was due to the way the method chose a good starting point. By modifying the way in which a starting point is obtained for their descent method, they enhanced its performance and made a case for its superiority for data fitting problems.

All the approaches considered so far are essentially vertex-to-vertex methods. They exploit the polyhedral nature of the function to be minimized, and are of course (in the absence of degeneracy) finite. The recent interest in interior point methods for linear programming problems has, as in the l_1 case, extended to the special case of Chebyshev problems. Ruzinsky and Olsen [178] in 1989, Zhang [209] in 1993 and Duarte and Vanderbei [56] in 1994 all proposed interior point methods. An affine scaling algorithm analogous to that for the l_1 problem was given by Coleman and Li [47] in 1992. This is a descent method which involves a sequence of least-squares problems to define descent directions. It provides a smooth transition from guaranteed descent steps far from a solution, to steps close to a solution which are sufficiently accurate approximations to the Newton step to permit quadratic convergence under suitable nondegeneracy assumptions.

In contrast to the l_1 situation, detailed comparisons of other methods with simplex type methods for large problems do not yet seem to be available. It should in any event not be assumed that conclusions can be drawn from the l_1 case, because large Chebyshev problems normally arise from discretizations of continuous Chebyshev approximation problems on intervals or multidimensional regions, and the data are highly systematic. Indeed, the solution is then normally part of a method for the continuous problem, or exploits the connection: we will defer further consideration of this until the following Section.

2.5. Linear Chebyshev approximation in $C[a,b]$

Let $C[a,b]$ denote the set of continuous functions defined on the real interval $[a,b]$, and let $f(x)$, $\phi_1(x),\dots,\phi_n(x)$ be in $C[a,b]$. Then the usual Chebyshev approximation problem in $C[a,b]$ can be expressed as

$$\text{find } \boldsymbol{a} \in \mathbb{R}^n \text{ to minimize } ||f - \phi||_\infty, \tag{12}$$

where $\phi = \sum_{i=1}^n a_i\phi_i(x)$, and

$$||f||_\infty = \max_{a \leqslant x \leqslant b} |f(x)|.$$

This class of problems was systematically investigated by Chebyshev from the 1850s, although Chebyshev credits Poncelet with originating the problem. The "classical" case occurs when the set of functions forms a Chebyshev set (or is a Haar subspace) on $[a,b]$, that is any nontrivial linear combination has at most $(n-1)$ zeros; the model problem here is approximation by polynomials of degree $n-1$. The problem (12), with the interval $[a,b]$ replaced by m points in $[a,b]$, reduces to a problem of the form considered in the previous section. Indeed it is readily seen that the matrix A in this case satisfies the Haar condition if and only if the set of functions $\phi_1(x),\dots,\phi_n(x)$ forms a Chebyshev set on $[a,b]$. Continuing this theme for a moment, arbitrarily good solutions to (12) can be obtained by choosing finer and finer discretizations; the main convergence results here are due to Motzkin and Walsh [132] in 1956. Although this observation by itself does not give practical algorithms, the use of a sequence of discretizations, where successive point sets are carefully chosen, is the key to the success of many good algorithms.

A general characterization result was obtained by Kirchberger [106] in 1903. Let

$$\bar{E} = \{x \in [a,b], |r(x,\boldsymbol{a})| = ||r(.,\boldsymbol{a})||_\infty\}.$$

Then $\boldsymbol{a}$ is a solution if and only if there exists $E \subset \bar{E}$ containing $t \leqslant n+1$ points $x_1,\dots,x_t$ and a nontrivial vector $\lambda \in \mathbb{R}^t$ such that

$$\sum_{i=1}^{t} \lambda_i\phi_j(x_i) = 0, \quad j = 1,\dots,n,$$

$$\lambda_i \,\text{sign}(r_i) \geqslant 0, \quad i = 1,\dots,t.$$

Borel [28] in 1905 established the well-known alternation result for approximation by degree $(n-1)$ polynomials, that $\boldsymbol{a}$ is a solution if and only if there are $n+1$ points in $[a,b]$ where the norm is attained with alternating sign as we move from left to right through the points; we can state this concisely in the form

$$\mathscr{A}(f - p_{n-1})_{[a,b]} \geqslant n + 1,$$

where p_n denotes the best degree n polynomial approximation. Uniqueness of solutions under these conditions is also due to Borel [28] in 1905. That this result extends to approximation by functions forming a Chebyshev set was shown by Young [208] in 1907, who also established uniqueness in this case. Haar [79] in 1918 showed that the solution is unique for all possible functions $f(x)$ if and only if $\phi_1(x), \ldots, \phi_n(x)$ forms a Chebyshev set on $[a, b]$.

Polya [156] in 1913 gave his algorithm for this problem, where a sequence of continuous L_p problems is solved with $p \to \infty$. A counterexample to a general convergence result for non-Chebyshev set problems was given by Descloux [54] in 1963. As in the discrete case this is anyway not a particularly effective approach.

Two important algorithms for solving the Chebyshev problem were given by Remes [165,166] in the 1930s. The method traditionally known as the "Second Algorithm" applies to Chebyshev set problems, exploiting the alternation property. It solves a sequence of discrete problems in $\mathbb{R}^{n+1}$ defined by sets of $n+1$ points in $[a, b]$: each of these is just the solution of a system of $n+1$ equations for $n + 1$ unknowns, using the fact the solutions have an alternation property. By exchanging the current set of $n + 1$ points for $n + 1$ local maxima of the modulus of the error function, subject to some simple rules, an ascent process is obtained. Under mild conditions this converges to the (unique) Chebyshev approximation, and at a second-order rate: the result, due to Veidinger [191] in 1960, is based on showing that the method is asymptotically Newton's method for solving the characterization conditions. A comparatively modern implementation of the method was given by Golub and Smith [73] in 1971. Note that if only one point is exchanged at each iteration (bringing in a point where the norm of the error is attained), then an equivalence can be drawn between a step of the method and a step of the Stiefel exchange method. An analysis of the one-point exchange method is given by Powell [158] in his 1980 book, where it is shown that this method also converges at a second-order rate.

The "First Algorithm of Remes" applies to general problems. Again it corresponds to the solution of a sequence of discrete problems, but of increasing size. Starting with a solution on $m_1 \geqslant (n + 1)$ discrete points in $[a, b]$, a point where the error function attains the norm is added, and a new solution obtained on $m_1 + 1$ points. If the matrix A of the initial problem has rank n, then successive matrices also have rank n and so linear programming techniques, for example, can be used, and implemented efficiently using postoptimality theory. This is an "implicit" exchange method, since every solution corresponds to a vertex defined on the current set of points. In fact since much of the work in implementing such a method lies in finding a global maximum of the error function, and this would normally involve calculating all the local maxima, it is sensible to add in all such local maxima: the method is then an implicit multiple exchange method. For Chebyshev set problems this is equivalent to the second algorithm of Remes. Modifications of the first algorithm of Remes to allow multiple exchanges have been considered by Carasso and Laurent [34] in 1978, and Blatt [25] in 1984, based on constructing "chains of references".

Unfortunately there are examples where this kind of approach performs badly, when the solution to the continuous problem does not occur at a vertex, that is it attains the norm in fewer than $n + 1$ points: such problems were called singular problems by Osborne and Watson [151] in 1969. Note that this phenomenon is specific to problems on a continuum, and has no analogue in the (full rank) discrete case. Therefore, because each discrete problem has a solution at a vertex, the limiting situation in this case is obtained by some of these points coalescing, slowing down convergence and giving ill-conditioned simplex bases.

For multivariate problems (where x is a vector in $\mathbb{R}^s$, $s > 1$), singularity is very common. A partial explanation for this is that Chebyshev sets of more than one function do not exist in continuums of dimension higher than one: this was first pointed out by Mairhuber [119] in 1956. Nevertheless, a method of this type can be developed for multivariate problems, as demonstrated by Watson [194] in 1975.

Therefore, there are two main difficulties with such methods: (a) the calculation of the local and global maxima of the error function, (b) the problem of singularity. It is perhaps only recently that close attention has been paid to efficient calculation in (a), for example by Reemtsen [163] in 1991, and Price and Coope [159] in 1996: it is usually assumed that all local maxima can be calculated to sufficient accuracy, and so the relevant algorithms are always implementable. But attempts to avoid (a) have been made, for example by Dunham [58] in 1981, Hettich [85] in 1986 and Reemtsen [162] in 1990. The main idea is to only require maxima of the error at each step on a grid, where the discretization error tends to zero as the method progresses. In particular, Reemtsen proved the convergence of a modified version of the first algorithm of Remes, in which the maximum of the kth error function was computed on a grid, with the grid density tending to zero. The method of Hettich is also based on successive grid refinement (and using a numerically stable simplex algorithm) and applies to one- and two-dimensional problems; solutions have been successfully obtained for problems with n up to 37.

An alternative approach which tries to avoid both (a) and (b) is through the use of *two-phase methods*. The first phase involves the solution of a single discretization of the problem, on a sufficiently dense set to enable identification of the number of points (with signs) where the norm is attained and good approximations of these. In the second phase, the characterization conditions, together with zero derivative conditions at points identified as extrema in (a, b), can then be solved (for example by Newton's method). This main idea for an approach of this type (in a more general context) is due to Gustafson [77] in 1970. Its application to Chebyshev approximation problems was considered by a number of people in the mid-1970s, among them Gustafson, Hettich, Andreassen and Watson [5,78,84]. The approach can be successful, but while the difficulty (a) above is essentially removed, (b) can still emerge in the first phase, and there is also the (new) difficulty of having to decide what level of discretization to use, or when to enter the second phase, and also when the information provided at that point is completely reliable. It may be necessary to permit re-entry to phase 1 with a more stringent exit criterion.

The second phase can be considered in two ways, depending on whether or not the local maxima are considered as differentiable functions of the unknown parameters, and whether or not this is exploited. If it is, then the zero derivative conditions can be used to eliminate these maxima in terms of the other unknowns, and there is a consequent reduction in the size of the linear system to be solved for the Newton step.

Of course the second phase applies equally to nonlinear problems, so we will return to some of these ideas in Section 3.4. Indeed, continuous Chebyshev approximation problems (both linear and nonlinear) are special cases of semi-infinite programming problems, that is problems with a finite number of variables and an infinite number of constraints, and many algorithms which have been developed for the more general problem class may be adapted for the Chebyshev approximation problem. Semi-infinite programming is an active research area – the recent survey paper of Reemtsen and Görner [164] in 1998 has 233 references, 96 of them dated 1990 or later. Algorithmic development has encompassed methods based on the ideas considered above, but also other approaches, for example the use of interior point methods. These are of comparatively recent origin, their usefulness (certainly as far as continuous Chebyshev approximation is concerned) does not appear to have been established, and we will not consider them further here.

2.6. Chebyshev approximation by splines with fixed knots

Approximation by splines is considered in some detail elsewhere, so we will not go into the history of the origins of this class of function. The main focus of approximation by splines has been on interpolation; however, Chebyshev approximation by splines has also attracted a lot of attention. Because we are concerned at present with linear problems, we assume in the present section that the knots are fixed a priori, and we will consider approximation from the space of spline functions defined as follows. Let integers m and k be given, and let $a = x_0 < x_1 < \cdots < x_{k+1} = b$. Then

$$S_m = \{s \in C^{m-1}[a,b] : s(x) \in \Pi_m \text{ on } [x_i, x_{i+1}],\ i = 0, \ldots, k\},$$

where Π_m denotes the space of polynomials of degree m, is the space of polynomial splines of degree m with k fixed knots. S_m is a linear space with dimension $m + k + 1$. The first results on Chebyshev approximation by splines seem to be due to Johnson [96] in 1960.

The theory of approximation by Chebyshev sets does not apply to approximation from S_m. However, S_m is an example of a family of functions forming a *weak Chebyshev set*: any linear combination of such a set of n functions has at most $(n - 1)$ changes of sign. For such sets Jones and Karlowitz [100] showed in 1970 that there exists at least one best Chebyshev approximation ϕ to any continuous function f which has the classical alternation property

$$\mathscr{A}(f - \phi)_{[a,b]} \geqslant n + 1,$$

(although there may be others which do not).

The theory of Chebyshev approximation by splines with fixed knots is fully developed, and a characterization of best approximation goes back to the Ph.D. dissertation of Schumaker in 1965, and his publications over the next few years, e.g. [180]. Results were also given by Rice [173] in 1967. What is required is the existence of an interval $[x_p, x_{p+q}] \subset [a,b]$, with $q \geqslant 1$ such that there are at least $q + m + 1$ alternating extrema on $[x_p, x_{p+q}]$, or in the notation previously introduced

$$\mathscr{A}(f - s)_{[x_p, x_{p+q}]} \geqslant q + m + 1,$$

where $s \in S_m$. In addition to characterization of solutions, there has been interest in conditions for uniqueness (and strong uniqueness) of best approximations. In general of course, best approximations are not unique. However, the uniqueness (and strong uniqueness) of best spline approximations is characterized by the fact that all knot intervals contain sufficiently many alternating extrema as shown by Nürnberger and Singer [143] in 1982.

The solution of a discretized problem by linear programming techniques was suggested by Barrodale and Young [14] in 1966 and also by Esch and Eastman in an 1967 technical report (see their 1969 paper [62]). These methods do not make explicit use of characterization results, in contrast to the (explicit) Remez exchange method of Schumaker presented again in technical reports about the same time (see his 1969 paper [181]). The latter method also solved the discretized problem, but had no convergence results.

Of course any methods for best Chebyshev approximation by linear functions may be used, but a special iterative algorithm for computing best Chebyshev approximations from spline spaces was given by Nürnberger and Sommer [144] in 1983. As in the classical Remes method, a substep at each iteration is the computation of a spline $s \in S_m$ such that

$$(-1)^i(f(\xi_i) - s(\xi_i)) = h, \quad i = 1, \ldots, m+k+2,$$

for some real number h, and given points $\xi_1, \ldots, \xi_{m+k+2}$ in $[a,b]$. The number of equations reflects the fact that S_m has dimension $m+k+1$. Then one of the points ξ_i is replaced by a point where $||f-s||$ is attained in $[a,b]$ to get a new set of points $\{\xi_i\}$. The usual Remes exchange rule can result in a singular system of equations, so a modified exchange rule is needed. Such a rule was given by Nürnberger and Sommer [144], which ensures that the new system has a unique solution. Because of possible nonuniqueness of best approximations, the proof of convergence is fairly complicated. However, convergence can be established.

A multiple exchange procedure can also be implemented, and quadratic convergence is possible. The above results can be extended to more general spline spaces, where the polynomials are replaced by linear combinations of functions forming Chebyshev sets: this was considered by Nürnberger et al. [141] in 1985.

To permit the full power of splines, one should allow the knots to vary, rather than be fixed in advance. The corresponding approximation problem is then a difficult nonlinear problem and we say more about this in Section 3.7.

2.7. Linear L_1 approximation in $C[a,b]$

Given the same setting as at the start of Section 2.5, we consider here the problem

$$\text{find } \boldsymbol{a} \in \mathbb{R}^n \text{ to minimize } \int_a^b \left| f(x) - \sum_{i=1}^n a_i\phi_i(x) \right| \mathrm{d}x. \tag{13}$$

This problem was apparently first considered by Chebyshev in 1889.

Characterization results go back to James [93] in 1947. A convenient form is the analogue of that available in the discrete case: a vector $\boldsymbol{a}$ solves the L_1 problem if and only if there exists a function v with $||v||_\infty \leqslant 1$ such that

$$\int_a^b v(x)\phi_j(x)\,\mathrm{d}x = 0, \quad j = 1, \ldots, n,$$

$$v(x) = \operatorname{sign} r(x), \quad r(x) \neq 0.$$

If the set $\{\phi_1(x), \ldots, \phi_n(x)\}$ forms a Chebyshev set in $[a,b]$, then Jackson [92] in 1921 showed that the solution is unique. For polynomial approximation, perhaps the first "algorithm" was given by Hoel [89] in 1935, who showed that the polynomials of best L_p approximation converge to the best L_1 approximation as $p \to 1$. This is the analogue of the Polya algorithm for Chebyshev approximation. A more general convergence result, and a characterization of the limiting element, was given by Landers and Rogge [109] in 1981.

The L_1 problem is greatly simplified if it can be assumed that the zeros of $f(x) - \sum_{i=1}^{n} a_i\phi_i(x)$ form a set of measure zero in the interval $[a,b]$ (for example the zeros just consist of a finite set of points). Then the function to be minimized in (13) is differentiable, and necessary and sufficient conditions for a solution are that

$$\int_a^b g(x,\boldsymbol{a})\phi_j(x) = 0, \quad j = 1,\ldots,n,$$

where $g(x,\boldsymbol{a})$ denotes the sign of $f(x) - \sum_{i=1}^{n} a_i\phi_i(x)$. This was known to Laasonen [107] in 1949. This means that great store is placed on the points where there are sign changes, or equivalently where the approximation interpolates f. If these points were known, and were exactly n in number, then we could compute the best approximation by interpolation, *provided that there were no other changes of sign in the error of the resulting approximation.* The points $x_1 < \cdots < x_t \in (a,b) = (x_0, x_{t+1})$, where $1 \leqslant t \leqslant n$, are called *canonical points* if

$$\sum_{i=0}^{t}(-1)^i \int_{x_i}^{x_{i+1}} \phi_j(x)\,\mathrm{d}x = 0, \quad j = 1,\ldots,n. \tag{14}$$

For the Chebyshev set case, Laasonen [107] in 1949 showed that there is a unique sign function and further $t = n$. This was extended to weak Chebyshev sets by Micchelli [129] in 1977. Existence of a set of $t \leqslant n$ canonical points for the general problem was shown by Hobby and Rice [87] in 1965.

For the special case when $\phi_i(x) = x^{i-1}$, $i = 1,\ldots,n$, then the location of the n canonical points is known – they lie at the zeroes of the Chebyshev polynomial of the second kind of degree n (shifted if necessary). This result is due to Bernstein [23] in 1926. Thus their location is independent of f. Interpolation at these points can quite frequently result in the best polynomial approximation, for example, if the set

$$\{f(x), \phi_1(x), \ldots, \phi_n(x)\}$$

forms a Chebyshev set in $[a,b]$. However, this is not usually the case, and so this is not a reliable method in general.

An algorithm of descent type seems first to have been given by Usow [188] in 1967, who gave an analysis applicable to problems with Chebyshev sets, and some numerical results for polynomial approximation. However, Marti [120] in 1975 gave an example where the method converges to a point which is not a solution. He gave an alternative descent method, valid when the functions $\{\phi_i\}$ form a Markov set (any rearrangement is a Chebyshev set).

The first general method seems due to Glashoff and Schultz [72] in 1979, based on using Newton's method to solve the characterization conditions (14) together with the corresponding interpolation conditions. A variant of this, which is globally convergent, was given by Watson [200] in 1981. It is essentially of exchange type, based on the calculation of the zeroes of the error at each iteration and the construction of descent directions. It is also of Newton type, since it constructs the Hessian matrix of the error when it exists, and therefore can have a second-order convergence rate. In a sense, it can be thought of as analogous to the second algorithm of Remes for Chebyshev problems, where here a sequence of sets of zeroes plays the role of a sequence of sets of extreme points in that problem; the connection with Newton's method under appropriate circumstances is also something the methods have in common. A method for L_1 problems based on Newton's method was also given by Blatt [25] in 1984.

3. Nonlinear approximation

There are two major differences which arise in moving from linear to nonlinear best approximation problems. Firstly, existence of solutions cannot generally be guaranteed. Secondly, there is normally a gap between conditions which are necessary and conditions which are sufficient for a best approximation. This reflects the loss of convexity. From an algorithmic pont of view, it is usual to seek to satisfy first-order conditions which are necessary for a solution to the best approximation problem, and such a point is conventionally referred to as a *stationary point*. At best this can be expected to be a local minimum of the norm. Assuming that the members of the approximating family are differentiable with respect to the free parameters at least in the region of interest, then a characterization of stationary points is straightforward: it is appropriate simply to replace in the linear case the basis elements (either vectors making up the columns of A or functions $\phi_i, i=1,\ldots,n$) by the partial derivatives of the approximating function with respect to the free parameters at the relevant points.

3.1. Nonlinear approximation in $\mathbb{R}^m$

Consider now the discrete problem

$$\text{find } \boldsymbol{x} \in \mathbb{R}^n \text{ to minimize } \|\boldsymbol{f}(\boldsymbol{x})\|,$$

where $\boldsymbol{f} \in \mathbb{R}^m$ depends nonlinearly on the components of $\boldsymbol{x}$, and where the norm is any norm on $\mathbb{R}^m$.

A general approach to this problem is through a sequence of linear subproblems. Assume that $\boldsymbol{f}$ is continuously differentiable in the region of interest, and at a given point $\boldsymbol{x}$, let A denote the $m \times n$ matrix of partial derivatives of the components of $\boldsymbol{f}$ with respect to the components of $\boldsymbol{x}$. Then consider the iterative method based on computing an updated $\boldsymbol{x}$ as follows:
(i) find $\boldsymbol{d} \in \mathbb{R}^n$ to minimize $\|\boldsymbol{f} + A\boldsymbol{d}\|$,
(ii) replace $\boldsymbol{x}$ by $\boldsymbol{x} + \gamma\boldsymbol{d}$, where $\gamma > 0$ is suitably chosen.
The problem in (i) is just a linear approximation problem in the given norm (a linear subproblem), and (ii) involves choosing γ so that

$$\|\boldsymbol{f}(\boldsymbol{x} + \gamma\boldsymbol{d})\| < \|\boldsymbol{f}(\boldsymbol{x})\|, \tag{15}$$

if this is possible: for example we may try to minimize the expression on the left-hand side with respect to γ.

When the norm is the least-squares norm, this kind of method (with $\gamma = 1$) most probably dates back to Gauss and is now known as the Gauss–Newton method. For the Chebyshev problem, this kind of approach was suggested by Zuhovickii et al. [212] in 1963, by Ishizaki and Watanabe [91] in 1968, and by Osborne and Watson [150] in 1969. Unless $\boldsymbol{x}$ is a stationary point, improvement can always be obtained via step (ii) since (15) holds for $\gamma > 0$ small enough. The theory given in the Osborne and Watson paper required that successive matrices A satisfied the Haar condition, and in that case convergence to a stationary point was established. The method was extended to the l_1 norm by Osborne and Watson [152] in 1971. Also in 1971, Osborne [146] was able to relax the Haar condition assumption for the l_∞ algorithm, and showed that the method was quadratically convergent if the maximum error at the limit point of the iteration was attained at $n + 1$ points. In that case, unit length steps were ultimately possible, and a ready connection could be drawn with Newton's method applied to the nonlinear equations satisfied at the stationary point. Osborne contrasted this with the behaviour of the method in the l_2 case, when good performance was dependent on the goodness of fit of the model, rather than on properties of the data.

The behaviour of the algorithm in a completely general setting was considered in 1978 by Osborne and Watson [153]; in particular, (15) was always shown to hold for $\gamma > 0$ small enough away from a stationary point. It was also pointed out that the above behaviour typified the situation for polyhedral norms on the one hand, and smooth strictly convex monotonic norms on the other.

A common basis for a convergence analysis which included this kind of algorithm was given by Cromme [51] in 1978: he showed that for second-order convergence, it was sufficient for the best approximation to be strongly unique. This criterion was also studied for the above algorithms in 1980 by Jittorntum and Osborne [95], who showed that strong uniqueness was not always necessary.

Meantime, (at least) two developments were taking place. The fact that the solution of the linear subproblem could be such that very small step lengths were sometimes required led to the idea of explicitly incorporating bounds. This Levenberg–Marquardt or trust region idea was finding favour in descent methods for more general optimization calculations. Another development was to do with the line search. Trying to find the value of γ to minimise $||\boldsymbol{f}||$ is clearly impractical, and the idea of inexact, but sufficiently good, line searches was again imported from contemporary optimization algorithms. These modifications were used by Madsen [115] in an algorithm for the Chebyshev problem, and by Anderson and Osborne [4] in 1977 for polyhedral norm problems (which include l_1 and l_∞). While this could improve things in certain cases, slow convergence could, however, still occur for many problems.

For fast local convergence in general, it was recognized that second derivative information had to be incorporated. Two stage methods for Chebyshev problems were given independently in 1979 by Watson [198] and by Hald and Madsen [80]. These methods solved a sequence of linear subproblems to build up information about the limit point (in particular, the number of points where the norm was attained, with signs). This information could then (if necessary) be used as input to a second (locally convergent) phase such as Newton's method applied to the nonlinear system of equations characterizing a stationary point. Thus they extended fast local convergence to a much wider range of problems.

It had long been recognized that the Chebyshev approximation problem could be posed as a nonlinearly constrained optimization problem, analogous to the way in which the linear problem could, although it seemed at one time that treating the problem in this way was likely to be less efficient than using linear subproblems. However, following advances in techniques for constrained optimization problems, and a recognition that there was much structure in the Chebyshev problem which could be exploited, Conn [48] in 1979, Murray and Overton [135] in 1980, Han [82] in 1981, and Womersley and Fletcher [207] in 1986 all proposed methods. These are all variants of a technique based on the solution of a sequence of quadratic programming problems, involving a Lagrangian function and linearizations of $r_i = h$, for i in a set which estimates the set of indices where the extrema are attained at a stationary point. They all incorporate second derivative information, and involve exploiting the structure and giving descent with respect to the norm. A line search descent method due to Conn and Li [49] in 1989 is claimed to make more explicit use of the structure: in addition to giving descent, it attempts to force satisfaction of the stationary point characterization conditions at the same time.

This general approach now seems the most effective for small problems with dense matrices A. However, for large problems with sparse structure in A, solving linear rather than quadratic programming problems is preferable, as the structure may be exploited. Therefore, for such problems, there has been some recent re-interest in methods of trust region type which use sequential linear programming. Some work of Jonasson and Madsen [97,98] from the mid-1990s is of relevance here.

As in the linear case, large problems may arise as discretizations of continuous problems; therefore we will return to this in Section 3.4.

There were analogous developments for the solution of the nonlinear l_1 problem. The first attempt to incorporate second derivative information into general classes of problems was probably by McLean and Watson [122] in 1980. This method was of two-phase type which used the solution of a sequence of bounded variable linear subproblems to provide information about Z at the desired stationary point, and then used Newton's method to get an accurate point. The exact Jacobian matrix was used for the Newton step. A similar method by Hald and Madsen [81] in 1985 used quasi-Newton approximations, and allowed several switches between phases. Meantime, (single phase) methods based on solving a sequence of quadratic programming problems were being developed, analogous to those mentioned before for Chebyshev approximation problems. In the main, these constructed the quadratic programming problems by defining a Lagrangian function, and by involving linear approximations to $r_i = 0$ for $i \in Z^k$, where Z^k was an estimate at iteration k to Z at the solution. Methods of this type which used line searches were proposed by Murray and Overton [136] in 1981 and Bartels and Conn [15] in 1982, and trust region methods were given by Fletcher [67,68] in 1981 and 1985.

Perhaps because there is no simple connection analogous to that between continuous and discrete Chebyshev approximation problems, the nonlinear l_1 problem has attracted much less recent interest.

3.2. Rational Chebyshev approximation in $\mathbb{R}^t$

Approximation by rational functions goes back to Chebyshev in 1859. The basic (discrete) problem is as follows. Let x_i, $i = 1, \ldots, t$ be in $[a, b]$. Then a best approximation is sought from the set

$$\mathbb{R}^D_{nm} = \left\{ P(x)/Q(x)\colon P(x) = \sum_{j=0}^{n} a_j p_j(x), \quad Q(x) = \sum_{j=0}^{m} b_j q_j(x),\ Q(x_i) > 0,\ i = 1, \ldots, t \right\},$$

to the set of values $f_1, \ldots, f_t$, in the sense that

$$\max_{1 \leqslant i \leqslant t} |\mathbb{R}(x_i) - f_i|$$

is minimized over all $\mathbb{R} \in \mathbb{R}^D_{nm}$. For this problem, existence of best approximations is not guaranteed, even in the case of quotients of polynomials, and characterization and uniqueness results are not available, although of course necessary conditions for a solution may be obtained. In fact necessary conditions based on alternations may be derived analogous to the characterization conditions which are available in the case of approximation to a continuous function on an interval: see Section 3.6. Because of this it is possible to implement an algorithm equivalent to the second algorithm of Remes, although for the discrete problem there are better approaches which do not explicitly use alternations.

The quest for algorithms for rational Chebyshev approximation appears to go back at least as far as Wenzl [201] in 1954. In the late 1950s Loeb considered some approaches which formed the basis for what was perhaps the first really effective algorithm for the discrete problem, the differential correction algorithm, given by Cheney and Loeb [38] in 1961. At that time, the convergence properties were uncertain, and a modified version was subsequently considered by the same authors in 1962 [39], and also by Cheney and Southard [42] in 1963, which was shown to have sure convergence properties, and drew attention away from the original method. However, in 1972 Barrodale et al. [9] studied both approaches, and showed that the method in its original form had not only guaranteed convergence from any starting approximation in $\mathbb{R}^D_{nm}$, but usually had a second-order convergence rate. Further, their comparisons of the methods showed that the performance of the original method was better. Some further analysis was given by Cheney and Powell [41] in 1987.

The differential correction algorithm is an iterative method where successive approximations from $\mathbb{R}^D_{nm}$ are computed by solving a linear programming subproblem, where one variable is minimized subject to $2t$ linear constraints involving also variables representing the coefficients of the new approximation, and bound constraints on the coefficients of the denominator. Each step of the method may be interpreted as working with an approximation to the original problem which is correct up to first order, and this "Newton method" connection gives a partial explanation of the second-order convergence rate. In fact, from the point of view of implementation, it is more efficient to solve the dual of the original linear programming subproblem.

A potentially unsatisfactory feature of approximation from $\mathbb{R}^D_{nm}$ is that the denominator, although positive, can become arbitrarily close to zero at certain points. It is not sufficient simply to impose a lower bound on Q, because of the possibility of multiplying both numerator and denominator by an arbitrary constant. A modification of the differential correction algorithm which applies to problems with a lower bound on the denominator and upper bounds on the absolute values of the coefficients b_j was given by Kaufmann and Taylor [104] in 1981.[5] It is more natural, however, to impose upper and lower bounds on the denominators themselves ("constrained denominators"). A modified differential correction algorithm for this problem has been given by Gugat [75] in 1996. This involves constraints of the form

$$\mu(x_i) \leqslant Q(x_i) \leqslant v(x_i), \quad i = 1, \ldots, t, \tag{16}$$

where μ and v are continuous functions, which replace the constraints on $Q(x_i)$ in the definition of $\mathbb{R}^D_{nm}$.

The linear programming subproblem corresponding to (16) above differs in that the additional conditions are imposed on the denominators. However, Gugat's method differs also in that there is greater flexibility in choice of initial values, and this turns out to be important. The original algorithm starts with an approximation $\mathbb{R}_1$ in $\mathbb{R}^D_{nm}$ and a value Δ_1 which is the maximum modulus error of this approximation on the discrete set. The method of Gugat starts with $\mathbb{R}_1$ as usual, but with an arbitrary number Δ_1 that is allowed to be smaller than the current maximum error. This flexibility turns out to be an important advantage: for example numerical results show that the choice $\Delta_1 = 0$ is a good one. It is shown by Gugat that convergence results for the original version carry over.

---

[5] This is an example of a constrained problem, which arises in a natural way from the rational approximation problem: it is not our intention to consider constrained problems per se.

It has been pointed out that the quadratic convergence of the differential correction algorithm is a consequence of a connection which it may be shown to have with Newton's method. Methods which set out deliberately to use variants of Newton's method are given by Hettich and Zenke [86] in 1990 and Gugat [76] in 1996. However, in contrast to the methods based on the differential correction algorithm, these do not generate a monotonically decreasing sequence of maximum modulus errors on successive approximations.

3.3. Nonlinear approximation in $C[a,b]$

Consider now the problem

$$\text{find } \boldsymbol{a} \in \mathbb{R}^n \text{ to minimize } \|f(.,\boldsymbol{a})\|, \tag{17}$$

where the norm is a given norm on $C[a,b]$ and where $\boldsymbol{a}$ occurs nonlinearly in f. It was shown by Watson [197] in 1978 that, provided that f was differentiable in the region of interest, methods of Gauss–Newton type (the continuous analogues of the methods introduced in Section 3.1) can be applied to this class of problems. However, while this may be of some theoretical interest, it does not lead to practical algorithms. Indeed, such problems cannot really be considered in any generality, and we will in fact restrict attention to the Chebyshev norm, and some important special cases.

3.4. Nonlinear Chebyshev approximation in $C[a,b]$

Here we consider (17) when the norm is the Chebyshev norm

$$\|f\|_\infty = \max_{x\in[a,b]} |f(x)|.$$

Some general problems of this type were considered by Chebyshev in 1859 [36], with particular reference to rational approximation.

Aside from some special cases (for example see below) it is not possible to say very much about the number of points where the norm is attained at a stationary point. In common with other general nonlinear problems, characterization results are not available, and numerical methods set out to find a stationary point.

The first practical numerical methods seem to have been of two-phase type (see Section 2.5), and these were proposed independently by Hettich [83] and Watson [195] in 1976. The basic idea is similar to that used for linear problems: a first phase is to solve a discretized problem, whose solution identifies the number and associated signs, along with good approximations, of the points where the norm is attained at a stationary point, and a second phase corresponding to the solution of a nonlinear system comprising the equations to be satisfied there. Only the first-phase calculation needs a method which is tailored to whether the problem is linear or not. If a single discretized problem is to be solved, then any of the methods for solving discrete Chebyshev problems can of course be used.

The second phase calculation is a Newton type method, whose steps may be interpreted as quadratic programming problems. The approach can be globalized, thus extending the domain of convergence. This idea was central to the single phase method given by Jonasson and Watson [99] in the mid-1980s, based on the use of a Lagrangian function, and solving a sequence of quadratic programming problems defined on the current set of local maxima of the modulus of the error function. Descent directions were defined, and both line search and trust region algorithms were developed. Second-order convergence is normal, and there is a nice connection with the second algorithm of Remes; however, the requirement to calculate exact local extrema at each step is a major disadvantage, and there can be sometimes slow progress far from a stationary point. A similar method was given by Jing and Fam [94] in 1987.

The connection between continuous Chebyshev approximation problems and semi-infinite programming problems has already been drawn, and the earlier comments apply to nonlinear problems. It may be that more recent methods being devised for nonlinear semi-infinite programming problems may also improve on these earlier methods for Chebyshev approximation problems. For example, a method by Görner [74] in 1997 consists of the solution of a finite set of discretized problems by sequential quadratic programming methods, following on from similar ideas used by Zhou and Tits [210] in 1996. These methods can lead into a second phase for accurate solution of the continuous problem: a feature of the method of Görner is that the same superlinearly convergent sequential quadratic programming method is used in both phases.

In any event, it would appear that this much at least can be said: a two-phase method with a discretization technique as first phase, and a variant of Newton's method as second phase, seems to be the most reliable and efficient method for solving small to medium size continuous Chebyshev set problems. However, the difficulties referred to near the end of Section 2.5 are still relevant for larger problems.

3.5. Nonlinear Chebyshev approximation in $C[a,b]$ – some special cases

In order to close the gap between conditions which are necessary and conditions which are sufficient, it is necessary to restrict the class of approximating functions, and the point at which this process converges may conveniently be described in terms of alternation conditions, analogous to those which apply in the (linear) Chebyshev set case. This clearly has implications for numerical methods, and so it is appropriate to look briefly at some of this theory. In the linear case, the Chebyshev set condition simultaneously implies the existence of an interpolation function with a certain (fixed) number of zeros. In nonlinear cases, these become two conditions which have to be assumed separately: the interpolation property is a *local* one (which depends on the approximation), but in addition we require a global property on the zeros.

The concept of unisolvency was introduced in 1949 by Motzkin [130]. Let $\phi(x,\boldsymbol{a}):\mathbb{R}^n \to C[a,b]$. Then given any $\boldsymbol{d} \in \mathbb{R}^n$ and n distinct points x_i, $i=1,\dots,n$ in $[a,b]$, this family is *unisolvent* if there exists a unique vector $\boldsymbol{a} \in \mathbb{R}^n$ such that

$$\phi(x_i,\boldsymbol{a}) = d_i, \quad i=1,\dots,n.$$

This particular generalization of the Chebyshev set property in the linear case leads to the existence of a unique best approximation ϕ which is characterized by

$$\mathscr{A}(f-\phi)_{[a,b]} \geqslant n+1,$$

as shown by Tornheim [187] in 1950. Unfortunately this is an extremely restrictive property, possessed by a small number of approximating functions, and Rice in his Ph.D. thesis in 1959, and in papers published in the next few years, suggested a more general property of *varisolvency*, which (provided the error is not constant) leads to the best approximation ϕ being characterized by

$$\mathscr{A}(f-\phi)_{[a,b]} \geqslant m(\phi)+1,$$

where $m(\phi)$ is the degree of local solvency [168,170]. Rice also showed that there is at most one best approximation. If ϕ is formed from a linear combination of n functions forming a Chebyshev set in $[a,b]$, then this is in fact a varisolvent family of constant degree n.

A related theory for nonlinear Chebyshev approximation on an interval was established by Meinardus and Schwedt [126] in 1964, valid for approximating functions differentiable with respect to their parameters. It essentially replaces the local condition required in varisolvency by a local Chebyshev set condition on the tangent space. An alternation characterization condition was established, along with an uniqueness result. Braess [31] in 1974 demonstrated the precise relationship between these various results.

Attempting to define a general class of nonlinear approximating functions which would be varisolvent, and so satisfy this kind of characterization result, Hobby and Rice [88] in 1967 defined *γ-polynomials*,

$$\phi(x,\boldsymbol{a}) = \sum_{i=1}^{n} a_i \gamma(a_{i+n}x),$$

where γ is a continuous function of its parameters. This class is of interest because it includes some important special cases, for example exponentials and spline functions. Subject to an additional assumption (Descartes' rule of signs), Hobby and Rice [88] established that the theory of varisolvent families applied. This condition is satisfied if the set

$$\{\gamma(t_1,x),\ldots,\gamma(t_n,x)\}$$

forms a Chebyshev set in $[a,b]$ for distinct t_i's. A best approximation ϕ is then characterized by

$$\mathscr{A}(f-\phi)_{[a,b]} \geqslant n+l(\phi)+1$$

where $l(\phi)$ is the length of the γ-polynomial ϕ, defined by the restriction that ϕ cannot be expressed by a sum of fewer terms. The closure of the set of γ-polynomials is in fact required for existence of best approximations, but then the alternating characterization is lost.

An important special case is given by taking

$$\gamma(t,x) = \mathrm{e}^{tx},$$

when we have approximation by sums of exponentials. This was studied first by Rice [169] in 1960 ($n=1$), and in 1962 (general n) [171]. Because the set $\{\mathrm{e}^{t_1x},\ldots,\mathrm{e}^{t_nx}\}$ forms a Chebyshev set in $[a,b]$ for distinct t_i's, then a Descartes' rule of signs holds (this result seems to go back to Laguerre [108] in 1898), and it follows that the approximating family is varisolvent. This was shown by Rice [171] in 1962, who also showed that a best approximation ϕ is characterized by

$$\mathscr{A}(f-\phi)_{[a,b]} \geqslant n+k(\phi)+1,$$

where the gradient vector of ϕ with respect to a_i, $i=1,\ldots,2n$ has $n+k(\phi)$ nonzero components. There is at most one best approximation. Existence of best approximations from the closure of the set of exponential functions was proved by Rice [171] in 1962 and Werner [204,205] in 1969.

As Bellman [21] wrote in 1970, "exponential approximation is a notoriously delicate enterprise", mainly because widely varying parameter values can give nearly optimal results. Therefore, the calculation of best Chebyshev approximations (or indeed any approximations) by sums of exponentials can be difficult. If an assumption is made about the number of alternations (that $k(\phi) = n$), then a method of Remes type can be applied with a nonlinear system of equations to be solved for the new coefficients at each iteration. This is considered by Dunham [57] in 1979, and in subsequent work with Zhu: it was necessary to have very good starting approximations.

The fact that n of the parameters occur linearly means that if the parameters $a_{n+1}, \dots, a_{2n}$ (the frequencies) are fixed, then the remaining parameters can be obtained by applying a linear solution method; this gives a problem which is essentially in the frequencies alone, and which could be tackled by iteration on the frequencies to obtain optimal values. Local descent methods were suggested by Braess [29] and Werner [205] in the late 1960s, and related methods were implemented in the 1970s by others such as Cromme, Kammler, Robitzsch and Schaback [50,101,177]. A method due to Dunham [59] in 1988 worked well with one frequency, but had difficulties with two or more. Nearly equal frequencies, or coalescing frequencies, are generally a problem.

One feature is that good initial approximations are necessary: in particular it is important to estimate the positions of the frequencies, before applying an optimal method, and this has led to interest in "suboptimal approximations". Prony's method of "approximate interpolation" may be applied, although the method is not generally stable. An alternative is Bellman's 1970 [21] method of differential approximation. These methods were considered in detail by Robitzsch and Schaback [177] in 1978 and by Schaback [179] in 1979. Any suboptimal method may be considered as a first phase method which can lead into a second phase based on Newton's method to satisfy the nonlinear system characterizing the solution.

But it would seem that in practice additional constaints are both natural physically, and necessary mathematically and computationally – for example, to bound frequencies, or to prevent frequencies from crossing each other. The computational approach then depends on precisely what is being assumed, and we will not pursue this further.

3.6. Rational Chebyshev approximation in $C[a,b]$

The continuous analogue of the class of problems considered in Section 3.2 is based on the approximating set $\mathbb{R}_{nm}$ defined by

$$\mathbb{R}_{nm} = \left\{ P(x)/Q(x) : P(x) = \sum_{j=0}^{n} a_j p_j(x), \right.$$
$$\left. Q(x) = \sum_{j=0}^{m} b_j q_j(x),\ Q(x) > 0 \text{ on } [a,b] \right\},$$

where the $p_j(x)$ and $q_j(x)$ are given sets of functions. Then given $f(x) \in C[a,b]$, we require to determine $\mathbb{R} \in \mathbb{R}_{nm}$ to minimize $||f - \mathbb{R}||$, where the norm is the Chebyshev norm on $[a,b]$. For the special case when $P(x)$ and $Q(x)$ are polynomials of degree n and m, respectively, existence of a best approximation is guaranteed, as shown by Walsh [193] in 1931. Achieser in 1947 (see his 1956 book [3]) showed that the best approximation is unique (up to a normalization), and earlier, in 1930 (again see his 1956 book [3]), he showed that a best approximation $\mathbb{R} = P/Q \in \mathbb{R}_{nm}$ is characterized by

$$\mathscr{A}(f - \mathbb{R})_{[a,b]} \geqslant n + m + 2 - d(\mathbb{R}),$$

where $d(\mathbb{R})$ is the *defect* of the approximation: the defect is just the minimum difference between the *actual* degree of $P(x)$ and $Q(x)$ and n and m respectively. If $d(\mathbb{R}) > 0$, the best approximation is said to be degenerate. These results also follow from the fact that the approximating family is varisolvent. (The necessary conditions referred to in Section 3.2 correspond to this alternation result defined *on the points of the set* $x_1 < \cdots < x_t$ introduced there.)

For more general quotients (of linear combinations of functions), existence is no longer guaranteed, although characterization results are available (not necessarily of alternation type), and uniqueness results may be extended. The main contributions here are from Cheney and Loeb [39,40] in the mid-1960s.

For rational approximation by quotients of polynomials on an interval, the analogue of the Remes exchange method may be applied, using sets of $m+n+2$ points. It, therefore, requires nondegeneracy of the best approximation, and can converge at a second-order rate if started from close enough to the solution: the analysis is primarily due to Werner in a series of papers in the early 1960s [202,203]. The system of linear equations which needs to be solved in the linear problem is replaced by a nonlinear system in the rational problem, equivalent to an eigenvalue problem. Werner [203] in 1963 showed that the eigenvalues are always real, and there is at most one pole free solution, that is a rational approximation with $Q(x) > 0$ on $[a,b]$. Maehly in 1963 [118] gave an example which showed that in fact no pole free solution need exist; even if it is does exist, it need not be associated with the smallest eigenvalue. Despite these potential problems, the second algorithm of Remes has been successfully used for rational approximation. Fraser and Hart [70] in 1962, Werner [202] in 1962 and Stoer [186] in 1964 gave methods based on solving the system of nonlinear equations directly. In 1966, Curtis and Osborne [52] gave an algorithm which used the eigenvalue connection explicitly, solving the eigenvalue problem by inverse iteration with zero as an initial estimate for the eigenvalue; they also established quadratic convergence. Breuer [32] in 1987 suggested a different direct approach to this subproblem which used continued fraction interpolation, and which it was claimed can lead to a considerable increase in efficiency, and also accuracy and robustness.

Variants of the second algorithm of Remes apply to rational Chebyshev approximation problems which incorporate a generalized weight function. Important work involving rational approximation on an interval to provide optimal starting values for computing $\sqrt{x}$ by the Newton Raphson method was done, for example, by Moursand [133] in the late 1960s.

The algorithms fail if the solution is degenerate, and indeed for problems which are nearly degenerate, extremely good starting approximations are required. Ralston [160,161] in a series of papers in the late 1960s and early 1970s considered degeneracy in detail. The computation of nearly degenerate approximations should if possible be avoided, as equally good results can be obtained through the use of smaller m and n.

It is possible to make the second algorithm of Remes more robust, by combining its merits with the differential correction algorithm. In particular the discrete subproblems can be solved by that method, and if no pole-free solution is obtained, additional points can be included. If sufficiently many points are taken in $[a,b]$, and always assuming that the continuous problem is not degenerate, then a pole-free solution can be obtained so that the algorithm can be continued. Methods based on this idea were given by Belogus and Liron [22] and also Kaufman et al. [103] both in 1978. Numerical evidence is that such an approach can be successful for problems which give difficulties with the traditional Remes method.

The differential correction algorithm may be applied to problems defined on an interval, although the subproblems are no longer finite. Dua and Loeb [55] in 1973 established a second order convergence rate if the best approximation is normal. The potentially unsatisfactory feature referred to in Section 3.2 where the deniminator, although positive, can become arbitrarily close to zero, also applies to $\mathbb{R}_{nm}$. The algorithm of Gugat referred to there also may be applied to intervals, although the numerical performance is unclear.

3.7. Chebyshev approximation by spline functions with free knots

To permit the full power of splines, one should allow the knots to vary, rather than be fixed in advance. The corresponding approximation problem is then a difficult nonlinear problem. This problem can be considered in terms of γ polynomials. However, the structure of the problem, and the way in which degeneracies can be introduced makes an attempt to make a straightforward application unhelpful.

To guarantee existence of best approximations, multiple knots have to be allowed. There may be local solutions; a characterization of best approximations is not known. For the case of k free knots, necessary and (different) sufficient conditions of the alternation kind given above may be proved. Let q' denote the sum of the knot multiplicities at the points $x_{p+1},\ldots,x_{p+q-1}$. Then it is *necessary* for $s \in S_m$ to be a best Chebyshev approximation with k free knots to f in $[a,b]$ that there exists an interval $[x_p, x_{p+q}] \subset [a,b]$ with $q \geqslant 1$ such that

$$\mathscr{A}(f-s)_{[x_p,x_{p+q}]} \geqslant m+q+q'+1,$$

as shown by Nürnberger et al. [142] in 1989; it is *sufficient* for $s \in S_m$ to be a best Chebyshev approximation with k free knots to f in $[a,b]$ that there exists an interval $[x_p, x_{p+q}] \subset [a,b]$ with $q \geqslant 1$ such that

$$\mathscr{A}(f-s)_{[x_p,x_{p+q}]} \geqslant m+k+q'+2,$$

as shown by Braess [30] in 1971. The necessary condition was strengthened to a possibly longer alternant by Mulansky [134] in 1992. Although a characterization of best spline approximations with free knots is not known, a characterization of strongly unique best spline approximations with free simple knots is available: what is required is that *all* knot intervals contain sufficiently many alternating extrema. The relevant work here is by Nürnberger [137,138] in 1987 and 1994.

Since approximation by splines with free knots is a nonlinear Chebyshev approximation problem, of course general methods can be used. However, the way in which the knots enter as free parameters makes this a particularly awkward problem and makes it important that the special structure be exploited.

For a discretization of the problem, a descent method based on Newton's method was given by Esch and Eastman [62] in 1969. Most algorithmic work has been concerned with uniform approximation from a space of piecewise polynomials where the continuity conditions at the knots are relaxed. A standard algorithmic approach is based on so-called segment approximation, originating from work of Lawson in 1964 [112], and methods were proposed by Pavlidis and Maika [155] in

1974, and McLaughlin and Zacharski [121] in 1984. Because pieces were fitted separately, continuity could be lost between segments. A recent method of this type is due to Nürnberger et al. [145] in 1986 (see also [124]). The algorithm converges through sequences of knot sets from an arbitrary set of knots. For each set of k knots, best Chebyshev degree m polynomial approximations to f are obtained on each subinterval using the classical Remes algorithm. The knots are then adjusted by a "levelling" process, so that the maximum errors of the polynomial best approximations are equalized. The result of this is a piecewise polynomial which is usually discontinuous. However, the procedure is augmented by the application of the method for fitting splines with fixed knots to the optimal knot positions obtained from the first part. The outcome of this is a differentiable spline approximation, which numerical results show to be a good one. Note that at present there is no algorithm for computing (global) best Chebyshev spline approximations with free knots. At best a local approximation can be expected, so producing a "good" spline approximation may be the most sensible strategy.

Generalizations to multivariate splines have mainly been concerned with interpolation problems. But consider bivariate splines on $[a_1,b_1]\times[a_2,b_2]$. This region can be divided into rectangles by knot lines $x=x_i$, $y=y_i$, $i=1,\dots,s$, and a tensor product spline space can be defined. As in the univariate problem, partitions can be defined and improved systematically in such a way that best Chebyshev approximations are obtained in the limit. Some recent work on this problem is given by Meinardus et al. [125] in 1996, and by Nürnberger [140] in 1997. However, there are many unsolved problems, as pointed out by Nürnberger [139] in 1996.

Acknowledgements

Charles Dunham attributes to P. Whippey, on the effort of writing a history: "It's a complex problem: the costs are real, the benefits imaginary". In any event, I am grateful to the many people who took the trouble to answer my questions about aspects of this work, or provided me with references. I am especially grateful to Mike Osborne, because he read through a draft and provided me with many helpful comments and suggestions. Of course, the responsibility for what is set down here is entirely mine.

Mike Osborne in fact has to shoulder quite a lot of responsibility for the existence of this paper, because he introduced me to the subject of approximation in 1964, he supervised my postgraduate study, and generally he has remained since that time a strong and guiding influence on my career. I dedicate this work to him, on the occasion of his 65th birthday, with my gratitude and my affection.

References

[1] N.N. Abdelmalik, Linear L_1 approximation for a discrete point set and L_1 solutions of overdetermined linear equations, J. Assoc. Comput. Mach. 18 (1971) 41–47.

[2] N.N. Abdelmalik, An efficient method for the discrete linear L_1 approximation problem, Math. Comp. 29 (1975) 844–850.

[3] N.I. Achieser, Theory of Approximation, Ungar, New York, 1956.

[4] D.H. Anderson, M.R. Osborne, Discrete, nonlinear approximation problems in polyhedral norms, Numer. Math. 28 (1977) 143–156.

[5] D.O. Andreassen, G.A. Watson, Linear Chebyshev approximation without Chebyshev sets, BIT 16 (1976) 349–362.

[6] R.D. Armstrong, J. Godfrey, Two linear programming algorithms for the linear discrete L_1 norm problem, Math. Comp. 33 (1979) 289–300.
[7] R.D. Armstrong, D.S. Kung, A dual method for discrete Chebyshev curve fitting, Math. Programming 19 (1979) 186–199.
[8] I. Barrodale, C. Phillips, Algorithm 495: solution of an overdetermined system of linear equations in the Chebyshev norm, ACM Trans. Math. Software 1 (1975) 264–270.
[9] I. Barrodale, M.J.D. Powell, F.D.K. Roberts, The differential correction algorithm for rational l_∞ approximation, SIAM J. Numer. Anal. 9 (1972) 493–504.
[10] I. Barrodale, F.D.K. Roberts, Applications of mathematical programming to l_p approximation, in: J.B. Rosen, O.L. Mangasarian, K. Ritter (Eds.), Nonlinear Programming, Academic Press, New York, 1970, pp. 447–464.
[11] I. Barrodale, F.D.K. Roberts, An improved algorithm for discrete l_1 linear approximation, SIAM J. Numer. Anal. 10 (1973) 839–848.
[12] I. Barrodale, F.D.K. Roberts, Algorithm 478: solution of an overdetermined system of equations in the L_1 norm, Comm. ACM 17 (1974) 319–320.
[13] I. Barrodale, A. Young, Algorithms for best L_1 and L_∞ linear approximation on a discrete set, Numer. Math. 8 (1966) 295–306.
[14] I. Barrodale, A. Young, A note on numerical procedures for approximation by spline functions, Comput. J. 9 (1966) 318–320.
[15] R. Bartels, A.R. Conn, An approach to nonlinear l_1 data fitting, in: J.P. Hennart (Ed.), Numerical Analysis, Cocoyoc, 1981, Springer, Berlin, 1982, pp. 48–58.
[16] R. Bartels, A.R. Conn, C. Charalambous, On Cline's direct method for solving overdetermined linear systems in the l_∞ sense, SIAM J. Numer. Anal. 15 (1978) 255–270.
[17] R. Bartels, A.R. Conn, Y. Li, Primal methods are better than dual methods for solving overdetermined linear systems in the l_∞ sense, SIAM J. Numer. Anal. 26 (1989) 693–726.
[18] R. Bartels, A.R. Conn, J.W. Sinclair, Minimization techniques for piecewise differentiable functions: the l_1 solution to an overdetermined linear system, SIAM J. Numer. Anal. 15 (1978) 224–241.
[19] R. Bartels, G.H. Golub, Stable numerical methods for obtaining the Chebyshev solution to an overdetermined system of equations, Comm. ACM. 11 (1968) 401–406.
[20] A.E. Beaton, J.W. Tukey, The fitting of power series, meaning polynomials, illustrated on band-spectrographic data, Technometrics 16 (1974) 147–185.
[21] R. Bellmann, Methods of Nonlinear Analysis I, Academic Press, New York, 1970.
[22] D. Belogus, N. Liron, DCR2: an improved algorithm for l_∞ rational approximation, Numer. Math. 31 (1978) 17–29.
[23] S.N. Bernstein, Lecons sur les Propriétés Extrémales et la Meillure Approximation des Fonctions Analytiques d'une Variable Réelle, Gauthier-Villars, Paris, 1926.
[24] L. Bittner, Das Austauschverfahren der linearen Tschebyscheff-Approximation bei nicht erfullter Haarscher Bedingung, Z. Angew. Math. Mech. 41 (1961) 238–256.
[25] H.-P. Blatt, Exchange algorithms, error estimates and strong unicity in convex programming and Chebyshev approximation, in: S.P. Singh, J.W.H. Burry, B. Watson (Eds.), Approximation Theory and Spline Functions, Reidel, Dodrecht, 1984, pp. 1–41.
[26] P. Bloomfield, W.L. Steiger, Least Absolute Deviations, Birkhaüser, Boston, 1983.
[27] P.T. Boggs, A new algorithm for the Chebyshev solution of overdetermined linear systems, Math. Comp. 28 (1974) 203–217.
[28] E. Borel, Lecons sur les Fonctions de Variables Reélles, Gauthier-Villars, Paris, 1905.
[29] D. Braess, Approximation mit Exponentialsummen, Computing 2 (1967) 309–321.
[30] D. Braess, Chebyshev approximation by spline functions with free knots, Numer. Math. 17 (1971) 357–366.
[31] D. Braess, Geometrical characterization for nonlinear uniform approximation, J. Approx. Theory 11 (1974) 260–274.
[32] P.T. Breuer, A new method for real rational uniform approximation, in: J.C. Mason, M.G. Cox (Eds.), Algorithms for Approximation, Clarendon Press, Oxford, 1987, pp. 265–283.
[33] P. Butzer, F. Jongmans, P.L. Chebyshev (1821–1894). A guide to his life and work, J. Approx. Theory 96 (1999) 11–138.

[34] C. Carasso, P.J. Laurent, An algorithm of successive minimization in convex programming, R. A. I. R. O. Numer. Anal. 12 (1978) 377–400.
[35] A. Charnes, W.W. Cooper, R.O. Ferguson, Optimal estimation of executive compensation by linear programming, Management Sci. 1 (1955) 138–151.
[36] P.L. Chebyshev, Sur les questions de minima qui se rattachent à la representation approximative des fonctions, Oeuvres I (1859) 273–378.
[37] E.W. Cheney, A.A. Goldstein, Newton's method for convex programming and Tchebycheff approximation, Numer. Math. 1 (1959) 253–268.
[38] E.W. Cheney, H.L. Loeb, Two new algorithms for rational approximation, Numer. Math. 3 (1961) 72–75.
[39] E.W. Cheney, H.L. Loeb, On rational Chebyshev approximation, Num. Math. 4 (1962) 124–127.
[40] E.W. Cheney, H.L. Loeb, Generalized rational approximation, SIAM J. on Num. Anal. 1 (1964) 11–25.
[41] E.W. Cheney, M.J.D. Powell, The differential correction algorithm for generalized rational functions, Constr. Approx. 3 (1987) 249–256.
[42] E.W. Cheney, T.H. Southard, A survey of methods for rational approximation, with particular reference to a new method based on a formula of Darboux, SIAM Rev. 5 (1963) 219–231.
[43] J.F. Claerbout, F. Muir, Robust modelling with erratic data, Geophysics 38 (1973) 826–844.
[44] D.I. Clark, The mathematical structure of Huber's M-estimator, SIAM J. Sci. Statist. Comput. 6 (1985) 209–219.
[45] A.K. Cline, A descent method for the uniform solution to overdetermined systems of linear equations, SIAM J. Numer. Anal. 13 (1976) 293–309.
[46] T.F. Coleman, Y. Li, A globally and quadratically convergent affine scaling algorithm for l_1 problems, Math. Programming 56 (1992) 189–222.
[47] T.F. Coleman, Y. Li, A globally and quadratically convergent method for linear l_∞ problems, SIAM J. Numer. Anal. 29 (1992) 1166–1186.
[48] A.R. Conn, An efficient second order method to solve the (constrained) minimax problem, Report CORR 79-5, University of Waterloo, 1979.
[49] A.R. Conn, Y. Li, An efficient algorithm for nonlinear minimax problems, Research Report CS-88-41, University of Waterloo, 1989.
[50] L. Cromme, Eine Klasse von Verfahren zur Ermittlung bester nichtlinearen Tschebyscheff-Approximationen, Numer. Math. 25 (1976) 447–459.
[51] L. Cromme, Strong uniqueness: a far reaching criterion for the convergence of iterative processes, Numer. Math. 29 (1978) 179–193.
[52] A.R. Curtis, M.R. Osborne, The construction of minimax rational approximations to functions, Comput. J. 9 (1966) 286–293.
[53] M. Davies, Linear approximation using the criterion of least total deviations, J. Roy. Statist. Soc. Ser. B 29 (1967) 101–109.
[54] J. Descloux, Approximations in L^p and Chebyshev approximations, SIAM J. 11 (1963) 1017–1026.
[55] S.N. Dua, H.L. Loeb, Further remarks on the differential correction algorithm, SIAM J. Numer. Anal. 10 (1973) 123–126.
[56] A.M. Duarte, R.J. Vanderbei, Interior point algorithms for lsad and lmad estimation, Technical Report SOR-94-07, Programs in Operations Research and Statistics, Princeton University, 1994.
[57] C.B. Dunham, Chebyshev approximation by exponential-polynomial sums, J. Comput. Appl. Math. 5 (1979) 53–57.
[58] C.B. Dunham, The weakened first algorithm of Remes, J. Approx. Theory 31 (1981) 97–98.
[59] C.B. Dunham, Approximation with one (or few) parameters nonlinear, J. Comput. Appl. Math. 21 (1988) 115–118.
[60] R. Dutter, Robust regression: different approaches to numerical solutions and algorithms, Research Report No 6, Technische Hochschule, Zurich, 1975.
[61] F.Y. Edgeworth, On a new method of reducing observations relating to several quantities, Philos. Mag. 5 (1888) 185–191.
[62] R.E. Esch, W.L. Eastman, Computational methods for best spline function approximation, J. Approx. Theory 2 (1969) 85–96.
[63] R.W. Farebrother, The historical development of the L_1 and L_∞ estimation procedures 1793–1930, in: Y. Dodge (Ed.), Statistical Data Analysis Based on the L_1 Norm and Related Methods, North-Holland, Amsterdam, 1987, pp. 37–63.

[64] R.W. Farebrother, Fitting Linear Relationships: A History of the Calculus of Observations 1750–1900, Springer, Berlin, 1999.
[65] J. Fischer, The convergence of the best discrete linear L_p approximation as $p \to 1$, J. Approx. Theory 39 (1983) 374–385.
[66] W.D. Fisher, A note on curve fitting with minimum deviations by linear programming, J. Amer. Statist. Assoc. 56 (1961) 359–362.
[67] R. Fletcher, Numerical experiments with an exact l_1 penalty function method, in: O.L. Mangasarian, R.R. Meyer, S.M. Robinson (Eds.), Nonlinear Programming 4, Academic Press, New York, 1981, pp. 99–129.
[68] R. Fletcher, An l_1 penalty method for nonlinear constraints, in: P.T. Boggs, R.H. Byrd, R.B. Schnabel (Eds.), Numerical Optimization 1984, SIAM Publications, Philadelphia, 1985, pp. 26–40.
[69] R. Fletcher, J.A. Grant, M.D. Hebden, Minimax approximation as the limit of best L_p approximation, SIAM J. Numer. Anal. 11 (1974) 123–136.
[70] W. Fraser, J.F. Hart, On the computation of rational approximations to continuous functions, Comm. ACM 5 (1962) 401–403.
[71] K. George, M.R. Osborne, The efficient computation of linear rank statistics, J. Comput. Simul. 35 (1990) 227–237.
[72] K. Glashoff, R. Schultz, Uber die genaue Berechnung von besten L^1-Approximierenden, J. Approx. Theory 25 (1979) 280–293.
[73] G.H. Golub, L.B. Smith, Algorithm 414: Chebyshev approximation of continuous functions by a Chebyshev system, Comm. ACM 14 (1971) 737–746.
[74] S. Görner, Ein Hybridverfahren zur Losung nichtlinearer semi-infiniter Optimierungsprobleme, Ph.D. Thesis, Technical University of Berlin, 1997.
[75] M. Gugat, An algorithm for Chebyshev approximation by rationals with constrained denominators, Constr. Approx. 12 (1996) 197–221.
[76] M. Gugat, The Newton differential correction algorithm for uniform rational approximation with constrained denominators, Numer. Algorithms 13 (1996) 107–122.
[77] S.-A. Gustafson, On the computational solution of a class of generalized moment problems, SIAM J. Numer. Anal. 7 (1970) 343–357.
[78] S.-A. Gustafson, K. Kortanek, Numerical treatment of a class of semi-infinite programming problems, Naval. Res. Logist. Quart. 20 (1973) 477–504.
[79] A. Haar, Die Minkowskische Geometrie und die Annaherung an stetige Funktionen, Math. Ann. 78 (1918) 294–311.
[80] J. Hald, K. Madsen, A two-stage algorithm for minimax optimization, in: A. Bensoussan, J. Lions (Eds.), International Symposium on Systems Optimization and Analysis, Springer, Berlin, 1979, pp. 225–239.
[81] J. Hald, K. Madsen, Combined LP and quasi-Newton methods for nonlinear l_1 optimization, SIAM J. Numer. Anal. 22 (1985) 68–80.
[82] S.-P. Han, Variable metric methods for minimizing a class of nondifferentiable functions, Math. Prog. 20 (1981) 1–13.
[83] R. Hettich, A Newton method for nonlinear Chebyshev approximation, in: R. Schaback, K. Scherer (Eds.), Approximation Theory, Springer, Berlin, 1976, pp. 222–236.
[84] R. Hettich, Numerical methods for nonlinear Chebyshev approximation in: G. Meinardus (Ed.), Approximation in Theorie und Praxis, B.I.-Wissenschaftsverlag, Mannheim, 1979, pp. 139–156.
[85] R. Hettich, An implementation of a discretization method for semi-infinite programming, Math. Programming 34 (1986) 354–361.
[86] R. Hettich, P. Zenke, An algorithm for general restricted rational Chebyshev approximation, SIAM J. Numer. Anal. 27 (1990) 1024–1033.
[87] C.R. Hobby, J.R. Rice, A moment problem in L_1-approximation, Proc. Amer. Math. Soc. 65 (1965) 665–670.
[88] C.R. Hobby, J.R. Rice, Approximation from a curve of functions, Arch. Rational Mech. Anal. 27 (1967) 91–106.
[89] P.G. Hoel, Certain problems in the theory of closest approximation, Amer. J. Math. 57 (1935) 891–901.
[90] M.J. Hopper, M.J.D. Powell, A technique that gains speed and accuracy in the minimax solution of overdetermined linear equations, in: J.R. Rice (Ed.), Mathematical Software III, Academic Press, New York, 1977.
[91] Y. Ishizaki, H. Watanabe, An iterative Chebyshev approximation method for network design, IEEE Trans. Circuit Theory 15 (1968) 326–336.
[92] D. Jackson, Note on a class of polynomials of approximation, Trans. Amer. Math. Soc. 13 (1921) 320–326.

[93] R.C. James, Orthogonality and linear functionals in normed linear spaces, Trans. Amer. Math. Soc. 61 (1947) 265–292.
[94] Z. Jing, A.T. Fam, An algorithm for computing continuous Chebyshev approximations, Math. Comp. 48 (1987) 691–710.
[95] K. Jittorntrum, M.R. Osborne, Strong uniqueness and second order convergence in nonlinear discrete approximation, Numer. Math. 34 (1980) 439–455.
[96] R.S. Johnson, On monosplines of least deviation, Trans. Amer. Math. Soc. 96 (1960) 458–477.
[97] K. Jonasson, A projected conjugate gradient method for sparse minimax problems, Numer. Algorithms 5 (1993) 309–323.
[98] K. Jonasson, K. Madsen, Corrected sequential linear programming for sparse minimax optimization, BIT 34 (1994) 372–387.
[99] K. Jonasson, G.A. Watson, A Lagrangian method for multivariate continuous Chebyshev approximation problems, in: W. Schempp, K. Zeller (Eds.), Multivariate Approximation Theory 2, I.S.N.M. 61, Birkhaüser, Basel, 1982, pp. 211–221.
[100] R.C. Jones, L.A. Karlovitz, Equioscillation under nonuniqueness in the approximation of continuous functions, J. Approx. Theory 3 (1970) 138–145.
[101] D.W. Kammler, Chebyshev approximation of completely monotonic functions by sums of exponentials, SIAM J. Numer. Anal. 13 (1976) 761–774.
[102] N. Karmarker, A new polynomial time algorithm for linear programming, Combinatorica 4 (1984) 373–395.
[103] E.H. Kaufmann, D.J. Leeming, G.D. Taylor, A combined Remes-Differential Correction algorithm for rational approximation, Math. Comp. 32 (1978) 233–242.
[104] E.H. Kaufmann, G.D. Taylor, Uniform approximation by rational functions having restricted denominators, J. Approx. Theory 32 (1981) 9–26.
[105] J.E. Kelley Jr., An application of linear programming to curve fitting, SIAM J. 6 (1958) 15–22.
[106] P. Kirchberger, Uber Tchebychefsche Annaherungsmethoden, Math. Ann. 57 (1903) 509–540.
[107] P. Laasonen, Einige Satze uber Tschebyscheffsche Funktionensysteme, Ann. Acad. Sci. Fenn. Ser. AI 52 (1949) 3–24.
[108] E. Laguerre, Ouvres I, Gauthier-Villars, Paris, 1898.
[109] D. Landers, L. Rogge, Natural choice of L_1-Approximants, J. Approx. Theory 33 (1981) 268–280.
[110] P.S. Laplace, Mechanique Celeste, Tome 111, No 39, 1799.
[111] C.L. Lawson, Contributions to the Theory of Linear Least Maximum Approximation, Ph.D. Dissertation, University of California, Los Angeles, 1961.
[112] C.L. Lawson, Characteristic properties of the segmented rational minimax approximation problem, Num. Math. 6 (1964) 293–301.
[113] W. Li, J.J. Swetits, The linear l_1 estimator and Huber M-estimator, SIAM J. on Optim. 8 (1998) 457–475.
[114] Y. Li, A globally convergent method for L_p problems, SIAM J. Optim. 3 (1993) 609–629.
[115] K. Madsen, An algorithm for minimax solution of over-determined systems of non-linear equations, J. Inst. Math. Appl. 16 (1975) 321–328.
[116] K. Madsen, H.B. Neilsen, A finite smoothing algorithm for linear l_1 estimation, SIAM J. Optim. 3 (1993) 223–235.
[117] K. Madsen, H.B. Nielson, M.C. Pinar, New characterizations of l_1 solutions of overdetermined linear systems, Operations Research Letters 16 (1994) 159–166.
[118] H.J. Maehly, Methods of fitting rational approximations, Part II, J. Assoc. Comput. Mach. 10 (1963) 257–266.
[119] J.C. Mairhuber, On Haar's theorem concerning Chebyshev approximation problems having unique solutions, Proc. Amer. Math. Soc. 7 (1956) 609–615.
[120] J.T. Marti, A method for the numerical computation of best L_1 approximations of continuous functions, Proceedings of Oberwolfach Conference, ISNM, Vol. 26, Birkhauser, Basel, 1975.
[121] H.W. McLaughlin, J.J. Zacharski, Segmented approximation, in: E.W. Cheney (Ed.), Approximation Theory III, Academic Press, New York, 1980, pp. 647–654.
[122] R.A. McLean, G.A. Watson, Numerical methods for nonlinear discrete L_1 approximation problems, Proceedings of Oberwolfach Conference on Numerical Methods in Approximation Theory 1979, ISNM, Vol. 52, Birkhauser, Basel, 1980.
[123] S. Mehrotra, On the implementation of a primal-dual interior point method, SIAM J. Optim. 2 (1992) 575–601.

[124] G. Meinardus, G. Nürnberger, M. Sommer, H. Strauss, Algorithms for piecewise polynomials and splines with free knots, Math. Comp. 53 (1989) 235–247.

[125] G. Meinardus, G. Nürnberger, G. Walz, Bivariate segment approximation and splines, Adv. Comput. Math. 6 (1996) 25–45.

[126] G. Meinardus, D. Schwedt, Nicht-lineare Approximationen, Arch. Rational Mech. Anal. 17 (1964) 297–326.

[127] M.S. Meketon, Least absolute value regression, Technical Report, AT&T Bell Laboratories, Murray Hill, New Jersey, 1987.

[128] G. Merle, H. Späth, Computational experience with discrete L_p approximation, Computing 12 (1974) 315–321.

[129] C.A. Micchelli, Best L^1 approximation by weak Chebyshev systems and the uniqueness of interpolating perfect splines, J. Approx. Theory 19 (1977) 1–14.

[130] T.S. Motzkin, Approximation by curves of a unisolvent family, Bull. Amer. Math. Soc. 55 (1949) 789–793.

[131] T.S. Motzkin, J.L. Walsh, Least pth power polynomials on a real finite point set, Trans. Amer. Math. Soc. 78 (1955) 67–81.

[132] T.S. Motzkin, J.L. Walsh, Least pth power polynomials on a finite point set, Trans. Amer. Math. Soc. 83 (1956) 371–396.

[133] D.G. Moursund, Optimal starting values for Newton-Raphson calculation of $\sqrt{x}$, Comm. ACM 10 (1967) 430–432.

[134] B. Mulansky, Chebyshev approximation by spline functions with free knots, IMA J. Numer. Anal. 12 (1992) 95–105.

[135] W. Murray, M.L. Overton, A projected Lagrangian algorithm for nonlinear minimax optimization, SIAM J. Sci. Statist. Comput. 1 (1980) 345–370.

[136] W. Murray, M.L. Overton, A projected Lagrangian algorithm for nonlinear l_1 optimization, SIAM J. Sci. Statist. Comput. 2 (1981) 207–214.

[137] G. Nürnberger, Strongly unique spline approximation with free knots, Constr. Approx. 3 (1987) 31–42.

[138] G. Nürnberger, Approximation by univariate and bivariate splines, in: D. Bainov, V. Covachev (Eds.), Second International Colloquium on Numerical Analysis, VSP, Utrecht, 1994, pp. 143–153.

[139] G. Nürnberger, Bivariate segment approximation and free knot splines: research problems 96-4, Constr. Approx. 12 (1996) 555–558.

[140] G. Nürnberger, Optimal partitions in bivariate segment approximation, in: A. Le Méhauté, C. Rabut, L.L. Schumaker (Eds.), Surface Fitting and Multiresolution Methods, Vanderbilt University Press, Nashville, 1997, pp. 271–278.

[141] G. Nürnberger, L.L. Schumaker, M. Sommer, H. Strauss, Approximation by generalized splines, J. Math. Anal. Appl. 108 (1985) 466–494.

[142] G. Nürnberger, L.L. Schumaker, M. Sommer, H. Strauss, Uniform approximation by generalized splines with free knots, J. Approx. Theory 59 (1989) 150–169.

[143] G. Nürnberger, I. Singer, Uniqueness and strong uniqueness of best approximations by spline subspaces and other spaces, J. Math. Anal. Appl. 90 (1982) 171–184.

[144] G. Nürnberger, M. Sommer, A Remez type algorithm for spline functions, Numer. Math. 41 (1983) 117–146.

[145] G. Nürnberger, M. Sommer, H. Strauss, An algorithm for segment approximation, Numer. Math. 48 (1986) 463–477.

[146] M.R. Osborne, An algorithm for discrete, nonlinear best approximation problems, Proceedings of Oberwolfach Conference on ISNM, Vol. 16, Birkhauser, Basel, 1971.

[147] M.R. Osborne, Finite Algorithms in Optimisation and Data Analysis, Wiley, Chichester, 1985.

[148] M.R. Osborne, The reduced gradient algorithm, in: Y. Dodge (Ed.), Statistical Data Analysis Based on the L_1 Norm and Related Methods, North-Holland, Amsterdam, 1987, pp. 95–107.

[149] M.R. Osborne, G.A. Watson, On the best linear Chebyshev approximation, Comput. J. 10 (1967) 172–177.

[150] M.R. Osborne, G.A. Watson, An algorithm for minimax approximation in the nonlinear case, Comput. J. 12 (1969) 63–68.

[151] M.R. Osborne, G.A. Watson, A note on singular minimax approximation problems, J. Math. Anal. Appl. 25 (1969) 692–700.

[152] M.R. Osborne, G.A. Watson, On an algorithm for discrete nonlinear L_1 approximation, Comput. J. 14 (1971) 184–188.

[153] M.R. Osborne, G.A. Watson, Nonlinear approximation problems in vector norms, in: G.A. Watson (Ed.), Proceedings of Dundee Numerical Analysis Conference, Springer, Berlin, 1978.

[154] M.R. Osborne, G.A. Watson, Aspects of M-estimation and l_1 fitting problems, in: D.F. Griffiths, G.A. Watson (Eds.), Numerical Analysis: A R Mitchell 75th Birthday Volume, World Scientific Publishing Co, Singapore, 1996, pp. 247–261.
[155] T. Pavlidis, A.P. Maika, Uniform piecewise polynomial approximation with variable joints, J. Approx. Theory 12 (1974) 61–69.
[156] G. Polya, Sur une algorithme toujours convergent pour obtenir les polynomes de meillure approximation de Tchebysheff pour une fonction continue quelconque, Comptes Rendues 157 (1913) 840–843.
[157] S. Portnoy, R. Koenker, The Gaussian hare and the Laplacian tortoise: computability of squared-error versus absolute-error estimators, Statist. Sci. 12 (1997) 279–296.
[158] M.J.D. Powell, Approximation Theory and Methods, Cambridge University Press, Cambridge, 1980.
[159] C.J. Price, I.D. Coope, Numerical experiments in semi-infinite programming problems, Comput. Optim. Appl. 6 (1996) 169–189.
[160] A. Ralston, Rational Chebyshev approximation by Remes algorithms, Numer. Math. 7 (1965) 322–330.
[161] A. Ralston, Some aspects of degeneracy in rational approximation, JIMA 11 (1973) 157–170.
[162] R. Reemtsen, Modifications of the first Remez algorithm, SIAM J. Numer. Anal. 27 (1990) 507–518.
[163] R. Reemtsen, Discretization methods for the solution of semi-infinite programming problems, J. Optim. Theory Appl. 71 (1991) 85–103.
[164] R. Reemtsen, S. Görner, Numerical methods for semi-infinite programming: a survey, in: R. Reemtsen, J.-J. Ruckman (Eds.), Semi-Infinite Programming, Kluwer Academic Publishers, Boston, 1998, pp. 195–275.
[165] E.YA. Remes, Sur un procédé convergent d'approximation sucessives pour déterminer les polynomes d'approximation, Comptes Rendues 198 (1934) 2063–2065.
[166] E.YA. Remes, Sur le calcul effectif des polynomes d'approximation de Tchebichef, Comptes Rendues 199 (1934) 337–340.
[167] E.C. Rhodes, Reducing observations by the method of minimum deviations, Philos. Mag. 9 (1930) 974–992.
[168] J.R. Rice, The characterization of best nonlinear Tchebycheff approximation, Trans. Amer. Math. Soc. 96 (1960) 322–340.
[169] J.R. Rice, Chebyshev approximation by $ab^x + c$, J. SIAM 10 (1960) 691–702.
[170] J.R. Rice, Tchebycheff approximation by functions unisolvent of variable degree, Trans. Amer. Math. Soc. 99 (1961) 298–302.
[171] J.R. Rice, Chebyshev approximation by exponentials, J. SIAM 10 (1962) 149–161.
[172] J.R. Rice, The Approximation of Functions, Vol. I, Addison-Wesley, Reading, MA, 1964.
[173] J.R. Rice, Characterization of Chebyshev approximation by splines, SIAM J. Math. Anal. 4 (1967) 557–567.
[174] F. Riesz, Uber lineare Funktionalgleichungen, Acta. Math. 41 (1918) 71–98.
[175] P.D. Robers, A. Ben-Israel, An interval programming algorithm for discrete linear L_1 approximation problems, J. Approx. Theory 2 (1969) 323–331.
[176] P.D. Robers, S.S. Robers, Algorithm 458: discrete linear L_1 approximation by interval linear programming, Comm. ACM 16 (1973) 629–631.
[177] H. Robitzsch, R. Schaback, Die numerische Berechnung von Startnaherungen bei der Exponentialapproximation, in: L. Collatz, G. Meinardus, H. Werner (Eds.), Numerical Methods of Approximation Theory, ISNM, Vol. 42, Birkhauser, Basel, 1978, pp. 260–280.
[178] S.A. Ruzinsky, E.T. Olsen, l_1 and l_∞ minimization via a variant of Karmarkar's algorithm, IEEE Trans. Acoust. Speech Signal Process. 37 (1989) 245–253.
[179] R. Schaback, Suboptimal exponential approximation, Report Nr 23, University of Gottingen, 1979.
[180] L.L. Schumaker, Uniform approximation by Tchebycheffian spline functions, J. Math. Mech. 18 (1968) 369–378.
[181] L.L. Schumaker, Some algorithms for the computation of interpolating and approximating spline functions, in: T.N.E. Greville (Ed.), Theory and Applications of Spline Functions, Academic Press, New York, 1969, pp. 87–102.
[182] R.R. Singleton, A method for minimizing the sum of absolute values of deviations, Ann. Math. Statist. 11 (1940) 301–310.
[183] K. Spyropoulos, E. Kiountouzis, A. Young, Discrete approximation in the L_1 norm, Comput. J. 16 (1973) 180–186.
[184] E.L. Stiefel, Uber diskrete und lineare Tschebyscheff- Approximationen, Numer. Math. 1 (1959) 1–28.
[185] E.L. Stiefel, Note on Jordan elimination, linear programming and Tchebyscheff approximation, Numer. Math. 2 (1960) 1–17.

[186] J. Stoer, A direct method for Chebyshev approximation by rational functions, J. Assoc. Comput. Mach. 1 (1964) 59–69.

[187] L. Tornheim, On n-parameter families of functions and associated convex functions, Trans. Amer. Math. Soc. 69 (1950) 457–467.

[188] K.N. Usow, On L_1 approximation I: computation for continuous functions and continuous dependence, SIAM J. Numer. Anal. 4 (1967) 70–88.

[189] K.N. Usow, On L_1 approximation II: computation for discrete functions and discretization effects, SIAM J. Numer. Anal. 4 (1967) 233–244.

[190] C.J. de la Vallée Poussin, Sur la methode de l'approximation minimum, Societe Scientifique de Bruxelles, Annales, Memoires, Vol. 35, 1911, pp. 1–16.

[191] L. Veidinger, On the numerical determination of the best approximations in the Chebyshev sense, Num. Math. 2 (1960) 99–105.

[192] H.M. Wagner, Linear programming techniques for regression analysis, J. Amer. Statist. Assoc. 54 (1959) 206–212.

[193] J.L. Walsh, The existence of rational functions of best approximation, Trans. Amer. Math. Soc. 33 (1931) 668–689.

[194] G.A. Watson, A multiple exchange algorithm for multivariate Chebyshev approximation, SIAM J. Numer. Anal. 12 (1975) 46–52.

[195] G.A. Watson, A method for calculating best nonlinear Chebyshev approximations, JIMA 18 (1976) 351–360.

[196] G.A. Watson, On two methods for discrete L_p approximation, Computing 18 (1977) 263–266.

[197] G.A. Watson, On a class of methods for nonlinear approximation problems, in: D.C. Handscomb (Ed.), Multivariate Approximation, Academic Press, London, 1978, pp. 219–227.

[198] G.A. Watson, The minimax solution of an overdetermined system of nonlinear equations, JIMA 23 (1979) 167–180.

[199] G.A. Watson, Approximation Theory and Numerical Methods, Wiley, Chichester, 1980.

[200] G.A. Watson, An algorithm for linear L_1 approximation of continuous functions, IMA J. Numer. Anal. 1 (1981) 157–167.

[201] F. Wenzl, Uber Gleichungssysteme der Tschebysheffschen Approximation, Z. Angew. Math. Mech. 34 (1954) 385–391.

[202] H. Werner, Die konstruktive Errmittlung der Tschebyscheff-Approximation in Bereich der rationalen Funktionen, Arch. Rational Mech. Anal. 11 (1962) 368–384.

[203] H. Werner, Rationale-Tschebyscheff-Approximation, Eigenwerttheorie und Differenzenrechnung, Arch. Rational Mech. Anal. 13 (1963) 330–347.

[204] H. Werner, Der Existenzsatz fur das Tschebyscheffsdhe Approximationsproblem mit Exponenttialsummen, in: L. Collatz, H. Unger (Eds.), Funktionalanalytische Methoden der numerischen Mathematik, ISNM, Vol. 12, Birkhauser, Basel, 1969, pp. 133–143.

[205] H. Werner, Tschebysheff-Approximation with sums of exponentials, in: A. Talbot (Ed.), Approximation Theory, Academic Press, London, 1969, pp. 109–134.

[206] J.M. Wolfe, On the convergence of an algorithm for discrete L_p approximation, Numer. Math. 32 (1979) 439–459.

[207] R.S. Womersley, R. Fletcher, An algorithm for composite nonsmooth optimization problems, J. Optim. Theory Appl. 48 (1986) 493–523.

[208] J.W. Young, General theory of approximation by functions involving a given number of arbitrary parameters, Trans. Amer. Math. Soc. 8 (1907) 331–344.

[209] Y. Zhang, A primal-dual interior point approach for computing the l_1 and l_∞ solutions of overdetermined linear systems, J. Optim. Theory and Appl. 77 (1993) 323–341.

[210] J.J. Zhou, A.L. Tits, An SQP algorithm for finitely discretized continuous minimax optimization problems and other minimax problems with many objective functions, SIAM J. Optim. 6 (1996) 461–487.

[211] S.I. Zuhovickii, L.I. Avdeyeva, Linear and Convex Programming, (Translated from the 1964 Russian edition), Saunders Co., Philadelphia, 1966.

[212] S.I. Zuhovickii, R.A. Poljak, M.E. Primak, An algorithm for the solution of the problem of Cebysev approximation, Soviet. Math. Dokl. 4 (1963) 901–904.

A tutorial history of least squares with applications to astronomy and geodesy ☆

Yves Nievergelt *

Department of Mathematics, Eastern Washington University, MS-32, Cheney, WA 99004-2431, USA

Abstract

This article surveys the history, development, and applications of least squares, including ordinary, constrained, weighted, and total least squares. The presentation includes proofs of the basic theory, in particular, unitary factorizations and singular-value decompositions of matrices. Numerical examples with real data demonstrate how to set up and solve several types of problems of least squares. The bibliography lists comprehensive sources for more specialized aspects of least squares.

Keywords: History; Constrained; Ordinary; Total; Weighted; Least squares

1. Introduction

The mathematical concept of least squares is the basis for several methods to fit certain types of curves and surfaces to data. Problems of fitting curves and surfaces have a history spanning several millenia, which is outlined in Section 2 to set in perspective the contribution of least squares to their solution. The citations provided here include page numbers from Dreyer's book [13] to identify the original texts. Examples of such problems include the determination of the shape and size of celestial bodies and of their trajectories.

These problems were still without satisfactory solutions near the end of the eighteenth century A.D., at the time of the development of the concepts of problems of least squares and their solution with normal equations; see Section 3. (For greater detail, see Stewart's translation [16] of Gauss's work.)

☆ Work supported by a faculty research grant from Eastern Washington University.

* Tel.: +1-509-359-4259; fax: +1-509-359-4700.

E-mail address: ynievergelt@ewu.edu (Y. Nievergelt).

For computations with floating-point or other approximate arithmetics, normal equations can exhibit a sensitivity to errors in data or in rounding larger than the sensitivity of methods with unitary factorizations. These factorizations also provide methods to solve problems of constrained and total least squares, as explained in Sections 4 and 5.

For the state-of-the-art in computing with least squares near the end of the second millenium A.D., Björk [1], Dennis Jr., and Schnabel [12], and Lawson and Hanson [32] present algorithms to solve least-squares problems, and Higham [22] also treats the analysis of sensitivity to errors. Van Huffel and Vandewalle [50] focus on total least-squares. These references also contain extensive bibliographies. To compute solutions of practically significant problems, the usual recommendation is to use one of the professionally maintained libraries of computer programs, for instance, `netlib` (`http://www.netlib.org/lapack/`).

2. An ancient history of curve and surface fitting

2.1. Fitting surfaces: the shapes of the earth and of the moon

One fitting problem consists in estimating the shape of the earth. Early in the first millenium B.C., several shapes were fitted to various combinations of religious canons, philosophical doctrines, and observations of the rôles of air, earth, fire, and water. Types of surfaces fitted to such ideas included a *circular disc* (Thales of Miletus, about 640–562 B.C. [13, p. 11]), an *infinite plane* (Xenophanes of Kolophon, about 570–475 B.C. [13, p. 18]), and a *sphere* (Parmenides of Elea, early in the fifth century B.C. [13, p. 20]). The type of surface was also fitted to observations of inequalities reported by travelers. For example, the star Canopus remained invisible to a traveler in Greece, became just visible above the horizon at Rhodes, and then appeared higher and higher above the horizon as the traveler went further and further south [13, p. 20]. Also while sailing toward the setting sun, mariners in the north saw the sun on their left, but mariners in the south saw the sun on their right [13, p. 39]. From the fifth century B.C., in Greece and India, the type of surface fitted to such observations was a sphere [13, pp. 39, 242].

Similarly, for the shape of the moon, a sphere fitted the observation that the lighted side of the moon always faces the sun (Parmenides [13, p. 21]; Anaxagoras of Klazomenæ, about 500–428 B.C. [13, p. 32]).

With the shape settled to be a sphere arises the problem of estimating its size.

To estimate the circumference of the earth, Posidonius of Apameia (about 135–50 B.C.) referred to a result attributed to Archimedes (287–212 B.C.) and Dikæarchus of Messana (about 285 B.C.), using two stars seen from two cities; see Fig. 1. The cities are Lysimachia in Thrace, and Syene in Upper Egypt, which lie 20 000 stadia apart from each other. The first star, γ Draconis, appears at the zenith (vertical direction) above Lysimachia. The second star, in the constellation Cancer, appears at the zenith above Syene. The difference between the declinations (angular elevations from the celestial equator) of the two stars is $1/15$ of a full circle, which is thus the difference between the vertical directions at the two cities. Therefore, the circumference of the earth is $15 * 20\,000 = 300\,000$ stadia, corresponding to approximately 100 000 stadia for the earth's diameter [13, pp. 173–174]. (Though Archimedes and Apollonius already knew the approximations $\pi \approx 22/7$ and $\pi \approx 3.1416$

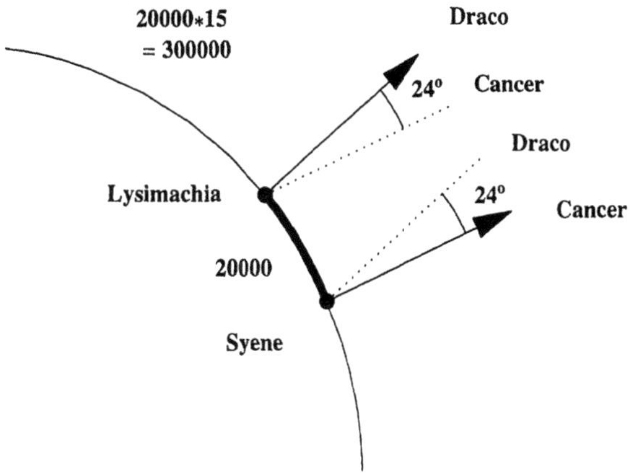

Fig. 1. Posidonius's estimate of the earth's circumference. Stars appear in the same direction from every point on earth. Two stars make an angle $2\pi/15$. One of them is at the zenith above Syene, the other is at the zenith at Lysimachia. Therefore, 15 times the distance from Syene to Lysimachia equals the earth's circumference.

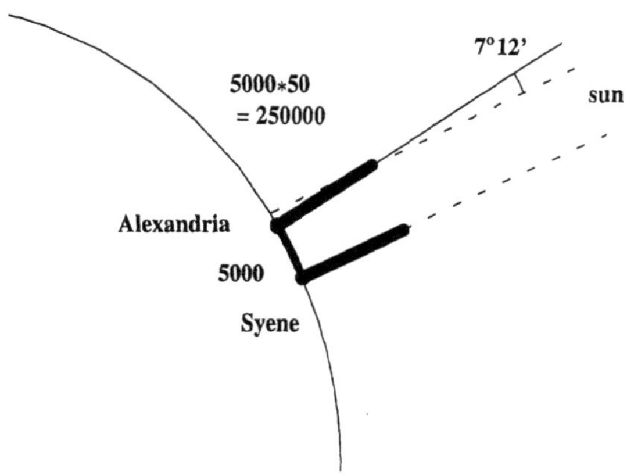

Fig. 2. Eratosthenes's estimate of the earth's circumference. The sun rays appear parallel on earth. They are vertical at Syene. At Alexandria, with a vertical stick they make an angle 1/50 of a full circle. Therefore, 50 times the distance from Syene to Alexandria equals the earth's circumference.

[49, pp. 185–186], the approximation $\pi \approx 3$ was then common for practical purposes not only in Greece but also in Babylon, in Egypt [49, p. 173], and in China [49, p. 196].)

With a different procedure, Eratosthenes of Alexandria (276–194 B.C.) used the shadows of vertical rods in two cities; see Fig. 2. At the summer soltice, at Syene the rod casts no shadow, so that the sun rays fall vertically, while at Alexandria the sun rays and the vertical rod make and angle equal to 1/50 of a full circle. (According to van der Waerden, the computation of this angle from measurements of the lengths of the rod and of its shadow proceeded through the Theorem of Pythagoras and tables of sines [49, p. 214].) Because Syene lies 5000 stadia away from Alexandria, it follows that the circumference of the earth is about $50*5000=250\,000$ stadia. Kleomedes corroborated this results through the same procedure at the same locations but at the winter soltice. Table 1 shows comparisons with the World Geodetic System WGS-84 [23].

Table 1
Comparisons of estimates of the earth's polar circumference and radius

Source	Circumference	Radius
Archimedes, Dikæarchus,	300 000 stadia	50 000 stadia
Posidonius, 3rd century B.C.	(47 250 000 m)	(7 875 000 m)
Eratosthenes, 2nd century B.C.	250 000 stadia	
(1 stade ≈ 157.5 m)	(39 375 000 m)	(6 562 500 m)
WGS-84, 1984 A.D.		$a = 6\,378\,137.000\,00$ m
		$b = 6\,356\,752.314\,25$ m
		$e^2 = 0.006\,694\,379\,990\,13$
Mathematica 4aEllipticE[e^2]	40 007 862.917 27 m	

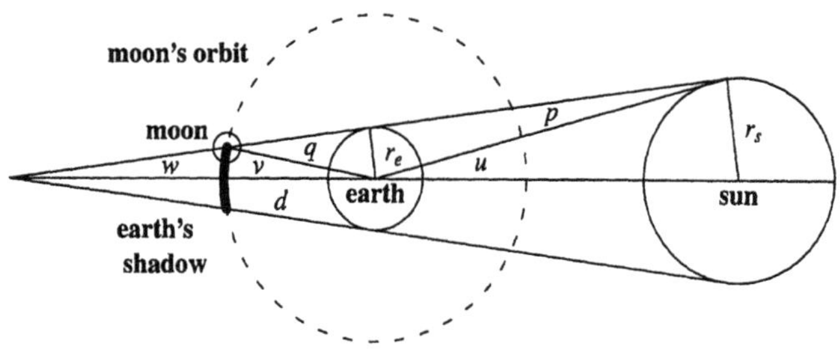

Fig. 3. Hipparchus's estimate of the radii of the moon and its orbit.

The estimate of the same circumference by different observers through different procedures or through repeated measurements hints at some attempts to detect errors, but no records of such attempts appear to remain [13, p. 177].

2.2. Fitting curves: the radii of the moon and its orbit

Another fitting problem consists in estimating the trajectories of celestial bodies. For example, rectilinear motions fitted the poetical ideas of Xenophanes in the sixth century B.C. [13, p. 18]. A century later, Philolaus of Thebes proposed circular orbits for the earth, the moon, the planets, and the sun, all around a "central fire" reflected by the sun toward the earth; such orbits fitted coarse observations of planetary motions [13, pp. 40–49]. In the third century B.C., Aristarchus of Samos outlined a heliocentric system with a circular orbit for the earth around the sun [13, p. 137].

With the orbits settled as circles arises the problem of estimating their size.

To estimate simultaneously the distance from the earth to the moon and the radius of the moon, Hipparchus of Nicæa (second century B.C.) used a full lunar eclipse [13, pp. 183–184]; see Fig. 3. Within the measurement accuracy available then, the sun's parallax p is nearly zero. Seen from the earth, the sun sustends an angle $u = 16'36''55'''$, and the path of the moon across the earth's shadow sustends and angle $v = 41'32''17.5'''$. The ratio $180^\circ/v \approx 260$ can also be calculated as the ratio t_1/t_2 of the time t_1 of a full revolution of the moon (29.5 days) and the time t_2 taken by the moon to cross the earth's shadow. Consequently, the parallax of the earth's shadow on the moon is nearly $q = u + v = 58'09''12.5'''$. Therefore, the ratio d/r_e of the distance d from the earth to

Table 2
Comparisons of estimates of the radii of the moon and its orbit

Source	Moon's radius	Orbit's radius
Hipparchus, 2nd century B.C.	$r_m = r_e/3.5$ (1 875 000 m)	$d = 59.1r_e$ (387 843 750 m)
Hipparchus, 2nd century B.C., reported by Kleomedes.	$r_m = r_e/3.4$ (1 930 147 m)	$d = 60\frac{5}{6}r_e$ (399 218 750 m)
[26, p. 476], 1984 A.D.	1 738 000 m ($b/r_m \approx 3.658$)	384 400 000 m ($d/b \approx 60.47$)

the moon and the earth's radius r_e is $1/\sin(q) = 1/\sin(u+v) = 59.1$. Moreover, the diameter of the earth's shadow at distance d from the earth equals about $d * v/180^\circ = d * t_2/t_1$. A measurement of the time t_3 from the moment the moon touches the earth's shadow to the moment it disappears in it then gives an estimate of the radius of the moon r_m in the form $2r_m/(d * t_2/t_1) = t_3/t_2$, whence $r_m = (d/2) * (t_3/t_1) = (59.1r_e/2) * (t_3/t_1) = r_e/3.5$.

According to Ptolemy's account, Hipparchus attempted to measure a lower bound and an upper bound for the sun's parallax p. The results just presented correspond to the lower bound 0. Kleomedes's report of another result from Hipparchus, $d = 60\frac{5}{6}r_e$ [13, pp. 183–184], corresponds to the upper bound $2'44''$. Such bounds hint at attempts to detect the maximum error.

With Eratosthenes's measure of the earth's radius, Hipparchus's results give 387 843 750 m for the distance of the moon, and 1 875 000 m for the radius of the moon. Table 2 shows comparisons with textbook values [26, p. 476].

2.3. *Fitting curves and surfaces: planetary orbits and earth's geoid*

It was also considerations of maximum errors, of the order of $8'$ between Tycho Brahe's observations of Mars and Copernicus's heliocentric model, which led Johann Kepler to abandon circles for the orbits, and finally (about 18 December 1604 A.D.) to substitute *ellipses* with a focus at the sun, along which planets sweep equal areas in equal times [13, pp. 389–392]. In 1687, Isaac Newton outlines in the *Philosophiae Naturalis Principia Mathematica* a proof that Kepler's laws are mathematically equivalent to the action of an attraction from the sun and inversely proportional to the square of the distance from the sun to the planet [39].

From Newton's law of gravitational attraction, it follows (from mathematical derivations by Newton, Ivory, Huygens, Clairaut, and Laplace [31, Book III, Section 18]) that a rotating mass of a homogenous and incompressible fluid can have the shape of an *ellipsoid* rotating around its shortest axis [20, pp. 172–175]. From 1700 through 1733, three surveys in France all suggested that the earth was an ellipsoid rotating around its *largest* axis [4, pp. 250–251]; such a surface failed to fit Newton's mathematical theory, based on Kepler's physics, itself based on Tycho Brahe's measurements. Ordered by Louis XV, a survey in Lapland and a survey in Peru in 1735 reversed the earlier results and confirmed that the earth was an ellipsoid rotating around it shortest axis [4, pp. 251–252].

The foregoing historical outline shows that for nearly three millenia, curves and surfaces were fitted to ideologies and theories. Yet errors — discrepancies between the fitted curve or surface and

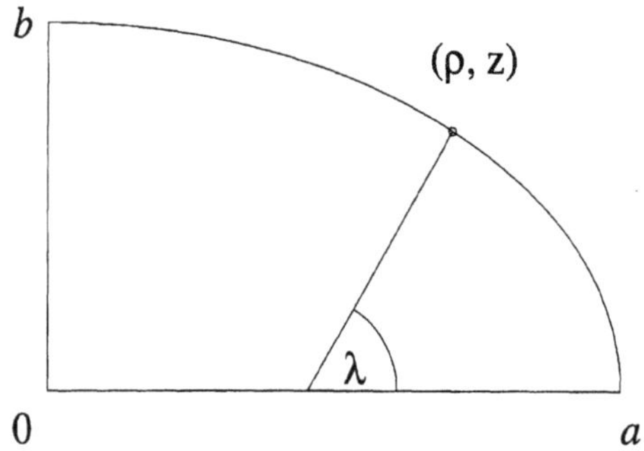

Fig. 4. The geodetic latitude of a point (ρ, z) is the angle λ between the normal to the surface through (ρ, z) and the equatorial plane.

observations — drew attention through gross departures from the theory or through unacceptable *maximum* values.

3. Weighted ordinary least squares and geodesy

3.1. Precursors: minimax and minimum average modulus

By the end of the 18th century A.D., the Marquis Pierre Simon de Laplace (1749–1827) was using a sequence of several methods to fit curves and surfaces to measurements in geodesy and astronomy. Each of his methods minimizes either the maximum residual, the average absolute residual, or the average squared residual, of a linearized model.

For example, consider the problem of fitting an ellipse to a polar cross section of the earth, with principal semi-axes of lengths $a \geqslant b > 0$. Let $e^2 := 1 - (b/a)^2$ be its squared eccentricity, and set $\sigma^2 := 1 - e^2 = (b/a)^2$. For each point $x := (x, y, z)$ on the earth's surface, the geodetic latitude of x is the angle λ between the normal to the surface at x and the equatorial plane, as in Fig. 4. With the cylindrical coordinate $\rho := \sqrt{x^2 + y^2}$, calculus gives

$$\rho = \frac{a\cos(\lambda)}{\sqrt{1 - e^2[\sin(\lambda)]^2}}, \qquad z = \frac{a\sigma^2\sin(\lambda)}{\sqrt{1 - e^2[\sin(\lambda)]^2}}.$$

Hence, the differential of the arclength s along a meridian becomes

$$\begin{aligned} ds &= \frac{a\sigma^2}{\{1 - e^2[\sin(\lambda)]^2\}^{3/2}}\, d\lambda \\ &= a\sigma^2\{1 + \tfrac{3}{2}e^2[\sin(\lambda)]^2 + \frac{3 * 5}{2 * 2 * 2!}e^4[\sin(\lambda)]^4 + \cdots\}\, d\lambda. \end{aligned}$$

Crude approximations indicate that $e^2 < 0.01$. Beyond the first two terms,

$$\sum_{k=2}^{\infty}\prod_{\ell=1}^{k}(2\ell + 1)\frac{|e\sin(\lambda)|^{2k}}{2^k * (k!)} < \frac{15e^4}{8}\left(1 + \frac{7e^2}{6}\right)\sum_{k=o}^{\infty} e^k < 0.000\,25.$$

Thus, with a relative error less than 0.000 25 uniformly over the earth's surface, the length Δs of an arc $\Delta\lambda$ of meridian at the geodetic latitude λ takes the following form, with $c_0 := a\sigma^2$ and $c_1 := \frac{3}{2}a\sigma^2 e^2$:

$$\frac{\Delta s}{\Delta\lambda} = c_0 + c_1[\sin(\lambda)]^2.$$

Thus, measurements of the lengths of n arcs of a meridian produce n equations.

Example 1. With lengths in double toises (1/0.256 537 m) and angles in grads ($2\pi = 400°$), Laplace considered the following system [31, Book III, Section 41]:

$\Delta s/\Delta\lambda = c_0 + c_1[\sin(\lambda)]^2$;	location;	latitude λ;	arc $\Delta\lambda$;
$25538.85 = c_0 + c_1 * 0.00000$;	Peru;	00.0000°;	3.4633°;
$25666.65 = c_0 + c_1 * 0.30156$;	Good Hope;	37.0093°;	1.3572°;
$25599.60 = c_0 + c_1 * 0.39946$;	Pennsylvania;	43.5556°;	1.6435°;
$25640.55 = c_0 + c_1 * 0.46541$;	Italy;	47.7963°;	2.4034°;
$25658.28 = c_0 + c_1 * 0.52093$;	France;	51.3327°;	10.7487°;
$25683.30 = c_0 + c_1 * 0.54850$;	Austria;	53.0926°;	3.2734°;
$25832.25 = c_0 + c_1 * 0.83887$;	Lapland;	73.7037°;	1.0644°.

The problem then consisted in fitting c_0 and c_1 to this linear system.

Laplace's first method aimed at determining the ellipsoid that *minimizes the maximum error* between the fitted ellipsoid and the measurements [31, Book III, Section 39]. From this first method he concluded that the earth's surface was not exactly an ellipsoid but the maximum error was within the measurement accuracy, with a flattening $f := 1 - (b/a) = 1/277$ [31, Book III, Section 41], which corresponds to a squared eccentricity $e^2 < 0.007\,207 < 0.01$.

Laplace's second method aimed at determining the ellipsoid that *minimizes the average absolute values of the errors* subject to the constraint that *the sum of the errors equal zero*; the result yielded what he considered the most probable ellipsoid [31, Book III, Section 40].

The second method presented several difficulties. Firstly, the "most probable" estimate depends on the probability distribution of the errors and can fail to coincide with the minimum average absolute error [24, pp. 400–401]. Secondly, Laplace's method did not lend itself to the methods of power series, and no efficient algorithm existed to determine the solutions (until George B. Dantzig's simplex algorithm in the 1950s [6,10,11]). Finally, for an overdetermined system of linear equations with a matrix of any rank, Laplace's method can lead to multiple solutions filling an entire polytope [6, p. 219].

Example 2. Consider the following system $Ax = b$ with maximal rank:

$$x + y = 4,$$

$$x - y = 0,$$

$$x - y = 2,$$

$$x + y = 6.$$

The residuals $r = Ax - b$ add to zero at $x = 3$. Setting $x = 3$ gives

$$\begin{aligned} &\{|(3+y)-4| + |(3-y)| + |(3-y)-2| + |(3+y)-6|\}/4 \\ &= \{|y-1| + |y-3|\}/2 \\ &= \begin{cases} 2-y > 1 & \text{if } y < 1, \\ 1 = 1 & \text{if } 1 \leqslant y \leqslant 3, \\ y-2 > 1 & \text{if } 3 < y. \end{cases} \end{aligned}$$

The average reaches its minimum everywhere on the segment $\{3\} \times [1,3]$.

Each of Laplace's numerical examples of a minimization of the average absolute error consists of an *odd* number of equations [31, Book III, Sections 41–42]. In contrast, for the determination of orbits of celestial bodies, Laplace used an *even* number of linearized equations, corresponding to measurements at times scattered symmetrically about a central time t_0:

$$t_0 - t_k, t_0 - t_{k-1}, \ldots, t_0 - t_1, t_0 + t_1, \ldots, t_0 + t_{k-1}, t_0 + t_k.$$

This produces a peculiar type of linear system, where the first column of coefficients $A(\,;1)$ is perpendicular to the second column of coefficients $A(\,;2)$, as in Example 2. For such systems, Laplace did not minimize the average absolute error. Instead, in effect, he computed the dot product of the system with the transposed column $A(\,;1)^*$ and solved for x, and then computed the dot product of the system with $A(\,;2)^*$ and solved for y [31, Book II, Section 37].

Example 3. Consider the system $Ax = b$ from Example 2:

$$\begin{pmatrix} 1 & 1 \\ 1 & -1 \\ 1 & -1 \\ 1 & 1 \end{pmatrix} \begin{pmatrix} x \\ y \end{pmatrix} = \begin{pmatrix} 4 \\ 0 \\ 2 \\ 6 \end{pmatrix},$$

$$\begin{pmatrix} 4x \\ 4y \end{pmatrix} = \begin{pmatrix} 1 & 1 & 1 & 1 \\ 1 & -1 & -1 & 1 \end{pmatrix} \begin{pmatrix} 1 & 1 \\ 1 & -1 \\ 1 & -1 \\ 1 & 1 \end{pmatrix} \begin{pmatrix} x \\ y \end{pmatrix} = \begin{pmatrix} 1 & 1 & 1 & 1 \\ 1 & -1 & -1 & 1 \end{pmatrix} \begin{pmatrix} 4 \\ 0 \\ 2 \\ 6 \end{pmatrix} = \begin{pmatrix} 12 \\ 8 \end{pmatrix}.$$

Thus, $x = 3$ and $y = 2$.

For the peculiar type of linear system in Examples 2 and 3, Laplace's method amounts to solving the normal equations for the least squares solution.

3.2. Weighted ordinary least squares

Around 1800, Laplace, Legendre, and Gauss were fitting functional forms to data through various types of least squares. Laplace's method applied to systems with mutually orthogonal columns. Legendre (1752–1833) published the method of normal equations in 1805 [33]. In 1821–1823, Gauss published the method of weighted least squares to solve linear systems $Ax = b$ with a matrix A with

n linearly independent columns and $m \geqslant n$ rows [16]. Though Gauss did not employ a notation with matrices, a derivation of weighted least squares with matrices can proceed as follows [47, p. 144].

The problem consists in determining a linear function u of n variables $a_1, \ldots, a_n$, which amounts to determining coefficients $x_1, \ldots, x_n$ so that

$$u(a_1, \ldots, a_n) = a_1 x_1 + \cdots + a_n x_n.$$

The data consist of m points $A(i;) = (a_{i,1}, \ldots, a_{i,n})$, arranged as the rows of the matrix A, and of the value b_i if u at each point. The problem then amounts to fitting coefficients $x_1, \ldots, x_n$ to the linear system $Ax = b$:

$$\begin{aligned} a_{1,1}x_1 + \cdots + a_{1,n}x_n &= b_1, \\ &\vdots \\ a_{m,1}x_1 + \cdots + a_{m,n}x_n &= b_{\mathrm{m}}. \end{aligned}$$

The data can also include estimates of the precision of the measurements, in the form of the reciprocal of the variance of each measurement, as investigated by Gauss, or, more generally, in the form of the inverse V^{-1} of the covariance matrix V of the measurements, as investigated by Aiken [2]. Specifically, if b_i represents the average $E(B_i)$ of a random variable B_i, estimated by the average of several observations, then $V_{i,j} = E[(B_i - b_i)(B_j - b_j)]$ is the covariance of B_i and B_j. The solution X of the linear system $AX = B$ is then also a random variable. The problem solved by Gauss consists in finding a linear transformation L such that $LA = I$, to solve for $\tilde{x} = I\tilde{x} = LA\tilde{x} = Lb$, such that $\tilde{x} = Lb$ minimizes the covariance

$$U = E[(X - \tilde{x})(X - \tilde{x})^*].$$

Gauss showed that $\tilde{x}$ is also the solution of the weighted least-squares system

$$WAx = Wb$$

with a matrix of weights W such that $W^*W = V^{-1}$, and then

$$L = (A^*V^{-1}A)^{-1}A^*V^{-1}.$$

Indeed, for every matrix K such that $KA = I$,

$$\begin{aligned} U &= E[(X - \tilde{x})(X - \tilde{x})^*] \\ &= E[(X - KB)(X - KB)^*] \\ &= E\{[X - KAX - K(B - AX)][X - KAX - K(B - AX)]^*\} \\ &= E\{[K(B - AX)][K(B - AX)]^*\} \\ &= KE\{(B - AX)(B - AX)^*\}K^* \\ &= KVK^* \\ &= LVL^* + (K - L)VL^* + LV(K - L)^* + (K - L)V(K - L)^*. \end{aligned}$$

The two middle terms equal zero, because of the condition $KA = I$ and the definition of L. The last term, $(K - L)V(K - L)^*$, is hermitian positive semidefinite. Hence, for each vector z,

$$z^*Uz = z^*LVL^*z + z^*(K - L)V(K - L)^*z \geqslant z^*LVL^*z,$$

with $z^* Uz$ minimum for $K:=L$. Moreover, the formula for U simplifies to

$$U = (A^* V^{-1} A)^{-1},$$

which is thus the covariance matrix of the weighted least squares solution X.

Example 4. For the system in Example 1, Laplace weighted each equation by the number of degrees $\Delta\lambda$ in the corresponding measured arc. There was no known correlation between the measurements from the different teams assigned to measure different arcs. Thus the weight matrix W is diagonal with the corresponding values of $\Delta\lambda$ on its diagonal:

$$W = \text{diagonal}\,(3.4633, 1.3572, 1.6435, 2.4034, 10.7487, 3.2734, 1.0644).$$

The weighted least-squares solution (computed through the command `LSQ` on a Hewlett–Packard's HP48GX calculator [21, pp. 14,15]) is

$$c_0 = 25534.47,$$
$$c_1 = 242.81.$$

Hence

$$e^2 = \tfrac{2c_1}{3c_0} = 0.006\,339\ldots,$$
$$\sigma^2 = 1 - e^2 = 0.993\,661\ldots,$$
$$f = 1 - b/a = 1 - \sigma = 0.003\,175\ldots,$$
$$a = \tfrac{c_0}{\sigma^2} = 25697.38^R \ldots = 100\,170.25\ m.$$

Laplace gives $f = 1/277 = 0.003\,610$, though Bowditch's calculations of Laplace's method lead to $f = 1/250 = 0.004$ [31, Book III, Section 41]. The values from WGS-84 are $f = 0.003\,352\,810\,664\,74$ and $a = 6\,378\,137$ m [23, p. xxiii]. Finally,

$$U = (A^* V^{-1} A)^{-1} = (A^* W^* W A)^{-1} = \begin{pmatrix} 0.007\,701 & -0.147\,686 \\ -0.147\,686 & 0.309\,706 \end{pmatrix}.$$

The method of weighted least squares assigns weights only to the measured values b of the function u, but not to be coordinates of the points $(a_{i,1}, \ldots, a_{i,n})$. In Laplace's application, this would correspond to treating the measurements of the lengths of arcs of meridians as random variables, but considering the determinations of the geodetic latitudes as exact. Allowances for adjustments of all data require different methods, as explained below in Section 6.

4. Unitary factorizations and constrained least squares

4.1. Householder symmetries and unitary factorizations

To solve linear systems, Gaussian elimination performs a linear transformation known as a shear that maps a column of coefficients $r = A(\ ;j)$ to a multiple of a canonical basis vector e_j, which "eliminates" the coefficients below the jth row. Yet shears alter Euclidean distances, in particular,

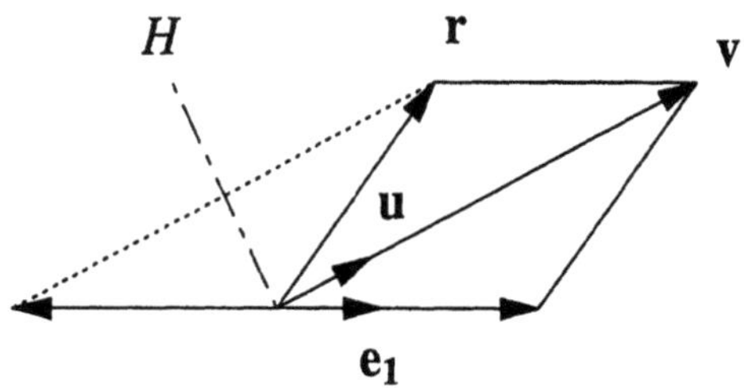

Fig. 5. A Householder symmetry maps r to $-\text{sign}(r_1)\|r\|_2 e_1$.

they do not reveal which vector lies closest to the "right-hand side" of the system. In contrast, one of the strategies for solving least squares problems consists in replacing Gaussian elimination by a type of linear elimination that preserves Euclidean distances, for instance, Modified Gram-Schmidt (MGS) orthogonalization [22, Section 19.3, 43], Givens rotations, or Householder symmetries [1,32].

Householder symmetries involves the function sign: $\mathbb{C} \to \mathbb{C}$ defined by

$$\text{sign}(z) := \begin{cases} z/|z| & \text{if } z \neq 0, \\ 1 & \text{if } z = 0. \end{cases}$$

For each non-zero vector $r \in \mathbb{C}^m \setminus \{0\}$, a Householder symmetry reflects r onto a multiple $-\text{sign}(r_1)\|r\|_2 e_1$ of the basis vector e_1 across the hyperplane $H \subset \mathbb{C}^m$ that passes through the origin 0 perpendicularly to the bisectrix of the angle formed by r and $\text{sign}(r_1)e_1$, which lies in the direction of $v := r + \text{sign}(r_1)\|r\|_2 \cdot e_1$, as in Fig. 5. The choice of $\text{sign}(r_1)$ minimizes rounding inaccuracies, so that if $r \neq 0$ then $v := r + \text{sign}(r_1)\|r\|_2 e_1 \neq 0$, because $|v_1| = |r_1 + \text{sign}(r_1)\|r\|_2| \geqslant \|r\|_2 > 0$. The hyperplane H is then perpendicular to the unit vector $u := (1/\|v\|_2)v$.

A Householder symmetry S thus amounts to subtracting from r twice its projection along u, so that $S(r) = r - 2\langle r, u\rangle u$, which leads to Algorithm 1.

Algorithm 1. Data: *any non-zero vector* $r \in \mathbb{C}^m \setminus \{0\}$
(1) $s := sign(r_1)$.
(2) $v := r + s\|r\|_2 e_1$.
(3) $\nu := 1/\{\|r\|_2(\|r\|_2 + |r_1|)\}$.

Result. $S(Z) = Z - \nu v(v^* Z)$ *for every* $Z \in \mathbb{M}_{m\times n}(\mathbb{C})$.

Proposition 5 verifies that Algorithm 1 produces a Householder symmetry.

Proposition 5. *The transformation S defined by algorithm 1 reflects r onto $S(r) = -s \cdot \|r\|_2 \cdot e_1$. Moreover, the matrix S of S is hermitian and unitary.*

Proof. With $S(r) = r - \nu v(v^* r)$ defined as in Algorithm 1,

$$v^* r = \sum_{j=1}^{m} \overline{v_j} r_j = \overline{v_1} r_1 + \sum_{j=2}^{m} \overline{v_j} r_j = (\overline{r_1} + \overline{s}\|r\|_2) r_1 + \sum_{j=2}^{m} \overline{r_j} r_j$$

$$= \|r\|_2^2 + \overline{s}\|r\|_2 r_1 = \|r\|_2(\|r\|_2 + |r_1|) = 1/\nu,$$

$$S(r) = r - v \cdot v \cdot (1/v) = r - v = r - (r + s \cdot \|r\|_2 \cdot e_1) = -s\|r\|_2 e_1.$$

Moreover, $\|v\|_2^2 = 2/v$:

$$\|v\|_2^2 = \sum_{j=1}^{m} |v_j|^2 = |r_1 + s\|r\|_2|^2 + \sum_{j=2}^{m} |r_j|^2$$

$$= |s|^2\|r\|_2^2 + (r_1\bar{s} + \overline{r_1}s)\|r\|_2 + |r_1|^2 + \sum_{j=2}^{m} |r_j|^2$$

$$= \|r\|_2^2 + 2|r_1| \cdot \|r\|_2 + \|r\|_2^2 = 2\|r\|_2(\|r\|_2 + |r_1|) = 2v.$$

Consequently, the Householder symmetry S has a hermitian matrix, $S^* = S$:

$$S^* = (I - vvv^*)^* = I^* - v(v^*)^*v^* = S.$$

Finally, the Householder symmetry S is a unitary transformation, $S^*S = I$:

$$S^*S = (I - vvv^*) \cdot (I - vvv^*) = I - 2Ivvv^* + vvv^*vvv^*$$

$$= I - 2Ivvv^* + v^2 v\|v\|_2^2 v^* = I - 2vvv^* + 2vvv^* = I. \qquad \square$$

Applied to the first column $r := A(\, ;1)$ of any rectangular matrix $A \in \mathbb{M}_{m\times n}(\mathbb{C})$, the Householder symmetry S produces zeros under the first entry $(SA)_{1,1} = -s\|A(\, ;1)\|_2$, and transforms the subsequent columns into $(SA(\, ;2), \ldots, SA(\, ;n))$. By induction, Householder symmetries $S_1, \ldots, S_n$ (such that each S_k modifies only entries on or below the kth row) produce a unitary — but not necessarily hermitian — matrix $Q = S_n^* \cdots S_1^*$, and an upper-triangular matrix R, with

$$A = QR.$$

4.2. Solving least-squares problems with orthogonal factorizations

Consider a linear system $Ax = b$ with $n \leqslant m$ linearly independent columns in $A \in \mathbb{M}_{m\times n}(\mathbb{C})$. If $A = QR$ with Q unitary and R upper triangular, then

$$Ax = b,$$

$$Q^*Ax = Q^*b,$$

$$Rx = Q^*b,$$

where multiplication Q preserves Euclidean distances, whence

$$\|Ax - b\|_2 = \|Rx - Q^*b\|_2.$$

Because R has n linearly independent columns and has only zeros below the rth row, $\|Rx - Q^*b\|_2$ reaches a minimum if and only if x is the unique solution $\tilde{x}$ of the first n equations. Moreover,

$$\|R\tilde{x} - Q^*b\|_2 = \|((Q^*b)_{n+1}, \ldots, (Q^*b)_{\mathrm{m}})\|_2.$$

For a matrix $A \in \mathbb{M}_{m\times n}(\mathbb{C})$ with rank $r \leqslant \min\{m, n\}$ and columns that *need not* be linearly independent, there exists a unitary factorization

$$AP = QR.$$

The matrix $P \in \mathbb{M}_{n\times n}(\mathbb{C})$ permutes the columns of A so that the first r columns of AP are linearly independent. Householder symmetries then yield a unitary factorization of the first r columns,

$$[AP(\ ;1),\dots,AP(\ ;r)] = Q[R(\ ;1),\dots,R(\ ;r)],$$

and $R = Q^*(AP)$ contains only zeros below the rth row. With the change of coordinates $z{:=}P^{-1}x$, there is then an affine subspace of dimension $n - r$ of least-squares solutions to the system

$$Ax = b,$$

$$(Q^*AP)(P^{-1}x) = Q^*b,$$

$$Rz = Q^*b.$$

One solution z results from setting $z_{r+1}{:=}\cdots{:=}z_n{:=}0$ and solving the $z_1,\dots,z_r$. the shortest least-squares solution x is then the orthogonal projection of any solution z on the orthogonal complement of the null space of R, in other words, on the row space of R.

Such a projection can employ a unitary factorization of R^*,

$$R^* = WT$$

with W unitary and T upper triangular. Because $T = W^*R^*$ has only zeros below its rth row, it follows that the last $n - r$ columns $w_{r+1},\dots,w_n$ of W form an orthonormal basis of Kernel (R), while the first r columns $w_1,\dots,w_r$ form an orthonormal basis of its row space. Consequently,

$$\tilde{x}{:=}(w_1 \dots w_r)\begin{pmatrix} w_1^* \\ \vdots \\ w_r^* \end{pmatrix} z$$

minimizes $\|A\tilde{x} - b\|_2$ with the smallest norm $\|\tilde{x}\|_2$; see also [32, Chapter 14].

In principle, the permutations P can be generated during the computation of each symmetry S_k, by swapping columns $A(\ ;k)$ and $A(\ ;\ell)$ for some $\ell > k$ if $A(\ ;k)$ lies in the subspace spanned by $A(\ ;1),\dots,A(\ ;k-1)$. However, detecting such linear dependencies and selecting a permutation amounts to computing the ranks of submatrices, which is not reliable with floating-point or other approximate arithmetics [12, p. 66]. The singular value decomposition will provide some information on the reliability of such computations.

4.3. Constrained least squares and geodesy

Such practical situations as geodesy lead to problems of least squares with linear constraints. The outline presented here expands on that of Lawson and Hanson [32, Chapter 20]. Specifically, for matrices

$$C \in \mathbb{M}_{k\times n}(\mathbb{C}),$$

$$E \in \mathbb{M}_{\ell\times n}(\mathbb{C}),$$

$$d \in \mathbb{C}^k,$$

$$f \in \mathbb{C}^\ell,$$

the problem consists in determining a vector $x \in \mathbb{C}^n$ that minimizes

$$\|Ex - f\|_2$$

subject to the constraint

$$Cx = d.$$

The strategy for solving such a problem uses an orthonormal basis $(q_1, \ldots, q_k;\ q_{k+1}, \ldots, q_n)$, where $(q_{k+1}, \ldots, q_n)$ is an orthonormal basis on the null space of C. The basis $(q_{k+1}, \ldots, q_n)$ provides a parametrization of the solution space of the system $Cx = d$, which reduces the problem to an unconstrained least squares problem in the subspace of $\mathbb{C}^n$ spanned by $(q_1, \ldots, q_k)$.

In the generic situation where C has k linearly independent rows and E has n linearly independent columns, C^* factors in the form

$$C^* = QR,$$

$$C = LQ^*,$$

where $Q \in \mathbb{M}_{n\times n}(\mathbb{C})$ is unitary, $R \in \mathbb{M}_{n\times k}(\mathbb{C})$ is upper triangular, and $L = R^*$ is lower triangular with linearly independent rows. Because $R = Q^*C^*$ has only zeros below the kth row, it follows that in Q^* all the rows $q_{k+1}^*, \ldots, q_n^*$ are perpendicular to all the columns of C^*. Hence, the rows $q_1^*, \ldots, q_k^*$ span the column space of C^*. Thus Q performs the required change of basis. With

$$w := Q^*x,$$

the system becomes

$$\begin{pmatrix} L \\ EQ \end{pmatrix} \begin{pmatrix} w_1 \\ w_2 \end{pmatrix} = \begin{pmatrix} C \\ E \end{pmatrix} Q(Q^*x) = \begin{pmatrix} d \\ f \end{pmatrix}.$$

Therefore, there exists exactly one solution $w_1 \in \mathbb{C}^k$ to the system

$$L \begin{pmatrix} w_1 \\ 0 \end{pmatrix} = d.$$

The initial problem thus reduces to determining $w_2 \in \mathbb{C}^\ell$ minimizing

$$\left\| \begin{pmatrix} L \\ EQ \end{pmatrix} \begin{pmatrix} w_1 \\ w_2 \end{pmatrix} - \begin{pmatrix} d \\ f \end{pmatrix} \right\|_2 = \left\| EQ \begin{pmatrix} 0 \\ w_2 \end{pmatrix} - \left[f - EQ \begin{pmatrix} w_1 \\ 0 \end{pmatrix} \right] \right\|_2.$$

The following application uses the Gauss-Bonnet Theorem.

Theorem 6 (Gauss-Bonnet). *Let D be a compact oriented domain with Euler characteristic χ on a Riemannian surface M in $\mathbb{R}^3$. Let $C = \partial D$ be the boundary of D in M, and let $\alpha_1, \ldots, \alpha_L$ be the oriented internal angles at the vertices (if any) of C. Moreover, let K be the Gaussian curvature of M, and let k_g be the geodesic curvature of C. Then*

$$\sum_{\ell=1}^{L} (\alpha_\ell - \pi) = \int\int_D K \, d\sigma + \int_C k_g \, ds - 2\pi\chi.$$

For a proof, see Chern's book [5, pp. 125–126].

For a triangle $D = \Delta$, with $v = 3$ vertices, $s = 3$ sides, and $f = 1$ facet, $\chi_\Delta = v - s + f = 1$. If each side lies on a geodesic on M, then $k_g = 0$, whence

$$\sum_{\ell=1}^{3} \alpha_\ell = \pi + \int\int_D K \, d\sigma.$$

Example 7. With the geodetic latitude λ and the longitude φ, the parametrization of the spheroidal earth surface takes the form

$$\rho = \frac{a\cos(\lambda)}{\sqrt{1 - e^2[\sin(\lambda)]^2}},$$

$$x = \rho\cos(\varphi),$$

$$y = \rho\sin(\varphi),$$

$$z = \frac{a\sigma^2\sin(\lambda)}{\sqrt{1 - e^2[\sin(\lambda)]^2}}.$$

Hence, calculus gives the surface area

$$d\sigma = \frac{a^2\sigma^2\cos(\lambda)}{\{1 - e^2[\sin(\lambda)]^2\}^2} d\varphi \, d\lambda$$

and the Gaussian curvature

$$K(\varphi, \lambda) = \frac{\{1 - e^2[\sin(\lambda)]^2\}^2}{a^2\sigma^2},$$

which is the reciprocal of the product of the radii of curvature R' in the plane of the meridian and N in the perpendicular plane [44, pp. 24, 25]:

$$R' = \frac{a(1 - e^2)}{\{1 - e^2[\sin(\lambda)]^2\}^{3/2}}, \quad N = \frac{a}{\{1 - e^2[\sin(\lambda)]^2\}^{1/2}}.$$

Thus, with Ω being the domain of the parametrization of Δ,

$$\int\int_\Delta K \, d\sigma = \int\int_\Omega \frac{\{1 - e^2[\sin(\lambda)]^2\}^2}{a^2\sigma^2} \frac{a^2\sigma^2\cos(\lambda)}{\{1 - e^2[\sin(\lambda)]^2\}^2} d\varphi \, d\lambda$$

$$= \int\int_\Omega \cos(\lambda) \, d\varphi \, d\lambda.$$

Example 8. Gauss investigated triangulations measured by De Krayenhof, for instance, the following internal angles of a spheroidal triangle [16, Section 23, p. 149]:

$\alpha = 50^\circ 58' 15.238''$ at Harlingen,

$\beta = 82^\circ 47' 15.351''$ at Leeuwarden,

$\gamma = 46^\circ 14' 27.202''$ at Ballum.

In the plane, no such triangle exists, because the sum of the three angles $\alpha+\beta+\gamma=179^\circ59'57.791''$ fails to equal 180°. On an ellipsoid, the sum of the internal angles in a geodesic triangle Δ exceeds 180° by the integral of the Gaussian curvature K over the triangle, which Gauss computed to be $1.749''$ for this example, so that $\alpha+\beta+\gamma=180^\circ0'1.749''$. In either case, it is impossible to place the three cities on a map without altering the data. One strategy consists in making the "smallest" adjustment while preserving $\alpha+\beta+\gamma=180^\circ$, in other words, minimizing

$$\begin{pmatrix}1&0&0\\0&1&0\\0&0&1\end{pmatrix}\begin{pmatrix}\alpha\\\beta\\\gamma\end{pmatrix}-\begin{pmatrix}50^\circ58'15.238''\\82^\circ47'15.351''\\46^\circ14'27.202''\end{pmatrix}$$

subject to the linear constraint

$$(1\quad 1\quad 1)\begin{pmatrix}\alpha\\\beta\\\gamma\end{pmatrix}=180^\circ+\frac{180^\circ}{\pi}\int\!\!\int_\Delta K\,d\sigma=180^\circ0'1.749''.$$

More generally, with n measurements $f_1,\ldots,f_n$ of quantities $\alpha_1,\ldots,\alpha_n$ subject to a constraint $\alpha_1+\cdots+\alpha_n=d$, the system becomes

$$\begin{pmatrix}1^*\\I\end{pmatrix}\begin{pmatrix}\alpha_1\\\vdots\\\alpha_n\end{pmatrix}=\begin{pmatrix}d\\f_1\\\vdots\\f_n\end{pmatrix}.$$

For the unitary factorization of the constraint equation,

$$r=C^*(\,;1)=1^*=(1,\ldots,1)^*,$$

$$v=r+\|r\|_2e_1=\begin{pmatrix}\sqrt{n}+1\\1\\\vdots\\1\end{pmatrix},$$

$$\nu=\frac{1}{\|r\|_2(\|r\|_2+|r_1|)}=\frac{1}{\sqrt{n}(\sqrt{n}+1)},$$

$$S=I-\nu vv^*,$$

$$L=CS^*=-\sqrt{3}1^*=(-\sqrt{n},0,0)=-\sqrt{n}e_1.$$

Consequently,

$$Lw_1=d,$$

$$-\sqrt{n}w_1=d,$$

$$w_1 = \frac{d}{-\sqrt{n}} = \frac{180^\circ 0' 1.749''}{-\sqrt{3}} = -103^\circ 55' 33.072''.$$

For the least-squares system, $E = I$. Consequently,

$$EQ = IS = S = (q_1; q_2, q_3, \ldots q_n),$$

and the least-squares system takes the form

$$(q_1;\ q_2\ q_3\ \ldots\ q_n) \begin{pmatrix} 0 \\ w_2 \end{pmatrix} = f - q_1 w_1.$$

Hence,

$$\begin{pmatrix} 0 \\ w_2 \end{pmatrix} = \begin{pmatrix} q_1^*; \\ q_2^* \\ q_3^* \\ \vdots \\ q_n^* \end{pmatrix} (f - q_1 w_1).$$

Because $q_j^* q_1 = 0$ for every $j > 1$, the least-squares solution is

$$w_2 = \begin{pmatrix} q_2^* \\ q_3^* \\ \vdots \\ q_n^* \end{pmatrix} (f - q_1 w_1) = \begin{pmatrix} q_2^* \\ q_3^* \\ \vdots \\ q_n^* \end{pmatrix} (f),$$

and the first coordinate (in this example) gives the least-squares error

$$\|Ex - f\|_2 = q_1^*(f - q_1 w_1) = q_1^* f - w_1 = -(1/\sqrt{n}) 1^* f + d/\sqrt{n}$$

$$= \frac{1}{\sqrt{n}} \left[180^\circ + \frac{180^\circ}{\pi} \int\!\!\int_\Delta K \mathrm{d}\sigma - (f_1 + f_2 + \cdots + f_n) \right].$$

Reverting to the canonical basis through the inverse change of basis gives the solution

$$x = Sw = (q_1; q_2, q_3, \ldots, q_n) \begin{pmatrix} w_1 \\ w_2 \end{pmatrix}$$

$$= (w_1 - q_1^* f) q_1 + (q_1; q_2, q_3, \ldots, q_n) \begin{pmatrix} q_1^* f \\ \begin{pmatrix} q_2^* \\ q_3^* \vdots \\ q_n^* \end{pmatrix} (f) \end{pmatrix}$$

$$= (w_1 - q_1^* f) q_1 + f$$

$$= \begin{pmatrix} 50^\circ 58'16.557\,333'' \\ 82^\circ 47'16.670\,333'' \\ 46^\circ 14'28.521\,333'' \end{pmatrix} = \begin{pmatrix} \tilde{\alpha} \\ \tilde{\beta} \\ \tilde{\gamma} \end{pmatrix},$$

which add to $180^\circ 0'1.749''$. The formula

$$x = (w_1 - q_1^* f)q_1 + f = \left(\frac{d}{\sqrt{n}} - \frac{1^* f}{\sqrt{n}}\right)\frac{1}{\sqrt{n}}1 + f$$

shows that the measurements f are all adjusted by the average discrepancy

$$\frac{d - 1^* f}{n} = \frac{180^\circ 0'1.749'' - 179^\circ 59'57.791''}{3} = \frac{0^\circ 0'3.958''}{3} = 1.319\,333\ldots''.$$

5. The singular-value decomposition and error analysis

5.1. The singular-value decomposition

Ordinary least-squares problems consist in determining the shortest vector $\tilde{x}$ that minimizes $\|A\tilde{x} - b\|_2$, perhaps subject to linear constraints. If $\tilde{b}:=A\tilde{x}$, then the solution minimizes the discrepancy in the right-hand side, $\|\tilde{b}-b\|_2$, but it does *not* adjust the matrix A. In other words, $\tilde{x}=(\tilde{x}_1,\ldots,\tilde{x}_n)$ is the gradient of the linear function $u: \mathbb{C}^n \to \mathbb{C}$ that minimizes the average squared discrepancy between the measurement b_j and the value $u[A(j;\,)]$, but it does *not* minimize the Euclidean distance from the graph of u (a hyperplane) to the data $(a_{j,1},\ldots,a_{j,n};b_j)$ in $\mathbb{C}^{n+1}$. Such more general problems of least squares admit solutions in term of a matrix factorization called the "singular-value decomposition" that was published independently by Eugenio Beltrami in 1873 and Camille Jordan in 1874, and extended by Erhard Schmidt in 1907 and Hermann Weyl in 1912. (For greater detail on the history of the singular value decomposition consult Stewart's account [45].)

Theorem 9. *For each matrix $A \in \mathbb{M}_{m\times n}(\mathbb{C})$ of rank r, there exist unitary matrices U, V, and a diagonal matrix Σ, such that*

$$\begin{aligned} A &= U\Sigma V^* \\ &= u_1\sigma_1 v_1^* + \cdots + u_r\sigma_r v_r^* \\ &= \tilde{U}\tilde{\Sigma}\tilde{V}^* \end{aligned}$$

with $\sigma_j:=\Sigma_{j,j}$ and with the following features.

(U) *The matrix $U \in \mathbb{M}_{m\times m}(\mathbb{C})$ is unitary. The first r columns $(u_1,\ldots,u_r)$ of U form an orthonormal basis for the range (column space) of A. The last $m-r$ columns $(u_{r+1},\ldots,u_m)$ of U form an orthonormal basis for the null space (kernel) of A^*.*

(V) *The matrix $V \in \mathbb{M}_{n\times n}(\mathbb{C})$ is unitary. The first r columns $(v_1,\ldots,v_r)$ of V form an orthonormal basis for the row space of A ($[\text{Kernel}(A)]^\perp$). The last $n-r$ columns $(v_{r+1},\ldots,v_n)$ of V form an orthonormal basis for the null space (kernel) of A.*

(Σ) *The matrix $\Sigma \in \mathbb{M}_{m\times n}(\mathbb{C})$ is diagonal: $\Sigma_{k,\ell} = 0$ for all $k \neq \ell$, with*

$$\sigma_1 \geqslant \sigma_2 \geqslant \cdots \geqslant \sigma_r > 0 = \sigma_{r+1} = \cdots = \sigma_{\min\{m,n\}}.$$

Moreover,

$$Av_j = \sigma_j u_j,$$
$$A^* u_j = \sigma_j v_j,$$

for every $j \in \{1,\ldots,r\}$, *and* $Av_j = 0$ *for every* $j \in \{r+1,\ldots,n\}$. *Finally,*

$$\tilde{U} = (u_1,\ldots,u_r) \in \mathbb{M}_{m\times r}(\mathbb{C}),$$
$$\tilde{V} = (v_1,\ldots,v_r) \in \mathbb{M}_{n\times r}(\mathbb{C}),$$
$$\tilde{\Sigma} = \text{diagonal}(\sigma_1,\ldots,\sigma_r) \in \mathbb{M}_{r\times r}(\mathbb{C}).$$

Proof. Let $V = (v_1,\ldots,v_r,v_{r+1},\ldots,v_n)$ be an orthonormal basis of eigen vectors for the hermitian positive semi-definite matrix $A^*A \in \mathbb{M}_{n\times n}(\mathbb{C})$, corresponding to its eigenvalues in nondecreasing order $\lambda_1 \geqslant \lambda_2 \geqslant \cdots \geqslant \lambda_r > 0 = \lambda_{r+1} = \cdots = \lambda_n$. Define $\sigma_j := \sqrt{\lambda_j}$, and $u_j := (1/\sigma_j)Av_j$ for every $j \in \{1,\ldots,r\}$. The remainder of the proof consists of straightforward verifications [30, Section 5.4; 46, Section 6.4]. □

Definition 10. The factorization $A = U\Sigma V^*$ is the *singular-value decomposition* of A. The scalars $\sigma_1,\ldots,\sigma_r$ are the *singular values* of A. The vectors $v_1,\ldots,v_n$ are the *right singular vectors* of A. The vectors $u_1,\ldots,u_m$ are the *left singular vectors* of A.

The singular-value decomposition also provides a means to solve ordinary least-squares problems. Firstly, the product $\tilde{U}^* b$ projects b orthogonally on the column space of A, whence

$$\|Ax - \tilde{U}^* b\|_2 \leqslant \|Ax - b\|_2$$

for every $x \in \mathbb{C}^n$. Because $\tilde{U}^* b$ lies in the column space of A, there exists a solution $x \in \mathbb{C}^n$ such that $Ax = \tilde{U}^* b$. Secondly, every solution to this system differs from x by a vector in the null space of A. Consequently, the shortest solution is the orthogonal projection $x^\dagger = \tilde{V}^* x$ of x on the orthogonal complement of the null space of A. A derivation of a formula for $x^\dagger$ can proceed as follows:

$$Ax = b,$$

$$(\tilde{U}\tilde{\Sigma}\tilde{V}^*)x = b,$$

$$\tilde{\Sigma}(\tilde{V}^* x) = \tilde{U}^* b,$$

$$\tilde{V}^* x = \tilde{\Sigma}^{-1}\tilde{U}^* b,$$

$$x^\dagger = (\tilde{V}\tilde{\Sigma}^{-1}\tilde{U}^*)b.$$

Definition 11. The *pseudoinverse* of A is the matrix

$$A^\dagger := \tilde{V}\tilde{\Sigma}^{-1}\tilde{U}^*.$$

Thus, the shortest least-squares solution of $Ax = b$ is $x^\dagger := A^\dagger b$.

5.2. *Norms and condition numbers of matrices*

5.2.1. *Norms of matrices*

The concepts of norms and condition numbers for matrices provide means to estimate the propagation of errors from the data and during computations through the solutions of linear systems, as developed by Gastinel [15].

Definition 12. For each norm $\| \ \|_p$ on $\mathbb{C}^n$ and each norm $\| \ \|_q$ on $\mathbb{C}^m$, the *subordinate matrix norm* $\| \ \|_{p,q}$ on $\mathbb{M}_{m\times n}(\mathbb{C})$ is defined by

$$\begin{aligned}\|A\|_{p,q} &:= \max\{\|Au\|_q \colon u \in \mathbb{C}^n,\ \|u\|_p = 1\}\\ &= \max\{\|Au\|_q/\|u\|_p \colon u \in \mathbb{C}^n,\ u \neq 0\}.\end{aligned}$$

Example 13. With $\|x\|_\infty := \max_j |x_j|$ on $\mathbb{C}^n$ and $\|Ax\|_\infty$ on $\mathbb{C}^m$,

$$\|A\|_{\infty,\infty} = \max_{1\leqslant i\leqslant m} \sum_{j=1}^{n} |A_{i,j}|.$$

With $\|x\|_1 := \sum_j |x_j|$ on $\mathbb{C}^n$ and $\|Ax\|_1$ on $\mathbb{C}^m$,

$$\|A\|_{1,1} = \max_{1\leqslant j\leqslant n} \sum_{i=1}^{m} |A_{i,j}|.$$

With $\|x\|_1$ on $\mathbb{C}^n$, and $\|Ax\|_\infty$ on $\mathbb{C}^m$,

$$\|A\|_{1,\infty} = \max_{1\leqslant i\leqslant m}\ \max_{1\leqslant j\leqslant n} |A_{i,j}|.$$

(For $p \in \{1, \infty\}$ the formulae for $\|x\|_p$ and $\|A\|_{p,p}$ coincide.)

The following considerations show that $\|A\|_{2,2} = \sigma_1$ and $\kappa_{2,2}(A) = \sigma_1/\sigma_n$ is the ratio of the largest to the smallest singular values of A.

Lemma 14. *For all real numbers* $\sigma_1 \geqslant \sigma_2 \geqslant \cdots \geqslant \sigma_{n-1} \geqslant \sigma_n \geqslant 0$,

$$\min_{\|x\|_2=1} \sum_{i=1}^{n} (\sigma_i x_i)^2 = \min_{i\in\{1,\ldots,n\}} \sigma_i^2 = \sigma_n^2,$$

$$\max_{\|x\|_2=1} \sum_{i=1}^{n} (\sigma_i x_i)^2 = \max_{i\in\{1,\ldots,n\}} \sigma_i^2 = \sigma_1^2.$$

Proof. Solving $\sum_{i=1}^{n} x_i^2 = 1$ for x_n^2 gives $x_n^2 = 1 - \sum_{i=1}^{n-1} x_i^2$. Hence

$$\sum_{i=1}^{n} (\sigma_i x_i)^2 = \sigma_n^2 x_n^2 + \sum_{i=1}^{n-1} (\sigma_i x_i)^2$$

$$= \sigma_n^2\left(1-\sum_{i=1}^{n-1} x_i^2\right) + \sum_{i=1}^{n-1} \sigma_i^2 x_i^2$$

$$= \sigma_n^2 + \sum_{i=1}^{n-1} (\sigma_i^2 - \sigma_n^2) x_i^2$$

$$\geqslant \sigma_n^2,$$

with equality if and only if $x_i = 0$ for $\sigma_i \neq \sigma_n$. Similarly, $\sum_{i=1}^{n}(\sigma_i x_i)^2 = \sigma_1^2 + \sum_{i=2}^{n}(\sigma_i^2 - \sigma_1^2)x_i^2 \leqslant \sigma_1^2$ with equality if and only if $x_i = 0$ for $\sigma_i \neq \sigma_1$.

Proposition 15. *For each matrix $A \in \mathbb{M}_{m\times n}(\mathbb{C})$, the subordinate Euclidean norm $\|A\|_{2,2} = \sigma_1$ is the largest singular value of A. Moreover, $\kappa_{2,2}(A) = \sigma_1/\sigma_n$ is the ratio of the largest to the smallest singular values of A.*

Proof. Consider a singular-value decomposition $A = U\Sigma V^*$. For each vector $x \in \mathbb{C}^n$ with $\|x\|_2 = 1$, let $w := V^* x$. Then $\|w\|_2 = \|x\|_2 = 1$. Hence,

$$\|Ax\|_2^2 = x^* A^* Ax = x^*(V\Sigma^* U^*)(U\Sigma V^*)x$$

$$= x^* V\Sigma^*\Sigma V^* x = \|\Sigma w\|_2^2 = \sum_{i=1}^{n} (\sigma_i w_i)^2 \leqslant \sigma_1^2$$

with the maximum value reached for $w = e_1$, or, equivalently, $x = v_1$. Hence, $\kappa_2(A) = \|A\|_2 \|A^{-1}\|_2 = (\max_i \sigma_i)(\max_i \sigma_i^{-1}) = \sigma_1/\sigma_n$. □

For norms of the type $\|x\|_p := (|x_1|^p + \cdots + |x_n|^p)^{1/p}$ with $p, q \notin \{1, 2, \infty\}$, no formula for the subordinate norm $\|A\|_{p,q}$ seems to be known [22, p. 124].

5.2.2. Condition numbers of matrices

For a square and invertible matrix $A \in \mathbb{M}_{n\times n}(\mathbb{C})$, the condition number provides lower and upper bounds on the discrepancy $\|\tilde{x} - x\|_p$ between the solution x of a linear system $Ax = b$ and any vector $\tilde{x}$. Such a vector $\tilde{x}$ can result, for instance, from an attempt at solving the system with floating-point or any other approximate arithmetic. To this end, let $\tilde{b} := A\tilde{x}$.

Definition 16. For each norm $\|\ \|_p$ on $\mathbb{C}^n$ and each norm $\|\ \|_q$ on $\mathbb{C}^m$, the *condition number* $\kappa_{p,q}$ is defined by

$$\kappa_{p,q}(A) := \|A\|_{p,q} \|A^\dagger\|_{q,p}.$$

Proposition 17. *For all b, $\tilde{b}$, x, $\tilde{x}$ and A invertible with $Ax = b$ and $A\tilde{x} = \tilde{b}$,*

$$\frac{1}{\kappa_{p,q}(A)} \frac{\|\tilde{b} - b\|_q}{\|b\|_q} \leqslant \frac{\|\tilde{x} - x\|_p}{\|x\|_p} \leqslant \kappa_{p,q}(A) \frac{\|\tilde{b} - b\|_q}{\|b\|_q}.$$

Proof. Use $\|b\|_q = \|Ax\|_q \leqslant \|A\|_{p,q} \cdot \|x\|_p$ and $\|\tilde{x} - x\|_p = \|A^{-1}(\tilde{b} - b)\|_p \leqslant \|A^{-1}\|_{q,p} \cdot \|\tilde{b} - b\|_q$ [30, Section 4.4; 46, Section 4.4]. □

Proposition 17 compares the solutions x and $\tilde{x}$ of two systems with right-hand sides b and $\tilde{b}$ but with the same matrix A. In contrast, with different matrices A and C the following result holds. For each invertible $A \in \mathbb{M}_{n\times n}(\mathbb{C})$, for each $C \in \mathbb{M}_{n\times n}(\mathbb{C})$. If $\|A - C\| < 1/\|A^{-1}\|$, then for each nonzero vector $b \in \mathbb{C}^n$ and for the solutions $x \in \mathbb{C}^n$ of $Ax = b$ and $w \in \mathbb{C}^n$ of $Cw = b$,

$$\frac{\|w - x\|}{\|x\|} \leqslant \frac{\kappa(A)}{1 - \kappa(A) \cdot \|A - C\|/\|A\|} \cdot \frac{\|A - C\|}{\|A\|}.$$

For a proof see [46, pp. 188–198], and for other similar error bounds see [22, Chapter 7]. Yet more generally, a theorem of Wedin for all matrices $A \in \mathbb{M}_{m\times n}(\mathbb{C})$ and $C \in \mathbb{M}_{m\times n}(\mathbb{C})$, with rank $r = n \leqslant m$, and for all vectors $b \in \mathbb{C}^m$ and $d \in \mathbb{C}^m$, if there exists a positive real ε for which

$$\kappa_2(A)\varepsilon < 1,$$

$$\|A - C\|_2 \leqslant \varepsilon \|A\|_2,$$

$$\|b - d\|_2 \leqslant \varepsilon \|b\|_2,$$

then the *least-squares* solutions $\tilde{x} \in \mathbb{C}^n$ and $\tilde{z} \in \mathbb{C}^n$ of the systems $Ax = b$ and $Cz = d$ satisfy the following inequalities [22, Chapter 19]:

$$\frac{\|x - z\|_2}{\|x\|_2} \leqslant \frac{\kappa_2(A)\varepsilon}{1 - \kappa_2(A)\varepsilon} \left\{ 2 + [1 + \kappa_2(A)] \frac{\|b - Ax\|_2}{\|A\|_2 \cdot \|x\|_2} \right\},$$

$$\frac{\|(b - Ax) - (d - Az)\|_2}{\|b\|_2} \leqslant [1 + 2\kappa_2(A)]\varepsilon.$$

The following theorem of Kahan [25, pp. 775,776], who credits Gastinel, shows that for each invertible matrix A the distance to the closest singular matrix is $1/\|A^{-1}\|$.

Theorem 18. *For every invertible matrix A and every subordinate norm*:

$$\min_{\det(S)=0} \|A - S\| = \frac{1}{\|A^{-1}\|}.$$

Proof. For each singular matrix S there exists a vector $z \neq 0$ with $Sz = 0$:

$$\|A - S\| \geqslant \frac{\|(A - S)z\|}{\|z\|} = \frac{\|Az\|}{\|z\|} = \frac{\|A^{-1}\|\|Az\|}{\|A^{-1}\|\|z\|} \geqslant \frac{\|A^{-1}Az\|}{\|A^{-1}\|\|z\|} = \frac{1}{\|A^{-1}\|}.$$

There exists a vector $y \neq 0$ with $\|A^{-1}y\| = \|A^{-1}\|\|y\|$. As in the Hahn–Banach theorem [48], choose a linear functional w dual to $A^{-1}y$, so that

$$w(A^{-1}y) = \|w\| \cdot \|A^{-1}y\| = 1,$$

let w^* be the matrix of w relative to the canonical basis, so that $w(z) = w^*z$ for every vector z, and define

$$S := A - yw^*.$$

Then S is singular, because

$$S(A^{-1}y) = (A - yw^*) \cdot (A^{-1}y) = y - y \cdot 1 = 0.$$

Moreover,

$$\begin{aligned}
\|A - S\| &= \max\{\|(yw^*)x\|: \|x\| = 1\} \\
&= \max\{\|y(w^*x)\|: \|x\| = 1\} \\
&= \|y\| \cdot \max\{w^*x: \|x\| = 1\} \\
&= \|y\| \cdot \|w^*\| \\
&= \|y\| \cdot \frac{1}{\|A^{-1}y\|} \\
&= \|y\| \cdot \frac{1}{\|A^{-1}\| \cdot \|y\|} \\
&= \frac{1}{\|A^{-1}\|}. \qquad \square
\end{aligned}$$

6. Matrix approximation and total least squares

6.1. The approximation theorems of Schmidt, Mirsky, and Weyl

A theorem of Schmidt [43], with later versions by Mirsky [34] and Weyl [52], approximates a matrix $C \in \mathbb{M}_{m\times n}(\mathbb{C})$ of rank r by a singular matrix $S \in \mathbb{M}_{m\times n}(\mathbb{C})$ of rank $s < r$ that minimizes the Frobenius norm $\|C - S\|_F$, defined by

$$\|A\|_F^2 := \sum_{i=1}^{m}\sum_{j=1}^{n} |A_{i,j}|^2 = \sum_{i=1}^{m} \|A(i;)\|_2^2 = \sum_{j=1}^{n} \|A(;j)\|_2^2.$$

All unitary matrices U and V preserve Euclidean and Frobenius norms:

$$\|UA\|_F^2 = \sum_{j=1}^{n} \|UA(;j)\|_2^2 = \sum_{j=1}^{n} \|A(;j)\|_2^2 = \|A\|_F,$$

$$\|AV\|_F = \sum_{i=1}^{m} \|A(i;)V\|_2^2 = \sum_{i=1}^{m} \|A(i;)\|_2^2 = \|A\|_F.$$

In particular, with a singular-value decomposition $A = U\Sigma V^*$,

$$\|A\|_F^2 = \|U\Sigma V^*\|_F^2 = \|\Sigma\|_F^2 = \sum_{i=1}^{r} \sigma_r^2 \geqslant \sigma_1^2 = \|A\|_2^2.$$

The following theorem follows Stewart's version [45, pp. 561, 562].

Theorem 19. *For each matrix* $C \in \mathbb{M}_{m\times n}(\mathbb{C})$ *with* $\sigma_1 \geqslant \cdots \geqslant \sigma_r > 0$ *and*

$$C = \sum_{i=1}^{r} \sigma_i u_i v_i^*$$

and for each matrix $S \in \mathbb{M}_{m\times n}(\mathbb{C})$ *of rank* $k \in \{0,\ldots,r\}$,

$$\|C - S\|_F^2 \geqslant \sum_{i=k+1}^{r} \sigma_i^2,$$

with the minimum $\sigma_{k+1}^2 + \cdots + \sigma_r^2$ *reached for*

$$S = \sum_{i=1}^{k} \sigma_i u_i v_i^*.$$

Proof. If $k = r$ then the theorem holds because $S = C$. Henceforth, assume that $k < r$. Also, for each matrix A, let $\sigma_i(A)$, $u_i(A)$, and $v_i(A)$ be the ith singular value and singular vectors of A, and define

$$A_k := \sum_{i=1}^{k} \sigma_i(A) u_i(A) v_i^*(A).$$

The following argument shows that $\sigma_1(C - S) \geqslant \sigma_{k+1}(C)$. If S has rank k then S has a singular value decomposition $S = \sum_{i=1}^{k} \tau_i w_i z_i^* = W\tau Z^*$. Moreover, the linear space $Z^\perp$ perpendicular to $z_1,\ldots,z_k$ has dimension $n - k > n - (k+1) \geqslant n - r$. Because the column space V spanned by $v_1,\ldots,v_{k+1}$ has dimension $k+1$, it follows that $Z^\perp \cap V \neq \{0\}$. Thus, there exists a non-zero vector of coefficients $\gamma \in \mathbb{C}^{k+1}$, for instance with $\|\gamma\|_2 = 1$, such that $x := V_\gamma = \sum_{i=1}^{k+1} \gamma_i v_i \in Z^\perp \cap V$, whence $0 = Zx$ and hence $Sx = 0$. Let $\tilde{\gamma} := (\gamma^*, 0^*)^* \in \mathbb{C}^n$:

$$\begin{aligned}
\sigma_1^2(C - S) &\geqslant x^*(C - S)^*(C - S)x \\
&= x^* C^* C x \\
&= \tilde{\gamma}^* V^* V \Sigma U^* U \Sigma V^* V \tilde{\gamma} \\
&= \tilde{\gamma}^* \Sigma^2 \tilde{\gamma} = \sum_{i=1}^{k+1} (\gamma_i \sigma_i)^2 \\
&\geqslant \sigma_{k+1}^2.
\end{aligned}$$

The next argument provides an upper bound on the change in the largest singular value caused by a change in a matrix. From the reverse triangle inequality for norms, it follows that

$$\sigma_1(A - B) = \|A - B\|_2 \geqslant \big| \|A\|_2 - \|B\|_2 \big| = |\sigma_1(A) - \sigma_1(B)|.$$

The following generalization provides inequalities for the other singular values. For each matrix G and each index ℓ, $\sigma_1(G - G_\ell) = \sigma_{\ell+1}(G)$. Consequently, for all matrices $G, H \in \mathbb{M}_{m\times n}(\mathbb{C})$, and for all indices k and ℓ, the foregoing result leads to

$$\begin{aligned}
&\sigma_{\ell+1}(G) + \sigma_{k+1}(H) \\
&\quad = \sigma_1(G - G_\ell) + \sigma_1(H - H_k) \geqslant \sigma_1([G - G_\ell] + [H - H_k])
\end{aligned}$$

$$= \sigma_1([G+H] - [G_\ell + H_k])$$

$$\geqslant \sigma_{\ell+k+1}(G+H)$$

because the rank of $G_\ell + H_k$ cannot exceed $\ell + k$. Equivalently, if $A{:=}G+H$ and $B{:=}H$, then

$$\sigma_{\ell+1}(A-B) \geqslant \sigma_{\ell+k+1}(A) - \sigma_{k+1}(B)$$

Finally, in the particular case where S has rank k, setting $G{:=}C-S$ and $H{:=}S$ gives

$$\sigma_{\ell+1}(C-S) + 0 = \sigma_{\ell+1}(C-S) + \sigma_{k+1}(S) \geqslant \sigma_{\ell+1+k}(C)$$

Finally,

$$\|C-S\|_F^2 = \sum_{i=1}^{r} \sigma_i^2(C-S) \geqslant \sum_{i=1}^{r} \sigma_{i+k}^2(C) = (\sigma_{k+1}^2 + \cdots + \sigma_r^2)(C).$$

Equality holds with $S = \sum_{i=1}^{k} \sigma_i u_i v_i^*$, for which $\|C-S\|_F^2 = \|\sigma_{k+1} u_{k+1} v_{k+1}^* + \cdots + \sigma_r u_r v_r^*\|_F^2 = \sigma_{k+1}^2 + \cdots + \sigma_r^2$. □

The approximation theorem of Schmidt, Mirsky, and Weyl amounts to identifying a matrix S minimizing a rotationally invariant norm $\|C-S\|$, for instance, the Euclidean norm, or the Frobenius norm, subject to the linear constraints $\sigma_{k+1}(S) = \cdots = \sigma_n(S) = 0$. There also exist other types of constraints, for example, with the vector of singular values $\sigma = (\sigma_1(S), \ldots, \sigma_n(S))$ subject to a linear system of constraints $K\sigma = d$ [37].

6.2. Total least squares

For a linear system $Ax = b$, the problem of *ordinary* least squares consists in determining the shortest vector $\tilde{x}$ that minimizes the Euclidean norm of the discrepancy between b and $\tilde{b}{:=}A\tilde{x}$, possibly subject to constraints. In other words, the *ordinary* least-squares solution $\tilde{x}$ solves *exactly* a related linear system $A\tilde{x} = \tilde{b}$ with $\|\tilde{b} - b\|_2$ minimum. In contrast, the problem of *total* least squares allows for minimal adjustments not only of b but also of A, also possibly subject to constraints. The problem of total least squares admit several mutually equivalent mathematical formulations. Their solutions in terms of singular value decompositions was published in 1980 by Golub and Van Loan [18;19, pp. 576–581]. Van Huffel and Vandewalle's monograph [50] describes further extensions and applications.

6.2.1. Geometric formulations of total least squares

Geometrically, the problem of total least squares amounts to fitting a hyperplane H minimizing the average squared Euclidean distance (measured perpendicularly to the fitted hyperplane) to data points $c_1, \ldots, c_m$ in $\mathbb{C}^{n+1}$. The problem then reduces to finding a point $c_0 \in H$ and a non-zero normal vector $x \perp H$ that minimize the sum D of the squared distances:

$$D(x, c_0; c_1, \ldots, c_m) {:=} \sum_{i=1}^{m} \frac{|\langle c_i - c_0, x\rangle|^2}{\langle x, x\rangle}.$$

To simplify notation, for every point $c_0 \in \mathbb{C}^{n+1}$, and for all data $c_1, \ldots, c_m$ in $\mathbb{C}^{n+1}$, define a matrix $C_{c_0} \in \mathbb{M}_{m\times(n+1)}(\mathbb{C})$ with ith row $c_i^* - c_0^*$:

$$C_{c_0} := \begin{pmatrix} c_1^* - c_0^* \\ \vdots \\ c_m^* - c_0^* \end{pmatrix}.$$

Consequently,

$$D(x, c_0; c_1, \ldots, c_m) = \frac{\|C_{c_0}x\|_2^2}{\|x\|_2^2}.$$

The following lemma reveals that an optimal hyperplane must pass through the centroid of the data,

$$\bar{c} = \frac{1}{m}\sum_{i=1}^{m} c_i,$$

which can thus serve as the point $c_0 \in H$.

Lemma 20. *For every normal vector $x \in \mathbb{C}^{n+1}\setminus\{0\}$, for every point $c_0 \in \mathbb{C}^{n+1}$, and for all data $c_1, \ldots, c_m$ in $\mathbb{C}^{n+1}$,*

$$D(x, c_0; c_1, \ldots, c_m) \geqslant D(x, \bar{c}; c_1, \ldots, c_m),$$

with equality if and only if $\langle x, (r - c_0)\rangle = \langle x, (r - \bar{c})\rangle$ for every r. Consequently, a hyperplane of total least squares must pass through the centroid $\bar{c}$.

Proof. Consider the vector $w := C_{c_0}x$, so that $w_i = \langle c_i - c_0, x\rangle$ and

$$D(x, c_0; c_1, \ldots, c_m) = \frac{\|w\|_2^2}{\|x\|_2^2}.$$

Also, consider the vector $z := C_{\bar{c}}x$, so that $z_i = \langle c_i - \bar{c}, x\rangle$ and

$$D(x, \bar{c}; c_1, \ldots, c_m) = \frac{\|z\|_2^2}{\|x\|_2^2}.$$

Moreover, define $1 := (1, \ldots, 1) \in \mathbb{C}^m$, and $h := \langle x, (\bar{c} - c_0)\rangle$, so that

$$w = z + h1.$$

Then $z \perp 1$:

$$\langle z, 1\rangle = 1^*(C_{\bar{c}}x) = (1^*C_{\bar{c}})x = \left(m\bar{c}^* - \sum_{j=1}^{m} c_j^*\right)x = 0^*x = 0.$$

Finally, the Pythagorean Theorem applied to $z \perp 1$ and $w = z + h1$ gives

$$\begin{aligned} D(x, c_0; c_1, \ldots, c_m) &= \|w\|_2^2/\|x\|_2^2 \\ &= (\|z\|_2^2 + h^2\|1\|_2^2)/\|x\|_2^2 \end{aligned}$$

$$= D(x, \bar{c}; c_1, \ldots, c_m) + h^2 m / \|x\|_2^2$$

$$\geqslant D(x, \bar{c}; c_1, \ldots, c_m),$$

with equality if and only if $0 = h = \langle x, (\bar{c} - c_0) \rangle$, which means that c_0 also lies in the hyperplane passing through $\bar{c}$ perpendicularly to x. □

The following lemma reveals that an optimal normal vector must be a right-singular vector corresponding to the smallest singular value of C_{c_0}.

Lemma 21. *For every point $c_0 \in \mathbb{C}^{n+1}$ and all data $c_1, \ldots, c_m$ in $\mathbb{C}^{n+1}$, let v be a right-singular vector corresponding to the smallest singular value σ of C_{c_0}. Then for every vector $x \in \mathbb{C}^{n+1} \setminus \{0\}$, the following inequality holds:*

$$D(x, c_0; c_1, \ldots, c_m) \geqslant D(v, c_0; c_1, \ldots, c_m),$$

with equality if, but only if, x is also a right-singular vector corresponding to the smallest singular value σ of C_{c_0}. Consequently, a hyperplane of total least squares must be perpendicular to such a singular vector. Moreover,

$$D(v, c_0; c_1, \ldots, c_m) = \sigma^2.$$

Proof. From $D(x, c_0; c_1, \ldots, c_m) = \|C_{c_0} x\|_2^2 / \|x\|_2^2$ it follows that D reaches its minimum at a unit vector $v = x / \|x\|_2$ that minimizes $\|C_{c_0} v\|_2$. The theory of the SVD shows that v coincides with any singular vector v for to the smallest singular value σ of C_{c_0}, with $D(v, c_0; c_1, \ldots, c_m) = \|C_{c_0} v\|_2^2 = \sigma^2$. □

Theorem 22. *For every set of data points $c_1, \ldots, c_m$ in $\mathbb{C}^{n+1}$, each hyperplane of total least-squares passes through the centroid of the data $\bar{c}$ perpendicularly to a right-singular vector v corresponding the smallest singular value σ of the matrix $C_{\bar{c}}$ with ith row $c_i^* - \bar{c}^*$. Moreover, for such a hyperplane, the sum of the squared distances to the data is σ^2:*

Proof. Combine the proofs of Lemmas 20 and 21. □

The matrix C can have a multiple smallest singular value $\sigma = \sigma_{k+1} = \cdots = \sigma_{k+\ell}$, corresponding to a linear subspace $V_\sigma \subseteq \mathbb{C}^{n+1}$ spanned by multiple singular vectors $v_{k+1}, \ldots, v_{k+\ell}$. In this situation, there exists a set $\mathscr{H}$ of hyperplanes of total least squares, with each hyperplane $H \in \mathscr{H}$ perpendicular to a vector $v \in V$ and containing the "axis" $\bar{c} + V^\perp$. In particular, if $\sigma = 0$, then the data lies at the intersection $\cap \mathscr{H}$ of all such hyperplanes, which is an affine subspace $\bar{c} + V^\perp$ of dimension $n + 1 - \ell$. For example, if $n + 1 = 3$ and $\ell = 2$, then $n + 1 - \ell = 1$ and all the data points lie on a common straight line in space.

With $x = v$ and σ computed, the vector

$$\hat{c}_i := c_i - \langle c_i - \bar{c}, v \rangle v$$

is the orthogonal projection of the data c_i onto H. Consequently,

$$\sum_{i=1}^{m} \|\hat{c}_i - c_i\|_2^2 = D(v, \bar{c}; c_1, \ldots, c_m) = \sigma^2.$$

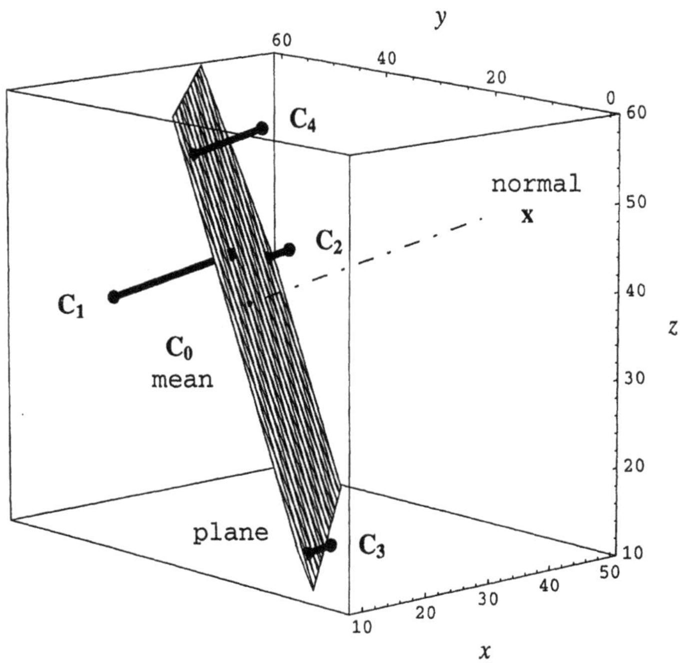

Fig. 6. The TLS plane minimizes the Euclidean distance to the data.

Example 23. Consider the four data points in space displayed in Fig. 6:

$c_1 = (11\ 45\ 38),$

$c_2 = (47\ 54\ 38),$

$c_3 = (17\ 12\ 14),$

$c_4 = (21\ 29\ 58).$

For these data, $\bar{c} = (\frac{1}{4})\sum_{i=1}^{4} c_i = (24, 35, 37)$, and

$$C_{\bar{c}} = \begin{pmatrix} c_1 - \bar{c} \\ c_2 - \bar{c} \\ c_3 - \bar{c} \\ c_4 - \bar{c} \end{pmatrix} = \begin{pmatrix} -13 & 10 & 1 \\ 23 & 19 & 1 \\ -7 & -23 & -23 \\ -3 & -6 & 21 \end{pmatrix}.$$

The smallest singular value of $C_{\bar{c}}$ is $\sigma = \sigma_3 = 18$, and the corresponding singular vector is $v_3 = (\frac{2}{3}, -\frac{2}{3}, \frac{1}{3})^*$. Thus, the hyperplane H passes through $c_0 = \bar{c} = (24, 35, 37)$ and lies perpendicularly to the vector $x = v_3 = (\frac{2}{3}, -\frac{2}{3}, \frac{1}{3})^*$, so that H satisfies the equation

$$\tfrac{2}{3}(x - 24) - \tfrac{2}{3}(y - 35) + \tfrac{1}{3}(z - 37) = 0.$$

Moreover, $C_{\bar{c}}v = (-15, 3, 3, 9)^*$ contains the signed distances $d(c_i, H)$ from the data points to the hyperplane H, here $\sum_{i=1}^{4} d(c_i, H)^2 = \|C_{\bar{c}}v\|_2^2 = \sigma^2 = 18^2$, which gives the orthogonal projections

$\hat{c}_1,\ldots,\hat{c}_4$ of the data on H:

$$\hat{C} = \begin{pmatrix} \hat{c}_1 \\ \vdots \\ \hat{c}_m \end{pmatrix} = \begin{pmatrix} c_1 \\ \vdots \\ c_m \end{pmatrix} - \begin{pmatrix} -15 \\ 3 \\ 3 \\ 9 \end{pmatrix} \begin{pmatrix} \frac{2}{3} & -\frac{2}{3} & \frac{1}{3} \end{pmatrix} = \begin{pmatrix} 21 & 35 & 43 \\ 45 & 56 & 37 \\ 15 & 14 & 13 \\ 15 & 35 & 55 \end{pmatrix}.$$

6.2.2. Algebraic formulations of total least squares

For a linear system $Ax = b$, the problem of *ordinary* least squares consists in determining a vector $\tilde{b}$ that minimizes $\|\tilde{b} - b\|_2$ subject to the constraint that the system $A\tilde{x} = \tilde{b}$ have a solution. More generally, the algebraic formulation of the problem of *total* least squares consists in determining a *matrix* $\hat{A}$ and a vector $\hat{b}$ that minimize the Frobenius norm

$$\|[\hat{A}; \hat{b}] - [A; b]\|_F$$

subject to the condition that the system

$$\hat{A}\hat{x} = \hat{b}$$

have a solution. Thus, with

$$C := [A; b],$$

$$\hat{C} := [\hat{A}; \hat{b}],$$

the problem reduces to determining a matrix $\hat{C}$ that minimizes $\|\hat{C} - C\|_F$ subject to the condition that Kernel $(\hat{C})$ contains a vector of the form $(\hat{x}^*, -1)^*$.

If the matrix $C = [A; b]$ has a singular-value decomposition

$$C = \sum_{i=1}^{n+1} \sigma_i u_i v_i^*$$

and the $n+1$ columns of C are linearly independent, then $\hat{C} = [\hat{A}, \hat{b}]$ must have rank at most n, and the Schmidt–Mirsky approximation theorem states that the closest such matrix is

$$\hat{C} = \sum_{i=1}^{n} \sigma_i u_i v_i^* = [A; b] - \sigma_{n+1} u_{n+1} v_{n+1}^*.$$

Then v_{n+1} spans Kernel $(\hat{C})$, and two cases arise. If $(v_{n+1})_{n+1} \neq 0$, then the problem admits the solution

$$\begin{pmatrix} \hat{x} \\ -1 \end{pmatrix} = \frac{-1}{(v_{n+1})_{n+1}} v_{n+1}.$$

If $(v_{n+1})_{n+1} = 0$, then the problem has no solution.

Some practical problems lead to problems of total least squares subject to the condition that the first k columns $A_1 := (A(\,;1),\ldots,A(\,;k))$ of the matrix $A \in \mathbb{M}_{m\times n}(\mathbb{C})$ be kept exact (unadjusted) while the last $n-k$ columns $A_2 := (A(\,;k+1),\ldots,A(\,;n))$ are subject to adjustments. Golub, Hoffman, and Stewart published the following method of solution [17]. If A_1 has rank $\ell \leqslant r$, if Q is the orthogonal projection on the column space of A_1 and $Q^{\perp}$ is the orthogonal projection on its orthogonal complement, then the matrix $[A_1; \hat{A}_2]$ of rank r that minimizes the Frobenius norm

$$\|[A_1; \hat{A}_2] - [A_1; A_2]\|_F$$

is defined in terms of a singular-value decomposition of $Q^{\perp}A_2$,

$$Q^{\perp}A_2 = \sum_{i=1}^{n-k} \tau_i w_i z_i^*,$$

by the formula

$$\hat{A}_2 := QA_2 + \sum_{i=1}^{r-\ell} \tau_i w_i z_i^*.$$

Example 24. To the data in Example 1, Laplace fitted an affine model for the length Δs of and arc of 1 grad along the meridian,

$$c_0 + c_1[\sin(\lambda)]^2 = \Delta s.$$

The coefficient 1 of c_0 is exact. Consequently, the first column $A_1 := 1$ remains fixed. The orthogonal projection on the space spanned by $1 \in \mathbb{C}^m$ has matrix $Q := 1/m\, 11^*$. Thus, $Q^{\perp} = I - Q$ subtracts from each column the mean of that column. Here, the matrix $C := Q^{\perp}A_2$ corresponds to a linear system for c_1:

$$\begin{pmatrix} 0.00\,000 - 0.43\,925 \\ 0.30\,156 - 0.43\,925 \\ 0.39\,946 - 0.43\,925 \\ 0.46\,541 - 0.43\,925 \\ 0.52\,093 - 0.43\,925 \\ 0.54\,850 - 0.43\,925 \\ 0.83\,887 - 0.43\,925 \end{pmatrix} (c_1) = \begin{pmatrix} 25\,538.85 - 25\,659.93 \\ 25\,666.65 - 25\,659.93 \\ 25\,599.60 - 25\,659.93 \\ 25\,640.55 - 25\,659.93 \\ 25\,658.28 - 25\,659.93 \\ 25\,683.30 - 25\,659.93 \\ 25\,832.25 - 25\,659.93 \end{pmatrix}.$$

The singular-value decomposition of C (computed with the command SVD on the HP 48GX [21, pp. 14–22]) shows two singular values,

$$221.279 = \sigma_1 > \sigma_2 = 0.266\,719,$$

the smallest of which corresponds to the right singular vector

$$\begin{pmatrix} \hat{x} \\ -1 \end{pmatrix} = \frac{-1}{(v_2)_2} v_2 = \frac{-1}{0.002\,562} \begin{pmatrix} -0.999\,997 \\ 0.002\,562 \end{pmatrix} = \begin{pmatrix} 390.356 \\ -1 \end{pmatrix}.$$

Adding QA_2 amounts to adding back the means, which yields

$$\hat{c}_1 = 390.356,$$

$$\hat{c}_0 = 25\,659.93 - 390.356 * 0.43\,925 = 25\,488.46.$$

These values lead to the estimate of the squared eccentricity $\hat{e}^2 = (\frac{2}{3})\hat{c}_1/\hat{c}_0 = 0.010\,210$, which is farther from the current estimate $e^2 = 0.006\,694\,379\,990\,13$ than the value obtained by ordinary least squares $\tilde{e}^2 = 0.006\,339$. The same values then lead to the estimate of the equatorial radius $\hat{a} = \hat{c}_0/(1-\hat{e}^2) = 99\,352$ m, which is closer to the current estimate $a = 6\,378\,137.000\,00$ m than the value obtained by ordinary least squares $\tilde{a}$=100 170.25 m, but still very inaccurate. The corresponding results for weighted total least squares, with $C = W[A; b]$, are $\hat{e}^2 = 0.008\,153$ and $\hat{a} = 100\,307$ m. Unweighted least squares, ordinary or total, give yet worse results.

Because the estimate of the eccentricity closest to the current estimate comes from ordinary least squares, rather than total least squares, such results suggest than most of the errors lie in the measurement of lengths along the meridian, rather than in the geodetic latitudes of the locations. Bowditch's comments corroborate such suggestions [31, Book II, Section 41].

For a system $AX = B$ with r right-hand sides, the problem of total least squares consists in determining matrices $\hat{A} \in \mathbb{M}_{m\times n}(\mathbb{C})$, $\hat{B} \in \mathbb{M}_{m\times r}(\mathbb{C})$, and $\hat{X} \in \mathbb{M}_{n\times r}(\mathbb{C})$, minimizing $\|[A; B] - [\hat{A}; \hat{B}]\|_{\mathrm{F}}$ subject to the constraint $\hat{A}\hat{X} = \hat{B}$. Equivalently, with $I \in \mathbb{M}_{r\times r}(\mathbb{C})$, there must exist a solution $[\hat{X}^*; -I]^*$ to the system $[\hat{A}; \hat{B}]\,[\hat{X}^*; -I]^* = O$. In particular, because the last r rows of $[\hat{X}^*; -I]^*$ are linearly independent, it follows that the rank of $S{:=}[\hat{A}; \hat{B}]$ cannot exceed $(n + r) - r = n$. Therefore, with

$$C{:=}[A; B] = \sum_{i=1}^{n+r} \sigma_i u_i v_i^*$$

the matrix S must be a matrix of rank at most n that minimizes $\|C - S\|_{\mathrm{F}}$. By the approximation theorem of Schmidt, Mirsky, and Weyl, it follows that

$$S = [\hat{A}; \hat{B}] = \sum_{i=1}^{n} \sigma_i u_i v_i^*.$$

The problem then admits a solution in the form $[\hat{X}^*; -I]^*$ if and only if in the matrix $(v_{n+1}, \ldots, v_{n+r})$ the last $n + r$ rows are linearly independent and hence form in invertible matrix $V_{n+1;n+r} \in \mathbb{M}_{r\times r}(\mathbb{C})$, so that

$$\begin{pmatrix} X \\ -I \end{pmatrix} = V_{n+1;n+r}^{-1}(v_{n+1}, \ldots, v_{n+r}).$$

6.3. Relations between the algebraic and geometric formulations

The matrix $\hat{C} = [\hat{A}; \hat{b}]$ in the algebraic formulation corresponds to the matrix C_0 in the geometric formulation. In other words, the algebraic formulation corresponds to the problem of fitting to the rows of $C = [A; b]$ a hyperplane constrained to pass through the origin $0 \in \mathbb{C}^{n+1}$ instead of through the centroid $\bar{c}$. Indeed, the system $\hat{A}\hat{x} = \hat{b}$ in the form

$$[\hat{A}; \hat{b}]\begin{pmatrix} \hat{x} \\ -1 \end{pmatrix} = 0$$

states that every row of $[\hat{A}; \hat{b}]$ lies on the hyperplane H passing through the origin perpendicularly to $(\hat{x}^*; -1)^*$. Moreover, the condition that $\hat{C}$ minimizes the Frobenius norm, or, equivalently, its square,

$$\|\hat{C} - C\|_F^2 = \sum_{i=1}^{m} \sum_{j=1}^{n+1} (c_{i,j} - \hat{c}_{i,j})^2 = \sum_{i=1}^{m} \|c_i - \hat{c}_i\|_2^2,$$

shows that $\hat{C}$ minimizes the sum of the squared distances from the rows of C in $\mathbb{C}^{n+1}$ to the rows of $\hat{C}$ on H. Therefore, the ith row $\hat{C}(i;)$ of $\hat{C}$ is the orthogonal projection of the ith row $C(i;)$ of C, for otherwise these orthogonal projections would lie on the same hyperplane H and would give a smaller total least squares, or squared Frobenius norm. However, Lemma 21 shows that v_{n+1} and hence $(\hat{x}^*; -1)^*$ is the normal direction of total least squares for all the hyperplanes passing through the origin.

7. Nonlinear least squares

7.1. Nonlinear least squares in astronomy and geodesy

The old problem of estimating the shape of the earth can be formulated as the total least-squares problem of fitting an ellipsoid by minimizing the sum of the squared distances to data points. The problem of reliably computing the distance from a point to an ellipse already causes difficulties, because it amounts to solving a quartic equation, and there does not seem to be any practical forward error bounds for the solutions by the quartic formulae. Nevertheless, for small eccentricities ($e^2 < 2 - \sqrt{2}$), there exists a provably reliable algorithm to compute the distance with a contracting map [28,38].

Similarly, the old problem of estimating the shape of the orbit of a celestial body can be formulated as the total least-squares problem of fitting a plane and a conic section in it by minimizing the sum of the squared distances to data points [41].

For both problems, the particular formulation depends on the type of data, for instance, azimuth and elevation only, or azimuth, elevation, and range (measured by radar, for instance) [41, Chapter 10; 51, pp. 302–305]. Despite the practicality of such problems, however, there does not yet seem to exist any theorem that guarantees the global convergence of any algorithm toward the globally optimum surface or orbit [41, p. 180].

7.2. Fitting circles by total least squares

Although the problem of fitting circles and spheres to observations can be traced to the first millenium B.C., the problem of designing an *algorithm* to calculate the center and the radius of a circle or a sphere fitted to a finite set of points can be traced to the 1970's A.D., through computer scientists' developments of algorithms [8] for medical devices [3] and typography [40], electrical engineers' adjustments of microwave calibrations [27], and particle physicists' writings on fitting circular trajectories to a large number of automated measurements of positions of electrically charged particles within uniform magnetic fields [9]. One method — called an *algebraic fit* — to fit a circle or a sphere to data points $c_1,\dots,c_m \in \mathbb{R}^n$ consists in computing the center $x \in \mathbb{R}^n$ and the radius $r \in \mathbb{R}$ that minimize the function $f : \mathbb{R}^{n+1} \to \mathbb{R}$ defined by

$$f(x,r;c_1,\dots,c_m):=\sum_{j=1}^{m}(\|x-c_j\|_2^2-r^2)^2=\sum_{j=1}^{m}(x^*x-r^2-2x^*c_j+c_j^*c_j)^2.$$

For each center $x \in \mathbb{R}^n$ the radius $r(x)$ that minimizes f is such that

$$[r(x)]^2=\frac{1}{m}\sum_{j=1}^{m}\|x-c_j\|_2^2.$$

Substituting $r(x)$ for r in f leads to a linear system $Ax=b$ for the center x, where A is $8m$ times the covariance matrix of the data, and the vector b is defined by $b_i:=\sum_{j=1}^{m}\{(c_j)_i-\bar{c}_i\}\|c_i\|_2^2$ [35]. With an approximate arithmetic, however, the computation of the entries of A and b can introduce errors.

An alternate method by Coope [7] performs the charge of coordinates

$$z:=2x, \quad z_{n+1}:=r^2-x^*x,$$

which leads to the following ordinary linear least-squares system $Cz=d$:

$$\begin{pmatrix} c_1^* & 1 \\ \vdots & \vdots \\ c_m^* & 1 \end{pmatrix}\begin{pmatrix} z \\ z_{n+1} \end{pmatrix}=\begin{pmatrix} \|c_1\|_2^2 \\ \vdots \\ \|c_m\|_2^2 \end{pmatrix}.$$

Thus with Coope's method forming the matrix C does not involve any computations and hence does not introduce any computational approximation.

However, Gander, Golub, and Strebel have demonstrated with examples that curves fitted by minimizing such algebraic objectives as f can lie farther from the data than curves fitted by total least squares of the Euclidean distances, called *geometric fits* [14]. Therefore, several authors have suggested using algebraic fits as initial estimates to start iterative methods for the computation of geometric fits [1, p. 357; 7, 14]. Several problems remain unsolved yet.

Example 25. Consider the following data in the plane:

$$c_2:=(0,2),$$

$$c_4:=(0,0),$$

$$c_3:=(-\sqrt{3},-1), \quad c_1:=(-\sqrt{3},-1)$$

with the sum of the squared distances to the circle with center x and radius r:

$$g(x,r;c_1,\dots,c_m):=\sum_{j=1}^{m}(\|x-c_j\|_2-r)^2.$$

Firstly, calculus yields three minima for g, corresponding to circles with centers x_k opposite to c_k, radius $r:=\frac{7}{4}$, and $g(x_k, r; c_1, \ldots, c_m) = 2$ for each k:

$$x_1 := e^{4\pi i/3} x_2, \qquad x_3 := e^{2\pi i/3} x_2,$$

$$x_2 := (0, -\tfrac{3}{4}).$$

Secondly, perturbations of any of the data can turn any of the local minima into a global minimum. Finally, the algebraic fit cannot serve as an initial estimate for Newton's methods to converge to any geometric fit. Indeed, the algebraically fitted circle has its center at the origin, which coincides with a data point, where the objective function is not differentiable.

7.3. Open problems in nonlinear least squares

Problems of fitting *affine* manifolds by minimizing a weighted sum of squared distances to data are extensively documented. Indeed, linear algebra yields affine parametrizations of their solutions and provides several methods of solution through orthogonal factorizations, for which there exist proven upper bounds on errors from the data or from approximate computations [1,19,22,32,50,51,53].

In contrast, problems of nonlinear least squares, for instance, problems of fitting nonaffine manifolds as simple as circles, remain mostly unsolved. There exist a substantial documentation of algorithms that converge globally (from every initial point) to a *local* minimum [12,29,42]. However, some of their shortcuts can succumb to rounding errors [36], and there does not yet seem to exist any theorem guaranteeing the convergence to a *global* minimum to fit curves as simple as conic sections and surfaces as simple as spheres.

References

[1] Å. Björk, Numerical Methods for Least Squares Problems, Society for Industrial and Applied Mathematics, Philadelphia, PA, 1996.

[2] A.C. Aiken, On least squares and linear combinations of observations, Proc. Roy. Soc. Edinburgh, Section A 55 (1934) 42–47.

[3] F.L. Bookstein, Fitting conic sections to scattered data, Comput. Graphics Image Process. 9 (1979) 56–71.

[4] L.A. Brown, The Story of Maps, Dover, New York, NY, 1979.

[5] S.-s. Chern, Curves and surfaces in Euclidean space, in: S.-s. Chern (Ed.), Global Differential Geometry, MAA Studies in Mathematics, Vol. 27, Mathematical Association of America, Washington, D.C, 1989, pp. 99–139.

[6] V. Chvátal, Linear Programming, W.H. Freeman, New York, NY, 1983.

[7] I.D. Coope, Circle fitting by linear and nonlinear least squares, J. Optim. Theory Appl. 76 (2) (1993) 381–388.

[8] M.G. Cox, H.M. Jones, An algorithm for least-squares circle fitting to data with specified uncertainty ellipses, IMA J. Numer. Anal. 9 (3) (1989) 285–298.

[9] J.F. Crawford, A non-iterative method for fitting circular arcs to measured points, Nucl. Instr. and Meth. 211 (2) (1983) 223–225.

[10] G.B. Dantzig, Inductive proof of the simplex method, IBM J. Res. Dev. 4 (1960) 505–506.

[11] G.B. Dantzig, Linear Programming and Extensions, Princeton University Press, Princeton, NJ, 1963.

[12] J.E. Dennis, Jr., R.B. Schnabel, Numerical Methods for Unconstrained Optimization and Nonlinear Equations, Classics In Applied Mathematics, Vol. 16, Society for Industrial and Applied Mathematics, Philadelphia, PA, 1996.

[13] J.L.E. Dreyer, A History of Astronomy from Thales to Kepler, 2nd Edition, Dover, New York, NY, 1953.

[14] W. Gander, G.H. Golub, R. Strebel, Least-squares fitting of circles and ellipses, BIT 34 (1994) 558–578.

[15] N. Gastinel, Matrices du Second Degré et Normes Générales en Analyse Numérique Linéaire, Thèse d'Etat, Université de Grenoble, 1960.
[16] C.F. Gauss, Theoria Combinationis Observationum Erroribus Minimis Obnoxiae, Society for Industrial and Applied Mathematics, Philadelphia, PA, 1995. (Theory of the Combination of Observations Least Subject to Errors, original with translation by Gilbert W. Stewart).
[17] G.H. Golub, A. Hoffman, G.W. Stewart, A generalization of the Eckart-Young-Mirsky matrix approximation theorem, Linear Algebra Appl. 88/89 (1987) 317–327.
[18] G.H. Golub, C.F.V. Loan, An analysis of the total least squares problem, SIAM J. Numer. Anal. 17 (6) (1980) 883–893.
[19] G.H. Golub, C.F.V. Loan, Matrix Computations, 2nd Edition, Johns Hopkins University Press, Baltimore, MD, 1989.
[20] N. Grossman, The Sheer Joy of Celestial Mechanics, Birkhäuser, Boston, Cambridge, MA, 1996.
[21] Hewlett-Packard Co., Corvallis Division, 1000 NE Circle Blvd., Corvallis, OR 97330, USA. HP 48G Series User's Guide, 7th edition, March 1994. (Part Number 00048-90126.).
[22] N.J. Higham, Accuracy and Stability of Numerical Algorithms, Society for Industrial and Applied Mathematics, Philadelphia, PA, 1996.
[23] B. Hofmann-Wellenhof, H. Lichteneggar, J. Collins, Global Positioning System: Theory and Practice, 4th revised Edition, Springer-Verlag, Wien, 1997.
[24] R.V. Hogg, A.T. Craig, Introduction to Mathematical Statistics, 4th Edition, Macmillan, New York, NY, 1978.
[25] W.M. Kahan, Numerical linear algebra, Canad. Math. Bull. 9 (6) (1966) 757–801.
[26] H. Karttunen, P. Kröger, H. Oja, M. Poutanen, K.J. Donner (Eds.), Fundamental Astronomy, 2nd enlarged Edition, Springer, New York, 1994.
[27] I. Kàsa, A circle fitting procedure and its error analysis, IEEE Trans. Instr. Meas. 25 (1) (1976) 8–14.
[28] S.P. Keeler, Y. Nievergelt, Computing geodetic coordinates, SIAM Rev. 40 (2) (1998) 300–309.
[29] C.T. Kelley, Iterative Methods for Linear and Nonlinear Equations, Frontiers in Applied Mathematics, Vol. 16, Society for Industrial and Applied Mathematics, Philadelphia, PA, 1995.
[30] D. Kincaid, W. Cheney, Numerical Analysis: Mathematics of Scientific Computing, Brooks/Cole, Pacific Grove, CA, 1996.
[31] P.S.d. Laplace, Traité de Mécanique Céleste, Durprat, Paris, 1798–1825. (Translated with commentary by Nathaniel Bowditch, Boston, MA, 1829; Chelsea, Bronx, NY, 1966).
[32] C.L. Lawson, R.J. Hanson, Solving Least Squares Problems, Classics In Applied Mathematics, Vol. 15, Society for Industrial and Applied Mathematics, Philadelphia, PA, 1995.
[33] A.M. Legendre, Nouvelle méthodes pour la déterminiation des orbites des comètes, Courcier, Paris, 1805.
[34] L. Mirsky, Symmetric gauge functions and unitarily invariant norms, Quart. J. Math. Oxford, Ser. 2 11 (41) (1960) 50–59.
[35] Y. Nievergelt, Computing circles and spheres of arithmetic least squares, Comput. Phys. Comm. 81 (3) (1994) 343–350.
[36] Y. Nievergelt, The condition of Steffensen's acceleration in several variables, J. Comput. Appl. Math. 58 (1995) 291–305.
[37] Y. Nievergelt, Schmidt-Mirsky matrix approximation with linearly constrained singular values, Linear Algebra Appl. 261 (1997) 207–219.
[38] Y. Nievergelt, S.P. Keeler, Computing geodetic coordinates in space, J. Spacecraft Rockets 37 (2000) 293–296.
[39] B. Pourciau, Reading the master: Newton and the birth of celestial mechanics, Amer. Math. Mon. 104 (1) (1997) 1–19.
[40] V. Pratt, Direct least-squares fitting of algebraic surfaces, ACM Comput. Graphics 21 (4) (1987) 145–152.
[41] J.E. Prussing, B.A. Conway, Orbital Mechanics, Oxford University Press, New York, NY, 1993.
[42] W.C. Rheinboldt, Methods for Solving Systems of Nonlinear Equations, CBMS-NSF Regional Conference Series in Applied Mathematics, Vol. 14, Society for Industrial and Applied Mathematics, Philadelphia, PA, 1974.
[43] E. Schmidt, Zur Theorie der linearen und nichtlinearen Integralgleichungen. 1. Teil: Entwicklung willkürlicher Funktionen nach Systemen vorgeschriebener, Math. Ann. 63 (1907) 433–476.
[44] J.P. Snyder, Map Projections — A Working Manual (U.S. Geological Survey Professional Paper 1395). United States Government Printing Office, Washington, D.C. 20402, 1987.
[45] G.W. Stewart, On the early history of the singular value decomposition, SIAM Rev. 35 (4) (1993) 551–566.

[46] J. Stoer, R. Bulirsch, Introduction to Numerical Analysis, 2nd Edition, Springer, New York, NY, 1993.
[47] G. Strang, Introduction to Applied Mathematics, Wellesley-Cambridge Press, Wellesley, MA, 1986.
[48] A.E. Taylor, Introduction to Functional Analysis, Wiley, New York, NY, 1958.
[49] B.L. van der Waerden, Geometry and Algebra in Ancient Civilizations, Springer, New York, NY, 1983.
[50] S. Van Huffel, J. Vandewalle, The Total Least Squares Problem: Computational Aspects and Analysis, Society for Industrial and Applied Mathematics, Philadelphia, PA, 1991.
[51] C.F. Van Loan, Introduction to Scientific Computing, Prentice-Hall, Upper Saddle River, NJ, 1997.
[52] H. Weyl, Das asymptotische Vertielungsgesetz der Eigenwert linearer partieller Differentialgleichungen (mit einer Anwendung auf der Theorie der Hohlraumstrahlung), Math. Ann. 71 (1912) 441–479.
[53] J.H. Wilkinson, The Algebraic Eigenvalue Problem, Oxford Science Publication, Monographs on Numerical Analysis, Oxford University Press, Oxford, UK, paperback edition, 1988.

Convergence acceleration during the 20th century

C. Brezinski

Laboratoire d'Analyse Numérique et d'Optimisation, UFR IEEA, Université des Sciences et Technologies de Lille, 59655–Villeneuve d'Ascq cedex, France

1. Introduction

In numerical analysis many methods produce sequences, for instance iterative methods for solving systems of equations, methods involving series expansions, discretization methods (that is methods depending on a parameter such that the approximate solution tends to the exact one when the parameter tends to zero), perturbation methods, etc. Sometimes, the convergence of these sequences is slow and their effective use is quite limited. Convergence acceleration methods consist of transforming a slowly converging sequence (S_n) into a new sequence (T_n) converging to the same limit faster than the initial one. Among such *sequence transformations*, the most well known are certainly Richardson's extrapolation algorithm and Aitken's Δ^2 process. All known methods are constructed by extrapolation and they are often called *extrapolation methods*. The idea consists of interpolating the terms $S_n, S_{n+1}, \ldots, S_{n+k}$ of the sequence to be transformed by a sequence satisfying a certain relationship depending on parameters. This set of sequences is called *kernel* of the transformation and every sequence of this set is transformed into a constant sequence by the transformation into consideration. For example, as we will see below, the kernel of Aitken's Δ^2 process is the set of sequences satisfying $\forall n, a_0(S_n - S) + a_1(S_{n+1} - S) = 0$, where a_0 and a_1 are parameters such that $a_0 + a_1 \neq 0$. If Aitken's process is applied to such a sequence, then the constant sequence $(T_n = S)$ is obtained. The parameters involved in the definition of the kernel are uniquely determined by the interpolation conditions and then the limit of the interpolating sequence of the kernel is taken as an approximation of the limit of the sequence to be accelerated. Since this limit depends on the index n, it will be denoted by T_n. Effectively, the sequence (S_n) has been transformed into a new sequence (T_n).

This paper, which is based on [31], but includes new developments obtained since 1995, presents my personal views on the historical development of this subject during the 20th century. I do not pretend to be exhaustive nor even to quote every important contribution (if a reference does not

E-mail address: claude.brezinski@univ-lille1.fr (C. Brezinski)

appear below, it does not mean that it is less valuable). I refer the interested reader to the literature and, in particular to the recent books [55,146,33,144]. For an extensive bibliography, see [28].

I will begin with scalar sequences and then treat the case of vector ones. As we will see, a sequence transformation able to accelerate the convergence of *all* scalar sequences cannot exist. Thus, it is necessary to obtain many different convergence acceleration methods, each being suitable for a particular class of sequences. Many authors have studied the properties of these procedures and proved some important classes of sequences to be accelerable by a given algorithm. Scalar sequence transformations have also been extensively studied from the theoretical point of view.

The situation is more complicated and more interesting for vector sequences. In the case of a sequence of vectors, it is always possible to apply a scalar acceleration procedure componentwise. However, such a strategy does not take into account connections which may exist between the various components, as in the important case of sequences arising from iterative methods for solving a system of linear or nonlinear equations.

2. Scalar sequences

Let (S_n) be a scalar sequence converging to a limit S. As explained above, an extrapolation method consists of transforming this sequence into a new one, (T_n), by a sequence transformation $T:(S_n) \to (T_n)$. The transformation T is said to *accelerate the convergence* of the sequence (S_n) if and only if

$$\lim_{n\to\infty} \frac{T_n - S}{S_n - S} = 0.$$

We can then say that (T_n) *converges* (to S) *faster than* (S_n).

The first methods to have been used were linear transformations

$$T_n = \sum_{i=0}^{\infty} a_{ni} S_i, \quad n = 0, 1, \ldots,$$

where the numbers a_{ni} are constants independent of the terms of the sequence (S_n). Such a linear transformation is usually called a *summation process* and its properties are completely determined by the matrix $A=(a_{ni})$. For practical reasons, only a finite number of the coefficients a_{ni} are different from zero for each n. Among such processes are those named after Euler, Cesaro and Hölder. In the case of linear methods, the convergence of the sequence (T_n) to S for any converging sequence (S_n) is governed by the Toeplitz summability theorem; see [115] for a review. Examples of such processes are

$$T_n = \frac{1}{n+1} \sum_{i=0}^{n} S_i$$

or

$$T_n = \frac{1}{k+1} \sum_{i=n}^{n+k} S_i.$$

In the second case, the sequence (T_n) also depends on a second index, k, and the convergence has to be studied either when k is fixed and n tends to infinity, or when n is fixed and k tends to infinity.

With respect to convergence acceleration, summation processes are usually only able to accelerate the convergence of restricted classes of sequences and this is why the numerical analysts of the 20th century turned their efforts to nonlinear transformations. However, there is one exception: Richardson's extrapolation process.

2.1. Richardson's process

It seems that the first appearance of a particular case of what is now called the Richardson extrapolation process is due to Christian Huygens (1629–1695). In 1903, Robert Moir Milne (1873) applied the idea of Huygens for computing π [101]. The same idea was exploited again by Karl Kommerell (1871–1948) in his book of 1936 [78]. As explained in [143], Kommerell can be considered as the real discoverer of Romberg's method although he used this scheme in the context of approximating π.

Let us now come to the procedures used for improving the accuracy of the trapezoidal rule for computing approximations to a definite integral. In the case of a sufficiently smooth function, the error of this method is given by the Euler–Maclaurin expansion. In 1742, Colin Maclaurin (1698–1746) [90] showed that its precision could be improved by forming linear combinations of the results obtained with various stepsizes. His procedure can be interpreted as a preliminary version of Romberg's method; see [49] for a discussion.

In 1900, William Fleetwood Sheppard (1863–1936) used an elimination strategy in the Euler–Maclaurin quadrature formula with $h_n = r_n h$ and $1 = r_0 < r_1 < r_2 < \cdots$ to produce a better approximation to the given integral [132].

In 1910, combining the results obtained with the stepsizes h and $2h$, Lewis Fry Richardson (1881–1953) eliminated the first term in a discretization process using central differences [119]. He called this procedure the *deferred approach to the limit* or *h^2-extrapolation*. The transformed sequence (T_n) is given by

$$T_n = \frac{h_{n+1}^2 S(h_n) - h_n^2 S(h_{n+1})}{h_{n+1}^2 - h_n^2}.$$

In a 1927 paper [120] he used the same technique to solve a 6th order differential eigenvalue problem. His process was called *(h^2, h^4)-extrapolation*. Richardson extrapolation consists of computing the value at 0, denoted by $T_k^{(n)}$, of the interpolation polynomial of the degree at most k, which passes through the points $(x_n, S_n), \ldots, (x_{n+k}, S_{n+k})$. Using the Neville–Aitken scheme for these interpolation polynomials, we immediately obtain

$$T_{k+1}^{(n)} = \frac{x_{n+k+1} T_k^{(n)} - x_n T_k^{(n+1)}}{x_{n+k+1} - x_n}$$

with $T_0^{(n)} = S_n$.

Let us mention that Richardson referred to a 1926 paper by Nikolai Nikolaevich Bogolyubov (born in 1909) and Nikolai Mitrofanovich Krylov (1879–1955) where the procedure (often called the *deferred approach to the limit*) can already be found [11].

In 1955, Werner Romberg (born in 1909) was the first to use repeatedly an elimination approach for improving the accuracy of the trapezoidal rule [121]. He himself refers to the book of Lothar Collatz (1910–1990) of 1951 [50]. The procedure became widely known after the rigorous error analysis given in 1961 by Friedrich L. Bauer [3] and the work of Eduard L. Stiefel (1909–1978) [138]. Romberg's derivation of his process was heuristic. It was proved by Pierre-Jean Laurent in 1963 [81] that the process comes out from the Richardson process by choosing $x_n = h_n^2$ and $h_n = h_0/2^n$. Laurent also gave conditions on the choice of the sequence (x_n) in order that the sequences $(T_k^{(n)})$ tend to S either when k or n tends to infinity. Weaker conditions were given by Michel Crouzeix and Alain L. Mignot in [52, pp. 52–55]. As we shall see below, extensions of Romberg's method to nonsmooth integrands leads to a method called the E-algorithm.

Applications of extrapolation to the numerical solution of ordinary differential equations were studied by H.C. Bolton and H.I. Scoins in 1956 [12], Roland Bulirsch and Josef Stoer in 1964–1966 [47] and William B. Gragg [65] in 1965. The case of difference methods for partial differential equations was treated by Guriĭ Ivanovich Marchuk and V.V. Shaidurov [91]. Sturm–Liouville problems are discussed in [117]. Finally, we mention that Heinz Rutishauser (1918–1970) pointed out in 1963 [122] that Romberg's idea can be applied to any sequence as long as the error has an asymptotic expansion of a form similar to the Euler–Maclaurin's.

For a detailed history of the Richardson method, its developments and applications, see [57,77,143].

2.2. *Aitken's process and the ε-algorithm*

The most popular nonlinear acceleration method is certainly Aitken's Δ^2 process which is given by

$$T_n = \frac{S_n S_{n+2} - S_{n+1}^2}{S_{n+2} - 2S_{n+1} + S_n} = S_n - \frac{(S_{n+1} - S_n)^2}{S_{n+2} - 2S_{n+1} + S_n}, \qquad n = 0, 1, \ldots$$

The method was stated by Alexander Craig Aitken (1895–1967) in 1926 [1], who used it to accelerate the convergence of Bernoulli's method for computing the dominant zero of a polynomial. Aitken pointed out that the same method was obtained by Hans von Naegelsbach (1838) in 1876 in his study of Furstenau's method for solving nonlinear equations [104]. The process was also given by James Clerk Maxwell (1831–1879) in his *Treatise on Electricity and Magnetism* of 1873 [95]. However, neither Naegelsbach nor Maxwell used it for the purpose of acceleration. Maxwell wanted to find the equilibrium position of a pointer oscillating with an exponentially damped simple harmonic motion from three experimental measurements. It is surprising that Aitken's process was known to Takakazu Seki (1642–1708), often considered the greatest Japanese mathematician. In his book *Katsuyō Sanpō*, Vol. IV, he used this process to compute the value of π, the length of a chord and the volume of a sphere. This book was written around 1680 but only published in 1712 by his disciple Murahide Araki. Parts of it can be found in [73]. Let us mention that the Japanese characters corresponding to Takakazu have another pronounciation which is Kōwa. This is the reason why this mathematician is often called, erroneously as in [29,31] Seki Kōwa.

What makes Aitken's process so popular is that it accelerates the convergence of all linearly converging sequences, that is sequences such that $\exists a \neq 1$

$$\lim_{n \to \infty} \frac{S_{n+1} - S}{S_n - S} = a.$$

It can even accelerate some logarithmic sequences (that is corresponding to $a = 1$) which are those with the slowest convergence and the most difficult to accelerate.

Aitken's Δ^2 process is exact (which means that $\forall n, T_n = S$) for sequences satisfying, $a_0(S_n - S) + a_1(S_{n+1} - S) = 0$, $\forall n$, $a_0 a_1 \neq 0$, $a_0 + a_1 \neq 0$. Such sequences form the kernel of Aitken's process. The idea naturally arose of finding a transformation with the kernel

$$a_0(S_n - S) + \cdots + a_k(S_{n+k} - S) = 0, \quad \forall n,$$

$a_0 a_k \neq 0$, $a_0 + \cdots + a_k \neq 0$. A particular case of $k = 2$ was already treated by Maxwell in his book of 1873 and a particular case of an arbitrary value of k was studied by T.H. O'Beirne in 1947 [107]. This last work remains almost unknown since it was published only as an internal report. The problem was handled in full generality by Daniel Shanks (1917–1996) in 1949 [130] and again in 1955 [131]. He obtained the sequence transformation defined by

$$T_n = e_k(S_n) = \frac{\begin{vmatrix} S_n & S_{n+1} & \cdots & S_{n+k} \\ S_{n+1} & S_{n+2} & \cdots & S_{n+k+1} \\ \vdots & \vdots & & \vdots \\ S_{n+k} & S_{n+k+1} & \cdots & S_{n+2k} \end{vmatrix}}{\begin{vmatrix} \Delta^2 S_n & \cdots & \Delta^2 S_{n+k-1} \\ \vdots & & \vdots \\ \Delta^2 S_{n+k-1} & \cdots & \Delta^2 S_{n+2k-2} \end{vmatrix}}.$$

When $k = 1$, Shanks transformation reduces to the Aitken's Δ^2 process. It can be proved that $e_k(S_n) = S$, $\forall n$ if and only if (S_n) belongs to the kernel of the transformation given above. The same ratios of determinants were obtained by R.J. Schmidt in 1941 [127] in his study of a method for solving systems of linear equations.

The determinants involved in the definition of $e_k(S_n)$ have a very special structure. They are called *Hankel determinants* and were studied by Hermann Hankel (1839–1873) in his thesis in 1861 [72]. Such determinants satisfy a five-term recurrence relationship. This relation was used by O'Beirne and Shanks to implement the transformation by computing separately the numerators and the denominators of the $e_k(S_n)$'s. However, numerical analysts know it is difficult to compute determinants (too many arithmetical operations are needed and rounding errors due to the computer often lead to a completely wrong result). A recursive procedure for computing the $e_k(S_n)$'s without computing the determinants involved in their definition was needed. This algorithm was obtained in 1956 by Peter Wynn. It is called the ε-algorithm [147]. It is as follows. One starts with

$$\varepsilon_{-1}^{(n)} = 0, \quad \varepsilon_0^{(n)} = S_n$$

and then

$$\varepsilon_{k+1}^{(n)} = \varepsilon_{k-1}^{(n+1)} + \frac{1}{\varepsilon_k^{(n+1)} - \varepsilon_k^{(n)}}.$$

Note that the numbers $\varepsilon_k^{(n)}$'s fill out a two-dimensional array. The ε-algorithm is related to Shanks transformation by

$$\varepsilon_{2k}^{(n)} = e_k(S_n) \quad \text{and} \quad \varepsilon_{2k+1}^{(n)} = 1/e_k(\Delta S_n).$$

Thus, the ε's with an odd lower index are only auxiliary quantities. They can be eliminated from the algorithm, thus leading to the so-called *cross rule* due to Wynn [153].

When implementing the ε-algorithm or using Wynn's cross rule, division by zero can occur and the algorithm must be stopped. However, if the singularity is confined, a term that will again be used in Section 1.6, that is if it occurs only for some adjacent values of the indexes k and n, one may jump over it by using *singular rules* and continue the computation. If a division by a number close to zero arises, the algorithm becomes numerically unstable due to the cancellation errors. A similar situation holds for the other convergence acceleration algorithms. The study of such problems was initiated by Wynn in 1963 [151], who proposed particular rules for the ε-algorithm which are more stable than the usual rules. They were extended by Florent Cordellier in 1979 [51,151]. Particular rules for the θ-algorithm were obtained by Redivo Zaglia [155].

The convergence and acceleration properties of the ε-algorithm have only been completely described only for two classes of sequences, namely totally monotonic and totally oscillating sequences [154,15,16].

Shanks' transformation and the ε-algorithm have close connections to Padé approximants, continued fractions and formal orthogonal polynomials; see, for example [18].

2.3. Subsequent developments

The Shanks transformation and the ε-algorithm sparked the rebirth of the study of nonlinear acceleration processes. They now form an independent chapter in numerical analysis with connections to other important topics such as orthogonal and biorthogonal polynomials, continued fractions, and Padé approximants. They also have applications to the solution of systems of linear and nonlinear equations, the computation of the eigenvalues of a matrix, the solution of systems of linear and nonlinear equations, and many other topics, see [40]. Among other acceleration methods which were obtained and studied, are the W-process of Samuel Lubkin [89], the method of Kjell J. Overholt [110], the ρ-algorithm of Wynn [148], the G-transformation of H.L. Gray, T.A. Atchison and G.V. McWilliams [70], the θ-algorithm of Claude Brezinski [14], the transformations of Bernard Germain–Bonne [63] and the various transformations due to David Levin [85]. To my knowledge, the only known acceleration theorem for the ρ-algorithm was obtained by Naoki Osada [108]. Simultaneously, several applications began to appear. For example, the ε-algorithm provides a quadratically convergent method for solving systems of nonlinear equations and its does not require the knowledge of any derivative. This procedure was proposed simultaneously by Brezinski [13] and Eckhart Gekeler [61]. It has important applications to the solution of boundary value problems for ordinary differential equations [44]. Many other algorithms are given in the work of Ernst Joachim Weniger [145], which also contains applications to physics, or in the book of Brezinski and Michela Redivo Zaglia [40] where applications to various domains of numerical analysis can be found. The authors of this book provide FORTRAN subroutines. The book of Annie Cuyt and Luc Wuytack must also be mentioned [53]. The ε-algorithm has been applied to statistics, see the work of Alain Berlinet [9], and to the acceleration of the convergence of sequences of random variables, considered by Hélène Lavastre [82]. Applications to optimization were proposed by Le Ferrand [84] and Bouchta Rhanizar [118].

Instead of using a quite complicated algorithm, such as the ε-algorithm, it can be interesting to use a simpler one (for instance, Aitken's Δ^2 process) iteratively. Such a use consists of applying the algorithm to (S_n) to produce a new sequence (T_n), then to apply the same algorithm to (T_n), and so on. For example, applying the iterated Δ^2 process to the successive convergents of a periodic continued fraction produces a better acceleration than using the ε-algorithm [24]. In particular, the iterated Δ^2 process transforms a logarithmic sequence into a sequence converging linearly and linear convergence into superlinear, to my knowledge the only known cases of such transformations.

The experience gained during these years lead to a deeper understanding of the subject. Research workers began to study more theoretical and general questions related to the theory of convergence acceleration. The first attempt was made by R. Pennacchi in 1968 [114], who studied rational sequence transformations. His work was generalized by Germain–Bonne in 1973 [62], who proposed a very general framework and showed how to construct new algorithms for accelerating some classes of sequences. However, a ground breaking discovery was made by Jean Paul Delahaye and Germain–Bonne in 1980 [56]. They proved that if a set of sequences satisfies a certain property, called *remanence* (too technical to be explained here), then a universal algorithm, i.e. one able to accelerate *all* sequences of this set, cannot exist. This result shows the limitations of acceleration methods. Many sets of sequences were proved to be remanent, for example, the sets of monotonic or logarithmic sequences. Even some subsets of the set of logarithmic sequences are remanent.

Moulay Driss Benchiboun [5] observed that all the sequence transformations found in the literature could be written as

$$T_n = \frac{f(S_n,\ldots,S_{n+k})}{Df(S_n,\ldots,S_{n+k})}$$

with $D^2 f \equiv 0$, where Df denotes the sum of the partial derivatives of the function f. The reason for that fact was explained by Brezinski [26], who showed that it is related to the translativity property of sequence transformations. Hassane Sadok [123] extended these results to the vector case. Abderrahim Benazzouz [7] proved that quasilinear transformations can be written as the composition of two projections.

In many transformations, such as Shanks', the quantities computed are expressed as a ratio of determinants. This property is related to the existence of a triangular recurrence scheme for their computation as explained by Brezinski and Guido Walz [46].

Herbert Homeier [74] studied a systematic procedure for constructing sequences transformations. He considered iterated transformations which are *hierarchically consistent*, which means that the kernel of the basic transformation is the lowest one in the hierarchy. The application of the basic transformation to a sequence which is higher in the hierarchy leads to a new sequence belonging to a kernel lower in the hierarchy. Homeier wrote several papers on this topics.

Thus, the theory of convergence acceleration methods has progressed impressively. The practical side was not forgotten and authors obtained a number of special devices for improving their efficiency. For example, when a certain sequence is to be accelerated, it is not obvious to know in advance which method will give the best result unless some properties of the sequence are already known. Thus, Delahaye [54] proposed using simultaneously several transformations and selecting, at each step of the procedure, one answer among the answers provided by the various algorithms. He proved that, under some assumptions, some tests are able to find automatically the best answer. The work of Delahaye was extended by Abdelhak Fdil [58,59]. The various answers could also be combined leading to *composite* transformations [23]. It is possible, in some cases, to extract a linear subsequence from the original one and then to accelerate it, for example, by Aitken's Δ^2 process [37]. Devices for controlling the error were also constructed [21].

When faced with the problem of accelerating the convergence of a given sequence, two approaches are possible. The first is to use a known extrapolation procedure and to try to prove that it accelerates the convergence of the given sequence. The second possibility is to construct an extrapolation procedure especially for that sequence. Convergence tests for sequences and series can be used for that purpose as explained by Brezinski [25]. This approach was mostly developed by Ana Cristina Matos [92]. Special extrapolation procedures for sequences such that $\forall n, S_n - S = a_n D_n$, where (D_n) is a known sequence and (a_n) an unknown one, can also be constructed from the asymptotic properties of the sequences (a_n) and (D_n). Brezinski and Redivo Zaglia did this in [39].

A.H. Bentbib [10] considered the acceleration of sequences of intervals. Mohammed Senhadji [129] defined and studied the condition number of a sequence transformation.

2.4. The E-algorithm

As we see above, the quantities involved in Shanks transformation are expressed as a ratio of determinants and the ε-algorithm allows one to compute them recursively. It is well known that an interpolation polynomial can be expressed as a ratio of determinants. Thus polynomial extrapolation also leads to such a ratio and the Neville–Aitken scheme can be used to avoid the computation of these determinants which leads to the Richardson extrapolation algorithm. A similar situation arises for many other transformations: in each case, the quantities involved are expressed as a ratio of special determinants and, in each case, one seeks for a special recursive algorithm for the practical implementation of the transformation. Thus, there was a real need for a general theory of such sequence transformations and for a single general recursive algorithm for their implementation. This work was performed independently between 1973 and 1980 by five different people. It is now known as the E-algorithm.

It seems that the first appearance of this algorithm is due to Claus Schneider in a paper received on December 21, 1973 [128]. The quantities $S(h_i)$ being given for $i = 0, 1, \ldots$, Schneider looked for $S'(h) = S' + a_1 g_1(h) + \cdots + a_k g_k(h)$ satisfying the interpolation conditions $S'(h_i) = S(h_i)$ for $i = n, \ldots, n+k$, where the g_j's are given functions of h. Of course, the value of the unknown S' thus obtained will depend on the indexes k and n. Assuming that $\forall j, g_j(0) = 0$, we have $S' = S'(0)$. Denoting by ϕ_k^n the extrapolation functional on the space of functions f defined at the points $h_0 > h_1 > \cdots > 0$ and at the point 0 and such that $\phi_k^n f = f(0)$, we have

$$\phi_k^n S' = c_0 S(h_n) + \cdots + c_k S(h_{n+k})$$

with $c_0 + \cdots + c_k = 1$. The interpolation conditions become

$$\phi_k^n E = 1, \quad \text{and} \quad \phi_k^n g_j = 0, \quad j = 1, \ldots, k$$

with $E(h) \equiv 1$. Schneider wanted to express the functional ϕ_k^n in the form $\phi_k^n = a\phi_{k-1}^n + b\phi_{k-1}^{n+1}$. He obtained the two conditions

$$\phi_k^n E = a + b = 1$$

and

$$\phi_k^n g_k = a\phi_{k-1}^n g_k + b\phi_{k-1}^{n+1} g_k = 0.$$

The values of a and b follow immediately and we have

$$\phi_k^n = \frac{[\phi_{k-1}^{n+1} g_k]\phi_{k-1}^n - [\phi_{k-1}^n g_k]\phi_{k-1}^{n+1}}{[\phi_{k-1}^{n+1} g_k] - [\phi_{k-1}^n g_k]}.$$

Thus, the quantities $\phi_k^n S'$ can be recursively computed by this scheme. The auxiliary quantities $\phi_k^n g_j$ needed in this formula must be computed separately by the same scheme using a different initialization. As we shall see below, this algorithm is just the E-algorithm. In a footnote, Schneider mentioned that this representation for ϕ_k^n was suggested by Börsch–Supan from Johannes Gutenberg Universität in Mainz.

In 1976, Günter Meinardus and G.D. Taylor wrote a paper [97] on best uniform approximation by functions from $\operatorname{span}(g_1,\ldots,g_N)\subset C[a,b]$. They defined the linear functionals L_n^k on $C[a,b]$ by

$$L_n^k(f)=\sum_{i=n}^{n+k} c_i f(h_i),$$

where $a\leqslant h_1<h_2<\cdots<h_{N+1}\leqslant b$ and where the coefficients c_i, which depend on n and k, are such that $c_n>0, c_i\neq 0$ for $i=n,\ldots,n+k$, sign $c_i=(-1)^{i-n}$ and

$$\sum_{i=n}^{n+k}|c_i|=1,$$

$$\sum_{i=n}^{n+k} c_i g_j(h_i)=0,\quad j=1,\ldots,k.$$

By using Gaussian elimination to solve the system of linear equations

$$\sum_{i=n}^{N} a_i g_i(h_j)+(-1)^j\lambda=f(h_j),\quad j=1,\ldots,k,$$

Meinardus and Taylor obtained a recursive scheme

$$L_i^k(f)=\frac{L_{i+1}^{k-1}(g_k)L_i^{k-1}(f)-L_i^{k-1}(g_k)L_{i+1}^{k-1}(f)}{L_{i+1}^{k-1}(g_k)-L_i^{k-1}(g_k)}$$

with $L_i^0(f)=f(h_i)$, $i=n,\ldots,n+k$. This is the same scheme as above.

Newton's formula for computing the interpolation polynomial is well known. It is based on divided differences. One can try to generalize these formulae to the case of interpolation by a linear combination of functions from a complete Chebyshev system (a technical concept which insures the existence and uniqueness of the solution). We seek

$$P_k^{(n)}(x)=a_0g_0(x)+\cdots+a_kg_k(x),$$

satisfying the interpolation conditions

$$P_k^{(n)}(x_i)=f(x_i),\quad i=n,\ldots,n+k,$$

where the x_i's are distinct points and the g_i's given functions. The $P_k^{(n)}$ can be recursively computed by an algorithm which generalizes the Neville–Aitken scheme for polynomial interpolation. This algorithm was obtained by Günter Mühlbach in 1976 [103] from a generalization of the notion of divided differences and their recurrence relationship. This algorithm was called the Mühlbach–Neville–Aitken algorithm, for short the MNA. It is as follows:

$$P_k^{(n)}(x)=\frac{g_{k-1,k}^{(n+1)}(x)P_{k-1}^{(n)}(x)-g_{k-1,k}^{(n)}(x)P_{k-1}^{(n+1)}(x)}{g_{k-1,k}^{(n+1)}(x)-g_{k-1,k}^{(n)}(x)}$$

with $P_0^{(n)}(x) = f(x_n)g_0(x)/g_0(x_n)$. The $g_{k,i}^{(n)}$'s can be recursively computed by a quite similar relationship

$$g_{k,i}^{(n)}(x) = \frac{g_{k-1,k}^{(n+1)}(x)g_{k-1,i}^{(n)}(x) - g_{k-1,k}^{(n)}(x)g_{k-1,i}^{(n+1)}(x)}{g_{k-1,k}^{(n+1)}(x) - g_{k-1,k}^{(n)}(x)}$$

with $g_{0,i}^{(n)}(x) = g_i(x_n)g_0(x)/g_0(x_n) - g_i(x)$. If $g_0(x) \equiv 1$, if it is assumed that $\forall i > 0, g_i(0) = 0$, the quantities $P_k^{(n)}(0)$ are the same as those obtained by the E-algorithm and the MNA reduces to it. Let us mention that, in fact, the MNA is closely related to the work of Henri Marie Andoyer (1862–1929) which goes back to 1906 [2]; see [30] for detailed explanations.

We now come to the work of Tore Håvie. We already mentioned Romberg's method for accelerating the convergence of the trapezoidal rule. The success of this procedure is based on the existence of the Euler–Maclaurin expansion for the error. This expansion only holds if the function to be integrated has no singularity in the interval. In the presence of singularities, the expansion of the error is no longer a series in h^2 (the stepsize) but a more complicated one depending on the singularity. Thus, Romberg's scheme has to be modified to incorporate the various terms appearing in the expansion of the error. Several authors worked on this question, treating several types of singularities. In particular, Håvie began to study this question under Romberg (Romberg emigrated to Norway and came to Trondheim in 1949). In 1978, Håvie wrote a report, published one year later [71], where he treated the most general case of an error expansion of the form

$$S(h) - S = a_1 g_1(h) + a_2 g_2(h) + \cdots,$$

where $S(h)$ denotes the approximation obtained by the trapezoidal rule with step size h to the definite integral S and the g_i are the known functions (forming an asymptotic sequence when h tends to zero) appearing in the expansion of the error. Let $h_0 > h_1 > \cdots > 0$, $S_n = S(h_n)$ and $g_i(n) = g_i(h_n)$. Håvie set

$$E_1^{(n)} = \frac{g_1(n+1)S_n - g_1(n)S_{n+1}}{g_1(n+1) - g_1(n)}.$$

Replacing S_n and S_{n+1} by their expansions, he obtained

$$E_1^{(n)} = S + a_2 g_{1,2}^{(n)} + a_3 g_{1,3}^{(n)} + \cdots$$

with

$$g_{1,i}^{(n)} = \frac{g_1(n+1)g_i(n) - g_1(n)g_i(n+1)}{g_1(n+1) - g_1(n)}.$$

The same process can be repeated for eliminating $g_{1,2}^{(n)}$ in the the expansion of $E_1^{(n)}$, and so on. Thus, once again we obtain the E-algorithm

$$E_k^{(n)} = \frac{g_{k-1,k}^{(n+1)} E_{k-1}^{(n)} - g_{k-1,k}^{(n)} E_{k-1}^{(n+1)}}{g_{k-1,k}^{(n+1)} - g_{k-1,k}^{(n)}}$$

with $E_0^{(n)} = S_n$ and $g_{0,i}^{(n)} = g_i(n)$. The auxiliary quantities $g_{k,i}^{(n)}$ are recursively computed by the quite similar rule

$$g_{k,i}^{(n)} = \frac{g_{k-1,k}^{(n+1)} g_{k-1,i}^{(n)} - g_{k-1,k}^{(n)} g_{k-1,i}^{(n+1)}}{g_{k-1,k}^{(n+1)} - g_{k-1,k}^{(n)}}$$

with $g_{0,i}^{(n)} = g_i(n)$.

Håvie gave an interpretation of this algorithm in terms of the Gaussian elimination process for solving the system

$$E_k^{(n)} + b_1 g_1(n+i) + \cdots + b_k g_k(n+i) = S_{n+i}, \quad i = 0,\ldots,k$$

for the unknown $E_k^{(n)}$.

In 1980, Brezinski took up the same problem, but from the point of view of extrapolation [19]. Let (S_n) be the sequence to be accelerated. Interpolating it by a sequence of the form $S'_n = S + a_1 g_1(n) + \cdots + a_k g_k(n)$, where the g_i's are known sequences which can depend on the sequence (S_n) itself, leads to

$$S_{n+i} = S'_{n+i}, \quad i = 0,\ldots,k.$$

Solving this system directly for the unknown S (which, since it depends on n and k, will be denoted by $E_k^{(n)}$) gives

$$E_k^{(n)} = \frac{\begin{vmatrix} S_n & \cdots & S_{n+k} \\ g_1(n) & \cdots & g_1(n+k) \\ \vdots & & \vdots \\ g_k(n) & \cdots & g_k(n+k) \end{vmatrix}}{\begin{vmatrix} 1 & \cdots & 1 \\ g_1(n) & \cdots & g_1(n+k) \\ \vdots & & \vdots \\ g_k(n) & \cdots & g_k(n+k) \end{vmatrix}}.$$

Thus $E_k^{(n)}$ is given as a ratio of determinants which is very similar to the ratios previously mentioned. Indeed, for the choice $g_i(n) = \Delta S_{n+i}$, the ratio appearing in Shanks transformation results while, when $g_i(n) = x_n^i$, we obtain the ratio expressing the quantities involved in the Richardson extrapolation process. Other algorithms may be similarly derived.

Now the problem is to find a recursive algorithm for computing the $E_k^{(n)}$'s. Applying Sylvester's determinantal identity, Brezinski obtained the two rules of the above E-algorithm. His derivation of the E-algorithm is closely related to Håvie's since Sylvester's identity can be proved by using Gaussian elimination. Brezinski also gave convergence and acceleration results for this algorithm when the $(g_i(n))$ satisfy certain conditions [19]. These results show that, for accelerating the convergence of a sequence, it is necessary to know the expansion of the error $S_n - S$ with respect to some asymptotic sequence $(g_1(n)), (g_2(n)), \ldots$. The $g_i(n)$ are those to be used in the E-algorithm. It can be proved that, $\forall k$

$$\lim_{n\to\infty} \frac{E_{k+1}^{(n)} - S}{E_k^{(n)} - S} = 0.$$

These results were refined by Avram Sidi [134–136]. Thus the study of the asymptotic expansion of the error of the sequences to be accelerated is of primary importance, see Walz [144]. For example, Mohammed Kzaz [79,80] and Pierre Verlinden [142] applied this idea to the problem of accelerating the convergence of Gaussian quadrature formulae [79] and Pedro Lima and Mario Graça to boundary value problems with singularities [88,87] (see also the works of Lima and Diogo [87], and Lima and Carpentier [86]). Other acceleration results were obtained by Matos and Marc Prévost [94], Prévost

[116] and Pascal Mortreux and Prévost [102]. An algorithm, more economical than the E-algorithm, was given by William F. Ford and Avram Sidi [60]. The connection between the E-algorithm and the ε-algorithm was studied by Bernhard Beckermann [4]. A general ε-algorithm connected to the E-algorithm was given by Carsten Carstensen [48]. See [27] for a more detailed review on the E-algorithm.

Convergence acceleration algorithms can also be used for predicting the unknowns terms of a series or a sequence. This idea, introduced by Jacek Gilewicz [64], was studied by Sidi and Levin [137], Brezinski [22] and Denis Vekemans [141].

2.5. *A new approach*

Over the years, a quite general framework was constructed for the theory of extrapolation algorithms. The situation was quite different for the practical construction of extrapolation algorithms and there was little systematic research in their derivation. However, thanks to a formalism due to Weniger [145], such a construction is now possible, see Brezinski and Matos [38]. It is as follows. Let us assume that the sequence (S_n) to be accelerated satisfies, $\forall n, S_n - S = a_n D_n$ where (D_n) is a known sequence, called a *remainder* (or error) *estimate* for the sequence (S_n), and (a_n) an unknown sequence. It is possible to construct a sequence transformation such that its kernel is precisely this set of sequences. For that purpose, we have to assume that a difference operator L (that is a linear mapping of the set of sequences into itself) exists such that $\forall n, L(a_n) = 0$. This means that the sequence obtained by applying L to the sequence (a_n) is identically zero. Such a difference operator is called an *annihilation* operator for the sequence (a_n). We have

$$\frac{S_n}{D_n} - \frac{S}{D_n} = a_n.$$

Applying L and using linearity leads to

$$L\left(\frac{S_n}{D_n}\right) - SL\left(\frac{1}{D_n}\right) = L(a_n) = 0.$$

We solve for S and designate it by the sequence transformation

$$T_n = \frac{L(S_n/D_n)}{L(1/D_n)}.$$

The sequence (T_n) is be such that $\forall n, T_n = S$ if and only if $\forall n, S_n - S = a_n D_n$. This approach is highly versatile.

All the algorithms described above and the related devices such as error control, composite sequence transformations, least squares extrapolation, etc., can be put into this framework. Moreover, many new algorithms can be obtained using this approach. The E-algorithm can also be put into this framework which provides a deeper insight and leads to new properties [41]. Matos [93], using results from the theory of difference equations, obtained new and general convergence and acceleration results when (a_n) has an asymptotic expansion of a certain form.

2.6. *Integrable systems*

The connection between convergence acceleration algorithms and discrete integrable systems is a subject whose interest is rapidly growing among physicists. When a numerical scheme is used for integrating a partial differential evolution equation, it is important that it preserves the quantities that are conserved by the partial differential equation itself. An important character is the *integrability* of the equation. Although this term has not yet received a completely satisfactory definition (see [66]), it can be understood as the ability to write the solution explicitly in terms of a finite number of functions or as the confinement of singularities in finite domains. The construction of integrable discrete forms of integrable partial differential equations is highly nontrivial. A major discovery in the field of integrability was the occurrence of a solitary wave (called a *soliton*) in the Korteweg–de Vries (KdV) equation. Integrability is a rare phenomenon and the typical dynamical system is nonintegrable. A test of integrability, called *singularity confinement*, was given by B. Grammaticos, A. Ramani and V. Papageorgiou [67]. It turns out that this test is related to the existence of singular rules for avoiding a division by zero in convergence acceleration algorithms (see Section 1.2).

The literature on this topic is vast and we cannot enter into the details of it. We only want to give an indication of the connection between these two subjects since both domains could benefit from it.

In the rule for the ε-algorithm, V. Papageorgiou, B. Grammaticos and A. Ramani set $m = k + n$ and replaced $\varepsilon_k^{(n)}$ by $u(n,m) + mp + nq$, where p and q satisfy $p^2 - q^2 = 1$. They obtained [111]

$$[p - q + u(n,m+1) - u(n+1,m)][p + q + u(n+1,m+1) - u(n,m)] = p^2 - q^2.$$

This is the discrete lattice KdV equation. Since this equation is integrable, one can expect integrability to hold also for the ε-algorithm, and, thanks to the singular rules of Wynn and Cordellier mentioned at the end of Subsection 1.2, this is indeed the case.

In the rule of the ε-algorithm, making the change of variable $k = t/\varepsilon^3$ and $n - 1/2 = x/\varepsilon - ct/\varepsilon^3$ and replacing $\varepsilon_k^{(n)}$ by $p + \varepsilon^2 u(x - \varepsilon/2, t)$ where c and p are related by $1 - 2c = 1/p^2$, A. Nagai and J. Satsuma obtained [105]

$$\varepsilon^2 u(x - \varepsilon/2 + c\varepsilon, t + \varepsilon^3) - \varepsilon^2 u(x + \varepsilon/2 - c\varepsilon, t - \varepsilon^3) = \frac{1}{p + \varepsilon^2 u(x + \varepsilon/2, t)} - \frac{1}{p + \varepsilon^2 u(x - \varepsilon/2, t)}.$$

We have, to terms of order ε^5, the KdV equation

$$u_t - \frac{1}{p^3} u u_x + \frac{1}{48 p^2}(1 - p^{-4}) u_{xxx} = 0.$$

Other discrete numerical algorithms, such as the *qd*, *LR*, and *ρ*-algorithms are connected to other discrete or continuous integrable equations (see, for example [112]). Formal orthogonal polynomials, continued fractions, Padé approximation also play a rôle in this topic [113].

By replacing the integer n in the ε-algorithm by a continuous variable, Wynn derived the *confluent form* of the ε-algorithm [149]

$$\varepsilon_{k+1}(t) = \varepsilon_{k-1}(t) + \frac{1}{\varepsilon_k'(t)}$$

with $\varepsilon_{-1}(t) \equiv 0$ and $\varepsilon_0(t) = f(t)$. This algorithm is the continuous counterpart of the ε-algorithm and its aim is to compute $\lim_{t\to\infty} f(t)$. Setting $N_k(t) = \varepsilon_k'(t)\varepsilon_{k+1}'(t)$, A. Nagai, T. Tokihiro and J. Satsuma [106] obtained

$$N_k'(t) = N_k(t)[N_{k-1}(t) - N_{k+1}(t)].$$

The above equation is the Bäcklund transformation of the discrete Toda molecule equation [139].

So, we see that some properties of integrable systems are related to properties of convergence acceleration algorithms. On the other hand, discretizing integrable partial differential equations leads to new sequence transformations which have to be studied from the point of view of their algebraic and acceleration properties. Replacing the second integer k in the confluent form of the ε-algorithm by a continuous variable, Wynn obtained a partial differential equation [152]. Its relation with integrable systems is an open question.

The connection between integrable systems and convergence acceleration algorithms needs to be investigated in more details to fully understand its meaning which is not clear yet.

3. The vector case

In numerical analysis, many iterative methods lead to vector sequences. To accelerate the convergence of such sequences, it is always possible to apply a scalar algorithm componentwise. However, vector sequence transformations, specially built for that purpose, are usually more powerful. The first vector algorithm to be studied was the vector ε-algorithm. It was obtained by Wynn [150] by replacing, in the rule of the scalar ε-algorithm, $1/\Delta\varepsilon_k^{(n)}$ by $(\Delta\varepsilon_k^{(n)})^{-1}$ where the inverse y^{-1} of a vector y is defined by $y^{-1} = y/(y,y)$. Thus, with this definition, the rule of the ε-algorithm can be applied to vector sequences. Using Clifford algebra, J.B. McLeod proved in 1971 [96] that $\forall n, \varepsilon_{2k}^{(n)} = S$ if the sequence (S_n) satisfies $a_0(S_n - S) + \cdots + a_k(S_{n+k} - S) = 0$, $\forall n$ with $a_0 a_k \neq 0$, $a_0 + \cdots + a_k \neq 0$. This result is only valid for real sequences (S_n) and real a_i's. Moreover, contrary to the scalar case, this condition is only sufficient. In 1983, Peter R. Graves–Morris [68] extended this result to the complex case using a quite different approach.

A drawback to the development of the theory of the vector ε-algorithm was that it was not known whether a corresponding generalization of Shanks transformation was underlying the algorithm, that is whether the vectors $\varepsilon_k^{(n)}$ obtained by the algorithm could be expressed as ratios of determinants (or some kind of generalization of determinants). This is why Brezinski [17], following the same path as Shanks, tried to construct a vector sequence transformation with the kernel $a_0(S_n - S) + \cdots + a_k(S_{n+k} - S) = 0$. He obtained a transformation expressed as a ratio of determinants. He then had to develop a recursive algorithm for avoiding their computation. This was the so-called topological ε-algorithm. This algorithm has many applications, in particular, to the solution of systems of linear equations (it is related to the biconjugate gradient algorithm [18, pp. 185ff]). In the case of a system of nonlinear equations, it gave rise to a generalization of Steffensen's method [13]. That algorithm has a quadratic convergence under some assumptions as established by Hervé Le Ferrand [83] following the ideas presented by Khalide Jbilou and Sadok [75]. The denominator of the vector $\varepsilon_{2k}^{(n)}$ obtained by the vector ε-algorithm was first written as a determinant of dimension $2k+1$ by Graves–Morris and Chris Jenkins in [69]. The numerator follows immediately by modifying the first row of the denominator, a formula given by Ahmed Salam and Graves–Morris [126]. However, the dimension of the corresponding determinants in the scalar case is only $k+1$. It was proved by Salam [124] that the vectors $\varepsilon_{2k}^{(n)}$ computed by the vector ε-algorithm can be expressed as a ratio of two *designants* of dimension $k+1$. A designant is a generalization of a determinant when solving a system of linear equations in a noncommutative algebra. An algebraic approach to this algorithm was given in [125]. This approach, which involves the use of a Clifford algebra, was used in [45] for extending the mechanism given in [41] to the vector and matrix cases. The vector generalization of the E-algorithm [19] can be explained similarly. This algorithm makes use of a fixed vector y. Jet Wimp [146, pp. 176–177] generalized it using a sequence (y_n) instead of y. Jeannette van Iseghem [140] gave an algorithm for accelerating vector sequences based on the vector orthogonal polynomials she introduced for generalizing Padé approximants to the vector case. Other vector sequence transformations are due to Osada [109] and Jbilou and Sadok [76]. Benchiboun [6] and Abderrahim Messaoudi [100] studied matrix extrapolation algorithms.

We have seen that, in the scalar case, the kernels of sequence transformations may be expressed as relationships with constant coefficients. This is also the case for the vector and the topological ε-algorithms and the vector E-algorithm. The first (and, to my knowledge, only) transformation treating a relationship with varying coefficients was introduced in [42]. The theory developed there also explains why the case of a relationship with non-constant coefficients is a difficult problem in the scalar case and why it could be solved, on the contrary, in the vector case. The reason is that the number of unknown coefficients appearing in the expression of the kernel must be strictly less than the dimension of the vectors. Brezinski in [34] proposed a general methodology for constructing vector sequence transformations. It leads to a unified presentation of several approaches to the subject and to new results. He also discussed applications to linear systems. In fact, as showed by Sidi [133], and Jbilou and Sadok [75], vector sequence transformations are closely related to projection methods for the solution of systems of equations. In particular, the RPA, a vector sequence transformation defined by Brezinski [20] was extensively studied by Messaoudi who showed its connections to direct and iterative methods for solving systems of linear equations [98,99].

Vector sequence transformations lead to new methods for the solution of systems of nonlinear equations. They also have other applications. First of all, it is quite important to accelerate the convergence of iterative methods for the solution of systems of linear equations, see [32,33,36]. Special vector extrapolation techniques were designed for the regularization of ill-posed linear systems in [43] and the idea of extrapolation was used in [35] to obtain estimates of the norm of the error when solving a system of linear equations by an arbitrary method, direct or iterative.

General theoretical results similar to those obtained in the scalar case are still lacking in the vector case although some partial results have been obtained. Relevant results on quasilinear transformations are in the papers by Sadok [123] and Benazzouz [8]. The present author proposed a mechanism for vector sequence transformations in [45,34].

4. Conclusions and perspectives

In this paper, I have tried to give a survey of the development of convergence acceleration methods for scalar and vector sequences in the 20th century. These methods are based on the idea of extrapolation. Since a universal algorithm for accelerating the convergence of all sequences cannot exist (and this is even true for some restricted classes of sequences), it was necessary to define and study a large variety of algorithms, each of them being appropriate for some special subsets of sequences.

It is, of course, always possible to construct other convergence acceleration methods for scalar sequences. However, to be of interest, such new processes must provide a major improvement over existing ones. For scalar sequence transformations, the emphasis must be placed on the theory rather than on special devices (unless a quite powerful one is found) and on the application of new methods to particular algorithms in numerical analysis and to various domains of applied sciences. In particular, the connection between convergence acceleration algorithms and continuous and discrete integrable systems brings a different and fresh look to both domains and could be of benefit to them.

An important problem in numerical analysis is the solution of large, sparse systems of linear equations. Most of the methods used nowadays are projection methods. Often the iterates obtained in such problems must be subject to acceleration techniques. However, many of the known vector convergence acceleration algorithms require the storage of too many vectors to be useful. New and cheaper acceleration algorithms are required. This difficult project, in my opinion, offers many opportunities for future research.

In this paper, I only briefly mentioned the confluent algorithms whose aim is the computation of the limit of a function when the variable tends to infinity (the continuous analog of the problem of convergence acceleration for a sequence). This subject and its applications will provide fertile ground for new discoveries.

Acknowledgements

I would like to thank Jet Wimp for his careful reading of the paper. He corrected my English in many places, he asked me to provide more explanations when needed, and suggested many improvements in the presentation. I am also indebted to Naoki Osada for his informations about Takakazu Seki.

References

[1] A.C. Aitken, On Bernoulli's numerical solution of algebraic equations, Proc. Roy. Soc. Edinburgh 46 (1926) 289–305.

[2] H. Andoyer, Interpolation, in: J. Molk (Ed.), Encyclopédie des Sciences Mathématiques Pures et Appliquées, Tome I, Vol. 4, Fasc. 1, I–21, Gauthier–Villars, Paris, 1904–1912, pp.127–160; (reprint by Editions Gabay, Paris, 1993).

[3] F.L. Bauer, La méthode d'intégration numérique de Romberg, in: Colloque sur l'Analyse Numérique, Librairie Universitaire, Louvain, 1961, pp. 119–129.

[4] B. Beckermann, A connection between the E-algorithm and the epsilon-algorithm, in: C. Brezinski (Ed.), Numerical and Applied Mathematics, Baltzer, Basel, 1989, pp. 443–446.

[5] M.D. Benchiboun, Etude de Certaines Généralisations du Δ^2 d'Aitken et Comparaison de Procédés d'Accélération de la Convergence, Thèse 3ème cycle, Université de Lille I, 1987.

[6] M.D. Benchiboun, Extension of Henrici's method to matrix sequences, J. Comput. Appl. Math. 75 (1996) 1–21.

[7] A. Benazzouz, Quasilinear sequence transformations, Numer. Algorithms 15 (1997) 275–285.

[8] A. Benazzouz, GL(E)-quasilinear transformations and acceleration, Appl. Numer. Math. 27 (1998) 109–122.

[9] A. Berlinet, Sequence transformations as statistical tools, Appl. Numer. Math. 1 (1985) 531–544.

[10] A.H. Bentbib, Acceleration of convergence of interval sequences, J. Comput. Appl. Math. 51 (1994) 395–409.

[11] N. Bogolyubov, N. Krylov, On Rayleigh's principle in the theory of differential equations of mathematical physics and upon Euler's method in the calculus of variation, Acad. Sci. Ukraine (Phys. Math.) 3 (1926) 3–22 (in Russian).

[12] H.C. Bolton, H.I. Scoins, Eigenvalues of differential equations by finite-difference methods, Proc. Cambridge Philos. Soc. 52 (1956) 215–229.

[13] C. Brezinski, Application de l'ε-algorithme à la résolution des systèmes non linéaires, C.R. Acad. Sci. Paris 271 A (1970) 1174–1177.

[14] C. Brezinski, Accélération de suites à convergence logarithmique, C. R. Acad. Sci. Paris 273 A (1971) 727–730.

[15] C. Brezinski, Etude sur les ε et ρ-algorithmes, Numer. Math. 17 (1971) 153–162.

[16] C. Brezinski, L'ε-algorithme et les suites totalement monotones et oscillantes, C.R. Acad. Sci. Paris 276 A (1973) 305–308.
[17] C. Brezinski, Généralisation de la transformation de Shanks, de la table de Padé et de l'ε-algorithme, Calcolo 12 (1975) 317–360.
[18] C. Brezinski, Padé-Type Approximation and General Orthogonal Polynomials, Birkhäuser, Basel, 1980.
[19] C. Brezinski, A general extrapolation algorithm, Numer. Math. 35 (1980) 175–187.
[20] C. Brezinski, Recursive interpolation, extrapolation and projection, J. Comput. Appl. Math. 9 (1983) 369–376.
[21] C. Brezinski, Error control in convergence acceleration processes, IMA J. Numer. Anal. 3 (1983) 65–80.
[22] C. Brezinski, Prediction properties of some extrapolation methods, Appl. Numer. Math. 1 (1985) 457–462.
[23] C. Brezinski, Composite sequence transformations, Numer. Math. 46 (1985) 311–321.
[24] C. Brezinski, A. Lembarki, Acceleration of extended Fibonacci sequences, Appl. Numer. Math. 2 (1986) 1–8.
[25] C. Brezinski, A new approach to convergence acceleration methods, in: A. Cuyt (Ed.), Nonlinear Numerical Methods and Rational Approximation, Reidel, Dordrecht, 1988, pp. 373–405.
[26] C. Brezinski, Quasi-linear extrapolation processes, in: R.P. Agarwal et al. (Eds.), Numerical Mathematics, Singapore 1988, International Series of Numerical Mathematics, Vol. 86, Birkhäuser, Basel, 1988, pp. 61–78.
[27] C. Brezinski, A survey of iterative extrapolation by the E-algorithm, Det Kong. Norske Vid. Selsk. Skr. 2 (1989) 1–26.
[28] C. Brezinski, A Bibliography on Continued Fractions, Padé Approximation, Extrapolation and Related Subjects, Prensas Universitarias de Zaragoza, Zaragoza, 1991.
[29] C. Brezinski, History of Continued Fractions and Padé Approximants, Springer, Berlin, 1991.
[30] C. Brezinski, The generalizations of Newton's interpolation formula due to Mühlbach and Andoyer, Electron Trans. Numer. Anal. 2 (1994) 130–137.
[31] C. Brezinski, Extrapolation algorithms and Padé approximations: a historical survey, Appl. Numer. Math. 20 (1996) 299–318.
[32] C. Brezinski, Variations on Richardson's method and acceleration, in: Numerical Analysis, A Numerical Analysis Conference in Honour of Jean Meinguet, Bull. Soc. Math. Belgium 1996, pp. 33–44.
[33] C. Brezinski, Projection Methods for Systems of Equations, North-Holland, Amsterdam, 1997.
[34] C. Brezinski, Vector sequence transformations: methodology and applications to linear systems, J. Comput. Appl. Math. 98 (1998) 149–175.
[35] C. Brezinski, Error estimates for the solution of linear systems, SIAM J. Sci. Comput. 21 (1999) 764–781.
[36] C. Brezinski, Acceleration procedures for matrix iterative methods, Numer. Algorithms, to appear.
[37] C. Brezinski, J.P. Delahaye, B. Germain-Bonne, Convergence acceleration by extraction of linear subsequences, SIAM J. Numer. Anal. 20 (1983) 1099–1105.
[38] C. Brezinski, A.C. Matos, A derivation of extrapolation algorithms based on error estimates, J. Comput. Appl. Math. 66 (1996) 5–26.
[39] C. Brezinski, M. Redivo Zaglia, Construction of extrapolation processes, Appl. Numer. Math. 8 (1991) 11–23.
[40] C. Brezinski, M. Redivo Zaglia, Extrapolation Methods, Theory and Practice, North-Holland, Amsterdam, 1991.
[41] C. Brezinski, M. Redivo Zaglia, A general extrapolation procedure revisited, Adv. Comput. Math. 2 (1994) 461–477.
[42] C. Brezinski, M. Redivo Zaglia, Vector and matrix sequence transformations based on biorthogonality, Appl. Numer. Math. 21 (1996) 353–373.
[43] C. Brezinski, M. Redivo Zaglia, G. Rodriguez, S. Seatzu, Extrapolation techniques for ill-conditioned linear systems, Numer. Math. 81 (1998) 1–29.
[44] C. Brezinski, A.C. Rieu, The solution of systems of equations using the vector ε-algorithm, and an application to boundary value problems, Math. Comp. 28 (1974) 731–741.
[45] C. Brezinski, A. Salam, Matrix and vector sequence transformation revisited, Proc. Edinburgh Math. Soc. 38 (1995) 495–510.
[46] C. Brezinski, G. Walz, Sequences of transformations and triangular recursion schemes with applications in numerical analysis, J. Comput. Appl. Math. 34 (1991) 361–383.
[47] R. Bulirsch, J. Stoer, Numerical treatment of ordinary differential equations by extrapolation methods, Numer. Math. 8 (1966) 1–13.
[48] C. Carstensen, On a general epsilon algorithm, in: C. Brezinski (Ed.), Numerical and Applied Mathematics, Baltzer, Basel, 1989, pp. 437–441.

[49] J.L. Chabert et al., Histoire d'Algorithmes, Belin, Paris, 1994.
[50] L. Collatz, Numerische Behandlung von Differentialgleichungen, Springer, Berlin, 1951.
[51] F. Cordellier, Démonstration algébrique de l'extension de l'identité de Wynn aux tables de Padé non normales, in: L. Wuytack (Ed.), Padé Approximation and its Applications, Lecture Notes in Mathematics, Vol. 765, Springer, Berlin, 1979, pp. 36–60.
[52] M. Crouzeix, A.L. Mignot, Analyse Numérique des Equations Différentielles, 2nd Edition, Masson, Paris, 1989.
[53] A. Cuyt, L. Wuytack, Nonlinear Methods in Numerical Analysis, North-Holland, Amsterdam, 1987.
[54] J.P. Delahaye, Automatic selection of sequence transformations, Math. Comp. 37 (1981) 197–204.
[55] J.P. Delahaye, Sequence Transformations, Springer, Berlin, 1988.
[56] J.P. Delahaye, B. Germain-Bonne, Résultats négatifs en accélération de la convergence, Numer. Math. 35 (1980) 443–457.
[57] J. Dutka, Richardson-extrapolation and Romberg-integration, Historia Math. 11 (1984) 3–21.
[58] A. Fdil, Sélection entre procédés d'accélération de la convergence, M2AN 30 (1996) 83–101.
[59] A. Fdil, A new technique of selection between sequence transformations, Appl. Numer. Math. 25 (1997) 21–40.
[60] W.F. Ford, A. Sidi, An algorithm for a generalization of the Richardson extrapolation process, SIAM J. Numer. Anal. 24 (1987) 1212–1232.
[61] E. Gekeler, On the solution of systems of equations by the epsilon algorithm of Wynn, Math. Comp. 26 (1972) 427–436.
[62] B. Germain-Bonne, Transformations de suites, RAIRO R1 (1973) 84–90.
[63] B. Germain-Bonne, Estimation de la Limite de Suites et Formalisation de Procédés d'Accélération de la Convergence, Thèse d'Etat, Université de Lille I, 1978.
[64] J. Gilewicz, Numerical detection of the best Padé approximant and determination of the Fourier coefficients of insufficiently sampled function, in: P.R. Graves-Morris (Ed.), Padé Approximants and their Applications, Academic Press, New York, 1973, pp. 99–103.
[65] W.B. Gragg, On extrapolation algorithms for initial-value problems, SIAM J. Numer. Anal. 2 (1965) 384–403.
[66] B. Grammaticos, A. Ramani, Integrability – and how to detect it, in: Y. Kosmann-Schwarzbach et al. (Eds.), Integrability of Nonlinear Systems, Springer, Berlin, 1997, pp. 30–94.
[67] B. Grammaticos, A. Ramani, V.G. Papageorgiou, Do integrable mappings have the Painlevé property? Phys. Rev. Lett. 67 (1991) 1825–1828.
[68] P.R. Graves-Morris, Vector valued rational interpolants I, Numer. Math. 42 (1983) 331–348.
[69] P.R. Graves-Morris, C.D. Jenkins, Vector valued rational interpolants III, Constr. Approx. 2 (1986) 263–289.
[70] H.L. Gray, T.A. Atchison, G.V. McWilliams, Higher order G – transformations, SIAM J. Numer. Anal. 8 (1971) 365–381.
[71] T. Håvie, Generalized Neville type extrapolation schemes, BIT 19 (1979) 204–213.
[72] H. Hankel, Ueber eine besondere Classe der symmetrischen Determinanten, Inaugural Dissertation, Universität Göttingen, 1861.
[73] A. Hirayama, K. Shimodaira, H. Hirose, Takakazu Seki's Collected Works Edited with Explanations, Osaka Kyoiku Tosho, Osaka, 1974.
[74] H.H.H. Homeier, A hierarchically consistent, iterative sequence transformation, Numer. Algorithms 8 (1994) 47–81.
[75] K. Jbilou, H. Sadok, Some results about vector extrapolation methods and related fixed-point iterations, J. Comput. Appl. Math. 36 (1991) 385–398.
[76] K. Jbilou, H. Sadok, Hybrid vector transformations, J. Comput. Appl. Math. 81 (1997) 257–267.
[77] D.C. Joyce, Survey of extrapolation processes in numerical analysis, SIAM Rev. 13 (1971) 435–490.
[78] K. Kommerell, Das Grenzgebiet der Elementaren und Höheren Mathematik, Verlag Köhler, Leipzig, 1936.
[79] M. Kzaz, Gaussian quadrature and acceleration of convergence, Numer. Algorithms 15 (1997) 75–89.
[80] M. Kzaz, Convergence acceleration of the Gauss–Laguerre quadrature formula, Appl. Numer. Math. 29 (1999) 201–220.
[81] P.J. Laurent, Un théorème de convergence pour le procédé d'extrapolation de Richardson, C.R. Acad. Sci. Paris 256 (1963) 1435–1437.
[82] H. Lavastre, On the stochastic acceleration of sequences of random variables, Appl. Numer. Math. 15 (1994) 77–98.
[83] H. Le Ferrand, Convergence of the topological ε-algorithm for solving systems of nonlinear equations, Numer. Algorithms 3 (1992) 273–284.

[84] H. Le Ferrand, Recherches d'extrema par des méthodes d'extrapolation, C.R. Acad. Sci. Paris, Sér. I 318 (1994) 1043–1046.
[85] D. Levin, Development of non-linear transformations for improving convergence of sequences, Int. J. Comput. Math. B3 (1973) 371–388.
[86] P.M. Lima, M.P. Carpentier, Asymptotic expansions and numerical approximation of nonlinear degenerate boundary-value problems, Appl. Numer. Math. 30 (1999) 93–111.
[87] P. Lima, T. Diogo, An extrapolation method for a Volterra integral equation with weakly singular kernel, Appl. Numer. Math. 24 (1997) 131–148.
[88] P.M. Lima, M.M Graça, Convergence acceleration for boundary value problems with singularities using the E-algorithm, J. Comput. Appl. Math. 61 (1995) 139–164.
[89] S. Lubkin, A method of summing infinite series, J. Res. Natl. Bur. Standards 48 (1952) 228–254.
[90] C. Maclaurin, Treatise of Fluxions, Edinburgh, 1742.
[91] G.I. Marchuk, V.V. Shaidurov, Difference Methods and their Extrapolations, Springer, Berlin, 1983.
[92] A.C. Matos, Acceleration methods based on convergence tests, Numer. Math. 58 (1990) 329–340.
[93] A.C. Matos, Linear difference operators and acceleration methods, IMA J. Numer. Anal., to appear.
[94] A.C. Matos, M. Prévost, Acceleration property for the columns of the E-algorithm, Numer. Algorithms 2 (1992) 393–408.
[95] J.C. Maxwell, A Treatise on Electricity and Magnetism, Oxford University Press, Oxford, 1873.
[96] J.B. McLeod, A note on the ε-algorithm, Computing 7 (1971) 17–24.
[97] G. Meinardus, G.D. Taylor, Lower estimates for the error of the best uniform approximation, J. Approx. Theory 16 (1976) 150–161.
[98] A. Messaoudi, Recursive interpolation algorithm: a formalism for solving systems of linear equations – I, Direct methods, J. Comput. Appl. Math. 76 (1996) 13–30.
[99] A. Messaoudi, Recursive interpolation algorithm: a formalism for solving systems of linear equations – II, Iterative methods, J. Comput. Appl. Math. 76 (1996) 31–53.
[100] A. Messaoudi, Matrix extrapolation algorithms, Linear Algebra Appl. 256 (1997) 49–73.
[101] R.M. Milne, Extension of Huygens' approximation to a circular arc, Math. Gaz. 2 (1903) 309–311.
[102] P. Mortreux, M. Prévost, An acceleration property for the E-algorithm for alternate sequences, Adv. Comput. Math. 5 (1996) 443–482.
[103] G. Mühlbach, Neville-Aitken algorithms for interpolating by functions of Čebyšev-systems in the sense of Newton and in a generalized sense of Hermite, in: A.G. Law, B.N. Sahney (Eds.), Theory of Approximation with Applications, Academic Press, New York, 1976, pp. 200–212.
[104] H. Naegelsbach, Studien zu Fürstenau's neuer Methode der Darstellung und Berechnung der Wurzeln algebraischer Gleichungen durch Determinanten der Coefficienten, Arch. Math. Phys. 59 (1876) 147–192; 61 (1877) 19–85.
[105] A. Nagai, J. Satsuma, Discrete soliton equations and convergence acceleration algorithms, Phys. Lett. A 209 (1995) 305–312.
[106] A. Nagai, T. Tokihiro, J. Satsuma, The Toda molecule equation and the ε-algorithm, Math. Comp. 67 (1998) 1565–1575.
[107] T.H. O'Beirne, On linear iterative processes and on methods of improving the convergence of certain types of iterated sequences, Technical Report, Torpedo Experimental Establishment, Greenock, May 1947.
[108] N. Osada, An acceleration theorem for the ρ-algorithm, Numer. Math. 73 (1996) 521–531.
[109] N. Osada, Vector sequence transformations for the acceleration of logarithmic convergence, J. Comput. Appl. Math. 66 (1996) 391–400.
[110] K.J. Overholt, Extended Aitken acceleration, BIT 5 (1965) 122–132.
[111] V. Papageorgiou, B. Grammaticos, A. Ramani, Integrable difference equations and numerical analysis algorithms, in: D. Levi et al. (Eds.), Symmetries and Integrability of Difference Equations, CRM Proceedings and Lecture Notes, Vol. 9, AMS, Providence, RI, 1996, pp. 269–279.
[112] V. Papageorgiou, B. Grammaticos, A. Ramani, Integrable lattices and convergence acceleration algorithms, Phys. Lett. A 179 (1993) 111–115.
[113] V. Papageorgiou, B. Grammaticos, A. Ramani, Orthogonal polynomial approach to discrete Lax pairs for initial-boundary value problems of the QD algorithm, Lett. Math. Phys. 34 (1995) 91–101.
[114] R. Pennacchi, Le trasformazioni razionali di una successione, Calcolo 5 (1968) 37–50.

[115] R. Powell, S.M. Shah, Summability Theory and its Applications, Van Nostrand Reinhold, London, 1972.
[116] M. Prévost, Acceleration property for the E-algorithm and an application to the summation of series, Adv. Comput. Math. 2 (1994) 319–341.
[117] J.D. Pryce, Numerical Solution of Sturm–Liouville Problems, Clarendon Press, Oxford, 1993.
[118] B. Rhanizar, On extrapolation methods in optimization, Appl. Numer. Math. 25 (1997) 485–498.
[119] L.F. Richardson, The approximate arithmetical solution by finite difference of physical problems involving differential equations, with an application to the stress in a masonry dam, Philos. Trans. Roy. Soc. London, Ser. A 210 (1910) 307–357.
[120] L.F. Richardson, The deferred approach to the limit. I: Single lattice, Philos. Trans. Roy. Soc. London, Ser. A 226 (1927) 299–349.
[121] W. Romberg, Vereinfachte numerische Integration, Kgl. Norske Vid. Selsk. Forsk. 28 (1955) 30–36.
[122] H. Rutishauser, Ausdehnung des Rombergschen Prinzips, Numer. Math. 5 (1963) 48–54.
[123] H. Sadok, Quasilinear vector extrapolation methods, Linear Algebra Appl. 190 (1993) 71–85.
[124] A. Salam, Non-commutative extrapolation algorithms, Numer. Algorithms 7 (1994) 225–251.
[125] A. Salam, An algebraic approach to the vector ε-algorithm, Numer. Algorithms 11 (1996) 327–337.
[126] P.R. Graves-Morris, D.E. Roberts, A. Salam, The epsilon algorithm and related topics, J. Comput. Appl. Math. 122 (2000).
[127] J.R. Schmidt, On the numerical solution of linear simultaneous equations by an iterative method, Philos. Mag. 7 (1941) 369–383.
[128] C. Schneider, Vereinfachte Rekursionen zur Richardson-Extrapolation in Spezialfällen, Numer. Math. 24 (1975) 177–184.
[129] M.N. Senhadji, On condition numbers of some quasi-linear transformations, J. Comput. Appl. Math. 104 (1999) 1–19.
[130] D. Shanks, An analogy between transient and mathematical sequences and some nonlinear sequence-to-sequence transforms suggested by it, Part I, Memorandum 9994, Naval Ordnance Laboratory, White Oak, July 1949.
[131] D. Shanks, Non linear transformations of divergent and slowly convergent sequences, J. Math. Phys. 34 (1955) 1–42.
[132] W.F. Sheppard, Some quadrature formulas, Proc. London Math. Soc. 32 (1900) 258–277.
[133] A. Sidi, Extrapolation vs. projection methods for linear systems of equations, J. Comput. Appl. Math. 22 (1988) 71–88.
[134] A. Sidi, On a generalization of the Richardson extrapolation process, Numer. Math. 57 (1990) 365–377.
[135] A. Sidi, Further results on convergence and stability of a generalization of the Richardson extrapolation process, BIT 36 (1996) 143–157.
[136] A. Sidi, A complete convergence and stability theory for a generalized Richardson extrapolation process, SIAM J. Numer. Anal. 34 (1997) 1761–1778.
[137] A. Sidi, D. Levin, Prediction properties of the t-transformation, SIAM J. Numer. Anal. 20 (1983) 589–598.
[138] E. Stiefel, Altes und neues über numerische Quadratur, Z. Angew. Math. Mech. 41 (1961) 408–413.
[139] M. Toda, Waves in nonlinear lattice, Prog. Theor. Phys. Suppl. 45 (1970) 174–200.
[140] J. van Iseghem, Convergence of vectorial sequences, applications, Numer. Math. 68 (1994) 549–562.
[141] D. Vekemans, Algorithm for the E-prediction, J. Comput. Appl. Math. 85 (1997) 181–202.
[142] P. Verlinden, Acceleration of Gauss–Legendre quadrature for an integral with an endpoint singularity, J. Comput. Appl. Math. 77 (1997) 277–287.
[143] G. Walz, The history of extrapolation methods in numerical analysis, Report No. 130, Universität Mannheim, Fakultät für Mathematik und Informatik, 1991.
[144] G. Walz, Asymptotics and Extrapolation, Akademie, Berlin, 1996.
[145] E.J. Weniger, Nonlinear sequence transformations for the acceleration of convergence and the summation of divergent series, Comput. Phys. Rep. 10 (1989) 189–371.
[146] J. Wimp, Sequence Transformations and their Applications, Academic Press, New York, 1981.
[147] P. Wynn, On a device for computing the $e_m(S_n)$ transformation, MTAC 10 (1956) 91–96.
[148] P. Wynn, On a procrustean technique for the numerical transformation of slowly convergent sequences and series, Proc. Cambridge Philos. Soc. 52 (1956) 663–671.
[149] P. Wynn, Confluent forms of certain non-linear algorithms, Arch. Math. 11 (1960) 223–234.

[150] P. Wynn, Acceleration techniques for iterated vector and matrix problems, Math. Comput. 16 (1962) 301–322.
[151] P. Wynn, Singular rules for certain non-linear algorithms, BIT 3 (1963) 175–195.
[152] P. Wynn, Partial differential equations associated with certain non-linear algorithms, ZAMP 15 (1964) 273–289.
[153] P. Wynn, Upon systems of recursions which obtain among the quotients of the Padé table, Numer. Math. 8 (1966) 264–269.
[154] P. Wynn, On the convergence and stability of the epsilon algorithm, SIAM J. Numer. Anal. 3 (1966) 91–122.
[155] M. Redivo Zaglia, Particular rules for the θ-algorithm, Numer. Algorithms 3 (1992) 353–370.

On the history of multivariate polynomial interpolation

Mariano Gasca[a,*], Thomas Sauer[b]

[a] *Department of Applied Mathematics, University of Zaragoza, 50009 Zaragoza, Spain*

[b] *Institute of Mathematics, University Erlangen-Nürnberg, Bismarckstr. $1\frac{1}{2}$, D-91054 Erlangen, Germany*

Abstract

Multivariate polynomial interpolation is a basic and fundamental subject in Approximation Theory and Numerical Analysis, which has received and continues receiving not deep but constant attention. In this short survey, we review its development in the first 75 years of this century, including a pioneering paper by Kronecker in the 19th century.

1. Introduction

Interpolation, by polynomials or other functions, is a rather old method in applied mathematics. This is already indicated by the fact that, apparently, the word "interpolation" itself has been introduced by J. Wallis as early as 1655 as it is claimed in [13]. Compared to this, polynomial interpolation in *several variables* is a relatively new topic and probably only started in the second-half of the last century with work in [6,22]. If one considers, for example, the *Encyklopädie der Mathematischen Wissenschaften* [13] (Encyclopedia of Math. Sciences), originated by the *Preußische Akademie der Wissenschaften* (Prussian Academy of Sciences) to sum up the "state of art" of mathematics at its time, then the part on interpolation, written by J. Bauschinger (Bd. I, Teil 2), mentions only one type of multivariate interpolation, namely (tensor) products of sine and cosine functions in two variables, however, without being very specific. The French counterpart, the *Encyclopédie de Sciences Mathematiques* [14], also contains a section on interpolation (Tome I, vol. 4), where Andoyer translated and extended Bauschinger's exposition. Andoyer is even more

* Corressponding author.

E-mail addresses: gasca@posta.unizar.es (M. Gasca), sauer@mi.uni-erlangen.de (T. Sauer).

explicit with his opinion on multivariate polynomial interpolation, by making the following statement which we think that time has contradicted:

> Il est manifeste que l'interpolation des fonctions de plusiers variables ne demande aucun principe nouveau, car dans tout ce qui précède le fait que la variable indépendante était unique n'a souvent joué aucun rôle. [1]

Nevertheless, despite of Andoyer's negative assessment, multivariate polynomial interpolation has received not deep but constant attention from one part of the mathematical community and is today a basic subject in Approximation Theory and Numerical Analysis with applications to many mathematical problems. Of course, this field has definitely been influenced by the availability of computational facilities, and this is one of the reasons that more papers have been published about this subject in the last 25 years than in the preceding 75 ones.

To our knowledge, there is not any paper before the present one surveying the early papers and books on multivariate polynomial interpolation. Our aim is a first, modest attempt to cover this gap. We do not claim to be exhaustive and, in particular, recognize our limitations with respect to the Russian literature. Moreover, it has to be mentioned that the early results on multivariate interpolation usually appear in the context of many different subjects. For example, papers on cubature formulas frequently have some part devoted to it. Another connection is Algebraic Geometry, since the solvability of a multivariate interpolation problem relies on the fact that the interpolation points do not lie on an algebraic surface of a certain type. So it is difficult to verify precisely if and when a result appeared somewhere for the first time or if it had already appeared, probably even in an implicit way, in a different context. We remark that another paper in this volume [25] deals, complementarily, with recent results in the subject, see also [16].

Along the present paper we denote by Π_k^d the space of d-variate polynomials of total degree not greater than k.

2. Kronecker, Jacobi and multivariate interpolation

Bivariate interpolation by the tensor product of univariate interpolation functions, that is when the variables are treated separately, is the classical approach to multivariate interpolation. However, when the set of interpolation points is not a Cartesian product grid, it is impossible to use that idea. Today, given any set of interpolation points, there exist many methods [2] to construct an adequate polynomial space which guarantees unisolvence of the interpolation problem. Surprisingly, this idea of constructing an appropriate interpolation space was already pursued by Kronecker [22] in a widely unknown paper from 1865, which seems to be the first treatment of multivariate polynomial interpolation with respect to fairly arbitrary point configurations. Besides the mathematical elegance of this approach, we think it is worthwhile to devote some detailed attention to this paper and to resolve its main ideas in today's terminology, in particular, as it uses the "modern" approach of connecting polynomial interpolation to the theory of polynomial ideals.

---

[1] It is clear that the interpolation of functions of several variables does not demand any new principles because in the above exposition the fact that the variable was unique has not played frequently any role.

[2] See [16,25] for exposition and references.

Kronecker's method to construct an interpolating polynomial assumes that the disjoint nodes $z_1,\ldots,z_N \in \mathbb{C}^d$ are given in *implicit* form, i.e., they are (all) the common *simple* zeros of d polynomials $f_1,\ldots,f_d \in \mathbb{C}[z]=\mathbb{C}[\zeta_1,\ldots,\zeta_d]$. Note that the nonlinear system of equations

$$f_j(\zeta_1,\ldots,\zeta_d)=0, \quad j=1,\ldots,d, \tag{1}$$

is a *square* one, that is, the number of equations and the number of variables coincide. We are interested in the *finite* variety V of solutions of (1) which is given as

$$V:=\{z_1,\ldots,z_N\}=\{z\in\mathbb{C}^d : f_1(z)=\cdots=f_d(z)=0\}. \tag{2}$$

The *primary decomposition* according to the variety V allows us to write the ideal $\mathcal{I}(V)=\{p : p(z)=0,\ z\in V\}$ as

$$\mathcal{I}(V)=\bigcap_{k=1}^{N}\langle \zeta_1-\zeta_{k,1},\ldots,\zeta_d-\zeta_{k,d}\rangle,$$

where $z_k=(\zeta_{k,1},\ldots,\zeta_{k,d})$. In other words, since $f_k\in\mathcal{I}(V)$, $k=1,\ldots,d$, any of the polynomials $f_1,\ldots,f_d$ can be written, for $k=1,\ldots,N$, as

$$f_j=\sum_{i=1}^{d} g_{i,j}^k(\cdot)(\zeta_i-\zeta_{k,i}), \tag{3}$$

where $g_{i,j}^k$ are appropriate polynomials. Now consider the $d\times d$ square matrices of polynomials

$$G_k=[g_{i,j}^k : i,j=1,\ldots,d], \quad k=1,\ldots,N$$

and note that, due to (3), and the assumption that $f_j(z_k)=0$, $j=1,\ldots,d$, $k=1,\ldots,N$, we have

$$0=\begin{bmatrix} f_1(Z_j)\\ \vdots \\ f_d(z_j)\end{bmatrix}=G_k(z_j)\begin{bmatrix}(\zeta_{j,1}-\zeta_{k,1})\\ \vdots \\ (\zeta_{j,d}-\zeta_{k,d})\end{bmatrix}, \quad k=1,\ldots,N. \tag{4}$$

Since the interpolation nodes are assumed to be *disjoint*, this means that for all $j\neq k$ the matrix $G_k(z_j)$ is *singular*, hence the determinant of $G_k(z_j)$ has to be zero. Moreover, the assumption that $z_1,\ldots,z_N$ are *simple* zeros guarantees that $\det G_k(z_k)\neq 0$. Then, Kronecker's interpolant takes, for any $f:\mathbb{C}^d\to\mathbb{C}$, the form

$$Kf=\sum_{j=1}^{N} f(z_j)\frac{\det G_k(\cdot)}{\det G_k(z_k)}. \tag{5}$$

Hence,

$$\mathcal{P}=\operatorname{span}\left\{\frac{\det G_k(\cdot)}{\det G_k(z_k)} : k=1,\ldots,N\right\}$$

is an interpolation space for the interpolation nodes $z_1,\ldots,z_N$. Note that this method does not give only *one* interpolation polynomial but in general *several different* interpolation spaces, depending on how the representation in (3) is chosen. In any way, note that for each polynomial $f\in\mathbb{C}[z]$ the difference

$$f-\sum_{j=1}^{N} f(z_j)\frac{\det G_k(z)}{\det G_k(z_k)}$$

belongs to the ideal $\langle f_1, \dots, f_d \rangle$, hence there exist polynomials $q_1, \dots, q_d$ such that

$$f - \sum_{j=1}^{N} f(z_j) \frac{\det G_k(z)}{\det G_k(z_k)} = \sum_{j=1}^{d} q_j f_j. \tag{6}$$

Moreover, as Kronecker points out, the "magic" polynomials $g_{i,j}^k$ can be chosen such that their leading homogeneous terms, say $G_{i,j}^k$, coincide with the leading homogeneous terms of $(1/\deg f_j)\partial f_j/\partial \zeta_i$. If we denote by F_j the leading homogeneous term of f_j, $j = 1, \dots, d$, then this means that

$$G_{i,j}^k = \frac{1}{\deg F_j} \frac{\partial F_j}{\partial \zeta_i}, \quad i,j = 1, \dots, d, \quad k = 1, \dots, N. \tag{7}$$

But this implies that the homogeneous leading term of the "fundamental" polynomials $\det G_k$ coincides, after this particular choice of $g_{i,j}^k$, with

$$g = \frac{1}{\deg f_1 \cdots \deg f_d} \det \left[\frac{\partial F_j}{\partial \zeta_i} : i,j = 1, \dots, d \right],$$

which is *independent of k* now; in other words, there exist polynomials $\hat{g}_k$, $k = 1, \dots, N$, such that $\deg \hat{g}_k < \deg g$ and $\det G_k = g + \hat{g}_k$. Moreover, g is a homogeneous polynomial of degree at most $\deg f_1 + \cdots + \deg f_d - d$. Now, let p be any polynomial, then

$$Kp = \sum_{j=1}^{N} p(z_j) \frac{\det G_j(\cdot)}{\det G_j(z_j)} = g \sum_{j=1}^{N} \frac{p(z_j)}{\det G_j(z_j)} + \sum_{j=1}^{N} \frac{p(z_j)}{\det G_j(z_j)} \hat{g}_j. \tag{8}$$

Combining (8) with (6) then yields the existence of polynomials $q_1, \dots, q_d$ such that

$$p = g \sum_{j=1}^{N} \frac{p(z_j)}{\det G_j(z_j)} + \sum_{j=1}^{N} \frac{p(z_j)}{\det G_j(z_j)} \hat{g}_j + \sum_{j=1}^{d} q_j f_j$$

and comparing homogeneous terms of degree $\deg g$ Kronecker realized that either, for any p such that $\deg p < \deg g$,

$$\sum_{j=1}^{N} \frac{p(z_j)}{\det G_j(z_j)} = 0 \tag{9}$$

or there exist homogeneous polynomials $h_1, \dots, h_d$ such that

$$g = \sum_{j=1}^{d} h_j \det F_j. \tag{10}$$

The latter case, Eq. (10), says (in algebraic terminology) that there is a *syzygy* among the leading terms of the polynomials F_j, $j = 1, \dots, d$, and is equivalent to the fact that $N < \deg f_1 \cdots \deg f_d$, while (9) describes and even characterizes the *complete intersection case* that $N = \deg f_1 \cdots \deg f_d$. In his paper, Kronecker also mentions that the condition (10) has been overlooked in [21]. Jacobi dealt there with the common zeros of two bivariate polynomials and derived *explicit* representations for the functional

$$[z_1, \dots, z_N] f := \sum_{j=1}^{N} \frac{f(z_j)}{\det G_j(z_j)}, \tag{11}$$

which behaves very much like a divided difference, since it is a combination of point evaluations which, provided that (9) hold true, annihilates $\Pi^d_{\deg g-1}$.

In addition, Kronecker refers to a paper [6] which he says treats the case of symmetric functions, probably elementary symmetric polynomials. Unfortunately, this paper is unavailable to us so far.

3. Bivariate tables, the natural approach

Only very few research papers on multivariate polynomial interpolation were published during the first part of this century. In the classical book *Interpolation* [45], where one section (Section 19) is devoted to this topic, the author only refers to two related papers, recent at that time (1927), namely [27,28]. The latter one [28], turned out to be inaccessible to us, unfortunately, but it is not difficult to guess that it might have pursued a tensor product approach, because this is the unique point of view of [45] (see also [31]).

The formulas given in [27] are Newton formulas for tensor product interpolation in two variables, and the author, Narumi, claims (correctly) that they can be extended to "many variables". Since it is a tensor product approach, the interpolation points are of the form (x_i, y_j), $0 \leqslant i \leqslant m$, $0 \leqslant j \leqslant n$, with x_i, y_j arbitrarily distributed on the axes OX and OY, respectively. Bivariate divided differences for these sets of points are obtained in [27], by recurrence, separately for each variable. With the usual notations, the interpolation formula from [27] reads as

$$p(x, y) = \sum_{i=0}^{m} \sum_{j=0}^{n} f[x_0, \ldots, x_i; y_0, \ldots, y_j] \prod_{h=0}^{i-1} (x - x_h) \prod_{k=0}^{j-1} (y - x_k), \tag{12}$$

where empty products have the value 1. Remainder formulas based on the mean value theorem are also derived recursively from the corresponding univariate error formulas in [27]. For f sufficiently smooth there exist values ξ, ξ', η, η' such that

$$R(x, y) = \frac{\partial^{m+1} f(\xi, y)}{\partial x^{m+1}} \frac{\prod_{h=0}^{m}(x - x_h)}{(m+1)!} + \frac{\partial^{n+1} f(x, \eta)}{\partial y^{n+1}} \frac{\prod_{k=0}^{n}(y - y_k)}{(n+1)!}$$

$$- \frac{\partial^{m+n+2} f(\xi', \eta')}{\partial x^{m+1} \partial y^{n+1}} \frac{\prod_{h=0}^{m}(x - x_h)}{(m+1)!} \frac{\prod_{k=0}^{n}(y - y_k)}{(n+1)!}. \tag{13}$$

The special case of equidistant points on both axes is particularly considered in [27], and since the most popular formulas at that time were based on finite differences with equally spaced arguments, Narumi shows how to extend Gauss, Bessel and Stirling univariate interpolation formulas for equidistant points to the bivariate case by tensor product. He also applies the formulas he obtained to approximate the values of bivariate functions, but he also mentions that some of his formulas had been already used in [49].

In [45], the Newton formula (12) is obtained in the same way, with the corresponding remainder formula (13). Moreover, Steffensen considers a more general case, namely when for each i, $0 \leqslant i \leqslant m$, the interpolation points are of the form $y_0, \ldots, y_{n_i}$, with $0 \leqslant n_i \leqslant n$. Now with a similar argument the interpolating polynomial becomes

$$p(x, y) = \sum_{i=0}^{m} \sum_{j=0}^{n_i} f[x_0, \ldots, x_i; y_0, \ldots, y_j] \prod_{h=0}^{i-1} (x - x_h) \prod_{k=0}^{j-1} (y - x_k) \tag{14}$$

with a slightly more complicated remainder formula. The most interesting particular cases occur when $n_i = n$, which is the Cartesian product considered above, and when $n_i = m - i$. This *triangular* case (triangular not because of the geometrical distribution of the interpolation points, but of the indices (i, j)), gives rise to the interpolating polynomial

$$p(x,y) = \sum_{i=0}^{m} \sum_{j=0}^{m-i} f[x_0,\dots,x_i; y_0,\dots,y_j] \prod_{h=0}^{i-1}(x - x_h) \prod_{k=0}^{j-1}(y - x_k), \tag{15}$$

that is

$$p(x,y) = \sum_{0 \leqslant i+j \leqslant m} f[x_0,\dots,x_i; y_0,\dots,y_j] \prod_{h=0}^{i-1}(x - x_h) \prod_{k=0}^{j-1}(y - x_k). \tag{16}$$

Steffensen refers for this formula to Biermann's lecture notes [4] from 1905, and actually it seems that Biermann has been the first who considered polynomial interpolation on the triangular grid in a paper [3] from 1903 (cf. [44]) in the context of cubature.

Since the triangular case corresponds to looking at the "lower triangle" of the tensor product situation only, this case can be resolved by tensor product methods. In particular, the respective error formula can be written as

$$R(x,y) = \sum_{i=0}^{m+1} \frac{\partial^{m+1} f(\xi_i, \eta_i)}{\partial x^i \partial y^{m+1-i}} \frac{\prod_{h=0}^{i-1}(x - x_h)}{i!} \frac{\prod_{k=0}^{m-i}(y - y_k)}{(m-i+1)!}. \tag{17}$$

In the case of Cartesian product Steffensen also provides the Lagrange formula for (12), which can be obviously obtained by tensor product of univariate formulas.

Remainder formulas based on intermediate points (ξ_i, η_i) can be written in many different forms. For them we refer to Stancu's paper [44] which also contains a brief historical introduction where the author refers, among others, to [3,15,27,40,41]. Multivariate remainder formulas with Peano (spline) kernel representation, however, have not been derived until very recently in [42] and, in particular, in [43] which treats the triangular situation.

4. Salzer's papers: from bivariate tables to general sets

In 1944, Salzer [33] considered the interpolation problem at points of the form $(x_1 + s_1 h_1, \dots, x_n + s_n h_n)$ where

(i) $(x_1, \dots, x_n)$ is a given point in $\mathbb{R}^n$,
(ii) $h_1, \dots, h_n$ are given real numbers,
(iii) $s_1, \dots, s_n$ are nonnegative integers summing up to m.

This is the multivariate extension of the triangular case (16) for equally spaced arguments, where finite differences can be used. Often, different names are used for the classical Newton interpolation formula in the case of equally spaced arguments using forward differences: Newton–Gregory, Harriot–Briggs, also known by Mercator and Leibnitz, etc. See [18] for a nice discussion of this issue. In [33], Salzer takes the natural multivariate extension of this formula considering the polynomial $q(t_1, \dots, t_n) := p(x_1 + t_1 h_1, \dots, x_n + t_n h_n)$ of total degree not greater than m in the variables $t_1, \dots, t_n$, which interpolates a function $f(x_1 + t_1 h_1, \dots, x_n + t_n h_n)$ at the points corresponding to $t_i = s_i$, $i = 1, \dots, n$, where the s_i are all nonnegative integers such that $0 \leqslant s_1 + \cdots + s_n \leqslant m$. The formula, which is called in [33] a *multiple Gregory–Newton formula*, is rewritten there in terms of the values of the function f at the interpolation points, i.e., in the form

$$q(t_1,\ldots,t_n) = \sum_{s_1+\cdots+s_n \leqslant m} \binom{t_1}{s_1} \cdots \binom{t_n}{s_n} \binom{m - t_1 - \cdots - t_n}{m - s_1 - \cdots - s_n} f(x_1 + s_1 h_1, \ldots, x_n + s_n h_n). \quad (18)$$

Note that (18) is the Lagrange formula for this interpolation problem. Indeed, each function

$$\binom{t_1}{s_1} \cdots \binom{t_n}{s_n} \binom{m - t_1 - \cdots - t_n}{m - s_1 - \cdots - s_n} \quad (19)$$

is a polynomial in $t_1,\ldots,t_n$ of total degree m which vanishes at all points $(t_1,\ldots,t_n)$ with t_i nonnegative integers $0 \leqslant t_1 + \cdots + t_n \leqslant m$, except at the point $(s_1,\ldots,s_n)$, where it takes the value 1. In particular, for $n = 1$ we get the well-known univariate Lagrange polynomials

$$\ell_s(t) = \binom{t}{s} \binom{m - t}{m - s} = \prod_{\substack{0 \leqslant i \leqslant m,\\ i \neq s}} \frac{t - i}{s - i}$$

for $s = 0,\ldots,m$.

Salzer used these results in [34] to compute tables for the polynomials (18) and, some years later in [35], he studied in a similar form how to get the Lagrange formula for the more general case of formula (16), even starting with this formula. He obtained the multivariate Lagrange polynomials by a rather complicated expression involving the univariate ones.

It should be noted that several books related to computations and numerical methods published around this time include parts on multivariate interpolation to some extent, surprisingly, more than most of the recent textbooks in Numerical Analysis. We have already mentioned Steffensen's book [45], but we should also mention Whittaker and Robinson [51, pp. 371–374], Mikeladze [26, Chapter XVII] and especially Kunz [23, pp. 248–274], but also Isaacson and Keller [20, pp. 294–299] and Berezin and Zhidkov [2, pp. 156–194], although in any of them not really much more than in [45] is told.

In [36,37], Salzer introduced a concept of bivariate divided differences abandoning the idea of iteration for each variable x and y taken separately. Apparently, this was the first time (in spite of the similarity with (11)), that bivariate divided differences were explicitly defined for irregularly distributed sets of points. Divided differences with repeated arguments are also considered in [37] by coalescence of the ones with different arguments. Since [36] was just a first attempt of [37], we only explain the latter one. Salzer considers the set of monomials $\{x^i y^j\}$, with i,j nonnegative integers, ordered in a graded lexical term order, that is,

$$(i,j) < (h,k) \Leftrightarrow i + j < h + k \quad \text{or} \quad i + j = h + k,\ i > h. \quad (20)$$

Hence, the monomials are listed as

$$\{1, x, y, x^2, xy, y^2, x^3, \ldots\}. \quad (21)$$

For any set of $n + 1$ points (x_i, y_i), Salzer defines the associated divided difference

$$[01\ldots n]f := \sum_{k=0}^{n} A_k f(x_k, y_k), \quad (22)$$

choosing the coefficients A_k in such a form that (22) vanishes when f is any of the first n monomials of list (21) and takes the value 1 when f is the $(n+1)$st monomial of that list. In other words, the coefficients A_k are the solution of the linear system

$$\sum_{k=0}^{n} A_k x_k^i y_k^j = 0, \quad x^i y^j \text{ any of the first } n \text{ monomials of (21)},$$

$$\sum_{k=0}^{n} A_k x_k^i y_k^j = 1, \quad x^i y^j \text{ the } (n+1)\text{th monomial of (21).} \tag{23}$$

These generalized divided differences share some of the properties of the univariate ones but not all. Moreover, they have some limitations, for example, they exist only if the determinant of the coefficients in (23) is different from zero, and one has no control of that property in advance. On the other hand, observe that for example the simple divided difference with two arguments (x_0, y_0) and (x, y), which is

$$\frac{f(x,y) - f(x_0,y_0)}{x - x_0},$$

gives, when applied to $f(x,y) = xy$, the rational function

$$\frac{xy - x_0 y_0}{x - x_0}$$

and not a polynomial of lower degree. In fact, Salzer's divided differences did not have great success. Several other definitions of multivariate divided differences had appeared since then, trying to keep as many as possible of the good properties of univariate divided differences, cf. [16].

5. Reduction of a problem to other simpler ones

Around the 1950s an important change of paradigm happened in multivariate polynomial interpolation, as several people began to investigate more general distributions of points, and not only (special) subsets of Cartesian products. So, when studying cubature formulae [32], Radon observed the following in 1948: if a bivariate interpolation problem with respect to a set $T \subset \mathbb{R}^2$ of $\binom{k+2}{2}$ interpolation points is unisolvent in Π_k^2, and U is a set of $k+2$ points on an arbitrary straight line $\ell \subset \mathbb{R}^2$ such that $\ell \cap T = \emptyset$, then the interpolation problem with respect to $T \cup U$ is unisolvent in Π_{k+1}^2. Radon made use of this observation to build up point sets which give rise to unisolvent interpolation problems for Π_m recursively by degree. Clearly, these interpolation points immediately yield interpolatory cubature formulae.

The well-known Bézout theorem, cf. [50], states that two planar algebraic curves of degree m and n, with no common component, intersect each other at exactly mn points in an algebraic closure of the underlying field, counting multiplicities. This theorem has many interesting consequences for bivariate interpolation problems, extensible to higher dimensions. For example, no unisolvent interpolation problem in Π_n^2 can have more than $n+1$ collinear points. Radon's method in [32] is a consequence of this type of observations, and some other more recent results of different authors can also be deduced in a similar form, as we shall see later.

Another example of a result which shows the more general point of view taken in multivariate interpolation at that time is due to Thacher Jr. and Milne [47] (see also [48]). Consider two univariate interpolation problems in Π^1_{n-1}, with T_1, T_2 as respective sets of interpolation points, both of cardinality n. Assume that $T_1 \cap T_2$ has cardinality $n-1$, hence $T = T_1 \cup T_2$ has cardinality $n+1$. The univariate Aitken–Neville interpolation formula combines the solutions of the two smaller problems based on T_1 and T_2 to obtain the solution in Π^1_n of the interpolation problem with T as the underlying set of interpolation points. The main idea is to find a *partition of unity*, in this case affine polynomials ℓ_1, ℓ_2, i.e., $\ell_1 + \ell_2 = 1$, such that

$$\ell_1(T_2 \setminus T_1) = \ell_2(T_1 \setminus T_2) = 0$$

and then combine the solutions p_1, p_2 with respect to T_1, T_2, into the solution $\ell_1 p_1 + \ell_2 p_2$ with respect to T. This method was developed in the 1930s independently by Aitken and Neville with the goal to avoid the explicit use of divided differences in the computation of univariate Lagrange polynomial interpolants.

It was exactly this idea which Thatcher and Milne extended to the multivariate case in [47]. Let us sketch their approach in the bivariate case. For example, consider an interpolation problem with 10 interpolation points, namely, the set $T = \{(i,j): 0 \leqslant i+j \leqslant 3\}$, where i,j are nonnegative integers, and the interpolation space Π^2_3. The solution p_T of this problem is obtained in [47] from the solutions $p_{T_k} \in \Pi^2_2$, $k = 1,2,3$, of the 3 interpolation problems with respect to the six-point sets $T_k \subset T$, $k = 1,2,3$, where

$$T_1 = \{(i,j): 0 \leqslant i+j \leqslant 2\},$$

$$T_2 = \{(i,j): (i,j) \in T,\ i > 0\},$$

$$T_3 = \{(i,j): (i,j) \in T,\ j > 0\}.$$

Then,

$$p_T = \ell_1 p_{T_1} + \ell_2 p_{T_2} + \ell_3 p_{T_3},$$

where ℓ_k, $k = 1,2,3$ are appropriate polynomials of degree 1. In fact, in this case these polynomials are the barycentric coordinates relative to the simplex (0, 0), (3, 0), (0, 3) and thus a partition of unity. In [47] the problem is studied in d variables and in that case $d+1$ "small" problems, with respective interpolation sets T_k, $k = 1, \ldots, d$, with a simplicial structure (the analogue of the triangular grid), are used to obtain the solution of the full problem with $T = T_1 \cap \cdots \cap T_{d+1}$ as interpolation points.

In 1970, Guenter and Roetman [19], among other observations, made a very interesting remark, which connects to the Radon/Bézout context and deserves to be explained here. Let us consider a set T of $\binom{m+d}{d}$ points in $\mathbb{R}^d$, where exactly $\binom{m+d-1}{d-1}$ of these points lie on a hyperplane H. Then $T \setminus H$ consists of $\binom{m-1+d}{d}$ points. Let us denote by $\Pi^m_{d,H}$ the space of polynomials of Π^m_d with the variables restricted to H, which is isomorphic to Π^m_{d-1}. If the interpolation problems defined by the sets $T \setminus H$ and $T \cap H$ are unisolvent in the spaces Π^{m-1}_d and $\Pi^m_{d,H}$, respectively, then the interpolation problem defined by T is unisolvent in Π^m_d. In other words, the idea is to decompose, whenever possible, a problem of degree m and d variables into two simpler problems, one of degree m and $d-1$ variables and the other one with degree $m-1$ and d variables.

6. The finite element approach

In 1943, Courant [11] suggested a finite difference method applicable to boundary value problems arising from variational problems. It is considered one of the motivations of the finite element method, which emerged from the engineering literature along the 1950s. It is a variational method of approximation which makes use of the Rayleigh–Ritz–Galerkin technique. The method became very successful, with hundreds of technical papers published (see, e.g., the monograph [52]), even before its mathematical basis was completely understood at the end of the 1960s.

Involved in the process of the finite element method there are local polynomial interpolation problems, generally for polynomials of low degree, thus, with only few interpolation data. The global solution obtained by solving all the local interpolation problems is a piecewise polynomial of a certain regularity, depending on the amount and type of interpolation data in the common boundary between pieces. Some of the interest in multivariate polynomial interpolation along the 1960/1970s was due to this method. Among the most interesting mathematical papers of that time in Finite Elements, we can mention [53,5], see also the book [46] by Strang and Fix, but, in our opinion, the most relevant papers and book from the point of view of multivariate polynomial interpolation are due to Ciarlet et al., for example [7–9].

In 1972, Nicolaides [29,30] put the classical problem of interpolation on a simplicial grid of $\binom{m+d}{d}$ points of $\mathbb{R}^d$, regularly distributed, forming what he called a *principal lattice*, into the finite element context. He actually used barycentric coordinates for the Lagrange formula, and moreover gave the corresponding error representations, see also [7]. However, much of this material can already be found in [3]. In general, taking into account that these results appeared under different titles, in a different context and in journals not accessible everywhere, it is not so surprising any more, how often the basic facts on the interpolation problem with respect to the simplicial grid had been rediscovered.

7. Hermite problems

The use of partial or directional derivatives as interpolation data in the multivariate case had not received much attention prior to the finite element method, where they were frequently used. It seems natural to approach partial derivatives by coalescence, as in univariate Hermite interpolation problems. However, things are unfortunately much more complicated in several variables. As it was already pointed out by Salzer and Kimbro [39] in 1958, the Hermite interpolation problem based on the values of a bivariate function $f(x,y)$ at two distinct points $(x_1,y_1),(x_2,y_2)$ and on the values of the partial derivatives $\partial f/\partial x$, $\partial f/\partial x$ at each of these two points is not solvable in the space Π_2^2 for any choice of points, although the number of interpolation conditions coincides with the dimension of the desired interpolation space. Some years later, Ahlin [1] circumvented some of these problems by using a tensor product approach: k^2 derivatives $\partial^{p+q}f/\partial x^p \partial y^q$ with $0 \leqslant p,q \leqslant k-1$ are prescribed at the n^2 points of a Cartesian product. The interpolation space is the one spanned by $x^\alpha y^\beta$ with $0 \leqslant \alpha, \beta \leqslant nk-1$ and a formula for the solution is easily obtained.

We must mention that Salzer came back to bivariate interpolation problems with derivatives in [38] studying *hyperosculatory interpolation* over Cartesian grids, that is, interpolation problems where all partial derivatives of first and second order and the value of the function are known at the interpolation points. Salzer gave some special configurations of points which yield solvability of this type of interpolation problem in an appropriate polynomial space and also provided the corresponding remainder formulae.

Nowadays, Hermite and Hermite–Birkhoff interpolation problems have been studied much more systematically, see [16,25] for references.

8. Other approaches

In 1966, Coatmelec [10] studied the approximation of functions of several variables by linear operators, including interpolation operators. At the beginning of the paper, he only considered interpolation operators based on values of point evaluations of the function, but later he also used values of derivatives. In this framework he obtained some qualitative and quantitative results on the approximation order of polynomial interpolation. At the end of [10], Coatmelec also includes some examples in $\mathbb{R}^2$ of points which are distributed irregularly along lines: $n+1$ of the points on a line r_0, n of them on another line r_1, but not on r_0, and so on until 1 point is chosen on a line r_n but not on $r_0 \cup \cdots \cup r_{n-1}$. He then points out the unisolvence of the corresponding interpolation problem in Π_n^2 which is, in fact, again a consequence of Bézout's theorem as in [32].

In 1971, Glaeser [17] considers Lagrange interpolation in several variables from an abstract algebraic/analytic point of view and acknowledges the inconvenience of working with particular systems of interpolation points due to the possibility of the nonexistence of a solution, in contrast to the univariate case. This is due to the nonexistence of polynomial spaces of dimension $k > 1$ in more than one variable such that the Lagrange interpolation problem has a unique solution for any system of k interpolation points. In other words, there are no nontrivial Haar (or Chebychev) spaces any more for two and more variables, cf. [12] or [24]. In [17], polynomial spaces with dimension greater than the number of interpolation conditions are considered in order to overcome this problem. Glaeser investigated these *underdetermined* systems which he introduced as *interpolation schemes* in [17] and also studied the problem of how to particularize the affine space of all solutions of a given interpolation problem in order to obtain a unique solution. This selection process is done in such a way that it controls the variation of the solution when two systems of interpolation points are very "close" to each other, with the goal to obtain a continuous selection process.

Acknowledgements

1. We thank Carl de Boor for several references, in particular, for pointing out to us the paper [32] with the result mentioned at the beginning of Section 4, related to Bezout's theorem. We are also grateful the help of Elena Ausejo, from the group of History of Sciences of the University of Zaragoza, in the search of references.
2. M. Gasca has been partially supported by the Spanish Research Grant PB96 0730.
3. T. Sauer was supported by a DFG Heisenberg fellowship, Grant SA 627/6.

References

[1] A.C. Ahlin, A bivariate generalization of Hermite's interpolation formula, Math. Comput. 18 (1964) 264–273.
[2] I.S. Berezin, N.P. Zhidkov, Computing Methods, Addison-Wesley, Reading, MA, 1965 (Russian version in 1959).
[3] O. Biermann, Über näherungsweise Kubaturen, Monatshefte Math. Phys. 14 (1903) 211–225.
[4] O. Biermann, Vorlesungen über Mathematische Näherungsmethoden, Vieweg, Braunschweig, 1905.
[5] G. Birkhoff, M.H. Schultz, R.S. Varga, Piecewise Hermite interpolation in one and two variables with applications to partial differential equations, Numer. Math. 11 (1968) 232–256.
[6] W. Borchardt, Über eine Interpolationsformel für eine Art symmetrischer Funktionen und deren Anwendung, Abh. d. Preuß. Akad. d. Wiss. (1860) 1–20.
[7] P.G. Ciarlet, The Finite Element Method for Elliptic Problems, North-Holland, Amsterdam, 1978.
[8] P.G. Ciarlet, P.A. Raviart, General Lagrange and Hermite interpolation in R^n with applications to finite element methods, Arch. Rational Mech. Anal. 46 (1972) 178–199.
[9] P.G. Ciarlet, C. Wagschal, Multipoint Taylor formulas and applications to the finite element method, Numer. Math. 17 (1971) 84–100.
[10] C. Coatmelec, Approximation et interpolation des fonctions differentiables de plusieurs variables, Ann. Sci. Ecole Norm. Sup. 83 (1966) 271–341.
[11] R. Courant, Variational methods for the solution of problems of equilibrium and vibrations, Bull. Amer. Math. Soc. 49 (1943) 1–23.
[12] P.J. Davis, Interpolation and Approximation, Blaisdell, Walthan, MA, 1963 (2nd Edition, Dover, New York, 1975).
[13] Encyklopädie der mathematischen Wissenschaften, Teubner, Leipzig, pp. 1900–1904.
[14] Encyclopédie des Sciences Mathematiques, Gauthier-Villars, Paris, 1906.
[15] I.A. Ezrohi, General forms of the remainder terms of linear formulas in multidimensional approximate analysis I, II Mat. Sb. 38 (1956) 389–416 and 43 (1957) 9–28 (in Russian).
[16] M. Gasca, T. Sauer, Multivariate polynomial interpolation, Adv. Comput. Math., 12 (2000) 377–410.
[17] G. Glaeser, L'interpolation des fonctions differentiables de plusieurs variables, in: C.T.C. Wall (Ed.), Proceedings of Liverpool Singularities Symposium II, Lectures Notes in Mathematics, Vol. 209, Springer, Berlin, 1971, pp. 1–29.
[18] H.H. Goldstine, A History of Numerical Analysis from the 16th Through the 19th Century, Springer, Berlin, 1977.
[19] R.B. Guenter, E.L. Roetman, Some observations on interpolation in higher dimensions, Math. Comput. 24 (1970) 517–521.
[20] E. Isaacson, H.B. Keller, Analysis of Numerical Methods, Wiley, New York, 1966.
[21] C.G.J. Jacobi, Theoremata nova algebraica circa systema duarum aequationum inter duas variabiles propositarum, Crelle J. Reine Angew. Math. 14 (1835) 281–288.
[22] L. Kronecker, Über einige Interpolationsformeln für ganze Funktionen mehrerer Variabeln. Lecture at the academy of sciences, December 21, 1865, in: H. Hensel (Ed.), L. Kroneckers Werke, Vol. I, Teubner, Stuttgart, 1895, pp. 133–141. (reprinted by Chelsea, New York, 1968).
[23] K.S. Kunz, Numerical Analysis, McGraw-Hill, New York, 1957.
[24] G.G. Lorentz, Approximation of Funtions, Chelsea, New York, 1966.
[25] R. Lorentz, Multivariate Hermite interpolation by algebaic polynomials: a survey, J. Comput. Appl. Math. (2000) this volume.
[26] S.E. Mikeladze, Numerical Methods of Mathematical Analysis, Translated from Russian, Office of Tech. Services, Department of Commerce, Washington DC, pp. 521–531.
[27] S. Narumi, Some formulas in the theory of interpolation of many independent variables, Tohoku Math. J. 18 (1920) 309–321.
[28] L. Neder, Interpolationsformeln für Funktionene mehrerer Argumente, Skandinavisk Aktuarietidskrift (1926) 59.
[29] R.A. Nicolaides, On a class of finite elements generated by Lagrange interpolation, SIAM J. Numer. Anal. 9 (1972) 435–445.
[30] R.A. Nicolaides, On a class of finite elements generated by Lagrange interpolation II, SIAM J. Numer. Anal. 10 (1973) 182–189.
[31] K. Pearson, On the construction of tables and on interpolation, Vol. 2, Cambridge University Press, Cambridge, 1920.
[32] J. Radon, Zur mechanischen Kubatur, Monatshefte Math. Physik 52 (1948) 286–300.

[33] H.E. Salzer, Note on interpolation for a function of several variables, Bull. AMS 51 (1945) 279–280.
[34] H.E. Salzer, Table of coefficients for interpolting in functions of two variables, J. Math. Phys. 26 (1948) 294–305.
[35] H.E. Salzer, Note on multivariate interpolation for unequally spaced arguments with an application to double summation, J. SIAM 5 (1957) 254–262.
[36] H.E. Salzer, Some new divided difference algorithms for two variables, in: R.E. Langer (Ed.), On Numerical Approximation, University of Wisconsin Press, Madison, 1959, pp. 61–98.
[37] H.E. Salzer, Divided differences for functions of two variables for irregularly spaced arguments, Numer. Math. 6 (1964) 68–77.
[38] H.E. Salzer, Formulas for bivariate hyperosculatory interpolation, Math. Comput. 25 (1971) 119–133.
[39] H.E. Salzer, G.M. Kimbro, Tables for Bivariate Osculatory Interpolation over a Cartesian Grid, Convair Astronautics, 1958.
[40] A. Sard, Remainders: functions of several variables, Acta Math. 84 (1951) 319–346.
[41] A. Sard, Remainders as integrals of partial derivatives, Proc. Amer. Math. Soc. 3 (1952) 732–741.
[42] T. Sauer, Yuan Xu, On multivariate Lagrange interpolation, Math. Comput. 64 (1995) 1147–1170.
[43] T. Sauer, Yuan Xu, A case study in multivariate Lagrange interpolation, in: S.P. Singh (Ed.), Approximation Theory, Wavelets and Applications, Kluwer Academic Publishers, Dordrecht, 1995, pp. 443–452.
[44] D.D. Stancu, The remainder of certain linear approximation formulas in two variables, J. SIAM Numer. Anal. 1 (1964) 137–163.
[45] I.F. Steffensen, Interpolation, Chelsea, New York, 1927 (2nd Edition, 1950).
[46] G. Strang, G.J. Fix, An Analysis of the Finite Element Method, Prentice-Hall, Englewood Cliffs, NJ, 1973.
[47] H.C. Thacher Jr., W.E. Milne, Interpolation in several variables, J. SIAM 8 (1960) 33–42.
[48] H.C. Thacher Jr., Derivation of interpolation formulas in several independent variables, Ann. N.Y. Acad. Sci. 86 (1960) 758–775.
[49] T.N. Thiele, Interpolationsrechnung, Teubner, Leipzig, 1909.
[50] R.S. Walker, Algebraic Curves, Springer, Berlin, 1978.
[51] E.T. Whittaker, G. Robinson, Calculus of Observations, 4th Edition, Blackie and Sons, London, 1944.
[52] O.C. Zienkiewicz, The Finite Element Method in Structural and Continuum Mechanics, McGraw-Hill, London, 1967.
[53] M. Zlamal, On the finite element method, Numer. Math. 12 (1968) 394–409.

Numerical linear algebra algorithms and software

Jack J. Dongarra[a,b,*], Victor Eijkhout[a]

[a] *Department of Computer Science, University of Tennessee, 107 Ayres Hall, Knoxville, TN 37996-1301, USA*
[b] *Mathematical Sciences Section, Oak Ridge National Laboratory, P.O. Box 2008, Bldg. 6012, Oak Ridge, TN 37831-6367, USA*

Abstract

The increasing availability of advanced-architecture computers has a significant effect on all spheres of scientific computation, including algorithm research and software development in numerical linear algebra. Linear algebra – in particular, the solution of linear systems of equations – lies at the heart of most calculations in scientific computing. This paper discusses some of the recent developments in linear algebra designed to exploit these advanced-architecture computers. We discuss two broad classes of algorithms: those for dense, and those for sparse matrices.

1. Introduction

The increasing availability of advanced-architecture computers has a significant effect on all spheres of scientific computation, including algorithm research and software development in numerical linear algebra. Linear algebra – in particular, the solution of linear systems of equations – lies at the heart of most calculations in scientific computing. This article discusses some of the recent developments in linear algebra designed to exploit these advanced-architecture computers. We discuss two broad classes of algorithms: those for dense, and those for sparse matrices. A matrix is called sparse if it has a substantial number of zero elements, making specialized storage and algorithms necessary.

Much of the work in developing linear algebra software for advanced-architecture computers is motivated by the need to solve large problems on the fastest computers available. In this article, we focus on four basic issues: (1) the motivation for the work; (2) the development of standards

* Corresponding author.
E-mail address: dongarra@cs.utk.edu (J.J. Dongarra).

for use in linear algebra and the building blocks for libraries; (3) aspects of algorithm design and parallel implementation; and (4) future directions for research.

As representative examples of dense matrix routines, we will consider the Cholesky and LU factorizations, and these will be used to highlight the most important factors that must be considered in designing linear algebra software for advanced-architecture computers. We use these factorization routines for illustrative purposes not only because they are relatively simple, but also because of their importance in several scientific and engineering applications that make use of boundary element methods. These applications include electromagnetic scattering and computational fluid dynamics problems, as discussed in more detail in Section 2.1.2.

For the past 15 years or so, there has been a great deal of activity in the area of algorithms and software for solving linear algebra problems. The goal of achieving high performance on codes that are portable across platforms has largely been realized by the identification of linear algebra kernels, the basic linear algebra subprograms (BLAS). We will discuss the EISPACK, LINPACK, LAPACK, and ScaLAPACK libraries which are expressed in successive levels of the BLAS.

The key insight of our approach to designing linear algebra algorithms for advanced architecture computers is that the frequency with which data are moved between different levels of the memory hierarchy must be minimized in order to attain high performance. Thus, our main algorithmic approach for exploiting both vectorization and parallelism in our implementations is the use of block-partitioned algorithms, particularly in conjunction with highly tuned kernels for performing matrix–vector and matrix–matrix operations (the Level 2 and 3 BLAS).

2. Dense linear algebra algorithms

2.1. Overview of dense algorithms

Common operations involving dense matrices are the solution of linear systems

$$Ax = b,$$

the least-squares solution of over- or underdetermined systems

$$\min_x ||Ax - b||$$

and the computation of eigenvalues and -vectors

$$Ax = \lambda x.$$

Although these problems are formulated as matrix–vector equations, their solution involves a definite matrix–matrix component. For instance, in order to solve a linear system, the coefficient matrix is first factored as

$$A = LU$$

(or $A = U^{\mathrm{T}}U$ in the case of symmetry) where L and U are lower and upper triangular matrices, respectively. It is a common feature of these matrix–matrix operations that they take, on a matrix of size $n \times n$, a number of operations proportional to n^3, a factor n more than the number of data elements involved.

Thus, we are led to identify three levels of linear algebra operations:

- *Level* 1: vector–vector operations such as the update $y \leftarrow y + \alpha x$ and the inner product $d = x^T y$. These operations involve (for vectors of length n) $O(n)$ data and $O(n)$ operations.
- *Level* 2: matrix–vector operations such as the matrix–vector product $y = Ax$. These involve $O(n^2)$ operations on $O(n^2)$ data.
- *Level* 3: matrix–matrix operations such as the matrix–matrix product $C = AB$. These involve $O(n^3)$ operations on $O(n^2)$ data.

These three levels of operations have been realized in a software standards known as the basic linear algebra subprograms (BLAS) [17,18,46]. Although BLAS routines are freely available on the net, many computer vendors supply a tuned, often assembly coded, BLAS library optimized for their particular architecture, see also Section 4.3.

The relation between the number of operations and the amount of data is crucial for the performance of the algorithm. We discuss this in detail in Section 3.1.

2.1.1. Loop rearranging

The operations of BLAS levels 2 and 3 can be implemented using doubly and triply nested loops, respectively. With simply modifications, this means that for level 2 each algorithms has two, and for level 3 six different implementations [20]. For instance, solving a lower triangular system $Lx = y$ is mostly written

for $i = 1 \ldots n$
  $t = 0$
  for $j = 1 \ldots i-1$
    $t \leftarrow t + \ell_{ij} x_j$
  $x = \ell_{ii}^{-1}(y_i - t)$

but can also be written as

for $j = 1 \ldots n$
  $x_j = \ell_{jj}^{-1} y_j$
  for $i = j+1 \ldots n$
    $y_i \leftarrow y_i - \ell_{ij} x_j$

(The latter implementation overwrites the right-hand side vector y, but this can be eliminated.)

While the two implementations are equivalent in terms of number of operations, there may be substantial differences in performance due to architectural considerations. We note, for instance, that the inner loop in the first implementation uses a row of L, whereas the inner loop in the second traverses a column. Since matrices are usually stored with either rows or columns in contiguous locations, column storage the historical default inherited from the FORTRAN programming language, the performance of the two can be radically different. We discuss this point further in Section 3.1.

2.1.2. Uses of LU factorization in science and engineering

A major source of large dense linear systems is problems involving the solution of boundary integral equations [26]. These are integral equations defined on the boundary of a region of interest. All examples of practical interest compute some intermediate quantity on a two-dimensional boundary and then use this information to compute the final desired quantity in three-dimensional space. The price one pays for replacing three dimensions with two is that what started as a sparse problem in $O(n^3)$ variables is replaced by a dense problem in $O(n^2)$.

Dense systems of linear equations are found in numerous applications, including:

- airplane wing design;
- radar cross-section studies;
- flow around ships and other off-shore constructions;
- diffusion of solid bodies in a liquid;
- noise reduction; and
- diffusion of light through small particles.

The electromagnetics community is a major user of dense linear systems solvers. Of particular interest to this community is the solution of the so-called radar cross-section problem. In this problem, a signal of fixed frequency bounces off an object; the goal is to determine the intensity of the reflected signal in all possible directions. The underlying differential equation may vary, depending on the specific problem. In the design of stealth aircraft, the principal equation is the Helmholtz equation. To solve this equation, researchers use the *method of moments* [37,62]. In the case of fluid flow, the problem often involves solving the Laplace or Poisson equation. Here, the boundary integral solution is known as the *panel method* [38,39], so named from the quadrilaterals that discretize and approximate a structure such as an airplane. Generally, these methods are called *boundary element methods*.

Use of these methods produces a dense linear system of size $O(N) \times O(N)$, where N is the number of boundary points (or panels) being used. It is not unusual to see size $3N \times 3N$, because of three physical quantities of interest at every boundary element.

A typical approach to solving such systems is to use LU factorization. Each entry of the matrix is computed as an interaction of two boundary elements. Often, many integrals must be computed. In many instances, the time required to compute the matrix is considerably larger than the time for solution.

The builders of stealth technology who are interested in radar cross-sections are using direct Gaussian elimination methods for solving dense linear systems. These systems are always symmetric and complex, but not Hermitian.

For further information on various methods for solving large dense linear algebra problems that arise in computational fluid dynamics, see the report by Alan Edelman [26].

2.2. Block algorithms and their derivation

It is comparatively straightforward to recode many of the dense linear algebra algorithms so that they use level 2 BLAS. Indeed, in the simplest cases the same floating-point operations are done, possibly even in the same order: it is just a matter of reorganizing the software. To illustrate this point, we consider the Cholesky factorization algorithm, which factors a symmetric positive-definite matrix as $A = U^{\mathrm{T}}U$. We consider Cholesky factorization because the algorithm is simple, and no pivoting is required on a positive-definite matrix.

Suppose that after $j-1$ steps the block A_{00} in the upper left-hand corner of A has been factored as $A_{00} = U_{00}^{\mathrm{T}}U_{00}$. The next row and column of the factorization can then be computed by writing

$A = U^{\mathrm{T}}U$ as

$$\begin{pmatrix} A_{00} & b_j & A_{02} \\ . & a_{jj} & c_j^{\mathrm{T}} \\ . & . & A_{22} \end{pmatrix} = \begin{pmatrix} U_{00}^{\mathrm{T}} & 0 & 0 \\ v_j^{\mathrm{T}} & u_{jj} & 0 \\ U_{02}^{\mathrm{T}} & w_j & U_{22}^{\mathrm{T}} \end{pmatrix} \begin{pmatrix} U_{00} & v_j & U_{02} \\ 0 & u_{jj} & w_j^{\mathrm{T}} \\ 0 & 0 & U_{22} \end{pmatrix},$$

where b_j, c_j, v_j, and w_j are column vectors of length $j-1$, and a_{jj} and u_{jj} are scalars. Equating coefficients on the jth column, we obtain

$$b_j = U_{00}^{\mathrm{T}} v_j, \qquad a_{jj} = v_j^{\mathrm{T}} v_j + u_{jj}^2.$$

Since U_{00} has already been computed, we can compute v_j and u_{jj} from the equations

$$U_{00}^{\mathrm{T}} v_j = b_j, \qquad u_{jj}^2 = a_{jj} - v_j^{\mathrm{T}} v_j.$$

The computation of v_j is a triangular system solution, a BLAS level 2 operation. Thus, a code using this will have a single call replacing a loop of level 1 calls or a doubly nested loop of scalar operations.

This change by itself is sufficient to result in large gains in performance on a number of machines – for example, from 72 to 251 megaflops for a matrix of order 500 on one processor of a CRAY Y-MP. Since this is 81% of the peak speed of matrix–matrix multiplication on this processor, we cannot hope to do very much better by using level 3 BLAS.

We can, however, restructure the algorithm at a deeper level to exploit the faster speed of the level 3 BLAS. This restructuring involves recasting the algorithm as a *block algorithm* – that is, an algorithm that operates on *blocks* or submatrices of the original matrix.

2.2.1. Deriving a block algorithm

To derive a block form of Cholesky factorization, we partition the matrices as shown in Fig. 1, in which the diagonal blocks of A and U are square, but of differing sizes. We assume that the first block has already been factored as $A_{00} = U_{00}^{\mathrm{T}} U_{00}$, and that we now want to determine the second block column of U consisting of the blocks U_{01} and U_{11}. Equating submatrices in the second block of columns, we obtain

$$A_{01} = U_{00}^{\mathrm{T}} U_{01},$$

$$A_{11} = U_{01}^{\mathrm{T}} U_{01} + U_{11}^{\mathrm{T}} U_{11}.$$

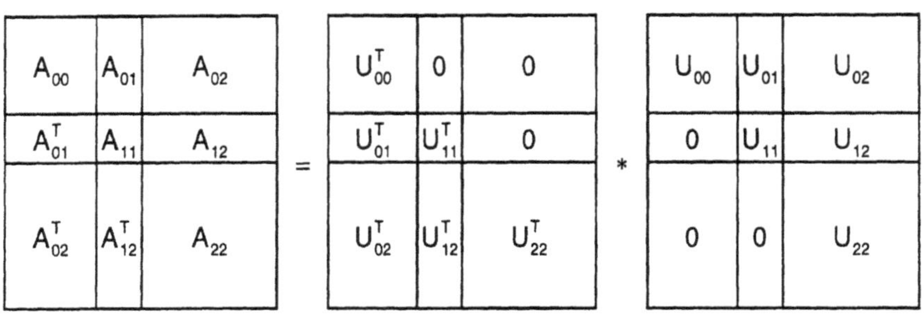

Fig. 1. Partitioning of A, U^{T}, and U into blocks. It is assumed that the first block has already been factored as $A_{00} = U_{00}^{\mathrm{T}} U_{00}$, and we next want to determine the block column consisting of U_{01} and U_{11}. Note that the diagonal blocks of A and U are square matrices.

Table 1
Speed (Megaflops) of Cholesky factorization $A = U^T U$ for $n = 500$

	CRAY T-90 1 proc.	CRAY T-90 4 proc.
j-variant: LINPACK	376	392
j-variant: using level 3 BLAS	1222	2306
i-variant: using level 3 BLAS	1297	3279

Hence, since U_{00} has already been computed, we can compute U_{01} as the solution to the equation

$$U_{00}^T U_{01} = A_{01}$$

by a call to the level 3 BLAS routine STRSM; and then we can compute U_{11} from

$$U_{11}^T U_{11} = A_{11} - U_{01}^T U_{01}.$$

This involves first updating the symmetric submatrix A_{11} by a call to the level 3 BLAS routine SSYRK, and then computing its Cholesky factorization. Since Fortran does not allow recursion, a separate routine must be called, using level 2 BLAS rather than level 3. In this way, successive blocks of columns of U are computed.

But that is not the end of the story, and the code given above is not the code actually used in the LAPACK routine SPOTRF. We mentioned earlier that for many linear algebra computations there are several algorithmic variants, often referred to as i-, j-, and k-variants, according to a convention introduced in [15,20] and explored further in [53,54]. The same is true of the corresponding block algorithms.

It turns out that the j-variant chosen for LINPACK, and used in the above examples, is not the fastest on many machines, because it performs most of the work in solving triangular systems of equations, which can be significantly slower than matrix–matrix multiplication. The variant actually used in LAPACK is the i-variant, which relies on matrix–matrix multiplication for most of the work.

Table 1 summarizes the results.

3. The influence of computer architecture on performance

3.1. Discussion of architectural features

In Section 2.1.1 we noted that for BLAS levels 2 and 3 several equivalent implementations of the operations exist. These differ, for instance, in whether they access a matrix operand by rows or columns in the inner loop. In FORTRAN, matrices are stored by columns, so accessing a column corresponds to accessing consecutive memory elements. On the other hand, as one proceeds across a row, the memory references jump across memory, the length of the jump being proportional to the length of a column.

We will now give a simplified discussion on the various architectural issues that influence the choice of algorithm. The following is, of necessity, a simplified account of the state of affairs for any particular architecture.

At first, we concentrate only on 'nonblocked' algorithms. In blocked methods, discussed in more detail below, every algorithm has two levels on which we can consider loop arranging: the block level, and the scalar level. Often, the best arrangement on one level is not the best on the other. The next two subsections concern themselves with the scalar level.

3.1.1. Using consecutive elements

The decision how to traverse matrix elements should usually be taken so as to use elements that are consecutive in storage. There are at least three architectural reasons for this.

Page swapping: By using consecutive memory elements, instead of ones at some stride distance of each other, the amount of memory page swapping is minimized.

Memory banks: If the processor cycle is faster than the memory cycle, and memory consists of interleaved banks, consecutive elements will be in different banks. By contrast, taking elements separated a distance equal to the number of banks, all elements will come from the same bank. This will reduce the effective performance of the algorithm to the memory speed instead of the processor speed.

Cache lines: Processors with a memory cache typically do not bring in single elements from memory to cache, but move them one 'cache line' at a time. A cache line consists of a small number of consecutive memory elements. Thus, using consecutive memory storage elements means that a next element will already be in cache and does not have to be brought into cache. This cuts down on memory traffic.

Whether consecutive elements correspond to rows or columns in a matrix depends on the programming language used. In Fortran, columns are stored consecutively, whereas C has row elements contiguous in memory.

The effects of column orientation are quite dramatic: on systems with virtual or cache memories, the LINPACK library codes (Section 4.4.2), which are written in FORTRAN and which are column-oriented, will significantly outperform FORTRAN codes that are not column-oriented. In the C language, however, algorithms should be formulated with row-orientation. We note that textbook examples of matrix algorithms are usually given in a row-oriented manner.

3.1.2. Cache reuse

In many contemporary architectures, memory bandwidth is not enough to keep the processor working at its peak rate. Therefore, the architecture incorporates some cache memory, a relatively small store of faster memory. The memory bandwidth problem is now shifted to bringing the elements into cache, and this problem can be obviated almost entirely if the algorithm can re-use cache elements.

Consider for instance a matrix–vector product $y = Ax$. The doubly nested loop has an inner statement

$$y_i \leftarrow y_i + a_{ij}a_j$$

implying three reads and one write from memory for two operations. If we write the algorithm as

$$y_* = x_1 a_{1*} + x_2 a_{2*} + \cdots,$$

we see that, keeping y in cache [1] and reusing the elements of x, we only need to load the column of A, making the asymptotic demand on memory one element load once x and y have been brought into cache.

3.1.3. Blocking for cache reuse

Above, we saw in the Cholesky example how algorithms can naturally be written in terms of level 2 operations. In order to use level 3 operations, a more drastic rewrite is needed.

Suppose we want to perform the matrix–matrix multiplication $C = AB$, where all matrices are of size $n \times n$. We divide all matrices in subblocks of size $k \times k$, and let for simplicity's sake k divide n: $n = km$. Then the triply nested scalar loop becomes, in one possible rearrangement

for $i = 1 \ldots m$
  for $k = 1 \ldots m$
    for $j = 1 \ldots m$
      $C_{ij} \leftarrow C_{ij} + A_{ik}B_{kj}$

where the inner statement is now a size k matrix–matrix multiplication.

If the cache is now large enough for three of these smaller matrices, we can keep C_{ij} and A_{ik} in cache, [2] while successive blocks B_{kj} are being brought in. The ratio of memory loads to operations is then (ignoring the loads of the elements of C and A, which is amortised) k^2/k^3, that is, $1/k$.

Thus, by blocking the algorithm, and arranging the loops so that blocks are reused in cache, we can achieve high performance in spite of a low-memory bandwidth.

3.2. Target architectures

The EISPACK and LINPACK software libraries were designed for supercomputers used in the 1970s and early 1980s, such as the CDC-7600, Cyber 205, and Cray-1. These machines featured multiple functional units pipelined for good performance [41]. The CDC-7600 was basically a high-performance scalar computer, while the Cyber 205 and Cray-1 were early vector computers.

The development of LAPACK in the late 1980s was intended to make the EISPACK and LINPACK libraries run efficiently on shared memory, vector supercomputers. The ScaLAPACK software library will extend the use of LAPACK to distributed memory concurrent supercomputers. The development of ScaLAPACK began in 1991 and is had its first public release of software by the end of 1994.

The underlying concept of both the LAPACK and ScaLAPACK libraries is the use of block-partitioned algorithms to minimize data movement between different levels in hierarchical memory. Thus, the ideas discussed in this chapter for developing a library for dense linear algebra computations are applicable to any computer with a hierarchical memory that (1) imposes a sufficiently large startup cost on the movement of data between different levels in the hierarchy, and for which (2) the cost of a context switch is too great to make fine grain size multithreading worthwhile. Our target machines are, therefore, medium and large grain size advanced-architecture computers. These

---

[1] Since many level-1 caches are write-through, we would not actually keep y in cache, but rather keep a number of elements of it in register, and reuse these registers by unrolling the '$*$' loop.

[2] Again, with a write-through level-1 cache, one would try to keep C_{ij} in registers.

include “traditional” shared memory, vector supercomputers, such as the Cray C-90 and T-90, and MIMD distributed memory concurrent supercomputers, such as the SGI Origin 2000, IBM SP, Cray T3E, and HP/Convex Exemplar concurrent systems.

Future advances in compiler and hardware technologies are expected to make multithreading a viable approach for masking communication costs. Since the blocks in a block-partitioned algorithm can be regarded as separate threads, our approach will still be applicable on machines that exploit medium and coarse grain size multithreading.

4. Dense linear algebra libraries

4.1. Requirements on high-quality, reusable, mathematical software

In developing a library of high-quality subroutines for dense linear algebra computations the design goals fall into three broad classes:

- performance,
- ease-of-use,
- range-of-use.

4.1.1. Performance

Two important performance metrics are *concurrent efficiency* and *scalability*. We seek good performance characteristics in our algorithms by eliminating, as much as possible, overhead due to load imbalance, data movement, and algorithm restructuring. The way the data are distributed (or decomposed) over the memory hierarchy of a computer is of fundamental importance to these factors. Concurrent efficiency, ε, is defined as the concurrent speedup per processor [32], where the concurrent speedup is the execution time, T_{seq}, for the best sequential algorithm running on one processor of the concurrent computer, divided by the execution time, T, of the parallel algorithm running on N_{p} processors. When direct methods are used, as in LU factorization, the concurrent efficiency depends on the problem size and the number of processors, so on a given parallel computer and for a fixed number of processors, the running time should not vary greatly for problems of the same size. Thus, we may write

$$\varepsilon(N, N_{\mathrm{p}}) = \frac{1}{N_{\mathrm{p}}} \frac{T_{\mathrm{seq}}(N)}{T(N, N_{\mathrm{p}})}, \tag{1}$$

where N represents the problem size. In dense linear algebra computations, the execution time is usually dominated by the floating-point operation count, so the concurrent efficiency is related to the performance, G, measured in floating-point operations per second by

$$G(N, N_{\mathrm{p}}) = \frac{N_{\mathrm{P}}}{t_{\mathrm{calc}}} \varepsilon(N, N_{\mathrm{p}}), \tag{2}$$

where t_{calc} is the time for floating-point operation. For iterative routines, such as eigensolvers, the number of iterations, and hence the execution time, depends not only on the problem size, but also on other characteristics of the input data, such as condition number. A parallel algorithm is said to be scalable [34] if the concurrent efficiency depends on the problem size and number of processors only through their ratio. This ratio is simply the problem size per processor, often referred to as the granularity. Thus, for a scalable algorithm, the concurrent efficiency is constant as the number of processors increases while keeping the granularity fixed. Alternatively, Eq. (2) shows that this is equivalent to saying that, for a scalable algorithm, the performance depends linearly on the number of processors for fixed granularity.

4.1.2. Ease-of-use

Ease-of-use is concerned with factors such as portability and the user interface to the library. Portability, in its most inclusive sense, means that the code is written in a standard language, such as Fortran or C, and that the source code can be compiled on an arbitrary machine to produce a program that will run correctly. We call this the "mail-order software" model of portability, since it reflects the model used by software servers such as *netlib* [19]. This notion of portability is quite demanding. It requires that all relevant properties of the computer's arithmetic and architecture be discovered at runtime within the confines of a compilable Fortran code. For example, if it is important to know the overflow threshold for scaling purposes, it must be determined at runtime *without overflowing*, since overflow is generally fatal. Such demands have resulted in quite large and sophisticated programs [24,44], which must be modified frequently to deal with new architectures and software releases. This "mail-order" notion of software portability also means that codes generally must be written for the worst possible machine expected to be used, thereby often degrading performances on all others. Ease-of-use is also enhanced if implementation details are largely hidden from the user, for example, through the use of an object-based interface to the library [22].

4.1.3. Range-of-use

Range-of-use may be gauged by how numerically stable the algorithms are over a range of input problems, and the range of data structures the library will support. For example, LINPACK and EISPACK deal with dense matrices stored in a rectangular array, packed matrices where only the upper- or lower-half of a symmetric matrix is stored, and banded matrices where only the nonzero bands are stored. In addition, some special formats such as Householder vectors are used internally to represent orthogonal matrices. In the second half of this paper we will focus on sparse matrices, that is matrices with many zero elements, which may be stored in many different ways.

4.2. Portability, scalability, and standards

Portability of programs has always been an important consideration. Portability was easy to achieve when there was a single architectural paradigm (the serial von Neumann machine) and a single programming language for scientific programming (Fortran) embodying that common model of computation. Architectural and linguistic diversity have made portability much more difficult, but no less important, to attain. Users simply do not wish to invest significant amounts of time to create large-scale application codes for each new machine. Our answer is to develop portable software libraries that hide machine-specific details.

In order to be truly portable, parallel software libraries must be *standardized*. In a parallel computing environment in which the higher-level routines and/or abstractions are built upon lower-level computation and message-passing routines, the benefits of standardization are particularly apparent.

Furthermore, the definition of computational and message-passing standards provides vendors with a clearly defined base set of routines that they can implement efficiently.

From the user's point of view, portability means that, as new machines are developed, they are simply added to network, supplying cycles where they are most appropriate.

From the mathematical software developer's point of view, portability may require significant effort. Economy in development and maintenance of mathematical software demands that such development effort be leveraged over as many different computer systems as possible. Given the great diversity of parallel architectures, this type of portability is attainable to only a limited degree, but machine dependences can at least be isolated.

LAPACK is an example of a mathematical software package whose highest-level components are portable, while machine dependences are hidden in lower-level modules. Such a hierarchical approach is probably the closest one can come to software portability across diverse parallel architectures. And the BLAS that are used so heavily in LAPACK provide a portable, efficient, and flexible standard for applications programmers.

Like portability, *scalabililty* demands that a program be reasonably effective over a wide range of number of processors. Maintaining scalability of parallel algorithms, and the software libraries implementing them, over a wide range of architectural designs and numbers of processors will likely require that the fundamental granularity of computation be adjustable to suit the particular circumstances in which the software may happen to execute. Our approach to this problem is block algorithms with adjustable block size. In many cases, however, polyalgorithms[3] may be required to deal with the full range of architectures and processor multiplicity likely to be available in the future.

Scalable parallel architectures of the future are likely to be based on a distributed memory architectural paradigm. In the longer term, progress in hardware development, operating systems, languages, compilers, and communications may make it possible for users to view such distributed architectures (without significant loss of efficiency) as having a shared memory with a global address space. For the near term, however, the distributed nature of the underlying hardware will continue to be visible at the programming level; therefore, efficient procedures for explicit communication will continue to be necessary. Given this fact, standards for basic message passing (send/receive), as well as higher-level communication constructs (global summation, broadcast, etc.), become essential to the development of scalable libraries that have any degree of portability. In addition to standardizing general communication primitives, it may also be advantageous to establish standards for problem-specific constructs in commonly occurring areas such as linear algebra.

The basic linear algebra communication subprograms (BLACS) [16,23] is a package that provides the same ease of use and portability for MIMD message-passing linear algebra communication that the BLAS [17,18,46] provide for linear algebra computation. Therefore, we recommend that future software for dense linear algebra on MIMD platforms consist of calls to the BLAS for computation and calls to the BLACS for communication. Since both packages will have been optimized for a particular platform, good performance should be achieved with relatively little effort. Also, since both packages will be available on a wide variety of machines, code modifications required to change platforms should be minimal.

---

[3] In a polyalgorithm the actual algorithm used depends on the computing environment and the input data. The optimal algorithm in a particular instance is automatically selected at runtime.

4.3. The BLAS as the key to portability

At least three factors affect the performance of compilable code:

1. *Vectorization/cache reuse*: Designing vectorizable algorithms in linear algebra is usually straightforward. Indeed, for many computations there are several variants, all vectorizable, but with different characteristics in performance (see, for example, [15]). Linear algebra algorithms can approach the peak performance of many machines – principally because peak performance depends on some form of chaining of vector addition and multiplication operations or cache reuse, and this is just what the algorithms require. However, when the algorithms are realized in straightforward Fortran77 or C code, the performance may fall well short of the expected level, usually because Fortran compilers fail to minimize the number of memory references – that is, the number of vector load and store operations or effectively reuse cache.
2. *Data movement*: What often limits the actual performance of a vector, or scalar, floating-point unit is the rate of transfer of data between different levels of memory in the machine. Examples include the transfer of vector operands in and out of vector registers, the transfer of scalar operands in and out of a high speed cache, the movement of data between main memory and a high-speed cache or local memory, paging between actual memory and disk storage in a virtual memory system, and interprocessor communication on a distributed memory concurrent computer.
3. *Parallelism*: The nested loop structure of most linear algebra algorithms offers considerable scope for loop-based parallelism. This is the principal type of parallelism that LAPACK and ScaLAPACK presently aim to exploit. On shared memory concurrent computers, this type of parallelism can sometimes be generated automatically by a compiler, but often requires the insertion of compiler directives. On distributed memory concurrent computers, data must be moved between processors. This is usually done by explicit calls to message passing routines, although parallel language extensions such as and Coherent Parallel C [30] and Split-C [13] do the message passing implicitly.

These issues can be controlled, while obtaining the levels of performance that machines can offer, through use of the BLAS, introduced in Section 2.1.

Level 1 BLAS are used in LAPACK, but for convenience rather than for performance: they perform an insignificant fraction of the computation, and they cannot achieve high efficiency on most modern supercomputers. Also, the overhead entailed in calling the BLAS reduces the efficiency of the code. This reduction is negligible for large matrices, but it can be quite significant for small matrices. Fortunately, level 1 BLAS can be removed from the smaller, more frequently used LAPACK codes in a short editing session.

Level 2 BLAS can achieve near-peak performance on many vector processors, such as a single processor of a CRAY X-MP or Y-MP, or Convex C-2 machine. However, on other vector processors such as a CRAY-2 or an IBM 3090 VF, the performance of level 2 BLAS is limited by the rate of data movement between different levels of memory.

Level 3 BLAS overcome this limitation. Level 3 of BLAS performs $O(n^3)$ floating-point operations on $O(n^2)$ data, whereas level 2 BLAS perform only $O(n^2)$ operations on $O(n^2)$ data. Level 3 BLAS also allow us to exploit parallelism in a way that is transparent to the software that calls them. While Level 2 BLAS offer some scope for exploiting parallelism, greater scope is provided by Level 3 BLAS, as Table 2 illustrates.

Table 2
Speed in Mflop/s of level 2 and level 3 BLAS operations on a CRAY C90 (all matrices are of order 1000; U is upper triangular)

Number of processors	1	2	4	8	16
Level 2: $y \leftarrow \alpha Ax + \beta y$	899	1780	3491	6783	11207
Level 3: $C \leftarrow \alpha AB + \beta C$	900	1800	3600	7199	14282
Level 2: $x \leftarrow Ux$	852	1620	3063	5554	6953
Level 3: $B \leftarrow UB$	900	1800	3574	7147	13281
Level 2: $x \leftarrow U^{-1}x$	802	1065	1452	1697	1558
Level 3: $B \leftarrow U^{-1}B$	896	1792	3578	7155	14009

The BLAS can provide portable high performance through being a standard that is available on many platforms. Ideally, the computer manufacturer has provided an assembly coded BLAS tuned for that particular architecture, but there is a standard implementation available that can simply be compiled and linked. Using this standard BLAS may improve the efficiency of programs when they are run on nonoptimizing compilers. This is because doubly subscripted array references in the inner loop of the algorithm are replaced by singly subscripted array references in the appropriate BLAS. The effect can be seen for matrices of quite small order, and for large orders the savings are quite significant.

4.4. Overview of dense linear algebra libraries

Over the past 25 years, we have been directly involved in the development of several important packages of dense linear algebra software: EISPACK, LINPACK, LAPACK, and the BLAS. Most recently, we have been involved in the development of ScaLAPACK, a scalable version of LAPACK for distributed memory concurrent computers. In this section, we give a brief review of these packages – their history, their advantages, and their limitations on high-performance computers.

4.4.1. EISPACK

EISPACK is a collection of Fortran subroutines that compute that eigenvalues and eigenvectors of nine classes of matrices: complex general, complex Hermitian, real general, real symmetric, real symmetric banded, real symmetric tridiagonal, special real tridiagonal, generalized real, and generalized real symmetric matrices. In addition, two routines are included that use singular value decomposition to solve certain least-squares problems.

EISPACK is primarily based on a collection of Algol procedures developed in the 1960s and collected by J.H. Wilkinson and C. Reinsch in a volume entitled *Linear Algebra* in the *Handbook for Automatic Computation* [64] series. This volume was not designed to cover every possible method of solution; rather, algorithms were chosen on the basis of their generality, elegance, accuracy, speed, or economy of storage.

Since the release of EISPACK in 1972, over 10 000 copies of the collection have been distributed worldwide.

4.4.2. LINPACK

LINPACK is a collection of Fortran subroutines that analyze and solve linear equations and linear least-squares problems. The package solves linear systems whose matrices are general, banded, symmetric indefinite, symmetric positive-definite, triangular, and tridiagonal square. In addition, the package computes the QR and singular-value decompositions of rectangular matrices and applies them to least-squares problems.

LINPACK is organized around four matrix factorizations: LU factorization, pivoted Cholesky factorization, QR factorization, and singular value decomposition. The term LU factorization is used here in a very general sense to mean the factorization of a square matrix into a lower triangular part and an upper triangular part, perhaps with pivoting. Next, we describe the organization and factors influencing LINPACK's efficiency.

LINPACK uses column-oriented algorithms to increase efficiency by preserving locality of reference. By column orientation we mean that the LINPACK codes always reference arrays down columns, not across rows. This works because Fortran stores arrays in column major order. This means that as one proceeds down a column of an array, the memory references proceed sequentially in memory. Thus, if a program references an item in a particular block, the next reference is likely to be in the same block. See further Section 3.1.1. LINPACK uses level 1 BLAS; see Section 4.3.

Since the release of LINPACK, over 20 000 copies of the collection have been distributed worldwide.

4.4.3. LAPACK

LAPACK [14] provides routines for solving systems of simultaneous linear equations, least-squares solutions of linear systems of equations, eigenvalue problems, and singular-value problems. The associated matrix factorizations (LU, Cholesky, QR, SVD, Schur, generalized Schur) are also provided, as are related computations such as reordering of the Schur factorizations and estimating condition numbers. Dense and banded matrices are handled, but not general sparse matrices. In all areas, similar functionality is provided for real and complex matrices, in both single and double precision.

The original goal of the LAPACK project was to make the widely used EISPACK and LINPACK libraries run efficiently on shared-memory vector and parallel processors. On these machines, LINPACK and EISPACK are inefficient because their memory access patterns disregard the multilayered memory hierarchies of the machines, thereby spending too much time moving data instead of doing useful floating-point operations. LAPACK addresses this problem by reorganizing the algorithms to use block matrix operations, such as matrix multiplication, in the innermost loops [2,14]. These block operations can be optimized for each architecture to account for the memory hierarchy [1], and so provide a transportable way to achieve high efficiency on diverse modern machines. Here, we use the term "transportable" instead of "portable" because, for fastest possible performance, LAPACK requires that highly optimized block matrix operations be already implemented on each machine. In other words, the correctness of the code is portable, but high performance is not – if we limit ourselves to a single Fortran source code.

LAPACK can be regarded as a successor to LINPACK and EISPACK. It has virtually all the capabilities of these two packages and much more besides. LAPACK improves on LINPACK and EISPACK in four main respects: speed, accuracy, robustness and functionality. While LINPACK and EISPACK are based on the vector operation kernels of level 1 BLAS, LAPACK was designed at the outset to exploit level 3 BLAS – a set of specifications for Fortran subprograms that do various types of matrix multiplication and the solution of triangular systems with multiple right-hand sides. Because of the coarse granularity of level 3 BLAS operations, their use tends to promote high efficiency on many high-performance computers, particularly if specially coded implementations are provided by the manufacturer.

LAPACK is designed to give high efficiency on vector processors, high-performance "superscalar" workstations, and shared memory multiprocessors. LAPACK in its present form is less likely to give good performance on other types of parallel architectures (for example, massively parallel SIMD machines, or MIMD distributed memory machines), but the ScaLAPACK project, described in Section 4.4.4, is intended to adapt LAPACK to these new architectures. LAPACK can also be used satisfactorily on all types of scalar machines (PCs, workstations, mainframes).

LAPACK, like LINPACK, provides LU and Cholesky factorizations of band matrices. The LINPACK algorithms can easily be restructured to use level 2 BLAS, though restructuring has little effect on performance for matrices of very narrow bandwidth. It is also possible to use level 3 BLAS, at the price of doing some extra work with zero elements outside the band [21]. This process becomes worthwhile for large matrices and semi-bandwidth greater than 100 or so.

4.4.4. ScaLAPACK

The ScaLAPACK software library extends the LAPACK library to run scalably on MIMD, distributed memory, concurrent computers [10,11]. For such machines the memory hierarchy includes the off-processor memory of other processors, in addition to the hierarchy of registers, cache, and local memory on each processor. Like LAPACK, the ScaLAPACK routines are based on block-partitioned algorithms in order to minimize the frequency of data movement between different levels of the memory hierarchy. The fundamental building blocks of the ScaLAPACK library are distributed memory versions of levels 2 and 3 BLAS, and a set of BLACS [16,23] for communication tasks that arise frequently in parallel linear algebra computations. In the ScaLAPACK routines, all interprocessor communication occurs within the distributed BLAS and BLACS, so the source code of the top software layer of ScaLAPACK looks very similar to that of LAPACK.

5. Future research directions in dense algorithms

Traditionally, large, general-purpose mathematical software libraries have required users to write their own programs that call library routines to solve specific subproblems that arise during a computation. Adapted to a shared-memory parallel environment, this conventional interface still offers some potential for hiding underlying complexity. For example, the LAPACK project incorporates parallelism in level 3 BLAS, where it is not directly visible to the user.

But when going from shared-memory systems to the more readily scalable distributed memory systems, the complexity of the distributed data structures required is more difficult to hide from the user. Not only must the problem decomposition and data layout be specified, but different phases of the user's problem may require transformations between different distributed data structures.

These deficiencies in the conventional user interface have prompted extensive discussion of alternative approaches for scalable parallel software libraries of the future. Possibilities include:

1. Traditional function library (i.e., minimum possible change to the status quo in going from serial to parallel environment). This will allow one to protect the programming investment that has been made.
2. Reactive servers on the network. A user would be able to send a computational problem to a server that was specialized in dealing with the problem. This fits well with the concepts of a networked, heterogeneous computing environment with various specialized hardware resources (or even the heterogeneous partitioning of a single homogeneous parallel machine).
3. General interactive environments like Matlab or Mathematica, perhaps with "expert" drivers (i.e., knowledge-based systems). With the growing popularity of the many integrated packages based on this idea, this approach would provide an interactive, graphical interface for specifying and solving scientific problems. Both the algorithms and data structures are hidden from the user, because the package itself is responsible for storing and retrieving the problem data in an efficient, distributed manner. In a heterogeneous networked environment, such interfaces could provide seamless access to computational engines that would be invoked selectively for different parts of the user's computation according to which machine is most appropriate for a particular subproblem.
4. Domain-specific problem solving environments, such as those for structural analysis. Environments like Matlab and Mathematica have proven to be especially attractive for rapid prototyping of new algorithms and systems that may subsequently be implemented in a more customized manner for higher performance.
5. Reusable templates (i.e., users adapt "source code" to their particular applications). A template is a description of a general algorithm rather than the executable object code or the source code more commonly found in a conventional software library. Nevertheless, although templates are general descriptions of key data structures, they offer whatever degree of customization the user may desire.

Novel user interfaces that hide the complexity of scalable parallelism will require new concepts and mechanisms for representing scientific computational problems and for specifying how those problems relate to each other. Very high level languages and systems, perhaps graphically based, not only would facilitate the use of mathematical software from the user's point of view, but also would help to automate the determination of effective partitioning, mapping, granularity, data structures, etc. However, new concepts in problem specification and representation may also require new mathematical research on the analytic, algebraic, and topological properties of problems (e.g., existence and uniqueness).

We have already begun work on developing such templates for sparse matrix computations. Future work will focus on extending the use of templates to dense matrix computations.

We hope the insight we gained from our work will influence future developers of hardware, compilers and systems software so that they provide tools to facilitate development of high quality portable numerical software.

The EISPACK, LINPACK, and LAPACK linear algebra libraries are in the public domain, and are available from *netlib*. For example, for more information on how to obtain LAPACK, send the following one-line email message to `netlib@ornl.gov`:

```
send index from lapack
```

or visit the web site at *http://www.netlib.org/lapack/*. Information for EISPACK, LINPACK, and ScaLAPACK can be similarly obtained.

6. Sparse linear algebra methods

6.1. Origin of sparse linear systems

The most common source of sparse linear systems is the numerical solution of partial differential equations. Many physical problems, such as fluid flow or elasticity, can be described by partial differential equations. These are implicit descriptions of a physical model, describing some internal relation such as stress forces. In order to arrive at an explicit description of the shape of the object or the temperature distribution, we need to solve the PDE, and for this we need numerical methods.

6.1.1. Discretized partial differential equations

Several methods for the numerical solution of PDEs exist, the most common ones being the methods of finite elements, finite differences, and finite volumes. A common feature of these is that they identify discrete points in the physical object, and give a set of equations relating these points.

Typically, only points that are physically close together are related to each other in this way. This gives a matrix structure with very few nonzero elements per row, and the nonzeros are often confined to a 'band' in the matrix.

6.1.2. Sparse matrix structure

Matrices from discretized partial differential equations contain so many zero elements that it pays to find a storage structure that avoids storing these zeros. The resulting memory savings, however, are offset by an increase in programming complexity, and by decreased efficiency of even simple operations such as the matrix–vector product.

More complicated operations, such as solving a linear system, with such a sparse matrix present a next level of complication, as both the inverse and the LU factorization of a sparse matrix are not as sparse, thus needing considerably more storage. Specifically, the inverse of the type of sparse matrix we are considering is a full matrix, and factoring such a sparse matrix fills in the band completely.

Example. Central differences in d dimensions, n points per line, matrix size $N = n^d$, bandwidth $q = n^{d-1}$ in natural ordering, number of nonzero $\sim n^d$, number of matrix elements $N^2 = n^{2d}$, number of elements in factorization $N^{1+(d-1)/d}$.

6.2. Basic elements in sparse linear algebra methods

Methods for sparse systems use, like those for dense systems, vector–vector, matrix–vector, and matrix–matrix operations. However, there are some important differences.

For iterative methods, discussed in Section 8, there are almost no matrix–matrix operations. See [43] for an exception. Since most modern architectures prefer these level 3 operations, the performance of iterative methods will be limited from the outset.

An even more serious objection is that the sparsity of the matrix implies that indirect addressing is used for retrieving elements. For example, in the popular row-compressed matrix storage format, the matrix–vector multiplication looks like

for $i = 1 \ldots n$
  $p \leftarrow$ pointer to row i
  for $j = 1, n_i$
    $y_i \leftarrow y_i + a(p + j)x(c(p + j))$

where n_i is the number of nonzeros in row i, and $p(\cdot)$ is an array of column indices. A number of such algorithms for several sparse data formats are given in [6].

Direct methods can have a BLAS 3 component if they are a type of dissection method. However, in a given sparse problem, the more dense the matrices are, the smaller they are on average. They are also not general full matrices, but only banded. Thus, we do not expect very high performance on such methods either.

7. Direct solution methods

For the solution of a linear system one needs to factor the coefficient matrix. Any direct method is a variant of Gaussian elimination. As remarked above, for a sparse matrix, this fills in the band in which the nonzero elements are contained. In order to minimize the storage needed for the factorization, research has focused on finding suitable orderings of the matrix. Re-ordering the equations by a symmetric permutation of the matrix does not change the numerical properties of the system in many cases, and it can potentially give large savings in storage. In general, direct methods do not make use of the numerical properties of the linear system, and thus their execution time is affected in a major way by the structural properties of the input matrix.

7.1. Matrix graph theory

The most convenient way of talking about matrix orderings or permutations is to consider the matrix 'graph' [55]. We introduce a node for every physical variable, and nodes i and j are connected in the graph if the (i,j) element of the matrix is nonzero. A symmetric permutation of the matrix then corresponds to a numbering of the nodes, while the connections stay the same. With these permutations, one hopes to reduce the 'bandwidth' of the matrix, and thereby the amount of fill generated by the factorization.

7.2. Cuthill–McKee ordering

A popular ordering strategy is the Cuthill–McKee ordering, which finds levels or wavefronts in the matrix graph. This algorithm is easily described:

1. Take any node as starting point, and call that 'level 0'.
2. Now successively take all nodes connected to the previous level, and group them into the next level.

3. Iterate this until all nodes are grouped into some level; the numbering inside each level is of secondary importance.

This ordering strategy often gives a smaller bandwidth than the natural ordering and there are further advantages to having a level structure, e.g., for out-of-core solution or for parallel processing. Often, one uses the 'reverse Cuthill–Mckee' ordering [50].

7.3. Minimum degree

An explicit reduction of bandwidth is effected by the minimum degree ordering, which at any point in the factorization chooses the variable with the smallest number of connections. Considering the size of the resulting fill-in is used as a tie breaker.

7.4. Nested dissection

Instead of trying to minimize fill-in by reducing the bandwidth, one could try a direct approach. The 'nested dissection' ordering recursively splits the matrix graph in two, thus separating it into disjoint subgraphs. Somewhat more precisely, given a graph, this algorithm relies on the existence of a 'separator': a set of nodes such that the other nodes fall into two mutually unconnected subgraphs. The fill from first factoring these subgraphs, followed by a factorization of the separator, is likely to be lower than for other orderings.

It can be shown that for PDEs in two space dimensions this method has a storage requirement that is within a log factor of that for the matrix itself, that is, very close to optimal [33]. This proof is easy for PDEs on rectangular grids, but with enough graph theory it can be generalized [48,49]. However, for problems in three space dimensions, the nested dissection method is no longer optimal.

An advantage of dissection-type methods is that they lead to large numbers of uncoupled matrix problems. Thus, to an extent, parallelization of such methods is easy. However, the higher levels in the tree quickly have fewer nodes than the number of available processors. In addition to this, they are also the larger subproblems in the algorithm, thereby complicating the parallelization of the method.

Another practical issue is the choice of the separator set. In a model case this is trivial, but in practice, and in particular in parallel, this is a serious problem, since the balancing of the two resulting subgraphs depends on this choice. Recently, the so-called 'second eigenvector methods' have become popular for this [56].

8. Iterative solution methods

Direct methods, as sketched above, have some pleasant properties. Foremost is the fact that their time to solution is predictable, either a priori, or after determining the matrix ordering. This is due to the fact that the method does not rely on numerical properties of the coefficient matrix, but only on its structure. On the other hand, the amount of fill can be substantial, and with it the execution time. For large-scale applications, the storage requirements for a realistic size problem can simply be prohibitive.

Iterative methods have far lower storage demands. Typically, the storage, and the cost per iteration with it, is of the order of the matrix storage. However, the number of iterations strongly depends on properties of the linear system, and is at best known up to an order estimate; for difficult problems the methods may not even converge due to accumulated round-off errors.

8.1. Basic iteration procedure

In its most informal sense, an iterative method in each iteration locates an approximation to the solution of the problem, measures the error between the approximation and the true solution, and based on the error measurement improves on the approximation by constructing a next iterate. This process repeats until the error measurement is deemed small enough.

8.2. Stationary iterative methods

The simplest iterative methods are the 'stationary iterative methods'. They are based on finding a matrix M that is, in some sense, 'close' to the coefficient matrix A. Instead of solving $Ax = b$, which is deemed computationally infeasible, we solve $Mx_1 = b$. The true measure of how well x_1 approximates x is the error $e_1 = x_1 - x$, but, since we do not know the true solution x, this quantity is not computable. Instead, we look at the 'residual': $r_1 = Ae_1 = Ax_1 - b$, which is a computable quantity. One easily sees that the true solution satisfies $x = A^{-1}b = x_1 - A^{-1}r_1$, so, replacing A^{-1} with M^{-1} in this relation, we define $x_2 = x_1 - M^{-1}r_1$.

Stationary methods are easily analyzed: we find that $r_i \to 0$ if all eigenvalues $\lambda = \lambda(I - AM^{-1})$ satisfy $|\lambda| < 1$. For certain classes of A and M this inequality is automatically satisfied [36,61].

8.3. Krylov space methods

The most popular class of iterative methods nowadays is that of 'Krylov space methods'. The basic idea there is to construct the residuals such that nth residual r_n is obtained from the first by multiplication by some polynomial in the coefficient matrix A, that is,

$$r_n = P_{n-1}(A)r_1.$$

The properties of the method then follow from the properties of the actual polynomial [3,7,9].

Most often, these iteration polynomials are chosen such that the residuals are orthogonal under some inner product. From this, one usually obtains some minimization property, though not necessarily a minimization of the *error*.

Since the iteration polynomials are of increasing degree, it is easy to see that the main operation in each iteration is one matrix–vector multiplication. Additionally, some vector operations, including inner products in the orthogonalization step, are needed.

8.3.1. The issue of symmetry

Krylov method residuals can be shown to satisfy the equation

$$r_n \in \text{span}\{Ar_{n-1}, r_{n-1}, \ldots, r_1\}.$$

This brings up the question whether all $r_{n-1},\dots,r_1$ need to be stored in order to compute r_n. The answer is that this depends on the symmetry of the coefficient matrix. For a symmetric problem, the r_n vectors satisfy a three-term recurrence. This was the original conjugate gradient method [40].

For nonsymmetric problems, on the other hand, no short recurrences can exist [29], and therefore, all previous residuals need to be stored. Some of these methods are OrthoDir and OrthoRes [65].

If the requirement of orthogonality is relaxed, one can derive short-recurrence methods for nonsymmetric problems [31]. In the biconjugate gradient method, two sequences $\{r_n\}$ and $\{s_n\}$ are derived that are mutually orthogonal, and that satisfy three-term recurrences.

A disadvantage of this latter method is that it needs application of the transpose of the coefficient matrix. In environments where the matrix is only operatively defined, this may exclude this method from consideration. Recently developed methods, mostly based on the work of [59,60], obviate this consideration.

8.3.2. True minimization

The methods mentioned so far minimize the error (over the subspace generated) in some matrix-related norm, but not in the Euclidean norm. We can effect a true minimization by collecting the residuals generated so far, and finding a minimizing convex combination. This leads to one of the most popular methods nowadays: GMRES [58]. It will always generate the optimal iterate, but for this it requires storage of all previous residuals. In practice, truncated or restarted version of GMRES are popular.

8.4. Preconditioners

The matrix M that appeared in the section on stationary iterative methods can play a role in Krylov space methods too. There, it is called a 'preconditioner', and it acts to improve spectral properties of the coefficient matrix that determine the convergence speed of the method. In a slight simplification, one might say that we replace the system $Ax = b$ by

$$(AM^{-1})(Mx) = b.$$

(Additionally, the inner product is typically changed.) It is generally recognized that a good preconditioner is crucial to the performance of an iterative method.

The requirements on a preconditioner are that it should be easy to construct, a system $Mx = b$ should be simple to solve, and in some sense M should be an approximation to A. These requirements need to be balanced: a more accurate preconditioner is usually harder to construct and more costly to apply, so any decrease in the number iterations has to be set against a longer time per iteration, plus an increased setup phase.

The holy grail of preconditioners is finding an 'optimal' preconditioner: one for which the number of operations for applying it is of the order of the number of variables, while the resulting number of iterations is bounded in the problem size. There are very few optimal preconditioners.

8.4.1. Simple preconditioners

Some preconditioners need no construction at all. For instance, the Jacobi preconditioner consists of simply the matrix diagonal D_A. Since in PDE applications the largest elements are on the diagonal, one expects some degree of accuracy from this. Using not just the diagonal, but the whole lower triangular part $D_A + L_A$ of the coefficient matrix, an even more accurate method results. Since this triangular matrix is nonsymmetric, it is usually balanced with the upper triangular part as $(D_A + L_A)D_A^{-1}(D_A + U_A)$.

8.4.2. *Incomplete factorizations*

A successful strategy for preconditioners results from mimicking direct methods, but applying some approximation process to them. Thus, the so-called 'incomplete factorization' methods ignore fill elements in the course of the Gaussian elimination process. Two strategies are to ignore elements in fixed positions, or to drop elements that are deemed small enough to be negligible. The aim is here to preserve at least some of the sparsity of the coefficient matrix in the factorization, while giving something that is close enough to the full factorization.

Incomplete factorizations can be very effective, but there are a few practical problems. For the class of M-matrices, these methods are well defined [52], but for other, even fairly common classes of matrices, there is a possibility that the algorithm breaks down [42,45,51].

Also, factorizations are inherently recursive, and coupled with the sparseness of the incomplete factorization, this gives very limited parallelism in the algorithm using a natural ordering of the unknowns. Different orderings may be more parallel, but take more iterations [25,27,43].

8.4.3. *Analytically inspired preconditioners*

In recent years, a number of preconditioners have gained in popularity that are more directly inspired by the continuous problem. First of all, for a matrix from an elliptic PDE, one can use a so-called 'fast solver' as preconditioner [12,28,63].

A particularly popular class of preconditioners based on the continuous problem, is that of 'domain decomposition' methods. If the continuous problem was elliptic, then decomposing the domain into simply connected pieces leads to elliptic problems on these subdomains, tied together by internal boundary conditions of some sort.

For instance, in the Schur complement domain decomposition method [8], thin strips of variables are assigned a function as interface region, and the original problem reduces to fully independent problems on the subdomains, connected by a system on the interface that is both smaller and better conditioned, but more dense, than the original one. While the subdomains can trivially be executed in parallel, the interface system poses considerable problems.

Choosing overlapping instead of separated subdomains leads to the class of Schwarz method [47]. The original Schwarz method on two domains proposed solving one subdomain, deriving interface conditions from it for the other subdomain, and solving the system there. Repetition of this process can be shown to converge. In a more parallel variant of this method, all subdomains solve their system simultaneously, and the solutions on the overlap regions are added together.

Multilevel methods do not operate by decomposing the domain. Rather, they work on a sequence of nested discretization, solving the coarser ones as a starting point for solving the finer levels. Under certain conditions such methods can be shown to be close to optimal [4,35]. However, they require explicit knowledge of the operator and boundary conditions. For this reason, people have investigated algebraic variants [5,57]. In both cases, these methods can be parallelised by distributing each level over the processors, but this may not be trivial.

9. Libraries and standards in sparse methods

Unlike in dense methods, there are few standards for iterative methods. Most of this is due to the fact that sparse storage is more complicated, admitting of more variation, and therefore less standardised. Whereas the (dense) BLAS has been accepted for a long time, sparse BLAS is not more than a proposal under research.

9.1. Storage formats

As is apparent from the matrix–vector example in Section 6.2, storage formats for sparse matrices include not just the matrix elements, but pointer information describing where the nonzero elements are placed in the matrix. A few storage formats are in common use (for more details see [6]):

Aij format: In the 'Aij' format, three arrays of the same length are allocated: one containing the matrix elements, and the other two containing the i and j coordinates of these elements. No particular ordering of the elements is implied.

Row/column-compressed: In the row-compressed format one array of integers is allocated in addition to the matrix element, giving the column indices of the nonzero elements. Since all elements in the same row are stored contiguously, a second, smaller, array is needed giving the start points of the rows in the two larger arrays.

Compressed diagonal: If the nonzero elements of the matrix are located, roughly or exactly, along subdiagonals, one could use contiguous storage for these diagonals. There are several diagonal storage formats. In the simplest, describing a contiguous block of subdiagonals, only the array of matrix elements is needed; two integers are sufficient to describe which diagonals have been stored.

There exist blocked versions of these formats, for matrices that can be partitioned into small square subblocks.

9.2. Sparse libraries

Since sparse formats are more complicated than dense matrix storage, sparse libraries have an added level of complexity. This holds even more so in the parallel case, where additional indexing information is needed to specify which matrix elements are on which processor.

There are two fundamentally different approaches for handling this complexity. Some sparse libraries require the user to set up the matrix and supply it to the library, while all handling is performed by the library. This requires the user to store data in a format dictated by the library, which might involve considerable work.

On the other hand, the library might do even the matrix setup internally, hiding all data from the user. This gives total freedom to the user, but it requires the library to supply sufficient access functions so that the user can perform certain matrix operations, even while not having access to the object itself.

References

[1] E. Anderson, J. Dongarra, Results from the initial release of LAPACK, Technical Report LAPACK Working Note 16, Computer Science Department, University of Tennessee, Knoxville, TN, 1989. `http://www.netlib.org/lapack/lawns/lawn16.ps`.

[2] E. Anderson, J. Dongarra, Evaluating block algorithm variants in LAPACK, Technical Report LAPACK Working Note 19, Computer Science Department, University of Tennessee, Knoxville, TN, 1990. http://www.netlib.org/lapack/lawns/lawn19.ps.
[3] O. Axelsson, A.V. Barker, Finite Element Solution of Boundary Value Problems, Theory and Computation, Academic Press, Orlando, FL, 1984.
[4] O. Axelsson, P. Vassilevski, Algebraic multilevel preconditioning methods, I, Numer. Math. 56 (1989) 157–177.
[5] O. Axelsson, V. Eijkhout, The nested recursive two-level factorization method for nine-point difference matrices, SIAM J. Sci. Statist. Comput. 12 (1991) 1373–1400.
[6] R. Barrett, M. Berry, T.F. Chan, J. Demmel, J. Donato, J. Dongarra, V. Eijkhout, R. Pozo, C. Romine, H. van der Vorst, Templates for the Solution of Linear Systems: Building Blocks for Iterative Methods, SIAM, Philadelphia, PA, 1994. http://www.netlib.org/templates/templates.ps.
[7] G. Birkhoff, R.E. Lynch, Numerical Solution of Elliptic Problems, SIAM, Philadelphia, PA, 1984.
[8] P. Bjørstad, O. Widlund, Iterative methods for the solution of elliptic problems on regions partitioned into substructures, SIAM J. Numer. Anal. 23 (1986) 1097–1120.
[9] T. Chan, Henk van der Vorst, Linear system solvers: sparse iterative methods, in: D. Keyes et al (Eds.), Parallel Numerical Algorithms, Proceedings of the ICASW/LaRC Workshop on Parallel Numerical Algorithms, May 23–25, 1994, Kluwer Academic Publishers, Dordrecht, The Netherlands, 1997, pp. 91–118.
[10] J. Choi, J.J. Dongarra, R. Pozo, D.W. Walker, Scalapack: a scalable linear algebra library for distributed memory concurrent computers, Proceedings of the Fourth Symposium on the Frontiers of Massively Parallel Computation, IEEE Computer Society Press, Silver Spring, MD, 1992, pp. 120–127.
[11] J. Choi, J.J. Dongarra, D.W. Walker, The design of scalable software libraries for distributed memory concurrent computers, in: J.J. Dongarra, B. Tourancheau (Eds.), Environments and Tools for Parallel Scientific Computing, Elsevier Science Publishers, Amsterdam, 1993.
[12] P. Concus, Gene H. Golub, Use of fast direct methods for the efficient numerical solution of nonseparable elliptic equations, SIAM J. Numer. Anal. 10 (1973) 1103–1120.
[13] D.E. Culler, A. Dusseau, S.C. Goldstein, A. Krishnamurthy, S. Lumetta, T. von Eicken, K. Yelick, Introduction to Split-C: Version 0.9, Technical Report, Computer Science Division – EECS, University of California, Berkeley, CA 94720, February 1993.
[14] J. Demmel, LAPACK: a portable linear algebra library for supercomputers, Proceedings of the 1989 IEEE Control Systems Society Workshop on Computer-Aided Control System Design, December 1989.
[15] J.J. Dongarra, Increasing the performance of mathematical software through high-level modularity, Proceedings of the Sixth International Symposium Comp. Methods in Engineering & Applied Sciences, Versailles, France, North-Holland, Amsterdam, 1984, pp. 239–248.
[16] J.J. Dongarra, LAPACK Working Note 34: Workshop on the BLACS, Computer Science Department, Technical Report CS-91-134, University of Tennessee, Knoxville, TN, May 1991. http://www.netlib.org/lapack/lawns/lawn16.ps.
[17] J.J. Dongarra, J. Du Croz, S. Hammarling, I. Duff, A set of level 3 basic linear algebra subprograms, ACM Trans. Math. Software 16 (1) (1990) 1–17.
[18] J.J. Dongarra, J. Du Croz, S. Hammarling, R. Hanson, An extended set of Fortran basic linear algebra subroutines, ACM Trans. Math. Software 14 (1) (1988) 1–17.
[19] J.J. Dongarra, E. Grosse, Distribution of mathematical software via electronic mail, Comm. ACM 30 (5) (1987) 403–407.
[20] J.J. Dongarra, F.C. Gustavson, A. Karp, Implementing linear algebra algorithms for dense matrices on a vector pipeline machine, SIAM Rev. 26 (1984) 91–112.
[21] J.J. Dongarra, P. Mayes, Giuseppe Radicati di Brozolo, The IBM RISC System/6000 and linear algebra operations, Supercomputer 44 (VIII-4) (1991) 15–30.
[22] J.J. Dongarra, R. Pozo, D.W. Walker, An object oriented design for high performance linear algebra on distributed memory architectures, Proceedings of the Object Oriented Numerics Conference, 1993.
[23] J.J. Dongarra, R.A. van de Geijn, Two-dimensional basic linear algebra communication subprograms, Technical Report LAPACK Working Note 37, Computer Science Department, University of Tennessee, Knoxville, TN, October 1991.

[24] J. Du Croz, M. Pont, The development of a floating-point validation package, in: M.J. Irwin, R. Stefanelli (Eds.), Proceedings of the Eighth Symposium on Computer Arithmetic, Como, Italy, May 19–21, 1987, IEEE Computer Society Press, Silver Spring, MD, 1987.
[25] I.S. Duff, G.A. Meurant, The effect of ordering on preconditioned conjugate gradients, BIT 29 (1989) 635–657.
[26] A. Edelman, Large dense numerical linear algebra in 1993: the parallel computing influence, Int. J. Supercomput. Appl. 7 (1993) 113–128.
[27] V. Eijkhout, Analysis of parallel incomplete point factorizations, Linear Algebra Appl. 154–156 (1991) 723–740.
[28] H.C. Elman, M.H. Schultz, Preconditioning by fast direct methods for non self-adjoint nonseparable elliptic equations, SIAM J. Numer. Anal. 23 (1986) 44–57.
[29] V. Faber, T. Manteuffel, Orthogonal error methods, SIAM J. Numer. Anal. 24 (1987) 170–187.
[30] E.W. Felten, S.W. Otto, Coherent parallel C, in: G.C. Fox (Ed.), Proceedings of the Third Conference on Hypercube Concurrent Computers and Applications, ACM Press, New York, 1988, pp. 440–450.
[31] R. Fletcher, Conjugate gradient methods for indefinite systems, in: G.A. Watson (Ed.), Numerical Analysis Dundee, 1975, Springer, New York, 1976, pp. 73–89.
[32] G.C. Fox, M.A. Johnson, G.A. Lyzenga, S.W. Otto, J.K. Salmon, D.W. Walker, Solving Problems on Concurrent Processors, Vol. 1, Prentice-Hall, Englewood Cliffs, NJ, 1988.
[33] A. George, H.-W. Liu, Computer Solution of Large Sparse Positive Definite Systems, Prentice-Hall, Englewood Cliffs, NJ, 1981.
[34] A. Gupta, V. Kumar, On the scalability of FFT on parallel computers, Proceedings of the Frontiers 90 Conference on Massively Parallel Computation, IEEE Computer Society Press, 1990. Also available as Technical Report TR 90-20 from the Computer Science Department, University of Minnesota, Minneapolis, MN 55455.
[35] W. Hackbusch, Multi-Grid Methods and Applications, Springer, Berlin, 1985.
[36] L.A. Hageman, D.M. Young, Applied Iterative Methods, Academic Press, New York, 1981.
[37] W. Croswell, Origin and development of the method of moments for field computation, IEEE Antennas Propagation Mag. 32 (1990) 31–34.
[38] J.L. Hess, Panel methods in computational fluid dynamics, Annu. Rev. Fluid Mech. 22 (1990) 255–274.
[39] J.L. Hess, M.O. Smith, Calculation of potential flows about arbitrary bodies, in: D. Küchemann (Ed.), Progress in Aeronautical Sciences, Vol. 8, Pergamon Press, Oxford, 1967.
[40] M.R. Hestenes, E. Stiefel, Methods of conjugate gradients for solving linear systems, Nat. Bur. Stand. J. Res. 49 (1952) 409–436.
[41] R.W. Hockney, C.R. Jesshope, Parallel Computers, Adam Hilger, Bristol, UK, 1981.
[42] A. Jennings, G.M. Malik, Partial elimination, J. Inst. Math. Appl. 20 (1977) 307–316.
[43] M.T. Jones, P.E. Plassmann, Parallel solution of unstructured, sparse systems of linear equations, in: R.F. Sincovec, D.E. Keyes, M.R. Leuze, L.R. Petzold, D.A. Reed (Eds.), Proceedings of the Sixth SIAM Conference on Parallel Processing for Scientific Computing, SIAM, Philadelphia, PA, pp. 471–475.
[44] W. Kahan, Paranoia, Available from netlib [19]: `http://www.netlib.org/paranoia`.
[45] D.S. Kershaw, The incomplete cholesky-conjugate gradient method for the iterative solution of systems of linear equations, J. Comput. Phys. 26 (1978) 43–65.
[46] C. Lawson, R. Hanson, D. Kincaid, F. Krogh, Basic linear algebra subprograms for Fortran usage, ACM Trans. Math. Software 5 (1979) 308–323.
[47] P.L. Lions, On the Schwarz alternating method. i., in: R. Glowinski, G.H. Golub, G. Meurant, J. Periaux (Eds.), Domain Decomposition Methods for Partial Differential Equations, Proceedings of the First Internation Symposium, Paris, January 7–9, 1987, SIAM, Philadelphia, PA, 1988, pp. 1–42.
[48] R.J. Lipton, D.J. Rose, R.E. Tarjan, Generalized nested dissection, SIAM J. Numer. Anal. 16 (1979) 346–358.
[49] R.J. Lipton, R.E. Tarjan, A separator theorem for planar graphs, SIAM J. Appl. Math. 36 (1979) 177–189.
[50] J.W-H. Liu, A.H. Sherman, Comparative analysis of the Cuthill–McKee and the reverse Cuthill–McKee ordering algorithms for sparse matrices, SIAM J. Numer. Anal. 13 (1973) 198–213.
[51] T.A. Manteuffel, An incomplete factorization technique for positive definite linear systems, Math. Comp. 34 (1980) 473–497.
[52] J.A. Meijerink, H.A. van der Vorst, An iterative solution method for linear systems of which the coefficient matrix is a symmetric m-matrix, Math. Comp. 31 (1977) 148–162.
[53] J.M. Ortega, The *ijk* forms of factorization methods I, Vector computers, Parallel Comput. 7 (1988) 135–147.

[54] J.M. Ortega, C.H. Romine, The *ijk* forms of factorization methods II, Parallel systems, Parallel Comput. 7 (1988) 149–162.

[55] S.V. Parter, The use of linear graphs in Gaussian elimination, SIAM Rev. 3 (1961) 119–130.

[56] A. Pothen, H.D. Simon, Kang-Pu Liou, Partitioning sparse matrices with eigenvectors of graphs, SIAM J. Matrix Anal. Appl. 11 (3) (1990) 430–452.

[57] J.W. Ruge, K. Stüben, Algebraic multigrid, in: S.F. McCormick (Ed.), Multigrid Methods, SIAM, Philadelphia, PA, 1987, (Chapter 4).

[58] Y. Saad, M.H. Schultz, GMRes: a generalized minimal residual algorithm for solving nonsymmetric linear systems, SIAM J. Sci. Statist. Comput. 7 (1986) 856–869.

[59] P. Sonneveld, CGS, a fast Lanczos-type solver for nonsymmetric linear systems, SIAM J. Sci. Statist. Comput. 10 (1989) 36–52.

[60] H. van der Vorst, Bi-CGSTAB: a fast and smoothly converging variant of Bi-CG for the solution of nonsymmetric linear systems, SIAM J. Sci. Statist. Comput. 13 (1992) 631–644.

[61] R.S. Varga, Matrix Iterative Analysis, Prentice-Hall, Englewood Cliffs, NJ, 1962.

[62] J.J.H. Wang, Generalized Moment Methods in Electromagnetics, Wiley, New York, 1991.

[63] O. Widlund, On the use of fast methods for separable finite difference equations for the solution of general elliptic problems, in: D.J. Rose, R.A. Willoughby (Eds.), Sparse Matrices and their Applications, Plenum Press, New York, 1972, pp. 121–134.

[64] J. Wilkinson, C. Reinsch, Handbook for Automatic Computation: Vol. II – Linear Algebra, Springer, New York, 1971.

[65] D.M. Young, K.C. Jea, Generalized conjugate-gradient acceleration of nonsymmetrizable iterative methods, Linear Algebra Appl. 34 (1980) 159–194.

Iterative solution of linear systems in the 20th century

Yousef Saad[a,1], Henk A. van der Vorst[b,*]
[a] *Department of Computer Science and Engineering, University of Minnesota, Minneapolis, USA*
[b] *Department of Mathematics, Utrecht University, P.O. Box 80.010, 3508 TA Utrecht, Netherlands*

Abstract

This paper sketches the main research developments in the area of iterative methods for solving linear systems during the 20th century. Although iterative methods for solving linear systems find their origin in the early 19th century (work by Gauss), the field has seen an explosion of activity spurred by demand due to extraordinary technological advances in engineering and sciences. The past five decades have been particularly rich in new developments, ending with the availability of large toolbox of specialized algorithms for solving the very large problems which arise in scientific and industrial computational models. As in any other scientific area, research in iterative methods has been a journey characterized by a chain of contributions building on each other. It is the aim of this paper not only to sketch the most significant of these contributions during the past century, but also to relate them to one another.

Keywords: ADI; Krylov subspace methods; Multigrid; Polynomial acceleration; Preconditioning; Relaxation methods; SOR; Sparse approximate inverse

1. Introduction

Numerical linear algebra is an exciting field of research and much of this research has been triggered by a problem that can be posed simply as: given $A \in \mathscr{C}^{m\times n}$, $b \in \mathscr{C}^m$, find solution vector(s) $x \in \mathscr{C}^n$ such that $Ax = b$. Many scientific problems lead to the requirement to solve linear systems of equations as part of the computations. From a pure mathematical point of view, this problem can be considered as being solved in the sense that we explicitly know its solution in terms of determinants. The actual computation of the solution(s) may however lead to severe complications, when carried out in finite precision and when each basic arithmetic operation takes finite time. Even

* Corresponding author.
E-mail addresses: saad@cs.umn.edu (Y. Saad), vorst@math.uu.nl (H.A. van der Vorst).
[1] Work supported by NSF/CCR and by the Minnesota Supercomputer Institute.

the "simple" case when $n = m$ and A is nonsingular, which is a trivial problem from a mathematical point of view, may become very complicated, from a computational point of view, and may even turn out to be impossible.

The traditional way to solve a nonsingular linear system is to employ Gaussian elimination, and, with all its enhancements, to overcome numerical instabilities. This process can be carried out in $\mathcal{O}(n^3)$ basic floating point operations (additions and multiplications, assuming $n = m$). Many applications lead to linear systems with a large n (where the notion of "large" depends, of course, on the capacity of the available computer), and it became soon evident that one has to exploit specific properties of the A at hand in order to make solution of the system feasible. This has led to variants of Gaussian elimination in which the nonzero structure of A is exploited, so that multiplications with zero result are avoided and that savings in computer storage could be realized.

Another direction of approach was based on the solution of a nearby linear system, with a matrix that admits a computationally inexpensive process (in terms of computing time and computer storage), and to embed this in an iterative process. Both approaches aim at making the impossible possible, and for the novice in this field this may seem to be just a collection of clever programming tricks: "in principle solving the problem is well understood but one has to be well organized to make the computational process a little faster". For this novice it will certainly come as a big surprise that a whole, still incomplete, mathematical framework had to be developed with deep and elegant results. As a result, relevant systems could be solved many orders of magnitude faster (and also often more accurate) than by a straightforward Gaussian elimination approach. In this paper, we will sketch the developments and progress that has taken place in the 20th century with respect to iterative methods alone. As will be clear, this subfield could not evolve in isolation, and the distinction between iterative methods and Gaussian elimination methods is sometimes artificial – and overlap between the two methodologies is significant in many instances. Nevertheless, each of the two has its own dynamics and it may be of interest to follow one of them more closely.

It is likely that future researchers in numerical methods will regard the decade just passed as the beginning of an era in which iterative methods for solving large linear systems of equations started gaining considerable acceptance in real-life industrial applications. In looking at past literature, it is interesting to observe that iterative and direct methods have often been in competition for solving large systems that arise in applications. A particular discovery will promote a given method from one camp only to see another discovery promote a competing method from the other camp. For example, the 1950s and 1960s saw an enormous interest in relaxation-type methods – prompted by the studies on optimal relaxation and the work by Young, Varga, Southwell, Frankel and others. A little later, sparse direct methods appeared that were very competitive – both from the point of view of robustness and computational cost. To this day, there are still applications dominated by direct solvers and others dominated by iterative solvers. Because of the high memory requirement of direct solvers, it was sometimes thought that these would eventually be replaced by iterative solvers, in all applications. However, the superior robustness of direct solvers prevented this. As computers have become faster, very large problems are routinely solved by methods from both camps.

Iterative methods were, even halfway in the 20th century, not always viewed as promising. For instance, Bodewig [23, p. 153], in 1956, mentioned the following drawbacks of iterative methods: nearly always too slow (except when the matrix approaches a diagonal matrix), for most problems they do not converge at all, they *cannot easily be mechanised*[2] and so they are *more appropriate for computing by hand than with machines*, and *do not take advantage of the situation when the equations are symmetric*. The only potential advantage seen was the observation that *Rounding errors do not accumulate, they are restricted to the last operation*. It is noteworthy that Lanczos' method was classified as a direct method in 1956.

The penetration of iterative solvers into applications has been a slow process that is still ongoing. At the time of this writing for example, there are applications in structural engineering as well as in circuit simulation, which are dominated by direct solvers.

This review will attempt to highlight the main developments in iterative methods over the past century. It is clear that a few pages cannot cover an exhaustive survey of 100 years of rich developments. Therefore, we will emphasize the ideas that were successful and had a significant impact.

Among the sources we used for our short survey, we would like to mention just a few that are notable for their completeness or for representing the thinking of a particular era. The books by Varga [188] and Young [205] give a complete treatise of iterative methods as they were used in the 1960s and 1970s. Varga's book has several excellent historical references. These two masterpieces remained the handbooks used by academics and practitioners alike for three decades. Householder's book [102] contains a fairly good overview of iterative methods – specifically oriented towards projection methods. Among the surveys we note the outstanding booklet published by the National Bureau of Standards in 1959 which contains articles by Rutishauser [150], Engeli [68] and Stiefel [170]. Later Birkhoff [21], who supervised David Young's Ph.D. thesis in the late 1940s, wrote an excellent historical perspective on the use of iterative methods as he experienced them himself from 1930 to 1980. The more recent literature includes the books by Axelsson [7], Brezinski [29], Greenbaum [88], Hackbusch [97], and Saad [157], each of which has a slightly different perspective and emphasis.

2. The quest for fast solvers: a historical perspective

Iterative methods have traditionally been used for the solution of large linear systems with diagonally dominant sparse matrices. For such systems the methods of Gauss–Jacobi and Gauss–Seidel could be used with some success, not so much because of the reduction in computational work, but mainly because of the limited amount of memory that is required. Of course, reduction of the computational work was also a serious concern, and this led Jacobi (1846) to apply plane rotations to the matrix in order to force stronger diagonal dominance, giving up sparsity. Jacobi had to solve many similar systems in the context of eigenvalue computations; his linear systems were rather small: of order 7.

In this century, simple iterative methods were predominantly applied for solving discretized elliptic self-adjoint partial differential equations, together with a local parameter for accelerating the iteration process. The first and simplest of these methods in Richardson's method [146]. Actually, this method

[2] This remark was removed from the second edition (in 1959); instead Bodewig included a small section on methods for automatic machines [24, Chapter 9]. The earlier remark was not as puzzling as it may seem now, in view of the very small memories of the available electronic computers at the time. This made it necessary to store intermediate data on punched cards. It required a regular flow of the computational process, making it cumbersome to include techniques with row interchanging.

was later viewed as a polynomial method and many authors have sought to optimize it by selecting its parameters so that the iteration polynomials became the Chebyshev polynomials; this was work done in the period 1950–1960 by Young, Lanczos and others. In the second half of this decade it became apparent that using the explicit three-term recurrence relation between Chebyshev polynomials, which led to three-term recurrence iteration methods (rather than the classical methods that are two-term iterations), were numerically superior in terms of stability [87].

The acceleration of the slightly more difficult to analyze Gauss–Seidel method led to point successive overrelaxation techniques introduced simultaneously by Frankel [78] and by Young [203]. It was shown, for rather simple Dirichlet problems, that a suitably chosen relaxation parameter could lead to drastic improvements in convergence. Young showed that these improvements could be expected for a larger class of matrices, characterized by his *property A* [203]. Successive overrelaxation methods, and numerous variants, became extremely popular and were the methods of choice in computer codes for large practical problems, such as nuclear reactor diffusion, oil reservoir modeling and weather prediction. Although their popularity has been overshadowed later, around after 1980, by more powerful techniques, they are still used in some applications either as the main iterative solution method or in combination with recent techniques (e.g. as smoothers for multigrid or as preconditioners for Krylov methods). The successive over-relaxation (SOR) methods made it possible to solve efficiently systems within the order of 20,000 unknowns by 1960 [188], and by 1965 systems of the order of 100,000 could be solved in problems related to eigenvalue computations in nuclear diffusion codes. The success of the SOR methods has led to a rich theory for iterative methods; this could be used fruitfully for the analysis of later methods as well. In particular, many methods, including SOR, could be viewed as simple Richardson iterations for specific splittings of the matrix of the linear system.

In 1955, Peaceman and Rachford [141] suggested a splitting that was motivated by the observation that the matrix for a three-point finite difference stencil for a one-dimensional second-order PDE is tridiagonal and this system can easily be solved. Their suggestion was to view the five-point finite difference approximation for a two-dimensional problem as the direct sum of two one-dimensional approximations. This led to an iteration in which alternatingly a tridiagonal associated with one of the two directions was split off, and this became popular as the alternating direction iteration (ADI). With the inclusion of iteration parameters, that steered the inclusion of a diagonal correction to the iteration matrices, the resulting ADI iterations could be tuned into a very effective method. Varga [188] gives a good overview of the theory for understanding ADI methods. He, as well as Birkhoff [21] mentions that ADI was initially derived as a by-product of numerical methods for parabolic equations (the correction to the diagonal was motivated by the effect of the time derivative in these methods). Sheldon and Wachspress, in 1957, gave an early proof for the convergence of ADI for fixed parameters [192]. Wachspress discusses these ADI methods in his book [193] and considers also other grid-oriented acceleration techniques. One of these techniques exploits approximations obtained on coarser grids and can be viewed as a primitive predecessor to multigrid.

The first half of the century begins also with simple local projection methods, in which one attempts to solve a set of equations by solving each separate equation by a correction that is small in some norm. These methods could be used for over- or underdetermined linear systems, such as those that arise in tomography problems. This has led to the methods of Cimmino [44] and Kaczmarz [106], which were later identified as instances of Gauss–Jacobi and or Gauss–Seidel for related systems with $A^{\mathrm{T}}A$ or AA^{T}. Modern variants of these methods, under the name of ART and

SIRT are very popular, for instance in medical and seismic tomography. ART and SIRT can be related to SOR and Block SOR. Spakman and Nolet [168] report on the solution of 292, 451 by 20,070 systems related to structures of the upper earth mantle, with these methods (and with LSQR).

The second half of the century was marked by the invention (paper published in 1952) of the conjugate gradient method by Hestenes and Stiefel [101] and the Lanczos algorithm for linear systems [117]. This started the era of Krylov iterative methods. Initially, these were not viewed as truly iterative techniques, but rather as direct solution algorithms since they terminated in exact arithmetic in fewer than n steps, if n is the order of the matrix (see, for instance, Householder's book where conjugate gradients is discussed in the chapter on direct methods [102, Chapter 5.7]). Hestenes and Stiefel already recognized that the method behaves as an iterative method, in the sense that the norm of the residual often decreases quite regularly, and that this might lead for some systems to acceptable approximations for the solution within n steps. A little earlier, papers by Lanczos [115] and by Arnoldi [2] had addressed the issue of transforming a matrix into simpler form for the purpose of diagonalizing it. These four papers together set the foundations of many methods that were developed later.

A famous publication by Engeli et al. [69] considered the method as a truly iterative process and showed that in rounding precision arithmetic, the conjugate gradient method did not terminate in the expected number of iteration steps (equal to at most the order of the matrix). This was shown for a matrix of order 64, a discretized biharmonic problem. Convergence occurred only after a few hundred steps. Notwithstanding this apparent failure, the method appeared later in the famous Wilkinson and Reinsch collection [202] as a kind of memory-friendly direct technique. It was mentioned that actual convergence might occur only after m iterations, where m could be 3 up to five times the order of the matrix. Because of this not well-understood behavior in rounded arithmetic, the method did not make it to the first universal linear algebra package LINPACK (mid-1970s). In the early to mid-1960s it became clear that the convergence of the conjugate gradient method depends on the distribution of the eigenvalues of the matrix, and not so much on the order of the matrix, as was, for example, explained in a paper by Kaniel [109]. Daniel [50,51] studied the conjugate gradient method as an iterative method for the minimization of functionals in (infinite dimensional) Hilbert spaces. This is a natural consequence of the observation that conjugate gradients, like other Krylov subspace methods, requires the action of the matrix as a linear operator and does not exploit the actual representation of the matrix (that is, the method does not require knowledge of the individual entries of the matrix). Also, Daniel expressed concerns about the convergence behavior of the method in finite precision, and he discussed modifications with guaranteed convergence [51, p. 134]. Note also that much of the convergence theory developed for the conjugate gradient and the Lanczos methods was almost invariably set in the context of operators on infinite-dimensional spaces, see, for example [109].

It was Reid [145] who suggested to use the conjugate gradient method again as an iterative technique, but now for large sparse linear systems arising in the discretization of certain PDEs. Soon after this, the notion of preconditioning (already proposed in the Hestenes and Stiefel paper) became quite popular. Thus, the incomplete Choleski decompositions of Meijerink and van der Vorst [125] led to the ICCG process, which became the de facto iterative solver for SPD systems.

Hence, it took about 25 years for the conjugate gradient method to become the method of choice for symmetric positive-definite matrices (the incomplete Choleski decompositions were shown to exist for M matrices). A good account of the first 25 years of the history of the CG method was given by Golub and O'Leary [86].

The unsymmetric variants of the Krylov methods required a similar amount of time to mature. The late 1960s and early 1970s, saw the roots for such methods. Techniques named ORTHODIR, ORTHOMIN, FOM, and others, were introduced but in their original formulations, these methods suffered from breakdowns and numerical instabilities. The GMRES variant, introduced by Saad and Schultz [158], was designed to avoid these undesirable features and became the de facto standard for unsymmetric linear systems. However, it suffered from the disadvantage of requiring increasing computational resources for increasing numbers of iterations. Bi-CG, the unsymmetric variant of conjugate gradients, did not have these disadvantages. The method, based on the unsymmetric Lanczos method (1952), was introduced by Fletcher in 1976 [76], but it is mathematically equivalent to a technique that had already been described in Lanczos' paper. Bi-CG, however, suffered from other practical problems, known as breakdowns of the first and second kind, which prevented early success. Moreover, the occurrence of nonorthogonal transformations led to much suspicion among numerical analysts. Nevertheless, the method became quite popular in a variant known as CGS (Sonneveld, 1984) [166] which, for virtually equal cost could essentially apply Bi-CG twice, leading often to a twice as fast convergence, but also amplifying the problems of Bi-CG. In the 1980s, Parlett and co-authors [140] and later Freund and Nachtigal [81] have shown how to repair the deficiencies in the Bi-CG method so that rather reliable software could be constructed. More recently, we have seen hybrids of the Bi-CG and GMRES approaches, with Bi-CGSTAB [186] as one of the most popular ones.

Originally, the usage of iterative methods was restricted to systems related to elliptic partial differential equations, discretized with finite difference techniques. Such systems came from oil reservoir engineering, weather forecasting, electronic device modeling, etc. For other problems, for instance related to various finite element modeling, practitioners preferred the usage of direct solution techniques, mainly efficient variants of Gaussian elimination, because of the lack of robustness of iterative methods for large classes of matrices. Until the end of the 1980s almost none of the big commercial packages for finite element problems included iterative solution techniques. Simon [164] presented results, obtained for matrices of the order of 55,000, for direct solution techniques. On the then fastest supercomputers, this required in the order of a few minutes of computing time. He claimed that direct sparse solvers would remain the method of choice for irregularly structured problems. Although this is certainly true if the structure of the matrix allows for an efficient elimination process, it became clear that for many PDE-related problems, the complexity of the elimination process increased too much to make realistic three-dimensional modeling feasible. Irregularly structured finite element problems of order 1,000,000, as foreseen by Simon, may be solved by direct methods – given a large enough computer (memory wise) but at tremendous cost and difficulty. However, *some* of them can be solved with iterative techniques, *if* an adequate preconditioning can be constructed. In the last decade of this century, much effort was devoted to the identification of effective preconditioners for classes of matrices. For instance, Pomerell [142] in 1994 reports on successful application of preconditioned Krylov methods for very ill-conditioned unstructured finite element systems of order up to 210,000 that arise in semiconductor device modeling.

While using iterative methods still requires know-how, skill, and insight, it can be said that enormous progress has been made for their integration in real-life applications. Still, linear systems arising from many relevant problems, for instance large electric and electronic circuits, are not easy to solve in an efficient and reliable manner by iterative methods. Steady progress is being made but the field as a whole can still be viewed as being in its infancy.

3. Relaxation-based methods

The Gauss–Seidel iteration was the starting point for the successive over-relaxation methods which dominated much of the literature on iterative methods for a big part of the second half of this century. The method was developed in the 19th century, originally by Gauss in the mid-1820s and then later by Seidel in 1874 (see references in [102]). In fact, according to Varga [188], the earliest mention on iterative methods is by Gauss (1823).

Indeed, on December 26, 1823, Gauss writes a letter to Gerling, in which he describes an iterative technique for the accurate computation of angles occurring in geodesy [84, p. 278]. The corrections for the four angles in a quadrangle, determined by four church towers, were computed from a singular linear systems of four equations with four unknowns (the singularity comes from the observation that the four angles sum up to 360°). The technique that Gauss describes is what we now know as the Gauss–Seidel algorithm. The order of processing of the equations was determined by the unknown that helped to reduce the residual most. Gauss recognized that the singularity of the system led to convergence to the solution modulo a vector in the null space, for which he could easily make a correction. The three pages of his letter are full of clever tricks. He concludes by recommending the new method to Gerling, arguing that the method is self correcting, and that one can easily determine how far to go and then ends his letter with the remark that the computations were a pleasant entertainment for him. He said that one could do this even half asleep, or one could think of other things during the computations. In view of this remark it may hardly be a surprise that the method became so popular in the era of electronic computing.

The method as it was developed in the 19th century was a relaxation technique, in which relaxation was done by "hand". It was therefore natural to eliminate the largest components, see for example [55,118]. This method is referred to as Nekrasov's method in the Russian literature [130]. Referring to the more modern method in which relaxation was done in a cyclic manner, Forsythe is quoted as having stated that "the Gauss–Seidel method was not known to Gauss and not recommended by Seidel", see [102, p. 115].

However, the blossoming of overrelaxation techniques seems to have been initiated by the Ph.D. work of David Young [203]. Young introduced important notions such as *consistent ordering* and *property A*, which he used for the formulation of an elegant theory for the convergence of these methods. Generalizations of Young's results to other relevant classes of matrices were due to Varga, who published his book on Matrix Iterative Analysis in 1962. For decades to come this book was the standard text for iterative methods for linear systems. It covered important notions such as *regular splittings*, a rather complete theory on *Stieltjes* and *M-matrices*, and a treatment of semi-iterative methods, including the Chebyshev semi-iteration method. The latter method, analyzed by Golub and Varga [87], also became more widely known, especially in the period when inner products were relatively expensive.

The accelerated Gauss–Seidel methods have motivated important developments in the theory of matrix linear algebra. In particular, relevant properties for M-matrices, introduced by Ostrowski [135], were uncovered and convergence results for so-called *regular splittings*, introduced by Varga [189] were established. A cornerstone in the convergence theory was the theorem of Stein–Rosenberg (1948) [169] which proved relations between the asymptotic rates of convergence for the successive overrelaxation methods, including the Gauss–Seidel method, and the Gauss–Jacobi method. The concept of *irreducibility* of a matrix, a natural property for grid-oriented problems, helped to extend results for strongly diagonally dominant matrices to matrices for which the strict diagonal dominance inequality is required to hold only for one single equation at least. Another important notion is the concept of cyclic matrices: an irreducible matrix with k eigenvalues of modulus $\rho(A)$ is said to be of index k. Varga [188] gives a good overview of the relevant theory and the implications of this concept for iterative methods. It has a close relationship with Young's property A [188, p. 99], and provides the basis for the convergence theory of the SOR methods. Sufficient conditions for the convergence of the SOR methods were given by theorems of Ostrowski [136] and Reich [144]. Lower bounds for the spectral radius of the SOR iteration matrix were derived by Kahan [107]. This together provided the basis for a theory for iterative methods, published in Varga's book [188] from which many new methods emerged. Later, in the 1970s major part of this theory served well in the development of preconditioners for Krylov methods.

The following is a quotation from Varga's book (page 1) "As an example of the magnitude of problems that have been successfully solved on digital computers by cyclic iterative methods, the Bettis Atomic Power laboratory of the Westinghouse Electric Corporation had in daily use in 1960 a two-dimensional program which would treat as a special case, Laplacean-type matrix equations of order 20,000". So the state of the art in 1960 was a $20,000 \times 20,000$ Laplace equation.

In the late 1960s and early 1970s a number of methods appeared in which the order of relaxation was not prescribed or even deterministic. These were appropriately termed "chaotic" or "asynchronous" relaxations. It was established that if a variable is relaxed an infinite number of times, the global method would always converge for any order in which the relaxation takes place. A few of the main contributions were by Chazan and Miranker [41], Miellou [128], Robert [147] and Robert et al. [148]. These methods were motivated by parallelism and were essentially ahead of their time for this reason.

4. Richardson and projection methods

Another line of development started with Richardson's method [146].

$$x_{k+1} = x_k + \omega r_k = (I - \omega A)x_k + \omega b,$$

which can be viewed as a straightforward iteration associated with the splitting $A = K - R$, with $K = (1/\omega)I$, $R = (1/\omega)I - A$. Here r_k is the residual vector of the current iterate:

$$r_k = b - Ax_k.$$

For the residual at the $(k+1)$th step, one obtains

$$r_{k+1} = (I - \omega A)^{k+1} r_0 = P_{k+1}(A) r_0,$$

where $P_{k+1}(A)$ is a $k+1$ degree polynomial in A, with $P_{k+1}(t) = (1-t)^{k+1}$. It is easy to see that for symmetric positive-definite matrices the process will converge for ω in the open interval $0 < \omega < 2/\lambda_{\max}$ where $\lambda_{\max}$ is the largest eigenvalue of A. In addition the best ω is known to be $2/(\lambda_{\min} + \lambda_{\max})$, see, e.g., [188,157] for details.

The original Richardson iteration is readily generalized by taking a different $\omega = \omega_k$ for each iteration, which leads to the generalized Richardson iteration

$$x_{k+1} = x_k + \omega_k r_k. \tag{1}$$

The sequence of ω_ks can be selected in a number of different ways. Note that the residual vector after step $k+1$ is given by

$$r_{k+1} = (I - \omega_k A)(I - \omega_{k-1} A) \cdots (I - \omega_0 A) r_0, \tag{2}$$

which shows that we can obtain any desired polynomial P_{k+1} with the property that $P_{k+1}(0) = 1$, by selecting its roots as the iteration parameters in (1). This process is referred to as polynomial acceleration of the simple splitting for A that we gave above. It was studied by, among others, Young [204], Shortley [162], and Lanczos [116]. By 1962 it was not considered competitive, since, as quoted from [188, p. 159]: "*Richardson's method has the disadvantage of being numerically unstable*".

In fact, the Chebyshev semi-iteration method [87] can, in exact arithmetic, be obtained from these polynomial accelerated methods, by choosing the acceleration parameters in successive Richardson iterations properly, but this approach is unstable. In the Chebyshev semi-iteration method one exploits the three term recurrence relation for Chebyshev polynomials, which leads to a stable three term recurrence iterative method. The main problem with these Chebyshev methods is that one needs fairly accurate information about extremal eigenvalues, since these define the interval for the Chebyshev polynomials.

The method of steepest descent which is attributed to Cauchy (1847) is also of the form (1). Kantorovitch later considered the method in a 1945 paper [110] that appeared in the Russian Doklady Akademii Nauk SSSR. In this case, the scalar ω_k is selected so as to minimize the quadratic form

$$J(x) = \tfrac{1}{2}(Ax, x) - (b, x)$$

in the direction of r_k.

In the 1950s and 1960s other matrix splittings were suggested as basis for iteration methods. We mentioned before one such splitting, namely the ADI method of Peaceman and Rachford [141]. In 1968, Stone [171] proposed the *strongly implicit procedure*, which is, in fact, a simple Richardson iteration with a series of splittings of the matrix. The idea, for a five-point finite difference stencil, is to factor the matrix in a lower triangular matrix and an upper triangular matrix each with a three-point stencil (as in incomplete LU with no fill-in). The factors are chosen in such a way that the fill-in is spread over the seven-point stencil of the product matrix. This is different from the Kendall–Dupont–Rachford [60] decomposition, where the fill-in is compensated by a correction to the diagonal in such a way that the sum of the elements of the error matrix equals zero for each row. Convergence in the SIP method is achieved by a set of iteration parameters (up to 18) for different distributions of the elements in the error matrix. The choice of the parameters is motivated by a Fourier analysis. Each value of the iteration parameter kills some components in the error (but may lead to increase in other components). Successful application of SIP requires to apply a special order of the iteration parameters, and for each value the decomposition has to be carried out from top to bottom and next from bottom to top. The SIP method gained quite some popularity in oil reservoir simulation and groundwater simulation problems, but its usage seem to have declined in favor of the Krylov methods. The Dupont–Kendall–Rachford splitting was proposed to be used in combination with Chebyshev polynomial acceleration.

In 1937 and 1938 two papers were published on methods that can be termed 'row-projection methods' (or column projection methods). These methods proposed by Kaczmarz [106] and Cimmino

[44] were also based on one-dimensional corrections:

$$x_{k+1} = x_k + \omega_k a_{i,.}, \tag{3}$$

where $a_{i,.} = A^{\mathrm{T}} e_i$ is the ith row vector of A. These rows are cycled through from 1 to n. Here ω_k is selected so that the ith component of r_{k+1} becomes zero. Because $r_{k+1} = r_k - \omega_k A A^{\mathrm{T}} e_i$, it is easy to see that this method is mathematically equivalent to the Gauss–Seidel iteration applied to the normal equations

$$AA^{\mathrm{T}} y = b, \quad x = A^{\mathrm{T}} y.$$

The method proposed by Cimmino was the Jacobi equivalent of this approach. It is also possible to define similarly a Gauss–Seidel approach for the normal equations

$$A^{\mathrm{T}} A x = b$$

as was noted by Björck and Elfving [22], and this would correspond to taking directions d_k along the columns of the matrix. This class of methods regained interest in the 1970s and 1980s with the work of Björck and Elfving [22], Tanabe [174], and later Kamath and Sameh [108] and Bramley and Sameh [25].

However, one-dimensional projections methods of a different type, based on very general definitions of norms were very popular in the later 1950s. Here, we mention the work of Gastinel among others. Gastinel's approach [83] consisted of defining generating vectors for norms. Consider an arbitrary vector norm ϕ (for example the norm $\|.\|_1$). Gastinel defines the vector v which realizes the norm of a vector r in the sense

$$(v, r) = \phi(r).$$

For example, for the 1-norm, the components of v can be defined as $v_i = \mathrm{sign}(e_i^{\mathrm{T}} r)$, where $r = b - Ax$ is the current residual vector. This vector v is chosen to do an orthogonal projection step. The method can be shown to converge for any nonsingular matrix.

5. Second-order and polynomial acceleration

An important observation regarding all acceleration methods of the form (1) is that their residuals take the form (2), so there is room for improvement to the scheme if successive iterates are considered.

In 1950 Frankel [78] proposed an acceleration method which used a three-term recurrence of the form

$$x_{k+1} = x_k + \beta_k \delta_k, \quad \delta_k = r_k - \alpha_k \delta_{k-1}.$$

This "second-order Richardson process" is initialized by $\delta_{-1} = r_{-1} = 0$. Frankel's method [78] uses constant coefficients and results in a residual polynomial which is a combination of Chebyshev polynomials of the first and second kind.

Naturally Chebyshev polynomials should give rise to optimal-like behavior and a number of authors discovered, rediscovered, or enhanced the method at various times. The paper by Flanders and Shortley [75] showed how to use Chebyshev polynomials for eigenvalue calculations. Later

Shortley [162] adapted this method for the solution of linear systems. In the Russian literature, Gavurin [85] also introduced the idea independently in 1950. In 1954, Young [204] proposed a method which amounted to compounding (or restarted) Chebyshev iterations. However the method was in the form (1) – which is unstable. Young gave some remedies to the process which consisted of reordering the roots ω_k before applying the polynomial.

In the particular case where A is symmetric positive definite the eigenvalues are located in an interval $[\lambda_{\min}, \lambda_{\max}]$. The best residual polynomial $1 - \lambda s(\lambda)$ in this case is a shifted and scaled Chebyshev polynomial of the first kind, and its three-term recurrence results in a simple three-term recurrence for the approximate solution, see, e.g., [157].

Using a different approach altogether, Lanczos in a 1952 paper [116] discusses the use of certain polynomials that are optimal in a least-squares sense. This paper, which was by and large overlooked by researchers, made a number of contributions. Its focus is on symmetric positive-definite matrices – for general matrices the author resorts to the normal equations. One of the main ideas proposed is to consider the problem of approximating the delta function in the interval $[0,1]$ which contains all eigenvalues (after the matrix is scaled by its largest eigenvalue). He then transforms the variable using the change of variables $x = (1 - \cos\theta)/2$. Now a least-squares approximation to the delta function is sought in the trigonometric basis. This leads to the so-called Dirichlet kernel whose solution is well known

$$P_k(\theta) = \frac{\sin(k+\frac{1}{2})\theta}{(k+\frac{1}{2})\sin\theta}$$

$$= \frac{1}{k+\frac{1}{2}}\left[\tfrac{1}{2} + \cos\theta + \cos 2\theta + \cdots + \cos k\theta\right].$$

To avoid the high oscillations around discontinuities, the so-called Gibbs phenomenon, Lanczos suggested a strategy due to Fejer.

Later, a remarkable paper by Stiefel gave a fairly complete view on similar ideas revolving around least-squares polynomials [170]. The above paper by Lanczos was not referenced by Stiefel. It is only in 1983 that the idea of using least-squares polynomials resurfaced in force again, motivated essentially by parallelism and vector processing. Earlier in 1979 a paper by Dubois et al. [58] suggested using simple Neumann series expansion of the matrix. In 1976 Axelsson addressed the problem of computing good polynomials when the spectrum is located in two intervals, and he was followed later in 1980 by deBoor and Rice [54] who showed how to compute the best min–max polynomial in this situation and the more general situation of multiple intervals. The least-squares alternative considered by Johnson et al. in [104] was for a single interval, assuming that A is symmetric positive definite. In other words, we need to solve

Find $s \in \Pi_k$ that minimizes:

$$\|1 - \lambda s(\lambda)\|_w, \qquad (4)$$

where w is some weight function on the interval $(\lambda_{\min}, \lambda_{\max})$, and $\|.\|_w$ is the L_2-norm associated with the corresponding inner product. Because the distribution of eigenvalues matters more than condition numbers for the preconditioned conjugate gradient method, the authors observed in [104] that least-squares polynomials tend to perform better than those based on the uniform norm, because they result in a better clustering of the spectrum. Moreover, Lanczos [116] and Rutishauser [150] already noted that the eigenvalue estimates need not be accurate: in fact, it suffices to use the simple bounds that are provided by Gershgorin's theorem. Further experiments in [151] did confirm that in some cases the least-squares polynomial over the Gershgorin interval, may perform as well as the infinity norm polynomial over $[\lambda_{min}, \lambda_{max}]$. Note that this is only a minor advantage of least-squares polynomials since effective adaptive procedures exist to compute $\lambda_{min}, \lambda_{max}$; see [98] for symmetric problems and [123,66] for nonsymmetric problems. We should add that the observations made in [104,151], and the simplicity of a method that bypasses eigenvalue estimates, have made least-squares polynomials more popular for polynomial preconditionings.

In the more general nonsymmetric case the interval (or union of intervals) that contains the spectrum is to be replaced by a more complex continuum E in $\mathbb{C}$, which ideally would contain the eigenvalues of the matrix A. Several choices have been used for E. The first idea, proposed by Manteuffel in 1977–1978 [122,123], is to use an ellipse E that encloses an approximate convex hull of the spectrum, or more precisely, the field of values of A. Then the shifted and scaled Chebyshev polynomials are optimal or nearly optimal and the use of these polynomials leads again to an attractive three-term recurrence. He exploited the fact that an unaccurate guess of extremal eigenvalues leads to either divergence or very slow convergence, in which the eigenvectors corresponding to the unidentified extremal eigenvalues play a dominant role. After a few iterations these directions can be identified and the parameters for the Chebyshev iteration polynomials can be adjusted. Although superseded by the parameter-free Krylov iteration methods, the Chebyshev methods are still of interest on computer platforms where the inner products are relatively expensive. They can be used in combination with Krylov methods, either as polynomial-type preconditioners in order to damp dominating parts of the spectrum, or to continue the iteration with the eigenvalue guesses that can be obtained from the Krylov methods (the Ritz values).

A second alternative is to use a polygon H that contains $\sigma(A)$ [160,152]. A notable advantage of using polygons is that they may better represent the shape of an arbitrary spectrum. The polynomial is not explicitly known but it may be computed by a Remez algorithm. As in the symmetric case an alternative is to use some weighted L_2-norm instead of the infinity norm. Saylor and Smolarski used a discrete norm on the polygon [160]. Saad [152] used an L_2-norm associated with Chebyshev weights on the edges of the polygon and expressed the best polynomial as a linear combination of Chebyshev polynomials associated with the ellipse of smallest area containing H.

Yet another attractive possibility, with polygons instead of ellipses, proposed by Fischer and Reichel [74] is to avoid the problem of best approximation altogether and interpolate the function $1/z$ with a polynomial at the Fejer points of E, i.e., the points $e^{2ji\pi/k}$, $j=0,\ldots,k$ that are conformally mapped from the unit circle to H. This is known to be an asymptotically optimal process. There are numerous publications related to this approach and the use of Faber polynomials; see the references in [74].

6. Krylov subspace methods: the first period

In the early 1950s a number of new methods appeared that dramatically changed the landscape of iterative methods. In separate contributions Lanczos [117] and Hestenes and Stiefel [101] propose in effect different versions of what is now known as the conjugate gradient method. Amazingly, Hestenes and Stiefel actually discovered the same method independently.[3] The method proposed by Lanczos is, for symmetric positive-definite matrices, mathematically equivalent to the conjugate gradient method, but it was described for the general case of nonsymmetric matrices.

There is no doubt that the origin of this class of methods was deeply rooted in approximation theory and, in particular in orthogonal polynomials. The ideas behind "gradient methods" as this class of methods was referred to, are based on some kind of global minimization. For instance, for positive-definite symmetric A, the CG method minimizes the so-called A-norm: $\|x_i - x\|_A^2 \equiv (x_i - x, A(x_i - x))$, for x_i that are in the Krylov subspace $K_i(A, r_0) \equiv \{r_0, \ldots A^{i-1}r_0\}$. For some PDE problems this norm is known as the energy norm, which has physical relevance. Another interpretation of the gradient methods is that the residual is orthogonal to the space of previously generated residuals, or some related space. Both interpretations are useful for the formulation of methods as well as for the analysis. A very useful consequence from the Krylov subspace basis is that x_i can be expressed as a polynomial in A of degree $i-1$, acting on r_0. The minimization interpretation makes it possible to bound the error for CG by replacing the "CG-polynomial" by easier to analyze polynomials, for instance a Chebyshev polynomial. This leads to the well-known upper bound [109,50,47,4]

$$\|x_i - x\|_A \leqslant 2 \left(\frac{\sqrt{\kappa} - 1}{\sqrt{\kappa} + 1} \right)^i \|x_0 - x\|_A \tag{5}$$

for symmetric positive-definite matrices, in which $\kappa = \lambda_{\max}(A)/\lambda_{\min}(A)$. This upper bound describes well the convergence behavior for matrices A of which the eigenvalues are distributed rather homogeneously. For more uneven distributions one can obtain bounds by making more subtle choices for the approximating polynomials, for instance, products of suitably chosen Chebyshev polynomials [4]. These choices do not reveal the superlinear convergence that is often observed for CG and also for other Krylov subspace methods. The notion of superlinear convergence refers to the observation that the convergence of CG accelerates in the course of the process. Proofs for superlinear convergence had been given already in the early 1950s [111,99], but these did not reveal that the superlinear behavior may take place in early phases of the iteration process; they were qualitative rather than quantitative. Concus et al. [47] related this convergence behavior to the Krylov subspace approximations, by stating that "the extremal eigenvalues are approximated especially well (by the Ritz values corresponding to the Krylov subspace) as CG proceeds, the iteration then behaving as if the corresponding eigenvectors are not present (thus leading to a smaller "effective" condition number in (5), which might then explain the faster convergence". In 1986, this was proven in a quantitative

---

[3] The anecdote told at the recent "Conference on preconditioning methods for sparse matrix problems in industrial applications" held in Minneapolis, by Emer. Prof. Marvin Stein, the post-doc who programmed the algorithm for M. Hestenes the first time, is that Stiefel was visiting UCLA at the occasion of a conference in 1951. Hestenes, then a faculty member at UCLA, offered to demonstrate this effective new method to Stiefel, in the evening after dinner. Stiefel was impressed by the algorithm. After seeing the deck of cards he discovered that this was the same method as the one he had developed independently in Zurich. Stiefel also had an assistant, by the name of Hochstrasser, who programmed the method.

way [179] and it was shown that the relevant eigenvalues needed to be approximated only in a modest degree by Ritz values, for an acceleration to set in.

As was mentioned in Section 2, the actual behavior of CG, in finite precision arithmetic, was initially not quite well understood. Several attempts to analyze this have been made, but it was in the early 1990s that this problem was satisfactorily explained. Greenbaum and Strakoš showed that CG in finite precision could be regarded as the exact process applied to an expanded system that is closely related to the given system, and the matrix of the expanded system has almost multiple exterior eigenvalues when the orthogonality in the process is lost. This helps explain why finite precision leads to a delay in the iteration process, but does not prevent the algorithm to deliver accurate approximate solutions (for symmetric positive-definite systems). For details, see [89,172], and [88, Chapter 4].

Surprisingly, it took some time before the ideas of the CG were generalized to other classes of matrices. Paige and Saunders [137], in 1975, showed that the underlying Lanczos method, could also be used for efficient implementations of related approaches for symmetric indefinite matrices. In MINRES, the norm of the residual $\|Ax_i - b\|_2$ is minimized, and this required a more careful approach in order to avoid breakdowns. Conjugate gradient can be interpreted as a process in which A is projected to the Krylov subspace (in fact, the Lanczos approach), which leads to a tridiagonal matrix T of low dimension. The actual approximation x_i is determined by solving a small linear system with T, and this is done with LU without pivoting. This leads to the elegant short recurrences in CG. For indefinite matrices, the LU factorization could lead to numerical difficulties.

Paige and Saunders circumvented this by employing a QR decomposition of T, which leads again to regular short recurrences, making the MINRES method attractive because of its minimal overhead and its economy of storage. It may come as a small wonder that one can also minimize $\|x_i - x\|_2$, without knowledge of the solution x. Paige and Saunders accomplished this in their SYMMLQ, by restricting x_i to $AK_i(A, r_0)$. The advantage of SYMMLQ over MINRES appears to be in less sensitivity to ill-conditioning of A, the price one has to pay is that SYMMLQ often takes more steps. A slight disadvantage is also that although the method minimizes the norm of the error, the value of this norm is not known and the only practical information one has is the norm of the residual.

In 1976, Concus and Golub [45] and Widlund [200] came up with the idea of splitting a matrix into its symmetric and nonsymmetric parts and using the symmetric part as a preconditioner. With the proper inner product, the resulting algorithm corresponds to an iteration with a skew-Hermitian matrix – and therefore a three-term recurrence – CG-like – algorithm (called CGW) can be formulated.

The Bi-CG method, proposed in 1976 by Fletcher [76], is actually an implementation of the two-sided Lanczos process, which was suggested by Lanczos in 1952. In Bi-CG, the residual is constructed to be orthogonal to a Krylov subspace generated with A^{T} and some vector s_0. Initially, many numerical analysts were very skeptical of the Bi-CG, mainly because of the various breakdown situations that may occur. Also, Bi-CG did not minimize any norm of interest and the convergence behavior can be very irregular. The fact that the underlying two-sided Lanczos process works with nonorthogonal projections led to serious doubts on the usefulness of the method. A good example of this concern is in [201, pp. 394, 395], where the two-sided Lanczos method (viewed as a finite method for reducing a nonsymmetric matrix to tridiagonal form) is commented on: "... we may well have to pay a heavy price in terms of numerical instability..." and "... it is difficult to think of any reason why we should use Lanczos' method in preference of Householder's". Wilkinson's analysis is, of course, still valid. However in the context of solving large sparse linear systems, we have learned to make the two-sided Lanczos method and the Bi-CG into useful solution techniques thanks to a number of enhancements and a better understanding of these processes.

7. Krylov subspace methods: the second period

Of course, the success of Krylov methods for symmetric matrices has inspired the construction of similar methods for unsymmetric matrices. The classical Lanczos method leads to a tridiagonal system, the projection of A with respect to the Krylov subspace. Factorization of this tridiagonal matrix as the product of a lower and an upper bidiagonal matrix, leads to the coupled two-term recurrences as in Bi-CG. As said before, Bi-CG suffered from several breakdown conditions. One breakdown, associated with the indefiniteness of the implicitly generated projected tridiagonal system, can be cured by admitting 2×2 blocks along the diagonal of one of the factors. This requires the combination of two successive iteration steps, which explains the name *Composite step Bi-CG* [12]. A similar idea had been used much earlier by Luenberger, in order to make the conjugate gradient algorithm robust for symmetric-indefinite matrices [119]. The other breakdown, a more serious one, arises when the bi-orthogonalization process leads to a zero inner product of the two new vectors in the Krylov subspace and its adjoint space (that is the Krylov subspace, generated in the two-sided Lanczos process, with A^T). Likewise, a near breakdown should also be avoided since it may lead to inaccuracies in the approximated solution. This breakdown can be cured with a *look-ahead strategy*, first proposed by Parlett et al. [140]. The idea is to expand the Krylov subspaces by two vectors simultaneously, and to make the new vectors block bi-orthogonal with respect to the similarly expanded adjoint space. Parlett et al. considered only look-aheads of length two, but a few years later, around 1990, the idea was picked up almost simultaneously by a number of other researchers who generalized it to look-aheads of arbitrary length. The most well known of these approaches were those published by Gutknecht and co-authors [95], Joubert [105], Parlett [139], Freund and Nachtigal [80], and Brezinski and co-authors [31,32]. In the latter work, the look-ahead strategy was related to the theory of orthogonal polynomials and referred to as *recursive zoom* technique. The connection between orthogonal polynomials and the Lanczos algorithms (and also the ε-algorithm) is discussed in [28]. This has proved to be very useful for getting more insight in the Lanczos and two-sided Lanczos algorithms. It also has helped to construct breakdown free variants of the hybrid Bi-CG algorithms, for details on this see [29].

Curing the breakdowns in Bi-CG was important, but there were other aspects as well that motivated further research. The convergence behavior of Bi-CG is usually quite irregular, in the sense that the norms of successive residuals can behave erratically. This motivated Freund and Nachtigal [80] to propose an algorithm in which the projected overdetermined tridiagonal system is solved in a least-squares sense. Since the basis vectors for the Krylov subspace, generated by the two-sided Lanczos process, are in general not orthogonal, this approach does not lead to a minimum residual approximate solution (as with MINRES), and this inspired for the name quasi-minimum residual (QMR). The full QMR method includes a look-ahead strategy, but it became also popular without it, since the first breakdown condition is cured by the least-squares solution of the tridiagonal system. For a template for this simplified QMR, see [13].

The other clear disadvantage in the basic two-sided Lanczos method was the necessity to construct two Krylov subspaces: $K^i(A; r_0)$, and $K^i(A^T; s_0)$. Of the two vector bases generated only one is exploited for the solution, the other can be regarded as an auxiliary set used to generate the inner products needed to generate the bi-orthogonal basis. Sonneveld [166] made the clever observation that the operations with A^T could be reformulated to operations with A, and these operations can be used for a further reduction of the residual norm. Whereas the previous approaches, look-ahead, composite-step, and QMR, help to cure deficiencies of the Lanczos method, they do not lead to essential faster convergence. However, Sonneveld's trick can lead to much faster convergence, for practically the same costs per iteration. The idea is based on the observation that the residual r_i can be expressed formally as $r_i = p_i(A)r_0$, with p_i a polynomial of degree i. Likewise, the shadow residual in the adjoint space can be expressed as $s_i = p_i(A^T)s_0$. The iteration coefficients for Bi-CG are computed from inner-products such as (r_i, s_i), and such an inner product can be rewritten formally, as $(p_i(A)r_0, p_i(A^T)s_0) = (p_i^2(A)r_0, s_0)$. This observation leads to an algorithm that generates $\tilde{x}_i \in K^{2i}(A; r_0)$, for which the corresponding residual $\tilde{r}_i$ can be expressed as $\tilde{r}_i = p_i^2(A)r_0$. For the situation where Bi-CG delivers a residual r_i $(= p_i(A)r_0)$ that is small compared with r_0, one may conclude that $p_i(A)$ has transformed r_0 into a small vector, and hopefully, if we apply $p_i(A)$ twice, then this leads to a double reduction. Indeed, the resulting method, conjugate gradients squared (CGS), often leads to a convergence about twice as fast as Bi-CG. This algorithm is also referred to as Bi-CGS, which is actually more appropriate. The downside of the squaring of p_i is that the convergence of CGS is usually much more irregular than for Bi-CG. This can lead to serious accuracy problems in finite precision arithmetic; we will come back to this aspect later.

Soon after the discovery of the CGS method, it was recognized that the operations with A^T could also be transformed to other polynomials in A. The first idea in this direction was Bi-CGSTAB [186], in which Bi-CG was combined with minimum residual steps of degree one. This led to a convergence that is rather smooth as well as faster than Bi-CG and it gave rise to many other hybrids. Gutknecht suggested to combine $2i$ Bi-CG steps with i times a minimum residual method of degree 2. This was generalized by Sleijpen and Fokkema [165] to Bi-CGSTAB(ℓ). The same principles can also be applied to QMR, and the analogue of CGS led to TFQMR [79]. The analogue of Bi-CGSTAB is QMRSTAB, suggested by Chan et al. [38]. Zhang [207] describes more general product methods based on Bi-CG. His framework includes the previously described methods, but also admits hybrid variants in which one can shift from CGS to Bi-CGSTAB at some iteration step. This principle admits further possibilities for reducing the residual in some norm. An interesting variant of CGS has been suggested by Fokkema et al. [77]. Here, the polynomial p_i^2 that generates the residuals, is replaced by the product $p_i \tilde{p}_i$, where $\tilde{p}_i$ is the polynomial that corresponds to a 'nearby' Bi-CG process. The principle can be used to help reduce severe irregularities in the convergence, while the quadratically reduced errors in important eigenvector directions are still realized. According to the authors, this is an advantage in the context of iterative solutions of Jacobian systems in a Newton process for nonlinear systems of equations. Similar ideas were also considered by Brezinski and Redivo Zaglia [30]. Their approach is to compute approximations by two different methods and to combine the two results in an effort to get a better approximation. For some methods, the combined method can be executed at reduced costs, that is some of the matrix vector products can be used for both methods. For a detailed overview of this approach see [29, Chapter 5].

A different direction is to try to minimize the norm of the residual over all vectors in the Krylov subspace, similar to the MINRES approach for symmetric A. A number of methods were proposed that achieved this goal, among them ORTHODIR [103], GCR [62], ORTHOMIN [190], and Axelsson's method [5], but many of these methods suffered from some numerical instability. An early specimen of this approach was suggested by Khabaza in 1963 [112]. He proposed a method which, in exact arithmetic, leads to the same iterands as GMRES(m). However, he used the defining vectors $r_0, Ar_0, \ldots, A^{m-1}r_0$ as the basis vectors, without further orthogonalization. The observation that Khabaza considered matrices with a condition number of 200 as being ill-conditioned, may serve as an illustration for the numerical problems that were encountered.

GMRES [158] developed in 1986, was mathematically equivalent to these techniques but soon it came to be preferred because of its better numerical behavior and its lower cost, both in terms of memory and arithmetic. An advantage of GMRES is its guarantee to compute the approximate solution with minimum residual norm, but the price to be paid is that the overhead costs per iteration, that is the computations other than the matrix vector product, increase linearly with the iteration count. Also, all basis vectors for the Krylov subspace have to be stored. The obvious solution seems to be to restart after a convenient number of iterations, before the costs for computation and storage become prohibitive. This is known as restarted GMRES, or simply GMRES(m). The disadvantage of this approach is that it decreases the robustness of the method since convergence is no longer guaranteed. Moreover, by restarting the convergence behavior may become very slow and one runs the risk to miss the faster convergence that might have occurred for a larger value of m. For that reason, researchers have tried to find ways to reduce the number of iteration steps, other than by preconditioning, or even in addition to preconditioning. One idea is to try to improve the preconditioner with updates from the Krylov subspace. This has been suggested first by Eirola and Nevanlinna [61]. Their approach leads to iterative methods that are related to Broyden's method [35], which is a Newton-type method. For specific but obvious choices, one recovers a method that is equivalent to GMRES. The Broyden method can be obtained from this update-approach if we do not restrict ourselves to Krylov subspaces. See [191] for a discussion on the relation of these methods.

The idea of preconditioning is to approximate $A^{-1}p$ for vectors p generated by the iteration method. One could do this in a different way for every iteration step, for instance, by incorporating information from previous iteration steps in the preconditioner, or by approximating $A^{-1}p$ by some iteration method again. The updated preconditioners cannot be applied immediately to GMRES, since the preconditioned operator now changes from step to step, and we are not forming a regular Krylov subspace. However, we can still minimize the residual over the new subspace. The idea of variable preconditioning has been exploited in this sense, by different authors. Axelsson and Vassilevski [11] have proposed a Generalized Conjugate Gradient method with variable preconditioning, Saad [155] has proposed a scheme very similar to GMRES, called Flexible GMRES (FGMRES), and Van der Vorst and Vuik [187] have published a scheme called GMRESR. FGMRES has received more attention, possibly because it is fairly easy to implement: only the update directions in GMRES have to be preconditioned, and each update may be preconditioned differently. This means that only one line in the GMRES algorithm has to be adapted. The price to be paid is that the method is no longer robust; it may break down. GMRESR and the generalized conjugate gradient method produce, in exact arithmetic, the same results, but GMRESR is numerically more stable and more efficient. In GMRESR the residual vectors are preconditioned and if this gives a further reduction then GMRESR does not breakdown. This gives slightly more control over the method in comparison with FGMRES. In most cases though the results are about the same.

Other methods that proved to be very useful include the LSQR method, suggested in 1982 by Paige and Saunders [138]. LSQR is a clever implementation of the Lanczos method that leads to a factorization of the tridiagonal reduced matrix for $A^{T}A$. This is often the method of choice for overdetermined or underdetermined systems; it minimizes the norm of the residual over the Krylov subspace generated with $A^{T}A$. For square systems the method is not so effective, unless one finds a good preconditioner, since the convergence behavior of LSQR is dictated by the square of the condition number of the system involved. The condition number, however, does not always give a good indication for the behavior of Krylov methods; Nachtigal et al. describe examples of matrices for which the singular values may predict the convergence behavior (much) better [129]. In extreme cases, GMRES with A may converge much slower than LSQR, or in fact, any method based on the normal equations. LSQR may also be viewed, in exact arithmetic, as CG applied to the normal equations. Applying CG in this fashion, however, will result in poorer stability. Craig's method [138] is a Krylov method which also works with $A^{T}A$ and in which the error $\|x_i - x\|_2$ is minimized, without computing the value of this norm.

Finally, we mention that Weiss, in the early 1990s, gave generalizations of SYMMLQ for unsymmetric matrices. These methods are known as GMERR methods [196,198]. It may have an advantage to have a method in which the norm of the error is minimized, but since this is done over a different subspace, namely $A^{T}K_i(A^{T}, r_0)$, it is not clear yet when this leads to advantages over, for example, GMRES, in terms of efficiency or stability.

Convergence results for Krylov methods in the non-Hermitian case were established following essentially similar ideas as for the CG algorithm, see [157, Chapter 6.11]; [88, Chapter 3] for overviews. However, this simple analysis which was given, for example, for GMRES [158] was soon viewed as insufficient. The traditional bounds on the norm of the error or the residual are expressed in terms of eigenvalues of A and the condition number of the eigenvector matrix. For highly nonnormal matrices this does not always lead to informative results. Embree, in his thesis [67], describes situations for which the field of values of A, or its pseudospectra, are used for understanding the observed convergence behavior of GMRES (see also [88, Chapter 3]).

Many attempts to get a better understanding of the behavior of GMRES were made. Work by Brown [34] and later by Cullum and Greenbaum [49] established relations between certain methods (GMRES and FOM, and then BiCG and QMR). Greenbaum and Strakos [90] showed a number of interesting properties of GMRES – in particular they characterize all linear systems that have the same GMRES convergence behavior. In a later paper, Greenbaum et al. [91] established that essentially any convergence behavior is possible for the same spectrum.

One fundamental question that was asked in the early 1980s was whether a short recurrence iterative process could be found that was also optimal for non-Hermitian matrices. Indeed, it was known how to generate short-term recurrence algorithms for nonsymmetric matrices (e.g. the Bi-CG) but these do not verify obvious optimality properties. On the other hand the optimal processes that were known required long recurrences (e.g., GMRES). The answer to this question was given by the excellent paper by Faber and Manteuffel in 1984, and alas is was a negative one [71]: short-term solution algorithms that are also optimal can essentially be devised only for a restricted class of matrices; for all practical purposes, these matrices are either hermitian or skew-hermitian. An essentially equivalent result had been published by Voevodin, just one year before Faber and Manteuffel [180].

For the Bi-CG and related methods, very little theory is available on convergence. An attempt on this was to introduce variable metrics, see [14] and the survey by Weiss [197].

8. Accelerators are not enough: preconditioning methods

The convergence of iterative methods depends on spectral properties of the matrix of the linear system and in order to improve these properties one often transforms the linear system by a suitable linear transformation. This process is known as preconditioning.

We do not know for sure who coined the term 'preconditioning' first – it may have been Turing (according to Golub and O'Leary [86]) or Forsythe (see below). Regardless, the idea was known quite early on. Cesari, in 1937 [37], proposed to *multiply* a given system $Ax = b$ with a polynomial $P(A)$, in an attempt to speed up the convergence of the Richardson iteration (see also [23, p. 156] for a discussion on this; in this reference the Richardson process is referred to as *Mises*' iteration – Cesari calls it the *von Mises*' iteration). In the 1952 paper by Lanczos [116] the notion of polynomial preconditioning is clearly defined: "*The construction of the inverse matrix is equivalent to a linear transformation which transforms the given matrix into the unit matrix. The unit matrix can be conceived as the extreme case of a well-conditioned matrix whose eigenvalues are all 1. We will ask for much less if we merely demand the transformation of the original system whose dispersion is moderate*". Lanczos then states that the goal of the procedure is to "reduce the initial skewness" of the system, not bring about the exact solution. Forsythe in his report on this paper in the mathematical reviews (review MR 16-751, 1955) does employ the term "preconditioning" explicitly. Polynomial preconditioning is also clearly mentioned in the review paper in Stiefel [170] – in 1959. Hestenes in 1956 [100], viewed the conjugate gradient method as an acceleration technique for suitable matrix splittings. His formulation of the algorithm is equivalent with preconditioned conjugate gradients. Finally, we mention Faddeev and Faddeeva, who used the term "preparing" in their 1963 book [72, p. 188] (a translation of the Russian text of 1960) for transforming a system $Ax = b$ to $KAx = Kb$, with K such that KA is close to a unit matrix.

Modern preconditioning methods started in the late 1960s and early 1970s. Evans [70] used the term *preconditioning* explicitly for the acceleration of SSOR by the Chebyshev iteration. However, this combination had already been studied by Sheldon in 1955 [161]. In 1972, Axelsson [3] proposed to use the SSOR method as a preconditioner for the conjugate gradient method. The incomplete Cholesky decompositions (Meijerink and van der Vorst, 1974, 1977), became quite popular and led to the ICCG process [125]. Concus, Golub and O'Leary [47] wrote an influential paper on the usage and effect of preconditioning for the CG method.

8.1. Incomplete factorizations

Preconditioning as we know it today refers mostly to approximate or incomplete factorizations of the coefficient matrix. Some of the early publications on such factorizations that are often cited include Buleev [36], Varga [189] and Oliphant [132]. Later in the 1960s a few other procedures were developed specifically for matrices arising from finite difference approximations to elliptic operators, these include the work by Dupont et al. [60]. In 1977, Meijerink and Van der Vorst introduced the more general incomplete LU factorization [125]. The paper suggests that the combination of this "preconditioning" and the conjugate gradient method could lead to a robust and very fast combination.

Similar ideas were in existence before. However, the paper provided a much needed stimulus to the whole area of iterative methods.

The Dupont–Kendall–Rachford splitting can be viewed as an incomplete LU factorization with zero fill-in, in which the elimination errors are compensated by corrections to the diagonal of the decomposition. In 1977, this procedure was generalized by Gustafsson [94] in 1978, as a modified form of the incomplete LU factorizations: MILU.

Several developments marked the years that followed. Two distinct ways of developing incomplete factorization preconditioners with improved accuracy were developed. The first approach is based on a symbolic factorization view, i.e., it only requires the nonzero structure of the matrix to determine which fill-ins to drop. A method proposed by Watts [195] for irregular sparse matrices attributes a "level of fill" recursively to each fill-in element from the levels of fill-in of its parents, in the Gaussian elimination process. Then each fill-in that is introduced and whose level exceeds a certain threshold is dropped. In practice for M-matrices, the higher the fill-in the lower the level. The second common approach is to modify a given direct solver by including a dropping rule, based on the numerical size of the fill-ins introduced [82,134,53,52,208,206].

Although the relation between the size of the dropped elements and the number of iterations required to achieve convergence is far from being understood, *on the average* dropping small elements is more likely to produce a better quality preconditioner than dropping large elements. However, experience reveals that this is not always true. Another drawback of the level-of-fill approach is that it is difficult to predict the amount of fill-in that will be generated.

The number of variations that can be found on incomplete factorization preconditioner is truly astounding and we will not attempt to list them all. It suffices to say that there were variants developed for specific architectures. (e.g., Twisted Factorizations), or for specific applications (e.g., element-by-element preconditioners), or to exploit specific features of the equations (e.g., block factorizations), among other classes. See [57] for an overview of these preconditioners, specially in view of their implementation for high-speed computers.

One of the interesting recurring themes in preconditioning methods is whether or not reordering the matrix prior to applying the ILU factorization can be helpful. Two early papers examined this carefully and concluded rather negatively. The first is a paper by Simon [163] who considered large nonsymmetric linear systems. For the systems he considered he concluded that standard techniques used for sparse direct solvers were not too helpful for use in preconditioners based on level-of-fill. Immediately following this was a paper by Duff and Meurant [59] which concluded, similarly, that ICCG does not in general benefit in any significant manner form reordering. These studies were limited to certain types of reorderings and certain types of preconditioners. It is now known [19] that certain reorderings, such as Reverse Cuthill McKee are beneficial in preconditioning methods, in particular with some form of dropping strategy. The beneficial impact of well-chosen fill-ins was already demonstrated in [59] for some orderings. What seems to be also clear is that the best approaches for direct solvers (such as Nested Dissection and minimal degree ordering) are not the best for iterative solvers.

Since ILU and IC factorizations were the most popular preconditioners, at least in a sequential environment, many attempts have been made to improve them, for instance by including more fill [126], by modifying the diagonal of the ILU factorization in order to force rowsum constraints [94,8,7,131,181,64], or by changing the ordering of the matrix [183,184]. A set of experiments with respect to the effects of ordering is contained in [59].

Saad [153] proposed a few variants on the incomplete LU approach for the matrix A, one of which is in fact an incomplete LQ decomposition. In this approach it is not necessary to form the matrix Q explicitly, and it turns out that the lower triangular matrix L can be viewed as the factor of an incomplete Cholesky factorization of the matrix $A^{\mathrm{T}}A$. This can be exploited in the preconditioning step, avoiding the use of Q. The second approach was to introduce partial pivoting in ILU, which appears to have some advantages for convection-dominated problems. This approach was further improved by including a threshold technique for fill-in as is done in the ILUT algorithm, see [157, p. 287].

Another major step forward, for important classes of problems, was the introduction of block variants of incomplete factorizations [176,46,6], and modified variants of them [46,6,120]. It was observed, by Meurant, that these block variants were more successful for discretized two-dimensional problems than for three-dimensional problems, unless the two-dimensional' blocks in the latter case were solved accurately. For discussions and analysis on ordering strategies, in relation to modified (block) incomplete factorizations, see [127,121].

8.2. Parallel preconditioners

Parallel preconditioners were discussed as early as with the first appearance of vector and parallel computers. It soon became apparent that the standard ILU-based preconditioners which were just becoming quite popular, were also very sequential in nature and had either to be replaced or implemented differently. The first ideas that were promoted or discussed were based on approximating the LU-solves by means of Neuman expansions in the L and U solves [182] as well as from the start by approximating the inverse of the original matrix by the Neuman series expansion of its inverse [58]. This gave rise to a number of papers on "polynomial preconditioners". The survey paper [154] gives an account of the state of the art toward the end of the 1980s and it can be seen that polynomial preconditioners figured prominently in the article. Another approach – termed "level-scheduling" or "wavefront" approach, was to unravel parallelism from the forward and backward solves. Because of sparsity, many equations can be solved at the same time in several levels during the forward and the backward solves – and a technique known in graph theory as "topological sorting" allows to determine these levels [1,15,20,159,184,185].

However, these two methods were soon viewed as having a limited potential. Level scheduling has limited parallelism and the first and last (smallest) levels were small enough to cause bottlenecks. A number of strategies could be used to improve the situation however. Polynomial preconditioners faced more serious challenges. Their performance relative to existing alternatives was not too convincing, especially for small number of processors. In addition, it is difficult to find a good polynomial in the case of indefinite matrices. Current interest in these techniques has all but vanished. This is a case where good mathematically based methods are not enough to overcome an inherent limitation of a given approach.

Red-black ordering is an obvious approach to improve parallel properties for well-structured problems, but experimental results were disappointing [59] so it was avoided. If carefully done though, they can lead to significant gains in efficiency. Elman and Golub [65] suggested such an approach, in which Red-Black ordering was combined with a reduced system technique. The idea is simply to eliminate the red points, and construct an ILU for the reduced system of black points. Recently, DeLong and Ortega [56] and Saad [156] suggested carrying out a few steps of red-black ordered

SOR as a preconditioner for GMRES and Bi-CGSTAB. The key to success in these cases seems to be a combined effect of fast convergence of SOR for red-black ordering, and the ability of the Krylov subspace to remove stagnations in convergence behavior associated with a few isolated eigenvalues of the preconditioned matrix.

Another stream of ideas for deriving parallel preconditioning methods came from domain decomposition-type methods. Such methods were in existence in the partial differential equations (PDE) literature already in a different form, see, e.g., the survey paper [40]. Though domain decomposition methods were motivated by parallel computing it appeared that the approach could be used with success also for the construction of sequential preconditioners. Domain decomposition has been used for problems that arise from discretization of a PDE over a given domain. The idea is to split the given domain into subdomains, and to solve the discretized PDEs over each subdomain separately. The main problem is to find proper boundary conditions along the interior boundaries of the subdomains. Domain decomposition is used in an iterative fashion and usually the interior boundary conditions are based upon information on the approximate solution of neighboring subdomains that is available from a previous iteration step.

It was shown by Chan and Goovaerts [39] that domain decomposition can actually lead to improved convergence rates, provided the number of domains is not too large. A splitting of the matrix with overlapping sub-blocks along the diagonal, which can be viewed as a splitting of the domain, if the matrix is associated with a discretized PDE and has been ordered properly, was suggested by Radicati and Robert [143]. They suggested to construct incomplete factorizations for the sub-blocks. These sub-blocks are then applied to corresponding parts of the vectors involved, and some averaging was applied on the overlapping parts. A more sophisticated domain-oriented splitting was suggested in [194], for SSOR and MILU decompositions, with a special treatment for unknowns associated with interfaces between the sub-domains.

The isolation of sub-blocks was done by Tang [175] in such a way that the sub-blocks correspond to subdomains with proper internal boundary conditions. In this case it is necessary to modify the sub-blocks of the original matrix such that the sub-blocks could be interpreted as the discretizations for subdomains with Dirichlet and Neumann boundary conditions in order to force some smoothness of the approximated solution across boundaries. In [173] this was further improved by requiring also continuity of cross-derivatives of the approximate solution across boundaries. The local fine tuning of the resulting interpolation formulae for the discretizations was carried out by local Fourier analysis. It was shown that this approach could lead to impressive reductions in numbers of iterations for convection dominated problems.

Note that domain decomposition methods for general sparse linear systems became successful at the same time as the machines for which they were designed (distributed memory, MIMD computers) were gaining importance. Currently, most of the parallel iterative solvers packages utilize essentially DD-type preconditioners.

For an overview of parallel preconditioners, and guidelines for their efficient implementation, see [57].

8.3. Multilevel preconditioners

Methods based on multilevel techniques, such as multigrid, have been popular for solving certain types of PDEs [96]. They are often designed specifically for problems arising from PDEs with regular meshes. Algebraic multilevel solution methods were developed as an attempt to extend the scope of these methods [149]. Clearly, nothing can prevent the use of these techniques as preconditioners for Krylov solvers. Since the multigrid method is viewed as optimal, its users have often avoided to use an accelerator such as GMRES or BICGSTAB to accelerate it. A study by Oosterlee and Washio [133] did, however, indicate that such a combination could be beneficial and lead to a much more robust solver than a multigrid solver alone.

Recently, a class of preconditioners that tended to close the gap between multilevel methods and preconditioned Krylov methods drew much attention. It was discovered that a multigrid-inspired ordering can be very effective for discretized diffusion–convection equations, leading in some cases to almost grid-independent speeds of convergence [177,178], see also [52]. These preconditioners can be viewed also from the angle of ILU factorization combined with a reordering as in the ILUM strategy, see [157, p. 371]. This type of approach can be fairly robust and scale well with problem size, unlike other ILU preconditioners.

In earlier related work, Axelsson and Vassilevski developed a method which was later referred to as AMLI [9,10] that is based on a set of nested finite element grids. The equations associated with the finer mesh are reduced (approximately) and the process is repeated to a number of levels until the coarsest mesh is reached.

It is interesting to note that currently, this general approach offers an excellent potential for providing a global method that can encompass most of the successful approaches for solving linear systems. By restricting the number of levels to one and performing the factorization accurately, one obtains a direct solver. A standard ILU solver can also be obtained by dropping fill-in.

8.4. Sparse approximate inverses

Many researchers and practitioners became aware of an important and damaging phenomenon in ILU techniques. An ILU factorization can be an accurate approximation to the original matrix but it can yield a very ill-conditioned factorization [181]. This phenomenon of instability of the LU factors was analyzed in particular by Elman [63]. This weakness of ILU factorizations, coupled with their sequential nature, spurred researchers to consider radical alternatives. The approximate inverse methods which were first proposed in the late 1970s [16] were in this category. It is only with the advent of massive parallel processing that such methods were considered as serious contenders of the now standard ILU methods [93]. A flurry of publications followed this work and the work by Kolotilina and Yeremin [113,114]. To cite just a few, [48,92] define strategies for determining the best pattern for the inverse, [43,42,18,17] define alternative schemes. While at the beginning, these preconditioning methods were received with much skepticism, it is fair to say that substantial progress has been made and a number of recent papers reported that approximate inverse schemes can often be competitive with ILU factorization methods – even in a sequential environment.

One idea for constructing an approximate inverse is to find a sparse matrix M such that $\|AM - I\|$ is small for some convenient norm. Kolotilina and Yeremin [114] presented an algorithm in which the inverse was delivered in factored form, which has the advantage that singularity of M is avoided. In [48] an algorithm is presented which uses the 1-norm for the minimization. We also mention Chow and Saad [43], who use GMRES for the minimization of $\|AM - I\|_F$. Drop-tolerance strategies are applied to limit the amount of fill-in allowed. The approach can also be used to correct explicitly some given implicit approximation, such as a given ILU decomposition.

An elegant approach was suggested by Grote and Huckle [92]. They also attempt to minimize the F-norm, which is equivalent to the Euclidean norm for the errors in the columns m_i of M

$$\|AM - I\|_F^2 = \sum_{i=1}^{n} \|Am_i - e_i\|_2^2.$$

Based on this observation they derive an algorithm that produces the sparsity pattern for the most error-reducing elements of M. This is done in steps, starting with a diagonal approximation, each steps adds more nonzero entries to M, and the procedure is stopped when the norms are small enough or when memory requirements are violated.

9. Multigrid methods

As was mentioned above, among the earliest preconditioning methods were the simple relaxation schemes since these have, historically, been quite popular techniques. Thus Krylov subspace methods were viewed as methods for accelerating such techniques. Another powerful way of accelerating relaxation techniques is to use multigrid – or multilevel methods. Although we have given little emphasis to these methods in this survey, they are nevertheless important methods which can give rise to very efficient solvers, actually of optimal complexity in some cases. The main observation of multigrid techniques is based on a Fourier analysis of the residual (or error) vector of a sequence of iterates that are generated by a scheme such as Jacobi or Gauss–Seidel. This means that these residual vectors are analyzed in the eigen-basis associated with the iteration matrix M – assuming that M has a complete set of eigenvectors. In the case of Jacobi, the observation is that the components associated with the largest eigenvalues (in the original matrix) will decrease rapidly. However, those associated with the smallest eigenvalues will converge much more slowly. As a result after a few steps, the "high-frequency" components may have converged while the "low-frequency" components may have made very little progress in comparison. To correct this situation, researchers developed methods that used several grids. The simplest idea is to use two meshes one fine and one that is coarser, where the fine mesh can be viewed as the result of refining the coarse one. The iteration initially takes place on the fine mesh. After a few steps, the residual is *projected* onto the coarse mesh, by some form of restriction. Let A_{2h} be the matrix for the problem on the coarse mesh and r_{2h} this projected residual. The system $A_{2h}\delta = r_{2h}$ is then solved on the coarse mesh by means of a few steps of relaxation. This is called a correction step. The vector δ is then extrapolated into the finer mesh and the result is added as a correction to the iterate on the fine mesh.

An early paper describing essentially such an idea can be traced back to 1935, when Southwell [167] discusses a "group-relaxation" scheme for a two-level setting. It is clear that we do not have to stop at two levels of meshes. Much later Fedorenko [73] described the first true multigrid technique – which employs more than two grids. The idea laid dormant for some time until Achi Brandt published a series of articles, the first of which in 1972 [26]. The paper [27] provided the needed analysis to boost this class of techniques. Many variants of multigrid methods have been developed and the literature is perhaps richer than that of Krylov subspace methods. The excellent "frontiers in applied mathematics" [124] volume published in 1987 listed already 607 references. A number of excellent books have been written on multigrid. For a quick tutorial see [33]. More complete texts include Hackbusch [96], and Wesseling [199]. The volume [124] contains an excellent collection of articles on multigrid and algebraic multigrid.

It is often asked what is the best method to use: preconditioned Krylov subspace methods or a multigrid approach? Users of iterative techniques are really split in two camps: those who use exclusively multigrid methods and those who use exclusively (preconditioned) Krylov subspace methods. Combination of the two methods have been advocated however, see Section 8.3, but this is an exception rather than a rule. When multigrid techniques work, they can be extremely efficient – far more so than preconditioned Krylov methods. However, their efficiency relies essentially on the inter-level restriction, and prolongation operators, the choice of which will vary from one application to the next. Such efficiencies can be achieved for regular meshes and for smooth elliptic PDEs. Standard multigrid methods cannot be applied without the existence of an underlying mesh – hence its major limitation.

This led to the development of algebraic multi-grid (AMG) initiated by Ruge and Stuben [149]. AMG was defined for algebraic systems – in the same manner as general sparse linear systems solvers – by defining restriction and prolongation operators algebraically. The overall success of AMG, which is derived based on an underlying PDE problem, has been somewhat limited.

10. Outlook

It is rather difficult to predict what the future will bring in the area of iterative methods. However, it is almost certain that the usage of these methods will increase substantially in the application areas. This is partly due to the impact of parallel architectures. Direct methods are more complex to implement in parallel than are iterative methods. Also it is clear that problem sizes are increasing to the point of making direct solvers exceedingly expensive – both in terms of memory and arithmetic costs. One ray of hope for those problems that are hard to solve by iterative techniques, is to combine techniques from direct and iterative solution technologies. As the communities from direct and iterative solvers are getting to learn each other's tricks, the distinction between the two methodologies is getting to be blurred and this results in better, more robust, methods. Indeed, if memory is the only difficulty with direct solvers, it may be possible to find preconditioners that are far more accurate than current ones – but which use moderate amounts of memory.

Acknowledgements

We wish to thank our colleages Michele Benzi, Claude Brezinski, Mark Embree, Gerard Meurant, Nick Trefethen, and Gerard Sleijpen, for their careful reading of the text and for adding useful information. Michele Benzi brought the early (and not very well-known) paper of Cesari, on polynomial preconditioning, to our attention.

References

[1] E.C. Anderson, Y. Saad, Solving sparse triangular systems on parallel computers, Internat. J. High Speed Comput. 1 (1989) 73–96.

[2] W.E. Arnoldi, The principle of minimized iteration in the solution of the matrix eigenvalue problem, Quart. Appl. Math. 9 (1951) 17–29.
[3] O. Axelsson, A generalized SSOR method, BIT 12 (1972) 443–467.
[4] O. Axelsson, Solution of linear systems of equations: iterative methods, in: V.A. Barker (Ed.), Sparse Matrix Techniques, Springer, Berlin, 1977, Copenhagen, 1976, pp. 1–51.
[5] O. Axelsson, Conjugate gradient type-methods for unsymmetric and inconsistent systems of linear equations, Linear Algebra Appl. 29 (1980) 1–16.
[6] O. Axelsson, A general incomplete block-factorization method, Linear Algebra Appl. 74 (1986) 179–190.
[7] O. Axelsson, Iterative Solution Methods, Cambridge University Press, New York, 1994.
[8] O. Axelsson, G. Lindskog, On the eigenvalue distribution of a class of preconditioning methods, Numer. Math. 48 (1986) 479–498.
[9] O. Axelsson, P. Vassilevski, Algebraic multilevel preconditioning methods, I, Numer. Math. 56 (1989) 157–177.
[10] O. Axelsson, P. Vassilevski, Algebraic multilevel preconditioning methods, II, SIAM J. Numer. Anal. 27 (6) (1990) 1569–1590.
[11] O. Axelsson, P.S. Vassilevski, A black box generalized conjugate gradient solver with inner iterations and variable-step preconditioning, SIAM J. Matrix Anal. Appl. 12 (4) (1991) 625–644.
[12] R.E. Bank, T.F. Chan, An analysis of the composite step biconjugate gradient method, Numer. Math. 66 (1993) 295–319.
[13] R. Barrett, M. Berry, T.F. Chan, J. Demmel, J. Donato, J. Dongarra, V. Eijkhout, R. Pozo, C. Romine, H. van der Vorst, Templates for the Solution of Linear Systems: Building Blocks for Iterative Methods, SIAM, Philadelphia, PA, 1994.
[14] T. Barth, T. Manteuffel, Variable metric conjugate gradient methods, in: Advances in Numerical Methods for Large Sparse Sets of Linear Equations, Number 10, Matrix Analysis and Parallel Computing, PCG 94, Keio University, Yokohama, Japan, 1994, pp. 165–188.
[15] D. Baxter, J. Saltz, M.H. Schultz, S.C. Eisenstat, K. Crowley, An experimental study of methods for parallel preconditioned Krylov methods, in: Proceedings of the 1988 Hypercube Multiprocessors Conference, Pasadena, CA, January 1988, pp. 1698–1711.
[16] M.W. Benson, P.O. Frederickson, Iterative solution of large sparse linear systems arising in certain multidimensional approximation problems, Utilitas Math. 22 (1982) 127–140.
[17] M. Benzi, J. Marín, M. Tůma, A two-level parallel preconditioner based on sparse approximate inverses, in: D.R. Kincaid, A.C. Elster (Eds.), Iterative Methods in Scientific Computation, II, IMACS, New Brunswick, NJ, 1999, pp. 167–178.
[18] M. Benzi, C.D. Meyer, M. Tůma, A sparse approximate inverse preconditioner for the conjugate gradient method, SIAM J. Sci. Comput. 17 (1996) 1135–1149.
[19] M. Benzi, D.B. Szyld, A. van Duin, Orderings for incomplete factorization preconditioning of nonsymmetric problems, SIAM J. Sci. Comput. 20 (1999) 1652–1670.
[20] H. Berryman, J. Saltz, W. Gropp, R. Mirchandaney, Krylov methods preconditioned with incompletely factored matrices on the CM-2, J. Partial Distrib. Comput. 8 (1990) 186–190.
[21] G. Birkhoff, Solving elliptic problems: 1930–1980, in: M.H. Schultz (Ed.), Elliptic Problem Solver, Academic Press, New York, 1981, pp. 17–38.
[22] A. Björck, T. Elfving, Accelerated projection methods for computing pseudo-inverse solutions of systems of linear equations, BIT 19 (1979) 145–163.
[23] E. Bodewig, Matrix Calculus, North-Holland, Amsterdam, 1956.
[24] E. Bodewig, Matrix Calculus (second revised and enlarged edition), North-Holland, Amsterdam, 1959.
[25] R. Bramley, A. Sameh, Row projection methods for large nonsymmetric linear systems, SIAM J. Sci. Statist. Comput. 13 (1992) 168–193.
[26] A. Brandt, Multi-level adaptive technique (MLAT) for fast numerical solutions to boundary problems, in: H. Cabannes, R. Temam (Eds.), Proceedings of the Third International Conference on Numerical Methods in Fluid Mechanics, Paris, 1972, Springer, Berlin, 1973, pp. 82–89.
[27] A. Brandt, Multi-level adaptive solutions to boundary value problems, Math. Comp. 31 (1977) 333–390.
[28] C. Brezinski, Padé-Type Approximation and General Orthogonal Polynomials, Birkhäuser, Basel, 1980.
[29] C. Brezinski, Projection Methods for Systems of Equations, North-Holland, Amsterdam, 1997.

[30] C. Brezinski, M. Redivo-Zaglia, Look-ahead in bicgstab and other product methods for linear systems, BIT 35 (1995) 169–201.
[31] C. Brezinski, M. Redivo-Zaglia, H. Sadok, Avoiding breakdown and near-breakdown in Lanczos type algorithms, Numer. Algorithms 1 (1991) 261–284.
[32] C. Brezinski, M. Redivo-Zaglia, H. Sadok, A breakdown-free Lanczos' type algorithm for solving linear systems, Numer. Math. 63 (1992) 29–38.
[33] W.L. Briggs, A Multigrid Tutorial, SIAM, Philadelphia, PA, 1987.
[34] P.N. Brown, A theoretical comparison of the Arnoldi and GMRES algorithms, SIAM J. Sci. Statist. Comput. 12 (1991) 58–78.
[35] G.C. Broyden, A new method of solving nonlinear simultaneous equations, Comput. J. 12 (1969) 94–99.
[36] N.I. Buleev, A numerical method for the solution of two-dimensional and three-dimensional equations of diffusion, Math. Sb. 51 (1960) 227–238 (in Russian).
[37] L. Cesari, Sulla risoluzione dei sistemi di equazioni lineari per approssimazioni successive, Atti Accad. Naz. Lincei. Rend. Cl. Sci. Fis. Mat. Nat. Ser. 6a 25 (1937) 422–428.
[38] T.F. Chan, E. Gallopoulos, V. Simoncini, T. Szeto, C.H. Tong, A quasi-minimal residual variant of the Bi-CGSTAB algorithm for nonsymmetric systems, SIAM J. Sci. Comput. 15 (1994) 338–347.
[39] T.F. Chan, D. Goovaerts, A note on the efficiency of domain decomposed incomplete factorizations, SIAM J. Sci. Statist. Comput. 11 (1990) 794–803.
[40] T.F. Chan, D. Resasco, A framework for the analysis of domain decomposition preconditioners, in: Proceedings of the First International Symposium on Domain Decomposition Methods for Partial Differential Equations, Society for Industrial and Applied Mathematics, Philadelphia, PA, 1988, pp. 217–230.
[41] D. Chazan, M. Miranker, Chaotic relaxation, Linear Algebra Appl. 2 (1969) 199–222.
[42] E. Chow, Y. Saad, Approximate inverse techniques for block-partitioned matrices, SIAM J. Sci. Comput. 18 (1997) 1657–1675.
[43] E. Chow, Y. Saad, Approximate inverse preconditioners via sparse-sparse iterations, SIAM J. Sci. Comput. 19 (1998) 995–1023.
[44] G. Cimmino, Calcolo approssimato per le soluzioni dei sistemi di equazioni lineari, Ric. Sci. Progr. Tecn. Econom. Naz. 9 (1938) 326–333.
[45] P. Concus, G.H. Golub, A generalized conjugate gradient method for nonsymmetric systems of linear equations, in: R. Glowinski, J.L. Lions (Eds.), Computing Methods in Applied Sciences and Engineering, Springer, New York, 1976, pp. 56–65.
[46] P. Concus, G.H. Golub, G. Meurant, Block preconditioning for the conjugate gradient method, SIAM J. Sci. Statist. Comput. 6 (1985) 220–252.
[47] P. Concus, G.H. Golub, D.P. O'Leary, A generalized conjugate gradient method for the numerical solution of elliptic partial differential equations, in: J.R. Bunch, D.J. Rose (Eds.), Sparse Matrix Computations, Academic Press, New York, 1976, pp. 309–332.
[48] J.D.F. Cosgrove, J.C. Díaz, A. Griewank, Approximate inverse preconditioning for sparse linear systems, Internat. J. Comput. Math. 44 (1992) 91–110.
[49] J. Cullum, A. Greenbaum, Relations between Galerkin and norm-minimizing iterative methods for solving linear systems, SIAM J. Matrix Anal. Appl. 17 (1996) 223–247.
[50] J.W. Daniel, The conjugate gradient method for linear and nonlinear operator equations, SIAM J. Numer. Anal. 4 (1967) 10–26.
[51] J.W. Daniel, The Approximate Minimization of Functionals, Prentice-Hall, Englewood Cliffs, NJ, 1971.
[52] E.F. D'Azevedo, F.A. Forsyth, W.P. Tang, Ordering methods for preconditioned conjugate gradient methods applied to unstructured grid problems, SIAM J. Matrix Anal. Appl. 13 (1992) 944–961.
[53] E.F. D'Azevedo, F.A. Forsyth, W.P. Tang, Towards a cost effective ILU preconditioner with high level fill, BIT 31 (1992) 442–463.
[54] C. de Boor, J.R. Rice, Extremal polynomials with applications to Richardson iteration for indefinite systems, SIAM J. Sci. Statist. Comput. 3 (1982) 47–57.
[55] R. Dedekind, Gauss in seiner Vorlesung über die Methode der kleisten Quadrate, Gesammelte Math. Werke 2 (1901) 293–306.
[56] M.A. DeLong, J.M. Ortega, SOR as a preconditioner, Appl. Numer. Math. 18 (1995) 431–440.

[57] J.J. Dongarra, I.S. Duff, D.C. Sorensen, H.A. van der Vorst, Numerical Linear Algebra for High-Performance Computers, SIAM, Philadelphia, PA, 1998.

[58] P.F. Dubois, A. Greenbaum, G.H. Rodrigue, Approximating the inverse of a matrix for use on iterative algorithms on vectors processors, Computing 22 (1979) 257–268.

[59] I.S. Duff, G.A. Meurant, The effect of ordering on preconditioned conjugate gradients, BIT 29 (1989) 635–657.

[60] T. Dupont, R. Kendall, H. Rachford, An approximate factorization procedure for solving self-adjoint elliptic difference equations, SIAM J. Numer. Anal. 5 (1968) 559–573.

[61] T. Eirola, O. Nevanlinna, Accelerating with rank-one updates, Linear Algebra Appl. 121 (1989) 511–520.

[62] S.C. Eisenstat, H.C. Elman, M.H. Schultz, Variational iterative methods for nonsymmetric systems of linear equations, SIAM J. Numer. Anal. 20 (1983) 345–357.

[63] H.C. Elman, A stability analysis of incomplete LU factorizations, Math. Comp. 47 (1986) 191–217.

[64] H.C. Elman, Relaxed and stabilized incomplete factorizations for non-self-adjoint linear systems, BIT 29 (1989) 890–915.

[65] H.C. Elman, G.H. Golub, Line iterative methods for cyclically reduced discrete convection–diffusion problems, SIAM J. Sci. Statist. Comput. 13 (1992) 339–363.

[66] H.C. Elman, Y. Saad, P. Saylor, A hybrid Chebyshev Krylov subspace algorithm for solving nonsymmetric systems of linear equations, SIAM J. Sci. Statist. Comput. 7 (1986) 840–855.

[67] M. Embree, Convergence of Krylov subspace methods for non-normal matrices, Ph.D. Thesis, Oxford University Computing Laboratory, Oxford, UK, 1999.

[68] M. Engeli, Overrelaxation and related methods, Refined Iterative Methods for Computation of the Solution and the Eigenvalues of Self-Adjoint Boundary Value Problems, Institute of Applied Mathematics, Zürich, Birkhäuser, Basel, 1959, pp. 79–107.

[69] M. Engeli, T. Ginsburg, H. Rutishauser, E. Stiefel, Refined Iterative Methods for Computation of the Solution and the Eigenvalues of Self-Adjoint Boundary Value Problems, Birkhäuser, Basel, 1959.

[70] D.J. Evans, The use of pre-conditioning in iterative methods for solving linear equations with symmetric positive definite matrices, J. Inst. Math. Appl. 4 (1968) 295–314.

[71] V. Faber, T. Manteuffel, Necessary and sufficient conditions for the existence of a conjugate gradient method, SIAM J. Numer. Anal. 21 (1984) 352–361.

[72] D.K. Faddeev, V.N. Faddeeva, Computational Methods of Linear Algebra, Freeman and Company, San Francisco, 1963.

[73] R.P. Fedorenko, On the speed of convergence of an iteration process, USSR Comput. Math. Math. Phys. 4 (1964) 227–235.

[74] B. Fischer, L. Reichel, A stable Richardson iteration method for complex linear systems, Numer. Math. 54 (1988) 225–241.

[75] D.A. Flanders, G. Shortley, Numerical determination of fundamental modes, J. Appl. Phys. 21 (1950) 1322–1328.

[76] R. Fletcher, Conjugate gradient methods for indefinite systems, in: G.A. Watson (Ed.), Proceedings of the Dundee Biennal Conference on Numerical Analysis 1974, Springer, New York, 1975, pp. 73–89.

[77] D.R. Fokkema, G.L.G. Sleijpen, H.A. van der Vorst, Generalized conjugate gradient squared, J. Comput. Appl. Math. 71 (1994) 125–146.

[78] S. Frankel, Convergence rates of iterative treatments of partial differential equations, MTAC (1950) 65–75.

[79] R.W. Freund, A transpose-free quasi-minimal residual algorithm for non-Hermitian linear systems, SIAM J. Sci. Comput. 14 (2) (1993) 470–482.

[80] R.W. Freund, N.M. Nachtigal, An implementation of the look-ahead Lanczos algorithm for non-Hermitian matrices, Part 2, Technical Report 90.46, RIACS, NASA Ames Research Center, 1990.

[81] R.W. Freund, N.M. Nachtigal, QMR: a quasi-minimal residual method for non-Hermitian linear systems, Numer. Math. 60 (1991) 315–339.

[82] K. Gallivan, A. Sameh, Z. Zlatev, A parallel hybrid sparse linear system solver, Comput. Systems Eng. 1 (2–4) (1990) 183–195.

[83] N. Gastinel, Analyse Numérique Linéaire, Hermann, Paris, 1966.

[84] C.F. Gauss, Werke, Band IX, Teubner, Leipzig, 1903.

[85] M.K. Gavurin, Application of the best approximation polynomial to improving the convergence of iterative methods, Uspehi Mat. Nauk. 5(3) (1950) 156–160 (in Russian).

[86] G.H. Golub, D.P. O'Leary, Some history of the conjugate gradient and Lanczos algorithms: 1948–1976, SIAM Rev. 31 (1989) 50–102.
[87] G.H. Golub, R.S. Varga, Chebyshev semi iterative methods successive overrelaxation iterative methods and second order Richardson iterative methods, Numer. Math. 3 (1961) 147–168.
[88] A. Greenbaum, Iterative Methods for Solving Linear Systems, SIAM, Philadelphia, PA, 1997.
[89] A. Greenbaum, Z. Strakoš, Predicting the behavior of finite precision Lanczos and conjugate gradient computations, SIAM J. Matrix Anal. Appl. 13 (1992) 121–137.
[90] A. Greenbaum, Z. Strakoš, Matrices that generate the same Krylov residual spaces, in: G. Golub, M. Luskin, A. Greenbaum (Eds.), Recent Advances in Iterative Methods, IMA Volumes in Mathematics and Its Applications, Vol. 60, Springer, New York, 1994, pp. 95–119.
[91] A. Greenbaum, V. Ptak, Z. Strakoš, Any nonincreasing convergence curve is possible for GMRES, SIAM J. Matrix Anal. Appl. 17 (1996) 465–469.
[92] M. Grote, T. Huckle, Parallel preconditioning with sparse approximate inverses, SIAM J. Sci. Comput. 18 (1997) 838–853.
[93] M. Grote, H.D. Simon, Parallel preconditioning and approximate inverses on the connection machine, in: R.F. Sincovec, D.E. Keyes, L.R. Petzold, D.A. Reed (Eds.), Parallel Processing for Scientific Computing, Vol. 2, SIAM, Philadelphia, PA, 1992, pp. 519–523.
[94] I. Gustafsson, A class of first order factorization methods, BIT 18 (1978) 142–156.
[95] M.H. Gutknecht, A completed theory of the unsymmetric Lanczos process and related algorithms, Part I, SIAM J. Matrix Anal. Appl. 13 (1992) 594–639.
[96] W. Hackbusch, Multi-Grid Methods and Applications, Springer, New York, 1985.
[97] W. Hackbusch, Iterative Solution of Large Linear Systems of Equations, Springer, New York, 1994.
[98] A.L. Hageman, D.M. Young, Applied Iterative Methods, Academic Press, New York, 1981.
[99] R.M. Hayes, Iterative methods of solving linear systems on Hilbert space, National Bureau of Standards, Appl. Math. Ser. 39 (1954) 71–103.
[100] M.R. Hestenes, The conjugate-gradient method for solving linear systems, in: Sympos. Appl. Math., Numerical Analysis, Vol. VI, McGraw-Hill, New York, 1956, pp. 83–102.
[101] M.R. Hestenes, E.L. Stiefel, Methods of conjugate gradients for solving linear systems, J. Res. Nat. Bur. Standards Section B 49 (1952) 409–436.
[102] A.S. Householder, Theory of Matrices in Numerical Analysis, Blaisdell Publishing Company, Johnson, CO, 1964.
[103] K.C. Jea, D.M. Young, Generalized conjugate-gradient acceleration of nonsymmetrizable iterative methods, Linear Algebra Appl. 34 (1980) 159–194.
[104] O.G. Johnson, C.A. Micchelli, G. Paul, Polynomial preconditionings for conjugate gradient calculations, SIAM J. Numer. Anal. 20 (1983) 362–376.
[105] W.D. Joubert, Generalized conjugate gradient and Lanczos methods for the solution of nonsymmetric systems of linear equations, Ph.D. Thesis, University of Texas, Austin, Center for Numerical Analysis, Austin, TX, 1990
[106] S. Kaczmarz, Angenäherte Auflösung von Systemen linearer Gleichungen, Bull. Internat. Acad. Pol. Sci. Lett. A 35 (1937) 355–357.
[107] W. Kahan, Gauss–Seidel methods of solving large systems of linear equations, Ph.D. Thesis, University of Toronto, 1958.
[108] C. Kamath, A. Sameh, A projection method for solving nonsymmetric linear systems on multiprocessors, Parallel Comput. 9 (1988/89) 291–312.
[109] S. Kaniel, Estimates for some computational techniques in linear algebra, Math. Comp. 20 (1966) 369–378.
[110] L. Kantorovitch, On an efficient method for solving optimization problems for quadratic functionals, DAN SSSR 48(7) (1945) 7–8 (in Russian).
[111] W. Karush, Convergence of a method of solving linear problems, Proc. Amer. Math. Soc. 3 (1952) 839–851.
[112] I.M. Khabaza, An iterative least-square method suitable for solving large sparse matrices, Comput. J. 6 (1963) 202–206.
[113] L.Yu. Kolotilina, A.Yu. Yeremin, On a family of two-level preconditionings of the incomplete block factorization type, Soviet J. Numer. Anal. Math. Modelling 1 (1986) 293–320.
[114] L.Yu. Kolotilina, A.Yu. Yeremin, Factorized sparse approximate inverse preconditionings I, Theory, SIAM J. Matrix Anal. Appl. 14 (1993) 45–58.

[115] C. Lanczos, An iteration method for the solution of the eigenvalue problem of linear differential and integral operators, J. Res. Nat. Bur. Standards 45 (1950) 255–282.

[116] C. Lanczos, Chebyshev polynomials in the solution of large-scale linear systems, Toronto Symposium on Computing Techniques (1952) 124–133.

[117] C. Lanczos, Solution of systems of linear equations by minimized iterations, J. Res. Nat. Bur. Standards 49 (1952) 33–53.

[118] H. Liebmann, Die angenäherte harmonischer Funktionen und konformer Abbildungen (nach Ideen von Boltzmann und Jacobi), S. B. Math. Nat. Kl. Bayerischen Akad. Wiss. München (1918) 385–416.

[119] D.G. Luenberger, Hyperbolic pairs in the method of conjugate gradients, SIAM J. Appl. Math. 17 (1979) 1263–1267.

[120] M.M. Magolu, Modified block-approximate factorization strategies, Numer. Math. 61 (1992) 91–110.

[121] M.M. Magolu, Ordering strategies for modified block incomplete factorizations, SIAM J. Sci. Comput. 16 (1995) 378–399.

[122] T.A. Manteuffel, The Tchebychev iteration for nonsymmetric linear systems, Numer. Math. 28 (1977) 307–327.

[123] T.A. Manteuffel, Adaptive procedure for estimation of parameter for the nonsymmetric Tchebychev iteration, Numer. Math. 28 (1978) 187–208.

[124] S.F. Mc Cormick (Ed.), Multigrid Methods, SIAM, Philadelphia, PA, 1987.

[125] J.A. Meijerink, H.A. van der Vorst, An iterative solution method for linear systems of which the coefficient matrix is a symmetric M-matrix, Math. Comp. 31 (137) (1977) 148–162.

[126] J.A. Meijerink, H.A. van der Vorst, Guidelines for the usage of incomplete decompositions in solving sets of linear equations as they occur in practical problems, J. Comput. Phys. 44 (1981) 134–155.

[127] G. Meurant, Computer Solution of Large Linear Systems, North-Holland, Amsterdam, 1999.

[128] J.C. Miellou, Algorithmes de relaxation chaotiques á retard, RAIRO R-1 (1975) 55–82.

[129] N.M. Nachtigal, S.C. Reddy, L.N. Trefethen, How fast are nonsymmetric matrix iterations?, SIAM J. Matrix Anal. Appl. 13 (1992) 778–795.

[130] P.A. Nekrasov, Determining unknowns using the least squares method when the number of unknowns is large, Mat. Sb. 12 (1884) 189–204 (in Russian).

[131] Y. Notay, DRIC: a dynamic version of the RIC method, Numer. Linear Algebra Appl. 1 (1994) 511–532.

[132] T.A. Oliphant, An extrapolation process for solving linear systems, Quart. Appl. Math. 20 (1962) 257–267.

[133] C.W. Oosterlee, T. Washio, An evaluation of parallel multigrid as a solver and a preconditioner for singularly perturbed problems, SIAM J. Sci. Statist. Comput. 19 (1991) 87–110.

[134] O. Osterby, Z. Zlatev, Direct Methods for Sparse Matrices, Springer, New York, 1983.

[135] A.M. Ostrowski, Uber die Determinanten mit überwiegender Hauptdiagonale, Comment. Math. Helv. 10 (1937) 69–96.

[136] A.M. Ostrowski, On the linear iteration procedures for symmetirc matrices, Rend. Mat. Appl. 14 (1954) 140–163.

[137] C.C. Paige, M.A. Saunders, Solution of sparse indefinite systems of linear equations, SIAM J. Numer. Anal. 12 (1975) 617–629.

[138] C.C. Paige, M.A. Saunders, LSQR: an algorithm for sparse linear equations and sparse least squares, ACM Trans. Math. Software 8 (1982) 43–71.

[139] B.N. Parlett, Reduction to tridiagonal form and minimal realizations, SIAM J. Matrix Anal. Appl. 13 (1992) 567–593.

[140] B.N. Parlett, D.R. Taylor, Z.S. Liu, A look-ahead Lanczos algorithm for nonsymmetric matrices, Math. Comp. 44 (1985) 105–124.

[141] D. Peaceman, H. Rachford, The numerical solution of elliptic and parabolic differential equations, J. SIAM 3 (1955) 28–41.

[142] C. Pommerell, Solution of large unsymmetric systems of linear equations, Ph.D. Thesis, Swiss Federal Institute of Technology, Zürich, 1992.

[143] G. Radicati di Brozolo, Y. Robert, Parallel conjugate gradient-like algorithms for solving sparse nonsymmetric systems on a vector multiprocessor, Parallel Comput. 11 (1989) 223–239.

[144] E. Reich, On the convergence of the classical iterative method of solving linear simultaneous equations, Ann. Math. Statist. 20 (1949) 448–451.

[145] J.K. Reid, On the method of conjugate gradients for the solution of large sparse systems of linear equations, in: J.K. Reid (Ed.), Large Sparse Sets of Linear Equations, Academic Press, New York, 1971, pp. 231–254.

[146] L.F. Richardson, The approximate arithmetical solution by finite differences of physical problems involving differential equations with an application to the stresses to a masonry dam, Philos. Trans. Roy. Soc. London Ser. A 210 (1910) 307–357.

[147] F. Robert, Contraction en norme vectorielle: convergence d'itérations chaotiques, Linear Algebra Appl. 13 (1976) 19–35.

[148] F. Robert, M. Charnay, F. Musy, Iterations chaotiques série–parallèle pour les équations non linéaires de point fixe, Apl. Math. 20 (1975) 1–38.

[149] A. Ruge, K. Stüben, Algebraic multigrid, in: S. McCormick (Ed.), Multigrid Methods, Frontiers in Applied Mathematics, Vol. 3, SIAM, Philadelphia, PA, 1987 (Chapter 4).

[150] H. Rutishauser, Theory of gradient methods, Refined Iterative Methods for Computation of the Solution and the Eigenvalues of Self-Adjoint Boundary Value Problems, Institute of Applied Mathematics, Zurich, Birkhäuser, Basel, 1959, pp. 24–49.

[151] Y. Saad, Practical use of polynomial preconditionings for the conjugate gradient method, SIAM J. Sci. Statist. Comput. 6 (1985) 865–881.

[152] Y. Saad, Least squares polynomials in the complex plane and their use for solving sparse nonsymmetric linear systems, SIAM J. Numer. Anal. 24 (1987) 155–169.

[153] Y. Saad, Preconditioning techniques for indefinite and nonsymmetric linear systems, J. Comput. Appl. Math. 24 (1988) 89–105.

[154] Y. Saad, Krylov subspace methods on supercomputers, SIAM J. Sci. Statist. Comput. 10 (1989) 1200–1232.

[155] Y. Saad, A flexible inner-outer preconditioned GMRES algorithm, SIAM J. Sci. Statist. Comput. 14 (1993) 461–469.

[156] Y. Saad, Highly parallel preconditioners for general sparse matrices, in: G. Golub, M. Luskin, A. Greenbaum (Eds.), Recent Advances in Iterative Methods, IMA Volumes in Mathematics and Its Applications, Vol. 60, Springer, New York, 1994, pp. 165–199.

[157] Y. Saad, Iterative Methods for Sparse Linear Systems, PWS Publishing, New York, 1996.

[158] Y. Saad, M.H. Schultz, GMRES: a generalized minimal residual algorithm for solving a nonsymmetric linear systems, SIAM J. Sci. Statist. Comput. 7 (1986) 856–869.

[159] J. Saltz, R. Mirchandaney, K. Crowley, Run-time paralellization and scheduling of loops, IEEE Trans. Comput. 40 (1991) 603–612.

[160] P.E. Saylor, D.C. Smolarski, Computing the roots of complex orthogonal kernel polynomials, SIAM J. Sci. Statist. Comput. 9 (1988) 1–13.

[161] J.W. Sheldon, On the numerical solution of elliptic difference equations, MTAC 9 (1955) 101–112.

[162] G.H. Shortley, Use of Tschebyscheff-polynomial operators in the solution of boundary value problems, J. Appl. Phys. 24 (1953) 392–396.

[163] H.D. Simon, Incomplete LU preconditioners for conjugate gradient type iterative methods, in: Proceedings of the SPE 1985 Reservoir Simulation Symposium, Dallas, TX, Society of Petroleum Engineers of AIME, Paper No 13533, 1988, pp. 302–306.

[164] H.D. Simon, Direct sparse matrix methods, in: J.C. Almond, D.M. Young (Eds.), Modern Numerical Algorithms for Supercomputers, Center for High Performance Computing, The University of Texas, Austin, 1989, pp. 325–344.

[165] G.L.G. Sleijpen, D.R. Fokkema, BICGSTAB(ℓ) for linear equations involving unsymmetric matrices with complex spectrum, ETNA 1 (1993) 11–32.

[166] P. Sonnoveld, CGS: a fast Lanczos-type solver for nonsymmetric linear systems, SIAM J. Sci. Statist. Comput. 10 (1989) 36–52.

[167] R.V. Southwell, Stress calculation in frameworks by the method of systematic relaxation of constraints, Proc. Roy. Soc. London Ser. A 151 (1935) 56–95.

[168] W. Spakman, G. Nolet, Imaging algorithms, accuracy and resolution in delay time tomography, in: N.J. Vlaar, N. Nolet, M. Wortel, S. Cloetingh (Eds.), Mathematical Geophysics: A Survey of Recent Developments in Seismology and Geodynamics, Reidel, Dordrecht, 1987, pp. 155–188.

[169] P. Stein, R.L. Rosenberg, On the solution of linear simultaneous equations by iteration, J. London Math. Soc. 23 (1948) 111–118.

[170] E.L. Stiefel, Kernel polynomials in linear algebra and their applications, U.S. Nat. Bur. Standards Appl. Math. Ser. 49 (1958) 1–24.

[171] H.L. Stone, Iterative solution of implicit approximations of multidimensional partial differential equations, SIAM J. Numer. Anal. 5 (1968) 530–558.

[172] Z. Strakoš, On the real convergence rate of the conjugate gradient method, Linear Algebra Appl. 154/156 (1991) 535–549.

[173] K.H. Tan, Local coupling in domain decomposition, Ph.D. Thesis, Utrecht University, Utrecht, The Netherlands, 1995.

[174] K. Tanabe, Projection method for solving a singular system of linear equations and its applications, Numer. Math. 17 (1971) 203–214.

[175] W.P. Tang, Generalized Schwarz splitting, SIAM J. Sci. Statist. Comput. 13 (1992) 573–595.

[176] R.R. Underwood, An approximate factorization procedure based on the block Cholesky factorization and its use with the conjugate gradient method, Technical Report Tech Report, General Electric, San Jose, CA, 1976.

[177] A. van der Ploeg, Preconditioning for sparse matrices with applications, Ph.D. Thesis, Department of Mathematics, University of Groningen, 1995.

[178] A. van der Ploeg, E.F.F. Botta, F.W. Wubs, Nested grids ILU-decomposition (NGILU), J. Comput. Appl. Math. 66 (1996) 515–526.

[179] A. van der Sluis, H.A. van der Vorst, The rate of convergence of conjugate gradients, Numer. Math. 48 (1986) 543–560.

[180] V.V. Voevodin, The problem of a non-selfadjoint generalization of the conjugate gradient method has been closed, USSR Comput. Math. Math. Phys. 23 (1983) 143–144.

[181] H.A. van der Vorst, Iterative solution methods for certain sparse linear systems with a non-symmetric matrix arising from PDE-problems, J. Comput. Phys. 44 (1981) 1–19.

[182] H.A. van der Vorst, A vectorizable variant of some ICCG methods, SIAM J. Statist. Sci. Comput. 3 (1982) 350–356.

[183] H.A. van der Vorst, Large tridiagonal and block tridiagonal linear systems on vector and parallel computers, Parallel Comput. 5 (1987) 45–54.

[184] H.A. van der Vorst, High performance preconditioning, SIAM J. Sci. Statist. Comput. 10 (1989) 1174–1185.

[185] H.A. van der Vorst, ICCG and related method for 3-D problems on vector computers, in: D. Truhlar (Ed.), Workshop on Practical Iterative Methods for Large Scale Computations, Minneaopolis, MN, October 23–25, 1988, Comput. Phys. Comm. 53 (1989).

[186] H.A. van der Vorst, Bi-CDSTAB: a fast and smoothly converging variant of Bi-CG for the solution of non-symmetric linear systems, SIAM J. Sci. Statist. Comput. 13 (1992) 631–644.

[187] H.A. van der Vorst, C. Vuik, GMRESR: a family of nested GMRES methods, Numer. Linear Algebra Appl. 1 (1994) 369–386.

[188] R.S. Varga, Matrix Iterative Analysis, Prentice-Hall, Englewood Cliffs, NJ, 1962.

[189] R.S. Varga, Factorization and normalized iterative methods, in: R.E. Langer (Ed.), Boundary Problems in Differential Equations, University of Wisconsin Press, Madison, 1960, pp. 121–142.

[190] P.K.W. Vinsome, ORTHOMIN: an iterative method solving sparse sets of simultaneous linear equations, Proceedings of the Fourth Symposium on Reservoir Simulation, Society of Petroleum Engineers of AIME, 1976, pp. 149–159.

[191] C. Vuik, H.A. van der Vorst, A comparison of some GMRES-like methods, Linear Algebra Appl. 160 (1992) 131–162.

[192] E.L. Wachspress, CURE, a generalized two-space-dimension multigroup coding for the IBM-704, Technical Report KAPL-1724, Knolls Atomic Power Laboratory, Schenectady, New York, 1957.

[193] E.L. Wachspress, Iterative Solution of Elliptic Systems and Applications to the Neutron Equations of Reactor Physics, Prentice-Hall, Englewood Cliffs, NJ, 1966.

[194] T. Washio, K. Hayami, Parallel block preconditioning based on SSOR and MILU, Numer. Linear Algebra Appl. 1 (1994) 533–553.

[195] J.W. Watts III, A conjugate gradient truncated direct method for the iterative solution of the reservoir simulation pressure equation, Soc. Petroleum Eng. J. 21 (1981) 345–353.

[196] R. Weiss, Error-minimixing Krylov subspace methods, SIAM J. Sci. Comput. 15 (1994) 511–527.

[197] R. Weiss, A theoretical overview of Krylov subspace methods, in: W. Schönauer, R. Weiss (Eds.), Special Issue on Iterative Methods for Linear Systems, Applied Numerical Methods, 1995, pp. 33–56.
[198] R. Weiss, Parameter-Free Iterative Linear Solvers, Akademie, Berlin, 1996.
[199] P. Wesseling, An Introduction to Multigrid Methods, Wiley, Chichester, 1992.
[200] O. Widlund, A Lanczos method for a class of nonsymmetric systems of linear equations, SIAM J. Numer. Anal. 15 (1978) 801–812.
[201] J.H. Wilkinson, The Algebraic Eigenvalue Problem, Clarendon Press, Oxford, 1965.
[202] J.H. Wilkinson, C. Reinsch, Handbook of Automatic Computation, Linear Algebra, Vol. II, Springer, New York, 1971.
[203] D.M. Young, Iterative methods for solving partial differential equations of elliptic type, Ph.D. Thesis, Harvard University, Cambridge, MA, USA, 1950.
[204] D.M. Young, On Richardson's method for solving linear systems with positive definite matrices, J. Math. Phys. 32 (1954) 243–255.
[205] D.M. Young, Iterative Solution of Large Linear Systems, Academic Press, New York, 1971.
[206] D.P. Young, R.G. Melvin, F.T. Johnson, J.E. Bussoletti, L.B. Wigton, S.S. Samant, Application of sparse matrix solvers as effective preconditioners, SIAM J. Sci. Statist. Comput. 10 (1989) 1186–1199.
[207] S.-L. Zhang, GPBi-CG: generalized product-type methods based on Bi-CG for solving nonsymmetric linear systems, SIAM J. Sci. Comput. 18 (1997) 537–551.
[208] Z. Zlatev, Use of iterative refinement in the solution of sparse linear systems, SIAM J. Numer. Anal. 19 (1982) 381–399.

Eigenvalue computation in the 20th century

Gene H. Golub[a, 1], Henk A. van der Vorst[b, *]
[a] *SCCM, Stanford University, Stanford, USA*
[b] *Department of Mathematics, Utrecht University, P.O. Box 80.010, 3508 TA, Utrecht, The Netherlands*

Abstract

This paper sketches the main research developments in the area of computational methods for eigenvalue problems during the 20th century. The earliest of such methods dates back to work of Jacobi in the middle of the 19th century. Since computing eigenvalues and vectors is essentially more complicated than solving linear systems, it is not surprising that highly significant developments in this area started with the introduction of electronic computers around 1950. In the early decades of this century, however, important theoretical developments had been made from which computational techniques could grow. Research in this area of numerical linear algebra is very active, since there is a heavy demand for solving complicated problems associated with stability and perturbation analysis for practical applications. For standard problems, powerful tools are available, but there still remain many open problems. It is the intention of this contribution to sketch the main developments of this century, especially as they relate to one another, and to give an impression of the state of the art at the turn of our century.

MSC: 65F15; 65N25

Keywords: QR; QZ; Lanczos' method; Arnoldi's method; Jacobi–Davidson; Power iteration; RKS; Jacobi's method; SVD; Perturbation theory

1. Sources

Numerical linear algebra is a very active field of research. Many problems are challenging of themselves, and in addition, much of scientific computing depends critically in one way or another on numerical linear algebra algorithms. Not only do the more classical scientific computational models for physical or engineering problems depend on linear algebra kernels, but many modern applications, such as information retrieval and image restoration, profit from numerical linear algebra

* Corresponding author.
E-mail addresses: golub@sccm.stanford.edu (G.H. Golub), vorst@math.uu.nl (H.A. van der Vorst).
[1] The work of this author was completed under a grant from the DOE: DE-FG03-97ER35330.

results. These factors have motivated numerical linear algebra research throughout the entire 20th century.

The field has blossomed, especially since the introduction of the modern computer, roughly from the early 1950s. This is evident from the large number of scientific journals in which articles in this area appear: *SIAM J. on Matrix Analysis and Applications* (*SIMAX*), *Linear Algebra and its Applications* (*LAA*), *Numerical Linear Algebra with Applications* (*NLAA*), are completely devoted to this specialty. Articles on numerical linear algebra, theoretical as well as applied, regularly appear in journals such as *BIT*, *SIAM J. Numerical Analysis*, *SIAM J. on Scientific Computing*, *J. on Computational and Applied Mathematics*, *J. Applied Numerical Mathematics*, *Numerische Mathematik*, *Numerical Algorithms*, *Mathematics of Computation*, *Parallel Computing*, *ACM Transactions on Mathematical Software*, *Computing*, *J. Inst. Math. Applic.*, *SIAM Review*, *IMA J. Num. Anal.*, and several others in more application oriented directions, such as *J. Computational Physics* and engineering journals. And from, for instance, the bibliography in Golub and Van Loan's book [51], one can see how many papers are referenced from these and other sources. A quick glance through the contents of the average 60 papers per year in SIMAX shows that roughly 40% of the papers are associated with eigenvalue problem research, and it is likely that this holds more or less for the many papers per year that focus on numerical linear algebra.

This makes any attempt to write a complete overview on the research on computational aspects of the eigenvalue problem a hopeless task. It also serves as an excuse for the incompleteness in the current overview. We have tried to highlight what seemed most important from our point of view. We have included references to main sources, and we have made a personally colored selection of references to more specialized details. Instead of trying to give an overview of all sorts of different approaches that have been followed to solve aspects of eigenproblems, we will try to emphasize the history of those methods that, in our opinion, still play a role. Our aim is to consider the algorithmic developments from a historical point of view and to indicate how the recent powerful techniques are the result of many smaller steps. This will also help to show how many of the algorithms are interrelated; we hope not to get lost in sidesteps. The reader who is interested in methods that have played a role but that are at present no longer considered to be on the main track, is referred to Householder's and Wilkinson's books [64,154]. In addition, Parlett [100] gives interesting historical information on older methods that still have some significance from a theoretical point of view.

In order to be active in this area of research, or to be informed about special aspects, then one might be interested in our main sources:

- Wilkinson: The Algebraic Eigenvalue Problem [154].
- Householder: The Theory of Matrices in Numerical Analysis [64].
- Wilkinson and Reinsch: The Handbook [158].
- Parlett: The Symmetric Eigenvalue Problem [100].
- Stewart and Sun: Matrix Perturbation Theory [129].
- Watkins: Fundamentals of Matrix Computations [150].
- Golub and Van Loan: Matrix Computations [51].
- Chatelin: Spectral Approximation of Linear Operators [18].
- Saad: Numerical Methods for Large Eigenvalue Problems [116].
- Demmel: Applied Numerical Linear Algebra [28].
- Trefethen and Bau: Numerical Linear Algebra [137].

- Arpack Guide (Lehoucq, Sorensen and Yang) [81].
- Dongarra et al.: Numerical Linear Algebra for High Performance Computers [31].
- Bai's paper in Numerical Lin. Alg. with Appl. [6].
- Watkin's paper in SIAM Review [151].
- Dongarra and Walker on Software [33].
- Wilkinson: State of the Art overview [157].
- van der Vorst and Golub: State of the Art paper [50].

By regularly examining the dedicated numerical linear algebra journals, one should be able to trace most of the relevant and interesting papers for further investigations and research.

It should be noted that we have concentrated on algebraic eigenvalue problems in this paper. For eigenvalue problems related to, for instance, PDEs, one may use methods that exploit the nature of the PDE or the expected behaviour of the solution. We have not considered such specialized techniques (of which multigrid is a good example).

2. Introduction

The eigenvalue problem for square matrices A, that is the determination of nontrivial solutions of $Ax = \lambda x$, is a central topic in numerical linear algebra. It is inherently nonlinear and this leads to many computational problems. Computation of the eigenvalues λ via the explicit construction of the characteristic equation

$$\det(A - \lambda I) = 0$$

is, except for very special cases, not an option since the coefficients of the characteristic equation cannot be computed from determinant evaluations in a numerically stable way. And even if the characteristic equation could be determined accurately, then the computation of its roots, in finite precision, may be highly unstable since small perturbations in the coefficients may lead to large perturbations of the roots. The numerical computation of the associated eigenvectors and generalized eigenvectors is even more delicate, in particular when eigenvectors of A make small angles with each other. In the limiting case, when the matrix is defective, A can be reduced to the Jordan canonical form, but arbitrary small perturbations in A may yield a nondefective matrix. This leads to many challenging numerical questions, which give rise to the central problem: how can we compute eigenvalues and eigenvectors in an efficient manner and how accurate are they?

In fact, this was already recognized by Jacobi, who, in 1846, computed the eigenvalues of symmetric matrices by rotating the matrix to a strongly diagonally dominant one. We will return to this later, since Jacobi's techniques are still relevant and have led to popular and powerful algorithms.

Another longstanding method that is of great significance and serves as the basis for many algorithms is the Power iteration. The method is based on the idea that if a given vector is repeatedly applied to a matrix, and is properly normalized, then ultimately, it will lie in the direction of the eigenvector associated with the eigenvalues which are largest in absolute value. The rate of convergence for the Power iteration depends on the ratio of the second largest eigenvalue (in absolute value) to the largest eigenvalue (in absolute value) and for many applications this leads to unacceptably slow convergence. The method can be problematic if one wants to compute a number of extremal eigenvalues. The Power iteration is still in use, but most frequently as (implicit) part of more efficient techniques, e.g., Krylov methods, inverse iteration, QR-method.

What becomes clear is that all these methods are of an iterative nature, and this is necessarily the case, since if there were a method of computing the eigenvalues of an nth-order matrix in a finite number of computations, depending only on n, then this would be in contradiction with the fundamental theorem of Abel–Ruffini (and also a well-known result in Galois theory) that no such algorithm exists for the computation of the roots of a general polynomial of degree greater than 4. Hence, an algorithm for a matrix with a general structure (that is, not a diagonal matrix or a triangular matrix or alike) is necessarily iterative and the problem is to identify iterative algorithms which have a fast rate of convergence and lead to accurate results.

In solving an eigenvalue problem there are a number of properties that need be considered. These greatly affect the choice of algorithm. We list below a number of questions that an investigator needs to consider in solving a particular problem.

- Is the matrix real or complex?
- What special properties does the matrix have?
  - symmetric,
  - Hermitian,
  - skew symmetric,
  - unitary.
- Structure?
  - band,
  - sparse,
  - structured sparseness,
  - Toeplitz.
- Eigenvalues required?
  - largest,
  - smallest in magnitude,
  - real part of eigenvalues negative,
  - sums of intermediate eigenvalues.

As well as the standard eigenproblem, there are a variety of more complicated eigenproblems, for instance $Ax = \lambda Bx$, and more generalized eigenproblems like $Ax + \lambda Bx + \lambda^2 Cx = 0$, higher-order polynomial problems, and nonlinear eigenproblems. All these problems are considerably more complicated than the standard eigenproblem, depending on the operators involved. However, as the standard eigenproblem has become better understood, in a numerical sense, progress has been made in the other problems and we will consider developments in solving these problems.

3. Canonical forms

The standard approach for the numerical solution of the eigenproblem is to reduce the operators involved to some simpler form, that yields the eigenvalues and eigenvectors directly, for instance, diagonal form. The idea is that the transformation be made with orthogonal operators as often as possible, in order to reduce the effects of perturbations.

The easiest situation is the symmetric case: for a real symmetric matrix, there exists an orthogonal matrix Q, so that $Q^{\mathrm{T}}AQ = D$, where D is a diagonal matrix. The diagonal elements of D are the eigenvalues of A, the columns of Q are the corresponding eigenvectors of A.

Unsymmetric matrices do not in general have an orthonormal set of eigenvectors, and may not have a complete set of eigenvectors, but they can be transformed unitarily to Schur form:

$$Q^*AQ = R,$$

in which R is upper triangular. In fact, the symmetric case is a special case of this Schur decomposition, since a symmetric triangular matrix is clearly diagonal. Apart from the ordering of the eigenvalues along the diagonal of R and the sign of each column of Q, the matrix Q is unique. Van Dooren [146] has described an algorithm for the orthogonal transformation of Q, so that the eigenvalues appear in prescribed order along the diagonal of R. If the eigenvalues are distinct then there exists a nonsingular matrix X (in general not orthogonal), that transforms A to diagonal form

$$X^{-1}AX = D.$$

An unsymmetric matrix can be transformed to Jordan form by a nonsingular matrix X. This Jordan matrix may have upper bidiagonal blocks along the diagonal. Each of these blocks has identical eigenvalues and the upper bidiagonal elements are equal, and most often set to 1. The numerical computation of the Jordan form is highly unstable, since a small perturbation suffices to obtain a matrix with different eigenvalues (and possibly a complete eigenvector system). Small angles between (some of) the eigenvectors reveal that A is close to a matrix that is similar to a nondiagonal Jordan form. For a discussion on how to compute elementary Jordan blocks (with the help of the singular value decomposition), see the 1976 paper by Golub and Wilkinson [52].

Just as the Jordan canonical form describes the eigenstructure of a matrix, the Kronecker form does this for matrix pencil $A - \lambda B$, even for rectangular A and B. For details on this we refer to papers by Wilkinson [155,156,143], and Kågström [66]. The latter has also developed software for the computation of the Kronecker structure [67]. Wilkinson, in his discussion on the progress made in the period 1976–1986 in eigenvalue computations [157], noted that the Jordan canonical and Kronecker canonical forms were largely regarded as irrelevant by numerical analysts because of their ill-posedness. He even stated: “Many felt that I should have ignored the Jordan canonical form in the Algebraic Eigenvalue Problem [154] and I had misgivings about including a discussion of it”. Since the 1970s, this has changed, and contributions have been made by many, including Demmel, Beelen, Van Dooren, Chaitin-Chatelin, Edelman, Kågström, Nichols, Kautsky, Golub, and Wilkinson. Although serious attempts have been undertaken for the computation of the Kronecker canonical form, by for instance Kågström and Van Dooren, this still needs further research. Also the computation of invariant subspaces of highly nonnormal matrices is still in its infancy, notwithstanding useful contributions by, for instance, Chaitin-Chatelin et al. [17,12] and Lee [80]. For recent references, see [4]. Van Dooren described, in papers published in 1981, how the Kronecker form can be used in system control problems (input–output systems) [145,144].

Related to eigendecompositions is the singular value decomposition. Let A be a real $m \times n$ matrix, then there exists an orthogonal $m \times m$ matrix U and an orthogonal $n \times n$ matrix V, such that

$$U^{\mathrm{T}}AV = \mathrm{diag}(\sigma_1, \ldots, \sigma_p),$$

with $p = \min\{m, n\}$, and $\sigma_1 \geqslant \sigma_2 \geqslant \cdots \geqslant \sigma_p \geqslant 0$. The values σ_i are the *singular values* of A, the columns v_i of V are the *right singular vectors* and the columns u_i of U are the *left singular vectors*. The number of nonzero singular values is equal to the rank of the matrix A. The singular value decomposition (SVD) plays an important role in numerical linear algebra, for instance in the solution of underdetermined or overdetermined linear systems.

4. Perturbation theorems

Perturbation theorems play a very essential role in computational processes for eigenproblems. As we have remarked, eigensolvers are essentially iterative processes and many of them rest on the principle of reducing a matrix to a special/simpler form, either diagonal or upper triangular. One has then to decide when a matrix is sufficiently close to the appropriate ultimate form. It is important to know what the approximation errors imply about the desired eigen information. A modern treatment of perturbation theory for a variety of eigenproblems is given in the book by Stewart and Sun [129]. We will restrict ourselves here to what we regard as some of the most relevant results in this area.

Many theoretical results rest on the famous Gershgorin Disc Theorem, which states that the eigenvalues of a matrix $A = (a_{ij})$ are located in the union of circles with centre a_{ii} and radius $\sum_{j \neq i} |a_{ij}|$. This theorem first appeared in a classic paper by Gerschgorin in 1931 [44]. A very useful refinement shows that if a group of s discs D_i is isolated from the others, then this group contains precisely s eigenvalues. In particular, if one disc is isolated then this disc contains one eigenvalue. This particular case is of great interest, since it can be used for stopping criteria in actual computations.

Wilkinson [154] discussed the application of Gerschgorin's theorem to various situations. He mentioned the discussion of useful extensions by Taussky in her 1949 paper [131]: Varga acknowledged Taussky's paper in his work on solving systems of linear equations by iterative methods (cf. [148]). An important extension, using block matrices, of the Gerschgorin Theorem was given by Feingold and Varga [38].

The eigenvalues depend continuously on the elements of A and if the ith eigenvalue λ_i is distinct, then it is even differentiable. In this case one can carry through a first-order perturbation analysis (cf. [129, p. 185]). This leads to the observation that if a matrix A is perturbed by δA, then the perturbation to λ_i is in first order of terms of δA given by

$$\delta\lambda_i = \frac{1}{y_i^* x_i} y_i^* \delta A x_i,$$

where x_i, and y_i are the normalized right and left eigenvectors, respectively, corresponding to λ_i, and y_i^* denotes the complex conjugate of y_i. The factor $1/y_i^* x_i$ is referred to as the *condition number* of the ith eigenvalue. The Bauer–Fike result (1960) [9], which is actually one of the more famous refinements of Gershgorin's theorem, makes this more precise: the eigenvalues $\tilde{\lambda}_j$ of $A + \delta A$ lie in discs B_i with centre λ_i, and radius $n(\|\delta A\|_2 / |y_i^* x_i|)$ (for normalized x_i and y_i).

The Courant–Fischer minimax theorem is the basis of many useful results. For a symmetric matrix A with ordered eigenvalues $\lambda_n \leqslant \cdots \leqslant \lambda_2 \leqslant \lambda_1$ it states that the eigenvalues are the stationary values of the *Rayleigh quotients*:

$$\lambda_k = \max_{dim(S)=k} \min_{0 \neq y \in S} \frac{y^* A y}{y^* y},$$

for $k = 1,2,\dots,n$. Some important consequences are the following.

For symmetric matrices, Weyl in 1912 proved an important property for symmetric perturbations δA:

$$|\tilde{\lambda}_i - \lambda_i| \leqslant \|\delta A\|_2.$$

In fact, Weyl gave even more detailed results in terms of the eigenvalues of A and of δA: let the eigenvalues of δA be denoted by $\eta_n \leqslant \cdots \leqslant \eta_2 \leqslant \eta_1$, then the eigenvalues $\tilde{\lambda}_i$ of $A+E$ satisfy

$$\lambda_k + \eta_n \leqslant \tilde{\lambda}_k \leqslant \lambda_k + \eta_1.$$

These inclusion formulas were later refined to, in Parlett's terminology, a blizzard of results, independently obtained by Kahan (1957)[2] and Weinberger (1974) [153]. See Parlett's book [100, Chapter 10.6] for an exposé of these results. It is interesting to note that by 1990 the theory had evolved to such an extent that Weyl's result and the Kato–Temple results could be left as exercises in Stewart and Sun's book [129, p. 210–211]. This illustrates a rich and powerful framework.

Another important property that plays a big role in iterative (projection type) algorithms, is the interlace property. Let A_r denote the leading $r \times r$ minor of A, with eigenvalues $\lambda_j^{(r)}$, then

$$\lambda_{r+1}^{(r+1)} \leqslant \lambda_r^{(r)} \leqslant \lambda_r^{(r+1)} \leqslant \cdots \leqslant \lambda_2^{(r+1)} \leqslant \lambda_1^{(r)} \leqslant \lambda_1^{(r+1)}.$$

An important result, that underlies the powerful *divide and conquer* method, comes from rank-one perturbations. If $B = A + \tau cc^{\mathrm{T}}$, with $\|c\|_2 = 1$, and real $\tau \geqslant 0$, then the ith eigenvalue of B is in the interval $[\lambda_i,\ \lambda_{i-1}]$, for $\tau \leqslant 0$ it is in $[\lambda_{i+1}, \lambda_i]$. In either case, there exist nonnegative $\alpha_1,\dots,\alpha_n$ with $\sum_i \alpha_i = 1$, such that $\lambda_i(B) = \lambda_i + \alpha_i \tau$.

Further details and results can be found in most books on the (numerical) eigenproblem; in particular Wilkinson's book [154] is a great source. A good overview of these results and similar results for invariant subspaces is given in [51]. From a result formulated as the Kato–Temple theorem, one can obtain sharp bounds for the Rayleigh quotients for symmetric matrices. This rests on work of Temple (1933) [132] and Kato (1949) [70]; more extended work in this direction has been done by Davis and Kahan (1970) [27] (see also [18, p. 46], [28, Chapter 5.2]). In these results the *gap* for the ith eigenvalue plays an essential role: $gap(i,A) \equiv \min_{j\neq i}|\lambda_j - \lambda_i|$. A small gap indicates a sensitive eigenvector. In particular, let x denote a normalized vector with Rayleigh quotient $\beta = x^{\mathrm{T}}Ax$, and residual $r = Ax - \beta x$. Then there is a (normalized) eigenvector q_i, corresponding to λ_i, for which

$$|\lambda_i - \beta| \leqslant \frac{\|r\|_2^2}{\mathrm{gap}(i,A)}, \qquad \sin(\theta) \leqslant \frac{\|r\|_2}{\mathrm{gap}(i,A)},$$

where θ denotes the angle between x and q_i. These results show the superior quality of a Rayleigh quotient from a given subspace. It is exploited in modern iterative methods, such as the Lanczos algorithm, but it is also essential in the QR algorithm. Related to the perturbation analysis for Rayleigh quotients is work of Kaniel (1966) [69] for errors in the Ritz approximations computed in the Lanczos process. For a comprehensive discussion of this, see [100, Chapter 12].

---

[2] According to Parlett [100, p. 203], who also mentions unpublished (?) results of Weinberger (1959).

5. Jacobi's method

For our discussion of the Jacobi method, we have used the following sources: [154, Chapter 5], [100, Chapter 9], [51, Chapter 8.4]. The Jacobi method which was originally proposed in 1846 [65], reduces a real symmetric matrix to diagonal form by a sequence of plane rotations. Jacobi, however, did not use the method to full convergence, as was done in the 20th century, but combined it with an iterative process (Gauss–Jacobi iterations) for the missing component of the desired eigenvector (for which he took a unit vector as an initial guess). Actually, Jacobi's technique may be viewed as a form of preconditioning for the Gauss–Jacobi iteration that he also used to solve linear least-squares systems. This has escaped the attention of most researchers that were active with Jacobi's method; the exception seems to be Bodewig [2, pp. 280–287]. The preconditioning part of the method, as an iterative technique to diagonalize a matrix, was reinvented in 1949 by Goldstine et al. and published in a manuscript. After Ostrowski had pointed out that this was actually a rediscovery of Jacobi's method, [3] the adapted manuscript was published only in 1959 [46]. [4] According to Wilkinson [154, p. 343] Jacobi's method was already being used independently on desk computers at the National Physical Laboratory in 1947. From 1950 on, the method got much attention. In the classical process, the maximal off-diagonal element is annihilated, and this guarantees convergence. Since it is a time-consuming process to determine the maximal element after each rotation, cyclic procedures were suggested, (cf. [54]). Later, threshold strategies were developed in order to avoid wasting time in trying to annihilate tiny elements [105]. Quadratic convergence for the cyclic Jacobi algorithm was proven, under various assumptions, by Henrici (1958) [59], Schönhage (1961) [119], Wilkinson (1962) [154], and van Kempen (1966) [147]. This rate of convergence sets in after a number of sweeps (that is $(n-1)n/2$ elementary rotations), but there is no rigorous bound on the number of sweeps required to achieve a specified accuracy. Brent and Luk [14] argued that this number is proportional to $\log(n)$, which is in line with Parlett's remark [100, p. 181] that after three or four sweeps through all the off-diagonal elements convergence is usually very rapid.

The success of the Jacobi method for diagonalizing a symmetric matrix by orthogonal similarity transformations inspired many investigators to find a similar method for nonsymmetric matrices. It was quickly realized that the Schur form was the appropriate decomposition. John Greenstadt, in 1955 [53], was one of the earliest investigators to develop such a method (indeed, Greenstadt made von Neumann aware of this canonical form). Unfortunately, these earliest attempts were not successful. The QR method, that gained more popularity somewhat later, can be viewed, however, as a Jacobi like method, since it can produce the Schur form via a sequence of similarity transformations composed of rotations.

Rutishauser made an Algol60 implementation of Jacobi's process, as a contribution to the Wilkinson Reinsch collection [158]. In the 1960s, the popularity of the method declined, because of the growing popularity first of the Givens method and slightly later, the Householder method: these latter two methods first reduced the matrix to tridiagonal form and then used an efficient procedure for computing the eigenvalues of the tridiagonal matrix. Interest in the Jacobi returned with the advent

---

[3] Michele Benzi brought to our attention that this story is narrated on p. 294 in Goldstine's book *The Computer from Pascal to von Neumann*, Princeton University Press, 1972; it is also nicely covered in the Master Thesis of Anjet den Boer [30] on the history of Jacobi's method.

[4] Parlett [100, p. 184] dates the reinvention in 1946, by Bargmann et al. [7], but this is presumably a misprint.

of parallel computers, starting with a paper of Sameh (1971) [118], and followed by others in the 1980s. Variants of Jacobi's method were proposed; we mention the extensions for normal matrices, by Paardekooper in 1971 [96], and to nonnormal matrices by Eberlein, in 1970 [36]. The latter process was also part of the Wilkinson and Reinsch collection (the Algol60 procedure *eigen*). In order to improve data locality for distributed memory computers, block Jacobi methods were suggested, see Bischof [10] for a discussion on the influence of the solution of the subproblems on the overall process.

Another interesting feature of the Jacobi method is its superior behaviour with respect to accuracy. Wilkinson analysed this and showed that the relative error in the eigenvalue approximations is eventually reduced to the order of the condition number of A times machine precision. This was perfected in 1992, by Demmel and Veselič [29], who showed that for symmetric positive-definite matrices, the condition number of A could be replaced by that of the matrix symmetrically scaled by the diagonal. If one is satisfied with less accuracy, then for large-scale computations, Jacobi's method is no longer regarded as competitive, not even for modern parallel computers.

6. Power method

For our discussion of the Power method, we have borrowed material from Householder's book [64]. The Power method, for general square matrices, is the simplest of all the methods for solving for eigenvalues and eigenvectors. The basic idea is to multiply the matrix A repeatedly by a well-chosen starting vector, so that the component of that vector in the direction of the eigenvector with largest eigenvalue in absolute value is magnified relative to the other components. Householder called this *Simple Iteration*, and attributed the first treatment of it to Müntz (1913). Bodewig [2, p. 250] attributes the power method to von Mises [149], and acknowledges Müntz for computing approximate eigenvalues from quotients of minors of the *explicitly* computed matrix A^k, for increasing values of k. For a careful analytic treatment of the Power method, Householder acknowledged work by Ostrowski and Werner Gautschi; the reader can find a fairly complete treatment in Wilkinson's book [154] together with the proper references. The speed of convergence of the Power iteration depends on the ratio of the second largest eigenvalue (in absolute value) to the largest eigenvalue (in absolute value). In many applications this ratio can be close to 1 – this has motivated research to improve the efficiency of the Power method. It is interesting that the most effective variant is the inverse Power method, in which one works with the matrix $(A - \mu I)^{-1}$, and this variant was proposed as late as 1944 by Wielandt (Wielandt's *fractional iteration*). Wielandt also proposed continuing the process after the largest eigenvalue has converged, by working with the deflated matrix $A - \lambda vv^*$, for which λ, v is the computed eigenpair (with $\|v\|_2 = 1$), associated with the largest eigenvalue in magnitude. (The deflation procedure outlined here is for symmetric matrices. For unsymmetric matrices it is necessary to work with at least two vectors; the choice of one of the vectors may not be unique.) This is called *implicit deflation*; another possibility is to keep the iteration vectors orthogonal to the computed eigenvector(s): *explicit deflation*. A compact description and analysis of these deflation techniques was given by Parlett [100]. The Power method and the Inverse Power method, in their pure form are no longer competitive methods even for the computation of a few eigenpairs, but they are still of interest since they are explicitly or implicitly part of most modern methods such as the QR method, and the methods of Lanczos and Arnoldi. These methods evolved in some way or another from the Power method and some of the techniques that were suggested as improvements to the Power method are still in use as acceleration techniques for modern iterative methods. One of these ideas is to work with polynomials of A, with the purpose of damping unwanted parts of the spectrum.

Another possibility is working with properly updated shifts μ in the inverse process and, in particular, if one takes the Rayleigh quotient with the most recent vector as a shift, then one obtains the *Rayleigh quotient iteration*. According to Parlett [100, p. 71], Lord Rayleigh used in the 1870s a less powerful technique: he did a single shift-and-invert step with a Rayleigh quotient for an eigenvector approximation, but with a unit vector as the right-hand side. (This saves the refactoring of the matrix $(A - \mu I)$ at each iteration.) The modern RQI, in which one takes the most current eigenvector approximation as the right-hand side, leads to very fast convergence. Ostrowski, in a series of six papers [95], studied the convergence properties for variance of RQI for the symmetric and unsymmetric case. He was able to establish cubic convergence in both cases under various circumstances (in the unsymmetric case for a properly generalized Rayleigh quotient). These results are essential for the understanding of modern iterative techniques that are based on (approximate) shift-and-invert strategies (for example, the Jacobi–Davidson method, see below).

A step forward was to work with a set of independent vectors in order to find a number of eigenvectors, instead of the deflation procedure suggested by Wielandt. A problem with the Power method is the determination of eigenvalues that have equal modulus, for instance, finding a conjugate pair of eigenvalues of a real unsymmetric matrix. It is therefore quite natural to work with a couple of independent vectors: this was first suggested in 1937 by Horst [62]. The next step that seems logical, in hindsight, is to force the vectors to be independent. This was initially done (cheaply) by Gaussian transformations by Bauer [8] in 1957, and led to *Treppeniteration*. If the set of vectors is denoted as a matrix L_s (an n by s unit lower-triangular matrix), then one forms AL_s and factors the resulting matrix, by Gaussian transformations, as $L_{s+1}R_{s+1}$. If the eigenvalues are distinct, then the $s \times s$ upper triangular matrix R_{s+1} converges, for increasing s, to a matrix whose eigenvalues are those of the dominant subspace on its diagonal. Rutishauser [111] made the important observation that if we factor A as $A = LR$ (again L unit lower triangular), then the similar matrix $L^{-1}AL = L^{-1}LRL = RL$. He proposed decomposing RL again, and repeating this process in an iterative fashion. This R also converges to an upper triangular matrix, and L is a unit matrix. This is the *LR method* of Rutishauser. The correspondence between Treppeniteration and LR is that if we start Treppeniteration with a unit full matrix, then in exact arithmetic we obtain the same matrices R in the process. For an efficient implementation, the matrix A is first reduced to an upper Hessenberg matrix. The LR method maintains this form throughout the process, and this makes LR computationally very attractive. Rutishauser's observation that permuting the factors of the matrix is equivalent to performing a similarity transformation was a key step. Wilkinson [154, p. 485] commented: "*In my opinion its development* (*i.e. of LR*) *is the most significant advance which has been made in connection with the eigenvalue problem since the advent of automatic computers*". However, Bauer's technique could be applied to a smaller set of starting vectors and it does not modify the matrix A. For this reason, in the words of Householder [64, p. 195], it is self-correcting. This seems to imply that Treppeniteration leads to more accurate results.

Since orthogonal reduction techniques often evidence superior stability properties, it became apparent that the LR factorization should be replaced by a QR factorization. This leads to one of the most popular and powerful methods of our time for eigenvalue problems: the QR method for computing all of the eigenvalues and associated eigenvectors of a dense symmetric matrix. (In fact, the QR method has essential enhancements that make the method really powerful; we will discuss this in another section.)

With the number of vectors less than n, this Power method in combination with QR orthogonalization is known as the *Simultaneous Iteration method*; Rutishauser studied this method in 1969 [112], see also [113]. Its convergence behaviour for general unsymmetric matrices was studied by Stewart [126] in 1976. Stewart also developed a subroutine, based on simultaneous iteration, for the computation of a dominant invariant subspace. This routine, SRRIT [127], was further improved in 1992, and made available for general use through Netlib [5].

The collection of vectors generated by the Power method define Krylov subspaces of increasing dimension. This motivated Krylov to try to determine the characteristic polynomial of a matrix by inspecting the dependence of a full set of these vectors. This procedure may fail because the system of equations is highly ill-conditioned but this can be repaired by orthogonalizing each new vector to the previous vectors and applying A onto the last constructed vector. This iteration process is known as the Lanczos method for symmetric matrices, and Arnoldi's method for unsymmetric matrices. We will discuss these Krylov methods below.

Our presentation might suggest that the Krylov methods have overshadowed the Simultaneous Iteration method, and for most situations this is indeed the case. Parlett, however, [100, p. 289] described situations where Simultaneous Iteration is still competitive. For instance, if we can store only a limited number of n-vectors in fast memory, or if the relative gap between the desired eigenvalues and the others is great, then Simultaneous Iteration is very useful.

7. Reduction algorithms

Early computational techniques, other than Jacobi's famous but slowly converging diagonalization method, and the unsatisfactory Power method with its many restrictions, attempted to exploit the fact that every matrix satisfies its characteristic equation. To this end, Krylov suggested in 1931 [73], using the vectors $x, Ax, A^2x, \ldots$, generated by the Power method, to determine the coefficients of the characteristic equation. This was not successful, because, as we have learned from Wilkinson's analysis [154] the roots of a polynomial may vary widely with only tiny perturbations to the coefficients of the polynomial. Even rounding the exact coefficients in floating point arithmetic may destroy much accuracy in many of the roots. Although Krylov's method failed, his name is still attached to the subspace generated by the Power method.

There is yet another reason for the failure of Krylov's method in finite precision arithmetic: the vectors generated by the Power method tend to converge in the direction of the eigenvectors associated with the dominating eigenvalues. Hence, the computed vectors for the subspace necessarily yield a very ill-conditioned basis. Checking mutual dependence of this basis, as is required in order to construct the characteristic polynomial, is an almost impossible task.

An early attempt to reduce the matrix A to a form that lends itself better for solving the characteristic equation was suggested by Hessenberg [60]. He suggested to compute a modified Krylov basis by making a set of basis vectors for the Krylov subspace, orthogonal to a given test-set, for instance the canonical basis vectors. This led to a reduction of A to upper Hessenberg form. This technique is very close to the techniques by Lanczos and Arnoldi.

In 1950, Lanczos [78] suggested building a basis for the Krylov subspace in a more stable way by orthogonalizing the vectors as they are constructed. The idea was to immediately orthogonalize the new vector for the Krylov subspace with respect to the already existing orthonormal basis. The difference with Hessenberg's approach is that Lanczos (and slightly later also Arnoldi) took the Krylov vectors themselves for the test-set.

The new vector for expansion is created by applying A to the latest orthogonal basis vector. For symmetric matrices A, this leads to a three-term recurrence relation between the basis vectors v_j and in exact arithmetic this can be formulated as

$$AV_j = V_jT_j + \gamma_j v_{j+1} e_{j+1}^{\mathrm{T}}.$$

Obviously, this recursion must terminate for some $j \leqslant n$, in which case V_j forms the basis for an invariant subspace of A, and the eigenvalues of the tridiagonal matrix T_j are the eigenvalues of A with respect to this invariant subspace. This algorithm is equivalent to the well-known algorithm of the Dutch mathematician Stieltjes for generating orthogonal polynomials by a three-term recurrence relationship. Lanczos also proposed a reduction process for unsymmetric matrices A, the so-called two-sided Lanczos process. In this process two sets of basis vectors are constructed, one for the Krylov subspace with A and one for a Krylov subspace with A^{T}. By requiring biorthogonality of the two sets, the two bases can be used for reduction of A to tridiagonal form. This form has suffered from many drawbacks. Not only are the reduction matrices nonorthogonal, a suspect property, but the algorithm also suffers from various break-down conditions. (The basic problem lies in the fact that the measure generated by the initial vectors is not nonnegative). The symmetric variant did not become popular in the 1950s, since it was soon recognized that rounding errors could spoil the process dramatically. Wilkinson [154] showed that the Lanczos algorithm is highly (forward) unstable and there seemed to be no way of stabilizing the process other than re-orthogonalizing the generated vectors. He showed this process is comparable to the methods of Householder or Givens (proposed in the late fifties), but the latter are more economical. He then concluded: "*it is difficult to think of any reason why we should use Lanczos' method in preference to Householder's*". This illustrates that the Lanczos method was commonly viewed as a direct reduction method at that time, and from that point of view Wilkinson's remarks were quite correct.

At about the same time as Lanczos, Arnoldi (1951) [1] gave a reduction algorithm for unsymmetric matrices. This was basically the same algorithm as Lanczos' algorithm for symmetric matrices, with the difference that each new basis vector had to be orthogonalized with respect to all previous basis vectors. In this way A is reduced by an orthogonal similarity transformation to upper Hessenberg form. Arnoldi's method suffers far less from numerical instability, depending on how well the orthogonalization process is carried out. But the method is more expensive than the Householder reduction, making it less attractive, as a direct method, for large (dense) matrices.

A very important notion was the recognition that matrices could be reduced, by orthogonal transformations, in a finite number of steps, to some special reduced form that lends itself more efficiently to further computations. In particular, a symmetric matrix can be reduced to tridiagonal form by Jacobi-rotations, provided that these rotations are restricted to annihilate entries of A outside its tridiagonal part. This was suggested by Givens in 1954 [45], and in this connection the Jacobi-rotations are also called Givens rotations. A few years later, Householder, in 1958 [63], discovered that complete columns of A could be reduced to zero, outside the tridiagonal part, by the more efficient Householder reflections. These are well-chosen orthogonal rank-one updates of the form

$$\left(I - \frac{2}{v^{\mathrm{T}}v}vv^{\mathrm{T}}\right);$$

these are discussed in [139]. The Householder method has become the method of choice for the reduction of matrices to tridiagonal form on serial computers. Thus for eigenproblems, a symmetric matrix can be reduced by orthogonal similarity transformations to tridiagonal form and unsymmetric matrices can be transformed to upper Hessenberg form.

By 1960, the eigenvalue problem for a symmetric tridiagonal matrix was solved by using the Sturm sequence property for successive subdeterminants. [5] The corresponding eigenvectors were solved by inverse iteration. The whole process is described in Givens' papers. A complete and thorough analysis for the Givens and Householder reductions and for the use of the Sturm sequences, is given in Wilkinson's book, which was the numerical linear algebra bible (Old Testament) for a long time.

As we have already shown in the section on the Power method, the QR method is, for determining the complete set of eigenvalues and eigenvectors, a superior technique. At the time that Wilkinson's book appeared, the blossoming of the QR method had just begun. Wilkinson devoted much attention to this method, but not as the method of choice for symmetric problems. We quote from Parlett [100, p. 172]: "Yet it was not invented until 1958–1959 and was not appreciated until the mid-1960s. The key idea came from Rutishauser with his construction of a related algorithm called LR in 1958". Whereas Wilkinson's book was the reference for eigenvalue problems in the period 1960–1980, after 1980, Parlett's book *The Symmetric Eigenvalue Problem* became the main source, at least for symmetric problems. Comparison of the two books clearly shows the progress made in this field.

The use of the QR (a mathematical equivalent is the QL algorithm) algorithm began with the work of Francis, [40] who recognized in 1961–1962 that a QR iteration maps a Hessenberg matrix to a Hessenberg matrix again, and this makes the process economical and also adds to stability since the zero elements need not be computed. Furthermore, Francis cleverly implicitly used origin shifts, and these can be carried out very economically for Hessenberg matrices. Kublanovskaja, in 1961 [74], also independently discovered the same process, but did not employ the invariance of the Hessenberg form. She deeply understood the mathematical aspects of the algorithm but was less concerned with the important computational details. The inclusion of Wilkinson shifts eventually makes the process very efficient, and for these shifts it can be proved that, for symmetric matrices, the process does not fail. The order of convergence for symmetric matrices is cubic (see, for instance, [61]), while for unsymmetric matrices it is quadratic [51,28]. These results rest on work of Ostrowski carried out in connection with the shift-and-inverse Power method (the RQI method, see that section). For a treatment of modern implementations of the QR method see [51, Chapter 7] or Demmel's book [28, Section 4.4.5]. These implementations incorporate techniques developed in the 1990s, such as (multiple) implicit shifts. This implicit shift technique leads to a rank-one perturbation of the Hessenberg structure, and this perturbation can be removed in its turn by a technique that is known as *chasing the bulge*: the perturbation (bulge) is chased down (and out of) the Hessenberg matrix with (double) Givens transformations. These chasing techniques were analysed in 1991 by Watkins and Elsner [152]. An important and complete overview of the practical QR algorithm can be found in [151].

---

[5] This shows that working with the characteristic equation, if not explicitly constructed, is not a bad idea in some situations.

With respect to the relation between the QR and the Lanczos algorithms, we note the following. Lanczos' method focusses on one particular starting vector; QR starts with a full orthogonal basis and keeps it orthogonal through the Power iterations; the inclusion of shifts does not destroy the structure of the Hessenberg matrix. With Lanczos' method, a shift only makes sense in damping unwanted parts of the spectrum, but one cannot vary the shift during the process.

By 1970 the standard numerical eigenproblem, for dense matrices of not too large order, could be regarded as essentially solved and research shifted to larger problems and other eigenproblems. The next important problem to consider was the generalized eigenproblem $Ax - \lambda Bx = 0$. An obvious approach is to transform this to a standard eigenproblem by inverting either A or B, or to work with more complicated transformations, such as the *Cayley Transform*: $(A - \sigma B)^{-1}(A - \tau B)$. These approaches share the disadvantage that matrices A and B are not treated in the same way, which is most obvious from the simple transformation $B^{-1}A$. This leads to problems if B is singular or ill-conditioned, but as Stewart (1980) has pointed out, this does not necessarily mean that the given eigenproblem is ill-conditioned because if A is well-conditioned then the pencil $B - \mu A$ may be well behaved (small perturbations in A and B lead to small perturbations in μ. In Stewart's analysis the matrices are treated symmetrically; in particular he suggests considering the pencil $\sigma A - \tau B$, and regarding multiples of (σ, τ), for which the determinant $|\sigma A - \tau B|$ vanishes, as generalized eigenpairs; see [129, Chapter VI] for more details. This approach, of course, requires a different reduction and this is accomplished by the Generalized Schur Decomposition, proposed by Moler and Stewart in 1973. This says that for arbitrary square A and B there exist unitary Q and Z such that $Q^*AZ = T$ and $Q^*BZ = S$ are upper triangular. For real matrices, the arithmetic can be kept real, but then the reduced matrices are quasi-triangular (that is 2×2 nonzero blocks along the diagonal may occur). Moler and Stewart [91] also proposed a stable algorithm to accomplish the reduction to (quasi) triangular form and this is known as the *QZ algorithm.* Major modern software packages include software for the QZ algorithm. For perturbation analysis, we refer to Stewart and Sun [129].

After the 1970s, the eigenproblem for dense matrices of moderate order seemed to be solved and further improvements were not expected, especially for symmetric dense matrices. However, with the ever increasing demand for higher efficiency and/or better accuracy, things changed from time to time. In 1981, Cuppen [25] proposed a divide and conquer algorithm for the solution of the eigenproblem for symmetric tridiagonal matrices. The idea was to split the tridiagonal matrix in two blocks, each of half the original size, plus a rank-one update. Cuppen showed how the eigenproblems for each of the blocks could be combined for the original full problem by exploiting the rank-one update property, which led to the solution of a secular equation. Initially, this approach was not seen as a competitive algorithm by itself for general matrices of modest dimensions, although Cuppen recognized that his algorithm was asymptotically (much) faster than QR. Further investigations by others were made on account of promising parallel properties. A major problem was that the original algorithm suffered from instabilities, especially for the eigenvectors belonging to close eigenvalues. Some scaling problems were rectified by Dongarra and Sorensen in 1987 [32], but the "right" implementation, according to Demmel [28, Section 5.3.3] was not discovered until 1992 and published in 1995, by Gu and Eisenstat [55]. Meanwhile, software for this algorithm found its way into LAPACK and ScaLAPACK. As stated by Demmel again, the divide and conquer approach is now the fastest algorithm for computing all eigenvalues and eigenvectors of a symmetric matrix of order larger than 25; this also holds true for nonparallel computers. If the subblocks are of order greater than 25, then they are further reduced; else, the QR algorithm is used for computing the eigenvalues and eigenvectors of the subblock. For a full treatment of the modern variant of the divide and conquer method, we refer to Demmel's book [28]. A recent discussion on parallel implementation aspects of this method can be found in [134].

There are still niches for other methods for dense symmetric matrices. Wilkinson advocated the bisection method for tridiagonal matrices if only a small subset of the eigenvalues is wanted. Inverse iteration may then be used to determine the corresponding eigenvectors. For the bisection method, based on the Sturm sequence property, the reader may find classical material in [154], a modern treatment of the inverse iteration (considered suspect by many because of the ill-conditioning of the shifted matrix) can be found in [100].

8. Iterative methods

Soon after its introduction in 1952, it was recognized that the Lanczos method was not a panacea for eigenproblems. The method showed strange behaviour, because of rounding errors, and in an influential paper by Engeli et al. in 1959 [37], it was shown by careful experiments that the theoretical finite termination within n steps, had no practical meaning. For a discretized biharmonic problem of order 64, they observed that hundreds of steps where necessary in order to obtain the 64 eigenvalues (together with extraneous other values). Wilkinson also analyzed the method and showed that it was forward unstable, which seemed to mark more or less the end of the Lanczos method. It was Paige, who showed in 1971 [97] that the Lanczos method could be used in a truly iterative way in order to obtain correct eigenvalue information. The crux of his analysis is that the observed loss of orthogonality in the Lanczos process, the source of all problems in the method, marked the convergence of an eigenpair; and most remarkably, it did not prevent convergence of other eigenpairs. This loss of orthogonality, by the re-introduction of components of the converged eigenvector to the process, led to duplication of the converged eigenpair in the reduced tridiagonal matrix. The main effect on the convergence of the left eigenpairs seemed to be some delay in the process in exact computation. Paige's analysis spurred much activity in this field and eventually the Lanczos method became a powerful tool and the method of choice for large sparse symmetric matrices, from 1980 on.

We mention the following major steps that led to improvements in the method, and to a better understanding. Parlett and Scott [103] proposed removing the delaying effects on convergence by a *selective* orthogonalization process for the Lanczos vectors. Also, the determination of converged eigenvalues became easier to identify by efficiently computing upper bounds for the residual of an eigenpair. Kaniel (1966) [69] derived upper bounds for the error in an eigenvalue approximation (the so-called *Ritz values*, the eigenvalues of the reduced tridiagonal matrix). These upper bounds have no direct practical implication, since they are in terms of unknown quantities associated with the matrix, such as gaps between eigenvalues relative to the span of the eigenvalues. However, these upper bounds are, in a general sense, rather sharp and can be used for the study of the convergence behaviour of the Ritz values. Later, Saad (1980) [114] refined these upper bounds. The convergence behaviour of Ritz values can be quite irregular; a temporary (almost) stagnation of the process can take place (also called *misconvergence*). Many of these effects were studied carefully, and explained, in a paper by Van der Sluis and Van der Vorst [142], through a rather complicated model for the Ritz values. This model, however, seemed to be necessary in order to show all the intricate dependencies in the apparently simple Lanczos process. In their paper, it is also shown that the convergence behaviour of the Ritz values is superlinear. No matter how irregular the rate of convergence may seem, on average it becomes faster as the iteration proceeds. Parlett also studied the so-called (occasional) misconvergence of the Ritz values [101]. His model is simpler, but does not explain all effects, such as the possible length of the stagnation phase. An algorithm ready for implementation was published by Parlett and Reid [102]; this algorithm computes upper bounds for the errors in the Ritz values, and exploits the fact that we have to compute these for successive tridiagonal matrices. We are not aware of a publicly available implementation, although we know that the process has been implemented for local use.

Note that, strictly mathematically speaking, the Lanczos process is finite and thus it is not correct to use a notion such as "convergence" or even "superlinear convergence". However, in finite precision and for large values of the order of the matrix, the Ritz values will become close to an eigenvalue and for practical purposes the method behaves like a truly convergent process.

In our discussion of the Power method, we showed how the use of a set of starting vectors arose quite naturally. This is also possible for the Lanczos (and Arnoldi) method, and this approach leads to block and banded Lanczos (and Arnoldi) methods [47,141,108].

For unsymmetric matrices it took longer for similar methods to gain popularity. An influential paper, that helped to promote Arnoldi's method as a useful tool, was published by Saad [115]. The Arnoldi method, for orthogonal reduction to upper Hessenberg form, was not only too expensive if one wanted to know only a few eigenpairs, it also suffered from poor convergence for specific eigenvalue distributions. Well-known is the Saad–Schultz example [117], which is a permuted identity matrix. The method leads to trivial approximations after the first $n-1$ steps, and after n steps all eigenpairs suddenly appear. This however, is at a much higher cost than for Householder's reduction. For this reason, the unsymmetric Lanczos process, also referred to as the *two-sided Lanczos method*, received some attention. Initially, the method was notorious for its break-down possibilities, its behaviour in finite precision arithmetic, and the fact that the reduction operators to tridiagonal form are nonorthogonal. Cullum and Willoughby, in 1986 [24], presented a code based on the two-sided Lanczos method, in which they solved a number of practical problems; this included a clever trick for identifying the *spurious* eigenvalues due to rounding errors. The code gained some popularity, for instance for plasma-physics eigenvalue computations [21]. Parlett and co-workers [104] introduced the concept of "look-ahead", mainly in order to improve the numerical stability of the process. The look-ahead strategy, introduced in order to prevent breakdown, was further perfected by Freund and Nachtigal, in 1996 [41]. They published a code based on *quasi-minimization* of residuals, and included look-ahead strategies, in which most of the original Lanczos problems were repaired (but the non-orthogonal reductions were still there). Gutknecht [57] published a thorough theoretical overview of the two-sided Lanczos algorithm and exploited its relation to Padé approximations. This gave a better understanding of look-ahead strategies and the convergence behaviour of the method (in the context of solving linear systems). Block variants of the two-sided Lanczos process were discussed in Day's Ph.D. thesis in 1993; for a further description of the algorithms see [3].

Almost simultaneously, there were efforts to make the Arnoldi method more practical. We mention firstly polynomial preconditioning, discussed extensively in Saad's book [116], which damps unwanted parts of the spectrum, and secondly, sophisticated restarting strategies. The method becomes effective for matrices for which shift-and-invert operations can be applied for given vectors. But the many (increasingly expensive) iterations for relevant problems were a bottleneck. A real breakthrough for the Arnoldi method was realized by Sorensen [125], in 1991, with the so-called *Implicit Restart Technique*. This is a clever technique by which unwanted information can be filtered away from the process. This leads to a reduced subspace with a basis, for which the matrix still has a Hessenberg form, so that Arnoldi's process can be continued with a subspace (rather than with a single vector as with the more classical restart techniques).

The Arnoldi iteration procedure is often carried out with the shift-and-invert approach. For instance, when solving the generalized eigenproblem $Ax = \lambda Bx$, the method is applied to the operator $(A - \sigma B)^{-1}B$. One step further is the *Cayley* transform $(A - \sigma B)^{-1}(A - \tau B)$, which can be used for emphasizing eigenvalues (near σ) and for damping of eigenvalues (near τ). Both techniques require expensive operations with an inverted operator, but the advantage is much faster convergence. Meerbergen and Roose [89] considered the use of inexact Cayley transforms, realized by a few steps of an iterative method, for Arnoldi's method. This technique bears a close relation to polynomial preconditioning. Ruhe [110] considered a more general shift-and-invert transform, and so-called *Rational Krylov Subspace* (*RKS*) *Method*:

$$(\delta_j A - \gamma_j B)^{-1}(\sigma_j A - \rho_j B),$$

in which the coefficients may be different for each iteration step j. It has been shown that by generating a subspace with this operator, the given problem can be reduced to a small projected generalized system

$$(\zeta K_{j,j} - \eta L_{j,j})s = 0,$$

where $K_{j,j}$ and $L_{j,j}$ are upper Hessenberg matrices of dimension j. This small system may be solved by the QZ algorithm in order to obtain approximate values for an eigenpair. The parameters in RKS can be chosen to obtain faster convergence to interior eigenvalues. For a comparison of RKS and Arnoldi, see [110,109].

In 1975, Davidson, a chemist, suggested an iterative method that had the idea of projection on a subspace in common with the Arnoldi method, but with the subspace chosen differently. Motivated by the observation that the matrices in his relevant applications were (strongly) diagonally dominant, Davidson computed the Ritz pairs (the eigenpairs of the projected matrix), computed the residual $r = (A - \theta I)z$ for a pair of interest (θ, z), and proposed expanding the subspace with the vector $(D_A - \theta I)^{-1}r$ (after proper orthogonalization with respect to the current subspace). The matrix D_A denotes the diagonal of the matrix A. For diagonally dominant matrices this approximates, in some sense, inverse iteration with Rayleigh quotient shifts. The Davidson method [26] became quite popular for certain applications. Although other approximations were suggested (see, e.g., [93]), its convergence behaviour for nondiagonal dominant matrices, or for poor initial starting vectors, was far from guaranteed. It was also puzzling that the "optimal" expansion $(A - \theta I)^{-1}r$, optimal from the inverse iteration point of view, led to the vector z, so that the method stagnated. As a consequence, Davidson's method was not able to find the eigenvalues of a diagonal matrix, a very unsatisfactory situation. The method was not very well understood by numerical analysts and as a consequence we find very little reference to it in the numerical analysis literature. Only after 1990, is there some serious analysis [20], almost simultaneously with a successful improvement of the method.

In 1996, Sleijpen and van der Vorst [121] suggested restricting the expansion of the current subspace to vectors from the space orthogonal to z, which restored a largely forgotten technique used by Jacobi (in 1846). Jacobi took an appropriate unit vector for z, and attempted to find the missing component to make it an eigenvector from the space spanned by the remaining $n-1$ unit vectors. He did this by solving a correction equation for the matrix shifted by θ and then restricted the correction to the subspace orthogonal to the chosen unit vector. This correction equation was solved by Gauss–Jacobi iterations and after each two iterations, Jacobi updated the value for θ. Sleijpen and van der Vorst suggested updating z as well and using the update vector for a subspace. This new Jacobi–Davidson method combined Davidson's idea of taking a different subspace with Jacobi's idea of restricting the search of an update to $z^{\perp}$. We refer to Sleijpen and van der Vorst [121] for details. A variant of the technique, in which the correction is approximately solved with one single invert step with a preconditioner, was suggested by Olsen et al. in 1990 [94]. An exact solution of the correction equation, or an approximate solution of high accuracy, leads to cubic convergence for a properly selected sequence of θ's if A is symmetric, and to quadratic convergence in the unsymmetric case. In the following years, it became clear how to efficiently implement the method with preconditioning [122] and how it could be applied to various other eigenproblems, amongst which are generalized eigenprobelms [39] and quadratic eigenproblems [123]. Thus, the transformation of these generalized eigenproblems to standard forms can be avoided. The Jacobi–Davidson method is attractive for large sparse eigenproblems for which shift-and-invert operations are too expensive, and for more unusual eigenproblems. The problem of identifying effective preconditioners for the correction matrix for larger classes of matrices is still largely open.

It is well-known that subspace methods lead to eigenvalue approximations that tend to converge towards exterior eigenvalues and that approximations for interior eigenvalues are difficult to obtain. In principle, it is easy to obtain these by working with A^{-1}, but this may be expensive. It is also possible to obtain eigenvalue approximations that converge (slowly) to the eigenvalues of A closest to the origin, from the subspaces generated by A. We explain this for the Arnoldi process. The Arnoldi process leads to

$$AV_m = V_{m+1}H_{m+1,m},$$

where $H_{m+1,m}$ is an upper Hessenberg matrix with $m+1$ rows and m columns. This means that we have a basis for the space with basis vectors Av_j and, using the above relation, this basis can be easily transformed into an orthogonal basis. This orthogonal basis can be used for the projection of A^{-1}, and multiplication by the inverse can be avoided, since all basis vectors have a factor A in common. The exterior eigenvalues of the projected A^{-1}, that is the inverses of interior eigenvalues of A, converge (slowly) to the exterior eigenvalues of A^{-1}. This way of approximating interior eigenvalues has received some attention in the 1990s. In [98], these eigenvalue approximations, in connection with the related Lanczos process, were called Harmonic Ritz values, and some nice relations for Harmonic Ritz values for symmetric indefinite matrices are given in that paper. Harmonic Ritz values had already been studied from a different viewpoint by other authors. Freund [42] studied them as the zeros of the GMRES and MINRES iteration polynomials. Morgan [92] had observed that the Harmonic Ritz values and vectors are very suitable for restarting purposes if one wants to compute interior eigenvalues with subspaces of restricted dimension. In [121,120], the Harmonic Ritz values are considered in connection with the Jacobi–Davidson process for the selection of proper shifts.

9. Related topics

9.1. *The singular value decomposition*

The singular value decomposition plays an essential role in many situations in which we want to decide between relevant and less relevant information. In addition to traditional applications such as regularization of linear least-squares problems, and the determination of the numerical rank of a matrix, there are applications to the reduction of information for images and information retrieval from large data bases. A headline contribution to the SIAM News by Berry and Dongarra, summer 1999, showed that the SVD can even help to reduce work in the organization of a conference. Interesting and unusual applications are also shown in a paper by Moler and Morrison [90]. A modern treatment of the numerical aspects of the SVD can be found in Golub and Van Loan's textbook [51] and we have taken most of our information from that source. Demmel [28] gives a good treatment of the important implementational aspects of the SVD.

Since the square of the singular values and the right singular vectors are the eigenpairs for the matrix $A^{\mathrm{T}}A$ (for the left singular vectors this holds with AA^{T}), it is not surprising that the numerical treatment of the SVD has many relationships to algorithms for symmetric eigenproblems. This is most visible in Demmel's book, where the discussion on modern SVD algorithms almost parallels the symmetric eigenvalue problem discussion [28, p. 211 and 241]. We note here that working directly with either $A^{\mathrm{T}}A$ or AA^{T} is not satisfactory for stability and complexity reasons, and this makes separate treatment of the numerical SVD necessary.

The origins of the singular value decomposition go back to the late 19th century, with work of Beltrami in 1873. Stewart [128] gave a historic overview of the SVD. The important numerical developments on the SVD started with work of Golub and Kahan in 1965 [49]. This led first to a contribution to the famous ACM-collection as Algorithm 358 [15] and later to the basis of the EISPACK and LAPACK routines in the Wilkinson–Reinsch collection [158, pp. 1334–1351]. The key trick in the numerical computation of the SVD is, instead of tridiagonalizing $A^{\mathrm{T}}A$, to bidiagonalize the matrix A. Then, with the bidiagonal reduced forms obtained by orthogonal transformations, one can make variants of QR, divide and conquer, and bisection techniques. The choice between these techniques can be made as above for the symmetric eigenproblem. In particular, for matrices of order larger than 25, the divide and conquer approach is currently regarded as the fastest option [28, p. 241].

9.2. *Nonlinear eigenproblems and related problems*

Standard and generalized eigenproblems arise, for instance, in the study of conservative mechanical systems, governed by Lagrange's equations of small free motion. According to Rogers [106], Rayleigh (in 1873) could not apply his technique for nonconservative systems (systems with a damping term). The well-known technique for the numerical solution of the resulting quadratic eigenproblem is to rewrite it as a generalized eigenproblem [43,77]. That is

$$\lambda^2 Mx + \lambda Cx + Kx = 0$$

is equivalent to

$$Az = \lambda Bz$$

with

$$A = \begin{bmatrix} 0 & I \\ -K & -C \end{bmatrix}, \quad B = \begin{bmatrix} I & 0 \\ 0 & M \end{bmatrix}, \quad z = \begin{bmatrix} x \\ \lambda x \end{bmatrix}.$$

Of course, algorithms and theory for the generalized eigenproblem can be used directly, but the unsatisfactory aspect of this is that we have to double the dimension of our spaces. Also, if the generalized eigenproblem is solved approximately by some iteration technique, then it is not straight forward to reduce the information from the double-dimensioned space to the original space. Duffin [34] seemed to be the first to generalize the Rayleigh–Ritz principle for the quadratic eigenproblem for symmetric K, C, and M, and A positive definite. Since then, the quadratic eigenproblem, and higher-order polynomial problems, have received attention in the numerical literature. Rogers [106], in 1964, considered a more general quadratic eigenproblem and used minimax principles for the investigation of such problems. In the late 1960s and early 1970s algorithms that avoid the linearization step appeared. We mention the work of Lancaster [77], Kublanovskaja [75,76], and Ruhe [107]. The suggested algorithms are mostly variants of Newton's method. More recently, in 1995, Guo et al. [56] described several iterative methods, that can be regarded as a fixed point iteration combined with the Lanczos method and a (simplified) Newton iteration. A backward error analysis for more general polynomial eigenproblems was given by Tisseur (1998 [133]), and a perturbation analysis for quadratic problems was published by Sun in 1999 [130]. In a paper that appeared in 1996, Sleijpen et al. [120] showed that the Jacobi–Davidson method could be applied in order to reduce a given polynomial problem in an n-dimensional space to a similar problem in a much lower m-dimensional space. The problem in the lower dimensional space can then be attacked by any of the previously mentioned approaches. For a quadratic equation from an accoustic problem, it was shown how this approach led successfully and efficiently to the desired eigenpair, for matrices of the order of about 240,000 [123]. For higher polynomial eigenproblems there has not been much experience to date. Bai [6] mentioned the need for algorithms for fifth-order polynomial eigenproblems in a review paper that appeared in 1995. Quite recently, in 1998, Heeg [58] showed how the Jacobi–Davidson approach could be successfully applied to fourth-order polynomial eigenproblems with complex matrices in the study of instabilities of attachment-line flows for airfoils.

Other than these polynomial eigenproblems, there is a wide variety of problems that are associated with the determination of invariant subspaces. Since the standard and generalized eigenproblems became more familiar and more or less routinely solvable, the more difficult problems received more attention. Among these problems we mention the following; the *Procrustes problem*: minimize $\|AY - YB\|_F$ for given A and B over the manifold $Y^*Y = I$; the determination of a nearest Jordan structure, and the problem of determining a simultaneous Schur decomposition for a set of perturbed matrices (under the assumption that the unperturbed matrices have a simultaneous Schur form). A nice overview of such problems, as well as software for the numerical solution is described by Edelman and Lippert [83].

9.3. Pseudospectra

Eigenvalues are often used as a source of information on stability or convergence, and the question arises as to the validity of the information gained from these values. For example, during the 1990s it was realized (see, e.g., [138]) that eigenvalues alone do not govern the instability and transition to turbulence of high Reynolds number fluid flows as had previously been thought. Many authors have studied the problem of sensitivity of the eigenvalues with respect to perturbations, see for instance [154,129,19]. These studies are usually related to perturbations caused by rounding errors, and not so much by the relevance of the eigenvalues due to the particular representation of a given problem, for instance the choice of basis.

Around 1987, Trefethen [140] began to emphasize this aspect of eigencomputations. He propagated the idea of inspecting the pseudospectra of a matrix as a relatively simple means for determining the significance of a particular part of the spectrum, without getting involved in complicated matters such as angles between eigenvectors or eigenspaces. [6] The definition of the pseudospectrum $\Lambda_\varepsilon(A)$ for a matrix A is directly related to perturbations:

$$\Lambda_\varepsilon(A) \equiv \{z \in \mathbb{C}: z \in \Lambda(A+E) \text{ for some } E \text{ with } \|E\| \leqslant \varepsilon\}.$$

The pseudospectra are usually shown graphically as a set of level curves for various values of ε. The level curves, or contour integrals, are more apparent from the original definition of the ε-pseudospectrum, in terms of the norm of the resolvent $(zI-A)^{-1}$:

$$\Lambda_\varepsilon(A) \equiv \{z \in \mathbb{C}: \|(zI-A)^{-1}\| \geqslant \varepsilon^{-1}\}$$

with the convention $\|(zI-A)^{-1}\| = \infty$ for $z \in \Lambda(A)$.

For symmetric matrices, the pseudospectra of A is a collection of discs around the eigenvalues of A (note that the perturbation E need not be symmetric). For unsymmetric matrices the pseudospectra can be any collection of curves around the set of eigenvalues of A. These level curves may give information that is hidden by the information provided by the eigenvalues themselves. For instance, when studying stability of integration methods for systems of ODEs, or in bifurcation problems, the eigenvalues may be in a proper region, for instance, in the left-half plane, while the level curves, even for small values of ε, may intersect with the right-half plane. This suggests that it may be necessary to ask further questions about the problem. On the other hand, the pseudospectra may not tell the full story. For instance, the sensitivity problems may be due to a single pair of ill-conditioned eigenvectors for which the more global level curves are too pessimistic. It may be the case that it is not realistic to assume equal perturbations for all matrix entries, but nevertheless the pseudospectra focus attention on critical places in the spectrum. A nice introduction to the relevance of pseudospectra is given in [135], where the pseudospectra are actually computed and discussed for a number of matrices.

Due to the nature of computing pseudospectra, this useful tool is often restricted to matrices of relatively moderate size and one has to be careful in generalizing the insights gained from smaller problems to larger problems which are similar. More recently, tools have become available for computing the pseudospectrum of large sparse matrices. Carpraux et al. [16] proposed an algorithm for computing the smallest singular value of $zI-A$, that is based on Davidson's method using ILU preconditioning. Lui [84] (see also [13]) suggested using the Lanczos method in combination with continuation techniques. This is a plausible approach, since we need to do the computation for many values of z, well-distributed over the region of interest, in order to obtain a complete picture of the pseudospectra. We know that currently such tools are being used for the analysis of instability problems of large sets of ODEs, related to climate modelling, but results have not yet been published.

---

[6] The notion had been defined by others earlier, the earliest of them, according to Trefethen, being Landau [79].

Valuable information on pseudospectra can also be derived from the Arnoldi iteration obtained, for instance, with ARPACK [81]. For an up-to-date overview on pseudospectra results, see Trefethen's 1999 paper [136].

9.4. Homotopy methods

The subspace methods that we have discussed above are often applied in combination with shift-and-invert operations. This means that if one wants to compute eigenvalues close to a value σ, then the methods are applied to $(A - \sigma I)^{-1}$. As we have seen, the Jacobi–Davidson method can be interpreted as an inexact shift-and-invert method, since the invert step is usually approximated by a few steps of some convenient preconditioned inner iteration method.

Another method related to these inexact shift-and-invert approaches is the homotopy approach which has received some attention in the 1990s. The idea is to compute some of the eigenvalues of a perturbed matrix $A + E$, when the eigenvalues of A are known, or can be relatively easily computed. In order to this we use the homotopy $H(t) = A + tE$, $0 \leqslant t \leqslant 1$. If eigenpairs of $H(t_0)$ are known, then they are used as approximations for those of $H(t_0 + \Delta t)$. These approximations are improved by a convenient subspace iteration (cf. [85]). Rayleigh quotient iterations are used for symmetric A and E (see references in [85] for earlier work on homotopy for eigenproblems). For the Rayleigh quotient iteration one needs to solve systems like $(H(t_0 + \Delta t) - \lambda I)y = x$, where (λ, x) represents the current approximation for an eigenpair of $H(t_0 + \Delta t)$. In the context of large sparse matrices, it may be undesirable to do this with a direct solver, and in [85] the system is solved with SYMMLQ [99]. Of course, one could restrict oneself to only a few steps with SYMMLQ, and then try to accelerate the inexact Rayleigh quotient steps, as is done in the Jacobi–Davidson method. This indicates relations between these different approaches, but as far as we know, these relations have not yet been explored. In [85], it is observed that SYMMLQ may have difficulty in converging for the nearly singular system $(H(t_0+\Delta t)-\lambda I)y=x$, and it is suggested that the situation would be improved by applying the Rayleigh quotient iteration to the approximately deflated matrix $H(t_0 + \Delta t) + xx^{\mathrm{T}}$. (The term approximately deflated is used to indicate that x is only an approximation to the desired eigenvector.) Note that similar deflation procedures are incorporated in the Jacobi–Davidson process. The whole procedure is repeated for successive increments Δt, until the final value $t = 1$ is reached. In [85] an elegant approach is followed for the selection of the step size Δt.

The homotopy approach lends itself quite naturally to situations where the matrix A varies in time, or where it varies as a linearization of a nonlinear operator, as in bifurcation problems. Another example of an interesting problem is the Schrödinger eigenvalue problem [85]

$$-\Delta u + fu = \lambda u$$

in the unit square in two dimensions with homogeneous Dirichlet boundary conditions. With the usual finite difference approximations on a uniform grid, this leads to the discrete Laplacian for $-\Delta u$, for which we know the eigensystem.

In [159], the homotopy approach is used for symmetric generalized eigenproblems, very much along the lines sketched above. The application to real unsymmetric eigenproblems is considered in [86].

In [82], the homotopy approach is suggested as a means of realizing a divide and conquer method for unsymmetric eigenproblems, as an alternative for the solution of the secular equation for symmetric problems (which cannot be used for unsymmetric problems).

9.5. Miscellaneous

Subspace methods such as Simultaneous Iteration, RQI, Lanczos, Arnoldi, and Davidson for large sparse systems are in general more powerful in combination with shift-and-invert techniques. This requires the solution of large sparse linear systems, since these methods only need operations with, for instance, $(A-\sigma I)$ in order to compute matrix vector products $(A-\sigma I)^{-1}v$. This has led to the question of how accurately these operations have to be carried out in order to maintain a reasonable rate of convergence. For instance, for the Power method, this approximation technique, or preconditioning, can be described as follows for the standard eigenproblem $Ax=\lambda x$. For each iteration one computes, for the current iteration vector $x^{(i)}$ the Rayleigh quotient $\mu^{(i)}=x^{(i)*}Ax^{(i)}/x^{(i)*}x^{(i)}$. Instead of computing $x^{(i+1)}$ with $Ax^{(i)}$, the defect $r = Ax^{(i)} - \mu^{(i)}x^{(i)}$ is computed and this defect is multiplied by the preconditioner K. The new vector $x^{(i+1)}$ is then computed as a normalized linear combination of x^i and Kr. Note the resemblance to Davidson's approach.

The convergence of this basic algorithm, including its use for the generalized eigenproblem, has been studied quite extensively in the Russian literature, starting in about 1980 in the work of D'yakonov [35] and others. Knyazev [72] gave an excellent overview of the work in the Russian literature on *preconditioned eigensolvers*. The study of these preconditioned iteration techniques is relevant also to the understanding of inexact, or preconditioned, forms of shift-and-invert Lanczos (proposed by Morgan and Scott [93]), inexact Arnoldi (work by Meerbergen [87]) and Davidson variants, including the Jacobi–Davidson method. A presentation of various iteration techniques including preconditioning was given by Knyazev in [71]. Smit [124, Chapter 4] studied the effect of approximate inverses on the convergence of the RQI method. For a discussion on inexact Krylov methods, see Meerbergen and Morgan [88]. Preconditioning in relation to the Jacobi–Davidson method is discussed in [122]. Note that in all these algorithms, the preconditioner is used for the computation of a promising new direction vector; the given eigenproblem is untouched. This is different from the situation with linear solution methods, where preconditioning is used to transform the given system to one that can be handled more efficiently.

10. Software

The history of reliable high-quality software for numerical linear algebra started with the book edited by Wilkinson and Reinsch, the *Handbook for Automatic Computation, Vol. 2, Linear Algebra*, published in 1971. This book contained a number of articles that had appeared previously in Numerische Mathematik, which described state-of-the-art algorithms for the solution of linear systems and eigenproblems. All these articles contained implementations in Algol60. Most of these algorithms are still alive albeit in other languages. Algol60 was a computer language that gained some popularity in academia, mostly in Europe, but it was not as fast as Fortran on most machines and it did not gain a foothold in the slowly emerging large industrial codes (the majority of which were written in Fortran or even in assembler language). For this reason, groups in the USA started the development of two influential software packages LINPACK and EISPACK in the early 1970s. These packages started as transcriptions of the major part of the Wilkinson and Reinsch collection: LINPACK covered the numerical solution of linear systems; EISPACK concentrated on eigenvalue problems. The most prominent omissions from these packages were iterative solution methods: the conjugate gradient method was not included in LINPACK and the Jacobi method was not included in EISPACK. At that time the Lanczos and Arnoldi methods were not even considered as candidates, because they were viewed either as direct methods, and in that sense not competitive with the then available methods, or as iterative methods that could not be safely automated [68]. The Lanczos method was only considered, in that time, as a safe iterative method provided one did complete re-orthogonalization (cf. [48]).

The "Wilkinson and Reinsch" procedures can also be viewed as prototypes for eigenvalue routines in the bigger software packages NAG and IMSL, and in the widely available software package MATLAB. EISPACK was replaced in 1995 by LAPACK, in the words of Golub and Van Loan (1996): "*LAPACK stands on the shoulders of two other packages* (*viz*: *LINPACK and EISPACK*) *that are milestones in the history of software development*". A more recent development along these lines in ScaLAPACK [11] which aims to provide close to optimal software for modern parallel computers.

We have already mentioned MATLAB in passing; the impact of this computing environment in the scientific computing world has been tremendous. MATLAB provides state-of-the-art software for all sorts of numerical linear algebra computations and has become the de facto standard for coding and testing in the 1990s. Its impact on the development of this field can hardly be overestimated.

As already indicated, in 1970 there were few robust and well-understood iterative methods available, and mainly for that reason, these methods were not included in the packages constructed then. They are still not available in their successors, with the exception of MATLAB (since 1998, MATLAB has had iterative methods for eigenproblems available). It was soon clear that the powerful "direct" methods, based on reduction to some special form, had their limitations for the large sparse matrices that occur in the modelling of realistic stability problems, and there was a heavy demand for methods that could, at least partially, handle big problems. Cullum and Willoughby filled this gap, in 1985, with their software based on the two-sided Lanczos procedure [22–24]. Of course, they realized the intricacies of the Lanczos method, and they advocated the use of iterated Rayleigh quotient steps for improvement of the information of the Lanczos output. They also provided software for this updating step (their software was available some years before 1985, but its publication in book form took place in that year).

Freund and Nachtigal proposed in 1992, a variant of the two-sided Lanczos process that improved convergence properties using a quasi minimization step; they also included sophisticated look-ahead facilities. The QMR method could also be used for eigenvalue computations, and they provided software through Netlib for this purpose, see QMRPACK [41]. Algorithms and software for adaptive block-variants of the two-sided Lanczos algorithm (ABLE) have been described by Bai et al. [3].

Because of the improvements made to the Arnoldi method, in particular the implicit restart technique, it became feasible to exploit the attractive orthogonal reduction properties in an efficient manner. This was realized in the ARPACK software, for which the User's guide was published, by Lehoucq et al. in 1998 [81]. The package was in existence and available to users a few years earlier. At present, ARPACK seems to be the default choice for large sparse eigenproblems, provided that either it is possible to implement shift-and-invert operations efficiently, or that unwanted parts of the spectrum can be damped by a fixed (polynomial) preconditioner. The parallel implementation of ARPACK is referred to as P_ARPACK, it is portable across a wide range of distributed memory platforms. The parallelism is mainly in the matrix vector products and the user has full control over this trough the reverse communication principle. For more details, see [81,31].

Kågström et al. have provided software for the generalized upper staircase Schur form. The software, called GUPTRI, can be used for the analysis of singular pencils $A - \lambda B$. This reduction admits more general cases than the QZ decomposition, for instance A and B may be singular. This software has been described in [4]. See also this reference for pointers to recent work by Kågström.

For a thorough overview of modern algorithms, software and pointers to sources for further information see [4].

11. Epilogue

For symmetric matrices, the eigenproblem is relatively simple, due to the existence of a complete orthogonal eigensystem, and the fact that all eigenvalues are real. These properties are exploited in the most efficient numerical methods, and the symmetric eigenproblem may be considered as solved: for small matrices $n \leqslant 25$ we have the QR method, one of the most elegant numerical techniques produced in the field of numerical analysis; for larger matrices (but smaller than a few thousand), we have a combination of divide and conquer with QR techniques. For the largest matrices, there is the Lanczos method, which in its pure form is strikingly simple but which conceals so many nice and attractive properties. All methods have a relatively long history. The theoretical aspects of the computations (convergence and perturbation theory) are relatively well understood.

For unsymmetric matrices the picture is less rosy. Unfortunately, it is not always possible to diagonalize an unsymmetric matrix, and even if it is known that all eigenvalues are distinct, then it may be numerically undesirable to do this. The most stable methods seem to be based on the Schur factorization, that is for each $n \times n$ matrix A, there exists an orthogonal Q, so that

$$Q^*AQ = R,$$

in which R is upper triangular. Apart from permutations and signs, the matrix Q is unique. The Schur factorization reveals much of the eigenstructure of A: its diagonal elements are the eigenvalues of A and the off-diagonal elements of R indicate how small the angles between eigenvectors may be. For matrices not too large, QR is the method of choice, but for larger matrices the picture is less clear. Modern variants of the Arnoldi method seem to be the first choice at the moment, and, if approximations are available, the Jacobi–Davidson method may be attractive. There is still a lot to investigate: if the matrix is nonnormal, that is, if the eigenvectors do not form a unitary set, then what kind of meaningful information can we extract from a given matrix (invariant subspaces, angles between subspaces, distance to Jordan canonical form), and how can we compute this as accurately as possible? Much has been done, but even more remains to be done.

There are even more open problems as eigenproblems become more complicated: generalized eigenproblems, polynomial eigenproblems, nonlinear eigenproblems, etc.

Looking back over the past century, we see that the solution of the eigenproblem has given up some of its mysteries through the work of many devoted and highly talented researchers. Novices in the field should be aware that the modern algorithms, even the apparently simple ones, are the result of many independent "small" steps. The fact that many of these steps can now be regarded as "small" illustrates how theory has kept up the pace with computational practice, so that new developments can find their place in an expanding but still elegant framework. Astonishingly much has been achieved, both computationally and theoretically, in a concerted effort, but much more remains to be unravelled. It is our firm belief that eigenproblem research will remain a lively and useful area of research for a long time to come. We hope that this overview will help to motivate young researchers to make their contributions to solving pieces of the gigantic puzzles that remain.

Acknowledgements

We wish to thank our colleagues Michele Benzi, Jan Brandts, Mark Embree, Karl Meerbergen, Gerard Sleijpen, Nick Trefethen, and David Watkins, for their suggestions for improvements and corrections.

References

[1] W.E. Arnoldi, The principle of minimized iteration in the solution of the matrix eigenproblem, Quart. Appl. Math. 9 (1951) 17–29.

[2] Z. Bai, D. Day, Q. Ye, ABLE: An adaptive block Lanczos method for non-Hermitian eigenvalue problems, Technical Report Research Report 95-04, University of Kentucky, Lexington, KY, 1995.

[3] Z. Bai, J. Demmel, J. Dongarra, A. Ruhe, H. van der Vorst, Templates for the Solution of Algebraic Eigenvalue Problems: a Practical Guide, SIAM, Philadelphia, PA, 1999.

[4] Z. Bai, G.W. Stewart, SRRIT — A FORTRAN subroutine to calculate the dominant invariant subspace of a nonsymmetric matrix, Technical Report 2908, Department of Computer Science, University of Maryland, 1992.

[5] Z. Bai, Progress in the numerical solution of the nonsymmetric eigenvalue problem, Numer. Linear Algebra Appl. 2 (1995) 219–234.

[6] V. Bargmann, C. Montgomery, J. von Neumann, Solution of linear systems of high order, Technical Report, Princeton, Institute for Advanced Study, 1946.

[7] F.L. Bauer, Das Verfahren der Treppeniteration und verwandte Verfahren zur Lösung algebraischers Eigenwertprobleme, Z. Angew. Math. Phys. 8 (1957) 214–235.

[8] F.L. Bauer, C.T. Fike, Norms and exclusion theorems, Numer. Math. 2 (1960) 137–141.

[9] C.H. Bischof, The two-sided block Jacobi method on hypercube architectures, in: M.T. Heath (Ed.), Hypercube Multiprocessors, SIAM, Philadelphia, PA, 1987.

[10] L. Blackford, J. Choi, A. Cleary, E. D'Azevedo, J. Demmel, I. Dhillon, J. Dongarra, S. Hammarling, G. Henry, A. Petitet, K. Stanley, D. Walker, R. Whaley, ScaLAPACK User's Guide, SIAM, Philadelphia, PA, 1997.

[11] E. Bodewig, Matrix Calculus, North-Holland, Amsterdam, 1956.

[12] T. Braconnier, F. Chatelin, V. Frayssé, The influence of large nonnormality on the quality of convergence of iterative methods in linear algebra, Technical Report TR-PA-94/07, CERFACS, Toulouse, 1994.

[13] T. Braconnier, N.J. Higham, Computing the field of values and pseudospectra using the Lanczos method with continuation, BIT 36 (1996) 422–440.

[14] R.P. Brent, F.T. Luk, The solution of singular value and symmetric eigenvalue problems on multiprocessor arrays, SIAM J. Sci. Statist. Comput. 6 (1985) 69–84.

[15] P.A. Businger, G.H. Golub, Algorithm 358: Singular value decomposition of a complex matrix, Comm. Assoc. Comp. Mach. 12 (1969) 564–565.

[16] J.F. Carpraux, J. Erhel, M. Sadkane, Spectral portrait for non-Hermitian large sparse matrices, Computing 53 (1994) 301–310.

[17] F. Chaitin-Chatelin, Is nonnormality a serious difficulty? Technical Report TR-PA-94/18, CERFACS, Toulouse, 1994.

[18] F. Chatelin, Spectral Approximation of Linear Operators, Academic Press, New York, 1983.

[19] F. Chatelin, Eigenvalues of Matrices, Wiley, Chichester, 1993; Masson, Paris, 1988.

[20] M. Crouzeix, B. Philippe, M. Sadkane, The Davidson method, SIAM J. Sci. Comput. 15 (1994) 62–76.

[21] J. Cullum, W. Kerner, R. Willoughby, A generalized nonsymmetric Lanczos procedure, Comput. Phys. Comm. 53 (1989) 19–48.

[22] J. Cullum, R.A. Willoughby, Lanczos Algorithms for Large Symmetric Eigenvalue Computations, Vol. I, Birkhaüser, Boston, MA, 1985.
[23] J. Cullum, R.A. Willoughby, Lanczos Algorithms for Large Symmetric Eigenvalue Computations, Vol. II, Birkhaüser, Boston, MA, 1985.
[24] J. Cullum, R.A. Willoughby, A practical procedure for computing eigenvalues of large sparse nonsymmetric matrices, in: J. Cullum, R.A. Willoughby (Eds.), Large Scale Eigenvalue Problems, Amsterdam, North-Holland, 1986, pp. 193–240.
[25] J.J.M. Cuppen, A divide and conquer method for the symmetric eigenproblem, Numer. Math. 36 (1981) 177–195.
[26] E.R. Davidson, The iterative calculation of a few of the lowest eigenvalues and corresponding eigenvectors of large real symmetric matrices, J. Comput. Phys. 17 (1975) 87–94.
[27] C. Davis, W. Kahan, The rotation of eigenvectors by a perturbation-iii, Soc. Ind. Appl. Math. J. Num. Anal. 7 (1970) 1–46.
[28] J.W. Demmel, Applied Numerical Linear Algebra, SIAM, Philadelphia, PA, 1997.
[29] J.W. Demmel, K. Veselić, Jacobi's method is more accurate than QR, SIAM J. Matrix Anal. Appl. 13 (1992) 1204–1246.
[30] A. den Boer, De Jacobi-methode van 1845 tot 1990, Master Thesis, Department of Mathematics, Utrecht University, 1991.
[31] J.J. Dongarra, I.S. Duff, D.C. Sorensen, H.A. van der Vorst, Numerical Linear Algebra for High-Performance Computers, SIAM, Philadelphia, PA, 1998.
[32] J.J. Dongarra, D.C. Sorensen, A fully parallel algorithm for the symmetric eigenvalue problem, SIAM J. Sci. Statist. Comput. 8 (1987) 139–154.
[33] J.J. Dongarra, D.W. Walker, Software libraries for linear algebra computations on high performance computers, SIAM Rev. 37 (1995) 151–180.
[34] R.J. Duffin, A minimax theory for overdamped networks, J. Rational Mech. Anal. 4 (1955) 221–233.
[35] E.G. D'yakonov, Iterative methods in eigenvalue problems, Math. Notes 34 (1983) 945–953.
[36] P.J. Eberlein, Solution to the complex eigenproblem by a norm-reducing Jacobi-type method, Numer. Math. 14 (1970) 232–245.
[37] M. Engeli, T. Ginsburg, H. Rutishauser, E. Stiefel, Refined Iterative Methods for Computation of the Solution and the Eigenvalues of Self-Adjoint Boundary Value Problems, Birkhäuser, Basel/Stuttgart, 1959.
[38] D.G. Feingold, R.S. Varga, Block diagonally dominant matrices in generalizations of the Gerschgorin circle theorem, Pacific J. Math. 12 (1962) 1241–1250.
[39] D.R. Fokkema, G.L.G. Sleijpen, H.A. van der Vorst, Jacobi–Davidson style QR and QZ algorithms for the reduction of matrix pencils, SIAM J. Sci. Comput. 20 (1996) 94–125.
[40] J.G.F. Francis, The QR transformation: a unitary analogue to the LR transformation, Parts I and II Comput. J. 4 1961 265–272, 332–345.
[41] R.W. Freund, N.M. Nachtigal, QMRPACK: a package of QMR algorithms, ACM Trans. Math. Software 22 (1996) 46–77.
[42] R.W. Freund, Quasi-kernel polynomials and their use in non-Hermitian matrix iterations, J. Comput. Appl. Math. 43 (1992) 135–158.
[43] F. Gantmacher, The Theory of Matrices, Vols. I and II, Chelsea, New York, 1959.
[44] S. Gerschgorin, Über die Abgrenzung der Eigenwerte einer Matrix, Izv. Akad. Nauk SSSR, Ser. Fiz.-Mat. 6 (1931) 749–754.
[45] W. Givens, Numerical computation of the characteristic values of a real symmetric matrix, Oak Ridge National Lab., ORNL-1574, 1954.
[46] H.H. Goldstine, F.J. Murray, J. von Neumann, The Jacobi method for real symmetric matrices, J. Assoc. Comput. Mach. 6 (1959) 176–195.
[47] G.H. Golub, Some uses of the Lanczos algorithm in numerical linear algebra, in: J.J.H. Miller (Ed.), Topics in Numerical Analysis, Academic Press, New York, 1973, pp. 173–184.
[48] G.H. Golub, Sparse matrix computations: Eigenvalues and linear equations, Seminaries Iria, Inst. Rech. D'Informatique et d'Automatique, Rocquencourt, 1975, pp. 117–140.
[49] G.H. Golub, W. Kahan, Calculating the singular values and pseudo-inverse of a matrix, SIAM J. Numer. Anal. 2 (1965) 205–224.

[50] G.H. Golub, H.A. van der Vorst, 150 Years old and still alive: Eigenproblems, in: I.S. Duff, G.A. Watson (Eds.), The State of the Art in Numerical Analysis, Clarendon Press, Oxford, 1997, pp. 93–119.
[51] G.H. Golub, C.F. Van Loan, Matrix Computations, The Johns Hopkins University Press, Baltimore, 1996.
[52] G.H. Golub, J.H. Wilkinson, Ill-conditioned eigensystems and the computation of the Jordan canonical form, SIAM Rev. 18 (1976) 578–619.
[53] J. Greenstadt, A method for finding roots of arbitrary matrices, Math. Tables Aids Comput. 9 (1955) 47–52.
[54] R.T. Gregory, Computing eigenvalues and eigenvectors of a symmetric matrix on the ILLIAC, Math. Tables Aids Comput. 7 (1953) 215–220.
[55] M. Gu, S.C. Eisenstat, A divide-and-conquer algorithm for the symmetric tridiagonal eigenproblem, SIAM J. Matrix Anal. Appl. 16 (1995) 172–191.
[56] J.-S. Guo, W.-W. Lin, C.-S. Wang, Numerical solutions for large sparse eigenvalue problems, Linear Algebra Appl. 225 (1995) 57–89.
[57] M.H. Gutknecht, A completed theory of the unsymmetric Lanczos process and related algorithms, Part I, SIAM J. Matrix Anal. Appl. 13 (1992) 594–639.
[58] R. Heeg, Stability and transition of attachment-line flow, Ph.D. Thesis, Universiteit Twente, Enschede, The Netherlands, 1998.
[59] P. Henrici, On the speed of convergence of cyclic and quasicyclic Jacobi methods for computing the eigenvalues of Hermitian matrices, SIAM J. Appl. Math. 6 (1958) 144–162.
[60] K. Hessenberg, Auflösung linearer Eigenwertaufgaben met Hilfe der Hamilton-Cayleyschen Gleichung, Ph.D. Thesis, Technische Hochschule, Darmstadt, 1940.
[61] W. Hoffmann, B.N. Parlett, A new proof of global convergence for the tridiagonal QL algorithm, SIAM J. Numer. Anal. 15 (1978) 929–937.
[62] P. Horst, A method of factor analysis by means of which all coordinates of the factor matrix are given simultaneously, Psychometrika 2 (1937) 225–236.
[63] A.S. Householder, Unitary triangularization of a nonsymmetric matrix, J. ACM 5 (1958) 339–342.
[64] A.S. Householder, The Theory of Matrices in Numerical Analysis, Dover, New York, 1964.
[65] C.G.J. Jacobi, Über ein leichtes Verfahren, die in der Theorie der Säcularstörungen vorkommenden Gleichungen numerisch aufzulösen, J. Reine Angew. Math. 30 (1846) 51–94.
[66] B. Kågström, The generalized singular value decomposition and the general $A - \lambda B$ problem, BIT 24 (1985) 568–583.
[67] B. Kågström, RGSVD: an algorithm for computing the Kronecker structure and reducing subspaces of singular $A - \lambda B$ pencils, SIAM J. Sci. Statist. Comput. 7 (1986) 185–211.
[68] W. Kahan, B.N. Parlett, How far should you go with the Lanczos process, in: J.R. Bunch, D.J. Rose (Eds.), Sparse Matrix Computations, Academic Press, New York, 1976.
[69] S. Kaniel, Estimates for some computational techniques in linear algebra, Math. Comput. 20 (1966) 369–378.
[70] T. Kato, On the upper and lower bounds of eigenvalues, J. Phys. Soc. Japan 4 (1949) 334–339.
[71] A.V. Knyazev, Preconditioned eigensolvers — an oxymoron?, Electron. Trans. Numer. Anal. 7 (1998) 104–123.
[72] A. Knyazev, Preconditioned eigensolvers, in: Z. Bai, J. Demmel, J. Dongarra, A. Ruhe, H. van der Vorst (Eds.), Templates for the Solution of Algebraic Eigenvalue Problems: a Practical Guide, SIAM, Philadelphia, PA, 2000, in press.
[73] A.N. Krylov, O cislennom resenii uravnenija, kotorym v techniceskih voprasah opredeljajutsja castoty malyh kolebanii material'nyh sistem, Izv. Akad. Nauk SSSR. Ser. Fiz.-Mat. 4 (1931) 491–539.
[74] V.N. Kublanovskaja, On some algorithms for the solution of the complete eigenvalue problem, Zh. Vych. Mat. 1 (1961) 555–570.
[75] V.N. Kublanovskaja, On an application of Newton's method to the determination of eigenvalues of λ-matrices, Soviet Math. Dokl. 10 (1969) 1240–1241.
[76] V.N. Kublanovskaja, On an approach to the solution of the generalized latent value problem for λ-matrices, SIAM J. Numer. Anal. 7 (1970) 532–537.
[77] P. Lancaster, Lambda-Matrices and Vibrating Systems, Pergamon Press, Oxford, UK, 1969.
[78] C. Lanczos, An iteration method for the solution of the eigenvalue problem of linear differential and integral operators, J. Res. Natl. Bur. Stand 45 (1950) 225–280.

[79] H.J. Landau, On Szegö's eigenvalue distribution theory and non-Hermitian kernels, J. Anal. Math. 28 (1975) 335–357.
[80] S.L. Lee, A practical upper bound for departure from normality, SIAM J. Matrix Anal. Appl. 16 (1995) 462–468.
[81] R.B. Lehoucq, D.C. Sorensen, C. Yang, ARPACK User's Guide — Solution of Large-Scale Eigenvalue Problems with Implicitly Restarted Arnoldi Methods, SIAM, Philadelphia, PA, 1998.
[82] T.-Y. Li, Z. Zeng, Homotopy-determinant algorithm for solving nonsymmetric eigenvalue problems, Math. Comput. 59 (1992) 483–502.
[83] R. Lippert, A. Edelman, Nonlinear eigenvalue problems with orthogonality constraints, in: Z. Bai, J. Demmel, J. Dongarra, A. Ruhe, H. van der Vorst (Eds.), Templates for the Solution of Algebraic Eigenvalue Problems: a Practical Guide, SIAM, Philadelphia, PA, 2000, in press.
[84] S.H. Lui, Computation of pseudospectra by continuation, SIAM J. Sci. Comput. 18 (1997) 565–573.
[85] S.H. Lui, G.H. Golub, Homotopy method for the numerical solution of the eigenvalue problem of self-adjoint partial differential operators, Numer. Algorithms 10 (1995) 363–378.
[86] S.H. Lui, H.B. Keller, W.C. Kwok, Homotopy method for the large sparse real nonsymmetric eigenvalue problem, SIAM J. Matrix Anal. Appl. 18 (1997) 312–333.
[87] K. Meerbergen, Robust methods for the calculation of rightmost eigenvalues of nonsymmetric eigenvalue problems, Ph.D. Thesis, Katholieke Universiteit Leuven, Leuven, Belgium, 1996.
[88] K. Meerbergen, R. Morgan, Inexact methods, in: Z. Bai, J. Demmel, J. Dongarra, A. Ruhe, H. van der Vorst (Eds.), Templates for the Solution of Algebraic Eigenvalue Problems: a Practical Guide, SIAM, Philadelphia, PA, 1999.
[89] K. Meerbergen, D. Roose, The restarted Arnoldi method applied to iterative linear system solvers for the computation of rightmost eigenvalues, SIAM J. Matrix Anal. Appl. 18 (1997) 1–20.
[90] C.B. Moler, D. Morrison, Singular value analysis of cryptograms, Amer. Math. Monthly 90 (1983) 78–87.
[91] C.B. Moler, G.W. Stewart, An algorithm for generalized matrix eigenvalue problems, SIAM J. Numer. Anal. 10 (1973) 241–256.
[92] R.B. Morgan, Computing interior eigenvalues of large matrices, Linear Algebra Appl. 154–156 (1991) 289–309.
[93] R.B. Morgan, D.S. Scott, Generalizations of Davidson's method for computing eigenvalues of sparse symmetric matrices, SIAM J. Sci. Statist. Comput. 7 (3) (1986) 817–825.
[94] J. Olsen, P. Jørgensen, J. Simons, Passing the one-billion limit in full configuration-interaction (FCI) calculations, Chem. Phys. Lett. 169 (1990) 463–472.
[95] A.M. Ostrowski, On the convergence of the Rayleigh quotient iteration for the computation of characteristic roots and vectors I–VI, Arch. Rational. Mesh. Anal. I 1:233–241, II 2:423–428, III 3:325–340, IV 3:341–347, V 3:472–481, VI 4:153–165, 1958–1959.
[96] M.H.C. Paardekooper, An eigenvalue algorithm for skew symmetric matrices, Numer. Math. 17 (1971) 189–202.
[97] C.C. Paige, The computation of eigenvalues and eigenvectors of very large matrices, Ph.D. Thesis, University of London, 1971.
[98] C.C. Paige, B.N. Parlett, H.A. van der Vorst, Approximate solutions and eigenvalue bounds from Krylov subspaces, Numer. Linear Algebra Appl. 2 (2) (1995) 115–134.
[99] C.C. Paige, M.A. Saunders, Solution of sparse indefinite systems of linear equations, SIAM J. Numer. Anal. 12 (1975) 617–629.
[100] B.N. Parlett, The Symmetric Eigenvalue Problem, Prentice-Hall, Englewood Cliffs, NJ, 1980.
[101] B.N. Parlett, Misconvergence in the Lanczos algorithm, in: M.G. Cox, S. Hammarling (Eds.), Reliable Numerical Computation, Clarendon Press, Oxford, UK, 1990, pp. 7–24 (Chapter 1).
[102] B.N. Parlett, J.K. Reid, Tracking the progress of the Lanczos algorithm for large symmetric eigenproblems, IMA J. Numer. Anal. 1 (1981) 135–155.
[103] B.N. Parlett, D.S. Scott, The Lanczos algorithm with selective orthogonalization, Math. Comput. 33 (1979) 217–238.
[104] B.N. Parlett, D.R. Taylor, Z.A. Liu, A look-ahead Lanczos algorithm for unsymmetric matrices, Math. Comput. 44 (1985) 105–124.
[105] D.A. Pope, C. Tompkins, Maximizing functions of rotations: experiments concerning speed of diagonalization of symmetric matrices using Jacobi's method, J. ACM 4 (1957) 459–466.
[106] E.H. Rogers, A minimax theory for overdamped systems, Arch. Rational Mech. Anal. 19 (1964) 89–96.
[107] A. Ruhe, Algorithms for the nonlinear eigenvalue problem, SIAM J. Numer. Anal. 10 (1973) 674–689.

[108] A. Ruhe, Implementation aspects of band Lanczos algorithms for computation of eigenvalues of large sparse symmetric matrices, Math. Comput. 33 (1979) 680–687.
[109] A. Ruhe, The rational Krylov algorithm for nonsymmetric eigenvalue problems. iii. complex shifts for real matrices, BIT 34 (1994) 165–176.
[110] A. Ruhe, Rational Krylov algorithms for nonsymmetric eigenvalue problems. ii. Matrix pairs, Linear Algebra Appl. 197/198 (1994) 283–295.
[111] H. Rutishauser, Solution of eigenvalue problems with the LR-transformation, Nat. Bur. Standards Appl. Math. Ser. 49 (1958) 47–81.
[112] H. Rutishauser, Computational aspects of F.L. Bauer's simultaneous iteration method, Numer. Math. 13 (1969) 4–13.
[113] H. Rutishauser, Simultaneous iteration method for symmetric matrices, in: J.H. Wilkinson, C. Reinsch (Eds.), Handbook for Automatic Computation, Vol. 2, Linear Algebra, Springer Verlag, Heidelberg, 1971, pp. 284–302.
[114] Y. Saad, On the rates of convergence of the Lanczos and the block Lanczos methods, SIAM J. Numer. Anal. 17 (1980) 687–706.
[115] Y. Saad, Variations on Arnoldi's method for computing eigenelements of large unsymmetric matrices, Linear Algebra Appl. 34 (1980) 269–295.
[116] Y. Saad, Numerical Methods for Large Eigenvalue Problems, Manchester University Press, Manchester, UK, 1992.
[117] Y. Saad, M.H. Schultz, GMRES: A generalized minimal residual algorithm for solving nonsymmetric linear systems, SIAM J. Sci. Statist. Comput. 7 (1986) 856–869.
[118] A. Sameh, On Jacobi and Jacobi-like algorithms for a parallel computer, Math. Comput. 25 (1971) 579–590.
[119] A. Schönhage, Zur Konvergenz des Jacobi-Verfahrens, Numer. Math. 3 (1961) 374–380.
[120] G.L.G. Sleijpen, J.G.L. Booten, D.R. Fokkema, H.A. van der Vorst, Jacobi–Davidson type methods for generalized eigenproblems and polynomial eigenproblems, BIT 36 (1996) 595–633.
[121] G.L.G. Sleijpen, H.A. van der Vorst, A Jacobi–Davidson iteration method for linear eigenvalue problems, SIAM J. Matrix Anal. Appl. 17 (1996) 401–425.
[122] G.L.G. Sleijpen, H.A. van der Vorst, E. Meijerink, Efficient expansion of subspaces in the Jacobi–Davidson method for standard and generalized eigenproblems, Electron. Trans. Numer. Anal. 7 (1998) 75–89.
[123] G.L.G. Sleijpen, H.A. van der Vorst, M.B. van Gijzen, Quadratic eigenproblems are no problem, SIAM News 29 (1996) 8–9.
[124] P. Smit, Numerical analysis of eigenvalue algorithms based on subspace iterations, Ph.D. Thesis, Tilburg University, The Netherlands, 1997.
[125] D.C. Sorensen, Implicit application of polynomial filters in a k-step Arnoldi method, SIAM J. Matrix Anal. Appl. 13 (1) (1992) 357–385.
[126] G.W. Stewart, Simultaneous iteration for computing invariant subspaces of non-Hermitian matrices, Numer. Math. 25 (1976) 123–136.
[127] G.W. Stewart, SRRIT: a Fortran subroutine to calculate the dominant invariant subspace of a nonsymmetric matrix, Technical Report TR 154, Department of Computer Science, University of Maryland, 1978.
[128] G.W. Stewart, On the early history of the singular value decomposition, SIAM Rev. 35 (1993) 551–566.
[129] G.W. Stewart, Ji-Guang Sun, Matrix Perturbation Theory, Academic Press, San Diego, CA, 1990.
[130] J.G. Sun, Perturbation analysis of quadratic eigenvalue problems, Report UNINF 99.01, Dept. Comp. Science, Umeå University, Sweden, 1999.
[131] O. Taussky, A recurring theorem on determinants, Amer. Math. Monthly 56 (1949) 672–676.
[132] G. Temple, The accuracy of Rayleigh's method of calculating the natural frequencies of vibrating systems, Proc. Roy. Soc. A 211 (1952) 204–224.
[133] F. Tisseur, Backward error and condition of polynomial eigenvalue problems, Technical Report NA-332, University of Manchester, Manchester, UK, 1998.
[134] F. Tisseur, J. Dongarra, A parallel divide and conquer algorithm for the symmetric eigenvalue problem on distributed memory architectures, SIAM J. Sci. Comput. 20 (1999) 2223–2236.
[135] L.N. Trefethen, Pseudospectra of matrices, in: D.F. Griffiths, G.A. Watson (Eds.), Numerical Analysis 1991, Longman, 1992, pp. 234–266.
[136] L.N. Trefethen, Computation of pseudospectra, Acta Numer. 8 (1999) 247–295.
[137] L.N. Trefethen, D. Bau, III, Numerical Linear Algebra, SIAM, Philadelphia, PA, 1997.

[138] L.N. Trefethen, A.E. Trefethen, S.C. Reddy, T.A. Driscoll, Hydrodynamic stability without eigenvalues, Science 261 (1993) 578–584.
[139] L.N. Trefethen, M.R. Trummer, An instability phenomenon in spectral methods, SIAM J. Numer. Anal. 24 (1987) 1008–1023.
[140] H.W. Turnbull, A.C. Aitken, Introduction to the Theory of Canonical Matrices, Dover Publications, New York, 1961.
[141] R.R. Underwood, An iterative block Lanczos method for the solution of large sparse symmetric eigenproblems, Technical Report STAN-CS-75-495, Department of Computer Science, Stanford University, Stanford, CA, 1975.
[142] A. van der Sluis, H.A. van der Vorst, The convergence behavior of Ritz values in the presence of close eigenvalues, Linear Algebra Appl. 88/89 (1987) 651–694.
[143] P. Van Dooren, The computation of Kronecker's canonical form of a singular pencil, Linear Algebra Appl. 27 (1979) 103–140.
[144] P. Van Dooren, The generalized eigenstructure problem in linear system theory, IEEE Trans. Automat. Control AC 26 (1981) 111–128.
[145] P. Van Dooren, A generalized eigenvalue approach of solving Ricatti equations, SIAM J. Sci. Statist. Comput. 2 (1981) 121–135.
[146] P. Van Dooren, Algorithm 590, DUSBSP AND EXCHQZZ: FORTRAN subroutines for computing deflating subspaces with specified spectrum, ACM Trans. Math. Software 8 (1982) 376–382.
[147] H.P.M. van Kempen, On quadratic convergence of the special cyclic Jacobi method, Numer. Math. 9 (1966) 19–22.
[148] R.S. Varga, Matrix Iterative Analysis, Prentice-Hall, Englewood Cliffs, NJ, 1962.
[149] R. von Mises, H. Pollaczek-Geiringer, Praktische Verfahren der Gleichungsauflösung, Z. Angew. Math. Mech. 9 (1929) 58–77, 152–164.
[150] D.S. Watkins, Fundamentals of Matrix Computations, Wiley, New York, 1991.
[151] D.S. Watkins, Some perspectives on the eigenvalue problem, SIAM Rev. 35 (1993) 430–471.
[152] D.S. Watkins, L. Elsner, Chasing algorithms for the eigenvalue problem, SIAM J. Matrix Anal. Appl. 12 (1991) 374–384.
[153] H.F. Weinberger, Variational Methods for Eigenvalue Approximation, SIAM, Philadelphia, 1974.
[154] J.H. Wilkinson, The Algebraic Eigenvalue Problem, Clarendon Press, Oxford, 1965.
[155] J.H. Wilkinson, Linear differential equations and Kronecker's canonical form, in: C. de Boor, G.H. Golub (Eds.), Recent advances in Numerical Analysis, Academic Press, New York, 1978, pp. 231–265.
[156] J.H. Wilkinson, Kronecker's canonical form and the QZ algorithm, Linear Algebra Appl. 28 (1979) 285–303.
[157] J.H. Wilkinson, Eigenvalue problems, in: A. Iserles, M.J.D. Powell (Eds.), The State of the Art in Numerical Analysis, Clarendon Press, Oxford, 1987, pp. 1–39.
[158] J.H. Wilkinson, C. Reinsch (Eds.), Handbook for Automatic Computation, Vol. 2, Linear Algebra, Springer, Heidelberg, 1971.
[159] T. Zhang, K.H. Law, G.H. Golub, On the homotopy method for perturbed symmetric generalized eigenvalue problems, SIAM J. Numer. Anal. 35 (1) (1998) 300–319.

Historical developments in convergence analysis for Newton's and Newton-like methods

Tetsuro Yamamoto

Department of Mathematical Sciences, Faculty of Science, Ehime University, Matsuyama 790-8577, Japan

Abstract

Historical developments in convergence theory as well as error estimates for Newton's method and Newton-like methods for nonlinear equations are described, mainly for differentiable equations in Banach spaces.

MSC: 47H17; 65J15

Keywords: Nonlinear equations; Convergence theorems; Error estimates; Newton's method; Newton-like methods; Halley's method

1. Introduction

More than three hundred years have passed since a procedure for solving an algebraic equation was proposed by Newton in 1669 and later by Raphson in 1690 [10]. The method is now called Newton's method or the Newton–Raphson method and is still a central technique for solving nonlinear equations. Many topics related to Newton's method still attract attention from researchers. For example, the construction of globally convergent effective iterative methods for solving nondifferentiable equations in $\mathbb{R}^n$ or $\mathbb{C}^n$ is an important research area in the fields of numerical analysis and optimization.

The purpose of this paper is to trace historical developments in Kantorovich-type convergence theory as well as error estimates for Newton's method and Newton-like methods, mainly for differentiable equations in Banach spaces.

First, in Section 2, we state fundamental results in the history of convergence study as well as error estimates for Newton's method based on the Newton–Kantorovich theorem. Next, in Section 3, we

E-mail address: yamamoto@math.sci.ehime-u.ac.jp (T. Yamamoto).

state some convergence results for Newton-like methods which generalize the results for Newton's method.

In Section 4, we describe Dennis' result for convergence of an iterative method with his "recalculation sequence", which includes the forward and the backward secant methods as special cases. Furthermore, in Section 5, we trace historical developments in convergence analysis for Halley's method and for Chebyshev's method.

Convergence theorems for a class of iterative methods for not necessarily differentiable equations are summarized in Section 6. Finally, in Section 7, concluding remarks are given.

2. Newton's method

Let X and Y be Banach spaces and $F : D \subseteq X \to Y$ be an operator where D is a domain of F. If F is differentiable in an open convex set $D_0 \subseteq D$, then Newton's method for solving the equation

$$F(x) = 0 \tag{2.1}$$

with a solution x^* is defined by the following:

(N1) Let x_k be an approximation to x^*;
(N2) Solve the linear equation

$$F(x_k) + F'(x_k)h = 0 \tag{2.2}$$

with respect to h, provided that $F'(x_k)$ is nonsingular;
(N3) Set $x_{k+1} = x_k + h$, expecting for it to be an improvement to x_k, where $k = 0, 1, 2, \ldots$.

Since $0 = F(x^*) = F(x_k + h) = F(x_k) + F'(x_k)h + \xi$ with $\|\xi\| = o(\|h\|)$ ($\|\xi\| = O(\|h\|^2)$ if F' satisfies a Lipschitz condition), (2.2) is a linearization procedure for the operator F around x_k. The procedure first employed by Newton in 1669 for the cubic equation $3x^3 - 2x - 5 = 0$ is different from the (N1)–(N3) [10], but it is easily verified that both are mathematically equivalent. The procedure (N1)–(N3) can also be written

$$x_{k+1} = x_k - F'(x_k)^{-1}F(x_k), \quad k = 0, 1, 2, \ldots . \tag{2.3}$$

Since Raphson described in 1690 the formula (2.3) for a general cubic equation $x^3 - bx = c$, the procedure (N1)–(N3) or (2.3) is also called the Newton–Raphson method.

Later, in 1818, Fourier [34] proved the quadratic convergence of the method for the case $X = \mathbb{R}$. In 1829, Cauchy first proved a convergence theorem which does not assume any existence of a solution. It asserts that if x_0 satisfies certain conditions, then Eq. (2.1) has a solution and iteration (2.3) starting from x_0 quadratically converges to the solution.

Cauchy's results are summarized as follows:

Theorem 2.1 (Cauchy [13, pp. 575–600]). *Let* $X = \mathbb{R}$, $F = f \in C^2$, $x_0 \in X$, $f'(x_0) \neq 0$, $\sigma_0 = -f(x_0)/f'(x_0)$, $\eta = |\sigma_0|$,

$$I = \langle x_0, x_0 + 2\sigma_0 \rangle \equiv \begin{cases} [x_0, x_0 + 2\sigma_0] & \text{if } \sigma_0 \geq 0, \\ [x_0 + 2\sigma_0, x_0] & \text{if } \sigma_0 < 0 \end{cases}$$

and $|f''(x)| \leqslant K$ *in* I. *Then the following results hold*:

(i) *If*

$$2K\eta < |f'(x_0)|, \tag{2.4}$$

then Eq. (2.1) *has a unique solution* x^* *in* I.

(ii) *If* $|f'(x)| \geqslant m$ *in* I *and*

$$2K\eta < m, \tag{2.5}$$

then the Newton sequence $\{x_k\}$ *starting from* x_0 *satisfies the following*:

$$|x_{k+1} - x_k| \leqslant \frac{K}{2m}|x_k - x_{k-1}|^2, \quad k \geqslant 1$$

and

$$x^* \in \langle x_k, x_k + 2\sigma_k \rangle,$$

where $\sigma_k = -f(x_k)/f'(x_k) = x_{k+1} - x_k$, *so that*

$$\begin{aligned} |x^* - x_k| &\leqslant 2|x_{k+1} - x_k| \quad (k \geqslant 0) \\ &\leqslant \frac{K}{m}|x_k - x_{k-1}|^2 \quad (k \geqslant 1) \\ &\leqslant 2\eta \left(\frac{K\eta}{2m}\right)^{2^k - 1} \quad (k \geqslant 0). \end{aligned}$$

A convergence theorem for an iterative method is called a local convergence theorem if it asserts convergence by assuming that a solution x^* exists and an initial approximation is chosen sufficiently close to x^*. On the other hand, a convergence theorem like Theorem 2.1, which does not assume the existence of any solution a priori, but assumes that some conditions hold at the initial point x_0, is called a semilocal convergence theorem.

Ostrowski [62] improved Theorem 2.1 by replacing (2.5) by $2K\eta \leqslant |f'(x_0)|$ and showing

$$|x_{k+1} - x_k| \leqslant \frac{K}{2|f'(x_0)|}|x_k - x_{k-1}|^2, \quad k \geqslant 1.$$

In 1916, Fine [33] first established a semilocal convergence theorem for (2.3) when $X = \mathbb{R}^n$ or $\mathbb{C}^n$, $n \geqslant 1$. He assumed each component of $F = (f_1, \ldots, f_n)^t$ has continuous first and second derivatives in the domain D and that

$$\bar{S} = \bar{S}(x_1, \eta) = \{x \in X \mid \|x - x_1\|_2 \leqslant \eta\} \subseteq D, \tag{2.6}$$

with

$$\eta = \|F'(x_0)^{-1}F(x_0)\|_2,$$

where $\|\cdot\|_2$ denotes the Euclidean norm. He then proved that if

$$|[F'(x)^{-1}]_{ij}| \leqslant \mu, \qquad \left|\frac{\partial^2 f_i}{\partial x_j \partial x_k}\right| \leqslant \nu \quad \text{in } D$$

and

$$\|F(x_0)\|_2 < \frac{1}{n^{7/2}\mu^2 v}, \tag{2.7}$$

then Eq. (2.1) has one and only one solution in $\bar{S}$ and the solution is obtained by (2.3). Furthermore, if instead of x_0 any other point in $\bar{S}$ is chosen as the starting point, (2.3) will converge to the same solution as $k \to \infty$. For a comparison of Cauchy's condition with those of Fine and Ostrowski for the case of a single equation (i.e., $n = 1$), see a historical note in Ostrowski's book [65, pp. 400–401].

In 1922, Banach [7] introduced a notion of a space of the type (B), which is now called a Banach space, and developed the theory in his famous book [8]. As is widely recognized, results established for such an abstract space can usually be applied to different fields of mathematics including finite and infinite algebraic equations, differential equations, integral equations, etc.

In 1939, Kantorovich [43] published a paper on iterative methods for functional equations in a space of the type (B) and applied his theory developed there to derive a convergence theorem for Newton's method, on the basis of the contraction mapping principle of Banach.

Later, in 1948, he [44] established a semilocal convergence theorem for Newton's method in a Banach space, which is now called Kantorovich's theorem or the Newton–Kantorovich theorem. It may be considered as a generalization and an improvement of Theorem 2.1.

Theorem 2.2 (Kantorovich's theorem). *Let $F : D \subseteq X \to Y$ be twice differentiable in an open convex set $D_0 \subseteq D$ and suppose that, for some $x_0 \in D_0$, $F'(x_0)^{-1}$ exists, $\|F'(x_0)^{-1}\| \leqslant B$, $\|F''(x)\| \leqslant K$ in X. Assume that $F(x_0) \neq 0$ without loss of generality and that*

$$\|F'(x_0)^{-1}F(x_0)\| \leqslant \eta, \quad h = KB\eta \leqslant \tfrac{1}{2}$$

$$t^* = \frac{2\eta}{1 + \sqrt{1 - 2h}}, \quad \bar{S} = \bar{S}(x_0, t^*) \subseteq D_0.$$

Then:

(i) *The iterates* (2.3) *are well defined, $x_k \in S$ (interior of $\bar{S}$), $k \geqslant 0$ and $\{x_k\}$ converges to a solution $x^* \in \bar{S}$ of* (2.1).

(ii) *The solution is unique in*

$$\tilde{S} = \begin{cases} S(x_0, t^{**}) \cap D_0 & \text{if } 2h < 1, \\ \bar{S}(x_0, t^{**}) \cap D_0 & \text{if } 2h = 1, \end{cases} \tag{2.8}$$

where $t^{**} = (1 + \sqrt{1 - 2h})/KB$.

(iii) *Error estimates*

$$\|x^* - x_k\| \leqslant \frac{2\eta_k}{1 + \sqrt{1 - 2h_k}} \leqslant 2^{1-k}(2h)^{2^k - 1}\eta, \quad k \geqslant 0, \tag{2.9}$$

hold, where $\{\eta_k\}$, $\{h_k\}$ are defined by the recurrence relations

$$B_0 = B, \quad \eta_0 = \eta, \quad h_0 = h,$$
$$B_k = \frac{B_{k-1}}{1 - h_{k-1}}, \quad \eta_k = \frac{h_{k-1}\eta_{k-1}}{2(1 - h_{k-1})}, \quad h_k = KB_k\eta_k, \quad k \geqslant 1. \tag{2.10}$$

Three years later, Kantorovich [45] introduced a majorant principle to give a new proof for Theorem 2.2. In fact, he showed that under the same conditions as in Theorem 2.2, Newton's sequence $\{x_k\}$ satisfies

$$\|x_{k+1} - x_k\| \leqslant t_{k+1} - t_k, \quad k \geqslant 0 \tag{2.11}$$

and

$$\|x^* - x_k\| \leqslant t^* - t_k, \quad k \geqslant 0, \tag{2.12}$$

with the scalar Newton sequence $\{t_k\}$ defined by

$$t_0 = 0, \quad t_{k+1} = t_k - \frac{f(t_k)}{f'(t_k)}, \quad k = 0, 1, 2, \ldots,$$

where

$$f(t) = \tfrac{1}{2}KBt^2 - t + \eta. \tag{2.13}$$

At first glance, the estimate (2.12) appears to be sharper than (2.9). However, we have (cf. [98])

$$t_{k+1} - t_k = \eta_k, \quad t^* - t_k = \frac{1 - \sqrt{1 - 2h_k}}{KB_k} = \frac{2\eta_k}{1 + \sqrt{1 - 2h_k}},$$

$$t^{**} - t_k = \frac{1 + \sqrt{1 - 2h_k}}{KB_k},$$

so that there is no difference between the estimates (2.9) and (2.12).

The condition $h = KB\eta \leqslant \frac{1}{2}$ in Theorem 2.2 is often called the Kantorovich condition.

Remark 2.3 (Updating the Newton–Kantorovich theorem). It is known that the conditions "$\|F''(x)\| \leqslant K$ in X" and "$\bar{S}(x_0, t^*) \subseteq D_0$" of Theorem 2.2 may be replaced by weaker ones

$$\|F'(x) - F'(y)\| \leqslant K\|x - y\|, \quad x, y \in D_0 \quad \text{(Fenyö [32])} \tag{2.14}$$

and

$$\bar{S}(x_1, t^* - \eta) \subseteq D_0 \quad \text{(Ostrowski [65])}, \tag{2.15}$$

respectively. Observe that $\bar{S}(x_1, t^* - \eta) \subseteq \bar{S}(x_0, t^*) \cap \bar{S}(x_1, \eta)$ and (2.15) improves Fine's condition (2.6), too. Furthermore, Deuflhard–Heindl [27] asserted that (2.14) should be replaced by an affine invariant condition

$$\|F'(x_0)^{-1}(F'(x) - F'(y))\| \leqslant K\|x - y\|, \quad x, y \in D_0, \tag{2.16}$$

since Newton's sequence is invariant under an affine transformation. If (2.16) replaces $\|F''(x)\| \leqslant K$ in X, then the conditions $h = KB\eta$, $t^{**} = (1 + \sqrt{1 - 2h})/(KB)$ and $B_0 = B$ in Theorem 2.2 should be replaced by $h = K\eta$, $t^{**} = (1 + \sqrt{1 - 2h})/K$ and $B_0 = 1$, respectively. Under these assumptions and notation, it can be shown that $x_k \in S$, $k \geqslant 1$, $x_k \to x^* \in \bar{S}$ as $k \to \infty$ and the assertions (ii) and (iii) hold [98].

Remark 2.4. If we replace the condition $\|F'(x_0)^{-1}\| \leqslant B$ by a stronger condition $\|F'(x)^{-1}\| \leqslant B$ in D_0 in Theorem 2.2, then we can prove convergence of Newton's method under the condition $h < 2$, which is weaker than the Kantorovich condition $2h \leqslant 1$. This is known as Mysovskii's theorem or the Newton–Mysovskii theorem (cf. Ortega–Rheinboldt [61] or Mysovskii [58]).

Remark 2.5. After the appearance of the Newton–Kantorovich theorem, sharp upper and lower bounds for the errors of Newton's method have been derived by many authors, for example, by Dennis [24], Döring [29], Rall–Tapia [75], Tapia [88], Ostrowski [64], Gragg–Tapia [37], Kornstaedt [49], Miel [53–55], Potra–Pták [69], Potra [68], Pták [70], Moret [57], etc.

In a series of papers [93–95,97,98], Yamamoto showed that their results follow from the Kantorovich theorem and that detailed comparison of the bounds can be made, which put a stop to the race for finding sharp error bounds for Newton's method. The following is a part of the chart for the upper error bounds arranged in an order reflecting the power of the result:

$$\|x^* - x_k\| \leqslant \frac{2d_k}{1+\sqrt{1-2K(1-K\Delta_k)d_k}} \quad (k \geqslant 0) \qquad \text{(Moret [57])}$$

$$\leqslant \frac{2d_k}{1+\sqrt{1-2KB_k d_k}} \quad (k \geqslant 0) \qquad \text{(Kantorovich [44])}$$

$$= \begin{cases} \dfrac{2d_k}{1+\sqrt{1-\frac{4}{\Delta}\frac{1-\theta^{2^k}}{1+\theta^{2^k}}d_k}} & (2h<1) \\[2ex] \dfrac{2d_k}{1+\sqrt{1-\frac{2^k}{\eta}d_k}} & (2h=1) \end{cases} \quad (k \geqslant 0) \qquad \text{(Yamamoto [94])}$$

$$\leqslant \frac{t^* - t_k}{t_{k+1} - t_k} d_k \quad (k \geqslant 0) \qquad \text{(Yamamoto [96])}$$

$$= \frac{2d_k}{1+\sqrt{1-2h_k}} \quad (k \geqslant 0) \qquad \text{(Döring [29])}$$

$$\leqslant \frac{Kd_{k-1}^2}{\sqrt{1-2h}+\sqrt{1-2h+(Kd_{k-1})^2}} \quad (k \geqslant 1)$$

$$= \sqrt{a^2 + d_{k-1}^2} - a \quad (k \geqslant 1) \qquad \text{(Potra–Pták [69])}$$

$$\leqslant t^* - t_k \quad (k \geqslant 0) \qquad \text{(Kantorovich [45])}$$

$$= \begin{cases} e^{-2^{k-1}\varphi} \dfrac{\sinh\varphi}{\sinh 2^{k-1}\varphi}\eta & (2h<1) \\ 2^{1-k}\eta & (2h=1) \end{cases} \qquad \text{(Ostrowski [64])}$$

$$= \begin{cases} \dfrac{\Delta\theta^{2^k}}{1-\theta^{2^k}} & (2h<1) \\ 2^{1-k}\eta & (2h=1) \end{cases} \quad (k \geqslant 0), \qquad \text{(Gragg–Tapia [37])}$$

where $\varphi \geqslant 0$, $(1+\cosh\varphi)h=1$, $\Delta_k=\|x_k-x_0\|$, $d_k=\|x_{k+1}-x_k\|$, $\Delta=t^{**}-t^*$, $\theta=t^*/t^{**}$, $a=\sqrt{1-2h}/KB_0$. We remark that the famous Gragg–Tapia bound in the above chart is also obtained by Ostrowski [63, p. 301]. Hence it should perhaps be known as the Ostrowski–Gragg–Tapia bound.

Similarly, we have the following chart for lower error bounds:

$$\|x^*-x_k\| \geqslant \frac{2d_k}{1+\sqrt{1+2K(1-K\Delta_k)^{-1}d_k}} \quad (k\geqslant 0) \qquad \text{(Yamamoto [98])}$$

$$\geqslant \frac{2d_k}{1+\sqrt{1+2K(1-Kt_k)^{-1}d_k}} \quad (k\geqslant 0) \qquad \text{(Schmidt [85])}$$

$$= \frac{2d_k}{1+\sqrt{1+2KB_kd_k}} \quad (k\geqslant 0) \qquad \text{(Miel [54])}$$

$$= \begin{cases} \dfrac{2d_k}{1+\sqrt{1+\frac{4}{\Delta}\frac{1-\theta^{2^k}}{1+\theta^{2^k}}d_k}} & (2h<1) \\[2ex] \dfrac{2d_k}{1+\sqrt{1+\frac{2^k}{\eta}d_k}} & (2h=1) \end{cases} \quad (k\geqslant 0) \qquad \text{(Miel [55])}$$

$$\geqslant \frac{2d_k}{1+\sqrt{1+\frac{2d_k}{d_k+\sqrt{a^2+d_k^2}}}} \quad (k\geqslant 0)$$

$$= \frac{2d_k(d_k+\sqrt{a^2+d_k^2})}{d_k+\sqrt{a^2+d_k^2}+\sqrt{(d_k+\sqrt{a^2+d_k^2})(3d_k+\sqrt{a^2+d_k^2})}} \quad (k\geqslant 0).$$

(Potra–Pták [69])

3. Newton-like methods

The majorant principle due to Kantorovich is so powerful that many authors have applied it to establish convergence theorems for variants of Newton's method [86,87,77].

In particular, Rheinboldt applied his majorant theory to obtain a convergence theorem for Newton-like method

$$x_{k+1}=x_k-A(x_k)^{-1}F(x_k), \quad k=0,1,2,\ldots, \tag{3.1}$$

for solving Eq. (2.1) in the Banach space, where $F:D\subseteq X\to Y$ is differentiable in an open convex set $D_0\subseteq D$, $x_0\in D_0$ and $A(x)$ denotes an invertible, bounded linear operator which may be considered as an approximation to $F'(x)$. Dennis [25] generalized Rheinboldt's result. Miel [54] improved their error bounds. Furthermore, Moret [57] gave a sharper error bound than Miel's, but under a rather stronger assumption on $A(x)$. Mysovskii-type theorems can be found in Dennis [23].

To state an updated version of Rheinboldt's and Dennis' results, we employ an affine-invariant formulation due to Deuflhard–Heindl [27] and assume

$$\begin{aligned}
&\|A(x_0)^{-1}(F'(x)-F'(y))\| \leqslant K\|x-y\|, \quad x,y\in D_0,\ K>0\\
&\|A(x_0)^{-1}(A(x)-A(x_0))\| \leqslant L\|x-x_0\|+l, \quad x\in D_0,\ L\geqslant 0,\ l\geqslant 0 \qquad (3.2)\\
&\|A(x_0)^{-1}(F'(x)-A(x))\| \leqslant M\|x-x_0\|+m, \quad x\in D_0,\ M\geqslant 0,\ m\geqslant 0,
\end{aligned}$$

$$l+m<1, \quad \sigma=\max\left(1,\frac{L+M}{K}\right), \quad F(x_0)\neq 0,$$

$$\|A(x_0)^{-1}F(x_0)\|\leqslant\eta, \quad h=\frac{\sigma K\eta}{(1-l-m)^2}\leqslant\frac{1}{2},$$

$$t^*=\frac{(1-l-m)(1-\sqrt{1-2h})}{\sigma K}, \quad t^{**}=\frac{1-m+\sqrt{(1-m)^2-2K\eta}}{K},$$

$$\bar{S}=\bar{S}(x_1,t^*-\eta)\subseteq D_0.$$

Under these assumptions, define the sequence $\{t_k\}$ by

$$t_0=0, \quad t_{k+1}=t_k+\frac{f(t_k)}{g(t_k)}, \quad k=0,1,2,\ldots,$$

where

$$f(t)=\tfrac{1}{2}\sigma Kt^2-(1-l-m)t+\eta \quad \text{and} \quad g(t)=1-l-Lt,$$

and the sequence $\{p_k\}$, $\{q_k\}$, $\{B_k\}$, $\{\eta_k\}$ and $\{h_k\}$ by

$$p_0=1-l, \quad q_0=1-l-m, \quad B_0=\frac{p_0}{q_0}, \quad \eta_0=\frac{\eta}{1-l}, \quad h_1=\sigma KB_0\eta_0,$$

$$p_k=1-l-L\sum_{j=0}^{k-1}\eta_j, \quad q_k=1-l-m-\sigma\sum_{j=0}^{k-1}\eta_j, \quad B_k=\frac{p_k}{q_k^2},$$

$$\eta_k=\{\tfrac{1}{2}\sigma K\eta_{k-1}^2+(p_{k-1}-q_{k-1})\eta_{k-1}\}/p_k, \quad h_k=\sigma KB_k\eta_k, \quad k\geqslant 1.$$

Furthermore, put

$$\varphi(t)=1-l-m-(L+M)t, \quad \Delta_k=\|x_k-x_0\|, \quad d_k=\|x_{k+1}-x_k\|.$$

Then, we have the following result which is an updated version of the results due to Rheinboldt, Dennis, Miel, Moret and others.

Theorem 3.1 (Yamamoto [99]).

(i) *The iteration* (3.1) *is well defined for every* $k\geqslant 0$, $x_k\in S$ *(interior of* $\bar{S}$*) for* $k\geqslant 1$ *and* $\{x_k\}$ *converges to a solution* $x^*\in\bar{S}$ *of Eq.* (2.1).

(ii) *The solution is unique in*

$$\tilde{S} = \begin{cases} S(x_0, \tilde{t}^{**}) \cap D_0 & (\textit{if } 2K\eta < (1-m)^2), \\ \bar{S}(x_0, \tilde{t}^{**}) \cap D_0 & (\textit{if } 2K\eta = (1-m)^2). \end{cases}$$

(iii) *Let* $\bar{S}_0 = \bar{S}$, $\bar{S}_k = \bar{S}(x_k, t^* - t_k)$ $(k \geqslant 1)$, $K_0 = L_0 = K$,

$$K_k = \sup_{\substack{x,y \in \bar{S} \\ x \neq y}} \frac{\|A(x_k)^{-1}(F'(x) - F'(y))\|}{\|x - y\|} \quad (k \geqslant 1),$$

$$L_k = \sup_{\substack{x,y \in \bar{S} \\ x \neq y}} \frac{\|A(x_k)^{-1}(F'(x) - F'(y))\|}{\|x - y\|} \quad (k \geqslant 1). \tag{3.3}$$

Then

$$x^* \in \bar{S}_k \subseteq \bar{S}_{k-1} \subseteq \cdots \subseteq \bar{S}_0, \quad t_{k+1} - t_k = \eta_k, \quad 2h_k \leqslant 1$$

and

$$\|x^* - x_k\| \leqslant \alpha_k \equiv \frac{2g(\Delta_k)d_k}{\varphi(\Delta_k) + \sqrt{\varphi(\Delta_k)^2 - 2K_k g(\Delta_k)^2 d_k}} \quad (k \geqslant 0)$$

$$\leqslant \beta_k \equiv \frac{2g(\Delta_k)d_k}{\varphi(\Delta_k) + \sqrt{\varphi(\Delta_k)^2 - 2L_k g(\Delta_k)^2 d_k}} \quad (k \geqslant 0)$$

$$\leqslant \gamma_k \equiv \frac{2g(\Delta_k)d_k}{\varphi(\Delta_k) + \sqrt{\varphi(\Delta_k)^2 - 2K g(\Delta_k) d_k}} \quad (k \geqslant 0) \tag{3.4}$$

(Yamamoto [96], Moret [57])

$$\leqslant \delta_k \equiv \frac{2g(t_k)d_k}{\varphi(t_k) + \sqrt{\varphi(t_k)^2 - 2K g(t_k) d_k}} \quad (k \geqslant 0) \qquad \text{(Yamamoto [96])}$$

$$\leqslant \frac{t^* - t_k}{t_k - t_{k-1}} d_{k-1} \quad (k \geqslant 1)$$

$$= \frac{1}{\eta_{k-1}} \left(\frac{2(p_k/q_k)\eta_k}{1 + \sqrt{1 - 2h_k}} \right) d_{k-1} \quad (k \geqslant 1)$$

$$\leqslant t^* - t_k \quad (k \geqslant 0) \qquad \text{(Rheinboldt [77], Dennis [25])}$$

$$= \frac{2(p_k/q_k)\eta_k}{1 + \sqrt{1 - 2h_k}} \quad (k \geqslant 0).$$

Remark 3.2. Moret [57] obtained (3.4) under stronger assumptions that $K \geqslant L$, $M = K - L$ and replacing (3.2) by

$$\|A(x_0)^{-1}(A(x) - A(y))\| \leqslant L\|x - y\| \quad \text{for } x, y \in D_0.$$

Remark 3.3. If we put $A(x_k)=F'(x_k)$, then (3.1) reduces to the Newton method and Theorem 3.1 reduces to an updated version of Theorem 2.2.

Remark 3.4. If we put $A(x)=F'(x_0)$, then (3.1) is called the simplified (or modified) Newton method. Theorem 3.1 then reduces to:

Corollary 3.5. *Consider the simplified Newton method*

$$x_{k+1}=x_k-F'(x_0)^{-1}F(x_k), \quad k=0,1,2,\ldots, \tag{3.5}$$

where we assume $x_0\in D_0$ and $F'(x_0)^{-1}$ exists and the following conditions hold:

$$\|F'(x_0)^{-1}(F'(x)-F'(x_0))\|\leqslant K\|x-x_0\|, \quad x\in D_0$$

$$0<\|F'(x_0)^{-1}F(x_0)\|\leqslant \eta, \quad h=K\eta\leqslant \tfrac{1}{2},$$

$$t^*=\frac{1-\sqrt{1-2h}}{K}, \quad t^{**}=\frac{1+\sqrt{1-2h}}{K},$$

$$\bar{S}=\bar{S}(x_1,t^*-\eta)\subseteq D_0.$$

Then:

(i) *The iteration* (3.5) *is well-defined for every $k\geqslant 0$, $x_k\in S$ for $k\geqslant 1$ and $\{x_k\}$ converges to a solution of* (2.1).

(ii) *The solution is unique in*

$$\tilde{S}=\begin{cases} S(x_0,t^{**})\cap D_0 & (2h<1), \\ \bar{S}(x_0,t^{**})\cap D_0 & (2h=1). \end{cases}$$

(iii) *Define the sequence $\{t_k\}$ by*

$$t_0=0, \quad t_{k+1}=\tfrac{1}{2}Kt_k^2+\eta, \quad k=0,1,2,\ldots .$$

Let $\bar{S}_0=\bar{S}$, $\bar{S}_k=\bar{S}(x_k,t^*-t_k)$ $(k\geqslant 1)$ and

$$K_k=\sup_{\substack{x\in\bar{S}_k \\ x\neq x_0}}\frac{\|F'(x_0)^{-1}(F'(x)-F'(x_0))\|}{\|x-x_0\|} \quad (k\geqslant 0).$$

Then

$$\begin{aligned}\|x^*-x_k\| &\leqslant \frac{2d_k}{1-K_k\Delta_k+\sqrt{(1-K_k\Delta_k)^2-2K_kd_k}} \quad (k\geqslant 0)\\ &\leqslant \frac{2d_k}{1-K\Delta_k+\sqrt{(1-K_k\Delta_k)^2-2Kd_k}} \quad (k\geqslant 0)\\ &\leqslant \frac{2d_k}{1-Kt_k+\sqrt{(1-Kt_k)^2-2Kd_k}} \quad (k\geqslant 0)\end{aligned}$$

$$\leqslant \frac{t^* - t_k}{t_{k+1} - t_k} d_k \quad (k \geqslant 0)$$

$$\leqslant \frac{t^* - t_k}{t_k - t_{k-1}} d_{k-1} \quad (k \geqslant 1) \tag{3.6}$$

$$\leqslant t^* - t_k \quad (k \geqslant 0)$$

$$= \frac{2(t_{k+1} - t_k)}{1 - Kt_k + \sqrt{(1 - Kt_k)^2 - 2K(t_{k+1} - t_k)}} \quad (k \geqslant 0)$$

$$\leqslant \frac{1}{2K}(1 - \sqrt{1 - 2h})^{k+1} \quad (k \geqslant 1) \qquad \text{(Andrew [4])} \tag{3.7}$$

$$\leqslant \frac{1}{K}(1 - \sqrt{1 - 2h})^{k+1} \quad (k \geqslant 0), \qquad \text{(Kantorovich–Akilov [46])}$$

where the inequality in (3.7) *is replaced by the strict inequality if* $k > 1$.

4. Secant method

Consider a Newton-like method using a divided difference operator $\delta F(x, y) \in L(X, Y)$ (Banach space of bounded linear operators of X into Y) in place of $A(x)$ in (3.1) such that

$$\delta F(x, y)(x - y) = F(x) - F(y), \quad x, y \in D_0 \tag{4.1}$$

and

$$\|\delta F(x, y) - \delta F(y, u)\| \leqslant a\|x - u\| + b\|x - y\| + b\|y - u\|, \quad u \in D_0, \tag{4.2}$$

where $a \geqslant 0$ and $b \geqslant 0$ are constants independent of x, y, u. These conditions are due to Schmidt [83]. Later, Laarsonen (1969) put the condition

$$\|\delta F(x', x'') - \delta F(y', y'')\| \leqslant M(\|x' - y'\| + \|x'' - y''\|), \quad x, x', y, y' \in D_0 \tag{4.3}$$

in place of (4.2) (cf. [26]). Dennis [26] showed that (4.3) implies (4.2) and that the conditions (4.1) and (4.2) imply that F is Fréchet differentiable, $\delta F(x,x) = F'(x)$ and F' satisfies the Lipschitz condition in D_0 with the Lipschitz constant $2(a + b)$.

According to Dennis, we shall call the iteration

$$x_{k+1} = x_k - \delta F(x_k, x_{k-1})^{-1} F(x_k), \quad k = 0, 1, 2, \ldots \tag{4.4}$$

(considered by Schmidt) the backward secant method and the iteration

$$x_{k+1} = x_k - \delta F(x_{k-1}, x_k)^{-1} F(x_k), \quad k = 0, 1, 2, \ldots \tag{4.5}$$

the forward secant method, where x_0, x_{-1} are given.

Laarsonen considered the following iterations:

$$y_{k+1} = y_k - \delta F(y_k, \bar{y}_k)^{-1} F(y_k), \tag{4.6}$$

$$\bar{y}_{k+1} = y_{k+1} - \delta F(y_k, \bar{y}_k)^{-1} F(y_{k+1}), \quad k = 0, 1, 2, \ldots \tag{4.7}$$

where y_0 and $\bar{y}_0$ are given.

Dennis gave a convergence theorem for the iteration

$$x_{k+1} = x_k - A_{\alpha_k}^{-1} F(x_k), \quad k = 0, 1, 2, \ldots, \tag{4.8}$$

which is the most general possible mixture of the forward and backward secant methods, where

$$A_{\alpha_k} = \delta F(x_{\alpha_k}, x_{\alpha_k - 1}) \quad \text{or} \quad A_{\alpha_k} = \delta F(x_{\alpha_k - 1}, x_{\alpha_k})$$

and $\{\alpha_k\}$ is a recalculation sequence defined by Dennis, that is, $\alpha_0 = 0$, $\alpha_k = k$ or $\alpha_k = \alpha_{k-1}$. Iterations (4.4), (4.5) and Laarsonen's (4.6)–(4.7) are included in (4.8) (cf. [26]). Dennis' result is stated as follows:

Theorem 4.1 (Dennis [26]). *Let* (4.1) *and* (4.2) *hold and let* $x_{-1}, x_0 \in D_0$ *have the following properties: for either* $A_0 = \delta F(x_0, x_{-1})$ *or for* $A_0 = \delta F(x_{-1}, x_0)$, $\|A_0^{-1}\| \leqslant \beta$, $\|x_{-1} - x_0\| \leqslant \eta_{-1}$,

$$\|A_0^{-1} F(x_0)\| \leqslant \eta, \quad \beta(a+b)\eta_{-1} < 1,$$

$$h \equiv \frac{(a+b)\beta\eta}{(1 - \beta(a+b)\eta_{-1})^2} < \tfrac{1}{4}$$

and $S(x_0, r_0) \subset D_0$, *where*

$$r_0 = \frac{1 - \sqrt{1 - 4h}}{2\beta(a+b)}(1 - \beta(a+b)\eta_{-1}).$$

Then, for an arbitrary recalculation sequence $\{\alpha_k\}$, (4.8) *converges to a solution* x^* *of* (2.1), *according to*

$$\begin{aligned} \|x^* - x_{k+1}\| &\leqslant r_0 - t_{k+1} \\ &\equiv r_0 - t_k - \frac{\beta(a+b)t_k^2 - (1 - \beta(a+b)\eta_{-1})t_k + \eta}{1 - \beta(a+b)(t_{\alpha_k} + t_{\alpha_{k-1}} + \eta_{-1})}, \\ &\qquad k = 0, 1, 2, \ldots, \end{aligned}$$

$$t_{-1} = -\eta_{-1}, \quad t_0 = 0.$$

Independently of Theorem 4.1, Schmidt [84] established a semilocal convergence theorem for the backward secant iteration (4.4). (He called this iteration the regular-falsi iteration.)

We remark here that, in many cases, uniqueness assertions in Kantorovich-type convergence theorems follow from the Newton–Kantorovich theorem, which may simplify proofs of the theorems. This remark also applies to Schmidt's theorem for (4.4). See Yamamoto [99,102,103].

5. Halley's and Chebyshev's methods

Let $X = \mathbb{R}$ and $F = f$ be a single function of C^2-class with a zero x^*. Then, as was mentioned in Section 2, a linear approximation

$$f(x+h) \doteqdot f(x) + f'(x)h$$

leads to the Newton method whose convergence speed is of second order. If we use a second-order approximation

$$f(x+h) \doteqdot f(x) + f'(x)h + \tfrac{1}{2}f''(x)h^2,$$

then an iterative method

$$x_{k+1} = x_k + h, \quad k = 0, 1, 2, \ldots,$$

$$h = \frac{-\dfrac{2f(x_k)}{f'(x_k)}}{1 + \sqrt{1 - \dfrac{2f(x_k)f''(x_k)}{f'(x_k)^2}}}, \tag{5.1}$$

is obtained. Cauchy [13] first established a semilocal convergence result for (5.1) under some assumptions. As is shown there, the convergence speed of this method is cubic. A more detailed discussion on its convergence can be found in Hitotumatu [41].

To avoid the complexity of computation of the square root, we replace $\sqrt{1-x}$ by its approximation $1 - \frac{1}{2}x$ near $x = 0$. Then we obtain Halley's method (1694)

$$x_{k+1} = x_k - \frac{\dfrac{f(x_k)}{f'(x_k)}}{1 - \dfrac{1}{2}\dfrac{f(x_k)f''(x_k)}{f'(x_k)^2}}, \quad k = 0, 1, 2, \ldots, \tag{5.2}$$

which is also called the method of tangent hyperbolas. It is interesting to note that if we replace the denominator of (5.2) by

$$\sqrt{1 - \frac{f(x_k)f''(x_k)}{f'(x_k)^2}},$$

then we get the square root iteration due to Ostrowski [65]

$$x_{k+1} = x_k - K(x_k), \quad k = 0, 1, 2, \ldots,$$

$$K(x) = \frac{f(x)/f'(x)}{\sqrt{1 - f(x)f''(x)/f'(x)^2}}.$$

Convergence analysis of (5.2) can be found in Salehov [82], Brown [12], Alefeld [2], Gander [36], Hernandez [40], etc.

As in Newton's method, the procedure (5.2) is easily extended to nonlinear operators in Banach spaces: Let X and Y be Banach spaces and $F: D \subseteq X \to Y$ as before. If F is twice Fréchet differentiable in an open convex domain $D_0 \subseteq D$, then the Halley method applied to Eq. (2.1) in a Banach space is defined by

$$x_0 \in D_0, \quad x_{k+1} = x_k - A(x_k)^{-1}F(x_k), \quad k = 0, 1, 2, \ldots \tag{5.3}$$

with

$$A(x) = F'(x)[I - \tfrac{1}{2}F'(x)^{-1}F''(x)F'(x)^{-1}F(x)],$$

which is equivalent to the procedure:

(H1) Solve the linear equation $F(x_k) + F'(x_k)c_k = 0$ with respect to c_k.
(H2) Solve the linear equation $F(x_k) + F'(x_k)d_k + \frac{1}{2}F''(x_k)c_k d_k = 0$ with respect to d_k.
(H3) Set $x_{k+1} = x_k + d_k, \quad k = 0, 1, 2, \ldots$.

A convergence theorem for (5.3) as well as a uniqueness result was first given by Mertvecova [52], who extended Salehov's result (1952) for a single equation of a real or complex function. Since then, Kantorovich-type convergence theorems for (5.3) or its variants have been obtained by many authors (Altman [3], Safiev [80,81], Döring [30], Alefeld [1], Werner [91,92], Yamamoto [101], Chen–Argyros–Qian [14], Ezquerro–Hernandez [31], Hernandez [40], etc.)

According to Safiev, but slightly changing his notation, we assume

(I) The operator $\Gamma = F'(x_0)^{-1}$ exists. We put $\Phi(x) = \Gamma F(x)$;
(II) $\zeta = \|\Phi(x_0)\| > 0,\ M = \|\Phi''(x_0)\| > 0$;
(III) $\|\Phi''(x) - \Phi''(y)\| \leqslant N\|x - y\|,\ x, y \in D_0,\ N > 0$;
(IV) The equation $f(t) = \frac{1}{6}Nt^3 + \frac{1}{2}Mt^2 - t + \zeta = 0$ has one negative root and two positive roots t^*, t^{**} such that $t^* \leqslant t^{**}$. Equivalently,

$$\zeta \leqslant \frac{M^2 + 4N - M\sqrt{M^2 + 2N}}{3N(M + \sqrt{M^2 + 2N})}, \tag{5.4}$$

where the equality holds if and only if $t^* = t^{**}$.

Under these assumptions, define the scalar sequence $\{t_k\}$ by

$$t_0 = 0, \quad t_{k+1} = t_k - a(t_k)^{-1} f(t_k), \quad k = 0, 1, 2, \ldots,$$

where $a(t) = f'(t) - \frac{1}{2}f''(t)f'(t)^{-1}f(t)$. Then the following result holds, which improves Döring's:

Theorem 5.1 (Yamamoto [101]). *To the assumptions* (I)–(IV) *add*

$$\bar{S} = \bar{S}(x_1, t^* - t_1) \subseteq D_0.$$

Then:

(i) *The iteration* (5.3) *is well defined for every* $k \geqslant 0$, $\{x_{k+1}\}$ *lies in* S (*interior of* $\bar{S}$) *and converges to a solution* x^* *of* (2.1).
(ii) *The solution is unique in*

$$\tilde{S} = \begin{cases} S(x_0, t^{**}) \cap D_0 & \text{if } t^* < t^{**}, \\ \bar{S}(x_0, t^{**}) \cap D_0 & \text{if } t^* = t^{**}. \end{cases}$$

(iii) *Error estimates*

$$\sigma_k^* \leqslant \|x^* - x_k\| \leqslant \tau_k^* \leqslant d_k + (t^* - t_{k+1})\left(\frac{d_k}{t_{k+1} - t_k}\right)^3$$

$$\leqslant (t^* - t_k)\left(\frac{d_k}{t_{k+1} - t_k}\right) \leqslant t^* - t_k$$

and

$$\|x^* - x_k\| \leqslant (t^* - t_k)\left(\frac{d_{k-1}}{t_k - t_{k-1}}\right)^3$$

hold, where $d_k = \|x_{k+1} - x_k\|$, τ_k^* *and* σ_k^* *are the smaller positive root of the equation*

$$\frac{(t^* - t_{k+1})}{(t^* - t_k)^3}t^3 - t + d_k = 0,$$

and

$$\frac{(t^* - t_{k+1})}{(t^* - t_k)^3}t^3 + t - d_k = 0,$$

respectively.

Remark 5.2. Since

$$\lim_{N\to 0}\frac{M^2 + 4N - M\sqrt{M^2+2N}}{3N(M+\sqrt{M^2+2N})} = \frac{1}{2M},$$

condition (5.4) reduces to the Kantorovich condition $2M\zeta \leqslant 1$, if $N = 0$. Hence (III) admits the case $N = 0$.

Kanno [42] shows that Theorem 5.1 is better than Safiev's result. See also Cuyt [21], Cuyt–Rall [22] for computational implementation of the method.

Chebyshev's method (1951) (cf. [60]) is similarly defined by

$$x_{k+1} = x_k - F'(x_k)^{-1}F(x_k)$$
$$-\tfrac{1}{2}F'(x_k)^{-1}F''(x_k)[F'(x_k)^{-1}F(x_k)]^2, \quad k = 0, 1, 2, \ldots \tag{5.5}$$

or by the following procedure:

(C1) Solve $F(x_k) + F'(x_k)c_k = 0$ with respect to c_k.
(C2) Solve $F(x_k) + F'(x_k)d_k + \frac{1}{2}F''(x_k)c_k^2 = 0$ with respect to d_k.
(C3) Set $x_{k+1} = x_k + d_k, \quad k \geqslant 0$.

The method is also obtained if we replace

$$[I - \tfrac{1}{2}F'(x)^{-1}F''(x)F'(x)^{-1}F(x)]^{-1}$$

in the Halley method (5.3) by

$$I + \tfrac{1}{2}F'(x)F''(x)F'(x)^{-1}F(x).$$

Convergence theorems for Chebyshev's and Chebyshev-like methods have been obtained by Necepurenko [60], Safiev [81], Alefeld [2], Werner [92] and others. A numerical test by Alefeld shows that Halley's method gives a slightly sharper result than Chebyshev's, although convergence speed of both methods is of cubic order.

Remark 5.3. Two steps of an iterative method with cubic order correspond in general to three steps of a method with second order. However, computational cost depends on the form of F. Hence, it is difficult to give a general statement about the efficiency/computational cost of Halley's or Halley-like methods versus Newton's method. In Ostrowski's book [65] the efficiency index of a procedure is defined and discussed for some algorithms.

6. A class of iterative methods for not necessarily differentiable equations

Recently, much attention has been paid to iterative methods for solving (2.1) when F is not necessarily differentiable. In 1963, Zincenko [107] considered using a differentiable operator $f: D_f \subseteq X \to Y$, where X and Y are Banach spaces, and putting $g(x) = F(x) - f(x)$. Then $F(x) = f(x) + g(x)$ and he proved convergence theorems for the iterations

$$x_{k+1} = x_k - f'(x_k)^{-1}(f(x_k) + g(x_k)), \quad k = 0, 1, 2, \ldots \tag{6.1}$$

and

$$x_{k+1} = x_k - f'(x_0)^{-1}(f(x_k) + g(x_k)), \quad k = 0, 1, 2, \ldots. \tag{6.2}$$

In [107], Zincenko observed that (6.1) and (6.2) were suggested by Krasnoselskii. Rheinboldt used his majorant theory to prove Zincenko's results. His result for (6.1) is stated as follows:

Theorem 6.1. *Let $g: D_g \subseteq X \to Y$ and suppose that on some convex set $D_0 \subseteq D_f \cap D_g$,*

$$\|f'(x) - f'(y)\| \leqslant \gamma \|x - y\|,$$

$$\|g(x) - g(y)\| \leqslant \delta \|x - y\|, \quad x, y \in D_0.$$

Assume that for some $x_0 \in D_0$, $f'(x_0)^{-1} \in L(Y, X)$ exists and that $\|f'(x_0)^{-1}\| \leqslant \beta$, $\|f'(x_0)^{-1} f(x_0)\| \leqslant \alpha$ as well as $\beta\delta < 1$ and

$$h = \beta\gamma\alpha/(1 - \beta\delta)^2 \leqslant 1/2.$$

Define t^*, t^{**} by

$$t^* = \frac{2}{1 + \sqrt{1 - 2h}} \cdot \frac{\alpha}{1 - \beta\delta}, \qquad t^{**} = \frac{1 + \sqrt{1 - 2h}}{h} \cdot \frac{\alpha}{1 - \beta\delta}.$$

If $\bar{S}(x_0, t^*) \subseteq D_0$, then the sequence (6.1) remains in $S(x_0, t^*)$ and converges to the only solution x^* of $F(x) = 0$ in $S(x_0, t^{**}) \cap D_0$.

(If $g = 0$, then Theorem 6.1 reduces to the Kantorovich theorem.)

Later, in 1982, Zabrejko–Zlepko [106] proved a semilocal convergence theorem for (6.1) and

$$x_{k+1} = x_k - C^{-1}(f(x_k) + g(x_k)), \quad k = 0, 1, 2, \ldots,$$

where $C \in L(X, Y)$. To state their result for (6.1), let f and g be defined in the closed ball $\bar{S}(x_0, R)$ of X and f be differentiable in the open ball $S(x_0, R)$. They considered the auxiliary scalar equation

$$r = \varphi(r), \tag{6.3}$$

where

$$\varphi(r) = a + b\left(\int_0^r \omega(\tau)\,\mathrm{d}\tau + H(r)\right),$$

$a > 0$, $b > 0$, and $\omega(r)$, $H(r)$, $H(r+t) - H(r)$, $t \geqslant 0$ are nonnegative, monotonically increasing and continuous functions, vanishing for $r = 0$.

Assume that (6.3) has at least one positive solution. Denote by r^* the least of these solutions. If it is isolated and $\varphi(r) < r$ for r sufficiently close to r^* and larger than r^*, then we denote by R^* the least upper bound of the number ξ for which $\varphi(r) < r$ for $r \in (r^*, \xi)$; otherwise we put $R^* = r^*$. Then the iteration $r_{k+1} = \varphi(r_k)$, $k = 0, 1, 2, \ldots$, $r_0 = 0$ is monotonically increasing and converges to r^*, while the iteration $R_{k+1} = \varphi(R_k)$, $k = 0, 1, 2, \ldots$ starting from $R_0 \in [r^*, R^*)$ is monotonically decreasing and converges to r^*.

Similarly, the iteration

$$\rho_{k+1} = \rho_k - \frac{\rho_k - \varphi(\rho_k)}{1 - b\omega(\rho_k)}, \quad k = 0, 1, 2, \ldots, \quad \rho_0 = 0$$

is monotonically increasing and converges to r^*.

In particular, putting

$$a = \|f'(x_0)^{-1}(f(x_0) + g(x_0))\|, \quad b = \|f'(x_0)^{-1}\| \tag{6.4}$$

the following Theorem 6.2 can be proved by assuming

$$\|f'(x+h) - f'(x)\| \leqslant \omega(r + \|h\|) - \omega(r), \quad \|x - x_0\| \leqslant r, \quad r + \|h\| \leqslant R,$$

$$\|g(x+h) - g(x)\| \leqslant H(\|x - x_0\| + \|h\|) - H(\|x - x_0\|).$$

Theorem 6.2 (Zabrejko–Zlepko [106]). *Let $r^* \leqslant R < R^*$. Then*

$$f(x) + g(x) = 0 \tag{6.5}$$

has a solution $x^* \in \bar{S}(x_0, r^*)$, which is unique in $\bar{S}(x_0, R)$. The sequence $\{x_k\}$ from (6.1) satisfies $x_k \in \bar{S}(x_0, r_k)$ for $k \geqslant 0$ and $x_k \to x^*$ as $k \to \infty$; we have

$$\|x_{k+1} - x_k\| \leqslant \rho_{k+1} - \rho_k \quad \text{and} \quad \|x^* - x_k\| \leqslant r^* - \rho_k, \quad k \geqslant 0.$$

In 1987, Zabrejko–Nguen [105] reformulated Theorem 6.2 as follows: the operators f and g are defined in $\bar{S}(x_0, R)$, f is differentiable in $S(x_0, R)$ and

$$\|f'(x') - f'(x'')\| \leqslant \kappa(r)\|x' - x''\|, \quad x', x'' \in \bar{S}(x_0, r),$$

$$\|g(x') - g(x'')\| \leqslant \varepsilon(r)\|x' - x''\|, \quad x', x'' \in \bar{S}(x_0, r),$$

where $\kappa(r)$ and $\varepsilon(r)$ are nondecreasing functions on the interval $[0, R]$. Furthermore, assume that $f'(x_0)^{-1}$ exists and put

$$\omega(r) = \int_0^r \kappa(t)\,\mathrm{d}t,$$

$$\varphi(r) = a + b\int_0^r \omega(t)\,\mathrm{d}t - r = a + b\int_0^r (r - s)\kappa(s)\,\mathrm{d}s - r,$$

$$\psi(r) = b\int_0^1 \varepsilon(t)\,\mathrm{d}t.$$

Under these assumptions and notation, they proved the following:

Theorem 6.3 (Zabrejko–Nguen [105]). *Suppose that the function $\chi(r) = \varphi(r) + \psi(r)$ has a unique zero ρ^* in $[0, R]$ and $\chi(R) \leqslant 0$. Then Eq. (6.5) admits a solution x^* in $\bar{S}(x_0, \rho)$, which is unique in $\bar{S}(x_0, R)$.*

The iteration (6.1) is defined for all $k \geqslant 0$, $x_k \in \bar{S}(x_0, \rho^*)$, $k \geqslant 0$, and $\{x_k\}$ converges to x^* as $k \to \infty$. Error estimates

$$\|x_{k+1} - x_k\| \leqslant \rho_{k+1} - \rho_k, \quad k \geqslant 0$$

and

$$\|x^* - x_k\| \leqslant \rho^* - \rho_k, \quad k \geqslant 0$$

hold where ρ_k is defined by

$$\rho_{k+1} = \rho_k - \frac{\chi(\rho_k)}{\varphi'(\rho_k)}, \quad k \geqslant 0, \quad \rho_0 = 0$$

and

$$\rho_0 < \rho_1 < \cdots < \rho_k \to \rho^* \quad \text{as } k \to \infty.$$

They used this result to generalize Pták's error estimates for Newton's method obtained by the "nondiscrete induction technique" [70] and showed that his estimates are a simple consequence of the classical majorant method due to Kantorovich.

In the same paper, they mentioned the following error estimate without proof: Let $r_k = \|x_k - x_0\|$, $\kappa_k(r) = \kappa(r_k + r)$ and $\varepsilon_k(r) = \varepsilon(r_k + r)$ for $r \in [0, R - r_k]$ and put

$$a_k = \|x_{k+1} - x_k\|, \quad b_k = (1 - \omega(r_k))^{-1}.$$

Without loss of generality, we may assume that $a_k > 0$. Then the equation

$$r = a_k + b_k \int_0^r \{(r - t)\kappa_k(t) + \varepsilon_k(t)\}\,\mathrm{d}t$$

has the unique positive zero ρ_k^* in $[0, R - r_k]$ and

$$\|x^* - x_k\| \leqslant \rho_k^*, \quad k = 0, 1, 2, \ldots, \tag{6.6}$$

where we understand $\rho_0^* = \rho^*$. A proof of (6.6) is given in Yamamoto [100] together with the following error estimates:

$$\begin{aligned}
\|x^* - x_k\| &\leqslant \rho_k^* \quad (k \geqslant 0) \\
&\leqslant \frac{a_k}{\rho_{k+1} - \rho_k}(\rho^* - \rho_k) \quad (k \geqslant 0) \\
&\leqslant \frac{a_{k-1}}{\rho_k - \rho_{k-1}}(\rho^* - \rho_k) \quad (k \geqslant 1) \\
&\leqslant \rho^* - \rho_k \quad (k \geqslant 0).
\end{aligned}$$

Motivated by Zabrejko–Nguen's paper, Chen–Yamamoto [17] considered the iteration,

$$x_{k+1} = x_k - A(x_k)^{-1}(f(x_k) + g(x_k)), \quad k = 0, 1, 2, \ldots, \tag{6.7}$$

which includes (6.1) as a special case, where $A(x)$ is an approximation for $f'(x)$. Under the Zabrejko–Nguen type hypotheses, they determined a domain Ω such that the method (6.5) starting from any point of Ω converges to a solution of (6.5).

Furthermore, a study of the iteration (6.7) is found in Yamamoto–Chen [104] and Chen–Yamamoto [18], where ball convergence theorems as well as error estimates are given. The results generalize and deepen those of Kantorovich [44], Mysovskii [58], Rall [74], Rheinboldt [78], Dennis [25], Yamamoto [96,99,100], Zabrejko–Nguen [105], and others. See also the works of Argyros [5,6].

For the case $X = Y = \mathbb{R}^n$, Clark [20] proposed a generalized Newton method which uses an element in the generalized Jacobian $\partial F(x^k)$ in place of the Jacobian if F is locally Lipschitzian but not differentiable. Since then, there has been a growing interest in the study of nonsmooth equations, which is closely related to the study of Newton-like methods. Such equations arise, for examples, from (i) nonlinear complementarity problems, (ii) nonlinear constrained optimization problems, (iii) nonsmooth convex optimization problems, (iv) compact fixed point problems, (v) nonsmooth eigenvalue problem related to ideal MHD (magnetohydrodynamics), etc.

Reformulating problems (i)–(iii) to nonsmooth equations and global and superlinear convergence of algorithms for solving such nonsmooth equations can be found in the works of Chen [15], Qi [71], Qi–Sun [73], Pang [66], Pang–Qi [67], Qi–Chen [72], Robinson [79], Fukushima–Qi [35] and others. Heinkenschloss et al. [39] discuss (iv) and Rappaz [76] and Kikuchi [47,48] discuss (v).

Finally, we remark that the general Gauss–Newton method for solving singular or ill-posed equations is defined by

$$x_{k+1} = x_k - B(x_k)F(x_k), \quad k = 0, 1, 2 \dots \tag{6.8}$$

where $F : D \subseteq X \to Y$ is differentiable and $B(x_k)$ is a linear operator which generalizes the Moore–Penrose pseudo-inverse. If $X = \mathbb{R}^n$, $Y = \mathbb{R}^m$ and $B(x_k) = F'(x_k)^+$, then (6.8) reduces to the ordinary Gauss–Newton method for solving the least square problem: Find $x \in D$ which minimizes $F(x)^t F(x)$. Convergence analysis for (6.8) is given in the works of Ben-Israel [11], Lawson and Hanson [50], Meyn [56], Häußler [38], Walker [89], Walker and Watson [90], Martínez [51], Deuflhard and Potra [28], Chen and Yamamoto [19], Nashed and Chen [59], Chen et al. [16] and others.

7. Concluding remarks

In this paper, we have traced historical developments in convergence theory and error estimates for Newton (and Newton-like) methods based on the Newton–Kantorovich theorem.

In spite of its simple principle, Newton's method is applicable to various types of equations such as systems of nonlinear algebraic equations including matrix eigenvalue problems, differential equations, integral equations, etc., and even to random operator equations [9]. Hence, the method fascinates many researchers.

However, as is well known, a disadvantage of the methods is that the initial approximation x_0 must be chosen sufficiently close to a true solution in order to guarantee their convergence. Finding a criterion for choosing x_0 is quite difficult and therefore effective and globally convergent algorithms are needed. This remark includes the important case of nonsmooth equations. It is expected that if such algorithms exist, then they too will be variants of Newton's method.

Acknowledgements

The author thanks Prof. Xiaojun Chen of Shimane University for her valuable comments during the preparation of this paper; and is also grateful to referees for their nice suggestions. By these suggestions this paper has been greatly improved.

References

[1] G. Alefeld, Zur Konvergenz des Verfahrens der tangierenden Hyperbeln und des Tschelyscheff-Verfahrens bei konvexen Abbildungen, Computing 11 (1973) 379–390.
[2] G. Alefeld, On the convergence of Halley's method, Amer. Math. Monthly 88 (1981) 530–536.
[3] M. Altman, Concerning the method of tangent hyperbolas for operator equations, Bull. Acad. Polon. Sci. Ser. Sci. Math. Astronom. Phys. 9 (1961) 633–637.
[4] A.L. Andrew, Error bounds for the modified Newton's method, Bull. Austral. Math. Soc. 14 (1976) 427–433.
[5] I.K. Argyros, The Theory and Applications of Iteration Methods, CRC Press, Boca Raton, 1993.
[6] I.K. Argyros, Improved á posteriori error bounds for Zincenko's iteration, Internat. J. Comput. Math. 51 (1994) 51–54.
[7] S. Banach, Sur les opérations dans les ensembles abstracts et leurs applications aux équations integrales, Fund. Math. 3 (1922) 133–181.
[8] S. Banach, Theorie des Opérations Linéaires, Warszawa, 1932. Reprinted by Chelsea, New York.
[9] A.T. Bharucha-Reid, R. Kannan, Newton's method for random operator equations, J. Nonlinear Anal. 4 (1980) 231–240.
[10] N. Bićanić, K.H. Johnson, Who was '-Raphson'?, Internat. J. Numer. Methods Eng. 14 (1979) 148–152.
[11] A. Ben-Israel, A Newton–Raphson method for the solution of systems of equations, J. Math. Anal. Appl. 15 (1966) 243–252.
[12] G.H. Brown, On Halley's variation of Newton's method, Amer. Math. Monthly 84 (1977) 726–728.
[13] A.L. Cauchy, Sur la détermination approximative des racines d'une équation algébrique ou transcendante, in: Leçons sur le Calcul Differentiel, Buré fréres, Paris, 1829, pp. 573–609.
[14] D. Chen, I.K. Argyros, Q.S. Qian, A note on the Halley method in Banach spaces, Appl. Math. Comp. 58 (1993) 215–224.
[15] X. Chen, Smoothing methods for complementarity problems and their applications: a survey, J. Oper. Res. Soc. Japan 43 (2000) 32–47.
[16] X. Chen, M.Z. Nashed, L. Qi, Convergence of Newton's method for singular smooth and nonsmooth equations using adaptive outer inverses, SIAM J. Optim. 7 (1997) 245–262.
[17] X. Chen, T. Yamamoto, Convergence domains of certain iterative methods for solving nonlinear equations, Numer. Funct. Anal. Optim. 10 (1989) 37–48.
[18] X. Chen, T. Yamamoto, A convergence ball for multistep simplified Newton-like methods, Numer. Funct. Anal. Optim. 14 (1993) 15–24.
[19] X. Chen, T. Yamamoto, Newton-like methods for solving undetermined nonlinear equations with nondifferentiable terms, J. Comput. Appl. Math. 55 (1994) 311–324.
[20] F.H. Clark, Optimization and Nonsmooth Analysis, Wiley, New York, 1983.
[21] A.A.M. Cuyt, Numerical stability of the Halley-iteration for the solution of a system of nonlinear equations, Math. Comp. 38 (1982) 171–179.
[22] A.A.M. Cuyt, L.B. Rall, Computational implementation of the multivariate Halley method for solving nonlinear systems of equations, ACM Trans. Math. Software 11 (1985) 20–36.
[23] J.E. Dennis Jr., On Newton-like methods, Numer. Math. 11 (1968) 324–330.
[24] J.E. Dennis Jr., On the Kantorovich hypothesis for Newton's method, SIAM J. Numer. Anal. 6 (1969) 493–507.
[25] J.E. Dennis Jr., On the convergence of Newton-like methods, in: P. Rabinowitz (Ed.), Numerical Methods for Nonlinear Algebraic Equations, Gordon & Breach, New York, 1970, pp. 163–181.

[26] J.E. Dennis Jr., Toward a unified convergence theory for Newton-like methods, in: L.B. Rall (Ed.), Nonlinear Functional Analysis and Applications, Academic Press, New York, 1971.
[27] P. Deuflhard, G. Heindl, Affine invariant convergence theorems for Newton's method and extensions to related methods, SIAM J. Numer. Anal. 16 (1979) 1–10.
[28] P. Deuflhard, F.A. Potra, A refined Gauss–Newton–Mysovskii theorem, ZIB SC 91-4 (1991) 1–9.
[29] B. Döring, Über das Newtonsche Näherungsverfahren, Math. Phys. Semi.-Ber. 16 (1969) 27–40.
[30] B. Döring, Einige Sätze über das Verfahren der tangierenden Hyperbeln in Banach-Räumen, Aplikace Mat. 15 (1970) 418–464.
[31] J.A. Ezquerro, M.A. Hernandez, Avoiding the computation of the second Fréchet-derivative in the convex acceleration of Newton's method, J. Comput. Appl. Math. 96 (1998) 1–12.
[32] I. Fenyö, Über die Lösung der in Banachschen Raume definierten nichtlinearen Gleichungen, Acta. Math. Acad. Sci. Hungar. 5 (1954) 85–93.
[33] H.B. Fine, On Newton's method of approximation, Proc. Nat. Acad. Sci. USA 2 (1916) 546–552.
[34] J.B.J. Fourier, Question d'analyse algébrique, in: Oeuvres Complétes (II), Gauthier-Villars, Paris, 1890, pp. 243–253.
[35] M. Fukushima, L. Qi, Reformulation: Nonsmooth, Piecewise Smooth, Semismooth and Smoothing Methods, Kluwer Academic Publishers, Dordrecht, 1989.
[36] W. Gander, On Halley's iteration method, Amer. Math. Monthly 92 (1985) 131–134.
[37] W.B. Gragg, R.A. Tapia, Optimal error bounds for the Newton–Kantorovich theorem, SIAM J. Numer. Anal. 11 (1974) 10–13.
[38] W.M. Häußler, A Kantorovich-type convergence analysis for the Gauss–Newton method, Numer. Math. 48 (1986) 119–125.
[39] M. Heinkenschloss, C.T. Kelley, H.T. Tran, Fast algorithms for nonsmooth compact fixed-point problems, SIAM J. Numer. Anal. 29 (1992) 1769–1792.
[40] M.A. Hernandez, A note on Halley's method, Numer. Math. 59 (1991) 273–276.
[41] S. Hitotumatu, A method of successive approximation based on the expansion of second order, Math. Japon. 7 (1962) 31–50.
[42] S. Kanno, Convergence theorems for the method of tangent hyperbolas, Math. Japon. 37 (1992) 711–722.
[43] L.V. Kantorovich, The method of successive approximations for functional analysis, Acta. Math. 71 (1939) 63–97.
[44] L.V. Kantorovich, On Newton's method for functional equations, Dokl Akad. Nauk SSSR 59 (1948) 1237–1240 (in Russian).
[45] L.V. Kantorovich, The majorant principle and Newton's method, Dokl. Akad. Nauk SSSR 76 (1951) 17–20 (in Russian).
[46] L.V. Kantorovich, G.P. Akilov, Functional Analysis, 2nd ed., Pergamon Press, Oxford, 1982.
[47] F. Kikuchi, An iteration scheme for a nonlinear eigenvalue problem, Theoret. Appl. Mech. 29 (1981) 319–333.
[48] F. Kikuchi, Finite element analysis of a nondifferentiable nonlinear problem related to MHD equilibria, J. Fac. Sci. Univ. Tokyo Sect. IA Math. 35 (1988) 77–101.
[49] H.J. Kornstaedt, Functional Ungleichungen und Iterations Verfahren, Aequationes Math. 13 (1975) 21–45.
[50] C.L. Lawson, R.J. Hanson, Solving Least Square Problems, Prentice-Hall, Englewood Cliffs, NJ, 1974.
[51] J.M. Martínez, Quasi-Newton methods for solving underdetermined nonlinear simultaneous equations, J. Comput. Appl. Math. 34 (1991) 171–190.
[52] M.A. Mertvecova, An analog of the process of tangent hyperbolas for general functional equations, Dokl. Akad. Nauk SSSR 88 (1953) 611–614 (in Russian).
[53] G.J. Miel, The Kantorovich theorem with optimal error bounds, Amer. Math. Monthly 86 (1979) 212–215.
[54] G.J. Miel, Majorizing sequences and error bounds for iterative methods, Math. Comp. 34 (1980) 185–202.
[55] G.J. Miel, An updated version of the Kantorovich theorem for Newton's method, Computing 27 (1981) 237–244.
[56] K.H. Meyn, Solution of underdetermined nonlinear equations by stationary iteration methods, Numer. Math. 42 (1983) 161–172.
[57] I. Moret, A note on Newton type iterative methods, Computing 33 (1984) 65–73.
[58] I. Mysovskii, On the convergence of L.V. Kantorovich's method of solution of functional equations and its applications, Dokl, Akad. Nauk SSSR 70 (1950) 565–585 (in Russian).

[59] M.Z. Nashed, X. Chen, Convergence of Newton-like methods for singular operator equations using outer inverses, Numer. Math. 66 (1993) 235–257.
[60] M.I. Necepurenko, On Chebyshev's method for functional equations, Uspehi Matem. Nauk 9 (1954) 163–170. English translation: L.B. Rall, MRC Technical Summary Report #648, 1966.
[61] J.M. Ortega, W.C. Rheinboldt, Iterative Solution of Nonlinear Equations in Several Variables, Academic Press, New York, 1970.
[62] A.M. Ostrowski, Sur la convergence et l'estimation des erreurs dans quelques procédés de résolution des équations numériques, in: Collection of Papers in Memory of D.A. Grave, Gosudarstv. Izdat. Tehn.-Teor. Lit., Moscow, 1940, pp. 213–234.
[63] A.M. Ostrowski, Solution of Equations and Systems of Equations, 2nd ed., Academic Press, New York, 1966.
[64] A.M. Ostrowski, La méthode de Newton dans les espaces de Banach, C.R. Acad. Sci. Paris Ser. A 272 (1971) 1251–1253.
[65] A.M. Ostrowski, Solution of Equations in Euclidean and Banach Space, Academic Press, New York, 1973.
[66] J.S. Pang, A *B*-differentiable equation-based, globally and locally quadratically convergent algorithm for nonlinear programs, complementarity and variational inequality problems, Math. Programming 51 (1991) 101–131.
[67] J.S. Pang, L. Qi, Nonsmooth equations: motivation and algorithms, SIAM J. Optim. 3 (1993) 443–465.
[68] F.A. Potra, On the a posteriori error estimates for Newton's method, Beiträge zur Numer. Math. 12 (1984) 125–138.
[69] F.A. Potra, V. Pták, Sharp error bounds for Newton's process, Numer. Math. 34 (1980) 63–72.
[70] V. Pták, The rate of convergence of Newton's process, Numer. Math. 25 (1976) 279–285.
[71] L. Qi, Convergence analysis of some algorithms for solving nonsmooth equations, Math. Oper. Res. 18 (1993) 227–244.
[72] L. Qi, X. Chen, A globally convergent successive approximation method for severely nonsmooth equations, SIAM J. Control. Optim. 33 (1995) 402–418.
[73] L. Qi, J. Sun, A nonsmooth version of Newton's method, Math. Programming 58 (1993) 353–367.
[74] L.B. Rall, A note on the convergence of Newton's method, SIAM J. Numer. Anal. 11 (1974) 34–36.
[75] L.B. Rall, R.A. Tapia, The Kantorovich theorem and error estimates for Newton's method, MRC Technical Summary Report No. 1043, Univ. of Wisconsin – Madison (1970).
[76] J. Rappaz, Approximation of a nondifferentiable nonlinear problem related to MHD equilibria, Numer. Math. 45 (1984) 117–133.
[77] W.C. Rheinboldt, A unified convergence theory for a class of iterative processes, SIAM J. Numer. Anal. 5 (1968) 42–63.
[78] W.C. Rheinboldt, An adaptive continuation process for solving systems of nonlinear equations, in: Mathematical Models and Numerical Methods, Banach Center Publications, Vol. 3, PWN–Polish Scientific Publishers, Warsaw, 1978, pp. 129–142.
[79] S.M. Robinson, Newton's method for a class of nonsmooth functions, Set-Valued Anal. 2 (1994) 291–305.
[80] R.A. Safiev, The method of tangent hyperbolas, Sov. Math. Dokl. 4 (1963) 482–485.
[81] R.A. Safiev, On some iterative processes, Ž . Vyčisl. Mat. i. Mat. Fiz. 4 (1964) 139–143 (in Russian).
[82] G.S. Salehov, On the convergence of the process of tangent hyperbolas, Dokl. A. N. SSSR 82 (1952) 525–528 (in Russian).
[83] J.W. Schmidt, Ein Übertragung der Regula falsi auf Gleichungen in Banachräumen, ZAMM 34 (1963) Part I, 1–8, Part II, 97–110.
[84] J.W. Schmidt, Regular-falsi-Verfahren mit konsistenter Steigung und Majorantenprinzip, Period. Math. Hungarica 5 (1974) 187–193.
[85] J.W. Schmidt, Untere Fehlerschranken für Regular-falsi-Verfahren, Period. Math. Hungarica 9 (1978) 241–247.
[86] J. Schröder, Nichtlineare Majoranten beim Verfahren der schrittweisen Näherung, Arch. Math. 7 (1956) 471–484.
[87] J. Schröder, Über das Newtonsche Verfahren, Arch. Rational Mech. Anal. 1 (1957) 154–180.
[88] R.A. Tapia, The Kantorovich theorem for Newton's method, Amer. Math. Monthly 78 (1971) 389–392.
[89] H.J. Walker, Newton-like methods for underdetermined systems, in: E.L. Allgower, K. Georg (Eds.), Computational Solution of Nonlinear Systems of Equations, Lectures in Appl. Math., Vol. 26, AMS, Providence, 1990.
[90] H.J. Walker, L.T. Watson, Least-change secant update methods for underdetermined systems, SIAM J. Numer. Anal. 27 (1990) 1227–1262.

[91] W. Werner, Some improvements of classical iterative methods for the solution of nonlinear equations, in: E.L. Allgower, K. Glashoft, H.O. Peitgen (Eds.), Numerical Solution of Nonlinear Equations, Lecture Notes in Math., Vol. 878, Springer, Heidelberg, 1981.

[92] W. Werner, Iterative solution of systems of nonlinear equations based upon quadratic approximations, Comput. Math. Appl. A 12 (1986) 331–343.

[93] T. Yamamoto, Error bounds for Newton's process derived from the Kantorovich theorem, Japan J. Appl. Math. 2 (1985) 258–292.

[94] T. Yamamoto, A unified derivation of several error bounds for Newton's process, J. Comput. Appl. Math. 12,13 (1985) 179–191.

[95] T. Yamamoto, Error bounds for Newton's iterates derived from the Kantorovich theorem, Numer. Math. 48 (1986) 91–98.

[96] T. Yamamoto, Error bounds for Newton-like methods under Kantorovich type assumptions, Japan J. Appl. Math. 3 (1986) 295–313.

[97] T. Yamamoto, A convergence theorem for Newton's method in Banach spaces, Japan J. Appl. Math. 3 (1986) 37–52.

[98] T. Yamamoto, A method for finding sharp error bounds for Newton's method under the Kantorovich assumptions, Numer. Math. 49 (1986) 203–220.

[99] T. Yamamoto, A convergence theorem for Newton-like methods in Banach spaces, Numer. Math. 51 (1987) 545–557.

[100] T. Yamamoto, A note on a posteriori error bound of Zabrejko and Nguen for Zincenko's iteration, Numer. Funct. Anal. Optim. 9 & 10 (1987) 987–994.

[101] T. Yamamoto, On the method of tangent hyerbolas in Banach spaces, J. Comput. Appl. Math. 21 (1988) 75–86.

[102] T. Yamamoto, Unified convergence analysis for a class of iterative methods for solving nonlinear equations, in: Th.M. Rassias (Ed.), Topics in Mathematical Analysis, World Scientific, Singapore, 1989, pp. 878–900.

[103] T. Yamamoto, Uniqueness of the solution in a Kantorovich-type theorem of Häußler for the Gauss–Newton method, Japan J. Appl. Math. 6 (1989) 77–81.

[104] T. Yamamoto, X. Chen, Ball convergence theorems and error estimates for certain iterative methods for nonlinear equations, Japan J. Appl. Math. 7 (1990) 131–143.

[105] P.P. Zabrejko, D.F. Nguen, The majorant method in the theory of Newton–Kantorovich approximations and the Pták error estimates, Numer. Funct. Anal. Optim. 9 (1987) 671–684.

[106] P.P. Zabrejko, P.P. Zlepko, On a generalization of the Newton–Kantorovich method for an equation with nondifferentiable operator, Ukr. Mat. Zhurn. 34 (1982) 365–369 (in Russian).

[107] A.I. Zincenko, Some approximate methods of solving equations with nondifferentiable operators, Dopovidi Akad. Nauk. Ukrain RSR (1963) 156–161 (in Ukrainian).

A survey of truncated-Newton methods

Stephen G. Nash[1]

Systems Engineering and Operations Research Department, George Mason University, Fairfax, VA 22030, USA

Abstract

Truncated-Newton methods are a family of methods for solving large optimization problems. Over the past two decades, a solid convergence theory has been derived for the methods. In addition, many algorithmic enhancements have been developed and studied, resulting in a number of publicly available software packages. The result has been a collection of powerful, flexible, and adaptable tools for large-scale nonlinear optimization.

1. Introduction

Truncated-Newton methods are a family of methods suitable for solving large nonlinear optimization problems. At each iteration, the current estimate of the solution is updated (i.e., a step is computed) by approximately solving the Newton equations using an iterative algorithm. This results in a doubly iterative method: an outer iteration for the nonlinear optimization problem, and an inner iteration for the Newton equations. The inner iteration is typically stopped or "truncated" before the solution to the Newton equations is obtained.

More generally, an "inexact" Newton method computes a step by approximately solving the Newton equations. This need not be done using an iterative method. These definitions, however, are not universal. In some papers, "inexact" Newton methods refer to methods for solving systems of nonlinear equations, and "truncated" Newton methods refer to methods for solving optimization problems. I focus here on truncated-Newton methods and optimization problems.

A truncated-Newton method will be effective if

- a small number of inner iterations is sufficient to produce a "good" step,
- each inner iteration can be performed efficiently,

[1] Partially supported by National Science Foundation grant DMI-9800544.

- the overall method is implemented with appropriate safeguards (a "globalization" strategy) to guarantee convergence to a stationary point or local optimum, in cases where the optimization problem satisfies appropriate assumptions.

These issues motivate much of my discussion.

Choices are available for the components of a truncated-Newton method:

- the globalization procedure (some form of line search or trust region strategy),
- the inner iterative algorithm,
- the preconditioner for the inner algorithm,
- the truncation rule for the inner algorithm,
- the technique for computing or estimating second-derivative information.

These choices provide a great deal of flexibility, and allow the method to be adapted to the optimization problem and the computing environment. "Black-box" software is available, but a sophisticated practitioner can enhance the basic method in many ways when faced with a difficult optimization problem.

In much of this paper I focus on the unconstrained problem

$$\min f(x) \tag{1}$$

since the ideas can be explained more simply in this setting, and many of the ideas carry over directly to the constrained case. The first-order optimality condition for this problem is

$$\nabla f(x) = 0,$$

which is a system of nonlinear equations. For this reason, results for nonlinear equations provide insight in the optimization setting.

Given some guess x_k of a solution x_*, Newton's method computes a step p_k as the solution to the linear system

$$\nabla^2 f(x_k)p = -\nabla f(x_k) \tag{2}$$

and then sets $x_{k+1} \leftarrow x_k + p_k$. In this simple form, Newton's method is not guaranteed to converge.

In a truncated-Newton method, an iterative method is applied to (2), and an approximate solution accepted. In [3], the rate of convergence of the outer iteration is proven to be related to the accuracy with which (2) is solved. The paper [3] focuses on nonlinear equations, but the results apply (with minor modification) to optimization problems. These results clarify the local convergence behavior of a truncated-Newton method (i.e., the behavior of the method when x_k is close to the solution x_*).

If the problem (1) satisfies appropriate assumptions, then global convergence (to a local solution) can be guaranteed in either a line search or a trust region framework by making adjustments to the inner algorithm (see Section 3). (In this paper, "global convergence" for an unconstrained problem means that the limit of the gradient norms is zero.) Building upon this foundation, many practical enhancements can be made to the overall method.

A basic question in a truncated-Newton method is the choice of an inner iterative algorithm for solving (2). Some variant of the linear conjugate-gradient method is almost always used. The conjugate-gradient method is an optimal iterative method for solving a positive-definite linear system $Ap = b$, in the sense that the ith iterate p_i minimizes the associated quadratic function $Q(p) = \frac{1}{2}p^{\mathrm{T}}Ap - p^{\mathrm{T}}b$ over the Krylov subspace spanned by $\{b, Ab, \ldots, A^{i-1}b\}$.

The Hessian matrix $\nabla^2 f(x_k)$ need not be positive definite, so the assumptions underlying the conjugate-gradient method may not be satisfied. However, the Hessian matrix is always symmetric. At a local minimizer of (1), the Hessian is guaranteed to be positive semi-definite; in nondegenerate cases it will be positive definite. Thus, as the solution is approached (and the Newton model for (1) is more accurate and appropriate) we can anticipate that the requirements for the conjugate-gradient method will be satisfied. If the Hessian matrix is not positive definite, then the techniques discussed in Section 5 should be used.

A truncated-Newton method will only be competitive if further enhancements are used. For example, a preconditioner for the linear system will be needed, and the stopping rule for the inner algorithm will have to be chosen so that it is effective both close to and far from the solution. With these enhancements, truncated-Newton methods are a powerful tool for large-scale optimization.

Because of all the choices that can be made in designing truncated-Newton methods, they form a flexible class of algorithms. For this reason, the method can be adapted to the problem being solved. Thus, if "black box" software is not able to solve a problem successfully, it is possible to modify the inner algorithm, the preconditioner, the stopping rule for the inner iteration, or a number of other details to enhance performance.

A constrained optimization problem

$$\begin{array}{ll} \min & f(x) \\ \text{subject to} & g(x) = 0, \\ & h(x) \geqslant 0, \end{array}$$

can be solved using a penalty-barrier method, in which one solves a sequence of unconstrained problems of the form

$$\min_x f(x) + \rho_j \sum g_i(x)^2 - \frac{1}{\rho_j} \sum \log(h_i(x))$$

for an increasing sequence of values of $\rho_j \to \infty$ [19]. Each of the unconstrained problems can be solved using a truncated-Newton method, and so all of the above comments apply in this case. (There are also some new issues; see Section 10.) This is not the only possible approach to constrained problems, but it does indicate the relevance of unconstrained optimization techniques in this setting.

Some applications where truncated-Newton methods have been effective include:

- weather modeling,
- potential-energy minimization,
- molecular geometry,
- multicommodity flow,
- medical imaging,
- molecular conformation.

Truncated-Newton methods have been extended to the infinite-dimensional case, at least in the setting of nonlinear equations. See, for example, [23].

Many of the above ideas are discussed in greater detail in the remainder of the paper. Here is an outline of the topics covered:

- controlling the convergence rate (Section 2),
- guaranteeing convergence (Section 3),

- computing second-derivative information (Section 4),
- handling nonconvex problems (Section 5),
- preconditioning (Section 6),
- parallel algorithms (Section 7),
- practical behavior (Section 8),
- software (Section 9),
- constrained problems (Section 10).

A version of this paper containing an expanded reference list can be obtained from `http://iris.gmu.edu/~snash/` under "New Papers", or by contacting the author.

1.1. Basics

The default norm $\|\cdot\|$ used in this paper is the 2-norm: for a vector $x=(x_1,\dots,x_n)$, $\|x\| = \sqrt{x_1^2+\cdots+x_n^2}$. All vectors are column vectors.

The conjugate-gradient method for solving a linear system $Ap=b$ is initialized with $p_0=0$, $r_0=b$ (r_i is the ith residual $b-Ap_i$), $v_{-1}=0$, and $\beta_0=0$. Then

For $i=0,1,\dots$
    If stopping rule satisfied, stop
    If $i>0$ set $\beta_i = r_i^{\mathrm T} r_i / r_{i-1}^{\mathrm T} r_{i-1}$
    Set $v_i = r_i + \beta_i v_{i-1}$
    Set $\alpha_i = r_i^{\mathrm T} r_i / v_i^{\mathrm T} A v_i$
    Set $p_{i+1} = p_i + \alpha_i v_i$
    Set $r_{i+1} = r_i - \alpha_i A v_i$

Stopping rules are discussed in Section 2. The algorithm requires the computation of the matrix–vector product Av_i, but other information about A need not be supplied.

A line-search method for solving (1) has the following basic form: Specify some initial guess of the solution x_0. Then

For $k=0,1,\dots$
    If stopping rule satisfied, stop
    Compute a search direction p_k
    Determine an improved estimate of the solution $x_{k+1} = x_k + \alpha_k p_k$
        [line search]

"Improvement" is often measured in terms of the function value $f(x_{k+1})$. For example, the new estimate x_{k+1} might be required to satisfy a "sufficient decrease" condition of the form

$$f(x_{k+1}) \leqslant f(x_k) + \mu\alpha p_k^{\mathrm T} \nabla f(x_k)$$

for some $0<\mu<1$ [19]. That is, there must be a decrease in the function value that is a fraction of the decrease predicted by the first-order Taylor series approximation to $f(x_k+\alpha p_k)$.

A trust-region method for solving (1) has the following basic form: Specify some initial guess of the solution x_0, and specify Δ_0, the bound on the size of the "trust region", i.e., the bound on the length of the allowable step at the current iteration. Then

For $k = 0, 1, \ldots$

If stopping rule satisfied, stop

Choose p_k so as to minimize some approximation $\psi_k(p) \approx f(x_k + p)$,
subject to the constraint $\|p\| \leqslant \Delta_k$

Compute x_{k+1} and Δ_{k+1} using p_k.

Algorithms for constrained problems can also be imbedded inside line search or trust region approaches, but the details are more complicated (both practically and theoretically). For more information on these topics, see [19].

2. Controlling the convergence rate

The basic local convergence theorem appeared in [3], in the context of nonlinear equations. Here is an adaptation of that theorem to unconstrained optimization. The definition of a q [strong] rate of convergence can be found in [3].

Theorem 1. *Assume that ∇f is continuously differentiable in a neighborhood of a local solution x_* of (1). In addition, assume that $\nabla^2 f(x_*)$ is nonsingular and that $\nabla^2 f$ is Lipschitz continuous at x_*. Assume that iteration k of the truncated-Newton method computes a step p_k that satisfies*

$$\|\nabla f(x_k) + \nabla^2 f(x_k) p_k\| \leqslant \eta_k \|\nabla f(x_k)\|$$

for a specified value of η_k; the new estimate of the solution is computed using $x_{k+1} \leftarrow x_k + p_k$. If x_0 is sufficiently close to x_* and $0 \leqslant \eta_k \leqslant \eta_{\max} < 1$ then $\{x_k\}$ converges to x_* q-linearly in the norm $\|\cdot\|_*$ defined by $\|v\|_* \equiv \|\nabla^2 f(x_*) v\|$, with asymptotic rate constant no greater than $\eta_{\max}$. If $\lim_{k\to\infty} \eta_k = 0$, then the convergence is q-superlinear. If $\eta_k = O(\|\nabla f(x_k)\|^r)$ for $0 < r \leqslant 1$, then the convergence is of order at least $(1 + r)$.

The sequence $\{\eta_k\}$ is referred to as a "forcing" sequence. The theorem shows that there is a direct relationship between the forcing sequence and the rate of convergence of the truncated-Newton method for (1). In [3] the authors suggest using

$$\eta_k = \min\{\tfrac{1}{2}, c\|\nabla f(x_k)\|^r\}$$

as a practical forcing sequence, where c is a positive constant, and $0 < r \leqslant 1$. This sequence leads to a method with a fast asymptotic convergence rate. However, it is not scale invariant, i.e., the behavior of the truncated-Newton method will change if the objective function $f(x)$ is multiplied by a positive constant.

If the conjugate-gradient method is used for the inner iteration, then the ith inner iteration finds a minimizer of the quadratic model

$$f(x_k + p) \approx f(x_k) + p^{\mathrm{T}} \nabla f(x_k) + \tfrac{1}{2} p^{\mathrm{T}} \nabla^2 f(x_k) p \equiv Q_k(p) \tag{3}$$

over the Krylov subspace spanned by $\{\nabla f(x_k), \ldots, [\nabla^2 f(x_k)]^{i-1} \nabla f(x_k)\}$. The model (3) has a global minimum when the residual of the Newton equations is zero. If p is not the minimizer of the quadratic model, however, the magnitudes of the residual and the quadratic model can be dramatically different [15], and the residual can be a deceptive measure of the quality of the search direction. For this reason, it may be preferable to base a stopping rule on the value of the quadratic model.

Let p_i be the search direction computed at the ith inner iteration, and let $Q_i = Q(p_i)$. The stopping rule suggested in [15] is to accept a search direction if

$$i(Q_i - Q_{i-1})/Q_i \leqslant \eta_k.$$

Quoting from [15]: "This criterion ... compares the reduction in the quadratic model at the current iteration $(Q_i - Q_{i-1})$ with the average reduction per iteration (Q_i/i). If the current reduction is small relative to the average reduction (with 'small' measured by η_k), then the inner iteration is terminated."

Newton's method is based on the Taylor series approximation (3). If this approximation is inaccurate then it may not be sensible to solve the Newton equations accurately. ("Over-solving" the Newton equations will not produce a better search direction.) In this circumstance, the inner algorithm should be truncated after a small number of iterations.

The conjugate-gradient method minimizes the quadratic model (3); in particular, it computes the value of the quadratic model. The quadratic model predicts the amount of decrease that will be obtained in the objective value. The outer iteration will typically compute $f(x_k + p)$, and hence determines the actual decrease in the objective value. By comparing these two quantities, the algorithm can determine if the quadratic model is accurate. If not, an alternative value of the forcing term η_{k+1} can be used at the next outer iteration. A simple rule of this type is used in [16]; more sophisticated approaches are analyzed in [5].

The paper [5] identifies two successful forcing sequences (in the context of solving nonlinear equations, but adapted here for optimization). Let p_k be the search direction at the kth outer iteration, and let α_k be the step length. The first stopping sequence uses $\eta_0 \in [0, 1)$, and then for $k = 0, 1, \ldots$

$$\eta_{k+1} = \frac{\big|\, \|\nabla f(x_{k+1})\| - \|\nabla f(x_k) + \alpha_k \nabla^2 f(x_k) p_k\| \,\big|}{\|\nabla f(x_k)\|}.$$

The second uses $\eta_0 \in [0, 1)$, $\gamma \in [0, 1]$, $\phi \in (1, 2)$, and then for $k = 0, 1, \ldots$

$$\eta_{k+1} = \gamma \left(\frac{\|\nabla f(x_{k+1})\|}{\|\nabla f(x_k)\|} \right)^{\phi}.$$

Both are designed to provide good asymptotic performance while at the same time preventing over-solving.

All of these results can be applied directly to the sequence of unconstrained problems that arise when a penalty-barrier method is used to solve a constrained problem. The convergence of the overall penalty-barrier method is discussed in [19]. Convergence results for a gradient-projection method for linearly-constrained problems can be found in [8].

3. Guaranteeing convergence

Convergence can be guaranteed by imbedding a truncated-Newton method in either a line-search or a trust-region framework. This is straightforward, although minor adjustments to the inner algorithm must be made. Basic convergence theorems for line-search and trust-region frameworks can be found (for example) in [19].

Global convergence results for the more general setting of inexact-Newton methods are developed in [4]. I assume here that a truncated-Newton method is used, with the conjugate-gradient method as the inner algorithm.

A variety of convergence results are available for line-search methods. In one such (from [19]), the line search method can be guaranteed to converge (in the sense that the limit of the gradient norms is zero) if the following assumptions are satisfied:

- the level set $S = \{x\colon f(x) \leqslant f(x_0)\}$ is bounded,
- ∇f is Lipschitz continuous for all $x \in S$,
- the search directions p_k satisfy a sufficient-descent condition

$$-\frac{p_k^{\mathrm{T}} \nabla f(x_k)}{||p_k|| \cdot ||\nabla f(x_k)||} \geqslant \varepsilon > 0$$

  for some fixed ε,
- the search directions are gradient related: $||p_k|| \geqslant m||\nabla f(x_k)||$ for some fixed $m > 0$,
- the search directions are bounded: $||p_k|| \leqslant M$ for some fixed M,
- an "appropriate" line search is used.

The first two conditions are assumptions on the optimization problem, and the final condition is independent of the inner algorithm.

Before discussing the other three conditions, it is useful to discuss the Lanczos method. The Lanczos method can be applied to any symmetric matrix A. It determines a sequence of orthogonal matrices V_i and tridiagonal matrices T_i such that

$$V_i^{\mathrm{T}} A V_i = T_i.$$

The Lanczos method is equivalent to the conjugate-gradient method. If $A = \nabla^2 f(x_k)$ and p_i is the result of the ith iteration of the conjugate-gradient method applied to (2), then

$$p_i = -V_i T_i(x_k)^{(-1)} V_i^{\mathrm{T}} \nabla f(x_k).$$

See [21] for further details. The other three conditions for convergence will be satisfied if the eigenvalues of the matrices T_i are uniformly bounded for all x_k:

$$0 < c_1 \leqslant \lambda_{\min}[T_i(x_k)] \leqslant \lambda_{\max}[T_i(x_k)] \leqslant c_2.$$

The upper bound can be guaranteed if the level set S is bounded, and if the Hessian is continuous on S. The lower bound can be guaranteed by making adjustments to the conjugate-gradient method. These ideas are discussed further in Section 5.

These convergence results are based on a "traditional" line search, i.e., the new estimate of the solution is obtained as $x_{k+1} \leftarrow x_k + \alpha p_k$, where α is chosen to ensure that the objective function decreases at every iteration. Convergence can also be proved for algorithms that use a curvilinear line search [9]; the new estimate of the solution is of the form

$$x_{k+1} = x_k + \alpha^2 p_k + \alpha d_k$$

where d_k is a direction of negative curvature (see Section 5). In addition, convergence can be proved for algorithms that use a non-monotone line search [7], where decrease in $f(x_k)$ is not required at every iteration.

When using a trust-region method, fewer adjustments need be made to the conjugate-gradient method. One significant issue, though, is to ensure that the output vector p_k satisfies the trust-region constraint:

$$||p_k|| \leqslant \Delta_k$$

where Δ_k is a parameter in the trust-region method. A technique for this was suggested by [25], and is outlined here. It is straightforward to prove that the iterates from the conjugate-gradient method increase monotonically in norm as long as the tridiagonal matrix T_i is positive definite:

$$||p_0|| < ||p_1|| < ||p_2|| < \cdots.$$

Thus, it is easy to determine at which iteration the trust-region constraint is violated, and to choose p_k as the point between p_i and p_{i+1} which exactly satisfies the constraint. Of course, if the termination rule is satisfied before the trust-region constraint is encountered, then the inner algorithm terminates before this occurs. If T_i becomes indefinite, then the quadratic model is unbounded below, and the next step in the inner iteration will cause the trust-region constraint to be violated. In [6] the authors examine more closely what happens in a truncated-Newton method when the trust-region boundary is encountered, and propose alternatives to simply truncating the inner iteration in this case.

Just as in Section 2, all of these results can be applied when a penalty-barrier method is used to solve a constrained problem. Global convergence results for a trust-region method for linearly constrained problems can be found in [8].

4. Computing second-derivative information

The conjugate-gradient method requires the computation or estimation of matrix-vector products involving the Hessian of the objective function

$$w = \nabla^2 f(x_k)v \tag{4}$$

for any vector v. This can be accomplished in a variety of ways.

If the Hessian $\nabla^2 f$ is explicitly available then (4) can be computed directly. This can be especially efficient if the Hessian is sparse. The user must be able to derive and program the formulas for the Hessian to use this technique. The remaining techniques require less effort on the part of the user.

An estimate of (4) can be obtained using finite differencing:

$$w \approx \frac{\nabla f(x_k + hv) - \nabla f(x_k)}{h}$$

for some "small" h. Each matrix–vector product requires one gradient evaluation, since $\nabla f(x_k)$ is already available as the right-hand side of (2). The choice of h is discussed in [19], as are alternative finite-difference formulas. This approach is widely used in practical truncated-Newton methods.

If it is possible to use complex arithmetic, then a more accurate finite-difference approximation to w can be obtained using

$$v_h = x_k + \sqrt{-1}hv, \quad g_h = \nabla f(v_h), \quad w \approx \mathrm{Im}(g_h)/h.$$

With this technique it is possible to choose a very small value of h (e.g., $h = 10^{-16}$) and obtain an estimate of w that is accurate to $\mathrm{O}(h)$.

A third alternative is to use automatic differentiation [19] to compute (4). This is an exact calculation (up to the limits of computer arithmetic). The computational cost is comparable to a gradient evaluation, and thus comparable to the finite-difference technique.

5. Nonconvex problems

As was mentioned in Section 3, the conjugate-gradient method is equivalent to the Lanczos method in the sense that

$$p_i = -V_i T_i(x_k)^{(-1)} V_i^{\mathrm{T}} \nabla f(x_k),$$

where $A = \nabla^2 f(x_k)$ and

$$V_i^{\mathrm{T}} A V_i = T_i.$$

Here $\{V_i\}$ is a sequence of orthogonal matrices and $\{T_i\}$ is a sequence of tridiagonal matrices.

If A is positive definite, then the formulas for the conjugate-gradient method correspond to computing the factorization

$$T_i = L_i D_i L_i^{\mathrm{T}}$$

where D_i is diagonal (with positive diagonal entries), and L_i is lower triangular (with ones along the diagonal). This factorization exists if and only if T_i is positive definite.

If T_i is not positive definite then this factorization cannot be computed. The algorithm will break down if a diagonal entry of D_i is zero, and will be numerically unstable if a diagonal entry of D_i is negative. To guarantee convergence (see Section 3) the diagonal entries of D_i must be positive and bounded away from zero.

The same situation occurs for certain implementations of Newton's method. In that setting a variety of proposals have been made that correspond to "modifying" the Hessian (or, equivalently, the factorization) to obtain a new, positive definite matrix that then replaces the Hessian in (2).

Any of these techniques could, in principle, be applied to the factorization of the tridiagonal matrix T_i. This is not usually done, however, because the components of the matrix T_i are generated iteratively, and the matrices T_i and V_i are not stored.

An alternative approach that uses information from two successive iterations of the conjugate-gradient method is developed in [12]. This "modified" conjugate-gradient method is iterative (like the regular conjugate-gradient method), and has many of the same theoretical and practical properties as modified-Newton methods.

It is possible to use a simpler technique, and develop a "modified" method using only information from the current iteration of the conjugate-gradient method. This approach is mentioned in [12]. The drawback to this approach is that the modification to the Hessian can be very large in norm, much larger than if information from two successive iterations is used.

If the tridiagonal matrix T_i is not positive semi-definite, then the matrix D_i must have a negative diagonal entry. This corresponds to a direction of negative curvature, i.e., a vector d satisfying

$$d^{\mathrm{T}}[\nabla^2 f(x_k)]d < 0.$$

Such a direction can be used as part of a search direction, since either d or $-d$ is a direction of nonascent. This idea is discussed in [9].

The trust-region techniques discussed in Section 3 provide an alternative way of handling non-convex problems. If a diagonal entry of D_i is negative, then the quadratic model can be decreased to $-\infty$ by following this direction of negative curvature. Thus, a sufficiently long step along such a direction (or any direction) is guaranteed to violate the trust-region constraint.

The application of these ideas to constrained problems is discussed in Section 10.

6. Preconditioning

The convergence of the conjugate-gradient method is strongly influenced by the condition number of the Hessian (i.e., its extreme eigenvalues), and by the number of distinct eigenvalues of the Hessian. Reducing either of these accelerates the convergence of the method.

Ideally, a preconditioner will be chosen based on the problem being solved. This can require considerable analysis and programming to accomplish, however, and is not suitable for routine cases.

If the Hessian matrix is available, a good "generic" choice of a preconditioner is an incomplete Cholesky factorization. The preconditioner is formed by factoring the Hessian, and ignoring some or all of the fill-in that occurs during Gaussian elimination. It may be necessary to modify the factorization (as discussed in Section 5) so that the preconditioner is positive definite. This idea is discussed in [24].

It is also possible to develop preconditioners based on partial separability in the objective function [2]. (A function $f(x)$ is partially separable if it can be written as the sum of functions $f_i(x)$, each of which has a large invariant subspace.)

If neither of these is possible, "automatic" preconditioners can be developed that do not require Hessian information. These preconditioners are based on quasi-Newton approximation to the Hessian. A quasi-Newton approximation is computed based on vector pairs (s_i, y_i). Traditionally, $s_i = x - \hat{x}$ for some pair of variable values, and $y_i = \nabla f(x) - \nabla f(\hat{x})$, the corresponding difference of gradient values. In the context of a truncated-Newton method, these might be $x = x_k$ and $\hat{x} = x_{k+1}$, i.e., two successive iterates.

It is also possible to use an arbitrary vector s_i with $y_i = \nabla^2 f(x_k)s_i$. At each iteration of the conjugate-gradient method, a matrix–vector product of this form is computed or estimated, and each of these matrix–vector products can be used to help construct a Hessian approximation.

In [13], both these ideas are combined to form a preconditioner. The matrix–vector products from the inner iteration are used to construct a diagonal approximation to the Hessian, using a BFGS update formula in which only the diagonal matrix entries are computed. This is in turn used to initialize a two-step limited-memory BFGS update formula which is the actual preconditioner. The limited-memory update is constructed using pairs in which s_i is the difference between a pair of x-vectors. Precise information is given in [14]. This preconditioner is implemented in the `TN/TNBC` software discussed in Section 9.

A more elaborate preconditioner is described in [11], based on an m-step limited memory BFGS update, with the (s_i, y_i) pairs chosen as a subset of the matrix–vector products in the inner iteration. Experiments are conducted with various choices of m. The authors propose an algorithm that "dynamically stores the correction pairs so that they are as evenly distributed as possible" among the set of pairs for a complete inner iteration.

The application of these ideas to constrained problems is discussed in Section 10.

7. Parallel algorithms

A parallel algorithm could be obtained by executing each of the steps of the truncated-Newton method in parallel. This would require converting the line search and the conjugate-gradient method so that they execute in parallel. By itself, this is not likely to be an effective strategy, since the steps in these algorithms consist of

- scalar operations,
- vector operations,
- function and gradient evaluations.

The scalar operations cannot be made parallel, and the vector operations do not offer much potential for speed-up on a parallel machine (since communication and synchronization delays could easily wipe out any computational savings obtained).

The function and gradient evaluations offer more hope, but this requires that the person solving the optimization problem be willing and able to compute these values effectively in parallel. For very large and difficult problems, however, this may be essential.

An alternative is to replace the line search and the inner algorithm with alternatives that are better able to exploit parallelism. Ideally, it should be possible to take advantage of both parallel linear algebra computations as well as parallel function and gradient evaluations (that is, simultaneous evaluations of the function and/or gradient on separate processors).

An approach of this type for unconstrained problems is discussed in [16]. In this work, the block conjugate-gradient method is used as the inner algorithm; this is a generalization of the conjugate-gradient method in which a block of vectors (rather than a single vector) is updated at every inner iteration. A simple parallel line search is used to compute $x_{k+1} \leftarrow x_k + \alpha p_k$. If the block size in the block conjugate-gradient method is equal to m, then each inner iteration requires the computation of m independent matrix–vector products (which can be approximated by m independent gradient evaluations). There is also considerable opportunity for parallel linear algebra computations. Each iteration of the line search requires m independent function evaluations.

A hybrid approach (combining parallelism in the algorithm with parallelism in the individual function evaluations) is also possible within the block conjugate-gradient method. The block size m need not be equal to the number of processors. This can be an advantage if the individual function and gradient evaluations can be performed in parallel. For example, suppose that a computer with 32 processors were available, and that each function or gradient evaluation could be spread over 4 processors. Then, if the block-size were chosen as $m = 8$, an inner iteration would require 8 gradient evaluations, each of which would require 4 processors. Thus a total of $4 \times 8 = 32$ processors would be used.

This algorithm is implemented in the software package `BTN`; see Section 9.

8. Practical behavior

Truncated-Newton methods use an inner iteration to compute a search direction, and thus expend considerable computational effort at each outer iteration. In contrast, nonlinear conjugate-gradient methods and limited-memory quasi-Newton methods use relatively few computations to obtain each search direction. (Precise operation counts can be found in [14].) A basic question is whether the effort per iteration for a truncated-Newton method can be worthwhile.

The tests in [14] compare the truncated-Newton method TN against the limited-memory quasi-Newton method L-BFGS. The tests imply that L-BFGS becomes more effective as the optimization problem (1) becomes more nonlinear. In a sense, the truncated-Newton method is more effective when the quadratic model (3) is more effective.

Attempts have been made to combine the best properties of both these methods. This consisted in

- monitoring the effectiveness of the quadratic model to avoid "over-solving" in cases where the quadratic model is poor,
- using limited-memory quasi-Newton formulas as preconditioners,
- combining both techniques in a single algorithm.

Testing of specific features of truncated-Newton software can be found in [12,13] and, for the parallel case, in [16]. The results of these tests have influenced the development of the software packages mentioned in Section 9. The paper [22] describes software for nonlinear equations, but many of the comments are also applicable to optimization. Tests of truncated-Newton methods for bound-constrained problems can be found in [8].

9. Software

Truncated-Newton software is available for unconstrained and bound constrained problems. The first three packages are available from the Netlib collection (`www.netlib.org`).

The package TN/TNBC solves both classes of problems. It requires that the user provide a subroutine to evaluate the function value and gradient of the objective function. The algorithm is described in [14]. This software is designed to be easy to use, and does not require or expect customization by the user.

The package TNPACK solves unconstrained minimization problems. In addition to function and gradient information, the user must provide formulas for the Hessian matrix and a user-supplied preconditioner. This software expects the user to supply information about the Hessian so that a preconditioner can be constructed. This can require considerable effort, but with the promise of improved performance.

The package BTN solves unconstrained problems on parallel computers (both shared and distributed memory). It requires that the user provide a (scalar) subroutine to evaluate the function value and gradient of the objective function. In addition BTN can take advantage of a parallel subroutine for the function and gradient if one is provided. The algorithm is described in [16]. This software comes with both easy-to-use and customizable top-level subroutines.

The TRON software [8] solves bound-constrained problems. It requires that the user supply function, gradient, and Hessian information. It uses an incomplete Cholesky factorization as a preconditioner. The software can be obtained from

`www.mcs.anl.gov/~more/tron/`

The Lancelot software [1] is a more general package, but a variety of truncated Newton algorithms can be used within it by appropriately selecting software parameters. Considerable customization is possible. Information about this software can be obtained from

`www.cse.clrc.ac.uk/Activity/LANCELOT`

A variety of software packages for solving nonlinear systems of equations are mentioned in [22].

Software for the quasi-Newton preconditioner in [11] is available from

`www.ece.nwu.edu/~nocedal/preqn.html`

10. Constrained problems

Many algorithms for constrained optimization problems are built upon algorithms, techniques, or principles from unconstrained optimization. It should not be surprising that truncated-Newton methods can be used in this setting.

One approach is to use a penalty-barrier method to solve the constrained problem [19]. The constrained problem is replaced by a sequence of unconstrained problems, where violations in the constraints are included as penalty terms in the objective function (see Section 1). A truncated-Newton method can then be applied to the sequence of unconstrained problems. Under appropriate assumptions, it is possible to derive complexity results for an algorithm of this type [20].

It has been known for decades that the unconstrained problems become increasingly ill-conditioned as the barrier parameter is increased (i.e., as the solution is approached). This ill-conditioning causes the behavior of the inner algorithm to deteriorate. This ill-conditioning is not inherent, however.

In [17], an approximation to the inverse of the Hessian matrix is derived that can be used within the conjugate-gradient method. With this approximation, the conjugate-gradient method is applied to a linear system whose conditioning reflects that of the underlying optimization problem, and not that of the penalty-barrier problem. The approximation formula requires, though, that an active set be identified (a prediction of the set of constraints that are binding at the solution to the optimization problem).

Similar techniques can be applied within augmented Lagrangian and modified barrier methods [1].

It is also possible to adapt truncated-Newton techniques to constrained methods based on sequential quadratic programming. At each iteration of such a method, the nonlinear constrained problem is approximated by a quadratic program. An inner iterative method can then be applied to the quadratic program. There are a number of choices in how this is done; for example, an interior-point method could be used to solve the quadratic program.

If the quadratic program is solved using a null-space approach, then the conjugate-gradient method would be applied to a linear system with a matrix of the form

$$Z^{\mathrm{T}}HZ$$

where H is an approximation to the Hessian, and Z is a null-space matrix for the Jacobian of the constraints. Optimality conditions for the constrained optimization problem imply that this matrix will be positive semi-definite at the solution of the optimization problem. Indefiniteness can be dealt with as in the unconstrained case.

The matrix $Z^{\mathrm{T}}HZ$ may be a dense matrix even if H and the constraint Jacobian are sparse. Matrix–vector products involving this matrix should be computed in stages:

$$w_1 = Zv, \quad w_2 = Hw_1, \quad w = Z^{\mathrm{T}}w_2.$$

The convergence of the conjugate-gradient method is enhanced if a preconditioner for $Z^T HZ$ is available. This is more challenging than in the unconstrained case. Since H is an approximation to the Hessian matrix, preconditioners for H will depend on properties of the optimization problem (i.e., this requires input from the user of the software). The null-space matrix Z depends on algorithmic details (i.e., this requires input from the developer of the optimization software). The paper [18] develops preconditioners for $Z^T HZ$ that combine preconditioning information from these two sources.

The quadratic program might be solved by looking at the combined linear system for the variables and multipliers of the quadratic program, which has a matrix of the form

$$\begin{pmatrix} H & A^T \\ A & 0 \end{pmatrix}. \tag{5}$$

H is an approximation to the Hessian of the Lagrangian, and A is the Jacobian matrix of the constraints. The matrix (5) is symmetric but indefinite. An iterative method can be applied to this system, derived from the Lanczos method (see Section 3). One possibility is to use SYMMLQ [21], which is designed for symmetric indefinite systems of equations.

It may be difficult to guarantee that the search direction that results is a descent direction. Straightforward tests require factorizations of $Z^T HZ$, where Z is a null-space matrix for A. In an iterative method, where matrices are not stored and matrix factorizations are not available, this is not usually possible.

It is also possible to derive preconditioners based on the structure of the system (5) [10].

11. Conclusions, recommendations

Truncated-Newton methods are a flexible set of methods for solving large optimization problems. They are built upon a sound theoretical foundation. They can be adjusted to achieve a desired asymptotic convergence rate, and they can be designed to limit the waste of over-solving at points far from the solution. They can also be customized to the problem being solved.

The easiest way to use a truncated-Newton method is via the software packages discussed in Section 9. If these are not adequate, then perhaps use of parallel software will help. Customization of the method may also be necessary. The greatest improvements in performance can be obtained by improving the preconditioner. In addition, the forcing sequence can be modified, as can the technique used to compute the matrix–vector product. The references in this paper provide much guidance in these areas.

Acknowledgements

I would like to thank an anonymous referee for many helpful and perceptive suggestions.

References

[1] A.R. Conn, N.M. Gould, P.L. Toint, LANCELOT: A Fortran Package for Large-Scale Nonlinear Optimization (Release A), Springer, Berlin, 1992.

[2] M.J. Daydé, J.-Y. L'Excellent, N.I.M. Gould, Element-by-element preconditioners for large partially separable optimization problems, SIAM J. Sci. Comput. 18 (1997) 1767–1787.
[3] R.S. Dembo, S.C. Eisenstat, T. Steihaug, Inexact Newton methods, SIAM J. Numer. Anal. 19 (1982) 400–408.
[4] S.C. Eisenstat, H.F. Walker, Globally convergent inexact Newton methods, SIAM J. Optim. 4 (1994) 393–422.
[5] S.C. Eisenstat, H.F. Walker, Choosing the forcing terms in an inexact Newton method, SIAM J. Sci. Comput. 17 (1996) 16–32.
[6] N.I.M. Gould, S. Lucidi, M. Roma, P.L. Toint, Solving the trust-region subproblem using the Lanczos method, SIAM J. Optim. 9 (1999) 504–525.
[7] L. Grippo, F. Lampariello, S. Lucidi, A truncated Newton method with nonmonotone line search for unconstrained optimization, J. Optim. Theory Appl. 609 (1989) 401–419.
[8] C.-J. Lin, J.J. Moré, Newton's method for large bound-constrained optimization problems, SIAM J. Optim. 9 (1999) 1100–1127.
[9] S. Lucidi, F. Rochetich, M. Roma, Curvilinear stabilization techniques for truncated Newton methods in large scale unconstrained optimization, SIAM J. Optim. 8 (1998) 916–939.
[10] L. Lukšan, J. Vlček, Indefinitely preconditioned inexact Newton method for large sparse equality constrained non-linear programming problems, Numer. Linear Algebra Appl. 5 (1998) 219–247.
[11] J.L. Morales, J. Nocedal, Automatic preconditioning by limited memory quasi-Newton updating, Rep. OTC 97/08, Opt. Tech. Ctr., Northwestern Univ., Evanston, IL, 1997.
[12] S.G. Nash, Newton-type minimization via the Lanczos method, SIAM J. Numer. Anal. 21 (1984) 770–788.
[13] S.G. Nash, Preconditioning of truncated-Newton methods, SIAM J. Sci. Statist. Comput. 6 (1985) 599–616.
[14] S.G. Nash, J. Nocedal, A numerical study of the limited memory BFGS method and the truncated-Newton method for large scale optimization, SIAM J. Optim. 1 (1991) 358–372.
[15] S.G. Nash, A. Sofer, Assessing a search direction within a truncated-Newton method, Oper. Res. Lett. 9 (1990) 219–221.
[16] S.G. Nash, A. Sofer, A general-purpose parallel algorithm for unconstrained optimization, SIAM J. Optim. 1 (1991) 530–547.
[17] S.G. Nash, A. Sofer, A barrier method for large-scale constrained optimization, ORSA J. Comput. 5 (1993) 40–53.
[18] S.G. Nash, A. Sofer, Preconditioning reduced matrices, SIAM J. Mat. Anal. Appl. 17 (1996) 47–68.
[19] S.G. Nash, A. Sofer, Linear and Nonlinear Programming, McGraw-Hill, New York, 1996.
[20] S.G. Nash, A. Sofer, On the complexity of a practical interior-point method, SIAM J. Optim. 8 (1998) 833–849.
[21] C.C. Paige, M.A. Saunders, Solution of sparse indefinite systems of linear equations, SIAM J. Numer. Anal. 12 (1975) 617–629.
[22] M. Pernice, H.F. Walker, NITSOL: A Newton iterative solver for nonlinear systems, SIAM J. Sci. Comput. 19 (1998) 302–318.
[23] E. Sachs, Rates of convergence for adaptive Newton methods, J. Optim. Theory Appl. 48 (1986) 175–190.
[24] T. Schlick, Modified Cholesky factorizations for sparse preconditioners, SIAM J. Sci. Comput. 14 (1993) 424–445.
[25] T. Steihaug, The conjugate gradient method and trust regions in large scale optimization, SIAM J. Numer. Anal. 20 (1983) 626–637.

Cubature formulae and orthogonal polynomials

R. Cools[a,*], I.P. Mysovskikh[b], H.J. Schmid[c]

[a] *Department of Computer Science, Katholieke Universiteit Leuven, Celestijnenlaan 200A, B-3001 Heverlee, Belgium*
[b] *Mathematical and Mechanical Faculty, Saint-Petersburg State University, Stary Peterhof, Bibliotechnaya pl. 2, 198904 Saint Petersburg, Russia*
[c] *Math. Institut, Universität Erlangen-Nürnberg, Bismarckstr. 1 1/2, D-91054 Erlangen, Germany*

Abstract

The connection between orthogonal polynomials and cubature formulae for the approximation of multivariate integrals has been studied for about 100 yr. The article J. Radon published about 50 yr ago (J. Radon, Zur mechanischen Kubatur, Monatsh. Math. 52 (1948) 286–300) has been very influential. In this text we describe some of the results that were obtained during the search for answers to questions raised by his article.

Keywords: Cubature; Multivariate integrals; Orthogonal polynomials

1. Introduction

The connection between orthogonal polynomials and algebraic integration formulae in higher dimension was already studied about 100 yr ago (early papers are, e.g., [1,6]). The problem became widely noticed after the second edition of Krylov's book "On the approximate calculation of integrals" [43], published in 1967, wherein Mysovskikh introduced Radon's construction of a formula of degree 5 published in 1948 [72].

Though no final solution – similar to the one-dimensional case – has been found up to now, the work in this field has been tremendous. In the textbooks by Krylov [43], Stroud [90], Sobolev [84], Engels [20], Mysovskikh [66], Davis and Rabinowitz [17], Xu [94] and Sobolev and Vaskevich [85], and in the survey article of Cools [10], the growth of knowledge in the field is documented. In this text we will only try to describe some relevant results – following Radon's ideas – that have been found in the meantime.

* Corresponding author.
E-mail address: ronald.cools@cs.kuleuven.ac.be (R. Cools).

The one-dimensional algebraic case, i.e., interpolatory quadrature formulae and their relation to orthogonal polynomials, is well known. In dimension 2 and beyond, things look worse: there are more questions than answers. Nevertheless, some progress has been made. Though several essential problems – important in applications – are still open, e.g., how minimal formulae of an arbitrary degree of exactness look like for the integral over the square with constant weight function, several results of some generality have been found. They make transparent why answers to important questions must be quite complex. We leave aside many particular results, in spite of their importance for applications. For these we refer to the surveys in [90,14,11].

2. Basic concepts and notations

We would have liked to preserve the flair of the old papers; however, we finally decided to use modern notations in order to achieve an easy and consistent way of presenting the results.

We denote by $\mathbb{N}$ the nonnegative integers. The monomials of degree m in n variables are written in the short notation

$$\boldsymbol{x}^{\boldsymbol{m}} = x_1^{m_1} x_2^{m_2} \cdots x_n^{m_n} \quad \text{with } |\boldsymbol{m}| = m,$$

where $\boldsymbol{x} = (x_1, x_2, \ldots, x_n)$, $\boldsymbol{m} = (m_1, m_2, \ldots, m_n) \in \mathbb{N}^n$, and

$$|\boldsymbol{m}| = \sum_{i=1}^{n} m_i \text{ is the length of the multi-index } \boldsymbol{m}.$$

A polynomial $f(\boldsymbol{x}) = f(x_1, x_2, \ldots, x_n)$ of (total) degree m can be represented as

$$f(\boldsymbol{x}) = \sum_{s=0}^{m} \sum_{|\boldsymbol{k}|=s} c_{\boldsymbol{k}} \boldsymbol{x}^{\boldsymbol{k}}, \quad c_{\boldsymbol{k}} \in \mathbb{C},$$

and the summation in

$$g_s(\boldsymbol{x}) = \sum_{|\boldsymbol{k}|=s} c_{\boldsymbol{k}} \boldsymbol{x}^{\boldsymbol{k}}$$

is done over all multi-indices $\boldsymbol{k}$ of length s. The polynomial g_s is called a *homogeneous component* of degree s. Hence f is an element of the ring of polynomials with complex coefficients, which will be denoted by $\mathbb{C}[\boldsymbol{x}] = \mathbb{C}[x_1, x_2, \ldots, x_n]$. The degree of a polynomial f will be denoted by $\deg(f)$.

The number of linearly independent polynomials of degree $\leqslant m$ is

$$M(n,m) = \binom{m+n}{n},$$

the number of pairwise distinct monomials of degree m is $M(n-1, m)$. When the linearly independent monomials are needed as an ordered sequence, we will represent them by

$$\{\varphi_j(\boldsymbol{x})\}_{j=1}^{\infty},$$

where $j < k$ whenever $\deg(\varphi_j(\boldsymbol{x})) < \deg(\varphi_k(\boldsymbol{x}))$. Hence

$$\{\varphi_j(\boldsymbol{x})\}_{j=1}^{\mu}, \quad \mu = M(n,m),$$

contains all monomials of degree $\leqslant m$. Most of the results in the sequel will be stated in the ring of polynomials with real coefficients, $\mathbb{R}[x_1, x_2, \ldots, x_n] = \mathbb{R}[\boldsymbol{x}]$, which will be denoted by $\mathbb{P}^n$. The polynomials in $\mathbb{P}^n$ of degree $\leqslant m$ will be denoted by $\mathbb{P}^n_m$. The elements of $\mathbb{P}^n_m$ form a vector space over $\mathbb{R}$ with dimension $\dim \mathbb{P}^n_m = M(n,m)$.

We consider integrals of the type

$$\mathscr{I}^n[f] = \int_\Omega f(\boldsymbol{x})\omega(\boldsymbol{x})\,\mathrm{d}\boldsymbol{x}, \quad f \in \mathscr{C}(\Omega), \tag{1}$$

where $\Omega \subseteq \mathbb{R}^n$ is a region with inner points and the real weight function ω is chosen such that the moments

$$\mathscr{I}^n[\boldsymbol{x}^{\boldsymbol{m}}], \quad \boldsymbol{m} \in \mathbb{N}^n,$$

exist. In many applications $\omega(\boldsymbol{x})$ is nonnegative. Hence $\mathscr{I}^n$ is a positive linear functional. In most of the results presented, $\mathscr{I}^n$ will be even strictly positive, i.e.,

$$\mathscr{I}^n[f] > 0 \text{ whenever } 0 \neq f \geqslant 0 \text{ on } \Omega,$$

so orthogonal polynomials with respect to $\mathscr{I}^n$ are defined.

The type of integrals we consider includes integrals over the so-called *standard regions*, for which we follow Stroud's notation [90].

C_n: the n-dimensional cube

$$\Omega = \{(x_1,\ldots,x_n):\ -1 \leqslant x_i \leqslant 1, i = 1,\ldots,n\}$$

with weight function $\omega(\boldsymbol{x}) = 1$,

S_n: the n-dimensional ball

$$\Omega = \left\{(x_1,\ldots,x_n):\ \sum_{i=1}^n x_i^2 \leqslant 1\right\}$$

with weight function $\omega(\boldsymbol{x}) = 1$,

T_n: the n-dimensional simplex

$$\Omega = \left\{(x_1,\ldots,x_n):\ \sum_{i=1}^n x_i \leqslant 1 \text{ and } x_i \geqslant 0, i = 1,\ldots,n\right\}$$

with weight function $\omega(\boldsymbol{x}) = 1$,

$E_n^{r^2}$: the entire n-dimensional space $\Omega = \mathbb{R}^n$ with weight function

$$\omega(\boldsymbol{x}) = \mathrm{e}^{-r^2}, \quad r^2 = \sum_{i=1}^n x_i^2,$$

E_n^r: the entire n-dimensional space $\Omega = \mathbb{R}^n$ with weight function

$$\omega(\boldsymbol{x}) = \mathrm{e}^{-r},$$

H_2: the region bounded by the regular hexagon with vertices $(\pm 1, 0)$, $(\pm\frac{1}{2}, \pm\frac{1}{2}\sqrt{3})$ and weight function $\omega(\boldsymbol{x}) = 1$.

A *cubature formula* for (1) is of the form

$$\mathscr{I}^n[f] = \mathscr{Q}[f] + R[f], \tag{2}$$

where

$$\mathcal{Q}[f] = \sum_{j=1}^{N} w_j f(\boldsymbol{x}^{(j)}) \tag{3}$$

is called a *cubature sum*. The $\boldsymbol{x}^{(j)}$'s are called *nodes*, the w_j's *weights* or *coefficients*, and $R[f]$ is the error. The shorthand notation

$$\mathcal{I}^n[f] \cong \sum_{j=1}^{N} w_j f(\boldsymbol{x}^{(j)})$$

is often used.

A nonnegative integer d is called *degree of exactness* or *degree of precision* or simply *degree* of formula (2), if $R[f] = 0$ for all polynomials f with $\deg(f) \leqslant d$ and if a polynomial g with $\deg(g) = d + 1$ exists such that $R[g] \neq 0$.

Let $f \in \mathbb{C}[\boldsymbol{x}]$ be given and $\deg(f) = m$; the algebraic manifold of degree m generated by f will be denoted by

$$\mathcal{H}_m = \{\boldsymbol{x} \in \mathbb{C}^n\colon\ f(\boldsymbol{x}) = 0\}.$$

A cubature formula (2) with $N = M(n, m)$ which is exact for all polynomials of degree $\leqslant m$ is called *interpolatory* if the nodes do not lie on an algebraic manifold of degree m and the coefficients are uniquely determined by the nodes.

If $n = 1$, then $N = m + 1$ and the converse is true, too. If the degree of exactness of the quadrature formula is m, then it is interpolatory. For $n \geqslant 2$ this does not hold in general. The number of nodes might be lower than $M(n, m)$ since some of the coefficients may vanish.

Theorem 1. *Let* (2) *be given such that* $R[f] = 0$ *for all polynomials of degree* $\leqslant m$ *and* $N \leqslant \mu = M(n, m)$. *The formula is interpolatory if and only if*

$$\operatorname{rank}([\varphi_1(\boldsymbol{x}^{(j)}), \varphi_2(\boldsymbol{x}^{(j)}), \ldots, \varphi_\mu(\boldsymbol{x}^{(j)})]_{j=1}^{N}) = N.$$

We are specifying $\mathcal{Q}$ by $\mathcal{Q}(n, m, N)$ if we refer to a cubature sum in n dimensions with a degree of exactness m and N nodes. We only consider interpolatory cubature formulae. A noninterpolatory formula can be transformed to an interpolatory formula by deleting nodes. In particular, minimal formulae (N is minimal for fixed m) are interpolatory.

A polynomial P with $\deg(P) = m$ is called *orthogonal* with respect to the underlying $\mathcal{I}^n$ if $\mathcal{I}^n[PQ] = 0$ for all Q, $\deg(Q) \leqslant m - 1$. It is called *quasi-orthogonal* if $\mathcal{I}^n[PQ] = 0$ for all $Q, \deg(Q) \leqslant m - 2$.

A set of polynomials in $\mathbb{R}[x_1, \ldots, x_n]$ is called a *fundamental system of degree m* whenever it consists of $M(n - 1, m)$ linearly independent polynomials of the form

$$\boldsymbol{x}^{\boldsymbol{m}} + Q_{\boldsymbol{m}}, \quad \boldsymbol{m} \in \mathbb{N}^n, \quad \deg(Q_{\boldsymbol{m}}) \leqslant |\boldsymbol{m}| - 1.$$

A set $\mathcal{M}$ of polynomials is called a *fundamental set of degree m* if a fundamental system of degree m is contained in $\operatorname{span}\{\mathcal{M}\}$.

In the two-dimensional case we drop the superscript n and use x and y as variables, i.e., $\mathbb{P}, \mathbb{P}_m$, and,

$$\mathscr{I}[f] = \int_\Omega f(x,y)\omega(x,y)\,\mathrm{d}x\,\mathrm{d}y, \quad f \in \mathscr{C}(\Omega), \ \Omega \subseteq \mathbb{R}^2.$$

Whenever possible, one tries to find cubature sums (3) such that the following constraints are satisfied:

(i) $w_j > 0$,

(ii) $\boldsymbol{x}^{(j)} \in \Omega$,

(iii) N is minimal for fixed degree m.

If $n = 1$, Gaussian quadrature formulae satisfy all constraints. These formulae are closely connected to orthogonal polynomials. The zeros of a quasi-orthogonal polynomial of degree k, $P_k^1 + \gamma P_{k-1}^1$ with free parameter γ, are the nodes of a minimal quadrature formula of degree $2k-2$ with all weights positive. The parameter γ can be chosen such that all nodes are inside the domain of integration, and, if $\gamma = 0$ one obtains a uniquely determined formula of degree $2k-1$ satisfying (i), (ii), (iii). Nonminimal interpolatory quadrature formulae have been characterised by Sottas and Wanner, Peherstorfer, and many others, most recently by Xu [86,68,69,96].

3. Radon's formulae of degree 5

The paper by Johann Radon [72], which appeared in 1948, is not the oldest studying the application of orthogonal polynomials to cubature formulae (earlier papers are, e.g., [1] to which Radon refers, and [6]). However, Radon's contribution became fundamental for all research in that field. Although the word "cubature" appeared in the written English language already in the 17th century, this paper is probably the first that used the term "Kubaturformel" (i.e., German for "cubature formula") for a weighted sum of function values to approximate a multiple integral (in contrast to quadrature formula to approximate one-dimensional integrals). As an introduction to the survey which follows, we will briefly sketch its main ideas.

Radon discusses the construction of cubature formulae of degree 5 with 7 nodes for integrals over T_2, C_2, S_2. We are sure Radon knew the estimate (22) and knew that this bound will not be attained for classical (standard) regions in the case of degree 5. In order to construct a cubature formula of degree m he counted the number of monomials of degree $\leqslant m$ and used this divided by 3 as number of necessary nodes. He was aware that for degrees 2, 3 and 4 this will not lead to a solution and thus degree 5 is the first nontrivial case he could consider.

Assuming a formula of degree 5 with 7 nodes for an integral $\mathscr{I}$, a geometric consideration leads to polynomials of degree 3 vanishing at the nodes. These polynomials have to be orthogonal with respect to $\mathscr{I}$ to all polynomials of degree 2, and exactly three such polynomials, P_1, P_2, P_3, can vanish at the nodes. In the next step further necessary conditions are derived for the P_i. They must satisfy

$$\sum_{i=1}^{3} L_i P_i = 0 \tag{4}$$

for some linear polynomials $L_i \neq 0$. This can be reduced to

$$xK_1 + yK_2 = K_3, \tag{5}$$

where K_i are orthogonal polynomials of degree 3. Assuming the orthogonal basis of degree 3 in a form

$$P_i^3 = x^{3-i}y^i + Q_i, \quad Q_i \in \mathbb{P}_2, \; i = 0,1,2,3,$$

one obtains

$$K_1 = \alpha P_1^3 + \beta P_2^3 + \gamma P_3^3 \quad \text{and} \quad K_2 = -\alpha P_0^3 - \beta P_1^3 - \gamma P_2^3.$$

The parameters α, β, γ have to be chosen such that K_3 is also an orthogonal polynomial of degree 3. Thus, by setting

$$A = \mathscr{I}[P_0^3 P_2^3 - P_1^3 P_1^3], \quad B = \mathscr{I}[P_0^3 P_3^3 - P_1^3 P_2^3], \quad C = \mathscr{I}[P_1^3 P_3^3 - P_2^3 P_2^3], \tag{6}$$

one obtains the linear system

$$\begin{bmatrix} 0 & A & B \\ -A & 0 & C \\ -B & -C & 0 \end{bmatrix} \begin{bmatrix} \alpha \\ \beta \\ \gamma \end{bmatrix} = 0. \tag{7}$$

Two cases may occur. The parameters α, β, γ can be determined up to a common factor, if

$$A^2 + B^2 + C^2 > 0, \tag{8}$$

otherwise they can be chosen arbitrarily. Radon did not further pursue the last case. He just remarked that he did not succeed in proving that this case never occurs.

In case (8) the polynomials K_1, K_2, K_3 can be computed. If they are linearly independent, the desired equation (4) is given by $xK_1 + yK_2 = K_3$. If these polynomials vanish at 7 pairwise distinct nodes, the degree of exactness follows from the orthogonality property of the K_i's.

If the K_i are linearly dependent, it follows that

$$K_1 = yQ \quad \text{and} \quad K_2 = -xQ$$

for some $Q \in \mathbb{P}_2$. In this case it can be shown that there is a K_3 such that all K_i vanish at 7 pairwise distinct nodes. This construction again is based on geometric considerations and finally allows the conclusion that such K_3's can be computed.

Radon's article continues with the construction of formulae of degree 5 with 7 nodes for integrals over the standard regions with constant weight function T_2, C_2 and S_2. The amount of computational work – in a pre-computer time – is tremendous. The article finishes with an examination of the cubature error.

Though the results are limited to a special case, Radon's approach is the basis for fundamental questions that were studied in the years following the publication of his result:

(i) Can this constructive method be generalised to a higher degree of exactness?
(ii) Can this constructive method be generalised to more than two dimensions?
(iii) Are there integrals for which the second case occurs, i.e., $A = B = C = 0$?
(iv) Is 7 a lower bound for the number of nodes of cubature formulae of degree 5 if (8) holds?
(v) Are there lower bounds of some generality for the number of nodes?
(vi) What intrinsic tools were applied for the solution?

Fifty years after the publication of Radon's paper, it is still not possible to answer these questions completely. We will outline what is known in the sequel.

4. Multivariate orthogonal polynomials

We assume that $\mathscr{I}^n$ is strictly positive. Different methods of generating an orthogonal polynomial basis were discussed by Hirsch [34]. If the moments $\mathscr{I}^n[\boldsymbol{x}^{\boldsymbol{m}}]$, $\boldsymbol{m} \in \mathbb{N}^n$, are known, one can either orthogonalise the monomial basis by the Gram–Schmidt procedure, or compute step by step fundamental systems of orthogonal polynomials of degree m.

A third way is to find a partial differential equation with boundary conditions the polynomial solutions of which lead to orthogonal systems. This permits finding formulae for the coefficients of the polynomials and deriving recursion formulae. However, one has to find out if there is an integral for which the polynomials form an orthogonal system.

4.1. Simple properties

The strictly positive integral (1) defines a scalar product in $\mathbb{C}[\boldsymbol{x}]$ by

$$(\phi,\psi) = \mathscr{I}^n[\phi,\bar{\psi}] = \int_\Omega \phi(\boldsymbol{x})\overline{\psi(\boldsymbol{x})}\omega(\boldsymbol{x})\,\mathrm{d}\boldsymbol{x}, \quad \phi,\psi \in \mathbb{C}[\boldsymbol{x}]. \tag{9}$$

Consider the polynomial

$$P_{k+1}(\boldsymbol{x}) = g_{k+1}(\boldsymbol{x}) + \sum_{i=1}^{\kappa} a_i\varphi_i(\boldsymbol{x}), \quad \kappa = M(n,k), \tag{10}$$

where $g_{k+1}(\boldsymbol{x})$ is a given homogeneous component of degree $k+1$ and the a_i's are unknown coefficients. Assuming (10) to be orthogonal to $\varphi_j(\boldsymbol{x})$, $j = 1,2,\ldots,\kappa$, with respect to (9), we obtain

$$\sum_{i=1}^{\kappa} a_i(\varphi_i,\varphi_j) = -(g_{k+1}(\boldsymbol{x}),\varphi_j), \quad j = 1,2,\ldots,\kappa. \tag{11}$$

The matrix of this system is the Gram matrix of the linearly independent polynomials $\varphi_1(\boldsymbol{x})$, $\varphi_2(\boldsymbol{x}),\ldots,\varphi_\kappa(\boldsymbol{x})$. Hence, the a_i are uniquely determined. The polynomial (10) is uniquely determined by its homogeneous component of degree $k+1$ and by orthogonality to all polynomials of degree $\leqslant k$.

We state some simple properties of orthogonal polynomials, which will be of use later.

Theorem 2. *The following properties hold for an orthogonal polynomial* P_{k+1}.

(1) *If the homogeneous component of degree* $k+1$ *has real coefficients, then all coefficients are real. This follows from* (11).

(2) *A real polynomial* P_{k+1} *changes sign in* Ω. *In particular,*

$$\{\boldsymbol{x} \in \Omega\colon P_{k+1} > 0\} \quad \text{and} \quad \{\boldsymbol{x} \in \Omega\colon P_{k+1} < 0\}$$

are of positive measure, which follows from

$$\int_\Omega P_{k+1}(\boldsymbol{x})\omega(\boldsymbol{x})\,\mathrm{d}\boldsymbol{x} = 0.$$

(3) *Whenever* $P_{k+1} = UV$ *with polynomials* U *and* V *of degree at least* 1, *then* U *is orthogonal with respect to* Ω *and the weight function* $\omega(\boldsymbol{x})|V(\boldsymbol{x})|^2$. *This implies Properties* (1) *and* (2) *for the factors as well.*
(4) *If the coefficients belonging to the highest-degree terms in the homogeneous component of a factor are real, then the remaining coefficients are real, too.*
(5) *A real factor* U *of an orthogonal polynomial changes sign in* Ω. *In particular,*

$$\{\boldsymbol{x} \in \Omega\colon U > 0\} \quad \text{and} \quad \{\boldsymbol{x} \in \Omega\colon U < 0\}$$

are of positive measure. From this we obtain
(6) *An orthogonal polynomial has no real multiple factors.*

We normalise the orthogonal polynomials of degree k to

$$P_k = \boldsymbol{x}^k + Q_k, \quad \boldsymbol{k} \in \mathbb{N}^n, \ |\boldsymbol{k}| = k, \ \deg(Q_k) \leqslant k - 1.$$

This fundamental system of degree k will be gathered in a polynomial vector of dimension $M(n-1,k)$ and be written as $\boldsymbol{P}_k$. We refer to the common zeros of all P_k as zeros of $\boldsymbol{P}_k$. The known explicit expressions for these normalised orthogonal polynomials are collected in [10].

4.2. Recursion formulae

For $n=2$ the following results were found. Jackson [37] discusses a three-term recursion formula for a given orthogonal system, Gröbner [27] generates orthogonal systems by solving a variational problem under constraints; Krall and Sheffer [42] study in a class of second-order differential equations special cases the polynomial solutions of which generate classical orthogonal systems. Since their approach is closely related to recursion formulae and leads to concrete results we will outline the main ideas.

Let

$$P_j^k = x^{k-j}y^j + Q_j, \quad Q_j \in \mathbb{P}_{k-1}, \ j = 0, 1, \ldots, k, \ k \in \mathbb{N},$$

be a basis of $\mathbb{P}$. We can collect these fundamental systems of degree k in vectors

$$\boldsymbol{P}_k = (P_0^k, P_1^k, \ldots, P_k^k)^{\mathrm{T}}.$$

The basis $\boldsymbol{P}_k$, $k \in \mathbb{N}$, is said to be a *weak orthogonal system* if there exist matrices

$$C_k, \bar{C}_k \in \mathbb{R}^{k+1\times k+1} \quad \text{and} \quad D_k, \bar{D}_k \in \mathbb{R}^{k+1\times k},$$

such that

$$\begin{aligned} x\boldsymbol{P}_k &= L_{k+1}\boldsymbol{P}_{k+1} + C_k\boldsymbol{P}_k + D_k\boldsymbol{P}_{k-1}, \\ y\boldsymbol{P}_k &= F_{k+1}\boldsymbol{P}_{k+1} + \bar{C}_k\boldsymbol{P}_k + \bar{D}_k\boldsymbol{P}_{k-1}, \end{aligned} \tag{12}$$

with shift matrices L_{k+1} and F_{k+1} defined by $[E_k \ 0]$ and $[0 \ E_k]$, where E_k is the identity in $\mathbb{R}^{k+1\times k+1}$ and $\boldsymbol{P}_{-1} = 0$.

A polynomial basis is said to be orthogonal with respect to a linear functional $\mathscr{L} : \mathbb{P} \to \mathbb{R}$, if, for each $k \in \mathbb{N}$, $\mathscr{L}[\boldsymbol{P}_k\boldsymbol{P}_l^{\mathrm{T}}]=0$, $l=0,1,\ldots,k-1$, and if $\operatorname{rank}(\mathscr{L}[\boldsymbol{P}_k\boldsymbol{P}_k^{\mathrm{T}}])=k+1$. Here, $\boldsymbol{P}_k\boldsymbol{P}_l^{\mathrm{T}}$ is the tensor product of the vectors $\boldsymbol{P}_k$ and $\boldsymbol{P}_l$, and $\mathscr{L}[\boldsymbol{P}_k\boldsymbol{P}_l^{\mathrm{T}}]$ is the matrix whose elements are determined by the functional acting on the polynomial coefficients of the tensor product. The matrix $M_k = \mathscr{L}[\boldsymbol{P}_k\boldsymbol{P}_k^{\mathrm{T}}]$ is known as the kth moment matrix. The basis $\{\boldsymbol{P}_k\}_{k\in\mathbb{N}}$ is said to be a (*positive*) definite orthogonal system in case the matrices M_k, $k \in \mathbb{N}$, are (positive) definite.

A definite system $\{\boldsymbol{P}_k\}_{k\in\mathbb{N}}$ is a weak orthogonal system, i.e., it satisfies the recurrence relations (12). Conversely, it follows from work by Xu [93] that a weak orthogonal system is an orthogonal system with respect to some $\mathscr{L}$ if and only if

$$\operatorname{rank}(S_k) = k+1 \quad \text{where } S_k = [D_k \quad \bar{D}_k] \in \mathbb{R}^{k+1\times 2k}.$$

The associated moment problem consists in assigning a measure to the functional $\mathscr{L}$ defined by a definite system. In particular, assigning a positive measure in case the system is positive definite (Favard's theorem) is quite complicated. We refer to Fuglede [24] and Xu [95].

Krall and Sheffer [42] studied the orthogonal polynomial systems which are generated by the following second-order differential equation:

$$\mathscr{D}\omega = -\lambda_k \omega, \quad \lambda_k \in \mathbb{R}, \ k \in \mathbb{N}, \tag{13}$$

where

$$\mathscr{D}\omega = (ax^2 + d_1x + e_1y + f_1)\frac{\partial^2\omega}{\partial x^2} + (2axy + d_2x + e_2y + f_2)\frac{\partial^2\omega}{\partial x\partial y}$$
$$+ (ay^2 + d_3x + e_3y + f_3)\frac{\partial^2\omega}{\partial y^2} + (gx + h_1)\frac{\partial\omega}{\partial x} + (gy + h_2)\frac{\partial\omega}{\partial y}$$

for some real constants $a \neq 0, g, d_i, e_i, f_i, h_i$, and for

$$\lambda_k = -k((k-1)a + g), \quad g + ka \neq 0, \ k \in \mathbb{N}.$$

They determined all weak orthogonal systems which are generated from (13) and proved that they are definite or positive definite, finding the classical orthogonal systems which had been derived in [2] and some new definite systems.

In [4] the recursion formulae for all positive-definite systems have been computed in the following way. Let $\{\boldsymbol{P}_k\}_{k=0,1,\dots}$ be a definite orthogonal system with respect to $\mathscr{I}$. Multiplying (12) by $\boldsymbol{P}_{k-1}^{\mathrm{T}}$, $\boldsymbol{P}_k^{\mathrm{T}}$, and $\boldsymbol{P}_{k+1}^{\mathrm{T}}$, respectively, and applying $\mathscr{I}$, we obtain

$$C_kM_k = \mathscr{I}[x\boldsymbol{P}_k\boldsymbol{P}_k^{\mathrm{T}}], \quad D_kM_{k-1} = \mathscr{I}[x\boldsymbol{P}_k\boldsymbol{P}_{k-1}^{\mathrm{T}}] = M_kL_k^{\mathrm{T}},$$
$$\bar{C}_kM_k = \mathscr{I}[y\boldsymbol{P}_k\boldsymbol{P}_k^{\mathrm{T}}], \quad \bar{D}_kM_{k-1} = \mathscr{I}[y\boldsymbol{P}_k\boldsymbol{P}_{k-1}^{\mathrm{T}}] = M_kF_k^{\mathrm{T}}.$$

By means of these identities the moment matrices can be computed by induction. Indeed, let $G_k = \operatorname{diag}\{[2, E_{k-2}]\}$ and $\bar{G}_k = \operatorname{diag}\{[E_{k-2}, 2]\}$; then

$$2E_k = L_k^{\mathrm{T}}G_kL_k + F_k^{\mathrm{T}}\bar{G}_kF_k,$$

and consequently,

$$2M_k = M_kL_k^{\mathrm{T}}G_kL_k + M_kF_k^{\mathrm{T}}\bar{G}_kF_k = D_kM_{k-1}G_kL_k + \bar{D}_kM_{k-1}\bar{G}_kF_k.$$

If one sets $M_0 = 1$, the last equation allows us to compute M_k from M_{k-1}, $k \in \mathbb{N}$. Based on [2], Verlinden [91] has computed explicit recursion formulae for classical two-dimensional integrals, too. So we refer to [91,4], if explicit recursion formulae are needed for standard integrals.

Not all two-dimensional orthogonal systems of interest can be obtained from (13). For further systems we refer to Koornwinder [38] and the references given there.

Kowalski in [39] presented a n-dimensional recursion formula and characterised it in [40,41]; Xu [93] refined this characterisation by dropping one condition. We will briefly outline these results. For a more complete insight into this development of a general theory of orthogonal polynomials in n dimensions we refer to the excellent survey by Xu [98].

Let $\mathscr{I}^n$ be given, and let

$$M_k = \mathscr{I}^n[\boldsymbol{P}_k\boldsymbol{P}_k^{\mathrm{T}}] \in \mathbb{R}^{M(n-1,k)\times M(n-1,k)}$$

be the moment matrix for $\mathscr{I}^n$. Then the recursion formula can be stated as follows.

Theorem 3. *For $k=0,1,\ldots$ there are matrices*

$$A_{k,i} \in \mathbb{R}^{M(n-1,k)\times M(n-1,k+1)}, \quad B_{k,i} \in \mathbb{R}^{M(n-1,k)\times M(n-1,k)},$$

and

$$C_{k,i} \in \mathbb{R}^{M(n-1,k)\times M(n-1,k-1)},$$

such that

$$x_i\boldsymbol{P}_k = A_{k,i}\boldsymbol{P}_{k+1} + B_{k,i}\boldsymbol{P}_k + C_{k,i}\boldsymbol{P}_{k-1}, \quad i=1,2,\ldots,n, \ k=0,1,\ldots,$$

where $\boldsymbol{P}_{-1}=0$ and for all $i=1,2,\ldots,n$ and all k

$$A_{k,i}M_{k+1} = \mathscr{I}^n[x_i\boldsymbol{P}_k\boldsymbol{P}_{k+1}^{\mathrm{T}}],$$

$$B_{k,i}M_k = \mathscr{I}^n[x_i\boldsymbol{P}_k\boldsymbol{P}_k^{\mathrm{T}}],$$

$$A_{k,i}M_{k+1} = M_kC_{k+1,i}^{\mathrm{T}}.$$

Furthermore, there are matrices

$$D_{k,i}, G_k \in \mathbb{R}^{M(n-1,k+1)\times M(n-1,k)}, \quad H_k \in \mathbb{R}^{M(n-1,k+1)\times M(n-1,k)}$$

such that

$$\boldsymbol{P}_{k+1} = \sum_{i=1}^{n} x_iD_{k,i}\boldsymbol{P}_k + G_k\boldsymbol{P}_k + H_k\boldsymbol{P}_{k-1},$$

where

$$\sum_{i=1}^{n} D_{k,i}A_{k,i} = E_{M(n-1,k+1)\times M(n-1,k+1)}$$

and

$$\sum_{i=1}^{n} D_{k,i}B_{k,i} = -G_k, \quad \sum_{i=1}^{n} D_{k,i}C_{k,i} = -H_k.$$

We will denote the fundamental set of orthonormal polynomials (with respect to $\mathscr{I}^n$) of degree k by $\boldsymbol{p}_k$. The recursion for orthonormal polynomials is given by Xu [95]. We reuse the notations $A_{k,i}$, $B_{k,i}$. In the following, these matrices will refer to the recursion for orthonormal matrices.

Theorem 4. *For $k=0,1,\ldots$ there are matrices*

$$A_{k,i}\in\mathbb{R}^{M(n-1,k)\times M(n-1,k+1)},\quad B_{k,i}\in\mathbb{R}^{M(n-1,k)\times M(n-1,k)}$$

such that

$$x_i\boldsymbol{p}_k=A_{k,i}\boldsymbol{p}_{k+1}+B_{k,i}\boldsymbol{p}_k+A_{k-1,i}^{\mathrm{T}}\boldsymbol{p}_{k-1},\quad i=1,2,\ldots,n,\ k=0,1,\ldots,$$

where $\boldsymbol{p}_{-1}=0$, $A_{-1,i}=0$ and

$$\operatorname{rank}(A_k)=\operatorname{rank}([A_{k,1}^{\mathrm{T}}|A_{k,2}^{\mathrm{T}}|\cdots|A_{k,n}^{\mathrm{T}}]^{\mathrm{T}})=M(n-1,k+1).$$

For $i,j=1,2,\ldots,n$, $i\neq j$, and $k\geqslant 0$, the following matrix equations hold for the coefficient matrices:

(i) $A_{k,i}A_{k+1,j}=A_{k,j}A_{k+1,i}$,

(ii) $A_{k,i}B_{k+1,j}+B_{k,i}A_{k,j}=B_{k,j}A_{k,i}+A_{k,j}B_{k+1,i}$,

(iii) $A_{k-1,i}^{\mathrm{T}}A_{k-1,j}+B_{k,i}B_{k,j}+A_{k,i}A_{k,j}^{\mathrm{T}}=A_{k-1,j}^{\mathrm{T}}A_{k-1,i}+B_{k,j}B_{k,i}+A_{k,j}A_{k,i}^{\mathrm{T}}$.

In order to characterise Gaussian cubature formulae, see Section 7.1.4; the use of orthonormal systems gives more insight and often is easier to apply.

4.3. Common zeros

A direct analog of the Gaussian approach for $n\geqslant 2$ suggests considering the common zeros of all orthogonal polynomials of degree k as nodes of a formula of degree $2k-1$. So the behaviour of common zeros of all orthogonal polynomials of degree k is of interest.

The following theorem, due to Mysovskikh [60,66], holds for (not necessarily orthogonal or real) fundamental systems of polynomials; it turned out to be essential.

Theorem 5. *Let*

$$R_{\boldsymbol{m}}=\boldsymbol{x}^{\boldsymbol{m}}+Q_{\boldsymbol{m}},\quad \deg(Q_{\boldsymbol{m}})\leqslant m-1,\ |\boldsymbol{m}|=m,$$

be a fundamental system of degree m. Then the following is true.

(i) *The polynomials $R_{\boldsymbol{m}}$ have at most* $\dim\mathbb{P}_{m-1}^n$ *common zeros.*

(ii) *No polynomial of degree $m-1$ vanishes at the common zeros of the $R_{\boldsymbol{m}}$, if and only if the $R_{\boldsymbol{m}}$ have exactly* $\dim\mathbb{P}_{m-1}^n$ *common pairwise distinct zeros.*

We will briefly derive the main properties of the zeros of fundamental systems of orthogonal polynomials. Orthonormalising the monomials $\{\varphi_j(\boldsymbol{x})\}_{j=1}^{\infty}$ with respect to $\mathscr{I}^n$, e.g., by the Gram–Schmidt procedure, we obtain

$$\{F_j(\boldsymbol{x})\}_{j=1}^{\infty}\quad\text{where } \mathscr{I}^n[F_iF_j]=\delta_{ij}.$$

The reproducing kernel in $\mathbb{P}_m^n$ is a polynomial in $2n$ variables,

$$K_m(\boldsymbol{u},\boldsymbol{x})=\sum_{j=1}^{\mu}F_j(\boldsymbol{u})F_j(\boldsymbol{x}),\quad \mu=M(n,m), \tag{14}$$

having the property

$$\mathscr{I}^n[K_m(\boldsymbol{u},\boldsymbol{x})f(\boldsymbol{x})]=f(\boldsymbol{u})\quad\text{for all } f\in\mathbb{P}_m^n. \tag{15}$$

Lemma 6. *For an* $\boldsymbol{a} \in \mathbb{C}^n$ *let* l *be a linear polynomial such that* $l(\boldsymbol{a}) = 0$. *Then* $R = l(\boldsymbol{x})K_m(\boldsymbol{a}, \boldsymbol{x})$ *is quasi-orthogonal. Whenever* $\boldsymbol{a}$ *is a common zero of* $\boldsymbol{P}_{m+1}$, *then* R *is orthogonal.*

Proof. If $Q \in \mathbb{P}^n_{m-1}$ we obtain by (15)

$$\mathscr{I}^n[l(\boldsymbol{x})K_m(\boldsymbol{a}, \boldsymbol{x})Q(\boldsymbol{x})] = l(\boldsymbol{a})Q(\boldsymbol{a}) = 0.$$

If $\boldsymbol{a}$ is a common zero of $\boldsymbol{P}_{m+1}$, then

$$F_{M(n-1,m)+i}(\boldsymbol{a}) = 0, \quad i = 1, 2, \ldots, M(n-1, m+1),$$

and thus $K_m(\boldsymbol{a}, \boldsymbol{x}) = K_{m+1}(\boldsymbol{a}, \boldsymbol{x})$. Hence

$$R(\boldsymbol{x}) = l(\boldsymbol{x})K_{m+1}(\boldsymbol{a}, \boldsymbol{x}) = l(\boldsymbol{x})K_m(\boldsymbol{a}, \boldsymbol{x})$$

is orthogonal to $\mathbb{P}^n_m$. Assuming $\deg(l(\boldsymbol{x})K_m(\boldsymbol{a}, \boldsymbol{x})) \leqslant m$, we obtain that R is zero, in contradiction to

$$K_m(\boldsymbol{a}, \bar{\boldsymbol{a}}) = \sum_{i=1}^{\mu} |F_j(\boldsymbol{a})^2| > 0, \quad \mu = M(n, m). \qquad \square$$

The following theorem was proved in [61,65].

Theorem 7. *The zeros of* $\boldsymbol{P}_{m+1}$ *are real and simple, and they belong to the interior of the convex hull of* Ω. *Furthermore,* $\boldsymbol{P}_{m+1}$ *and* $\boldsymbol{P}_m$ *have no zeros in common.*

Proof. Let $\boldsymbol{a} \in \mathbb{C}^n$ be a common zero of $\boldsymbol{P}_{m+1}$. By Lemma 6 the polynomials

$$(x_i - a_i)K_m(\boldsymbol{a}, \boldsymbol{x}), \quad i = 1, 2, \ldots, n, \tag{16}$$

are orthogonal to all polynomials of degree m. Because of property (3) in Theorem 2, the linear factor $x_i - a_i$ is real, hence $\boldsymbol{a} \in \mathbb{R}^n$. The Jacobian matrix of (16) in $\boldsymbol{a}$ is diagonal with elements $K_m(\boldsymbol{a}, \boldsymbol{a}) > 0$. This implies that $\boldsymbol{a}$ is simple. If $\boldsymbol{a}$ is not an interior point of the convex hull of Ω, there is a separating hyperplane $l(\boldsymbol{x})$ through $\boldsymbol{a}$, e.g., $l(\boldsymbol{x}) \geqslant 0$ for all $\boldsymbol{x}$ in the interior of the convex hull. Since $l(\boldsymbol{x})$ is a real factor of $l(\boldsymbol{x})K_m(\boldsymbol{a}, \boldsymbol{x})$, this is a contradiction to property (5) in Theorem 2. Finally, $\boldsymbol{a}$ is no common zero of $\boldsymbol{P}_m$ since by Lemma 6 the degree of (16) is $m + 1$, hence $\deg(K_m(\boldsymbol{a}, \boldsymbol{x})) = m$. $\square$

Using the matrices presented in Theorem 4, Xu [97] defines infinite tridiagonal block Jacobi matrices of the form

$$T_i = \begin{bmatrix} B_{0,i} & A_{0,i} & 0 & 0 & \ldots & 0 \\ A_{0,i}^{\mathrm{T}} & B_{1,i} & A_{1,i} & 0 & \ldots & 0 \\ 0 & A_{1,i}^{\mathrm{T}} & B_{1,i} & A_{2,i} & \ldots & 0 \\ \vdots & \vdots & \vdots & \vdots & \vdots & \vdots \end{bmatrix}, \quad i = 1, 2, \ldots, n,$$

and truncated versions of these. He found a relation between an eigenvalue problem for these matrices and the zeros of all orthogonal polynomials of a fixed degree. These results and their relation to cubature formulae are further elaborated in [94].

5. Lower bounds

5.1. Numerical characteristics

Mysovskikh [56] proved that Radon's formulae are minimal whenever (8) holds. In [67] an integral is constructed where the matrix in (7) is zero and a formula of degree 5 with 6 nodes can be constructed. Applying Radon's method to formulae of degree 3, Mysovskikh [59] found that such formulae with 4 nodes exist if and only if

$$\mathscr{I}[P_0^2 P_2^2 - P_1^2 P_1^2] \neq 0; \tag{17}$$

otherwise the formula has only 3 nodes. This has been further studied by Günther [28,29]. Fritsch [23] gave an example of an integral for which $\mathscr{Q}(n, 3, n+1)$ exists. Cernicina [9] constructed a region in $\mathbb{R}^n$, $3 \leqslant n \leqslant 8$, admitting minimal formulae of type $\mathscr{Q}(n, 4, (n+1)(n+2)/2)$; for $n = 2$ a formula $\mathscr{Q}(2, 5, 6)$ is obtained. Stroud [90] extended (6) in the following way:

$$B = [\mathscr{I}[P_i^k P_j^k - P_\nu^k P_\mu^k]]_{i+j=\nu+\mu,\ \mu\neq i\neq\nu,\ 0\leqslant i,j,\mu,\nu\leqslant k}, \tag{18}$$

in order to obtain the lower bound in Theorem 10.

Mysovskikh [60] generalised (5) and (6) (in order to study the case which Radon did not further pursue) by defining for given k and $\mathscr{I}$ the following matrices:

$$M_{k-1}^{\star} = [\mathscr{I}[P_{j+1}^k P_i^k - P_j^k P_{i+1}^k]]_{i,j=0,1,\ldots,k-1}, \tag{19}$$

– note that $M_k^{\star}$ is skew-symmetric – and

$$A = \tfrac{1}{2}[\mathscr{I}[P_{i+1}^k P_{j-1}^k - 2P_i^k P_j^k - P_{i-1}^k P_{j+1}^k]]_{i,j=1,2,\ldots,k-1}.$$

The elements of these matrices characterise the behaviour of the orthogonal polynomials with respect to $\mathscr{I}$; so they were called *numerical characteristics*.

Theorem 8. *The following are equivalent*:

(i) *the matrix A vanishes,*
(ii) *the matrix $M_{k-1}^{\star}$ vanishes,*
(iii) *the orthogonal basis of degree k has $k(k+1)/2$ common pairwise distinct real zeros,*
(iv) *a cubature formula of degree $2k-1$ with the lowest possible number of nodes exists. Its nodes are the common zeros of the orthogonal basis of degree k.*

The proof in [66, p. 189], is based on Theorem 5 and the following considerations. The polynomials

$$Q_i = yP_i^k - xP_{i+1}^k, \quad i = 0, 1, \ldots, k-1, \tag{20}$$

are of degree k; this is Radon's equation (5). Hence if the common zeros of all P_i^k are the nodes of a formula of degree $2k-1$, then the Q_i are orthogonal polynomials of degree k. This implies $M_{k-1}^{\star} = 0$. Evidently, $M_{k-1}^{\star} = 0$ implies $A = 0$. On the other hand, if $A = 0$, then $M_{k-1}^{\star}$ is of Hankel type, and, since $M_k^{\star}$ is skew-symmetric, this implies $M_{k-1}^{\star} = 0$. The existence of integrals admitting the conditions of Theorem 8 was studied by Kuzmenkov in [44–46].

The articles based on Mysovskikh's results prefer to work with $M_{k-1}^{\star}$, and it turns out that this matrix in many ways characterises the behaviour of the associated orthogonal polynomials.

In order to generalise Theorem 8 to n dimensions, Eq. (20) has to be studied for all possible variables. By means of the recursion formulae for orthonormal systems, n-dimensional numerical characteristics can be defined. By condition (ii) in Theorem 4,

$$A_{k-1,j}x_i\boldsymbol{p}_k - A_{k-1,i}x_j\boldsymbol{p}_k$$

can be computed under the condition that these polynomials are orthogonal to $\mathbb{P}^n_{k-1}$. This leads to the matrices

$$M^\star_{k-1}(i,j) = A_{k-1,i}A^{\mathrm{T}}_{k-1,j} - A_{k-1,j}A^{\mathrm{T}}_{k-1,i}, \quad i,j = 1,2,\ldots,n, \ i \neq j, \tag{21}$$

which are representing the numerical characteristics of orthogonal polynomials in n dimensions.

5.2. Lower bounds

To settle the question of minimal formulae, lower bounds for the number of nodes are needed. The one-dimensional result can be generalised directly to find the following result, which seems to be folklore.

Theorem 9. *If $\mathcal{Q}(n,m,N)$ is a cubature sum for an integral $\mathcal{I}^n$, then*

$$N \geqslant \dim \mathbb{P}^n_{\lfloor m/2 \rfloor} = M(n, \lfloor m/2 \rfloor). \tag{22}$$

As we have seen in Section 3, this lower bound is not sharp for $n = 2$ and $m = 5$. A simple consequence of Theorem 8 is

Theorem 10. *If $\mathcal{Q}(2,2k-1,N)$ is a cubature sum for an integral $\mathcal{I}$ for which* $\mathrm{rank}(M^\star_{k-1}) > 0$, *then*

$$N \geqslant \dim \mathbb{P}_{k-1} + 1.$$

Stroud [90] showed this under the condition that B in (18) does not vanish.

Considerable progress was made by Möller [49], who improved the lower bound for $n = 2$.

Theorem 11. *If $\mathcal{Q}(2,2k-1,N)$ is a cubature sum for an integral $\mathcal{I}$, then*

$$N \geqslant \dim \mathbb{P}_{k-1} + \tfrac{1}{2}\mathrm{rank}(M^\star_{k-1}). \tag{23}$$

Proof. If a cubature sum $\mathcal{Q}(2,2k-1,N)$ is given, then no polynomial in $\mathbb{P}_{k-1}$ vanishes at all nodes. If no polynomial of degree k vanishes at all nodes, then $N \geqslant \dim \mathbb{P}_k > \dim \mathbb{P}_{k-1} + \frac{1}{2}\mathrm{rank}(M^\star_{k-1})$, since $M^\star_{k-1} \in \mathbb{R}^{k\times k}$. So Möller assumed the existence of s orthogonal polynomials Q_i of degree k which vanish at the nodes and first searched for a bound on s. Note that Q_i, xQ_i, yQ_i belong to an ideal which does not contain any polynomial of $\mathbb{P}_{k-1}$. Let

$$\mathcal{W} = \mathrm{span}\{Q_i, xQ_i, yQ_i, \ i = 1,2,\ldots,s\};$$

then

$$3s - \eta = \dim \mathcal{W} \leqslant k + 2 + s,$$

where η is the number of those xQ_i, yQ_i which can be dropped without diminishing the dimension of $\mathscr{W}$. These dependencies in $\mathscr{W}$ are of the form

$$x\sum_{i=1}^{k} a_i P_i^k - y\sum_{i=0}^{k-1} a_{i+1} P_i^k = \sum_{i=0}^{k} b_i P_i^k.$$

By orthogonality this leads to

$$\sum_{i=0}^{k-1} a_{i+1}\mathscr{I}[P_j^k P_{i+1}^k - P_{j+1}^k P_i^k] = 0, \quad j = 0, 1, \ldots, k-1,$$

i.e.,

$$M_{k-1}^{\star}(a_1, a_2, \ldots, a_k)^{\mathrm{T}} = 0.$$

Thus we get $\eta \leqslant k - \operatorname{rank}(M_{k-1}^{\star})$, from which we finally obtain

$$s \leqslant k + 1 - \tfrac{1}{2}\operatorname{rank}(M_{k-1}^{\star}). \qquad \square$$

For all classes of integrals for which the rank of $M_{k-1}^{\star}$ has been computed, it turned out that either the rank is zero (cf. [82]) or

$$\operatorname{rank}(M_{k-1}^{\star}) = 2\lfloor k/2 \rfloor.$$

Classes of integrals for which the second rank condition holds have been already given by Möller [49]. He showed this for product integrals and integrals enjoying central symmetry. This includes the standard regions $C_2, S_2, H_2, E_2^{r^2}$ and E_2^r. Further classes with the same rank were detected by Rasputin [73], Berens and Schmid [3]. These include the standard region T_2.

Another important fact was observed by Möller. If (23) is attained, then the polynomials xQ_i, yQ_i form a fundamental set of degree $k+1$.

The improved lower bound, in general, is not sharp. Based on a characterisation of cubature sums $\mathscr{Q}(2, 4k+1, 2(k+1)^2 - 1)$ for circularly symmetric integrals in [92], it was shown in [16] that for all $k \in \mathbb{N} \setminus \{1\}$ the integrals

$$\int_{\mathbb{R}^2} f(x,y)(x^2+y^2)^{\alpha-1} \mathrm{e}^{-x^2-y^2}\, \mathrm{d}x\, \mathrm{d}y, \quad \alpha > 0,$$

and for $\alpha, \beta > -1$ the integrals

$$\int_{S_2} f(x,y)(x^2+y^2)^{\alpha}(1-x^2-y^2)^{\beta}\, \mathrm{d}x\, \mathrm{d}y$$

admit cubature sums $\mathscr{Q}(2, 4k+1, N)$ where at least $N \geqslant 2(k+1)^2$. Note that this includes the standard regions S_2 and $E_2^{r^2}$. This result can however not be generalised to all circularly symmetric integrals. In [92] the existence of a circularly symmetric integral admitting a cubature sum $\mathscr{Q}(2, 9, 17)$ has been proven.

The n-dimensional version of Theorem 11 was stated in [51]. An explicit form of the matrices involved was given in [97], using (21), which allows us to formulate (23) as follows.

Theorem 12. *If $\mathscr{Q}(n, 2k-1, N)$ is a cubature sum for an integral $\mathscr{I}^n$, then*

$$N \geqslant \dim \mathbb{P}_{k-1}^n + \tfrac{1}{2}\max\{\operatorname{rank}(M_{k-1}^{\star}(i,j)) \colon i, j = 1, 2, \ldots, n\}. \tag{24}$$

Let us denote by $\mathscr{G}_{2m}$ the linear space of all even polynomials in $\mathbb{P}^n_{2m}$ and by $\mathscr{G}_{2m-1}$ the linear space of all odd polynomials in $\mathbb{P}^n_{2m-1}$. For integrals $\mathscr{I}^n$ which are centrally symmetric, i.e., for which

$$\mathscr{I}^n[Q] = 0 \quad \text{if } Q \in \mathscr{G}_{2m-1}, \quad m \in \mathbb{N},$$

holds, another lower bound is known, which is not based on orthogonality.

Theorem 13. *If $\mathscr{Q}(n, 2k-1, N)$ is a cubature sum for a centrally symmetric integral $\mathscr{I}^n$, then*

$$N \geqslant 2 \dim \mathscr{G}_{k-1} - \begin{cases} 1, & \textit{if } 0 \textit{ is a node and } k \textit{ is even}, \\ 0, & \textit{else}. \end{cases}$$

This bound was given for degree 3 by Mysovskikh [55]; the general case is due to Möller [47,51] and Mysovskikh [64]. Möller proved that cubature formulae attaining the bound of Theorem 13 (having the node 0, if k is even) are centrally symmetric, too. For $n \geqslant 3$ and $\mathscr{I}^n$ centrally symmetric, the bound of Theorem 13 is better than the one of Theorem 12. For $n=2$ and $\mathscr{I}^n$ centrally symmetric, they coincide.

To conclude this section, we remark that cubature formulae with all nodes real and attaining the bounds of Theorem 9 or Theorem 13 are known to have all weights positive [58,90,47,12].

6. Methods of construction

6.1. Interpolation

Let $\Omega \subseteq \mathbb{R}^n$ be given and assume that Ω contains interior points. By virtue of the linear independence of $\{\varphi_j(\boldsymbol{x})\}_{j=1}^{\infty}$ we can find for each m exactly $\mu = M(n,m)$ points from Ω such that they generate a regular Vandermonde matrix. We remark that ν points, $\nu < \mu$, are always contained in an algebraic manifold of degree m, hence μ is the minimal number of points which do not belong to such a manifold. We denote by

$$V_m = [\varphi_1(\boldsymbol{x}^{(j)}), \varphi_2(\boldsymbol{x}^{(j)}), \ldots, \varphi_\mu(\boldsymbol{x}^{(j)})]_{j=1}^{\mu}, \quad \mu = M(n,m), \tag{25}$$

the Vandermonde matrix defined by $\boldsymbol{x}^{(j)}$, $j = 1, 2, \ldots, \mu$.

Theorem 14. *The points $\boldsymbol{x}^{(j)}$, $j = 1, 2, \ldots, \mu$, do not lie on an algebraic manifold of degree m if and only if* $\det V_m \neq 0$.

A natural way to construct a cubature formula is interpolation. Choose μ points $\boldsymbol{x}^{(j)} \in \mathbb{R}^n$ which do not lie on a manifold of degree m. Because of the nonsingularity of the corresponding Vandermonde matrix we can construct the interpolating polynomial of f:

$$P_m(\boldsymbol{x}) = \sum_{j=1}^{\mu} L_j^{(m)}(\boldsymbol{x}) f(\boldsymbol{x}^{(j)}),$$

where

$$L_j^{(m)}(\boldsymbol{x}^{(i)}) = \delta_{ij}, \quad i,j = 1,2,\ldots,\mu.$$

Substituting P_m for f in (1), we obtain

$$\mathscr{I}[f] = \sum_{j=1}^{\mu} w_j f(\boldsymbol{x}^{(j)}) + R[f], \tag{26}$$

where

$$w_j = \int_{\Omega} L_j^{(m)}(\boldsymbol{x})\omega(\boldsymbol{x})\,\mathrm{d}\boldsymbol{x}. \tag{27}$$

A cubature formula obtained in this way is obviously interpolatory, see Theorem 1.

6.2. *Reproducing kernels*

The method of reproducing kernel was introduced in [58] in order to construct cubature formulae of degree $2k$ with a minimal number of nodes $N = M(n,k)$. Most often the method will produce cubature formulae with more nodes. By means of the orthonormal basis such cubature formulae may be constructed by inserting $f = F_l F_m$ in (2) and studying

$$\sum_{j=1}^{N} w_j F_l(\boldsymbol{x}^{(j)}) F_m(\boldsymbol{x}^{(j)}) = \delta_{lm}, \quad l,m = 1,2,\ldots,N. \tag{28}$$

Introducing the $N \times N$ matrices

$$F = [F_1(\boldsymbol{x}^{(j)}), F_2(\boldsymbol{x}^{(j)}),\ldots,F_N(\boldsymbol{x}^{(j)})]_{j=1}^{N}$$

and $C = \mathrm{diag}\{w_1, w_2, \ldots, w_N\}$, we can write Eq. (28) as

$$F^{\mathrm{T}} C F = E.$$

This can be written as $FF^{\mathrm{T}} = C^{-1}$, i.e.,

$$\sum_{i=1}^{N} F_i(\boldsymbol{x}^{(r)}) F_i(\boldsymbol{x}^{(s)}) = w_r^{-1}\delta_{rs}, \quad r,s = 1,2,\ldots,N.$$

If we are using (14), this can be rewritten as

$$K_k(x^{(r)}, x^{(s)}) = w_r^{-1}\delta_{rs}, \quad r,s,=1,2,\ldots,N. \tag{29}$$

If we assume that (28) will lead to a cubature formula, then the nodes and coefficients can be determined by (29).

Let $\boldsymbol{a}^{(i)}$, $i=1,2,\ldots,n$, be pairwise distinct nodes of such a formula. We denote by $\mathscr{H}_i$ the algebraic manifold defined by the polynomial $K_k(\boldsymbol{a}^{(i)}, \boldsymbol{x})$. From (29) we obtain

$$K_k(\boldsymbol{a}^{(i)}, \boldsymbol{a}^{(j)}) = b_i\delta_{ij}, \quad b_i = w_i^{-1}, \; i,j = 1,2,\ldots,n. \tag{30}$$

The remaining nodes of the formula belong to $\bigcap_{i=1}^{n} \mathscr{H}_i$ and can be computed by solving for the unknown variables $\boldsymbol{x}$ from

$$K_k(\boldsymbol{a}^{(i)}, \boldsymbol{x}) = 0, \quad i = 1,2,\ldots,n.$$

Since the nodes of the cubature sum $\mathscr{Q}(n,2k,M(n,k))$ are not known, we proceed in the following way. The nodes $\boldsymbol{a}^{(i)}$ are chosen (whenever possible) in Ω but different from any of the common zeros of the fundamental system of orthogonal polynomials of degree k. So the order of the manifold $\mathscr{H}_i$ generated by $K_k(\boldsymbol{a}^{(i)},\boldsymbol{x})$ is k.

If $\boldsymbol{a}^{(1)}$ is fixed, then $\boldsymbol{a}^{(2)}$ will be chosen on $\mathscr{H}_1$, and, if possible in Ω. If $\boldsymbol{a}^{(i)}$, $i=1,2,\ldots,t-1$, are fixed, the next node $\boldsymbol{a}^{(t)}$ is chosen on $\bigcap_{i=1}^{t-1}\mathscr{H}_i$, if possible in Ω. The $\boldsymbol{a}^{(i)}$, $i=1,2,\ldots,n$, constructed in this way satisfy (30). If all nodes are chosen in $\mathbb{R}^n$, then $b_i>0$; in fact,

$$b_i=\sum_{j=1}^{N}F_j^2(\boldsymbol{a}^{(i)})>0,$$

since the $\boldsymbol{a}^{(i)}$ are no zeros of the fundamental system of orthogonal polynomials of degree k.

If there are no points at infinity on $\mathscr{H}=\bigcap_{i=1}^{n}\mathscr{H}_i$, then $\mathscr{H}$ consists of r points $\boldsymbol{x}^{(j)}$. Thus we obtain

$$\int_\Omega f(\boldsymbol{x})\omega(\boldsymbol{x})\,\mathrm{d}\boldsymbol{x}\cong\mathscr{Q}(n,2k,n+r)=\sum_{j=1}^{n}A_jf(\boldsymbol{a}^{(j)})+\sum_{j=1}^{r}B_jf(\boldsymbol{x}^{(j)}).\tag{31}$$

The coefficients A_i can be computed from (31) since the formula is exact for $K_k(\boldsymbol{a}^{(i)},\boldsymbol{x})$, i.e.,

$$\int_\Omega K_k(\boldsymbol{a}^{(i)},\boldsymbol{x})\omega(\boldsymbol{x})\,\mathrm{d}\boldsymbol{x}=\sum_{j=1}^{n}A_jK_k(\boldsymbol{a}^{(i)},\boldsymbol{a}^{(j)})=A_iK_k(\boldsymbol{a}^{(i)},\boldsymbol{a}^{(i)}),$$

or, by using (15) with $f\equiv 1$,

$$A_i=\frac{1}{b_i}=\frac{1}{K_k(\boldsymbol{a}^{(i)},\boldsymbol{a}^{(i)})}.$$

If $n+r=N=M(n,k)$, the coefficients B_j can be computed in an analogous way; if $n+r>N$, the B_j are determined by the condition for (31) to be of degree $2k$.

The method of reproducing kernels can be applied to regions in $\mathbb{R}^n$ without inner points, see [36,52]. The method was applied in [58,7,8,25] to construct cubature formulae of degree 4 for a variety of regions and in [47] to construct a cubature formula of degree 9 for the region S_2.

Möller [47] and Gegel' [26] proved

Theorem 15. *If* $\boldsymbol{a}^{(i)}$, $i=1,2,\ldots,n$, *satisfy* (30) *where* $b_i\neq 0$, *and if* $\bigcap_{i=1}^{n}\mathscr{H}_i$ *consists of pairwise distinct nodes* $\boldsymbol{x}^{(j)}$, $j=1,2,\ldots,k^n$, *then*

$$\int_\Omega f(\boldsymbol{x})\omega(\boldsymbol{x})\,\mathrm{d}\boldsymbol{x}\cong\mathscr{Q}(m,2k,n+k^n)=\sum_{i=1}^{n}\frac{1}{b_i}f(\boldsymbol{a}^{(i)})+\sum_{j=1}^{k^n}w_jf(\boldsymbol{x}^{(j)}),$$

where $b_i=K_k(\boldsymbol{a}^{(i)},\boldsymbol{a}^{(i)})$.

Möller modified this for centrally symmetric integrals by using the following important observation. For these integrals, the orthogonal polynomials of degree m are even (odd) polynomials, if m is even (odd). For the linear space $\tilde{\mathbb{P}}_k^n$ of all even (odd) polynomials of degree $\leqslant k$ if k is even (odd) the reproducing kernel

$$\tilde{K}_k(\boldsymbol{u},\boldsymbol{x}) = {\sum_{j=t}^{N}}' F_j(\boldsymbol{u})F_j(\boldsymbol{x}), \quad t = k - 2\lfloor k/2 \rfloor + 1, \ N = M(n,k),$$

is considered. Here $\sum'$ denotes summation over all even (odd) polynomials F_j if j is even (odd).

Again, nodes $\boldsymbol{a}^{(i)}$, $i = 1,2,\ldots,n$, are chosen (whenever possible) in Ω but different from any of the common zeros of the fundamental system of orthogonal polynomials of degree k.

The manifolds corresponding to $\tilde{K}_k(\boldsymbol{a}^{(i)},\boldsymbol{x})$ will be denoted by $\tilde{\mathscr{H}}_i$. If $\boldsymbol{a}^{(i)}$, $i = 1,2,\ldots,t-1$, are already selected, the node $\boldsymbol{a}^{(t)}$ is chosen on $\bigcap_{i=1}^{t-1} \tilde{\mathscr{H}}_i$, if possible in Ω. These nodes satisfy

$$\tilde{K}_k(\boldsymbol{a}^{(i)},\boldsymbol{a}^{(j)}) = b_i\delta_{ij}, \quad i,j = 1,2,\ldots,n, \tag{32}$$

where $b_i > 0$ since $\boldsymbol{a}^{(i)} \in \mathbb{R}^n$.

We remark that the nodes in the modified method are chosen as $\boldsymbol{a}^{(i)}, -\boldsymbol{a}^{(i)}$, $i=1,2,\ldots,n$. By central symmetry it follows that

$$\tilde{K}_k(\boldsymbol{a}^{(i)},-\boldsymbol{a}^{(j)}) = (-1)^k\tilde{K}_k(\boldsymbol{a}^{(i)},\boldsymbol{a}^{(j)}),$$

hence by (32), if $b_i \neq 0$, we get $\boldsymbol{a}^{(i)} \neq -\boldsymbol{a}^{(j)}$ if $i \neq j$. So the $\boldsymbol{a}^{(i)}, -\boldsymbol{a}^{(i)}$ are pairwise distinct, if $\boldsymbol{a}^{(i)} \neq 0$, $i = 1,2,\ldots,n$. If k is odd, this is satisfied; if k is even, the number of pairwise distinct nodes $\boldsymbol{a}^{(i)}$ and $-\boldsymbol{a}^{(i)}$ may be $2n$ or $2n-1$. In [47] the following is derived.

Theorem 16. *Let the integral be centrally symmetric. If the nodes $\boldsymbol{a}^{(i)}$, $i = 1,2,\ldots,n$, satisfy* (32) *where $b_i \neq 0$, and $\bigcap_{i=1}^{n} \tilde{H}_i$ consists of pairwise distinct points $\boldsymbol{x}^{(j)}$, $j = 1,2,\ldots,k^n$, then*

$$\int_\Omega f(\boldsymbol{x})\omega(\boldsymbol{x})\,\mathrm{d}\boldsymbol{x} \cong \mathscr{Q}(n,2k+1,2n+k^n)$$

$$= \sum_{i=1}^{n} \frac{1}{2b_i}[f(\boldsymbol{a}^{(i)}) + f(-\boldsymbol{a}^{(i)})] + \sum_{j=1}^{k^n} w_j f(\boldsymbol{x}^{(j)}),$$

where $b_i = \tilde{K}_k(\boldsymbol{a}^{(i)},\boldsymbol{a}^{(i)})$.

6.3. *Ideal theory*

Let

$$X = \{\boldsymbol{x}^{(j)},\ j = 1,2,\ldots,N\} \subset \mathbb{R}^n$$

be a finite set of points, and define the subspace

$$\mathscr{W} = \{P \in \mathbb{P}^n_m\colon P(\boldsymbol{x}) = 0 \text{ for all } \boldsymbol{x} \in X\} \subset \mathbb{P}^n.$$

Sobolev [83] proved

Theorem 17. *The points X are the nodes of $\mathscr{Q}(n,m,N)$ for $\mathscr{I}^n$ if and only if*

$$P \in \mathscr{W} \quad \textit{implies } \mathscr{I}^n[P] = 0.$$

An English rendering of the proof can be found in [10].

The relationship of orthogonal polynomials and cubature formulae was studied since 1967 by Mysovskikh [57,59,61–63] and Stroud [87–90]; in particular they introduced elements from algebraic geometry.

Möller [47] recognised that this connection can be represented more transparently by using ideal theory, and that this theory will help in determining common zeros of orthogonal polynomials. E.g., Theorem 5 follows easily from this theory. Let $\mathcal{Q}(n,m,N)$ be given such that

$$X = \{x^{(j)},\ j=1,2,\dots,N\} \subset \mathbb{R}^n$$

is the finite set of nodes, and define the polynomial ideal

$$\mathfrak{A} = \{P \in \mathbb{P}^n\colon P(x) = 0 \text{ for all } x \in X\} \subset \mathbb{P}^n.$$

Then we obtain for each $P \in \mathfrak{A} \cap \mathbb{P}^n_m$ the orthogonality condition $\mathscr{I}^n[PQ] = 0$, whenever $PQ \in \mathbb{P}^n_m$. Möller introduced the notion of m-orthogonality. A set of polynomials is said to be m-orthogonal if for every element P we have $\mathscr{I}^n[PQ] = 0$, when $PQ \in \mathbb{P}^n_m$. Hence, orthogonal polynomials of degree m are $(2m-1)$-orthogonal, while quasi-orthogonal polynomials of degree m are $(2m-2)$-orthogonal.

The main problem is the selection of a suitable basis. It turns out that an H-basis suits best. $\{P_1, P_2, \dots, P_s\}$ is such a basis, if every $Q \in \mathfrak{A}$ can be written as

$$Q = \sum_{i=1}^{s} Q_i P_i \quad \text{where } \deg(Q_i P_i) \leqslant \deg(Q).$$

The ideal then is written as $\mathfrak{A} = (P_1, P_2, \dots, P_s)$.

Theorem 18. *Let Q_i, $i=1,2,\dots,s$, be an H-basis of a zero-dimensional ideal $\mathfrak{A}$. Then the following are equivalent:*

(i) *The Q_i are m-orthogonal with respect to $\mathscr{I}^n$.*

(ii) *There is a $\mathcal{Q}(n,m,N)$ using the N common zeros of $\mathfrak{A}$ as nodes if no multiple nodes appear.*

For multiple nodes, function derivatives can be used; this was proposed in [62,63]. Möller called formulae of this type *generalised cubature formulae* of algebraic degree. This was further developed in [48–50].

In this ideal-theoretic setting, condition (5) can be interpreted as syzygy. If an H-basis of $\mathfrak{A}$ is fixed, then syzygies of higher order will occur; e.g., if $P_1, P_2 \in \mathfrak{A}$ are of degree m, then it is possible that

$$Q_1 P_1 - Q_2 P_2 \in \mathbb{P}^n_{m-1} \quad \text{for } Q_i \in \mathbb{P}^n.$$

Möller found that such syzygies will occur in an H-basis and that they impose restrictions on the polynomials which can be used constructively to compute a suitable ideal. Furthermore, the Hilbert function can be used to study the number of common zeros. For the connection to Gröbner bases we refer to [53,13].

However, Theorem 18 allows nodes to be in $\mathbb{C}^n$. Schmid [78] proposed to avoid this by considering real ideals.

If the common zeros of an ideal $\mathfrak{A}$ are denoted by $\mathscr{V}(\mathfrak{A})$ and the ideal of all polynomials which vanish at a finite set $X \subset \mathbb{R}^n$ by $\mathfrak{A}_X$, then

$$X \subseteq \mathscr{V}(\mathfrak{A}_X).$$

An ideal $\mathfrak{A}$ is called real if

$$X = \mathscr{V}(\mathfrak{A}_X).$$

So m-orthogonal real ideals characterise cubature formulae, and real ideals are characterised by the following theorem due to Dubois et al. (cf. [18,19,75,76]).

Theorem 19. *The following are equivalent*:
(i) $\mathfrak{A}$ *is a real ideal,*
(ii) *the common zeros of* $\mathfrak{A}$ *are pairwise distinct and real,*
(iii) P *vanishes on* $\mathscr{V}(\mathfrak{A})$ *if and only if* $P \in \mathfrak{A}$,
(iv) *for all* $M \in \mathbb{N}$ *and all* $Q_i \in \mathbb{P}^n$, $i = 1, 2, \ldots, M$,

$$\sum_{i=1}^{M} Q_i^2 \in \mathfrak{A} \text{ implies } Q_i \in \mathfrak{A}, \quad i = 1, 2, \ldots, M.$$

By combining Möller's results and the conditions which can be derived from Theorem 19 it is possible to give a complete characterisation of cubature formulae. However, if the degree of the formula m is fixed, the conditions which have to be satisfied strongly depend on the number of nodes. Indeed, the number of nodes influences the number of polynomials in the ideal basis and their degree. The conditions derived from Theorem 19 depend on the structure of the ideal basis, and their complexity therefore increases with m.

In [79] Theorem 5 is extended by using Theorem 19 and applying it to ideals containing a fundamental set of an arbitrary degree. It was then applied to construct cubature formulae for the regions C_2, S_2, T_2.

Theorem 20. *Let* R_i, $i = 1, 2, \ldots, t$, *be linearly independent polynomials in* $\mathbb{P}_m^n$ *containing a fundamental system of degree* m. *If* $\mathfrak{A} = (R_1, R_2, \ldots, R_t)$, *then*
(i) $\mathscr{V}(\mathfrak{A}) \leqslant N = \dim \mathbb{P}_m^n - t$,
(ii) $\mathscr{V}(\mathfrak{A}) = N$ *if and only if* $\mathfrak{A}$ *is a real ideal.*

6.3.1. *Even-degree formulae*

By applying the syzygies of first order to quasi-orthogonal polynomials it is possible to characterise all even-degree formulae attaining the lower bound in (22).

Theorem 21. *Let*

$$R_i = P_i^k + \sum_{j=0}^{k-1} \gamma_{ij} P_j^{k-1}, \quad i = 0, 1, \ldots, k,$$

be quasi-orthogonal polynomials generating the ideal $\mathfrak{A}$. *A cubature formula* $\mathcal{Q}(2, 2k-2, \dim \mathbb{P}_{k-1})$ *with all weights positive exists if and only if the parameters* γ_{ij} *can be chosen such that*

$$yR_i - xR_{i+1} \in \operatorname{span}\{R_j,\ j = 0, 1, \ldots, k\}, \quad i = 0, 1, \ldots, k-1, \tag{33}$$

holds. If (33) *holds, then the nodes of the formula are given by* $\mathcal{V}(\mathfrak{A})$, *and* $\mathfrak{A}$ *is real.*

Morrow and Patterson [54] proved this by applying Möller's Theorem 18 and using the Hilbert function to count the common zeros, counting multiplicities.

Schmid [77] applied Theorem 19 to prove that (33) is necessary and sufficient for $\mathfrak{A}$ to be a real ideal. From the work in [54] a classical integral is known for which all even-degree minimal formulae can be computed, see Section 7.1.

The complexity of (33) can be realised by considering the equivalent matrix equation given in [81] for integrals having central symmetry. The quadratic matrix equation

$$0 = M_{k-1}^{\star} + \Gamma_k M_k^{-1} M_k^{\star} M_k^{-1} \Gamma_k^{\mathrm{T}}$$

has to be solved. Here $M_k = [\mathcal{I}[P_i^k P_j^k]]_{i,j=0,1,\ldots,k}$ is the moment matrix and $M_{k-1}^{\star}$ the matrix of the numerical characteristics. Γ_k is a $k \times k+1$ Hankel matrix, which has to be determined; from its coefficients the γ_{ij}'s can be computed.

The straightforward generalisation to the n-dimensional case has been studied in [79,74]; however, only moderate-degree formulae could be constructed for C_n, $n = 2, 3, 4, 5$.

6.3.2. *Odd-degree formulae*

Stroud and Mysovskikh [88,59] proved that $\mathcal{Q}(2, 2k-1, k^2)$ can be constructed if two orthogonal polynomials of degree k can be found having exactly k^2 common pairwise distinct real zeros. Franke [21] derived sufficient conditions implying the existence of $\mathcal{Q}(2, 2k-1, N)$, where $N < k^2$, for special integrals over planar regions. Further generalisations were obtained in [62,63] by admitting point evaluations of derivatives and preassigning nodes.

We recall Theorem 8 in the following form.

Theorem 22. $\mathcal{Q}(n, 2k-1, \dim \mathbb{P}_{k-1})$ *exists if and only if the nodes are the zeros of* $\boldsymbol{P}_k$.

For the standard regions, such formulae exist for $n = 1$ and $k = 1, 2, \ldots$ or $k = 1$ and $n = 1, 2, \ldots$; for $n \geqslant 2$, $k \geqslant 2$, such formulae do not exist. The existence of $M(n, k-1)$ common roots of the polynomials gathered in $\boldsymbol{P}_k$ can be reduced to the solution of a nonlinear system in n unknowns; however, the number of equations is larger than n, since for $n, k \geqslant 2$, we have

$$M(n-1, k) \geqslant M(n-1, 2) = \frac{n(n+1)}{2} \geqslant n+1.$$

The existence of special regions for which Theorem 22 holds for moderate k have been discussed in Section 5.2. A class of integrals for which Theorem 22 holds for arbitrary k was presented in [82] for $n = 2$, and in [5] for n arbitrary.

In order to find cubature sums $\mathscr{Q}(2,2k-1,N)$ where N attains the improved lower bound (23), Möller derived the following necessary conditions:

Theorem 23. *If $\mathscr{Q}(2,2k-1,N)$ exists where $N = \dim \mathbb{P}_{k-1} + \frac{1}{2}\mathrm{rank}(M^{\star}_{k-1})$, then there are $s = k + 1 - \frac{1}{2}\mathrm{rank}(M^{\star}_{k-1})$ orthogonal polynomials P_i of degree k vanishing at the nodes of the formula and satisfying the following conditions.*

(i) *Whenever orthogonal polynomials of degree k satisfy $xQ_1 - yQ_2 = Q_3$, then $Q_i \in \mathrm{span}\{P_i,\ i = 1,2,\ldots,s\}$.*

(ii) *xP_i, yP_i form a fundamental set of degree $k+1$.*

(iii) *There are $2k - \frac{3}{2}\mathrm{rank}\, M^{\star}_{k-1}$ linearly independent vectors $a \in \mathbb{R}^{3(k+1)}$ such that*

$$x^2 \sum_{i=0}^{k} a_i P_i + xy \sum_{i=0}^{k} a_{k+1+i} P_i + y^2 \sum_{i=0}^{k} a_{2k+2+i} P_i = \sum_{i=1}^{s} L_i P_i,$$

where L_i are linear polynomials.

These conditions are almost sufficient, too.

Theorem 24. *If there are $s = k+1-\frac{1}{2}\mathrm{rank}(M^{\star}_{k-1})$ orthogonal polynomials P_i of degree k satisfying the conditions* (i), (ii), *and* (iii) *in Theorem* 23, *then these polynomials have $N = \dim \mathbb{P}_{k-1} + \frac{1}{2}\mathrm{rank}(M^{\star}_{k-1})$ affine common zeros. If they are pairwise distinct and real, then a cubature sum $\mathscr{Q}(2,2k-1,N)$ exists.*

The surprising result from this theorem was the construction of $\mathscr{Q}(2,9,17)$ for C_2, the square with constant weight function. Franke [22] expected that 20 would be the lowest possible number of nodes for such a formula. Haegemans and Piessens [70,33] conjectured that 18 would be lowest possible.

Again, by applying Theorem 19 one can determine further conditions which guarantee that the polynomials P_i generate a real ideal, i.e., have pairwise distinct real zeros. To check this, choose U_i, $i = 1,2,\ldots,k+1-s$, such that P_i, U_i are a fundamental system of degree k. By virtue of condition (ii) of Theorem 23 there are polynomials R_{ij} and $P \in \mathrm{span}\{P_i,\ i = 1,2,\ldots,s\}$ such that $U_iU_j - R_{ij}P \in \mathbb{P}_{k+1}$. If, in addition, the P_i are chosen such that

$$\mathscr{I}[U^2 - RP] > 0$$

for all $U \in \mathrm{span}\{U_i,\ i=1,2,\ldots,k+1-s\}$, $P \in \mathrm{span}\{P_i,\ i=1,2,\ldots,s\}$, and R such that $U^2 - RP \in \mathbb{P}_{k+1}$, then the ideal $(P_1,P_2,\ldots,P_s)$ is real.

This holds in the n-dimensional case, too, even if we admit ideals with a fundamental system of maximal degree $m+1$ [79].

Theorem 25. *Let R_i, $i = 1,2,\ldots,t$, be an m-orthogonal fundamental set of degree $m+1$ of linearly independent polynomials in $\mathbb{P}^n$, and let $\mathfrak{A} = (R_1,R_2,\ldots,R_t)$ and $\mathscr{W} = \mathrm{span}\{R_1,R_2,\ldots,R_t\}$. Let $\mathscr{U}$, $\dim \mathscr{U} = N$, be an arbitrary, but fixed, complement of $\mathscr{W}$ such that $\mathbb{P}^n_{m+1} = \mathscr{W} \oplus \mathscr{U}$. Then the following are equivalent*:

(i) *A positive $\mathscr{Q}(n,m,N)$ for $\mathscr{I}^n$ exists with nodes in $\mathscr{V}(\mathfrak{A})$.*

(ii) $\mathfrak{A}$ *and* $\mathscr{U}$ *are characterised by*
   (a) $\mathfrak{A} \cap \mathscr{U} = (0)$,
   (b) $\mathscr{I}^n[Q^2 - R^+] > 0$ *for all* $Q \in \mathscr{U}$, *where* R^+ *is chosen such that* $Q^2 - R^+ \in \mathbb{P}^n_m$.

(iii) $\mathfrak{A}$ *is a real ideal with a zero-set of* N *pairwise distinct real points, which are the nodes of the cubature formula of degree* m.

6.4. Formulae characterised by three orthogonal polynomials

The nodes of a Gauss quadrature formula are the zeros of one particular polynomial. The nodes of a Gauss product cubature formula in n dimensions are the common zeros of n polynomials in n variables. Franke [21] derived conditions for planar product regions implying the existence of cubature sums $\mathscr{Q}(2, 2k-1, N)$ where $N < k^2$, see Section 6.3.2. This is based on the common zeros of two orthogonal polynomials.

Huelsman [35] proved that for fully symmetric regions, $\mathscr{Q}(2,7,10)$ cannot exist. Franke [22] proved that for these regions, and also for symmetric product regions, a cubature sum $\mathscr{Q}(2,7,11)$ cannot exist. He observed that from Stroud's characterisation [90] there follows that a cubature sum $\mathscr{Q}(2,7,12)$ is characterised by three orthogonal polynomials of degree 4, and he exploited this to construct some formulae.

In [70,71], Piessens and Haegemans observed that there are actually three orthogonal polynomials of degree k that vanish in the nodes of their cubature formulae of degree $2k-1$ for $k = 5, 6$. Following this observation, and using earlier results of Radon and Franke, in a series of articles [30,32,33] they constructed cubature formulae for a variety of planar regions whose nodes are the common zeros of three orthogonal polynomials in two variables. They restricted their work to regions that are symmetric with respect to both coordinate axes and noticed that Radon's cubature formulae for these regions have the same symmetry.

At first sight, it may look strange that Radon, Franke, Haegemans and Piessens characterised cubature formulae in two dimensions as the common zeros of three polynomial equations in two unknowns, i.e., as an overdetermined system of nonlinear equations. We now know, see Section 5.2, that for centrally symmetric regions there are $\lfloor \frac{k}{2} \rfloor + 1$ linearly independent orthogonal polynomials of degree k that vanish in the nodes of a cubature formula of degree $2k-1$ that attains the lower bound of Theorem 11. We thus know that formulae of degree 5 and 7 that attain this bound are fully characterised by three such polynomials. Formulas of higher degrees $2k-1$ that attain this bound will have even more than three linearly independent orthogonal polynomials of degree k that vanish in their nodes.

Franke, Haegemans and Piessens proceeded as follows. They assumed the existence of three linearly independent orthogonal polynomials of the form

$$\phi_i = \sum_{j=0}^{k} a_{ij} P_j^k, \quad i = 1, 2, 3.$$

The first set of conditions on the unknowns a_{ij} is obtained by demanding that whenever a node (α_i, β_i) is a common zero of ϕ_1, ϕ_2, and ϕ_3, then also $(\pm\alpha_i, \pm\beta_i)$ is. A second set of conditions is obtained by demanding that these three polynomials have sufficiently many common zeros. Obtaining these conditions requires much labour, and a computer algebra system was used to derive some of these. For higher degrees, the result contains some free parameters, and consequently a continuum

Table 1
Number of nodes in known cubature formulae [22,30,32,33][a]

Degree	C_2	S_2	$E_2^{r^2}$	E_2^r	H_2
7	12(∞)	12(∞)	12(∞)	12(∞)	12(∞)
9	19(∞) 18(2)	19(∞) 18(1)	19(∞) 18(1)	19(∞)	19(∞) 18(1)
11	28(∞) 26(∞) 25(2)	28(∞) 26(∞) 25(1)	28(∞) 26(∞) 25(1)	28(∞) 26(∞)	28(∞) 26(∞) 25(1)

[a] In parentheses the number of such cubature formulae is given.

of cubature formulae was obtained. In such a continuum Haegemans and Piessens searched for the formula with the lowest number of nodes, e.g., by searching for a formula with a weight equal to zero. An overview of the cubature formulae for the symmetric standard regions C_2, S_2, $E_2^{r^2}$, E_2^r, and H_2 obtained in this way is presented in Table 1.

This approach was also used to construct cubature formulae of degree 5 for the four standard symmetric regions in three dimensions [31]. A continuum of formulae with 21 nodes is obtained. It is mentioned that this continuum contains formulae with 17, 15, 14 and 13 nodes, the last being the lowest possible.

7. Cubature formulae of arbitrary degree

For an overview of all known minimal formulae, we refer to [10]. In this final section we present those integrals for which minimal cubature formulae for an arbitrary degree of exactness were constructed by using orthogonal polynomials. Though these examples are limited, they illustrate that all lower bounds which have been discussed will be attained for special integrals and that the construction methods based on orthogonal polynomials can be applied. Indirectly this shows that improving these bounds will require more information about the given integral to be taken into account. The symmetry of the region Ω and the weight function ω is not enough!

7.1. Minimal formulae for the square with special weight functions

Two-dimensional integrals with an infinite number of minimal cubature formulae have been presented by Morrow and Patterson [54]. They studied

$$\mathscr{I}_{1/2}[f] = \int_{-1}^{1}\int_{-1}^{1} f(x,y)(1-x^2)^{1/2}(1-y^2)^{1/2}\,\mathrm{d}x\,\mathrm{d}y.$$

The associated fundamental orthogonal system of degree k, U_i^k, $i=0,1,\dots,k$, is gathered in

$$\boldsymbol{U}_k = (U_0^k, U_1^k, \dots, U_k^k)^{\mathrm{T}}.$$

Similarly, they studied

$$\mathscr{I}_{-1/2}[f] = \int_{-1}^{1}\int_{-1}^{1} f(x,y)(1-x^2)^{-1/2}(1-y^2)^{-1/2}\,\mathrm{d}x\,\mathrm{d}y,$$

where the associated fundamental orthogonal system of degree k, T_i^k, $i=0,1,\ldots,k$, is gathered in

$$\boldsymbol{T}_k = (T_0^k, T_1^k, \ldots, T_k^k)^{\mathrm{T}}.$$

7.1.1. Even-degree formulae for $\mathscr{I}_{1/2}$

A minimal cubature sum $\mathscr{Q}(2, 2k-2, \dim \mathbb{P}_{k-1})$ has been derived in [54]; the nodes are the common zeros of

$$\boldsymbol{U}_k + 1/2F_k^{\mathrm{T}}\boldsymbol{U}_{k-1},$$

where $F_k = [0\ E_k]$. This is the special case $\sigma=1$ from the following result [80,81]:

For $k \geqslant 6$, up to symmetries, all minimal cubature sums $\mathscr{Q}(2, 2k-2, \dim \mathbb{P}_{k-1})$ are generated by the common zeros of

$$\boldsymbol{U}_k + 1/2\Gamma_k^{\mathrm{T}}\boldsymbol{U}_{k-1},$$

where Γ_k is a Hankel matrix of the form

$$\Gamma_k = \begin{bmatrix} \gamma_0 & \sigma\gamma_0 & \sigma^2\gamma_0 & \cdots & \sigma^{k-1}\gamma_0 & 1/\sigma \\ \sigma\gamma_0 & \sigma^2\gamma_0 & \sigma^3\gamma_0 & \cdots & 1/\sigma & 0 \\ \vdots & \vdots & \vdots & & \vdots & \vdots \\ \sigma^{k-1}\gamma_0 & 1/\sigma & 0 & \cdots & 0 & 0 \end{bmatrix}, \quad \gamma_0 = \frac{1-\sigma^2}{\sigma^{k+1}}, \quad 0 \neq \sigma \in \mathbb{R}.$$

7.1.2. Odd-degree formulae for $\mathscr{I}_{1/2}$

Up to symmetries, all minimal cubature sums $\mathscr{Q}(2, 2k-1, \dim \mathbb{P}_{k-1} + \lfloor k/2 \rfloor)$, k odd, for $\mathscr{I}_{1/2}$ are generated by the common zeros of

$$(E_k + \Gamma_k)\boldsymbol{U}_k,$$

where Γ_k is an orthogonal Hankel matrix of the form

$$\Gamma_k = \begin{bmatrix} \gamma_0 & \sigma\gamma_0 & \ldots & \sigma^{k-1}\gamma_0 & \sigma^k\gamma_0 - \sigma \\ \sigma\gamma_0 & \sigma^2\gamma_0 & \ldots & \sigma^k\gamma_0 - \sigma & \gamma_0 \\ \vdots & \vdots & & \vdots & \vdots \\ \sigma^{k-1}\gamma_0 & \sigma^k\gamma_0 - \sigma & \ldots & \sigma^{k-3}\gamma_0 & \sigma^{k-2}\gamma_0 \\ \sigma^k\gamma_0 - \sigma & \gamma_0 & \ldots & \sigma^{k-2}\gamma_0 & \sigma^{k-1}\gamma_0 \end{bmatrix}, \tag{34}$$

where

$$\gamma_0 = 2/(k+1), \quad \sigma^2 = 1, \quad \text{or } \gamma_0 = \frac{\sigma^2-1}{\sigma^{k+1}-1}, \quad \sigma^2 \neq 1, \ \sigma \in \mathbb{R}.$$

Note that there are redundancies in (34), $\operatorname{rank}(E_k + \Gamma_k) = \lfloor k/2 \rfloor + 2$. The general form is obtained in [81], special cases having been known long before: for $\sigma=0$, $\gamma_0=1$ see [78], for $\sigma=1$ and $\sigma=-1$ see [15].

For odd-degree formulae, k even, no general formula is known. However, there are minimal cubature sums $\mathscr{Q}(2, 2k-1, \dim \mathbb{P}_{k-1} + k/2)$, k even, for $\mathscr{I}_{1/2}$, generated by the common zeros of

$$(E_k + \Gamma_k)\boldsymbol{U}_k,$$

where Γ_k is an orthogonal Hankel matrix of the form (34), where

$$\gamma_0 = 2/(k+1), \quad \sigma = 1, \quad \text{or} \quad \gamma_0 = \frac{\sigma^2 - 1}{\sigma^{k+1} - 1}, \quad \sigma \neq 1, \ \sigma \in \mathbb{R}.$$

The case $\sigma = -1$, $\gamma_0 = 0$ was stated in [54]. The result for $\sigma = 1$ is obtained in [15], for $\sigma \neq 1$ it is obtained in [81].

7.1.3. Odd-degree formulae for $\mathscr{I}_{-1/2}$

If k is even, a minimal formula of degree $2k-1$ exists, the nodes being the common zeros of

$$T_i^k + T_{k-i}^k, \quad i = 0, 1, \ldots, k/2,$$

this result is due to [54]. Minimal formulae of degree $2k-1$, k odd or even, were derived in [15], the nodes are the common zeros of

$$T_i^k - T_{k-i}^k, \quad i = 0, 1, \ldots, \lfloor k/2 \rfloor, \quad T_0^k + T_1^k + \cdots + T_{k-1}^k + T_k^k.$$

A third formula of degree $2k-1$ for k even is given in [15], the nodes are the common zeros of

$$T_i^k - T_{k-i}^k, \quad i = 0, 1, \ldots, k/2 - 1, \quad T_0^k + T_2^k + \cdots + T_{k-2}^k + T_k^k.$$

7.1.4. Gaussian formulae

Cubature formulae attaining the lower bound (22) for even and odd degree are often called formulae of *Gaussian type* or *Gaussian formulae*. They exist for a class of (nonstandard) integrals, which will be shown in this section. This result is due to [82].

Let $\omega(x)$ be a nonnegative function on $I \subseteq \mathbb{R}$ and let $\{p_s\}$ be the orthonormal polynomials with respect to ω. Koornwinder [38] introduced bivariate orthogonal polynomials as follows.

For given $s \in \mathbb{N}$ let $u = x + y$ and $v = xy$ and define

$$P_i^{s,(-1/2)}(u, v) = \begin{cases} p_s(x)p_i(y) + p_s(y)p_i(x) & \text{if } i < s, \\ \sqrt{2}\, p_s(x)p_s(y) & \text{if } i = s, \end{cases}$$

and

$$P_i^{s,(1/2)}(u, v) = \frac{p_{s+1}(x)p_i(y) - p_{s+1}(y)p_i(x)}{x - y}.$$

Then $P_i^{s,(\pm 1/2)}$ are polynomials of total degree s. Koornwinder showed that they form a bivariate orthogonal system with respect to the weight function

$$(u^2 - 4v)^{\pm 1/2} W(u, v).$$

Let $x_{i,s}$ be the zeros of the quasi-orthogonal polynomial $p_s + \rho p_{s-1}$ where $\rho \in \mathbb{R}$ is arbitrary but fixed. The roots are ordered by $x_{1,s} < \cdots < x_{s,s}$. Let $u = x + y$ and $v = xy$, and define $W(u, v) = \omega(x)\omega(y)$.

Then we have the following Gaussian cubature formula of degree $2k-2$:

$$\iint_\Omega f(u, v) W(u, v)(u^2 - 4v)^{-1/2}\, du\, dv \cong \sum_{i=1}^{k} \sum_{j=1}^{i} \omega_{i,j} f(x_{i,k} + x_{j,k}, x_{i,k}x_{j,k}),$$

and

$$\iint_{\Omega} f(u,v)W(u,v)(u^2-4v)^{1/2}\,\mathrm{d}u\,\mathrm{d}v \cong \sum_{i=1}^{k+1}\sum_{j=1}^{i-1} \omega_{i,j} f(x_{i,k+1}+x_{j,k+1}, x_{i,k+1}x_{j,k+1}),$$

where

$$\Omega = \{(u,v)\colon (x,y) \in I \times I \text{ and } x < y\}.$$

If $\rho = 0$, then a uniquely determined formula of degree $2k-1$ will be obtained.

So there are classes of two-dimensional integrals for which the one-dimensional result of Gaussian quadrature formulae can be regained. The lower bound (22) will be attained for odd and even degree, the common zeros of

$$\boldsymbol{P}_k + \rho \Gamma_k \boldsymbol{P}_{k-1}, \quad \Gamma_k \in \mathbb{R}^{k+1 \times k}, \quad \rho \in \mathbb{R},$$

are the nodes of the formula, where Γ_k is determined from commuting properties in the orthonormal recursion formula in Theorem 4 and a matrix equation which follows from (33) in Theorem 21.

These examples have been extended to the n-dimensional case in [5].

Acknowledgements

The authors would like to thank Frau Jutta Zintchenko for her help in translating Russian texts. They are also indebted to Professor Walter Gautschi for removing the influences of Dutch and German from our version of English and for his suggestions to improve the presentation.

References

[1] P. Appell, Sur une classe de polynômes a deux variables et le calcul approché des intégrales doubles, Ann. Fac. Sci. Univ. Toulouse 4 (1890) H1–H20.
[2] P. Appell, J.K. de Fériet, Fonctions hypergéometriques et hypersphériques – Polynomes d'Hermite, Gauthiers-Villars et Cie., Paris, 1926.
[3] H. Berens, H.J. Schmid, On the number of nodes of odd degree cubature formulae for integrals with Jacobi weights on a simplex, in: T.O. Espelid, A. Genz (Eds.), Numerical Integration, Kluwer Academic Publishers, Dordrecht, 1992, pp. 37–44.
[4] H. Berens, H.J. Schmid, Y. Xu, On twodimensional definite orthogonal systems and on a lower bound for the number of nodes of associated cubature formulae, SIAM J. Math. Anal. 26 (1995) 468–487.
[5] H. Berens, H.J. Schmid, Y. Xu, Multivariate Gaussian cubature formulae, Arch. Math. 64 (1995) 26–32.
[6] H. Bourget, Sur une extension de la méthode de quadrature de Gauss, Acad. Sci. Paris 126 (1898) 634–636.
[7] T.M. Bykova, Cubature formulae with a minimal number of nodes in a plane region which are exact for polynomials of degree 4, Vestnik Leningrad. Univ. Math. 7 (1969) 145–147 (in Russian).
[8] T.M. Bykova, Cubature formulae for three-dimensional integrals of degree four with eleven nodes, Isw. AN BSSR. Ser. Phys.-Math. 10 (1) (1970) 51–54 (in Russian).
[9] V.Ja. Cernicina, Construction of domains for which there exist interpolation cubature formulae with the least number of nodes, Vestnik Leningrad. Univ. No. 1, Mat. Meh. Astronom. Vyp. 28 (1973) 144–147 (in Russian).

[10] R. Cools, Constructing Cubature Formulae: The Science Behind the Art, Acta Numerica, Vol. 6, Cambridge University Press, Cambridge, 1997, pp. 1–54.

[11] R. Cools, Monomial cubature rules since "Stroud": a compilation – Part 2, J. Comput. Appl. Math. 112 (1999) 21–27.

[12] R. Cools, A. Haegemans, Why do so many cubature formulae have so many positive weights?, BIT 28 (1988) 792–802.

[13] R. Cools, A. Haegemans, Construction of symmetric cubature formulae with the number of knots (almost) equal to Möller's lower bound, in: H. Braß, G. Hämmerlin (Eds.), Numerical Integration, ISNM, Vol. 85, Birkhäuser, Basel, 1988, pp. 25–36.

[14] R. Cools, P. Rabinowitz, Monomial cubature rules since "Stroud": a compilation, J. Comput. Appl. Math. 48 (1993) 309–326.

[15] R. Cools, H.J. Schmid, Minimal cubature formulae of degree $2k-1$ for two classical functionals, Computing 43 (1989) 141–157.

[16] R. Cools, H.J. Schmid, A new lower bound for the number of nodes in cubature formulae of degree $4n+1$ for some circularly symmetric integrals, in: H. Brass, G. Hämmerlin (Eds.), Integration IV, International Series of Numerical Mathematics, Vol. 112, Birkhäuser, Basel, 1993, pp. 57–66.

[17] P.J. Davis, P. Rabinowitz, Methods of Numerical Integration, 2nd Edition, Academic Press, New York, 1984.

[18] D.W. Dubois, G. Efroymson, Algebraic theory of real varieties I, in: Studies and Essays Presented to Yu-Why Chen on his Sixtieth Birthday, Academica Sinica, Tapei, 1970, pp. 107–135.

[19] G. Efroymson, Local reality on algebraic varieties, J. Algebra 29 (1974) 133–142.

[20] H. Engels, Numerical Quadrature and Cubature, Academic Press, London, 1980.

[21] R. Franke, Obtaining cubatures for rectangles and other planar regions by using orthogonal polynomials, Math. Comp. 25 (1971) 813–817.

[22] R. Franke, Minimal point cubatures of precision seven for symmetric planar regions, SIAM J. Numer. Anal. 10 (1973) 849–862.

[23] N.N. Fritsch, On the existence of regions with minimal third degree integration formulas, Math. Comp. 24 (1970) 855–861.

[24] B. Fuglede, The multidimensional moment problem, Exposition. Math. 1 (1983) 47–65.

[25] G.N. Gegel', On a cubature formula for the four-dimensional ball, Ž. Vyčisl. Mat. i Mat. Fiz. 15 (1) (1975) 234–236 (in Russian), USSR Math. Comp. 15 (1) (1975) 226–228.

[26] G.N. Gegel', Construction of cubature formulae exact for polynomials of degree $2m$, Voprosy Vychisl. i Prikl. Mat., Tashkent 32 (1975) 5–9 (in Russian).

[27] W. Gröbner, Über die Konstruktion von Systemen orthogonaler Polynome in ein- und zweidimensionalen Bereichen, Monatsh. Math. 52 (1948) 38–54.

[28] C. Günther, Über die Anzahl von Stützstellen in mehrdimensionalen Integrationsformeln mit positiven Gewichten, Ph.D., Universität Karlsruhe, 1973.

[29] C. Günther, Third degree integration formulas with four real points and positive weights in two dimensions, SIAM J. Numer. Anal. 11 (1974) 480–493.

[30] A. Haegemans, Tables of symmetrical cubature formulas for the two-dimensional hexagon, Department of Computer Science, K.U. Leuven, Report TW, 1976.

[31] A. Haegemans, Construction of known and new cubature formulas of degree five for three-dimensional symmetric regions, using orthogonal polynomials, in: G. Hämmerlin (Ed.), Numerical Integration, International Series of Numerical Mathematics, Vol. 57, Birkhäuser, Basel, 1982, pp. 119–127.

[32] A. Haegemans, R. Piessens, Construction of cubature formulas of degree eleven for symmetric planar regions, using orthogonal polynomials, Numer. Math. 25 (1976) 139–148.

[33] A. Haegemans, R. Piessens, Construction of cubature formulas of degree seven and nine symmetric planar regions, using orthogonal polynomials, SIAM J. Numer. Anal. 14 (1977) 492–508.

[34] P.M. Hirsch, Evaluation of orthogonal polynomials and relationship to evaluating multiple integrals, Math. Comp. 22 (1968) 280–285.

[35] C.B. Huelsman III, Quadrature formulas over fully symmetric planar regions, SIAM J. Numer. Anal. 10 (1973) 539–552.

[36] G.P. Ismatullaev, On the construction of formulae for hypercube and sphere, Voprosy Vycisl. i Prikl. Mat., Tashkent 51 (1978) 191–203 (in Russian).
[37] D. Jackson, Formal properties of orthogonal polynomials in two variables, Duke Math. J. 2 (1936) 423–434.
[38] T. Koornwinder, Two-variable analogues for the classical orthogonal polynomials, in: R.A. Askey (Ed.), Theory and Applications of Special Functions, Academic Press, New York, 1975.
[39] M.A. Kowalski, The recursion formulas for orthogonal polynomials in *n* variables, SIAM J. Math. Anal. 13 (1982) 309–315.
[40] M.A. Kowalski, Orthogonality and recursion formulas for polynomials in *n* variables, SIAM J. Math. Anal. 13 (1982) 316–323.
[41] M.A. Kowalski, Algebraic Characterization of Orthogonality in the Space of Polynomials, Lecture Notes in Mathematics, Vol. 1171, Springer, Berlin, 1985, pp. 101–110.
[42] H.L. Krall, I.M. Sheffer, Orthogonal polynomials in two variables, Ann. Mat. Pura Appl. (4) 76 (1967) 325–376.
[43] V.I. Krylov, Approximate Calculation of Integrals, 2nd Edition, Nauka, Moscow, 1967 (in Russian).
[44] V.A. Kuzmenkov, The finite moment problem in several variables, Vestnik Leningr. Univ. 19 (1975) 26–31 (in Russian).
[45] V.A. Kuzmenkov, On the existence of cubature formulae with a minimal number of nodes, Ž h. Vyčisl. Mat. i Mat. Fiz. 16(5) (1976) 1337–1334 (in Russian). USSR Math. Comp. 16(5) (1975) 242–245.
[46] V.A. Kuzmenkov, Gaussian type cubature formulae and the finite moment problem, Metody Vychisl. 11 (1978) 21–42 (in Russian).
[47] H.M. Möller, Polynomideale und Kubaturformeln, Ph.D. Thesis, Univ. Dortmund, 1973.
[48] H.M. Möller, Mehrdimensionale Hermite-Interpolation und numerische Integration, Math. Z. 148 (1976) 107–118.
[49] H.M. Möller, Kubaturformeln mit minimaler Knotenzahl, Numerische Mathematik 25 (1976) 185–200.
[50] H.M. Möller, Hermite Interpolation in Several Variables Using Ideal-Theoretic Methods, Lecture Notes in Mathematics, Vol. 571, Springer, Berlin, 1977, pp. 155–163.
[51] H.M. Möller, Lower bounds for the number of nodes in cubature formulae, in: G. Hämmerlin (Ed.), Numerische Integration, International Series of Numerical Mathematics, Vol. 45, Birkhäuser, Basel, 1979, pp. 221–230.
[52] H.M. Möller, The construction of cubature formulae and ideals of principal classes, in: G. Hämmerlin (Ed.), Numerische Integration, International Series of Numerical Mathematics, Vol. 45, Birkhäuser, Basel, 1979, pp. 249–263.
[53] H.M. Möller, On the construction of cubature formulae with few nodes using Gröbner bases, in: P. Keast, G. Fairweather (Eds.), Numerical Integration – Recent Developments, Software and Applications, Reidel, Dordrecht, 1987, pp. 177–192.
[54] C.R. Morrow, T.N.L. Patterson, Construction of algebraic cubature rules using polynomial ideal theory, SIAM J. Numer. Anal. 15 (1978) 953–976.
[55] I.P. Mysovskikh, Proof of the minimality of the number of nodes in the cubature formula for a hypersphere, Ž h. Vyčisl. Mat. i Mat. Fiz. 6 (4) (1966) 621–630 (in Russian), USSR Math. Comp. 6 (4) (1966) 15–27.
[56] I.P. Mysovskikh, Radon's paper on the cubature formula, Ž h. Vyčisl. Mat. i Mat. Fiz. 7(4) (1967) 889–891 (in Russian), USSR Comp. Math. 7(4) (1967) 232–236.
[57] I.P. Mysovskikh, Construction of cubature formulae and orthogonal polynomials, Ž h. Vyčisl. Mat. i Mat. Fiz. 7(1) (1967) 185–189 (in Russian), USSR Comp. Math. 7(1) (1967) 252–257.
[58] I.P. Mysovskikh, On the construction of cubature formulas with fewest nodes, Dokl. Akad. Nauk SSSR 178 (1968) 1252–1254 (in Russian), Soviet Math. Dokl. 9 (1968) 277–280 (in English).
[59] I.P. Mysovskikh, Cubature formulae and orthogonal polynomials, Ž h. Vyčisl. Mat. i Mat. Fiz. 9(2) (1969) 419–425 (in Russian), USSR Comp. Math. 9(2) (1969) 217–228.
[60] I.P. Mysovskikh, Numerical characteristics of orthogonal polynomials in two variables, Vestnik Leningrad. Univ. Mat. 25 (1970) 46–53 (in Russian).
[61] I.P. Mysovskih, A multidimensional analogon of quadrature formulae of Gaussian type and the generalised problem of Radon, Voprosy Vychisl. i Prikl. Mat., Tashkent 38 (1970) 55–69 (in Russian).
[62] I.P. Mysovskikh, On the paper: Cubature formulae and orthogonal polynomials, Ž h. Vyčisl. Mat. i Mat. Fiz. 10(2) (1970) 444–447 (in Russian), USSR Comp. Math. 10(2) (1970) 219–223.

[63] I.P. Mysovskikh, Application of orthogonal polynomials to construct cubature formulae, Ž. Vyčisl. Mat. i Mat. Fiz. 12(2) (1972) 467–475 (in Russian), USSR Comp. Math. 12(2) (1972) 228–239.

[64] I.P. Mysovskikh, Construction of cubature formulae, Voprosy Vycisl. i Prikl. Mat., Tashkent 32 (1975) 85–98 (in Russian).

[65] I.P. Mysovskikh, Orthogonal polynomials in several variables, Metody Vyčisl. 10 (1976) 26–35 (in Russian).

[66] I.P. Mysovskikh, Interpolatory Cubature Formulas, Nauka, Moscow, 1981 (in Russian), Interpolatorische Kubaturformeln, Institut für Geometrie und Praktische Mathematik der RWTH Aachen, Aachen, 1992, Bericht No. 74 (in German).

[67] I.P. Mysovskikh, V. Ja. Cernicina, The answer to a question of Radon, Dokl. Akad. Nauk SSSR 198(3) (1971) (in Russian), Soviet Math. Dokl. 12 (1971) 852–854 (in English).

[68] F. Peherstorfer, Characterization of positive quadrature formulas, SIAM J. Math. Anal. 12 (1981) 935–942.

[69] F. Peherstorfer, Characterization of quadrature formulas II, SIAM J. Math. Anal. 15 (1984) 1021–1030.

[70] R. Piessens, A. Haegemans, Cubature formulas of degree nine for symmetric planar regions, Math. Comp. 29 (1975) 810–815.

[71] R. Piessens, A. Haegemans, Cubature formulas of degree eleven for symmetric planar regions, J. Comput. Appl. Math. 1 (1975) 79–83.

[72] J. Radon, Zur mechanischen Kubatur, Monatsh. Math. 52 (1948) 286–300.

[73] G.G. Rasputin, On the question of numerical characteristics for orthogonal polynomials of two variables, Metody Vyčisl. 13 (1983) 145–154 (in Russian).

[74] G.G. Rasputin, Zur Konstruktion der Kubaturformel mit geradem agebraischen Grad und minimaler Knotenzahl, Wiss. Z. d. Päd. Hochsch. in Erfurt-Mühlhausen, Math.-Nat. Reihe 23 (1987) 158–165.

[75] J.J. Risler, Une charactérisation des idéaux des varietés algébriques réelles, Note aux CRAS, Paris 272 (1970) 522, 531.

[76] J.J. Risler, Un Théorème de Zéroes en Géometrie Algébrique et Analytique Réelles, in: Lecture Notes in Mathematics, Vol. 409, Springer, Berlin, 1974, pp. 522–531.

[77] H.J. Schmid, On cubature formulae with a minimal number of knots, Numer. Math. 31 (1978) 282–297.

[78] H.J. Schmid, Interpolatorische Kubaturformeln und reelle Ideale, Math. Z. 170 (1980) 267–280.

[79] H.J. Schmid, Interpolatorische Kubaturformeln, Diss. Math. CCXX (1983) 1–122.

[80] H.J. Schmid, On minimal cubature formulae of even degree, in: H. Brass, G. Hämmerlin (Eds.), Integration III, International Series of Numerical Mathematics, Vol. 85, Birkhäuser, Basel, 1988, pp. 216–225.

[81] H.J. Schmid, Two-dimensional minimal cubature formulas and matrix equations, SIAM J. Matrix Anal. Appl. 16 (1995) 898–921.

[82] H.J. Schmid, Y. Xu, On bivariate Gaussian cubature formulae, Proc. Amer. Math. Soc. 122 (1994) 833–841.

[83] S.L. Sobolev, Formulas of mechanical cubature on the surface of the sphere, Sibirsk. Math. 3 (1962) 769–796 (in Russian).

[84] S.L. Sobolev, Introduction to Cubature Formulae, Nauka Publishers, Moscow, 1974 (in Russian), Cubature Formulas and Modern Analysis, Gordon and Breach Science Publishers, Philadelphia, PA, 1992, xvi + 379 pp. (in English).

[85] S.L. Sobolev, V.L. Vaskevich, The Theory of Cubature Formulas, Moskau, Nauka, 1996 (in Russian), Kluwer Academic Publishers, Dordrecht, 1997 (in English).

[86] G. Sottas, G. Wanner, The number of weights of a quadrature formula, BIT 22 (1982) 339–352.

[87] A.H. Stroud, Integration formulas and orthogonal polynomials, SIAM J. Numer. Anal. 4 (1967) 381–389.

[88] A.H. Stroud, Integration formulas and orthogonal polynomials for two variables, SIAM J. Numer. Anal. 6 (1969) 222–229.

[89] A.H. Stroud, Integration formulas and orthogonal polynomials II, SIAM J. Numer. Anal. 7 (1967) 271–276.

[90] A.H. Stroud, Approximate Calculation of Multiple Integrals, Prentice-Hall, Englewood Cliffs, NJ, 1971.

[91] P. Verlinden, Expliciete uitdrukkingen en rekursie-betrekkingen voor meerdimensionale invariante orthogonale veeltermen, Master's Thesis, Katholieke Universiteit Leuven, 1986.

[92] P. Verlinden, R. Cools, On cubature formulae of degree $4k + 1$ attaining Möller's lower bound for integrals with circular symmetry, Numer. Math. 61 (1992) 395–407.

[93] Y. Xu, On multivariate orthogonal polynomials, SIAM J. Math. Anal. 24 (1993) 783–794.

[94] Y. Xu, Common zeros of polynomials in several variables and higher dimensional quadrature, Pitman Research Notes in Mathematics Series, Vol. 312, Longman Scientific & Technical, Harlow, 1994.
[95] Y. Xu, Recurrence formulas for multivariate orthogonal polynomials, Math. Comp. 62 (1994) 687–702.
[96] Y. Xu, A characterization of positive quadrature formulae, Math. Comp. 62 (1994) 703–718.
[97] Y. Xu, Block Jacobi matrices and zeros of multivariate orthogonal polynomials, Trans. Amer. Math. Soc. 342 (1994) 855–866.
[98] Y. Xu, On orthogonal polynomials in several variables, in: Special functions, q-series and related topics, The Fields Institute for Research in Mathematical Sciences, Communications Series, Vol. 14, American Mathematical Society, Providence, RI, 1997, pp. 247–270.

Computation of Gauss-type quadrature formulas

Dirk P. Laurie
School for Modelling Sciences, Potchefstroomse Universiteit vir CHO, Vaal Triangle Campus, P.O. Box 1174 Vanderbijlpark, South Africa

Abstract

Gaussian formulas for a linear functional L (such as a weighted integral) are best computed from the recursion coefficients relating the monic polynomials orthogonal with respect to L. In Gauss-type formulas, one or more extraneous conditions (such as pre-assigning certain nodes) replace some of the equations expressing exactness when applied to high-order polynomials. These extraneous conditions may be applied by modifying the same number of recursion coefficients. We survey the methods of computing formulas from recursion coefficients, methods of obtaining recursion coefficients and modifying them for Gauss-type formulas, and questions of existence and numerical accuracy associated with those computations.

Keywords: Gaussian quadrature; Numerical integration; Lobatto; Radau; Kronrod

1. Introduction

By n-point *Gaussian quadrature* we mean the approximation of a given linear functional L by a discrete linear functional G_n, the n-point *Gaussian quadrature formula* given by

$$G_n[f] = \sum_{j=1}^{n} w_j f(x_j),$$

such that $G_n[f] = L[f]$ whenever $f \in \mathscr{P}_{2n-1}$, where $\mathscr{P}_m$ is the space of polynomials of degree not exceeding m. We say that the formula has *degree* $2n-1$. It is convenient to think of L as defined on the space of all functions $f : \mathbb{R} \to \mathbb{R}$, because when f is defined on a smaller interval, one can simply define it to be zero elsewhere.

E-mail address: dlaurie@na-net.ornl.gov (D.P. Laurie).

The definition suggests an obvious brute force calculation method, namely:

(1) Select a basis for $\mathscr{P}_{2n-1}$ for which L can reliably be evaluated.
(2) Solve the system of $2n$ non-linear equations for the $2n$ unknowns $\{w_j\}$ and $\{x_j\}$ without taking any advantage of its structure, apart from possibly using a solver that can treat implicitly the "linear" variables $\{w_j\}$.

The brute force method is not totally devoid of theoretical merit, since the Jacobian of the system gives information about the condition of the calculation problem, but is intractable for large n. In some cases the brute force method, with ad hoc simplifying techniques that reflect great ingenuity, may be the only available method to find certain formulas of *Gauss-type*. This nonstandard term is here used to mean an n-point formula of degree at least n (i.e., of higher degree than an interpolatory formula) and defined by $2n$ equations, not all of which express exactness when applied to a polynomial. In general there can be at most $n-1$ such equations, the other $n+1$ equations being those that define the degree.

The best-known formulas of Gauss-type are the Radau, Lobatto and Kronrod formulas, respectively, of degrees $2n-2$, $2n-3$ and approximately $\frac{3}{2}n$. They are discussed together with some others in Section 4.

The problem is easiest when we know the coefficients of the three-term recursion

$$p_{l+1}(x) = (x - a_l)p_l(x) - b_l p_{l-1}(x) \tag{1}$$

satisfied by the monic orthogonal polynomials $\{p_l\}$ (i.e., monic polynomials for which $L[p_l q] = 0$ when $q \in \mathscr{P}_{l-1}$; by pq we mean the pointwise product of p and q, i.e., $(pq)(x) = p(x)q(x)$.) In order that (1) be valid for $l = 0, 1, 2, \ldots$, it is customary to define $p_{-1}(x) = 0$; of course, $p_0(x) = 1$. This allows freedom in the choice of b_0; a convenient value suggested by Gautschi is $b_0 = L[p_0]$, which leads to the useful relation

$$L[p_l^2] = b_0 b_1 b_2 \ldots b_l. \tag{2}$$

Unless stated otherwise, we will assume that all b_l are positive: this is for example the case when L denotes integration with a positive measure. In Section 6 we discuss what could happen if some b_l are not positive.

The purpose of this paper is to survey the calculation methods currently regarded as efficient and accurate for Gaussian and some Gauss-type formulas. To keep the paper concise and devoted to a single theme, any method that does not explicitly use the three-term coefficients for G_n is only mentioned when no method based on them is available. The state of the art in software for computing with three-term coefficients is Gautschi's Fortran package `ORTHPOL` [20].

There are at least two ways of using the recursion coefficients.

(1) The nodes $\{x_j\}$ are the zeros of p_n. The recursion (1) is a numerically stable way of computing p_n, and can easily be differentiated to give a recursion formula for the derivatives. One can therefore calculate the nodes by a method such as Newton's or Weierstrass'. Once a node has been found, any of several classical formulas, such as

$$w_j = \frac{L[p_{n-1}^2]}{p_{n-1}(x_j)p_n'(x_j)}, \tag{3}$$

Table 1
Two- and three-term recursion coefficients for the classical orthogonal polynomials. In the Jacobi case, $\kappa = 2k + \alpha + \beta$

	Jacobi over (0,2)	Laguerre over $(0,\infty)$
a_k	$1 + \dfrac{\beta^2 - \alpha^2}{\kappa(\kappa+2)}$	$2k + 1 + \alpha$
b_k	$\dfrac{4k(k+\alpha)(k+\beta)(k+\alpha+\beta)}{\kappa^2(\kappa+1)(\kappa-1)}$	$k(k+\alpha)$
e_k	$\dfrac{2k(k+\alpha)}{\kappa(\kappa+1)}$	k
q_k	$\dfrac{2(k+\beta)(k+\alpha+\beta)}{\kappa(\kappa-1)}$	$k+\alpha$

can be applied to calculate its weight. This method is used in the monumental work by Stroud and Secrest [47], which was the first to suggest that the use of tabulated formulas (of which the book has many) might be superseded by subroutines (also given there).

(2) The symmetric tridiagonal matrix (known as the *Jacobi matrix*)

$$T = \text{tridiag}\begin{pmatrix} & \sqrt{b_1} & & \sqrt{b_1} & \bullet & \sqrt{b_{n-1}} & \\ a_0 & & a_1 & & \bullet \quad \bullet & & a_{n-1} \\ & \sqrt{b_1} & & \sqrt{b_1} & \bullet & \sqrt{b_{n-1}} & \end{pmatrix}$$

has the nodes as eigenvalues, and when its eigenvectors are normalized to have length 1, the weights are just $w_j = b_0 u_{1,j}^2$ where $u_{1,j}$ is the first component of the eigenvector corresponding to the eigenvalue x_j. Golub and Welsch [23] show how to modify the QR algorithm so that only the required components are found.

In view of the Golub–Welsch result, the calculation of a Gaussian formula is considered to be essentially a solved problem in the case where the recursion coefficients are known, and all $b_j > 0$. Still, even a fully satisfactory algorithm might be improved a little; some alternatives to the original algorithm are discussed in Section 2. In particular, when some b_l are negative, the algebraic identities behind the Golub–Welsch algorithm still hold, but the algorithm itself would require complex arithmetic and its numerical stability and freedom from breakdown would become uncertain.

Analytical expressions for the recursion coefficients are only available in exceptional cases, including the classical orthogonal polynomials (see Table 1). In practice one has to calculate them from other information about the functional. Here two main general approaches have emerged.

(1) If L arises from integration with a weight function (by far the most common application), one can approximate L by a quadrature formula with $N \gg n$ points. This approach was first suggested by Gautschi [12,14]. The problem of recovering the recursion coefficients from a discrete linear functional is an important one in its own right, discussed in Section 3.1, although for our purpose it is not necessary to obtain great accuracy right up to a_{N-1} and b_{N-1}.
(2) In some cases it is possible to obtain *modified moments* $\mu_k = L[\pi_k]$, where the polynomials $\{\pi_k\}$ themselves are orthogonal with respect to another linear functional Λ. This approach is only useful when the recursion coefficients $\{\alpha_k\}$ and $\{\beta_k\}$ for the polynomials $\{\pi_k\}$ are known: the most common cases are the monic Chebyshev and Legendre polynomials (see Section 3.2).

It is essential to remember that any algorithm for calculating $\{a_l; b_l\}$ from $\{\alpha_k; \beta_k\}$ can in principle deliver no better accuracy than that implied by the condition of the map between the two polynomial bases, a question thoroughly discussed by Gautschi [16]. In particular, it is well known (to numerical analysts, at least) that the case $\alpha_k = \beta_k = 0$, when the polynomials $\{\pi_k\}$ reduce to the monomials and the modified moments to ordinary moments, is catastrophically ill-conditioned.

Finally, it is sometimes possible to view certain Gauss-type quadratures for L as Gaussian quadratures for a related linear functional L', so that the recursion coefficients for L' can be derived with more or less ease from those of L. These include Radau, Lobatto, anti-Gaussian and Kronrod formulas, and are discussed in Section 4.

For much of the paper, the questions of when the methods are applicable and how well they work is at best mentioned in passing. These issues are collected in Section 6. That section also contains a new idea on how to check the accuracy of a Gaussian formula.

Recent developments in eigenvalue methods indicate that when it is known that not only the weights but also the nodes must be positive, it is better to work with two-term recursion coefficients. Laurie [33] discusses this question fully, and therefore only a brief summary is given in Section 5.

2. What to do when the recursion coefficients are known

The timeless fact about the Golub–Welsch algorithm is that any advance in our ability to solve the symmetric tridiagonal eigenproblem immediately implies a corresponding advance in our ability to compute Gaussian formulas. Therefore, the fragment of Matlab code

```
% Given the number of nodes n —
a=zeros(1,n);
b=sqrt(1 ./(4-1 ./(1:n-1).^2));
b0=2;
[S,D]=eig(diag(b,1)+diag(a)+diag(b,-1));
x=diag(D); w= b0*S(1,:).^2;
% — we now have nodes x and weights w
```

that computes the n-point Gaussian quadrature formula for $Lf = \int_{-1}^{1} f(x)\,dx$ may be inefficient because the eigenvectors are found in their entirety, but as far as accuracy is concerned, it will remain as state-of-the-art as the Linpack or LAPACK or whatever subroutines that underlie one's available Matlab implementation. Only the lines defining a, b and $b0$ need be changed to make it work for other linear functionals with known recursion coefficients.

The above code fragment requires $O(n^3)$ operations, whereas the Golub–Welsch algorithm only requires $O(n^2)$, as follows:

(1) The inner loop of the algorithm consists of an implicit QR sweep, implemented as a sequence of plane rotations [44] in the $(i, i+1)$ plane, with $i = 1, 2, \ldots, n-1$.
(2) The same rotations are made to the first row of the eigenvector matrix, which starts at $[1, 0, 0, \ldots, 0]$ and ends as the square roots of the normalized weights.

(3) All decisions as to shift strategy and deflation are made as usual, uninfluenced by the weight calculation.

Since 1968, several improvements to the QR algorithm have been published. When eigenvalues only are needed, there exist ways to avoid taking square roots, e.g., Pal–Walker–Kahan [35] and Gates–Gragg [11] rational QR algorithms. The latter is slightly faster than the former, using one multiplication less per inner loop. Parlett has suggested [35] that it may be better to calculate the eigenvectors after the eigenvalues have already been found.

These ideas could lead to improvements to the Golub–Welsch algorithm if one could also calculate the weights in a square-root free way. Laurie [32] has presented one way of so doing. The idea is to calculate all the eigenvalues first, and then to do a QL instead of a QR sweep to find each weight. In that case the sequence of plane rotations runs $i = n-1, n-2, \ldots, 1$, so that the first row of the eigenvector matrix is unchanged until the very last rotation. The first element changes to $\cos\theta$. The required weight is $b_0 \cos^2\theta$. But the quantity $C = \cos^2\theta$ is available directly from the rational QL inner loop.

This idea should not be taken to its logical conclusion, as in algorithm `sqlGauss` of [32], where the matrix is deflated immediately, and the other weights are computed using the deflated matrix. It is well known that when 'ultimate' shifts are used in an attempt to converge to each eigenvalue in a single iteration, the accumulation of roundoff is such that deflation is not always justified. This may happen even when each shift is the correctly rounded machine representation of the corresponding eigenvalue. Instead, each weight should be computed on its own, starting from $[1, 0, 0, \ldots, 0]$.

The main recent alternative to the QR algorithm for the symmetric tridiagonal eigenproblem is the divide-and-conquer algorithm. This algorithm is faster than QR on large problems when all eigenvectors are required, and has been modified by Cuppen [4] to yield only those components necessary to the quadrature formula. A numerical study that compares the original Golub–Welsch algorithm to the square-root free QL-based version outlined above, and to the divide-and-conquer method, would be an interesting topic for a master's thesis. At the moment of writing the Golub–Welsch algorithm is the only one of the three that has been published in the form of actual Fortran code [20], but divide-and-conquer (in the numerically robust version of Gu and Eisenstat [25]) is in LAPACK (called via `dsyevd` in double precision) and it would be easy to make the abbreviated version that would be required for such a comparison.

3. Obtaining recursion coefficients

The subject of obtaining recursion coefficients is of importance in all applications of orthogonal polynomials, and in this paper we concentrate on methods that are necessary in order to compute Gauss-type quadratures.

3.1. *Recursion coefficients via quadrature*

Very often the functional L is given in the form

$$L[f] = \int_A^B f(x)w(x)\,dx$$

(or even more generally as a Stieltjes integral). When the integral can be evaluated analytically, a procedure commonly called the *Stieltjes algorithm* gives the recursion coefficients. The formulas

$$b_l = \frac{L[p_l^2]}{L[p_{l-1}^2]},$$

$$a_l = \frac{L[xp_l^2]}{L[p_l^2]},$$

are used in bootstrap fashion, alternating with the recursion (1): each coefficient is computed just before it is needed to generate the next p_l. (The notation x above stands for the identity polynomial: $x(t) = t$.)

The luxury of analytical evaluation is rarely possible, and therefore Gautschi [12,14] suggests replacing L by a suitable N-point quadrature formula Q with $N \gg n$, of sufficient accuracy that the first n polynomials orthogonal to Q will coincide to working accuracy with those orthogonal to L. The *discrete Stieltjes algorithm* is then obtained by running the above computation with Q in the place of L, stopping when enough coefficients have been found.

One might equally well say that the desired Gaussian quadrature is the n-point formula corresponding to the functional Q. The problem of calculating the coefficients $\{a_l\}$ and $\{b_l\}$ from a given Gaussian quadrature formula can be viewed as an inverse eigenvalue problem. This approach is taken by De Boor and Golub [5] and by Gragg and Harrod [24]. The latter paper gives examples, involving either very nearly coincident nodes or very nearly vanishing weights, where the discrete Stieltjes produces results with no significant figures. Reichel has given some other illustrative examples [41] showing that the Stieltjes algorithm may give inaccurate recursion coefficients a_l and b_l for $l < n$ even when $N \gg n$. So, although there are cases where the Stieltjes procedure is adequate, the proverbial 'cautious man' will prefer to use the Gragg–Harrod algorithm [24].

An algorithm with a further small theoretical edge is discussed in Section 5.

3.2. *Recursion coefficients from modified moments*

Suppose that it is possible to obtain in a numerically stable way the *modified moments*

$$\mu_k = L[\pi_k],$$

where the monic polynomials $\{\pi_k\}$ satisfy the recursion

$$\pi_{k+1}(x) = (x - \alpha_k)\pi_k(x) - \beta_k\pi_{k-1}(x). \tag{4}$$

To do so is seldom a trivial task, even in simple cases such as when the polynomials $\{\pi_k\}$ are monic Chebyshev or Legendre polynomials. For example, it is likely to be disastrous to try direct quadrature of $\{\pi_k\}$. When the map from the basis $\{\pi_k\}$ to the basis $\{p_l\}$ is well conditioned, π_k typically has k sign changes in the integration interval, thereby leading to severe the loss of significant digits by cancellation. The most promising approach is to look for a recursion formula connecting the $\{\mu_k\}$. Early attempts to do so needed inspiration, but Lewanowicz ([34] and many later papers) has shown how to obtain such recursions when the functional L is an integral involving a weight function that satisfies a differential equation with polynomial coefficients.

We leave the question as to the stable computation of modified moments and ask what can be done with them when they are available.

Sack and Donovan [45] point out that the *mixed moments*

$$\sigma_{k,l} = L[\pi_k p_l]$$

satisfy the five-term two-dimensional recursion relation

$$\begin{pmatrix} & -\beta_k & \\ b_l & a_l - \alpha_k & 1 \\ & -1 & \end{pmatrix} \cdot \begin{pmatrix} & \sigma_{k-1,l} & \\ \sigma_{k,l-1} & \sigma_{k,l} & \sigma_{k,l+1} \\ & \sigma_{k+1,l} & \end{pmatrix} = 0.$$

Here the dot is taken to mean, as in the case of vectors, that the quantities to its left and right are multiplied component by component and summed.

The matrix with entries $[\sigma_{k,l}]$ is lower triangular, and this fact, together with the known first column, allows us to generate the other entries along anti-diagonals running southwest to northeast. When the main diagonal is reached, the stencil is centred over known entries to give the values of b_l and a_l that are required to complete the next anti-diagonal.

The five-term stencil makes it easy to see that when two functionals are equal on the space P_m, then all mixed moments $\sigma_{k,l}$ with $k+l \leqslant m$ and $k \neq l$ are zero, and therefore the first $m+1$ members of the recursion coefficient sequences $b_0, a_0, b_1, a_1, \ldots$ and $\beta_0, \alpha_0, \beta_1, \alpha_1, \ldots$ are equal. For want of a better term, we call this the *degree-revealing* property of the recursion coefficients.

4. Gauss-type quadratures

Our main tool for computing Gauss-type quadratures is to obtain the recursion coefficients for the system of orthogonal polynomials that are generated by the quadrature formula itself. All the machinery of Section 2 can then be brought to bear.

When the formula has degree close to the maximum, only a few of these coefficients differ from those of the Gaussian formula itself, and the problem is easy. When the formula is of comparatively low degree, many coefficients differ and the problem is difficult.

In this section, we use notations such as a'_l, b''_l, etc., to indicate recursion coefficients for Gauss-type formulas that do not necessarily equal the a_l and b_l of the Gaussian formula G_n.

4.1. Radau, Lobatto, anti-Gaussian and other related formulas

The best-known Gauss-type quadratures are those with one or two preassigned nodes. We shall call them, respectively, Radau and Lobatto formulas, although strictly speaking these names apply to the case where the preassigned nodes are endpoints of the integration interval. A nice discussion on Radau and Lobatto formulas is provided by Golub [21].

It is possible to calculate Radau and Lobatto formulas just like Gaussian formulas, by specifying the recursion coefficients that define their orthogonal polynomials. By the degree-revealing property, only a_{n-1} in the case of a Radau formula, and both a_{n-1} and b_{n-1} in the case of a Lobatto formula, will differ from that for the Gaussian formula. We obtain

$$a'_{n-1} = a_{n-1} + p_n(\xi)/p_{n-1}(\xi)$$

for a formula with one preassigned node ξ, and

$$\begin{bmatrix} a''_{n-1} \\ b''_{n-1} \end{bmatrix} = \begin{bmatrix} a_{n-1} \\ b_{n-1} \end{bmatrix} + \begin{bmatrix} p_{n-1}(\xi) & p_{n-2}(\xi) \\ p_{n-1}(\eta) & p_{n-2}(\eta) \end{bmatrix}^{-1} \begin{bmatrix} p_n(\xi) \\ p_n(\eta) \end{bmatrix}$$

for a formula with two preassigned nodes ξ and η.

The anti-Gaussian formulas introduced by Laurie [30] are likewise easily calculated. The idea here is that G'_n should have error equal in magnitude but opposite in sign to G_{n-1}. The formula is therefore also of degree $2n-1$. By the degree-revealing property, only a_{n-1} and b_{n-1} need to be modified: in fact, it turns out that only the latter changes. We obtain

$$b'_{n-1} = 2b_{n-1}.$$

4.2. General modification algorithms

If the recursion coefficients for the Gaussian formula corresponding to a weight function $w(x)$ are known, then it is possible to derive the coefficients for the following weights in a reasonably straightforward manner:

(1) $(x-\xi)w(x)$;
(2) $((x-\xi)^2+\eta^2)w(x)$;
(3) $(x-\xi)^{-1}w(x)$;
(4) $((x-\xi)^2+\eta^2)^{-1}w(x)$.

Software for this calculation is included in ORTHPOL [20].

The $(n-1)$-point Gaussian rule G^{ξ}_{n-1} for the weight function $(x-\xi)w(x)$ is related to the n-point Radau formula R_n for $w(x)$ with prescribed node ξ as follows if:

$$R_n[f] = \rho f(\xi) + \sum_{j=1}^{n} r_j f(x_j),$$

then

$$G^{\xi}_{n-1}[f] = \sum_{j=1}^{n} (x_j-\xi) r_j f(x_j).$$

To go from R_n to G^{ξ}_{n-1} we keep the nodes and multiply the weights by $(x_j-\xi)$; to go the other way we divide the weights by $(x_j-\xi)$ and calculate the weight ρ for the fixed node from the condition that $f(x)=1$ is integrated exactly.

This approach to calculating the Radau formula differs from the one in Section 4.1 in that the fixed node is treated in different way to the others. By repeating the calculation, one can obtain a Gaussian formula for functions with any number of preassigned poles or zeros, as is done by Gautschi [19].

Similarly, the modification algorithms can be used to obtain the formulas with up to $n-1$ preassigned nodes. The details of the algorithms involved, including their extension to multiple nodes, are discussed by Elhay, Golub and Kautsky (variously collaborating) in [22,26,27].

4.3. Kronrod formulas

Formulas with m preassigned nodes, where $n=2m+1$, have been well studied. The best-known case is the formula K_{2m+1} whose preassigned nodes are those of G_m. Although theoretical results go back to Stieltjes and later Szegő, the quadrature formulas themselves are named after Kronrod [28], who computed them for $Lf=\int_{-1}^{1} f(x)\,\mathrm{d}x$ by a brute force method.

Many methods for computing Kronrod formulas are known (see Gautschi's survey [18]) but most of them separate the calculation of the new nodes and weights from the calculation of new weights for the old nodes. In the spirit of our theme we mention only methods based on the recurrence relation for the polynomials generated by K_{2m+1}.

By the degree-revealing property, the last m entries in the recursion coefficient sequence differ from those of G_{2m+1}. An algorithm to calculate them, based on the five-term recurrence of mixed moments, is given by Laurie [31]. The theoretical base for that algorithm is a theorem, proved in [31], that the leading and trailing $n\times n$ submatrices of the Jacobi matrix corresponding to K_{2m+1} have the same characteristic polynomial. This theorem is used by Calvetti et al. [3] to derive a method for calculating the Kronrod formula by a divide-and-conquer method applied to that Jacobi matrix, in which the new coefficients are never explicitly found.

The divide-and-conquer method seems on the basis of numerical experiments to be a little more accurate than the method based on finding the new recursion coefficients, but the latter also has some theoretical and practical advantages. From the theoretical point of view, it allows us to tell whether the formula is real with positive weights before calculating it: the question is equivalent to asking whether all $b'_l > 0$. For example, in [31], an analytic criterion is obtained for the existence of a 5-point Kronrod formula for the Jacobi weight $(1-x)^{\alpha}(1+x)^{\beta}$ in terms of the positivity of certain sixth-degree polynomials in α and β. From a practical point of view, Kronrod formulas with complex nodes and/or negative weights can also be calculated.

4.4. Other formulas with $m=(n-1)/2$ preassigned nodes

An easy alternative to the Kronrod formula is to require that not only should the nodes of G_m be preassigned, but their weights in the new formula should be the old weights multiplied by $\frac{1}{2}$. In that case, the required formula is simply the mean of the Gaussian formula G_m and the anti-Gaussian formula G'_{m+1}. For lack of a better term, we call the formula $\frac{1}{2}(G_m+G'_{m+1})$ an *averaged Gaussian formula*. This formula exists in many more cases than does the Kronrod formula (see [31]).

The general case with $m=(n-1)/2$ preassigned nodes is numerically very difficult. A software package has been published by Patterson [40], but one must be prepared to compute in much higher precision than that of the required formula.

The philosophy behind the Kronrod and anti-Gaussian formulas can be iterated to form sequences of embedded quadrature formulas in which each member is related to its predecessor in the same way as a Kronrod or averaged Gaussian formula is related to a Gaussian formula. In the first case, the Patterson formulas [38,39] are obtained: each Patterson formula is of Gauss-type under the restriction that all the nodes of the preceding formula are pre-assigned. In the second case, Laurie's stratified formulas [29] are obtained: each formula is of Gauss-type under the restriction that all the nodes of the preceding formula are pre-assigned, with their weights multiplied by $\frac{1}{2}$.

For both of these families of formulas, the best available calculation algorithms (too complicated to describe here) require very high precision. It is an open question whether the difficulty is inherent (which seems likely) or an ingenious algorithm that can calculate the formulas in a numerically stable way might one day be discovered.

5. Two-term recursion coefficients

The symmetric tridiagonal matrix T defined in Section 1 is positive definite if and only if all the nodes are positive. In that case, T can be factorized as $T = R^{\mathrm{T}}R$ where

$$R = \text{bidiag}\begin{pmatrix} & \sqrt{e_1} & & \sqrt{e_2} & \bullet & & \sqrt{e_{n-1}} & \\ \sqrt{q_1} & & \sqrt{q_1} & & \bullet & \bullet & & \sqrt{q_n} \end{pmatrix}.$$

The coefficients are related by the equations

$$a_l = q_{l+1} + e_l, \tag{5}$$

$$b_l = q_l e_l. \tag{6}$$

Although the matrix R is useful from a theoretical point of view, practical algorithms involving two-term coefficients all work with two bidiagonal matrices

$$L = \text{bidiag}\begin{pmatrix} 1 & & 1 & & \bullet & \bullet & & 1 \\ & e_1 & & e_2 & & \bullet & e_{n-1} & \end{pmatrix}, \tag{7}$$

$$U = \text{bidiag}\begin{pmatrix} & 1 & & 1 & \bullet & 1 & \\ q_1 & & q_2 & & \bullet & \bullet & q_n \end{pmatrix}, \tag{8}$$

which have the property that LU is similar to T.

The main computational tool is Rutishauser's qd algorithm [43], which requires only rational operations. It is slightly older, and computationally more efficient, than the QR algorithm, but was neglected for many years because the standard form of the algorithm is numerically suspect.

Recent work by Dhillon, Fernando and Parlett (variously collaborating) [6,7,36,37] has catapulted the qd algorithm back into the limelight. The point is that the 'differential' form of the qd algorithm (which Rutishauser knew but did not love, because it uses one multiplication more per inner loop than the standard form) has a remarkable stability property: when the two-term coefficients are known to full floating-point accuracy, the eigenvalues of T can also be calculated to high floating-point accuracy.

Laurie [33] surveys the theoretical and computational aspects of two-term versus three-term recursions. It is argued there that two-term recursions might become the method of choice, but at this stage of software development that is not yet the case.

5.1. Computing formulas from two-term recursion coefficients

Since the eigenvalues of $R^{T}R$ are the squares of the singular values of R, it is possible to give a 'timeless' algorithm also in the two-term case:

```
% Given the number of nodes n —
q=(1:n)./(1:2:2*n);
e=(1:n-1)./(3:2:2*n);
b0=2;
[U,D,V]=svd(diag(sqrt(q))+diag(sqrt(e),1);
x=diag(D). ^2; w=b0*V(1,:).^2;
% — we now have nodes x and weights w.
```

In practice, this algorithm might not be all that is desired, because high-floating-point accuracy of the smallest singular value is not relevant in the traditional applications of the SVD and the implementation might not deliver such accuracy. It is also annoying that one must take square roots before entering the svd routine, and square the result afterwards, but such is the price one pays for using general-purpose software.

Software for the two-term formulas is not as highly developed yet as in the case of three-term formulas, but the area is one of intense current reaserch, and it is already planned to add such methods to the standard linear algebra package LAPACK which is at present the engine driving higher-level languages like Matlab and Octave.

As for Gaussian quadrature, there exists an algorithm given by Rutishauser himself [43] that computes the weights in a single qd array, but the algorithm relies on 'ultimate' shifts and therefore the later weights may be badly contaminated. At this stage, there is no generally accepted 'best' way to calculate the weights.

Much work remains to be done: in particular, a two-term version of ORTHPOL would be very welcome.

5.2. Obtaining two-term recursion coefficients

The one thing that one should not do is to start from known three-term coefficients and solve Eqs. (5) and (6), because that requires subtraction of numbers with the same sign. The reverse procedure, to obtain a_l and b_l from q_l and e_l, is on the other hand quite a reasonable thing to do.

One should therefore obtain the required two-term coefficients directly. In the case of the classical orthogonal polynomials (if necessary shifted to move the left endpoint of their interval of orthogonality to the origin), this is easily done, and in fact the formulas for the two-term coefficients are simpler than those for the three-term ones (see Table 1). In the case of formulas such as those for the Jacobi weight, the shift back to $[-1,1]$ will of course lead to some cancellation if a node happens to lie close to 0. Such a node cannot be obtained in this way to high-floating-point accuracy.

In the case where the coefficients themselves are to be obtained by quadrature, it is easy to modify the discrete Stieltjes procedure. Alternatively, one may use an algorithm given by Laurie [32], which requires no subtractions and is therefore numerically stable in the sense of componentwise relative error.

An analogue of the Sack–Donovan algorithm is given in by Laurie [33], which allows us to calculate the two-term recursion coefficients directly from modified moments.

6. Existence and numerical considerations

We use the following terminology [30] when discussing questions about the existence of a Gaussian or Gauss-type formula:

- A quadrature formula *exists* if its defining equations have a (possibly complex) solution.
- The formula is *real* if the points and weights are all real.
- A real formula is *internal* if all the points belong to the (closed) interval of integration. A node not belonging to the interval is called an *exterior node.*
- The formula is *positive* if all the weights are positive.

The classical results are:

- The Gaussian formula G_n is real and positive if and only if all $b_l > 0,\ l = 0, 1, \ldots, n-1$.
- When $L[f] = \int_A^B f(x)\,\mathrm{d}\omega(x)$, and $\omega(x)$ is nondecreasing with at least n points of increase in the interval (A, B), then G_n is real, positive and internal. This condition covers the classical integrals with weight functions and also discrete linear functionals with positive coefficients.
- The nodes and weights are well-conditioned functions of the coefficients $\{b_l\}$ and $\{a_l\}$ when all $b_l > 0,\ l = 0, 1, \ldots, n-1$.

The main positive results about conditioning and numerical stability are

- When the Gaussian formula is real and positive, the computation of nodes and weights from recursion coefficients is well conditioned in a vector norm.
- When the Gaussian formula is symmetric around zero, the computation of nodes is well conditioned in the sense of componentwise relative error.
- The computation of recursion coefficients from positive weights and inter-node gaps $x_{j+1} - x_j$ is well conditioned in the sense of componentwise relative error [32].

There are also some negative results

- The computation of recursion coefficients from positive weights and distinct nodes is ill-conditioned in a vector norm [24].

The last two results seem to give contradictory answers to the question: is the map from a positive quadrature formula to its recursion coefficients well conditioned? The answer is generally thought to be 'yes', e.g., Gautschi [17] bluntly states: 'The map H_n is always well conditioned'. The negative result of Gragg and Harrod [24] arises because they measure the distance between two weight vectors $\mathbf{w}$ and $\hat{\mathbf{w}}$ as $\max|w_j - \hat{w}_j|$ and not as $\max|w_j - \hat{w}_j|/|w_j|$. The matter is fully discussed by Laurie [32].

To summarize: when starting from accurate three-term coefficients, with all b_l positive, a formula with positive weights is obtained and its absolute error will be small. When starting from accurate two-term coefficients, all positive, a formula with positive nodes and weights is obtained and its relative error will be small.

When starting from accurate modified moments, all depends on the condition of the map from the one polynomial base to the other. The classic paper of Gautschi [16] analyzes many cases, and further work has been done by Fischer [8–10].

In the case of a finite interval, it has been shown by Beckermann and Bourreau [2] that the auxiliary base must be orthogonal with respect to a measure that has the same minimal support as that of the polynomials we are trying to find, otherwise the condition number rises exponentially. This is only a sufficient condition for instability, and it is therefore still necessary to exhibit caution even when the condition is satisfied. Section 6.1 shows how to obtain an *a posteriori* numerical estimate for the error in such a case.

The temptation is great to use the methods even when the hypotheses required for the theorems do not hold. For example, some of the b_k values obtained by Laurie's algorithm [31] for computing a Kronrod formula via its the three-term coefficients may be negative. This does not mean that the Kronrod formula does not exist, only that it is not positive or not real (or both). In the case where the formula is still real, any method that only uses b_k and not its square root, such as those in [32], should work. Some methods that particularly cater for such a possibility have also been developed by Ammar et al. [1].

The case of a general indefinite measure can be written in terms of a symmetric (not Hermitian) tridiagonal matrix with complex entries. The inverse eigenvalue problem for such a matrix is discussed by Reichel [42]. In general, algorithms with look-ahead have to be used, but these tend to be fairly complicated.

As a general rule, the algebraic properties of the methods remain valid when the hypotheses fail, but the numerical stability might be adversely affected. Therefore, a method should still calculate the correct numbers in exact arithmetic as long as everything stays real, and may well deliver satisfactory results also in finite precision. Of course, as soon as quantities that should be real turn out to be complex, software that requires real numbers will fail, and software in a language that silently goes complex (like Matlab) may give wrong answers, since the original version may have tacitly relied on identities such as $z = \bar{z}$ that are no longer true.

6.1. Estimating accuracy of the computed formula

Gautschi's paper *How and how not to check the accuracy of Gaussian formulae* [15] is one of those little gems of numerical analysis that should be required reading for all postgraduate students. The main 'how not to' dictum is that it is meaningless to demonstrate that the formula integrates the required monomials to machine precision: some very bad formulas also do that. The 'how to' suggestions involve higher precision than was used to compute the formula itself.

If the only available data are the moments with respect to an ill-conditioned base (such as the monomials) then higher precision is indeed unavoidable, even when the moments are given as exact rational numbers. In the cases where we start from modified moments or recursion coefficients, it is relatively easy to calculate the sensitivity of the formula.

We give the argument for three-term recursions but it is equally well applicable to two-term ones. The Sack–Donovan algorithm for computing recursion coefficients from modified moments is noniterative and uses $\mathcal{O}(n^2)$ operations. One can vary each modified moment in turn by approximately $\sqrt{\varepsilon}$, where ε is the machine epsilon, to obtain in $\mathcal{O}(n^3)$ operations a numerical Jacobian J_1 for the dependence of recursion coefficients on modified moments.

The computation of the formula itself is iterative, and one would be squeamish to compute a numerical Jacobian by varying the recursion coefficients. But the Gragg–Harrod algorithm, which uses $\mathcal{O}(n^2)$ operations, is noniterative, and one could therefore obtain a numerical inverse Jacobian J_2^{-1} (where J_2 is the Jacobian for the dependence of the nodes and weights on recursion coefficients) by varying the nodes and weights.

Under the assumption (which may be a highly optimistic one) that the original modified moments are correct to machine precision one could compute an iterative correction to the formula as follows:

(1) Calculate the recursion coefficients $\hat{a}_k, \hat{b}_k$ for the computed formula by the Gragg–Harrod algorithm.
(2) Calculate the modified moments $\hat{\mu}_i$ for these recursion coefficients by the Salzer algorithm [46].
(3) Calculate the residual $\mu_i - \hat{\mu}_i$.
(4) Calculate the correction to the formula by applying $(J_2^{-1})^{-1}J_1$ to the residual.

Of course, without the aid of higher precision in the first three steps of the above procedure, one cannot hope to improve the formula, but the correction can be expected to be of approximately the same order of magnitude as the error in the formula.

6.2. *Underflow and overflow*

On computers with double-precision IEEE arithmetic, underflow and overflow is less of a problem than on some older machines. Nevertheless, in the case of calculating Gaussian quadrature formulas, it could be a difficulty if care is not taken. For example, the examples in [24] showing ill-conditioning in a vector norm arise because some weights underflow and are set to zero.

In the case of integration over a finite interval, it is easy to avoid underflow and overflow. The critical quantities of interest are the numbers $L[p_l^2]$. These appear in, e.g., the Sack–Donovan algorithm and in certain formulas for the weights. By the formula (2), overflow or underflow will arise if $b_l \to b$ for any positive value of b except 1. When the length of the interval is 4, it is fairly well known that $b_l \to 1$ for a very large family of weight functions. Any software with claims to robustness against overflow or underflow should therefore scale the interval to have length 4 before embarking on any computation.

On an infinite interval it is not typical for $\lim_{l\to\infty} b_l$ to exist. The best that one could achieve by scaling is to have some b_l less than and some greater than 1, a technique that has its limitations. Care must therefore be taken to detect the overflow and underflow in the computation, since their avoidance cannot be guaranteed.

6.3. *Floating-point computation and relative accuracy*

It is a common practice to perform the calculations for finite intervals on $[-1,1]$. The underflow and overflow argument above suggests that $[-2,2]$ is a much safer interval to work on. This is still not good enough. In practice, points of a Gaussian formula cluster near the endpoints of the interval and have very small weights there. When they are used to integrate functions with singularities at the endpoints (there are better ways to do that, we know, but a numerical analyst has no control over the abuses to which software will be put) the important quantity is the distance between the node and the endpoint, which is subject to cancellation when the endpoint is nonzero.

For careful work, one should store as floating-point numbers not the nodes, but the gaps between them. These are precisely the quantities for which the floating-point stability result in [32] holds. A challenge for future research is the development of software that can reliably compute those gaps. One approach may be to shift each endpoint in turn to the origin, and keep half the gaps and weights from each computation; the difficulty is then transferred to obtaining accurate floating-point values for the shifted coefficients, which is possible for the Jacobi polynomials but may require methods not yet known in the general case.

7. Conclusion

The calculation of Gauss-type quadrature formulas is a well-understood problem for which fully satisfactory methods are available when the recursion coefficients are known. Questions on which work remains to be done include:

- Do rational QR and divide-and-conquer methods really improve on the Golub–Welsch algorithm in practice?
- Does the QL algorithm give higher accuracy in practice than the QR algorithm?
- Does the qd algorithm really improve on the Golub–Welsch algorithm in practice?
- To what extent do the answers to the previous three questions depend on implementation?
- Is the use of very high precision for the Patterson and stratified sequences of embedded integration formulas inherently unavoidable?
- Two-term analogues for current three-term methods need to be developed, in particular in the form of reliable software, culminating in a package like `ORTHPOL` but based on two-term recursions.
- Can the inter-node gaps of a quadrature formula be accurately computed without using higher precision?

Acknowledgements

When starting work on this article, I asked NA-Net[1] members for references to the post-1980 literature, the pre-1980 literature being very well covered in Gautschi's comprehensive survey [13]. The following members were very helpful, in some cases e-mailing me extensive BiBTEX bibliographies: Bradley Alpert, John Crow, Kai Diethelm, Alex Fedoseyev, Daniele Funaro, Walter Gautschi, Gene Golub, Bill Gragg, Martin Gutknecht, Jaroslav Kautsky, Fred Krogh, Tom Patterson, Lothar Reichel, Van Snyder, Garry Tee. If I have missed anyone, please blame it on a hard disk crash I had in October 1999. The comments of an anonymous referee have led to many improvements in the paper. Lothar Reichel and Walter Gautschi, as editors of this special volume, have saved me from some embarrassing solecisms.

---

[1] `mailto://help@na-net.ornl.gov`

References

[1] G.S. Ammar, D. Calvetti, L. Reichel, Computation of Gauss–Kronrod quadrature rules with non-positive weights, Elec. Trans. Numer. Anal. 9 (1999) 26–38.

[2] B. Beckermann, E. Bourreau, How to choose modified moments?, J. Comput. Appl. Math. 98 (1998) 81–98.

[3] D. Calvetti, G.H. Golub, W.B. Gragg, L. Reichel, Computation of Gauss-Kronrod quadrature rules, Math. Comput. 69 (2000) 1035–1052.

[4] J.J.M. Cuppen, A divide and conquer method for the symmetric tridiagonal eigenproblem, Numer. Math. 36 (1981) 177–195.

[5] C. de Boor, G.H. Golub, The numerically stable reconstruction of a Jacobi matrix from spectral data, Linear Algebra Appl. 21 (1978) 245–260.

[6] I.S. Dhillon, A new $O(n^2)$ algorithm for the symmetric tridiagonal eigenvalue/eigenvector problem, Ph.D. Thesis, University of California, Berkeley, 1997.

[7] K.V. Fernando, B.N. Parlett, Accurate singular values and differential quotient-difference algorithms, Numer. Math. 67 (1994) 191–229.

[8] H.-J. Fischer, On the condition of orthogonal polynomials via modified moments, Z. Anal. Anwendungen 15 (1996) 223–244.

[9] H.-J. Fischer, Explicit calculation of some polynomials introduced by W. Gautschi, Z. Anal. Anwendungen 17 (1998) 963–977.

[10] H.-J. Fischer, On generating orthogonal polynomials for discrete measures, Z. Anal. Anwendungen 17 (1998) 183–205.

[11] K. Gates, W.B. Gragg, Notes on TQR algorithms, J. Comput. Appl. Math. 86 (1997) 195–203.

[12] W. Gautschi, Construction of Gauss–Christoffel quadrature formulae, Math. Comput. 22 (1968) 251–270.

[13] W. Gautschi, A survey of Gauss–Christoffel quadrature formulae, in: P.L. Butzer, F. Fehér (Eds.), E.B. Christoffel: The Influence of his Work on Mathematics and the Physical Sciences, Birkhäuser, Basel, 1981, pp. 72–147.

[14] W. Gautschi, On generating orthogonal polynomials, SIAM J. Sci. Statist. Comput. 3 (1982) 289–317.

[15] W. Gautschi, How and how not to check Gaussian quadrature formulae, BIT 23 (1983) 209–216.

[16] W. Gautschi, Questions of numerical condition related to polynomials, in: G.H. Golub (Ed.), Studies in Numerical Analysis, Math. Assoc. Amer. 1984, pp. 140–177.

[17] W. Gautschi, On the sensitivity of orthogonal polynomials to perturbations in the moments, Numer. Math. 48 (1986) 369–382.

[18] W. Gautschi, Gauss–Kronrod quadrature – a survey, in: G.V. Milovanović (Ed.), Numerical Methods and Approximation Theory III, University of Niš, 1987, pp. 39–66.

[19] W. Gautschi, Gauss-type quadrature rules for rational functions, in: H. Brass, G. Hämmerlin (Eds.), Numerical Integration IV, International Series of Numerical Mathematics, Vol. 112, Birkhäuser, Basel, 1993, pp. 111–130.

[20] W. Gautschi, Algorithm 726: ORTHPOL – a package of routines for generating orthogonal polynomials and Gauss-type quadrature rules, ACM Trans. Math. Software 20 (1994) 21–62.

[21] G.H. Golub, Some modified matrix eigenvalue problems, SIAM Rev. 15 (1973) 318–334.

[22] G.H. Golub, J. Kautsky, Calculation of Gauss quadratures with multiple free and fixed knots, Math. Comp. 41 (1983) 147–163.

[23] G.H. Golub, J.H. Welsch, Calculation of Gauss quadrature rules, Math. Comp. 23 (1969) 221–230.

[24] W.B. Gragg, W.J. Harrod, The numerically stable reconstruction of Jacobi matrices from spectral data, Numer. Math. 44 (1984) 317–335.

[25] M. Gu, S.C. Eisenstat, A divide-and-conquer algorithm for the bidiagonal svd, SIAM J. Matrix Anal. Appl. 16 (1995) 79–92.

[26] J. Kautsky, S. Elhay, Gauss quadratures and Jacobi matrices for weight functions not of one sign, Math. Comp. 43 (1984) 543–550.

[27] J. Kautsky, G.H. Golub, On the calculation of Jacobi matrices, Linear Algebra Appl. 52/53 (1983) 439–455.

[28] A.S. Kronrod, Nodes and Weights of Quadrature Formulas, Consultants Bureau, New York, 1965.

[29] D.P. Laurie, Stratified sequences of nested integration formulas, Quaest. Math. 15 (1992) 365–384.

[30] D.P. Laurie, Anti-Gaussian quadrature formulas, Math. Comp. 65 (1996) 739–747.

[31] D.P. Laurie, Calculation of Gauss–Kronrod quadrature formulas, Math. Comp. 66 (1997) 1133–1145.

[32] D.P. Laurie, Accurate recovery of recursion coefficients from Gaussian quadrature formulas, J. Comput. Appl. Math. 112 (1999) 165–180.
[33] D.P. Laurie, Questions related to Gaussian quadrature formulas and two-term recursions, in: W. Gautschi, G.H. Golub, G. Opfer (Eds.), Computation and Application of Orthogonal Polynomials, International Series of Numerical Mathematics, Vol. 133, Birkhäuser, Basel, 1999, pp. 133–144.
[34] S. Lewanowicz, Construction of a recurrence relation for modified moments, Technical Report, Institute of Computer Science, Wroclaw University, 1977.
[35] B.N. Parlett, The Symmetric Eigenvalue Problem, Prentice-Hall, Englewood Cliffs, NJ, 1980.
[36] B.N. Parlett, The new qd algorithms, in: A. Iserles (Ed.), Acta Numerica 1995, Cambridge University Press, Cambridge, 1995, pp. 459–491.
[37] B.N. Parlett, I.S. Dhillon, Fernando's solution to Wilkinson's problem: an application of double factorization, Linear Algebra Appl. 267 (1997) 247–279.
[38] T.N.L. Patterson, The optimal addition of points to quadrature formulae, Math. Comp. 22 (1968) 847–856.
[39] T.N.L. Patterson, On some Gauss and Lobatto based quadrature formulae, Math. Comp. 22 (1968) 877–881.
[40] T.N.L. Patterson, Algorithm 672: generation of interpolatory quadrature rules of the highest degree of precision with pre-assigned nodes for general weight-functions, ACM Trans. Math. Software 15 (2) (1989) 137–143.
[41] L. Reichel, Fast QR decomposition of Vandermonde-like matrices and polynomial least squares approximation, SIAM J. Matrix Anal. Appl. 12 (1991) 552–564.
[42] L. Reichel, Construction of polynomials that are orthogonal with respect to a discrete bilinear form, Adv. Comput. Math. 1 (1993) 241–258.
[43] H. Rutishauser, Der Quotienten-Differenzen-Algorithmus, Birkhäuser, Basel, 1957.
[44] H. Rutishauser, On Jacobi rotation patterns, in: Experimental Arithmetic, High Speed Computation and Mathematics, American Mathematical Society, Providence, RI, 1963, pp. 219–239.
[45] R.A. Sack, A.F. Donovan, An algorithm for Gaussian quadrature given modified moments, Numer. Math. 18 (1972) 465–478.
[46] H.E. Salzer, A recurrence scheme for converting from one orthogonal expansion into another, Comm. ACM 16 (1973) 705–707.
[47] A.H. Stroud, D. Secrest, Gaussian Quadrature Formulas, Prentice-Hall, Englewood Cliffs, NJ, 1966.

A review of algebraic multigrid

K. Stüben

German National Research Center for Information Technology (GMD), Institute for Algorithms and Scientific Computing (SCAI), Schloss Birlinghoven, D-53754 St. Augustin, Germany

Abstract

Since the early 1990s, there has been a strongly increasing demand for more efficient methods to solve large sparse, *unstructured* linear systems of equations. For practically relevant problem sizes, classical *one-level* methods had already reached their limits and new *hierarchical* algorithms had to be developed in order to allow an efficient solution of even larger problems. This paper gives a review of the first hierarchical and *purely matrix-based* approach, *algebraic multigrid* (AMG). AMG can directly be applied, for instance, to efficiently solve various types of elliptic partial differential equations discretized on unstructured meshes, both in 2D and 3D. Since AMG does not make use of any geometric information, it is a "plug-in" solver which can even be applied to problems without any geometric background, provided that the underlying matrix has certain properties.

Keywords: Algebraic multigrid

1. Introduction

The efficient numerical solution of large systems of discretized elliptic partial differential equations (PDEs) requires hierarchical algorithms which ensure a rapid reduction of both short- and long-range error components. A break-through, and certainly one of the most important advances during the last three decades, was due to the multigrid principle. Any corresponding method operates on a hierarchy of grids, defined a priori by coarsening the given discretization grid in a geometrically natural way ("geometric" multigrid). Clearly, this is straightforward for logically regular grids. However, the definition of a natural hierarchy may become very complicated for highly complex, unstructured meshes, if possible at all.

The first attempt to automate the coarsening process took place in the early 1980s [10,12,13], at the time when the so-called *Galerkin-principle* and *operator-dependent interpolation* were combined in geometric multigrid to increase its robustness (aiming at the efficient solution of diffusion

E-mail address: stueben@gmd.de (K. Stüben).

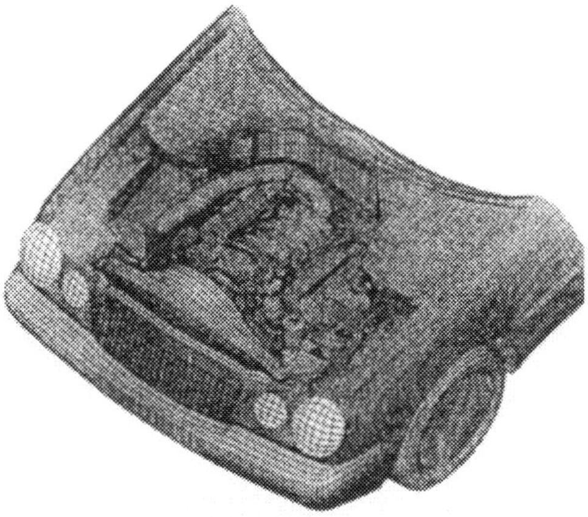

Fig. 1. Mesh for computing the underhood flow of a Mercedes-Benz E-Class.

problems with jumping coefficients [1,20]). This attempt was motivated by the observation that reasonable operator-dependent interpolation and the Galerkin operator can often be derived directly from the underlying matrices, without any reference to the grids. The result was a multigrid-like approach which did not merely allow an automatic coarsening process, but could be directly applied to (line sparse) algebraic equations of certain types, without any pre-defined hierarchy ("algebraic" multigrid,[1] AMG).

The first fairly general AMG program was described and investigated in [47,48,50]. Since this code was made publically available in the mid-1980s (AMG1R5), there had been no substantial further research and development in AMG for many years. However, since the early 1990s, and even more since the mid-1990s, there was a strong increase of interest in algebraically oriented multilevel methods. One reason for this was certainly the increasing geometrical complexity of applications which, technically, limited the immediate use of geometric multigrid. Another reason was the steadily increasing demand for efficient "plug-in" solvers. In particular, in commercial codes, this demand was driven by increasing problem sizes which made clear the limits of the classical one-level solvers still used in most packages.

For instance, CFD applications in the car industry involve very complicated flow regions. Flows through heating and cooling systems, complete vehicle underhood flows, or flows within passenger compartments are computed on a regular basis. Large complex meshes, normally unstructured, are used to model such situations. Requirements on the achievable accuracy are ever increasing, leading to finer and finer meshes. Locally refined grid patches are introduced to increase the accuracy with as few additional mesh points as possible. Fig. 1 shows an example.

In the recent past, several ways to realize concrete AMG algorithms have been investigated and there is still an ongoing rapid development of new AMG and AMG-like approaches and variants. Consequently, there is no unique and best approach yet. Whenever we talk about AMG in the

[1] We should actually use the term multi*level* rather than multi*grid*. It is just for historical reasons that we use the term multi*grid*.

context of concrete numerical results, we actually refer to the code RAMG05 [2] (described in detail in [51]), which is a successor of the original code AMG1R5 mentioned above. However, RAMG05 is completely new and, in particular, incorporates more efficient and more robust interpolation and coarsening strategies.

This paper gives a survey of the classical AMG idea [48], certain improvements and extensions thereof, and various new approaches. The focus in Sections 2–6 is on fundamental ideas and aspects, targeting classes of problems for which AMG is best-developed, namely, *symmetric positive-definite* (s.p.d.) problems of the type as they arise, for instance, from the discretization of *scalar* elliptic PDEs of second order. We want to point out, however, that the potential range of applicability is much larger. In particular, AMG has successfully been applied to various nonsymmetric (e.g. convection–diffusion) and certain indefinite problems. Moreover, important progress has been achieved in the numerical treatment of *systems* of PDEs (mainly Navier–Stokes and structural mechanics applications). However, major research is still ongoing and much remains to be done to obtain an efficiency and robustness comparable to the case of scalar applications. In particular in Section 7, we will set pointers to the relevant literature where one can find further information or more recent AMG approaches. Although we try to cover the most important references, the list is certainly not complete in this rapidly developing field of research.

2. Algebraic versus geometric multigrid

Throughout this paper, we assume the reader to have some basic knowledge of geometric multigrid. In particular, he should be familar with the fundamental principles (smoothing and coarse-grid correction) and with the recursive definition of multigrid cycles. This is because, for simplicity, we limit our main considerations to just two levels. Accordingly, whenever we talk about the efficiency of a particular approach, we implicitly always assume the underlying two-level method to be recursively extended to a real multi-level method, involving only a small number of variables (20–40, say) on the coarsest level. Regarding more detailed information on geometric multigrid, we refer to [61] and the extensive list of references given therein.

A two-level AMG cycle to solve (sparse) s.p.d. systems of equations

$$A_h u^h = f^h \quad \text{or} \quad \sum_{j \in \Omega^h} a_{ij}^h u_j^h = f_i^h \quad (i \in \Omega^h) \tag{1}$$

is formally defined in the same way as a Galerkin-based geometric two-grid cycle. The only difference is that, in the context of AMG, Ω^h is just an *index set* while it corresponds to a *grid* in geometric multigrid. Accordingly, a coarser level, Ω^H, just corresponds to a (much smaller) index set.

If we know how to map H-vectors into h-vectors by some (full rank) "interpolation" operator I_H^h, the (s.p.d.) coarse-level operator A_H is defined via

$$A_H := I_h^H A_h I_H^h \quad \text{with } I_h^H = (I_H^h)^{\mathrm{T}}.$$

One two-level correction step then runs as usual, that is

$$u_{\text{new}}^h = u_{\text{old}}^h + I_H^h e^H, \tag{2}$$

[2] The development of RAMG05 has partly been funded by Computational Dynamics Ltd., London.

where the correction e^H is the exact solution of

$$A_H e^H = r^H \quad \text{or} \quad \sum_{j \in \Omega^H} a_{ij}^H e_j^H = r_i^H \quad (i \in \Omega^H)$$

with $r^H = I_h^H(r_{\text{old}}^h)$ and $r_{\text{old}}^h = f^h - A_h u_{\text{old}}^h$. (Note that we normally use the letter u for *solution* quantities and the letter e for *correction* or *error* quantities.) For the corresponding errors $e^h = u_\star^h - u^h$ ($u_\star^h$ denotes the exact solution of (1)), this means

$$e_{\text{new}}^h = K_{h,H} e_{\text{old}}^h \quad \text{with } K_{h,H} := I_h - I_H^h A_H^{-1} I_h^H A_h, \tag{3}$$

being the *coarse-grid correction operator* (I_h denotes the identity).

We finally recall that – given any relaxation operator, S_h, for smoothing – the convergence of Galerkin-based approaches can most easily be investigated w.r.t. the *energy norm*, $\|e^h\|_{A_h} = (A_h e^h, e^h)^{1/2}$. Assuming ν relaxation steps to be performed for (pre-) smoothing, the following well-known variational principle holds (see, for instance [51]),

$$\|K_{h,H} S_h^\nu e^h\|_{A_h} = \min_{e^H} \|S_h^\nu e^h - I_H^h e^H\|_{A_h}. \tag{4}$$

As a trivial consequence, convergence of two-level cycles and, if recursively extended to any number of levels, the convergence of complete V-cycles is always ensured as soon as the relaxation method converges. This is true for any sequence of coarser levels and interpolation operators. More importantly, (4) indicates that the *speed* of convergence strongly depends on the efficient interplay between *relaxation* and *interpolation*. Based on (4), we want to outline the basic conceptual difference between geometric and algebraic multigrid.

2.1. Geometric multigrid

In geometric multigrid, fixed coarsening strategies are employed and interpolation is usually defined geometrically, typically by linear interpolation. Depending on the given problem, this necessarily imposes strong requirements on the smoothing properties of S_h (in order for the right-hand side in (4) to become small), namely, that the error after relaxation varies in a geometrically smooth way from the fine-level grid points to the neighboring coarse-level ones. In other words, the error after relaxation *has to be geometrically smooth*, relative to the coarse grid.

As an illustration, let us assume the coarser levels to be defined by standard geometric $h \to 2h$ coarsening in each spatial direction. It is well known that pointwise relaxation geometrically smooths the error in each direction only if the given problem is essentially isotropic. In case of anisotropic problems, however, smoothing is only "in the direction of strong couplings". In such cases, more complex smoothers, such as alternating line-relaxation or ILU-type schemes, are required in order to still achieve a good interplay between smoothing and interpolation and, thus, fast multigrid convergence.

While the construction of "robust smoothers" is not difficult in 2D model situations, for 3D applications on complex meshes their realization tends to become rather cumbersome. For instance, the robust 3D analog of alternating line relaxation is alternating *plane* relaxation (e.g., realized by 2D multigrid within each plane) which, in complex geometric situations, becomes very complicated to implement, if possible at all. ILU smoothers, on the other hand, loose much of their smoothing property in general 3D situations. The only way to loosen the requirements on the smoothing properties of the relaxation and still maintain an efficient interplay relaxation and interpolation is to use more sophisticated coarsening techniques. In geometric multigrid, steps in this direction have been done by, for example, employing more than one coarser grid on each multigrid level ("multiple semi-coarsening" [35,59,21,34]).

2.2. *Algebraic multigrid*

While geometric multigrid essentially relies on the availability of robust smoothers, AMG takes the opposite point of view. It assumes a simple relaxation process to be given (typically plain Gauss–Seidel relaxation) and then attempts to construct a suitable operator-dependent interpolation I_H^h (including the coarser level itself). According to (4), this construction has to be such that error of the form $S_h^\nu e^h$ is sufficiently well represented in the *range of interpolation*, $\mathscr{R}(I_H^h)$. The better this is satisfied, the faster the convergence can be. Note that it is *not* important here whether relaxation smooths the error in any geometric sense. What *is* important, though, is that the error after relaxation can be characterized algebraically to a degree which makes it possible to automatically construct coarser levels and define interpolations which are *locally* adapted to the properties of the given relaxation. This local adaptation is the main reason for AMG's flexibility in adjusting itself to the problem at hand and its robustness in solving large classes of problems *despite using very simple point-wise smoothers.*

3. The classical AMG approach

In classical AMG, we regard the coarse-level variables as a subset of the fine-level ones. That is, we assume the set of fine-level variables to be split into two disjoint subsets, $\Omega^h = C^h \cup F^h$, with C^h representing those variables which are to be contained in the coarse level (*C-variables*) and F^h being the complementary set (*F-variables*). Given such a C/F-splitting, we define $\Omega^H = C^h$ and consider (full rank) interpolations $e^h = I_H^h e^H$ of the form

$$e_i^h = (I_H^h e^H)_i = \begin{cases} e_i^H & \text{if } i \in C^h, \\ \sum_{k \in P_i^h} w_{ik}^h e_k^H & \text{if } i \in F^h, \end{cases} \tag{5}$$

where $P_i^h \subset C^h$ is called the set of *interpolatory variables*. (For reasons of sparsity of A_H, P_i^h should be a reasonably small subset of C-variables "near" i.) Clearly, $\mathscr{R}(I_H^h)$ strongly depends on both the concrete selection of the C-variables *and* the definition of the interpolation. In a given situation, one can easily imagine "bad" C/F-splittings which just do not allow any interpolation which is suitable in the sense that was outlined in the previous section. That is, the construction of concrete C/F-splittings and the definition of interpolation are closely related processes.

Concrete algorithms used in practice are largely heuristically motivated. In Section 3.1, we mainly summarize the basic ideas as described in [48] and some modifications introduced in [51]. In Section 3.2, we take a closer look at some theoretical and practical aspects in case that A_h contains only nonpositive off-diagonal entries (M-matrix). To simplify notation, we usually omit the index h in the following, for instance, we write S, A, K and e instead of S_h, A_h, $K_{h,H}$ and e^h. Moreover, instead of (5), we simply write

$$e_i = \sum_{k \in P_i} w_{ik} e_k \quad (i \in F). \tag{6}$$

3.1. The basic ideas

Classical AMG uses plain Gauss–Seidel relaxation for smoothing. From some heuristic arguments, one can see that the error e, obtained after a few relaxation steps, is characterized by the fact that the (scaled) residual is, on the average for each i, much smaller than the error itself, $|r_i| \ll a_{ii}|e_i|$. This implies that e_i can *locally* be well approximated by

$$e_i \approx -\left(\sum_{j \in N_i} a_{ij} e_j\right) \Big/ a_{ii} \quad (i \in \Omega), \tag{7}$$

where $N_i = \{j \in \Omega : j \neq i,\ a_{ij} \neq 0\}$ denotes the *neighborhood* of any $i \in \Omega$. Such an error is called *algebraically smooth.* According to the remarks at the end of Section 2, it is this kind of error which has to be well represented in $\mathscr{R}(I_H^h)$. That is, the general goal is to construct C/F-splittings and define sets of interpolatory variables $P_i \subset C$ $(i \in F)$ along with corresponding weights w_{ik} such that (6) yields a reasonable approximation for each algebraically smooth vector e.

Obviously, a very "accurate" interpolation in this sense is obtained by directly using (7), that is, by choosing $P_i = N_i$ and $w_{ik} = -a_{ik}/a_{ii}$. However, this would require selecting a C/F-splitting so that, for each $i \in F$, *all of its neighbors* are contained in C. Although any such selection can even be seen to yield a *direct* solver, this approach is of no real practical relevance since, in terms of computational work and memory requirement, the resulting method will generally be extremely inefficient if recursively extended to a hierarchy of levels [51].

In practice, we want to achieve rapid convergence with as small sets of interpolatory variables P_i as possible (in order to allow for a rapid coarsening and to obtain reasonably sparse Galerkin operators). Various approaches have been tested in practice which cannot be described in detail here. In the following, we just give an outline of some typical approaches.

3.1.1. Direct interpolation

We talk about *direct interpolation* if $P_i \subseteq N_i$. Such an interpolation can immediately be derived from (7) if we know how to approximate the "noninterpolatory part" (i.e. that part of the sum in (7) which refers to $j \in N_i \setminus P_i$) for an algebraically smooth error. This approximation is the most critical step in defining interpolation.

For M-matrices A, for instance, such an approximation can be obtained by observing that an algebraically smooth error varies slowly in the direction of *strong* (large) couplings. In particular, the more strong couplings of any variable i are contained in P_i, the better an algebraically smooth error satisfies

$$\frac{1}{\sum_{k \in P_i} a_{ik}} \sum_{k \in P_i} a_{ik} e_k \approx \frac{1}{\sum_{j \in N_i} a_{ij}} \sum_{j \in N_i} a_{ij} e_j \quad (i \in \Omega).$$

Inserting this into (7), we obtain an interpolation (6) with positive weights

$$w_{ik} = -\alpha_i a_{ik}/a_{ii} \quad \text{where } \alpha_i = \frac{\sum_{j \in N_i} a_{ij}}{\sum_{\ell \in P_i} a_{i\ell}}. \tag{8}$$

Practically, this means that we have to construct C/F-splittings so that each $i \in F$ has a sufficiently large number of *strongly* coupled C-neighbors which are then taken as the set of interpolatory variables P_i. (See Section 3.2 regarding some important additional aspects.)

In the case of (scalar) elliptic PDEs, the largest off-diagonal entries are usually negative. If there are also *positive* off-diagonal entries, a similar process as before can be applied as long as such entries are relatively small: Small positive couplings can simply be ignored by just considering them as *weak*. However, the situation becomes less obvious, if A contains *large* positive off-diagonal entries. In many such cases, an algebraically smooth error can be assumed to *oscillate* along such couplings (e.g. in case of *weakly diagonally dominant* s.p.d. matrices A [26,51]). This can be used to generalize the above approach by, for instance, a suitable separation of positive and negative couplings, leading to interpolation formulas containing both positive and negative weights. A corresponding approach has been proposed in [51] which can formally be applied to arbitrary s.p.d. matrices. However, the resulting interpolation is heuristically justified only as long as, for any i, those error components e_k which correspond to large positive couplings $a_{ik} > 0$, change slowly *among each other* (unless a_{ik} is very small in which case its influence can be ignored).

In practice, these simple algebraic approaches to construct an interpolation cover a large class of applications. However, there is no best way yet to automatically construct an interpolation which is good for *arbitrary* s.p.d. matrices, at least not by merely considering the size and sign of coefficients. For instance, in case of particular higher-order finite-element discretizations or discretizations by bilinear elements on quadrilateral meshes with large aspect ratios, the resulting matrices typically contain significant positive off-diagonal entries and are far from being weakly diagonally dominant. In such cases, the performance of AMG may be only suboptimal. If this happens, it often helps to exploit some information about the origin of these positive connections rather than to rely only on information directly contained in the matrix. For instance, one could try to structurally simplify the given matrix before applying AMG (see, e.g. [41]). Alternative approaches, defining the coarsening and interpolation are outlined in Section 7.

3.1.2. *More complex interpolations*

There are several ways to improve the direct interpolation of the previous section. To outline some possibilities, let us assume a C/F-splitting and, for each $i \in F$, a set of (strongly coupled) interpolatory variables $P_i \subseteq N_i \cap C$ to be given. Rather than to immediately approximate the noninterpolatory part of the ith equation (7) as done above, one may first (approximately) eliminate all strongly coupled e_j $(j \notin P_i)$ by means of the corresponding jth equation. The ideas outlined in the previous section can then analogously be applied to the resulting extended equation for e_i, leading to an interpolation with an increased set of interpolatory variables. The corresponding interpolation (called *standard interpolation* in [51]) is, in general, considerably more robust in practice. Alternatively, one may obtain an improved interpolation by applying one F-relaxation step (for more details, see "Jacobi-interpolation" in Section 5) to either the direct or the standard interpolation.

In both approaches, compared to the direct interpolation, the "radius" of interpolation increases which, in turn, will reduce the sparsity of the resulting Galerkin operator. However, interpolation weights corresponding to variables "far away" from variable i are typically much smaller than the largest ones. Before computing the Galerkin operator, one should therefore truncate the interpolation operator by ignoring all small entries (and re-scale the remaining weights so that the total sum remains unchanged). Note that, because of the variational principle, the truncation of interpolation is a "safe process"; in the worst case, overall convergence may slow down, but no divergence can occur. On the other hand, a truncation of the *Galerkin operators* can be dangerous since this destroys the validity of the variational principle and, if not applied with great care, may even cause strong divergence in practice.

Apart from other minor differences, the original AMG interpolation proposed in [48] (and realized in the code AMG1R5) can be regarded as a compromise between the direct interpolation and the standard interpolation described before. There, an attempt was made to replace all strongly coupled $e_j\,(j \notin P_i)$ in (7) by averages *involving only variables in* P_i. However, for this to be reasonable, based on certain criteria, new C-variables had to be added to a given splitting a posteriori ("final C-point choice" in [48]). Although this approach worked quite well in those cases treated in [48], typically too many additional C-variables are required in geometrically complex 3D situations, causing unacceptably high fill-in towards coarser levels (see Section 4.2 for examples). In practice, the standard interpolation outlined above (in combination with a reasonable truncation) has turned out to be more robust and often considerably more efficient.

The above improvements of interpolation generally lead to faster convergence but also increase the computational work per cycle and the required memory to some extent. Whether or not this finally pays, depends on the given application. If memory is an issue (as it is often in commercial environments), one may, instead, wish to *simplify* interpolation at the expense of a reduced convergence speed. One way to achieve this is to generally allow interpolation from variables which are *not* in the direct neighborhood. Such "long-range" interpolation [48] generally allows a much faster coarsening and drastically increases the sparsity on coarser levels. For details of a simple approach which has been tested in practice, see [51] ("aggressive coarsening" and "multi-pass interpolation").

3.2. *The M-matrix case*

In practice, it turns out that AMG V-cycle convergence is, to a large extent, independent of the problem size. Unfortunately, this cannot be proved based merely on algebraic arguments. Nevertheless, some important aspects can be investigated theoretically, in particular, matrix-independent two-level convergence can be proved for various classes of matrices if interpolation is defined properly. We consider here the class of M-matrices. Generalizations to other classes and the corresponding proofs can be found in [51].

3.2.1. *Two-level considerations*

The following theorem shows that direct interpolation based on (8) ensures matrix-independent two-level convergence if, for each $i \in F$, the connectivity represented in P_i is a *fixed fraction* of the total connectivity.

Theorem 1. *Let A be a symmetric, weakly diagonally dominant M-matrix. With fixed $0 < \tau \leqslant 1$ select a C/F-splitting so that, for each $i \in F$, there is a set $P_i \subseteq C \cap N_i$ satisfying*

$$\sum_{k \in P_i} |a_{ik}| \geqslant \tau \sum_{j \in N_i} |a_{ij}| \tag{9}$$

and define interpolation according to (8). *Then the two-level method, using one Gauss–Seidel relaxation step for* (*post-*) *smoothing, converges at a rate which depends only on* τ *but not on the*

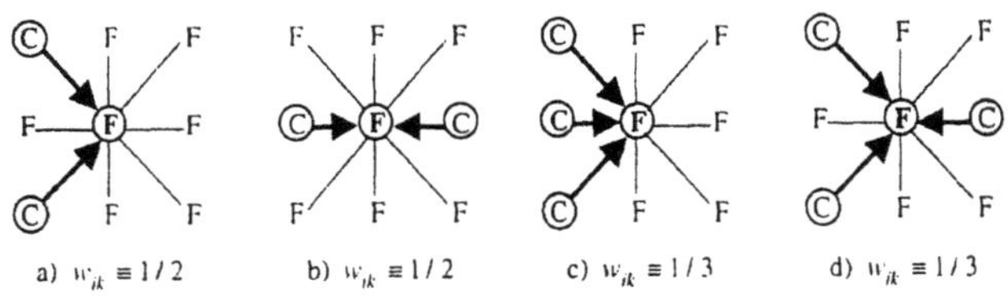

Fig. 2. Different C/F-arrangements and corresponding interpolation formulas.

given matrix,

$$\|SK\|_A \leqslant \sqrt{1 - \tau/4}.$$

The above theorem confirms that it is the *strong* couplings which are important to interpolate from, while the use of *weak* couplings would increase the computational work but hardly affect the convergence. The more strong connections are used for interpolation, the better the convergence can be. Note that this implicitly means that coarsening will be "in the direction of smoothness" which is the main reason for the fact that AMG's convergence does not sensitively depend on anisotropies. Moreover, AMG's interpolation can be regarded as an algebraic generalization of the operator-dependent interpolation introduced in [1,20], which explains why the performance of AMG does not sensitively depend on large, discontinuous variations in the coefficients of the given system of equations. For an illustration, see Section 4.1.

From a practical point of view, the above convergence estimate is a worst-case estimate, at least if the given problem has some kind of geometric background (which it typically does). The reason is that the algebraic requirement (9) does not take the location of the interpolatory C-points, relative to the F-points, into account. For an illustration, consider the 9-point discretization of the Poisson operator

$$\frac{1}{3h^2}\begin{bmatrix} -1 & -1 & -1 \\ -1 & 8 & -1 \\ -1 & -1 & -1 \end{bmatrix}_h . \tag{10}$$

From geometric multigrid, we know that linear interpolation yields fast convergence. The algebraic interpolation, however, cannot distinguish between geometrically "good" and "bad" C/F-splittings. For instance, in Fig. 2a and b we use the same total weight for interpolation but the second arrangement will clearly result in much better convergence. Similarly, the arrangement in Fig. 2d, although it does not give exactly second order, will be much better than the one in Fig. 2c.

This illustrates that the concrete arrangement of a C/F-splitting will have a substantial influence on the quality of interpolation, and, through this, on the final convergence. In order to *strictly ensure* an optimal interpolation, we would have to exploit the geometric location of (strongly coupled) points among each other. In practice, however, it turns out to be sufficient to base the construction of a C/F-splitting on the following additional objective. As a rule, one should arrange the C/F-splitting so that the set of C-variables builds (approximately) a *maximal* set with the property that the C-variables are not strongly coupled among each other ("maximally independent set") and that the F-variables are "surrounded" by their interpolatory C-variables. This can be ensured to a sufficient extent by merely exploiting the connectivity information contained in the matrix (for an algorithm, see [48,51]). Note that strong connectivity does not necessarily have to be via direct couplings.

Observing this objective will, in practice, substantially enhance convergence even if only small sets of interpolatory variables are used.

3.2.2. Extension to multi-level cycles

Unfortunately, the assumptions on interpolation in Theorem 1 are sufficient for uniform two-level, but not for uniform V-cycle convergence. Although, by choosing $\tau \geqslant \frac{1}{2}$, one can ensure that all recursively defined Galerkin operators remain weakly diagonally dominant M-matrices and, hence, the formal extension to complete V-cycles is straightforward, A-independent V-cycle convergence cannot be proved. The reason is the limited accuracy of purely algebraically defined interpolation as discussed in the previous section. We will return to this problem in Section 6 where we consider a worst-case scenario in the context of "aggregation-type" AMG approaches.

In practice, however, one can observe V-cycle convergence which is, to a large extent, independent of the problem size if we take the additional objective of the previous section into account. Furthermore, it turns out that it is not important to force the coarse-level matrices to exactly remain M-matrices. On the contrary, such a requirement puts too many restrictions on the coarsening process, in particular on lower levels, where the size of the Galerkin operators then may grow substantially.

In this context, we want to emphasize again that, for an efficient overall performance, convergence speed is only one aspect. An equally important aspect is the complexity (sparsity) of the coarser level matrices produced by AMG (which directly influences both the run time and the overall memory requirement). Only if both the convergence speed and the *operator complexity*

$$c_{\mathrm{A}} = \sum_{\ell} |m_{\ell}|/|m_1|, \tag{11}$$

(m_{ℓ} denotes the number of nonzero entries contained in the matrix on level ℓ) are bounded independent of the size of A, do we have an asymptotically optimal performance. The typical AMG performance in case of some complex problems is given in Section 4.2.

3.3. AMG as a pre-conditioner

In order to increase the robustness of geometric multigrid approaches, it has become very popular during the last years, to use multigrid not as a stand-alone solver but rather combine it with acceleration methods such as conjugate gradient. BI-CGSTAB or GMRES. This development was driven by the observation that it is often not only simpler but also more efficient to use accelerated multigrid approaches rather than to try to optimize the interplay between the various multigrid components in order to improve the convergence of stand-alone multigrid cycles.

This has turned out to be similar for AMG which, originally, was designed to be used stand-alone. Practical experience has clearly shown that AMG is also a very good pre-conditioner, much better than standard (one-level) ILU-type pre-conditioners, say. Heuristically, the major reason is due to the fact that AMG, in contrast to any one-level pre-conditioner, aims at the efficient reduction of *all* error components, short range as well as long range. However, although AMG tries to capture all relevant influences by proper coarsening and interpolation, its interpolation will hardly ever be optimal. It may well happen that error reduction is significantly less efficient for some very specific error components. This may cause a few eigenvalues of the AMG iteration matrix to be considerably closer to 1 than all the rest. If this happens, AMG's convergence factor is limited by the slow convergence of just a few exceptional error components while the majority of the error components is reduced very quickly. Acceleration by, for instance, conjugate gradient typically eliminates these particular frequencies very efficiently. The alternative, namely, to try to prevent such situations by putting more effort into the construction of interpolation, will generally be much more expensive. Even then, there is no final guarantee that such situations can be avoided. (We note that this even happens with "robust" geometric multigrid methods.)

4. Applications and performance

The flexibility of AMG and its simplicity of use, of course, have a price: A *setup phase*, in which the given problem is analyzed, the coarse levels are constructed and all operators are assembled, has to be concluded before the actual *solution phase* can start. This extra overhead is one reason for the fact that AMG is usually less efficient than geometric multigrid approaches (if applied to problems for which geometric multigrid *can* be applied efficiently). Another reason is that AMG's components can, generally, not be expected to be "optimal". They are always constructed on the basis of compromises between numerical work and overall efficiency. Nevertheless, if applied to standard elliptic test problems, the computational cost of AMG's solution phase (ignoring the setup cost) is typically comparable to the solution cost of a *robust* geometric multigrid solver [47].

However, AMG should not be regarded as a competitor of geometric multigrid. The strengths of AMG are its robustness, its applicability in complex geometric situations and its applicability to even solve certain problems which are out of the reach of geometric multigrid, in particular, problems with no geometric or continuous background at all. In such cases, AMG should be regarded as an efficient alternative to standard numerical methods such as conjugate gradient accelerated by typical (one-level) pre-conditioners. We will show some concrete performance comparisons in Section 4.2. Before, however, we want to illustrate the flexibility of AMG in adjusting its coarsening process locally to the smoothing properties of relaxation by means of a simple but characteristic model equation.

4.1. A model problem for illustration

We consider the model equation

$$-(au_x)_x-(bu_y)_y+cu_{xy}=f(x,y), \tag{12}$$

defined on the unit square with Dirichlet boundary conditions. We assume $a=b=1$ everywhere except in the upper left quarter of the unit square (where $b=10^3$) and in the lower right quarter (where $a=10^3$). The coefficient c is zero except for the upper right quarter where we set $c=2$. The resulting discrete system is isotropic in the lower left quarter of the unit square but strongly anisotropic in the remaining quarters. Fig. 3a shows what a "smooth" error looks like on the finest level after having applied a few Gauss–Seidel point relaxation steps to the homogeneous problem,

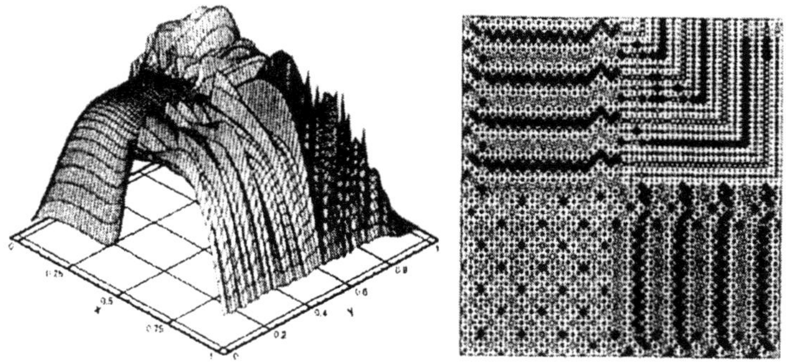

(a) (b)

Fig. 3. (a) "Smooth" error in case of problem (12). (b) The finest and three consecutive levels created by the standard AMG coarsening algorithm.

starting with a random function. The different anisotropies as well as the discontinuities across the interface lines are clearly reflected in the picture.

It is heuristically clear that such an error can only be effectively reduced by means of a coarser grid if that grid is obtained by essentially coarsening in the directions in which the error really changes smoothly in the geometric sense and if interpolation treats the discontinuities correctly. Indeed, see Section 3.2, this is exactly what AMG does. First, the operator-based interpolation ensures the correct treatment of the discontinuities. Second, AMG coarsening is in the direction of strong connectivity, that is, in the direction of smoothness.

Fig. 3b depicts the finest and three consecutive grids created by using standard AMG coarsening and interpolation. The smallest dots mark grid points which are contained *only* on the finest grid, the squares mark those points which are also contained on the coarser levels (the bigger the square, the longer the corresponding grid point stays in the coarsening process). The picture shows that coarsening is uniform in the lower left quarter where the problem is isotropic. In the other quarters, AMG adjusts itself to the different anisotropies by locally coarsening in the proper direction. For instance, in the lower right quarter, coarsening is in the x-direction only. Since AMG takes only *strong* connections in coarsening into account and since all connections in the y-direction are *weak*, the individual lines are coarsened *independently of each other*. Consequently, the coarsening of neighboring x-lines is not "synchronized"; it is actually a matter of "coincidence" where coarsening starts within each line. This has to be observed in interpreting the coarsening pattern in the upper right quarter: within each diagonal line, coarsening is essentially in the direction of this line.

4.2. Complex applications

For a demonstration of AMG's efficiency, we consider some complex problems of the type typically solved in two commercial codes designed for *oil reservoir simulation* and for *computational fluid dynamics*, respectively. In both codes, the numerical kernel requires the fast solution of scalar elliptic equations. While, in oil reservoir simulation, geometries are typically fairly simple but the underlying problems have strongly anisotropic and discontinuous coefficients (jumps by several orders of magnitude in a nearly random way), in computational fluid dynamics these problems are

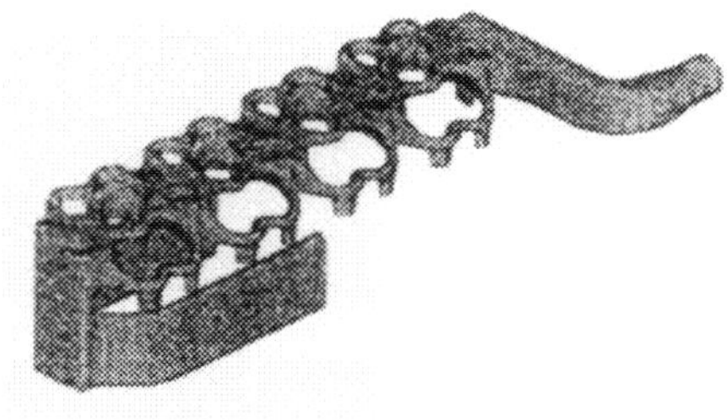

Fig. 4. Cooling jacket of a four-cylinder engine.

Poisson-like but defined on very complex, unstructured grids. For more details on these codes, see [51].

The following test cases are considered: [3]

(1) *Reservoir*. The underlying reservoir corresponds to a simple domain discretized by a mesh consisting of 1.16 million cells. The variation of absolute permeabilities results in a discontinuous variation of the coefficients by four orders of magnitude.
(2) *Cooling jacket*. Computation of the flow through the cooling jacket of a four-cylinder engine. The underlying mesh consists of 100 000 tetrahedra cells (see Fig. 4).
(3) *Coal furnace*. Computation of the flow inside the model of a coal furnace. The underlying mesh consists of 330 000 hexahedra and a few thousand pentahedra, including many locally refined grid patches.
(4) *Underhood*. Computation of the underhood flow of a Mercedes-Benz E-class model. The mesh is highly complex and consists of 910 000 cells (see Fig. 1).
(5) *E-Class*. Computation of the exterior flow over a Mercedes-Benz E-class model (see Fig. 5). The original mesh consists of 10 million cells. Due to memory restrictions, our test runs to two reduced mesh sizes consisting of 2.23 and 2.82 million cells, respectively. (Note that the underlying mesh also includes all modeling details of the previous underhood case.)

Memory requirement is a major concern for any commercial software provider. Industrial users of commercial codes always drive their simulations to the limits of their computers, shortage of memory being a serious one. For these reasons, in a commercial environment, low-memory AMG approaches are of primary interest, even if the reduction of the memory requirement is at the expense of a (limited) increase of the total computational time. Compared to standard one-level solvers, a memory overhead of some tens of percents is certainly acceptable. In any case, however, the operator complexity c_A (11) must not be larger than 2.0 say. Therefore, in the following test runs, we employ an aggressive coarsening strategy (cf. Section 3.1) and, in order to make up for the resulting reduced convergence speed, use AMG as a pre-conditioner rather than stand-alone.

Fig. 6 shows AMG's V-cycle convergence histories for each of the above cases, based on the code RAMG05 mentioned in the introduction. The results reflect the general experience that the convergence of AMG depends, to a limited extent, on the type of elements used as well as on the type of problem, but hardly on the problem size. In particular, the three Mercedes meshes are

[3] The first case has been provided by StreamSim Technologies Inc., the other ones by Computational Dynamics Ltd.

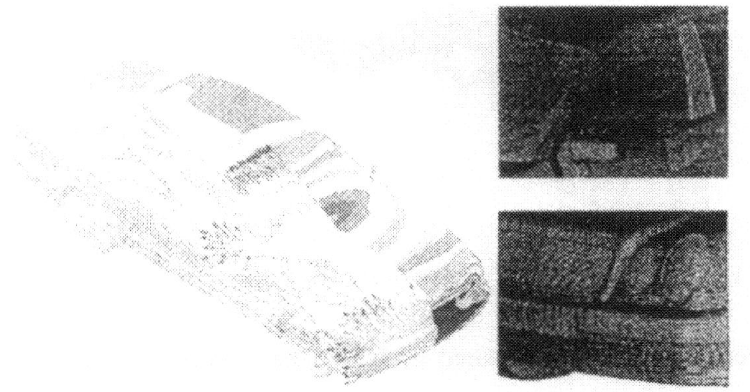

Fig. 5. Model of a Mercedes-Benz E-Class.

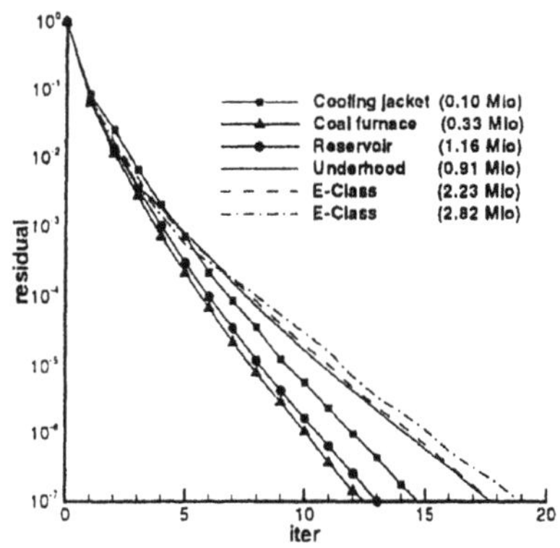

Fig. 6. RAMG05 convergence histories for various problems.

	RAMG05/cg			ILU(0)/cg		AMG1R5
problem	time	iter	c_A	time	iter	c_A
jacket	12.3	21	1.44	218.2	926	5.35
furnace	45.7	18	1.47	292.8	286	7.06
reservoir	165.0	18	1.41	2707.0	720	7.66
underhood	172.9	25	1.43	1364.0	461	5.64
eclass (2.23)	438.8	25	1.46	8282.0	1151	6.24

Fig. 7. Performance of RAMG05 vs. ILU(0)/cg.

comparable in their type but their size varies by more than a factor of three. Convergence, obviously, is influenced only marginally.

Fig. 7 compares the RAMG05 performance with that of ILU(0) pre-conditioned conjugated gradient. For both methods and for each of the above problems, the number of iterations as well as total run times (in s), required to reduce the residuals by nine digits, are shown. Compared to ILU(0)/cg.

AMG reduces the number of iterations by up to a factor of 46. In terms of run time, AMG is up to 19 times faster. The figure also shows that the industrial requirements in terms of memory, mentioned before, are fully met. In fact, the A-complexity (11) is very satisfactory for all cases, namely $c_A \approx 1.45$.

For a comparison, the last column in the figure shows the unacceptably high complexity values of RAMG05's forerunner, AMG1R5. As already mentioned in Section 3.1, AMG1R5 typically performs quite well in the case of 2D problems. In complex 3D cases as considered here, however, the results clearly demonstrate one of the advantages of the different coarsening and interpolation approaches used in RAMG05 (Fig. 7). (For more information on the differences in the two codes we refer to [51].)

5. AMG based on mere F-relaxation

In this section, we consider a very different approach [28,51] which can be used to *force* the right-hand side of (4) to become small. For a description, we assume vectors and matrices to be re-ordered so that, w.r.t a given C/F-splitting, the set of equations (1) can be written in block form

$$A_h u = \begin{pmatrix} A_{FF} & A_{FC} \\ A_{CF} & A_{CC} \end{pmatrix} \begin{pmatrix} u_F \\ u_C \end{pmatrix} = \begin{pmatrix} f_F \\ f_C \end{pmatrix} = f. \tag{13}$$

Correspondingly, the interpolation operator is re-written as $I_H^h = (I_{FC}, I_{CC})^{\mathrm{T}}$ with I_{CC} being the identity operator. Instead of $e^h = I_H^h e^H$, we simply write $e_F = I_{FC} e_C$.

5.1. The basic idea

The approach mentioned above is based on the fact that the sub-matrix A_{FF} is very well conditioned if we just select the C/F-splitting accordingly. For instance, for all problems we have in mind here, we can easily force A_{FF} to be *strongly diagonally dominant*,

$$a_{ii} - \sum_{j \in F, j \neq i} |a_{ij}| \geqslant \delta a_{ii} \quad (i \in F) \tag{14}$$

with some fixed $\delta > 0$. Assuming this to hold in the following, we can efficiently approximate the solution of the F-equations (with frozen e_C),

$$A_{FF} e_F + A_{FC} e_C = 0 \tag{15}$$

for instance, by relaxation (in the following called *F-relaxation*). Using this as the basis for both the definition of smoothing *and* interpolation, we can force the right-hand side of (4) to become as small as we wish.

To be more specific, given any e_C, interpolation is defined by applying μF-relaxation steps to approximately solve (15). In order to keep the resulting operator as "local" as possible, we only consider Jacobi-relaxation (below, we refer to this as *Jacobi-interpolation*). That is, we iteratively define a sequence of operators,

$$I_{FC}^{(\mu)} = P_{FF} I_{FC}^{(\mu-1)} - D_{FF}^{-1} A_{FC} \quad \text{where } P_{FF} = I_{FF} - D_{FF}^{-1} A_{FF}, \tag{16}$$

starting with some reasonable first-guess interpolation operator, $I_{FC}^{(0)}$. Because of (14), we have rapid convergence $(I_H^h)^{(\mu)}e_C \rightarrow \hat{e}\,(\mu \rightarrow \infty)$ at a rate which depends only on δ. Here $\hat{e} := (\hat{e}_F, e_C)^{\mathrm{T}}$ where $\hat{e}_F := -A_{FF}^{-1}A_{FC}e_C$ denotes the solution of (15).

Similarly, we also use F-relaxation for smoothing (referred to as *F-smoothing* below). That is, we define one smoothing step by $u \rightarrow \bar{u}$ where

$$Q_{FF}\bar{u}_F + (A_{FF} - Q_{FF})u_F + A_{FC}u_C = f_F, \quad \bar{u}_C = u_C. \tag{17}$$

In contrast to the interpolation, we here normally use Gauss–Seidel relaxation, i.e., Q_{FF} is the lower triangular part of A_{FF} (including the diagonal). The corresponding smoothing operator is easily seen to be

$$S_h^\nu e = \begin{pmatrix} S_{FF}^\nu(e_F - \hat{e}_F) + \hat{e}_F \\ e_C \end{pmatrix} \quad \text{where } S_{FF} = I_{FF} - Q_{FF}^{-1}A_{FF}. \tag{18}$$

As with the interpolation, for any given $e = (e_F, e_C)^{\mathrm{T}}$, we have rapid convergence $S_h^\nu e \rightarrow \hat{e}\ (\nu \rightarrow \infty)$.

5.2. Two-level convergence

For various classes of matrices A one can show that F-smoothing and Jacobi-interpolation can be used to obtain *matrix-independent* two-level convergence if the first-guess interpolation, $I_{FC}^{(0)}$, is selected properly. Moreover, two-level convergence becomes arbitrarily fast if ν, μ are chosen sufficiently large. As an example, we again consider the class of M-matrices (cf. Theorem 1).

Theorem 2. *Let A be a symmetric, weakly diagonally M-matrix. Define the interpolation by applying $\mu \geqslant 0$ Jacobi F-relaxation steps, using an interpolation as defined in Theorem 1 as the first-guess (with fixed $0 < \tau \leqslant 1$). Then, if $\nu \geqslant 1$ Gauss–Seidel F-relaxation steps are used for (pre-) smoothing, the following two-level convergence estimate holds,*

$$\|KS^\nu\|_A \leqslant \|S_{FF}\|_{A_{FF}}^\nu + \tilde{\tau}\|P_{FF}\|_{A_{FF}}^\mu,$$

where $\|S_{FF}\|_{A_{FF}} < 1$ and $\|P_{FF}\|_{A_{FF}} < 1$, and both norms as well as $\tilde{\tau}$ depend only on τ but not on the matrix A.

In this theorem, we have used the interpolation from Theorem 1 as a first guess. In particular, we assume the C/F-splitting to satisfy (9) which can easily be seen to ensure strong diagonal dominance (14) with $\delta = \tau$. Although one may think of various other ways to define the first-guess interpolation, we want to point out that a proper selection of the first-guess interpolation is important for obtaining matrix-independent two-level convergence (it is, for instance, not sufficient to simply select $I_{FC}^{(0)} = 0$). Generally, the first-guess interpolation has to be such that the Galerkin operator which corresponds to it, $A_H^{(0)}$, is spectrally equivalent to the Schur complement, $A_{CC} - A_{CF}A_{FF}^{-1}A_{FC}$, w.r.t. all matrices in the class under consideration. For more details and generalizations of the above theorem as well as the proofs, see [51].

Note that the AMG approach discussed here is not really in the spirit of standard multigrid since smoothing in the usual sense is not exploited. In fact, the role of F-smoothing is merely to *force* $S^\nu e \approx \hat{e}$ rather than to smooth the error of the full system. This, together with Jacobi-interpolation, is a "brute force" approach to make $\|S^\nu e - I_H^h e_C\|_A$ small for all $e = (e_F, e_C)^{\mathrm{T}}$.

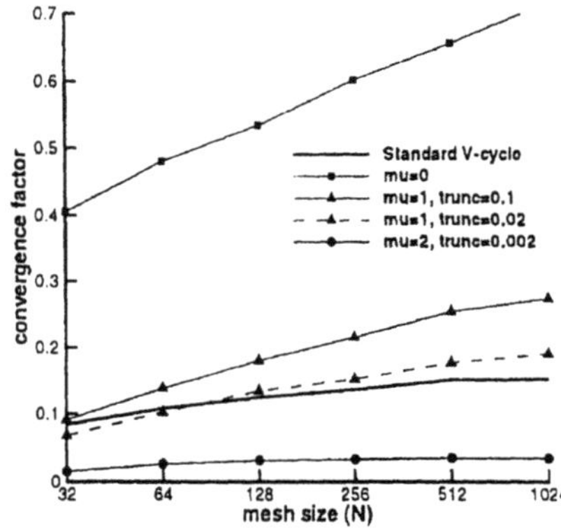

Fig. 8. Convergence factors of AMG based on F-relaxation.

5.3. *Practical remarks*

The mere fact that AMG can be *forced* to converge as fast as we wish, is only of little relevance in practice. Each F-relaxation step applied to the interpolation increases its "radius" by one additional layer of couplings, causing increased fill-in in the Galerkin operator. The resulting gain in convergence speed is, generally, more than eaten up for by a corresponding increase of matrix complexities towards coarser levels. Consequently, the main problem is the tradeoff between convergence and numerical work (which is directly related to the memory requirements). Note that this is, in a sense, just opposite to geometric multigrid where the numerical work per cycle is known and controllable but the convergence may not be satisfactory.

For a practical realization of Jacobi-interpolation, several things are important to observe. First, most of the new entries introduced by each additional relaxation step will be relatively small and can be truncated (before computing the Galerkin operator) without sacrificing convergence seriously (cf. also Section 3.1). Second, it is usually not necessary to perform F-relaxation with the complete matrix A_{FF}. Instead, one may well ignore all those entries of A_{FF} which are relatively small (and add them to the diagonal, say, in order to preserve the row sums of interpolation). Finally, we want to remark that, although Theorem 2 states fast convergence only if μ is sufficiently large, in practice, $\mu > 2$ is hardly ever required (at least if δ is not too small).

Fig. 8 shows some V-cycle convergence factors as a function of the mesh size for model equation

$$-((1+\sin(x+y))u_x)_x - (e^{x+y}u_y)_y = f(x,y), \tag{19}$$

discretized on the unit square with uniform mesh size $h=1/N$. We first observe the rapid h-independent convergence of the "standard" AMG V-cycle (corresponding to the approach outlined in Section 3, using one full Gauss–Seidel relaxation step both for pre- and post-smoothing). Convergence drastically drops, and becomes strongly h-dependent, if we just replace each full smoothing step by two F-smoothing steps and leave interpolation unchanged (case $\mu=0$). This has to be expected since the definition of interpolation in classical AMG is based on the assumption that the error after relaxation is *algebraically smooth* (cf. Section 3.1). This is, clearly, not true if only *partial* relaxation, such as F-relaxation, is performed. However, if we use just one Jacobi F-relaxation step to improve interpolation ($\mu=1$), convergence becomes comparable to that of the standard cycle. Results are shown using two different truncation parameters, 0.1 and 0.02, respectively. Finally, the case $\mu=2$ (and four partial relaxation steps for smoothing rather than two) gives a convergence which is about twice as fast as that of the standard cycle.

We note that, if computational time and memory requirement is taken into account in this example, the standard V-cycle is more efficient than the others. In particular, the cycle employing $\mu=2$ is substantially inferior, mainly due to the considerably higher setup cost. This seems typical for applications for which algebraically smooth errors, in the sense of Section 3.1, can be characterized sufficiently well. The heuristic reason is that then, using *full* smoothing steps, relatively simple interpolations of the type outlined in Section 3.1 are usually sufficient to approximate algebraically smooth errors and obtain fast convergence. This is no longer true if mere F-smoothing is employed and, generally, additional effort needs to be invested to "improve" interpolation by F-relaxation in order to cope with all those error components which are not affected by mere F-smoothing. (In particular, note that all error components of the form $\hat{e}$ are not reduced at all by F-smoothing.)

In general, however, when the characterization of algebraically smooth errors is less straightforward, the use of F-relaxation provides a means to enhance convergence. Further numerical experiments employing F-smoothing and Jacobi-interpolation can be found in [28,51]. F-relaxation is a special case of a "compatible relaxation" which, in a more general context, is considered in [11].

6. Aggregation-type AMG

In the previous sections, we have considered increasingly complex interpolation approaches. In this section, we go back and consider the most simple case that each F-variable interpolates from *exactly* one C-variable only. We have already pointed out in Section 3.2 that the use of such "one-sided" interpolations is not recommendable. In fact, one important goal of the additional objective introduced in Section 3.2 was just *to avoid* such extreme interpolations. On the other hand, the resulting method is so easy to implement that it, nevertheless, has drawn some attention. We will outline the fundamental problems with this approach in Section 6.2 and summarize three possibilities of improvement in Sections 6.3–6.5. Since we just want to highlight the main ideas, we restrict our motivations to very simple but characteristic (Poisson-like) problems.

6.1. The basic approach

Let us consider C/F-splittings and interpolations (6) where, for each $i \in F$, $w_{ik}=1$ for just one particular $k \in C$ and zero otherwise. Consequently, the total number of variables can be subdivided into "aggregates" I_k ($k \in C$) where I_k contains (apart from k itself) all indices i corresponding to F-variables which interpolate from variable k (see Fig. 9).

With this notation, the computation of the Galerkin operator becomes very simple. One easily sees that

$$I_h^H A_h I_H^h = (a_{kl}^H) \quad \text{where } a_{kl}^H = \sum_{i \in I_k} \sum_{j \in I_l} a_{ij}^h \quad (k,l \in C), \tag{20}$$

that is, the coefficient a_{kl}^H is just the sum of all cross-couplings between I_k and I_l. Obviously, regarding the coefficients a_{kl}^H, the particular role of the variables k and l (as being C-variables) is

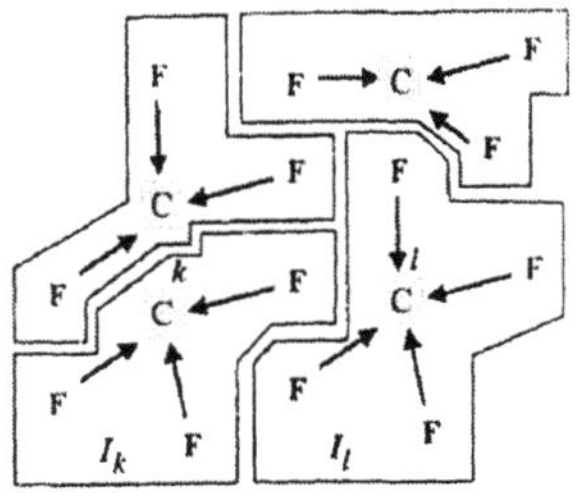

Fig. 9. Subdivision of fine-level variables into aggregates. The arrows indicate which C-variable an F-variable interpolates from.

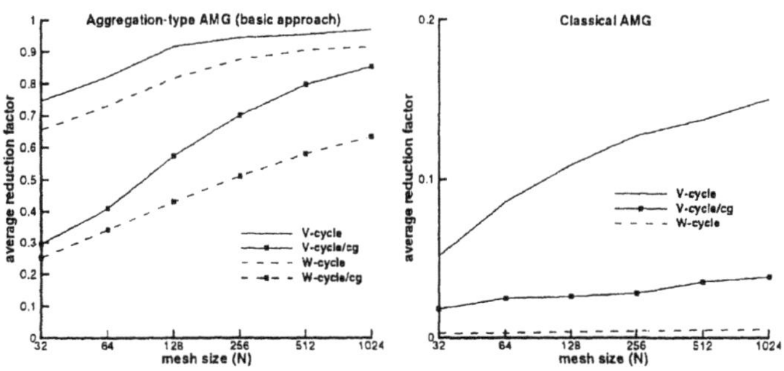

Fig. 10. Convergence of (a) aggregation-type AMG, (b) classical AMG.

not distinguished from the other variables; the Galerkin operator merely depends *on the definition of the aggregates*. Consequently, we might as well associate each aggregate I_k with some "new" coarse-level variable which has no direct relation to the C-variable k. The above interpolation then is nothing else than *piecewise constant interpolation* from these new coarse-level variables to the associated aggregates.

Originally, such *aggregation-type* AMG approaches [52,53,9] have been developed the other way around: Coarsening is defined by building aggregates (rather than constructing C/F-splittings), a new coarse-level variable is associated with each aggregate and interpolation I_H^h is defined to be piecewise constant. (That is, the set of coarse-level variables is generally *not* considered as a subset of the fine-level ones.) Clearly, for a given subdivision into aggregates to be reasonable, all variables in the same aggregate should strongly depend on each other. Otherwise, piecewise constant interpolation makes no real sense.

As expected, an immediate implementation of this simple coarsening and interpolation approach leads to very inefficient solvers, even if used only as a pre-conditioner. Fig. 10a shows the typical convergence of both the V- and the W-cycle, used as stand-alone and as pre-conditioner, in solving the model equation (19). Convergence is indeed very slow and exhibits a *strong h-dependency*. For a comparison, the much better performance of classical AMG is depicted in Fig. 10b.

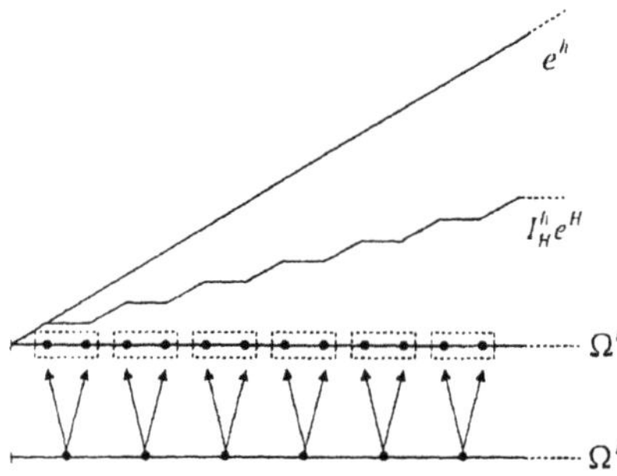

Fig. 11. Optimal approximation $I_H^h e^H$ of e^h w.r.t. the energy norm.

6.2. The reason for slow convergence

The main reason for this unsatisfactory convergence is that piecewise constant interpolation is not able to approximate the *values* of smooth error if approximation is based on the *energy norm* (cf. (4)). In fact, the approximation order becomes zero.

To illustrate this, let us consider the most simple case that A_h is derived from discretizing $-u''$ on the unit interval with meshsize h, i.e., the rows of A_h correspond to the difference stencil

$$\frac{1}{h^2}[-1 \quad 2 \quad -1]_h$$

with Dirichlet boundary conditions. Let e^h be an error satisfying the homogeneous boundary conditions. According to the variational principle, the corresponding two-level correction, $I_H^h e^H$, is optimal in the sense that it minimizes $\|e^h - I_H^h e^H\|_{A_h}$ w.r.t. all possible corrections in $\mathcal{R}(I_H^h)$. (At this point, the concrete choice of I_H^h is not relevant.) This implies that $I_H^h e^H$ minimizes

$$\|v^h\|_{A_h}^2 = (A_h v^h \cdot v^h) = \frac{1}{2h^2} \sum_{i,j}{}' (v_i^h - v_j^h)^2 + \sum_i s_i (v_i^h)^2, \tag{21}$$

where $v^h = e^h - I_H^h e^H$ and $s_i = \sum_j a_{ij}^h$. (The prime indicates that summation is only over neighboring variables i and j.) This, in turn, means that, away from the boundary (where we have $s_i \equiv 0$), the Euclidian norm of the *slope* of v^h is minimal. At the boundary itself we have $s_i \neq 0$, and v^h equals zero.

Assuming now the aggregates to be built by joining the pairs of neighboring variables, the result of this minimization is illustrated in Fig. 11 (see also [9,10]). We here consider a smooth error e^h in the neighborhood of the left boundary of the unit interval. On each aggregate, interpolation is constant and the slope of $I_H^h e^H$ necessarily vanishes. On the remaining intervals, the Euclidian norm of the slope of v^h becomes minimal if the slope of $I_H^h e^H$ equals that of e^h. Consequently, $I_H^h e^H$ has, on the average, *only half the slope of* e^h (independent of h). That is, the resulting approximation is off by a factor of approximately 0.5 if compared to the best approximation in the Euclidian sense. (Note that subsequent smoothing smooths out the "wiggles", but does not improve the quality of the correction.) Accordingly, the Galerkin operator, which can easily be computed, turns out to be too large by a factor of two compared to the "natural" $2h$-discretization of $-u''$.

If the same strategy is now used recursively to introduce coarser and coarser levels, the above arguments carry over to each of the intermediate levels and, in particular, each coarser-grid Galerkin operator is off by a factor of 2 compared to the previous one. A simple recursive argument shows that errors are accumulated from grid to grid and the asymptotic V-cycle convergence factor cannot be expected to be better than $1 - 2^{-m}$, where m denotes the number of coarser levels. That is, *the V-cycle convergence is strongly h-dependent.*

6.3. Improvement by re-scaling the Galerkin operators

The fact that piecewise constant interpolation produces badly scaled AMG components, was the basis for an improvement introduced in [9]. In that paper, it is demonstrated that convergence can be substantially improved by just multiplying corrections $I_H^h e^H$ by some suitable factor $\alpha > 1$ ("over-correction"). This is equivalent to re-scaling the Galerkin operator by $1/\alpha$

$$I_h^H A_h I_H^h \rightarrow \frac{1}{\alpha} I_h^H A_h I_H^h$$

and leaving everything else unchanged.

In case of the simple model equation $-u''$ considered in the previous section, $\alpha = 2$ would be the optimal choice. However, the main arguments carry over to the Poisson equation in 2D and 3D, assuming a uniform grid and the aggregates to be built by 2×2 and $2 \times 2 \times 2$ blocks of neighboring variables, respectively. In case of more general problems and/or different grids, the optimal weight is no longer $\alpha = 2$. Nevertheless, it has been demonstrated in [9] that a slightly reduced value of $\alpha = 1.8$ (in order to reduce the risk of "overshooting") yields substantially improved V-cycle convergence for various types of problems, *if the cycle is used as a pre-conditioner* and if the number of coarser levels is kept fixed (in [9] four levels are always used). Smoothing is done by symmetric Gauss–Seidel relaxation sweeps.

A comparison of Figs. 10a and 12a shows the convergence improvement if re-scaling by $\alpha = 1.8$ is applied to the model equation (19). (In contrast to [9], we here have not restricted the number of coarser levels.) Fig. 12a shows that there is indeed a risk of "overshooting": For larger meshes, the V-cycle starts to diverge. (Note that the above re-scaling destroys the validity of the variational principle and the iteration process may well diverge.) Using the V-cycle as a pre-conditioner, eliminates the problem.

We want to point out that the above comparison shows only the tendency of improvements due to re-scaling, the concrete gain depends on how the aggregates are chosen precisely (which is not optimized here and can certainly be improved to some extent). In any case, the gain in convergence, robustness and efficiency of this (very simple and easily programmable) approach are somewhat limited, one reason being that a good value of α depends on various aspects such as the concrete problem, the type of mesh and, in particular, the type and size of the aggregates. For instance, if the aggregates are composed of three neighboring variables (rather than two) in each spatial direction, the same arguments as in the previous section show that the best weight would be $\alpha \approx 3$ in case of Poisson's equation. If the size of the aggregates strongly varies over the domain, it becomes difficult to define a good value for α.

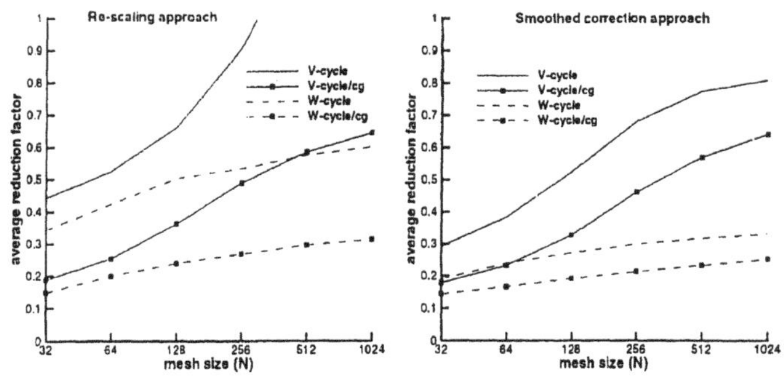

Fig. 12. (a) Re-scaling approach ($\alpha = 1.8$), (b) smoothed correction approach.

6.4. *Improvement by smoothing corrections*

Rather than explicitly prescribing a scaling factor α as before, a reasonable scaling can also be performed automatically. The idea is to modify the coarse-level correction step (2) by replacing the true (piece-wise constant) correction $e^h = I_H^h e^H$ by some approximation, e_0^h, and then compute u_{new}^h by

$$u_{\mathrm{new}}^h = u_{\mathrm{old}}^h + \alpha e_0^h \quad \text{with } \alpha = \frac{(f^h - A_h u_{\mathrm{old}}^h, e_0^h)}{(A_h e_0^h, e_0^h)}, \tag{22}$$

instead of (2). Note that α is defined so that the energy norm of the error of u_{new}^h becomes minimal.

Clearly, for this minimization to be meaningful, the selection of e_0^h is crucial. Most importantly, e_0^h should be some sufficiently *smooth* approximation to e^h. (The choice $e_0^h = e^h$ would not give any gain: The variational principle would just imply $\alpha = 1$.) One possible selection is

$$e_0^h = S_h^\nu e^h, \tag{23}$$

which requires the application of ν smoothing steps to the *homogeneous* equations (starting with e^h). Note that, loosely speaking, this process will leave the "smooth part" of e^h essentially unchanged; only its "high-frequency part" will be reduced. Consequently, the regular smoothing steps, applied to u_{new}^h after the coarse-grid correction, will effectively correct this.

The effect of this process of *smoothing corrections*, is demonstrated in Fig. 12b (using $\nu = 2$). Apart from the fact that, compared to the re-scaling approach (see Fig. 12a), convergence is slightly better here, there is no risk of "overshooting" as before since the process "controls itself". On the other hand, the additional smoothing steps increase the overall computing time.

Although smoothing of corrections is a simple means to automatically correct wrong scalings to some extent, its possibilities are limited. In any case, the resulting overall performance is generally worse than that of classical AMG.

6.5. *Improvement by smoothing the interpolation*

A more sophisticated (but also more costly) way to accelerate the basic aggregation-type AMG approach is developed and analyzed in [52–54]. Here, piecewise constant interpolation is only considered as a first-guess interpolation which is improved by some *smoothing process* ("smoothed

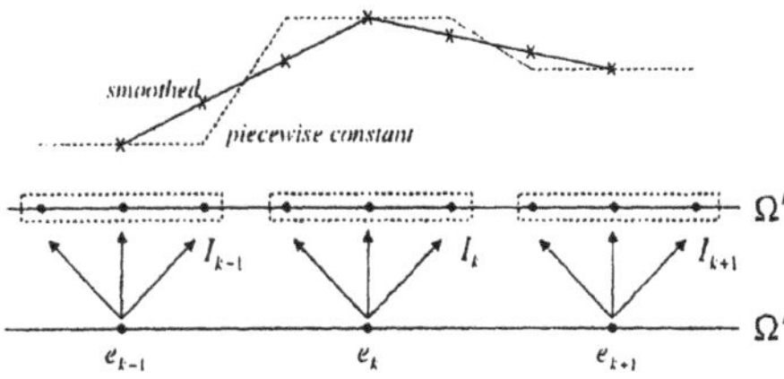

Fig. 13. Piecewise constant versus smoothed interpolation.

aggregation") before the Galerkin operator is computed. In [52,53], this smoothing is proposed to be done by applying one ω-Jacobi-relaxation step.

To be more specific, denote the operator corresponding to piecewise constant interpolation by $\tilde{I}_H^h$. Then the interpolation actually used is defined by

$$I_H^h = (I_h - \omega D_h^{-1} A_h^f)\tilde{I}_H^h,$$

where $D_h = \mathrm{diag}(A_h^f)$ and A_h^f is derived from the original matrix A_h by adding all weak connections to the diagonal ("filtered matrix"). That is, given some coarse-level vector e^H, $e^h = I_H^h e^H$ is defined by applying one ω-Jacobi relaxation step to the homogeneous equations $A_h^f v^h = 0$, starting with the piecewise constant vector $\tilde{I}_H^h e^H$. (Note that this process will increase the "radius" of interpolation and, hence, destroy the simplicity of the basic approach. Moreover, interpolation will generally *not* be of the special form (5) any more. Note also that here Jacobi relaxation serves a quite different purpose than Jacobi-F-relaxation as considered in Section 5. In particular, Jacobi-relaxation is here used as a smoother, applied to the full system of equations, which requires the use of an under-relaxation parameter, ω.)

To illustrate this process, we again consider the 1D case of $-u''$ and assume the aggregates to consist of three neighboring variables (corresponding to the typical size of aggregates used in [52,53] in each spatial direction). Note first that, since all connections are strong, we have $A_h^f = A_h$. Fig. 13 depicts both the piecewise constant interpolation (dashed line) and the smoothed interpolation obtained after the application of one Jacobi-step with $\omega = \frac{2}{3}$ (solid line). Obviously, the smoothed interpolation just corresponds to *linear* interpolation if the coarse-level variables are regarded as the fine-level analogs of those variables sitting in the center of the aggregates. Linear interpolation, however, does not exhibit a scaling problem as described in Section 6.2 for piecewise constant interpolation.

Of course, in more general situations, relaxation of piecewise constant interpolation will not give exact linear interpolation any more and a good choice of ω depends on the situation. Nevertheless, even if $\omega = \frac{2}{3}$ is kept fixed, smoothed interpolation will typically be much better than the piecewise constant one. (Actually, the real advantage of smoothed compared to piecewise constant interpolation is that errors, obtained after interpolation from the coarser level, have a much lower "energy"; see also Section 7.) This is demonstrated in [53] by means of various examples using a mixture of Gauss–Seidel and SOR sweeps for error smoothing. The tendency is to compose aggregates of three neighboring variables in each spatial direction. Note that a good value for ω depends not only on the problem and the underlying mesh, but also on the size of the aggregates. If, instead, only *two* neighbors would be aggregated in each spatial direction, one easily confirms that $\omega \approx 0.5$ should be chosen.

In general, classical AMG and AMG based on smoothed aggregation perform comparably if applied to relatively smooth (Poisson-like) problems. A certain advantage of aggregation-type AMG is that it, often, requires less memory than classical AMG (due to its very fast coarsening which causes a particularly low operator complexity c_A). On the other hand, this is payed for by a reduced robustness: The aggregation-type code seems to *require* acceleration by conjugate-gradient to maintain an acceptable efficiency and robustness in more complex situations. Since classical AMG puts more effort into the construction of interpolation and performs a slower coarsening, its performance generally depends on aspects such as strong discontinuities only to a lesser extent (for some examples, see [51]).

7. Further developments and conclusions

Algebraic multigrid provides a general approach to developing robust and efficient methods for solving large classes of matrix equations such as those typically arising in the numerical solution of elliptic PDEs, on structured as well as unstructured grids, in 2D and 3D. The construction of suitable interpolation operators (including the coarsening process) is crucial for obtaining fast and (nearly) h-independent convergence. Generally, the more effort is put into this construction, the faster the convergence can be, but, unfortunately, the required numerical work may increase even faster. That is, the main problem in designing efficient AMG algorithms is the tradeoff between convergence and numerical work, and keeping the balance between the two is the ultimate goal of any practical algorithm.

We have summarized and discussed various possibilities to realize concrete AMG algorithms. For most applications of the type discussed in this paper, methods based on the classical AMG approach still belong to the most efficient ones. An extensive list of experiments, based on the original code AMG1R5, can be found in [18]. Robustness and efficiency can substantially be improved, in particular in case of complex 3D meshes, by employing modifications as mentioned in Sections 3.1 and 5 (for more details, see [51]). AMG methods based on smoothed aggregation (see Section 6.5) are an efficient alternative to classical AMG, at least if employed as a pre-conditioner rather than stand-alone. Further developments and applications which are close to the classical AMG ideas are, for example, contained in [16,23,24,26,32,33,40,62,63]. Related approaches, but with a focus on different coarsening and interpolation strategies, are, for example, found in [22,27]. Applications of the aggregation-type approach in computational fluid dynamics are found in [30,39,46,60].

However, there are still many applications for which algebraically defined interpolation, and hence the resulting AMG performance, are not yet satisfactory. For instance, one of the major current research activities in AMG aims at its generalization to efficiently treat *systems* of PDEs such as linear elasticity problems. Although AMG has successfully been applied to various cases (see, e.g., [9,14,31,48,53]), its development has not yet reached a state where a particular approach is well settled. However, even for scalar applications, there are still questions about best ways to define coarsening and interpolation, for instance, if the given matrix is s.p.d., contains relatively large positive off-diagonal entries, and is far from being weakly diagonally dominant. In such cases, the performance of classical AMG may be only suboptimal.

It is often possible to avoid such situations by simplifying the given matrix before applying AMG [41]. One can also imagine situations where it would be advantageous (and easy) to provide AMG with some additional information on the problem at hand. For instance, information on the geometry (in terms of point locations) or more concrete descriptions on what an "algebraically smooth" error looks like (e.g. in the form of some user-provided "test-vector(s)"). This additional information can be used to fit AMG's interpolation in order to approximate certain types of error components particularly well. Straightforward possibilities have already been pointed out in [48].

In the following, we briefly summarize a few more recent approaches to define interpolation which aim at increasing the robustness in cases such as those mentioned above.

A new way to construct interpolation (AMGe, [14]) starts from the fact that an algebraically smooth error is nothing else but an error which is slow-to-converge w.r.t. the relaxation process. Hence, an algebraically smooth error, generally, corresponds to the eigenvectors of A belonging to the smallest eigen-values. Instead of defining interpolation by directly approximating (7), the goal in [14] is to define interpolation so that the smaller the associated eigenvalue is the better the eigenvectors are interpolated. To satisfy this by explicitly computing eigenvectors is, of course, much too expensive. However, in the case of finite element methods – assuming the element stiffness matrices to be known – one can derive measures (related to measures used in classical multigrid theory) whose minimization allows the determination of *local* representations of algebraically smooth error components in the above sense. The added robustness has been demonstrated in [14] by means of certain model applications. However, the approach is still in its infancy. In particular, significant development work still has to be invested to link the processes of coarsening and interpolation definition in order to obtain an optimal algorithm. In any case, it is an interesting new approach which has the potential of leading to more generally applicable AMG approaches.

Other algebraic approaches, designed for the solution of equations derived from finite-element discretizations, have been considered in [31,58]. Both approaches are aggregation based and the coarse space basis functions are defined so that their energy is minimized in some sense. (In the finite-element context it is natural to define interpolation implicitly by constructing the coarse space basis functions.) This does not require the element stiffness matrices to be known, but leads to a *global* (constraint) minimization problem the exact solution of which would be very expensive. However, iterative solution processes are proposed in both papers to obtain approximate solutions, indicating that the extra work (invested in the setup phase) is acceptable. While Wan et al. [58] concentrate on scalar applications, an extension to systems of PDEs (from linear elasticity) is one major aspect in [31]. Special attention is paid to the correct treatment of zero-energy modes (e.g. rigid-body modes in case of linear elasticity): such modes should be contained in the span of the coarse space basis functions, at least away from Dirichlet boundaries. (Note that, for typical scalar problems, this corresponds to the requirement that constants should be interpolated exactly away from Dirichlet boundaries, cf. (8).) It is interesting that the approach in [31] can be regarded as an extension of the earlier work [53] on smoothed aggregation: if only one iteration step is performed to approximately solve the energy minimization problem, the resulting method coincides with the smoothed aggregation approach. In contrast to the latter, however, further iterations will *not* increase the support of the basis functions (i.e., the radius of interpolation). Some test examples in [31] indicate the advantages of this new interpolation in terms of convergence speed. Unfortunately, however, this benefit is essentially offset by the expense of the minimization steps.

There are various other papers with focus on the development of multigrid methods to solve finite-element problems on unstructured grids. Although some of them are also based on algorithmical components which are, more or less, algebraically defined, most of them are not meant to be algebraic multigrid solvers in the sense as considered in this paper. We therefore do not want to discuss such approaches here further but rather refer, for example, to [15] and the references given therein.

In the approach of [56], A is not assumed to be asymmetric, and interpolation and restriction are constructed separately. Interpolation, for instance, is constructed so that a smooth error, $S_h e^h$, is interpolated particularly well w.r.t. the *Euclidian* norm, $\|\cdot\|_2$. More precisely, the attempt is to make

$$\|S_h e^h - I_H^h e^H\|_2,$$

where e^H denotes the straight injection of $S_h e^h$ to the coarse level, as small as possible (cf. (4)). In [56], this leads to certain local minimizations which are used to find, for each variable, pairs of neighboring variables which would allow a good interpolation in the above sense, and, at the same time, compute the corresponding weights (of both the interpolation and the restriction). Based on this information, a C/F-splitting is constructed which allows each F-variable to interpolate from one of the pairs found before. A heuristic algorithm is used to minimize the total number of C-variables.

In this context, we want to point out that, although classical AMG has been developed in the variational framework, it has successfully been applied to a large number of non-symmetric problems without any modification. This can be explained heuristically but no theoretical justification is available at this time. In the context of smoothed aggregation-based AMG, a theoretical analysis can be found in [25].

An important aspect which has not been addressed in this paper is the parallelization of AMG. An efficient parallelization of classical AMG is rather complicated and requires certain algorithmical modifications in order to limit the communication cost without sacrificing convergence significantly. Most parallelization approaches investigated up to now either refer to simple aggregation-based variants (e.g. [46]) or use straightforward domain decomposition techniques (such as Schwarz' alternating method) for parallelization. A parallelization strategy which stays very close to the classical AMG approach is presented in [29]. Results for complex 3D problems demonstrate that this approach scales reasonably well on distributed memory computers as long as the number of unknowns per processor is not too small. The method discussed in [56] is also available in parallel. There are several further ongoing parallelization activities, for instance, at the University of Bonn and the National Laboratories LLNL [17] and LANL, but no results have been published by now.

It is beyond the scope of this paper to also discuss the variety of hierarchical algebraic approaches which are not really related to the multigrid idea in the sense that these approaches are not based on the fundamental multigrid principles, smoothing and coarse-level correction. There is actually a rapid and very interesting ongoing development of such approaches. For completeness, however, we include some selected references. Various approaches based on approximate block Gauss elimination ("Schur-complement" methods) are found in [2–5,19,36–38,42,57]. Multi-level structures have also been introduced into ILU-type pre-conditioners, for example, in [49]. Very recently, some hybrid methods have been developed which use ideas from ILU and from multigrid [6–8,43–45]. For a further discussion, see also [55].

Summarizing, the development of hierarchically operating algebraic methods to efficiently tackle the solution of large sparse, unstructured systems of equations, currently belongs to one of the most active fields of research in numerical analysis. Many different methods have been investigated but, by now, none of them is really able to efficiently deal with *all* practically relevant problems. All methods seem to have their range of applicability but all of them may fail to be efficient in certain other applications. Hence, the development in this exciting area of research has to be expected to continue for the next years.

References

[1] R.E. Alcouffe, A. Brandt, J.E. Dendy, J.W. Painter, The multi-grid method for the diffusion equation with strongly discontinuous coefficients, SIAM J. Sci. Statist. Comput. 2 (1981) 430–454.

[2] O. Axelsson, The method of diagonal compensation of reduced matrix entries and multilevel iteration, J. Comput. Appl. Math. 38 (1991) 31–43.

[3] O. Axelsson, M. Neytcheva, Algebraic multilevel iteration method for Stieltjes matrices, Numer. Linear Algebra Appl. 1 (3) (1994) 213–236.

[4] O. Axelsson, P.S. Vassilevski, Algebraic multilevel preconditioning methods I, Numer. Math. 56 (1989) 157–177.

[5] O. Axelsson, P.S. Vassilevski, Algebraic multilevel preconditioning methods II, SIAM Numer. Anal. 27 (1990) 1569–1590.

[6] R.E. Bank, R.K. Smith, The incomplete factorization multigraph algorithm, SIAM J. Sci. Comput., to appear.

[7] R.E. Bank, R.K. Smith, The hierarchical basis multigraph algorithm, SIAM J. Sci. Comput., submitted.

[8] R.E. Bank, Ch. Wagner, Multilevel ILU decomposition, Numer. Math. 82 (1999) 543–576.

[9] D. Braess, Towards algebraic multigrid for elliptic problems of second order, Computing 55 (1995) 379–393.

[10] A. Brandt, Algebraic multigrid theory: the symmetric case, Appl. Math. Comput. 19 (1986) 23–56.

[11] A. Brandt, General highly accurate algebraic coarsening schemes, Proceedings of the Ninth Copper Mountain Conference on Multigrid Methods, Copper Mountain, April 11–16, 1999.

[12] A. Brandt, S.F. McCormick, J. Ruge, Algebraic multigrid (AMG) for automatic multigrid solution with application to geodetic computations, Institute for Computational Studies, POB 1852, Fort Collins, Colorado, 1982.

[13] A. Brandt, S.F. McCormick, J. Ruge, Algebraic multigrid (AMG) for sparse matrix equations, in: D.J. Evans (Ed.), Sparsity and its Applications, Cambridge University Press, Cambridge, 1984, pp. 257–284.

[14] M. Brezina, A.J. Cleary, R.D. Falgout, V.E. Henson, J.E. Jones, T.A. Manteuffel, S.F. McCormick, J.W. Ruge, Algebraic multigrid based on element interpolation (AMGe), LLNL Technical Report UCRL-JC-131752, SIAM J. Sci. Comput., to appear.

[15] T. Chan, J. Xu, L. Zikatanov, An agglomeration multigrid method for unstructured grids, Proceedings of the 10th International Conference on Domain Decomposition Methods, 1988.

[16] Q. Chang, Y.S. Wong, H. Fu, On the algebraic multigrid method, J. Comput. Phys. 125 (1996) 279–292.

[17] A.J. Cleary, R.D. Falgout, V.E. Henson, J.E. Jones, Coarse-grid selection for parallel algebraic multigrid, Proceedings of the "Fifth International Symposium on Solving Irregularly Structured Problems in Parallel", Lecture Notes in Computer Science, vol. 1457, Springer, New York, 1998, pp. 104–115.

[18] A.J. Cleary, R.D. Falgout, V.E. Henson, J.E. Jones, T.A. Manteuffel, S.F. McCormick, G.N. Miranda, J.W. Ruge, Robustness and scalability of algebraic multigrid, SIAM J. Sci. Comput., special issue on the "Fifth Copper Mountain Conference on Iterative Methods", 1998.

[19] W. Dahmen, L. Elsner, Algebraic Multigrid Methods and the Schur Complement, Notes on Numerical Fluid Mechanics, 23, Vieweg Verlag, Braunschweig, 1988.

[20] J.E. Dendy Jr., Black box multigrid, J. Comput. Phys. 48 (1982) 366–386.

[21] J.E. Dendy, S.F. McCormick, J. Ruge, T. Russell, S. Schaffer, Multigrid methods for three-dimensional petroleum reservoir simulation, Proceedings of 10th SPE Symposium on Reservoir Simulation, February 6–8, 1989.

[22] J. Fuhrmann, A modular algebraic multilevel method, Technical Report Preprint 203, Weierstrass-Institut für Angewandte Analysis und Stochastik, Berlin, 1995.

[23] T. Grauschopf, M. Griebel, H. Regler, Additive multilevel-preconditioners based on bilinear interpolation, matrix dependent geometric coarsening and algebraic multigrid coarsening for second order elliptic PDEs, Appl. Numer. Math. 23 (1997) 63–96.

[24] M. Griebel, T. Neunhoeffer, H. Regler, Algebraic multigrid methods for the solution of the Navier–Stokes equations in complicated geometries, SFB-Bericht Nr. 342/01/96 A, Institut für Informatik, Technische Universität München, 1996.
[25] H. Guillard, P. Vanek, An aggregation multigrid solver for convection-diffusion problems on unstructured meshes, Center for Computational Mathematics, University of Denver, Report 130, 1998.
[26] W.Z. Huang, Convergence of algebraic multigrid methods for symmetric positive definite matrices with weak diagonal dominance, Appl. Math. Comput. 46 (1991) 145–164.
[27] F. Kickinger, Algebraic multi-grid for discrete elliptic second order problems, Institutsbericht 513, Universität Linz, Institut für Mathematik, 1997.
[28] A. Krechel, K. Stüben, Operator dependent interpolation in algebraic multigrid, Proceedings of the Fifth European Multigrid Conference, Stuttgart, October 1–4, 1996; Lecture Notes in Computational Science and Engineering, vol. 3, Springer, Berlin, 1998.
[29] A. Krechel, K. Stüben, Parallel algebraic multigrid based on subdomain blocking, Parallel Comput., to appear.
[30] R.D. Lonsdale, An algebraic multigrid solver for the Navier–Stokes equations on unstructured meshes, Int. J. Numer. Methods Heat Fluid Flow 3 (1993) 3–14.
[31] J. Mandel, M. Brezina, P. Vanek, Energy optimization of algebraic multigrid bases, UCD/CCM Report 125, 1998.
[32] S. McCormick, J. Ruge, Algebraic multigrid methods applied to problems in computational structural mechanics, in: State-of-the-Art Surveys on Computational Mechanics, ASME, New York, 1989, pp. 237–270.
[33] R. Mertens, H. De Gersem, R. Belmans, K. Hameyer, D. Lahaye, S. Vandewalle, D. Roose, An algebraic multigrid method for solving very large electromagnetic systems, IEEE Trans. Magn. 34 (1998), to appear.
[34] W.A. Mulder, A new multigrid approach to convection problems, J. Comput. Phys. 83 (1989) 303–323.
[35] N.H. Naik, J. van Rosendale, The improved robustness of multigrid elliptic solvers based on multiple semicoarsened grids, SIAM Numer. Anal. 30 (1993) 215–229.
[36] Y. Notay, An efficient algebraic multilevel preconditioner robust with respect to anisotropies, in: O. Axelsson, B. Polman (Eds.), Algebraic Multilevel Iteration Methods with Applications, Department of Mathematics, University of Nijmegen, 1996, pp. 111–228.
[37] Y. Notay, Using approximate inverses in algebraic multilevel methods, Numer. Math. 80 (1998) 397–417.
[38] Y. Notay, Optimal V-cycle algebraic multilevel preconditioning, Numer. Linear Algebra Appl., to appear.
[39] M. Raw, A coupled algebraic multigrid method for the 3D Navier–Stokes equations, Report: Advanced Scientific Computing Ltd., 554 Parkside Drive, Waterloo, Ontario N2L 5Z4, Canada.
[40] H. Regler, Anwendungen von AMG auf das Plazierungsproblem beim Layoutentwurf und auf die numerische Simulation von Strömungen, Ph.D. thesis, TU München, 1997.
[41] S. Reitzinger, Algebraic multigrid and element preconditioning I, SFB-Report 98-15, University Linz, Austria, December 1998.
[42] A.A. Reusken, Multigrid with matrix-dependent transfer operators for a singular perturbation problem, Computing 50 (3) (1993) 199–211.
[43] A.A. Reusken, A multigrid method based on incomplete Gaussian elimination, Eindhoven University of Technology, Report RANA 95-13, ISSN 0926-4507, 1995.
[44] A.A. Reusken, On the approximate cyclic reduction preconditioner, Report 144, Institut für Geometrie und Praktische Mathematik, RWTH Aachen, 1997. SIAM J. Sci. Comput., to appear.
[45] A.A. Reusken, Approximate cyclic reduction preconditioning, in: W. Hackbusch, G. Wittum (Eds.), Multigrid Methods, Vol. 5, Lecture Notes in Computational Science and Engineering, Vol. 3, Springer, Berlin, 1998, pp. 243–259.
[46] G. Robinson, Parallel computational fluid dynamics on unstructured meshes using algebraic multigrid, in: R.B. Pelz, A. Ecer, J. Häuser (Eds.), Parallel Computational Fluid Dynamics, Vol. 92, Elsevier, Amsterdam, 1993.
[47] J.W. Ruge, K. Stüben, Efficient solution of finite difference and finite element equations by algebraic multigrid (AMG), in: D.J. Paddon, H. Holstein (Eds.), Multigrid Methods for Integral and Differential Equations, The Institute of Mathematics and its Applications Conference Series, New Series vol. 3, Clarendon Press, Oxford, 1985, pp. 169–212.
[48] J.W. Ruge, K. Stüben, Algebraic multigrid (AMG), in: S.F. McCormick (Ed.), Multigrid Methods, Frontiers in Applied Mathematics, Vol. 5, SIAM, Philadelphia, 1986.

[49] Y. Saad, ILUM: a multi-elimination ILU preconditioner for general sparse matrices, SIAM J. Sci. Comput. 17 (1996) 830–847.

[50] K. Stüben, Algebraic multigrid (AMG): Experiences and comparisons, Appl. Math. Comput. 13 (1983) 419–452.

[51] K. Stüben, Algebraic multigrid (AMG): an introduction with applications, in: U. Trottenberg, C.W. Oosterlee, A. Schüller (Eds.), Multigrid, Academic Press, New York, 2000. Also GMD Report 53, March 1999.

[52] P. Vanek, J. Mandel, M. Brezina, Algebraic multigrid on unstructured meshes, University of Colorado at Denver, UCD/CCM Report No. 34, 1994.

[53] P. Vanek, J. Mandel, M. Brezina, Algebraic multigrid by smoothed aggregation for second and fourth order elliptic problems, Computing 56 (1996) 179–196.

[54] P. Vanek, M. Brezina, J. Mandel, Convergence of algebraic multigrid based on smoothed aggregation, UCD/CCM Report 126, 1998. Numer. Math., submitted.

[55] C. Wagner, Introduction to algebraic multigrid, Course notes of an algebraic multigrid course at the University of Heidelberg in the Wintersemester 1998/99, http://www.iwr.uni-heidelberg.de/ ~ Christian.Wagner, 1999.

[56] C. Wagner, On the algebraic construction of multilevel transfer operators, IWR-Report, Universität Heidelberg, Computing (2000).

[57] C. Wagner, W. Kinzelbach, G. Wittum, Schur-complement multigrid – a robust method for groundwater flow and transport problems, Numer. Math. 75 (1997) 523–545.

[58] W.L. Wan, T.F. Chan, B. Smith, An energy minimization interpolation for robust multigrid methods, Department of Mathematics, UCLA, UCLA CAM Report 98-6, 1998.

[59] T. Washio, C.W. Oosterlee, Flexible multiple semicoarsening for 3D singularly perturbed problems, SIAM J. Sci. Comput. 19 (1998) 1646–1666.

[60] R. Webster, An algebraic multigrid solver for Navier–Stokes problems in the discrete second order approximation, Int. J. Numer. Methods Fluids 22 (1996) 1103–1123.

[61] P. Wesseling, C.W. Oosterlee, Geometric multigrid with applications to computational fluid dynamics, this volume, Comput. Appl. Math. 128 (2001) 311–334.

[62] L. Zaslavsky, An adaptive algebraic multigrid for multigroup neutron diffusion reactor core calculations, Appl. Math. Comput. 53 (1993) 13–26.

[63] L. Zaslavsky, An adaptive algebraic multigrid for reactor critically calculations, SIAM J. Sci. Comput. 16 (1995) 840–847.

From finite differences to finite elements A short history of numerical analysis of partial differential equations

Vidar Thomée

Department of Mathematics, Chalmers University of Technology, S-412 96 Göteborg, Sweden

Abstract

This is an account of the history of numerical analysis of partial differential equations, starting with the 1928 paper of Courant, Friedrichs, and Lewy, and proceeding with the development of first finite difference and then finite element methods. The emphasis is on mathematical aspects such as stability and convergence analysis.

MSC: 01A60; 65-03; 65M10; 65N10; 65M60; 65N30

Keywords: History; Finite difference methods; Finite element methods

0. Introduction

This article is an attempt to give a personal account of the development of numerical analysis of partial differential equations. We begin with the introduction in the 1930s and further development of the finite difference method and then describe the subsequent appearence around 1960 and increasing role of the finite element method. Even though clearly some ideas may be traced back further, our starting point will be the fundamental theoretical paper by Courant, Friedrichs and Lewy (1928)[1] on the solution of problems of mathematical physics by means of finite differences. In this paper a discrete analogue of Dirichlet's principle was used to define an approximate solution by means of the five-point approximation of Laplace's equation, and convergence as the mesh width tends to zero was established by compactness. A finite difference approximation was also defined for the

E-mail address: thomee@math.chalmers.se (V. Thomée).

[1] We refer to original work with publication year; we sometimes also quote survey papers and textbooks which are numbered separately.

wave equation, and the CFL stability condition was shown to be necessary for convergence; again compactness was used to demonstrate convergence. Since the purpose was to prove existence of solutions, no error estimates or convergence rates were derived. With its use of a variational principle for discretization and its discovery of the importance of mesh-ratio conditions in approximation of time-dependent problems this paper points forward and has had a great influence on numerical analysis of partial differential equations.

Error bounds for difference approximations of elliptic problems were first derived by Gerschgorin (1930) whose work was based on a discrete analogue of the maximum principle for Laplace's equation. This approach was actively pursued through the 1960s by, e.g., Collatz, Motzkin, Wasow, Bramble, and Hubbard, and various approximations of elliptic equations and associated boundary conditions were analyzed.

For time-dependent problems considerable progress in finite difference methods was made during the period of, and immediately following, the Second World War, when large-scale practical applications became possible with the aid of computers. A major role was played by the work of von Neumann, partly reported in O'Brien, Hyman and Kaplan (1951). For parabolic equations a highlight of the early theory was the important paper by John (1952). For mixed initial–boundary value problems the use of implicit methods was also established in this period by, e.g., Crank and Nicolson (1947). The finite difference theory for general initial value problems and parabolic problems then had an intense period of development during the 1950s and 1960s, when the concept of stability was explored in the Lax equivalence theorem and the Kreiss matrix lemmas, with further major contributions given by Douglas, Lees, Samarskii, Widlund and others. For hyperbolic equations, and particularly for nonlinear conservation laws, the finite difference method has continued to play a dominating role up until the present time, starting with work by, e.g., Friedrichs, Lax, and Wendroff.

Standard references on finite difference methods are the textbooks of Collatz [12], Forsythe and Wasow [14] and Richtmyer and Morton [28].

The idea of using a variational formulation of a boundary value problem for its numerical solution goes back to Lord Rayleigh (1894, 1896) and Ritz (1908), see, e.g., Kantorovich and Krylov [21]. In Ritz's approach the approximate solution was sought as a finite linear combination of functions such as, for instance, polynomials or trigonometrical polynomials. The use in this context of continuous piecewise linear approximating functions based on triangulations adapted to the geometry of the domain was proposed by Courant (1943) in a paper based on an address delivered to the American Mathematical Society in 1941. Even though this idea had appeared earlier, also in work by Courant himself (see Babuška [4]), this is often thought of as the starting point of the finite element method, but the further development and analysis of the method would occur much later. The idea to use an orthogonality condition rather than the minimization of a quadratic functional is attributed to Galerkin (1915); its use for time-dependent problems is sometimes referred to as the Faedo–Galerkin method, cf. Faedo (1949), or, when the orthogonality is with respect to a different space, as the Petrov–Galerkin or Bubnov–Galerkin method.

As a computational method the finite element method originated in the engineering literature, where in the mid 1950s structural engineers had connected the well established framework analysis with variational methods in continuum mechanics into a discretization method in which a structure is thought of as divided into elements with locally defined strains or stresses. Some of the pioneering work was done by Turner, Clough, Martin and Topp (1956) and Argyris [1] and the name of the finite element method appeared first in Clough (1960). The method was later applied to other classes of problems in continuum mechanics; a standard reference from the engineering literature is Zienkiewicz [43].

Independently of the engineering applications a number of papers appeared in the mathematical literature in the mid-1960s which were concerned with the construction and analysis of finite difference schemes by the Rayleigh–Ritz procedure with piecewise linear approximating functions, by, e.g., Oganesjan (1962, 1966), Friedrichs (1962), Céa (1964), Demjanovič (1964), Feng (1965), and Friedrichs and Keller (1966) (who considered the Neumann problem). Although, in fact, special cases of the finite element method, the methods studied were conceived as finite difference methods; they were referred to in the Russian literature as variational difference schemes.

In the period following this, the finite element method with piecewise polynomial approximating functions was analyzed mathematically in work such as Birkhoff, Schultz and Varga (1968), in which the theory of splines was brought to bear on the development, and Zlámal (1968), with the first stringent a priori error analysis of more complicated finite elements. The so called mixed finite element methods, which are based on variational formulations where, e.g., the solution of an elliptic equation and its gradient appear as separate variables and where the combined variable is a saddle point of a Lagrangian functional, were introduced in Brezzi (1974); such methods have many applications in fluid dynamical problems and for higher-order elliptic equations.

More recently, following Babuška (1976), Babuška and Rheinboldt (1978), much effort has been devoted to showing a posteriori error bounds which depend only on the data and the computed solution. Such error bounds can be applied to formulate adaptive algorithms which are of great importance in computational practice.

Comprehensive references for the analysis of the finite element method are Babuška and Aziz [5], Strang and Fix [34], Ciarlet [11], and Brenner and Scott [8].

Simultaneous with this development other classes of methods have arisen which are related to the above, and we will sketch four such classes: In a *collocation* method an approximation is sought in a finite element space by requiring the differential equation to be satisfied exactly at a finite number of collocation points, rather than by an orthogonality condition. In a *spectral* method one uses globally defined functions, such as eigenfunctions, rather than piecewise polynomials approximating functions, and the discrete solution may be determined by either orthogonality or collocation. A *finite volume* method applies to differential equations in divergence form. Integrating over an arbitrary volume and transforming the integral of the divergence into an integral of a flux over the boundary, the method is based on approximating such a boundary integral. In a *boundary integral* method a boundary value problem for a homogeneous elliptic equation in a d-dimensional domain is reduced to an integral equation on its $(d-1)$-dimensional boundary, which in turn can be solved by, e.g., the Galerkin finite element method or by collocation.

An important aspect of numerical analysis of partial differential equations is the numerical solution of the finite linear algebraic systems that are generated by the discrete equations. These are in general very large, but with sparse matrices, which makes iterative methods suitable. The development of convergence analysis for such methods has parallelled that of the error analysis sketched above. In the 1950s and 1960s particular attention was paid to systems associated with finite difference approximation of positive type of second-order elliptic equations, particularly the five-point scheme, and starting with the Jacobi and Gauss–Seidel methods techniques were developed such as the Frankel and Young successive overrelaxation and the Peaceman–Rachford (1955) alternating direction methods, as described in the influential book of Varga [39]. In later years systems with positive-definite matrices stemming from finite element methods have been solved first by the conjugate gradient method proposed by Hestenes and Stiefel (1952), and then making this more effective by preconditioning. The multigrid method was first introduced for finite difference methods in the 1960s by Fedorenko and Bahvalov and further developed by Brandt in the 1970s. For finite elements the multigrid method and the associated method of domain decomposition have been and are being intensely pursued by, e.g., Braess, Hackbusch, Bramble, and Widlund.

Many ideas and techniques are common to the finite difference and the finite element methods, and in some simple cases they coincide. Nevertheless, with its more systematic use of the variational approach, its greater geometric flexibility, and the way it more easily lends itself to error analysis, the finite element method has become the dominating approach both among numerical analysts and in applications. The growing need for understanding the partial differential equations modeling the physical problems has seen an increase in the use of mathematical theory and techniques, and has attracted the interest of many mathematicians. The computer revolution has made large-scale real-world problems accessible to simulation, and in later years the concept of computational mathematics has emerged with a somewhat broader scope than classical numerical analysis.

Our approach in this survey is to try to illustrate the ideas and concepts that have influenced the development, with as little technicalities as possible, by considering simple model situations. We emphasize the mathematical analysis of the discretization methods, involving stability and error estimates, rather than modeling and implementation issues. It is not our ambition to present the present state of the art but rather to describe the unfolding of the field. It is clear that it is not possible in the limited space available to do justice to all the many important contributions that have been made, and we apologize for omissions and inaccuracies; writing a survey such as this one is a humbling experience. In addition to references to papers which we have selected as important in the development we have quoted a number of books and survey papers where additional and more complete and detailed information can be found; for reason of space we have tried to limit the number of reference to any individual author.

Our presentation is divided into sections as follows. Section 1: The Courant–Friedrichs–Lewy paper; Section 2: Finite difference methods for elliptic problems; Section 3: Finite difference methods for initial value problems; Section 4: Finite difference methods for mixed initial–boundary value problems; Section 5: Finite element methods for elliptic problems; Section 6: Finite element methods for evolution equations; Section 7: Some other classes of approximation methods; and Section 8: Numerical linear algebra for elliptic problems.

1. The Courant–Friedrichs–Lewy paper

In this seminal paper from 1928 the authors considered difference approximations of both the Dirichlet problems for a second-order elliptic equation and the biharmonic equation, and of the initial–boundary value problem for a second-order hyperbolic equation, with a brief comment also about the model heat equation in one space variable. Their purpose was to derive existence results for the original problem by constructing finite-dimensional approximations of the solutions, for which the existence was clear, and then showing convergence as the dimension grows. Although the aim was not numerical, the ideas presented have played a fundamental role in numerical analysis. The paper appears in English translation, together with commentaries by Lax, Widlund, and Parter concerning its influence on the subsequent development, see the quotation in references.

The first part of the paper treats the Dirichlet problem for Laplace's equation,

$$-\Delta u = 0 \quad \text{in } \Omega, \ \text{with } u = g \text{ on } \partial\Omega, \tag{1.1}$$

where $\Omega \subset \mathbb{R}^2$ is a domain with smooth boundary $\partial\Omega$. Recall that by Dirichlet's principle the solution minimizes $\int\int_\Omega |\nabla\varphi|^2\,\mathrm{d}x$ over φ with $\varphi = g$ on $\partial\Omega$. For a discrete analogue, consider mesh-points $x_j = jh$, $j \in \mathbb{Z}^2$, and let Ω_h be the mesh-points in Ω for which the neighbors $x_{j\pm e_1}, x_{j\pm e_2}$ are in Ω ($e_1 = (1,0)$, $e_2 = (0,1)$), and let ω_h be those with at least one neighbor outside Ω. For $U_j = U(x_j)$ a mesh-function we introduce the forward and backward difference quotients

$$\partial_l U_j = (U_{j+e_l} - U_j)/h, \qquad \bar\partial_l U_j = (U_j - U_{j-e_l})/h, \quad l = 1,2. \tag{1.2}$$

By minimizing a sum of terms of the form $(\partial_1 U_j)^2 + (\partial_2 U_j)^2$ one finds a unique mesh-function U_j which satisfies

$$-\partial_1\bar\partial_1 U_j - \partial_2\bar\partial_2 U_j = 0 \quad \text{for } x_j \in \Omega_h, \ \text{with } U_j = g(x_j) \text{ on } \omega_h; \tag{1.3}$$

the first equation is the well-known five-point approximation

$$4U_j - (U_{j+e_1} + U_{j-e_1} + U_{j+e_2} + U_{j-e_2}) = 0. \tag{1.4}$$

It is shown by compactness that the solution of (1.3) converges to a solution u of (1.1) when $h \to 0$. By the same method it is shown that on compact subsets of Ω difference quotients of U converge to the corresponding derivatives of u as $h \to 0$. Also included are a brief discussion of discrete Green's function representation of the solution of the inhomogeneous equation, of discretization of the eigenvalue problem, and of approximation of the solution of the biharmonic equation.

The second part of the paper is devoted to initial value problems for hyperbolic equations. In this case, in addition to the mesh-width h, a time step k is introduced and the discrete function values $U_j^n \approx u(x_j, t_n)$, with $t_n = nk, n \in \mathbb{Z}_+$. The authors first consider the model wave equation

$$u_{tt} - u_{xx} = 0 \quad \text{for } x \in \mathbb{R}^2, \ t \geqslant 0, \ \text{with } u(\cdot,0), \ u_t(\cdot,0) \text{ given} \tag{1.5}$$

and the approximate problem, with obvious modification of notation (1.2),

$$\partial_t\bar\partial_t U_j^n - \partial_x\bar\partial_x U_j^n = 0 \quad \text{for } j \in \mathbb{Z}, \ n \geqslant 1, \ \text{with } U_j^n \text{ given for } n = 0,1.$$

When $k = h$ the equation may also be expressed as $U_j^{n+1} + U_j^{n-1} - U_{j+1}^n - U_{j-1}^n = 0$, and it follows at once that in this case the discrete solution at $(x,t) = (x_j, t_n)$ depends only on the initial data in the interval $(x - t, x + t)$. For a general time-step k the interval of dependence becomes $(x - t/\lambda, x + t/\lambda)$, where $\lambda = k/h$ is the mesh-ratio. Since the exact solution depends on data in $(x - t, x + t)$ it follows that if $\lambda > 1$, not enough information is used by the scheme, and hence a necessary condition for convergence of the discrete solution to the exact solution is that $\lambda \leqslant 1$; this is referred to as the Courant–Friedrichs–Lewy or CFL condition. By an energy argument it is shown that the appropriate sums over the mesh-points of positive quadratic forms in the discrete solution are bounded, and compactness is used to show convergence as $h \to 0$ when $\lambda = k/h$=constant $\leqslant 1$. The energy argument is a clever discrete analogue of an argument by Friedrichs and Lewy (1926): For the wave equation in (1.5) one may integrate the identity $0 = 2u_t(u_{tt} - u_{xx}) = (u_t^2 + u_x^2)_t - 2(u_x u_t)_x$ in x to show that $\int_R (u_t^2 + u_x^2)\,\mathrm{d}x$ is independent of t, and thus bounded. The case of two spatial variables is also briefly discussed.

In an appendix brief discussions are included concerning a first-order hyperbolic equation, the model heat equation in one space variable, and of wave equations with lower-order terms.

2. Finite difference methods for elliptic problems

The error analysis of finite difference methods for elliptic problems started with the work of Gerschgorin (1930). In contrast to the treatment in Courant, Friedrichs and Lewy (1928) this work was based on a discrete version of the maximum principle. To describe this approach we begin with the model problem

$$-\Delta u = f \quad \text{in } \Omega, \quad \text{with } u = 0 \text{ on } \partial\Omega, \tag{2.1}$$

where we first assume Ω to be the square $\Omega=(0,1)\times(0,1)\subset\mathbb{R}^2$. For a finite difference approximation consider the mesh-points $x_j = jh$ with $h = 1/M$, $j \in \mathbb{Z}^2$ and the mesh-function $U_j = U(x_j)$. With the notation (1.2) we replace (2.1) by

$$-\varDelta_h U_j := -\partial_1\bar{\partial}_1 U_j - \partial_2\bar{\partial}_2 U_j = f_j \quad \text{for } x_j \in \Omega, \ U_j = 0 \ \text{for } x_j \in \partial\Omega. \tag{2.2}$$

This problem may be written in matrix form as $AU=F$, where A is a symmetric $(M-1)^2\times(M-1)^2$ matrix whose elements are $4, -1$, or 0, with 0 the most common occurrence, cf. (1.4).

For the analysis one first shows a discrete maximum principle: If U is such that $-\varDelta_h U_j \leqslant 0$ ($\geqslant 0$) in Ω, then U takes its maximum (minimum) on $\partial\Omega$; note that $-\varDelta_h U_j \leqslant 0$ is equivalent to $U_j \leqslant (U_{j+e_1} + U_{j-e_1} + U_{j+e_2} + U_{j-e_2})/4$. Letting $W(x) = \frac{1}{2} - |x - x_0|^2$ where $x_0 = (\frac{1}{2}, \frac{1}{2})$ we have $W(x) > 0$ in Ω, and applying the discrete maximum principle to the function $V_j = \pm U_j - \frac{1}{4}|\varDelta_h U|_\Omega W_j$ one concludes easily that, for any mesh-function U on $\bar{\Omega}$,

$$|U|_{\bar{\Omega}} \leqslant |U|_{\partial\Omega} + C|\varDelta_h U|_\Omega, \quad \text{where } |U|_S = \max_{x_j\in S} |U_j|.$$

Noting that the error $z_j = U_j - u(x_j)$ satisfies

$$-\varDelta_h z_j = f_j + \varDelta_h u(x_j) = (\varDelta_h - \Delta)u(x_j) = \tau_j, \quad \text{with } |\tau_j| \leqslant Ch^2\|u\|_{C^4}, \tag{2.3}$$

one finds, since $z_j = 0$ on $\partial\Omega$, that

$$|U - u|_{\bar{\Omega}} = |z|_{\bar{\Omega}} \leqslant C|\tau|_\Omega \leqslant Ch^2\|u\|_{C^4}.$$

The above analysis uses the fact that all neighbors $x_{j\pm e_l}$ of the interior mesh-points $x_j \in \Omega$ are either interior mesh-points or belong to $\partial\Omega$. When the boundary is curved, however, there will be mesh-points in Ω which have neighbors outside $\bar{\Omega}$. If for such a mesh-point $x = x_j$ we take a point $b_{h,x} \in \partial\Omega$ with $|b_{h,x} - x| \leqslant h$ and set $U_j = u(b_{h,x}) = 0$, then it follows from Gerschgorin, loc. cit., that $|U - u|_{\bar{\Omega}} \leqslant Ch\|u\|_{C^3}$. To retain second-order accuracy Collatz (1933) proposed to use linear interpolation near the boundary: Assuming for simplicity that Ω is a convex plane domain with smooth boundary, we denote by Ω_h the mesh-points $x_j \in \Omega_h$ that are truly interior in the sense that all four neighbors of x_j are in $\bar{\Omega}$. (For the above case of a square, Ω_h simply consists of all mesh-points in Ω.) Let now $\tilde{\omega}_h$ be the mesh-points in Ω that are not in Ω_h. For $x_j \in \tilde{\omega}_h$ we may then find a neighbor $y = x_i \in \Omega_h \cup \tilde{\omega}_h$ such that the line through x and y cuts $\partial\Omega$ at a point $\bar{x}$ which is not a mesh-point. We denote by $\bar{\omega}_h$ the set of points $\bar{x} \in \partial\Omega$ thus associated with the points of $\tilde{\omega}_h$, and define for $x = x_j \in \tilde{\omega}_h$ the error in the linear interpolant

$$\ell_h u_j := u(x_j) - \alpha u(x_i) - (1 - \alpha)u(\bar{x}), \quad \text{where } \alpha = \gamma/(1+\gamma) \leqslant \tfrac{1}{2} \text{ if } |x - \bar{x}| = \gamma h.$$

As $u=0$ on $\partial\Omega$ we now pose the problem

$$-\Delta_h U_j = f_j \quad \text{in } \Omega_h, \qquad \ell_h U_j = 0 \quad \text{in } \tilde{\omega}_h, \quad U(\bar{x}) = 0 \text{ on } \bar{\omega}_h$$

and since $\alpha \leqslant \frac{1}{2}$ it is not difficult to see that

$$|U|_{\Omega_h \cup \tilde{\omega}_h} \leqslant C(|U|_{\bar{\omega}_h} + |\ell_h U|_{\tilde{\omega}_h} + |\Delta_h U|_{\Omega}). \tag{2.4}$$

Using again (2.3) together with $|\ell_h z|_{\tilde{\omega}_h} \leqslant Ch^2\|u\|_{C^3}$, $z=0$ on $\bar{\omega}_h$, one finds that $|U-u|_{\Omega_h\cup\tilde{\omega}_h} \leqslant Ch^2\|u\|_{C^4}$. Another approximation near $\partial\Omega$ was proposed by Shortley and Weller (1938). For $x_j \in \tilde{\omega}_h$ it uses the points defined by the intersections of $\partial\Omega$ with the horizontal and vertical mesh-lines through x_j, which together with the neighbors that are in Ω form an irregular five-point star. This gives an approximation to $-\Delta$ which is only first-order accurate, but, using it in a boundary operator ℓ_h similarly to the Collatz interpolation error it will yield a second-order error bound.

Consider more generally the variable coefficient Dirichlet problem

$$Au := -\sum_{j,k=1}^{d} a_{jk}\frac{\partial^2 u}{\partial x_j \partial x_k} - \sum_{j=1}^{d} b_j \frac{\partial u}{\partial x_j} f \quad \text{in } \Omega \subset \mathbb{R}^d, \;\; u = g \text{ on } \partial\Omega, \tag{2.5}$$

where the matrix (a_{jk}) is uniformly positive definite in $\bar{\Omega}$, and a corresponding finite difference operator with finitely many terms of the form

$$A_h u(x) = -h^{-2}\sum_j a_j u(x-jh), \quad a_j = a_j(x,h), \;\; j \in \mathbb{Z}^2 \tag{2.6}$$

which is consistent with A so that $A_h u(x) \to Au(x)$ as $h \to 0$. Following Motzkin and Wasow (1953) such an operator is said to be of positive type if $a_j \geqslant 0$ for $j \neq 0$, with $a_0 < 0$.

For mesh-points x, let $N(x)$ be the convex hull of the set of neighbors of x defined by (2.6), i.e., the mesh-points $x - jh$ with $a_j(x,h) \neq 0$, and let Ω_h denote the set of mesh-points with $N(x) \not\subset \bar{\Omega}$. The remaining mesh-points in $\bar{\Omega}$ form the set $\tilde{\omega}_h$ of boundary mesh-points. We set $\bar{\Omega}_h = \Omega_h \cup \tilde{\omega}_h$. For $x \in \Omega_h$ we want to use the equation $A_h U(x) = M_h f(x)$ as an analogue of the differential equation in (2.5), where M_h is a linear operator approximating the identity operator I (in most cases $M_h = I$). For $x \in \tilde{\omega}_h$, $A_h U(x)$ is not defined by the values of U on $\bar{\Omega}_h$, and at such points we therefore want to choose an equation of the form

$$\ell_h u(x) := \sum_{x_j \in \bar{\Omega}} \tilde{a}_j u(x-jh) = m_h(g,f), \quad \tilde{a}_j = \tilde{a}_j(x,h),$$

where m_h is a suitable linear operator. The values of u at points in $\bar{\omega}_h$ will now be included in the right-hand side by $u=g$ on $\partial\Omega$. Together these equations form our difference scheme, and we say (see Bramble, Hubbard and Thomée (1969)) that this is of essentially positive type if A_h is of positive type and $\tilde{a}_0 = 1$, $\sum_{j\neq 0}|\tilde{a}_j| \leqslant \gamma < 1$ for $x \in \omega_h$. A discrete maximum principle shows that the analogue of (2.4) remains valid in this case (with Δ_h replaced by A_h and without the term $|U|_{\bar{\omega}_h}$). The scheme is said to be accurate of order q if $A_h u - M_h A u = O(h^q)$ on Ω_h, and $\ell_h u - m_h(u|_{\partial\Omega}, Au) = O(h^q)$ on $\tilde{\omega}_h$. Under somewhat more precise natural assumptions one may now conclude from (2.4) that $|U-u|_{\Omega_h} \leqslant Ch^q\|u\|_{C^{q+2}}$. Error bounds may also be expressed in terms of data, and a $O(h^q)$ error bound holds if $f \in C^s(\bar{\Omega})$, $g \in C^s(\partial\Omega)$ with $s > q$. For homogeneous boundary conditions this follows easily using the Schauder estimate $\|u\|_{C^{q+2}} \leqslant C\|f\|_{C^s}$. For inhomogeneous boundary conditions a more precise analysis may be based on a representation using a nonnegative discrete Green's function $G_{jl} = G(x_j, x_l)$,

$$U_j = h^d \sum_{x_l \in \Omega_h} G_{jl} A_h U_l + \sum_{x_l \in \omega_h} G_{jl} U_l \quad \text{for } x_j \in \bar{\Omega}_h,$$

where $h^d \sum_{x_l \in \Omega_h} G_{jl} \leqslant C$ and $\sum_{x_l \in \bar{\omega}_h} G_{jl} \leqslant 1$, and where also a discrete analogue of the estimate $\int_{\Gamma_\delta} G(x,y)\,\mathrm{d}s \leqslant C\delta$ for the continuous problem holds, where $\Gamma_\delta = \{y \in \Omega;\ \mathrm{dist}(y, \partial\Omega) = \delta\}$. The latter is related to the important observation by Bahvalov (1959) that the regularity demands on the solution u of the continuous problem can be relaxed in some cases by essentially two derivatives at the boundary without loosing the convergence rate. For less regular data one can obtain correspondingly weaker convergence estimates: When $f \in C^s(\bar{\Omega})$, $g \in C^s(\partial\Omega)$ with $0 \leqslant s < q$, $O(h^s)$ order convergence may be shown by interpolation between C^s-spaces. The regularity demands of f may be further relaxed by choosing for M_h an averaging operator, see Tikhonov and Samarskii (1961); this paper also demonstrated how to construct finite difference approximations of elliptic equations with discontinuous coefficients by taking local harmonic averages. When the boundary itself is nonsmooth the exact solution may have singularities which make the above results not applicable. Laasonen (1957) showed that the presence of a corner with an accute inner angle does not affect the rate of convergence but if the angle is $\pi\alpha$ with $\alpha > 1$ he shows the weaker estimate $O(h^{1/\alpha - \varepsilon})$ for any $\varepsilon > 0$.

As an example of an operator of the form (2.6), consider the nine-point formula

$$-\Delta_h^{(9)} u(x) = \tfrac{1}{6} h^{-2} \left(20\, u(x) - 4 \sum_{|j|=1} u(x - jh) - \sum_{|j_1| = |j_2| = 1} u(x - jh) \right).$$

With $M_h f = f + \frac{1}{12} h^2 \Delta_h f$ one finds $\Delta_h^{(9)} u + M_h \Delta u = O(h^4)$ and Bramble and Hubbard (1962) showed that the operator ℓ_h can be chosen so that the corresponding scheme is of essentially positive type and accurate of order $q = 4$. Further, Bramble and Hubbard (1963) constructed second-order accurate schemes of essentially positive type in the case of a general A ($d = 2$), also with mixed derivative terms. Here the neighbors of x may be several mesh-widths away from x. Related results were also obtained in special cases by Wasow (1952), Laasonen (1958), and Volkov (1966), see also Bramble, Hubbard and Thomée (1969).

We shall now turn to some schemes that approximate elliptic equations containing mixed derivatives but are not generally of essentially positive type. Assume now that A is in divergence form, $A = -\sum_{i,k=1}^{d} (\partial/\partial x_i)(a_{ik} \partial u/\partial x_k)$. To our above notation ∂_i, $\bar{\partial}_i$ of (1.2) we add the symmetric difference quotient $\hat{\partial}_i = (\partial_i + \bar{\partial}_i)/2$ and set, with $a_{ik}^{(h)} = a_{ik}(x + \frac{1}{2} h e_i)$,

$$A_h^{(1)} u = -\sum_{i,k} \bar{\partial}_i (a_{ik}^{(h)} \partial_k u), \qquad A_h^{(2)} u = -\sum_i \bar{\partial}_i (a_{ii}^{(h)} \partial_i u) - \sum_{i \neq k} \hat{\partial}_i (a_{ik} \hat{\partial}_k u).$$

These operators are obviously consistent with A and second-order accurate. Except in special cases the $A_h^{(l)}$ are not of positive type and the above analysis does not apply. Instead one may use energy arguments to derive error estimates in discrete l_2-norms, see Thomée (1964). For $x \in \tilde{\omega}_h$ let as above $b_{h,x} \in \partial\Omega$, and consider the discrete Dirichlet problem

$$A_h U = f \quad \text{on } \Omega_h, \qquad U = g(b_{h,x}) \quad \text{on } \tilde{\omega}_h, \text{ with } A_h = A_h^{(1)} \text{ or } A_h^{(2)}. \tag{2.7}$$

With $\mathscr{D}_h$ the mesh-functions which vanish outside Ω_h, we define for $U, V \in \mathscr{D}_h$

$$(U,V) = h^d \sum_j U_j V_j, \quad \|U\| = (U,U)^{1/2}, \quad \text{and} \quad \|U\|_1 = \|U\| + \sum_{j=1}^{d} \|\partial_j U\|.$$

By summation by parts one easily derives that $\|V\|_1^2 \leqslant C(A_h V, V)$ for $V \in \mathscr{D}_h$, and this shows at once the uniqueness and hence the existence of a solution of (2.7). When $\tilde{\omega}_h \subset \partial\Omega$, application to $U - u$, using the second-order accuracy of A_h shows that $\|U - u\|_1 \leqslant C(u)h^2$. When $\tilde{\omega}_h \not\subset \partial\Omega$, h^2 has to be replaced by $\sqrt{h}$, but with $\|\cdot\|_1'$ a slightly weaker norm it was shown in Bramble, Kellogg and Thomée (1968) that $\|U - u\| \leqslant \|U - u\|_1' \leqslant C(u)h$.

Consider now a constant coefficient finite difference operator of the form (2.6), which is consistent with A of (2.5). Introducing the symbol of A_h, the trigonometric polynomial $p(\xi) = \sum_j a_j e^{ij\xi}$, we say that A_h is elliptic if $p(\xi) \neq 0$ for $|\xi_l| \leqslant \pi$, $l = 1,2$, $\xi \neq 0$. For the five-point operator $-\Delta_h$ we have $p(\xi) = 4 - 2\cos\xi_1 - 2\cos\xi_2$ and $-\Delta_h$ is thus elliptic. For such operators A_h we have the following interior estimate by Thomée and Westergren (1968). Set $\|U\|_S = (h^d \sum_{x_j \in S} U_j^2)^{1/2}$ and $\|U\|_{k,S} = \left(\sum_{|\alpha| \leqslant k} \|\partial_h^\alpha U\|_S^2\right)^{1/2}$ where $\partial_h^\alpha = \partial_1^{\alpha_1} \cdots \partial_d^{\alpha_d}$, $|\alpha| = \alpha_1 + \cdots + \alpha_d$, and let $\bar{\Omega}_0 \subset \Omega_1 \subset \bar{\Omega}_1 \subset \Omega$. Then for any α and h small we have

$$|\partial_h^\alpha U|_{\Omega_0} \leqslant C(\|A_h U\|_{l,\Omega_1} + \|U\|_{\Omega_1}) \quad \text{if } l \geqslant |\alpha| + [d/2] - 1.$$

Thus, the finite difference quotients of the solution of the equation $A_h U = f$ may be bounded in Ω_0 by the difference quotients of f in a slightly larger domain Ω_1 plus a discrete l_2-norm of U in Ω_1. Assuming u is a solution of (2.1) this may be used to show that if A_h is accurate of order q and Q_h is a finite difference operator which approximates the differential operator Q to order q, then

$$|Q_h U - Qu|_{\Omega_0} \leqslant C(u)h^q + C\|U - u\|_{\Omega_1}. \tag{2.8}$$

Thus, if we already know that $U - u$ is of order $O(h^q)$ in the l_2-norm, then $Q_h u - Qu$ is of the same order in maximum norm in the interior of the domain.

Finite difference approximations for elliptic equations of higher order were studied by, e.g., Saulev (1957); the work in Thomée (1964) concerns also such equations.

3. Finite difference methods for initial value problems

In this section we sketch the development of the stability and convergence theory for finite difference methods applied to pure initial value problems. We first consider linear constant coefficient evolution equations and then specialize to parabolic and hyperbolic equations.

We begin with the initial value problem for a general linear constant coefficient scalar equation

$$u_t = P(D)u \quad \text{for } x \in \mathbb{R}^d,\ t > 0, \qquad u(\cdot,0) = v, \quad \text{where } P(\xi) = \sum_{|\alpha| \leqslant M} P_\alpha \xi^\alpha \tag{3.1}$$

with $u = u(x,t)$, $v = v(x)$, $\xi^\alpha = \xi_1^{\alpha_1} \cdots \xi_d^{\alpha_d}$, and $D = (\partial/\partial x_1, \ldots, \partial/\partial x_d)$. Such an initial value problem is said to be well-posed if it has a unique solution that depends continuously on the initial data, in some sense that has to be specified. For example, the one-dimensional wave equation $u_t = \rho u_x$ has the unique solution $u(x,t)=v(x+\rho t)$ and since $\|u(\cdot,t)\|_{L_p}=\|v\|_{L_p}$, this problem is well posed in L_p for $1\leqslant p\leqslant\infty$. Similarly, for the heat equation $u_t=u_{xx}$ we have

$$u(x,t)=\frac{1}{\sqrt{4\pi t}}\int_{-\infty}^{\infty}e^{-(x-y)^2/4t}v(y)\,dy$$

and $\|u(\cdot,t)\|_{L_p}\leqslant\|v\|_{L_p}$, $1\leqslant p\leqslant\infty$. More precisely, (3.1) is well-posed in L_p if $P(D)$ generates a semigroup $E(t)=e^{tP(D)}$ in L_p which grows at most exponentially, so that the solution $u(t)=E(t)v$ satisfies $\|u(t)\|_{L_p}\leqslant Ce^{\kappa t}\|v\|_{L_p}$ for $t\geqslant 0$, for some κ. For $p=2$, which is the case we will concentrate on first, we see by Fourier transformation and Parseval's relation that this is equivalent to

$$|e^{tP(i\xi)}|\leqslant Ce^{\kappa t},\quad \forall\xi\in\mathbb{R}^d,\ t>0, \tag{3.2}$$

or $\operatorname{Re}P(i\xi)\leqslant\kappa$ for $\xi\in\mathbb{R}^d$; if only the highest-order terms are present in $P(D)$, this is equivalent to $\operatorname{Re}P(i\xi)\leqslant 0$ for $\xi\in\mathbb{R}^d$, in which case (3.2) holds with $\kappa=0$, so that $E(t)$ is uniformly bounded in L_2 for $t\geqslant 0$. In the above examples $P(i\xi)=i\rho\xi$ and $P(i\xi)=-\xi^2$, respectively, and $\kappa=0$.

Generalizing to systems of the form (3.1) with u and v N-vectors and the P_α $N\times N$ matrices, the condition for well-posedness in L_2 is again (3.2) where now $|\cdot|$ denotes any matrix norm. Here it is clear that a necessary condition for (3.2) is that $\operatorname{Re}\lambda_j(\xi)\leqslant\kappa$ for $\xi\in\mathbb{R}^d$, for any eigenvalue $\lambda_j(\xi)$ of $P(i\xi)$, and if $P(i\xi)$ is a normal matrix this is also sufficient. Necessary and sufficient conditions for (3.2) were given by Kreiss (1959b), and depend on the following lemma. Here, for any $N\times N$ matrix A with eigenvalues $\{\lambda_j\}_{j=1}^N$ we set $\Lambda(A)=\max_j\operatorname{Re}\lambda_j$ and $\operatorname{Re}A=(A+A^*)/2$, where A^* is the adjoint matrix. For Hermitian matrices $A\leqslant B$ means $Au\cdot u\leqslant Bu\cdot u$ for all $u\in\mathbb{R}^N$. With this notation, (3.2) holds if and only if the set $\mathcal{F}=\{P(i\xi)-\kappa;\ \xi\in\mathbb{R}^d\}$ satisfies the conditions of the following lemma:

Let $\mathcal{F}$ be a set of $N\times N$ matrices. Then the following conditions are equivalent:

(i) $|e^{tA}|\leqslant C$ for $A\in\mathcal{F}$, $t\geqslant 0$, for some $C>0$;

(ii) $\Lambda(A)\leqslant 0$ *and* $\operatorname{Re}(z\,|R(A;z)|)\leqslant C$ *for* $A\in\mathcal{F}$, $\operatorname{Re}z>0$, *for some* $C>0$;

(iii) $\Lambda(A)\leqslant 0$ *for* $A\in\mathcal{F}$ *and there exist* C_1 *and* C_2 *and for each* $A\in\mathcal{F}$ *a matrix* $S=S(A)$ *such that* $\max(|S|,|S^{-1}|)\leqslant C_1$ *and such that* $SAS^{-1}=B=(b_{jk})$ *is a triangular matrix with off-diagonal elements satisfying* $|b_{jk}|\leqslant C_2\min(|\operatorname{Re}\lambda_j|,|\operatorname{Re}\lambda_k|)$ *where* $\lambda_j=b_{jj}$;

(iv) *Let* $0\leqslant\gamma<1$. *There exists* $C>0$ *such that for each* $A\in\mathcal{F}$ *there is a Hermitian matrix* $H=H(A)$ *with* $C^{-1}I\leqslant H\leqslant CI$ *and* $\operatorname{Re}(HA)\leqslant\gamma\Lambda(A)H\leqslant 0$.

Equations of higher order in the time variable such as the wave equation $u_{tt}=u_{xx}$ may be written in system form (3.1) by introducing the successive time derivatives of u as dependent variables, and is therefore covered by the above discussion.

For the approximate solution of the initial value problem (3.1), let h and k be (small) positive numbers. We want to approximate the solution at time level $t_n=nk$ by U^n for $n\geqslant 0$ where $U^0=v$ and $U^{n+1}=E_kU^n$ for $n\geqslant 0$, and where E_k is an operator of the form

$$E_kv(x)=\sum_j a_j(h)v(x-jh),\quad \lambda=k/h^M=\text{constant} \tag{3.3}$$

with summation over a finite set of multi-indices $j=(j_1,\ldots,j_d)\in\mathbb{Z}^d$; such operators are called explicit. The purpose is to choose E_k so that E_k^nv approximates $u(t_n)=E(t_n)v=E(k)^nv$. In numerical applications we would apply (3.3) only for $x=x_l=lh$, $l\in\mathbb{Z}^d$, but for convenience in the analysis we shall think of $E_k v$ as defined for all x. As an example, for the heat equation $u_t = u_{xx}$ the simplest such operator is obtained by replacing derivatives by finite difference quotients, $\partial_t U_j^n = \partial_x \bar{\partial}_x U^n$. Solving for U^{n+1} we see that this defines $E_k v(x) = \lambda v(x+h) + (1-2\lambda)v(x) + \lambda v(x-h)$, $\lambda = k/h^2$. We shall consider further examples below.

We say that E_k is consistent with (3.1) if $u(x, t+k) = E_k u(x,t) + o(k)$ when $k \to 0$, for sufficiently smooth solutions $u(x,t)$ of (3.1), and accurate of order r if the term $o(k)$ may be replaced by $kO(h^r)$. By Taylor series expansion around (x,t) these conditions are seen to be equivalent to algebraic conditions between the coefficients $a_j(h)$. In our above example, $u(x, t+k) - E_k u(x,t) = ku_t + \frac{1}{2}k^2 u_{tt} - \lambda h^2 u_{xx} - \frac{1}{12}\lambda h^4 u_{xxxx} = kO(h^2)$ when $u_t = u_{xx}$, so that $r = 2$.

For the purpose of showing convergence of $E_k^n v$ to $E(t_n)v$ and to derive an error bound one needs some stability property of the operator E_k^n: This operator is said to be stable in L_2 if for any $T > 0$ there are constants C and κ such that $\|E_k^n v\| \leqslant C e^{\kappa n k}\|v\|$ for $v \in L_2$, $n \geqslant 0$, where $\|\cdot\| = \|\cdot\|_{L_2}$. If this holds, and if (3.1) is well posed in L_2 and E_k is accurate of order r, then it follows from the identity $E_k^n - E(t_n) = \sum_{j=0}^{n-1} E_k^{n-1-j}(E_k - E(k))E(t_j)$ that, with $\|\cdot\|_s = \|\cdot\|_{H^s(\mathbb{R}^d)}$,

$$\|(E_k^n - E(t_n))v\| \leqslant C \sum_{j=0}^{n-1} kh^r \|E(t_j)v\|_{M+r} \leqslant C_T h^r \|v\|_{M+r} \quad \text{for } t_n \leqslant T, \tag{3.4}$$

where we have also used the fact that spatial derivatives commute with $E(t)$.

The sufficiency of stability for convergence of the solution of the discrete problem to the solution of the continuous initial value problem was shown in particular cases in many places, e.g., Courant, Friedrichs and Lewy (1928), O'Brien, Hyman and Kaplan (1951), Douglas (1956). It was observed by Lax and Richtmyer (1959) that stability is actually a necessary condition for convergence to hold for all $v \in L_2$; the general Banach space formulation of stability as a necessary and sufficient condition for convergence is known as the Lax equivalence theorem, see Richtmyer and Morton [28]. We note that for individual, sufficiently regular v convergence may hold without stability; for an early interesting example with analytic initial data and highly unstable difference operator, see Dahlquist (1954). Without stability, however, roundoff errors will then overshadow the theoretical solution in actual computation.

We shall see that the characterization of stability of finite difference operators (3.3) is parallel to that of well-posedness of (3.1). Introducing the trigonometric polynomial $E_k(\xi) = \sum_j a_j(h) e^{ijh\xi}$, the symbol of E_k, Fourier transformation shows that E_k is stable in L_2 if and only if, cf. (3.2),

$$|E_k(\xi)^n| \leqslant C e^{\kappa n k}, \quad \forall \xi \in \mathbb{R}^d, \quad n \geqslant 0.$$

In the scalar case this is equivalent to $|E_k(\xi)| \leqslant 1 + Ck$ for $\xi \in \mathbb{R}^d$ and small k, or $|E_k(\xi)| \leqslant 1$ for $\xi \in \mathbb{R}^d$ when the coefficients of E_k are independent of h, as is normally the case when no lower-order terms occur in $P(D)$. For our above example $E_k(\xi) = 1 - 2\lambda + 2\lambda \cos h\xi$ and stability holds if and only if $\lambda = k/h^2 \leqslant \frac{1}{2}$. For the equation $u_t = \rho u_x$ and the scheme $\partial_t U_j^n = \hat{\partial}_x U_j^n$ we have $E_k(\xi) = 1 + i\rho\lambda \sin h\xi$ and this method is therefore seen to be unstable for any choice of $\lambda = k/h$.

Necessary for stability in the matrix case is the von Neumann condition

$$\rho(E_k(\xi)) \leqslant 1 + Ck, \quad \forall \xi \in \mathbb{R}^d, \tag{3.5}$$

where $\rho(A) = \max_j |\lambda_j|$ is the spectral radius of A, and for normal matrices $E_k(\xi)$ this is also sufficient. This covers the scalar case discussed above. Necessary and sufficient conditions are given in the following discrete version of the above Kreiss matrix lemma, see Kreiss (1962), where we denote $|u|_H=(Hu,u)^{1/2}$ and $|A|_H=\sup_{u\neq 0}|Au|_H/|u|_H$ for H positive definite:

Let $\mathscr{F}$ be a set of $N\times N$ matrices. Then the following conditions are equivalent:

(i) $|A^n|\leqslant C$ *for* $A\in\mathscr{F}$, $n\geqslant 0$, *for some* $C>0$;

(ii) $\rho(A)\leqslant 1$ *and* $(|z|-1)|R(A;z)|\leqslant C$ *for* $A\in\mathscr{F}$, $|z|>1$, *with* $C>0$;

(iii) $\rho(A)\leqslant 1$ *for* $A\in\mathscr{F}$ *and there are* C_1 and C_2 *and for each* $A\in\mathscr{F}$ *a matrix* $S=S(A)$ *such that* $\max(|S|,|S^{-1}|)\leqslant C_1$ *and such that* $SAS^{-1}=(b_{jk})$ *is triangular with off-diagonal elements satisfying* $|b_{jk}|\leqslant C_2\min(1-|\lambda_j|,1-|\lambda_k|)$ *where* $\lambda_j=b_{jj}$;

(iv) *let* $0\leqslant\gamma<1$. *There exists* $C>0$ *and for each* $A\in\mathscr{F}$ *a Hermitian matrix* $H=H(A)$ *with* $C^{-1}I\leqslant H\leqslant CI$ *and* $|A|_H\leqslant 1-\gamma+\gamma\rho(A)$.

Application shows that if E_k is stable in L_2, then there is a κ such that $\mathscr{F}=\{e^{-\kappa k}E_k(\xi);\ k\leqslant k_0,\ \xi\in\mathbb{R}^d\}$ satisfies conditions (i)–(iv). On the other hand, if one of these conditions holds for some κ, then E_k is stable in L_2.

Other related sufficient conditions were given in, e.g., Kato (1960), where it was shown that if the range of an $N\times N$ matrix is in the unit disc, i.e., if $|Av\cdot v|\leqslant 1$ for $|v|\leqslant 1$, then $|A^n v\cdot v|\leqslant 1$, and hence, by taking real and imaginary parts, $|A^n|\leqslant 2$ for $n\geqslant 1$.

Using the above characterizations one can show that a necessary and sufficient condition for the existence of an L_2-stable operator which is consistent with (3.1) is that (3.1) is well-posed in L_2, see Kreiss (1959a). It was also proved by Wendroff (1968) that for initial value problems which are well-posed in L_2 one may construct L_2-stable difference operators with arbitrarily high-order of accuracy.

It may be shown that von Neumann's condition (3.5) is equivalent to growth of at most polynomial order of the solution operator in L_2, or $\|E_k^n v\|_{L_2}\leqslant Cn^q\|v\|_{L_2}$ for $t_n\leqslant T$, for some $q\geqslant 0$. This was used by Forsythe and Wasow [14] and Ryabenkii and Filippov [29] as a definition of stability.

For variable coefficients it was shown by Strang (1965) that if the initial-value problem for the equation $u_t=\sum_{|\alpha|\leqslant M}P_\alpha(x)D^\alpha u$ is well-posed in L_2, then the one for the equation without lower-order terms and with coefficients fixed at $x\in\mathbb{R}^d$ is also well-posed, and a similar statement holds for the stability of the finite difference scheme $E_k v(x)=\sum_j a_j(x,h)v(x-jh)$. However, Kreiss (1962) showed that well-posedness and stability with frozen coefficients is neither necessary nor sufficient for well-posedness and stability of a general variable coefficient problem. We shall return below to variable coefficients for parabolic and hyperbolic equations.

We now consider the special case when system (3.1) is parabolic, and begin by quoting the fundamental paper of John (1952) in which maximum-norm stability was shown for finite difference schemes for second-order parabolic equations in one space variable. For simplicity, we restrict our presentation to the model problem

$$u_t=u_{xx}\quad\text{for } x\in R,\ t>0,\quad\text{with } u(\cdot,0)=v\quad\text{in } \mathbb{R} \tag{3.6}$$

and a corresponding finite difference approximation of the form (3.3) with $a_j(h)=a_j$ independent of h. Setting $a(\xi)=\sum_j a_j e^{-ij\xi}=E_k(h^{-1}\xi)$ one may write

$$U^n(x)=\sum_j a_{nj}v(x-jh),\quad\text{where } a_{nj}=\frac{1}{2\pi}\int_{-\pi}^{\pi}e^{-ij\xi}a(\xi)^n\,d\xi. \tag{3.7}$$

Here the von Neumann condition reads $|a(\xi)|\leqslant 1$, and is necessary and sufficient for stability in L_2. To show maximum-norm stability we need to estimate the a_{nj} in (3.7). It is easily seen that if the difference scheme is consistent with (3.6) then $a(\xi)=e^{-\lambda\xi^2}+o(\xi^2)$ as $\xi\to 0$, and if we assume that $|a(\xi)|<1$ for $0<|\xi|\leqslant\pi$ it follows that

$$|a(\xi)|\leqslant e^{-c\xi^2} \quad \text{for } |\xi|\leqslant\pi, \text{ with } c>0. \tag{3.8}$$

One then finds at once from (3.7) that $|a_{nj}|\leqslant Cn^{-1/2}$, and integration by parts twice, using $|a'(\xi)|\leqslant C|\xi|$, shows $|a_{nj}|\leqslant Cn^{1/2}j^{-2}$. Thus

$$\sum_j |a_{nj}|\leqslant C\sum_{j\leqslant n^{1/2}} n^{-1/2}+\sum_{j>n^{1/2}} n^{1/2}j^{-2}\leqslant C,$$

so that $\|U^n\|_\infty\leqslant C\|v\|_\infty$ by (3.7) where $\|\cdot\|_\infty=\|\cdot\|_{L_\infty}$. We remark that for our simple example above we have $|E_k v(x)|\leqslant(\lambda+|1-2\lambda|+\lambda)\|v\|_\infty=\|v\|_\infty$ for $\lambda\leqslant\frac{1}{2}$, which trivially yields maximum-norm stability.

In the general constant coefficient case system (3.1) is said to be parabolic in Petrowski's sense if $\Lambda(P(\xi))\leqslant-\delta|\xi|^M+C$ for $\xi\in\mathbb{R}^d$, and using (iii) of the Kreiss lemma one shows easily that the corresponding initial value problem is well-posed in L_2. For well-posedness in L_∞ we may write

$$u(x,t)=(E(t)v)(x)=\int_{\mathbb{R}^d} G(x-y,t)v(y)\,dy$$

where, cf., e.g., [15], with $D^\alpha=D_x^\alpha$,

$$|D^\alpha G(x,t)|\leqslant C_T t^{-(|\alpha|+d)/M} e^{(-\delta(|x|^M/t)^{1/(M-1)})} \quad \text{for } 0<t\leqslant T. \tag{3.9}$$

This implies that $\|D^\alpha u(\cdot,t)\|_\infty\leqslant Ct^{-|\alpha|/M}\|v\|_\infty$, so that the solution, in addition to being bounded, is smooth for positive time even if the initial data are only bounded. Consider now a difference operator of the form (3.3) which is consistent with (3.1). Generalizing from (3.8) we define this operator to be parabolic in the sense of John if, for some positive δ and C,

$$\rho(E_k(h^{-1}\xi))\leqslant e^{-\delta|\xi|^M}+Ck, \quad \text{for } \xi\in Q=\{\xi;\ |\xi_j|\leqslant\pi,\ j=1,\ldots,d\},$$

such schemes always exist when (3.1) is parabolic in Petrowski's sense. Extending the results of John (1952) to this situation Aronson (1963) and Widlund (1966) showed that if we write $U^n(x)=E_k^n v(x)=\sum_j a_{nj}(h)v(x-jh)$, then, denoting difference quotients corresponding to D^α by ∂_h^α, we have, cf. (3.9),

$$|\partial_h^\alpha a_{nj}(h)|\leqslant Ch^d t_n^{-(|\alpha|+d)/M} e^{(-c(|j|^M/n)^{1/(M-1)})} \quad \text{for } t_n\leqslant T,$$

which implies that $\|\partial_h^\alpha E_k^n v\|_\infty\leqslant Ct_n^{-|\alpha|/M}\|v\|_\infty$. In the work quoted also multistep methods and variable coefficients were treated.

From estimates of these types follow also convergence estimates such as, if D_h^α is a difference operator consistent with D^α and accurate of order r,

$$\|D_h^\alpha U^n-D^\alpha u(t_n)\|_\infty\leqslant Ch^r\|v\|_{W_\infty^{r+|\alpha|}} \quad \text{for } t_n\leqslant T.$$

Note in the convergence estimate for $\alpha=0$ that, as a result of the smoothing property for parabolic equations, less regularity is required of initial data than for the general well-posed initial value problem, cf. (3.4). For even less regular initial data lower convergence rates have to be expected; by interpolation between the W_∞^s spaces one may show (see Peetre and Thomée (1967)), e.g.,

$$\|U^n-u(t_n)\|_\infty\leqslant Ch^s\|v\|_{W_\infty^s} \quad \text{for } 0\leqslant s\leqslant r,\ t_n\leqslant T.$$

We remark that for nonsmooth initial data v it is possible to make a preliminary smoothing of v to recover full accuracy for t bounded away from zero: It was shown in Kreiss, Thomée and Widlund (1970) that there exists a smoothing operator of the form $M_h v = \Phi_h * v$, where $\Phi_h(x) = h^{-d}\Phi(h^{-1}x)$, with Φ an appropriate function, such that if $M_h v$ is chosen as initial data for the difference scheme, then

$$\|U^n - u(t_n)\|_\infty = \|E_h^n M_h v - E(t_n)v\|_\infty \leqslant Ch^r t_n^{-r/M}\|v\|_\infty.$$

Let us also note that from a known convergence rate for the difference scheme for fixed initial data v conclusions may be drawn about the regularity of v. For example, under the above assumptions, assume that for some p, s with $1 \leqslant p \leqslant \infty, 0 < s \leqslant r$, we know that $\|U^n - u(t_n)\|_{L_p} \leqslant Ch^s$ for $t_n \leqslant T$. Then v belongs to the Besov space $B_p^{s,\infty}$ ($\approx W_p^s$), and, if $s > r$ then $v = 0$. Such inverse results were given in, e.g., Hedstrom (1968), Löfström (1970), and Thomée and Wahlbin (1974).

We now turn our attention to hyperbolic equations, and consider first systems with constant real coefficients

$$u_t = \sum_{j=1}^d A_j u_{x_j} \quad \text{for } t > 0, \quad \text{with } u(0) = v. \tag{3.10}$$

Such a system is said to be strongly hyperbolic if it is well-posed in L_2, cf. Strang (1967). With $P(\xi) = \sum_{j=1}^d A_j\xi_j$ this holds if and only if for each $\xi \in \mathbb{R}^d$ there exists a nonsingular matrix $S(\xi)$, uniformly bounded together with its inverse, such that $S(\xi)P(\xi)S(\xi)^{-1}$ is a diagonal matrix with real elements. When the A_j are symmetric this holds with $S(\xi)$ orthogonal; the system is then said to be symmetric hyperbolic. The condition is also satisfied when the eigenvalues of $P(\xi)$ are real and distinct for $\xi \neq 0$; in this case the system is called strictly hyperbolic.

One important feature of hyperbolic systems is that the value $u(x,t)$ of the solution at (x,t) only depends on the initial values on a compact set $K(x,t)$, the smallest closed set such that $u(x,t) = 0$ when v vanishes in a neighborhood of $K(x,t)$. The convex hull of $K(x,t)$ may be described in terms of $P(\xi)$: if $K = K(0,1)$ we have for its support function $\phi(\xi) = \sup_{x\in K} x\xi = \lambda_{\max}(P(\xi))$.

Consider now the system (3.10) and a corresponding finite difference operator of the form (3.3) (with $M = 1$). Here we introduce the domain of dependence $\tilde{K}(x,t)$ of E_k as the smallest closed set such that $E_k^n v(x) = 0$ for all n, k with $nk = t$ when v vanishes in a neighborhood of $\tilde{K}(x,t)$. Corresponding to the above the support function of $\tilde{K}(x,t)$ satisfies $\tilde{\phi}(\xi) = \lambda^{-1}\max_{a_j\neq 0} j\xi$. Since clearly convergence, and hence stability, demands that $\tilde{K} = \tilde{K}(0,1)$ contains the continuous domain of dependence $K = K(0,1)$ it is necessary for stability that $\lambda_{\max}(P(\xi)) \leqslant \lambda^{-1}\max_{a_j\neq 0} j\xi$; this is the CFL-condition, cf. Section 1. In particular, $\min_{a_j\neq 0} j_l \leqslant \lambda\lambda_{\min}(A_l) \leqslant \lambda\lambda_{\max}(A_l) \leqslant \max_{a_j\neq 0} j_l$. For the equation $u_t = \rho u_x$ and a difference operator of the form (3.3) using only $j = -1, 0, 1$, this means $\lambda|\rho| \leqslant 1$: In this case $u(0,1) = v(\rho)$ so that $K = \{\rho\}$, and the condition is thus $\rho \in \tilde{K} = [-\lambda^{-1}, \lambda^{-1}]$.

We shall now give some sufficient conditions for stability. We first quote from Friedrichs (1954) that if a_j are symmetric and positive semidefinite, with $\sum_j a_j = I$, then $|E_k(h^{-1}\xi)| = |\sum_j a_j e^{ij\xi}| \leqslant 1$ so that the scheme is L_2-stable. As an example, the first order accurate Friedrichs operator

$$E_k v(x) = \frac{1}{2}\sum_{j=1}^d ((d^{-1}I + \lambda A_j)v(x + he_j) + (d^{-1}I - \lambda A_j)v(x - he_j)) \tag{3.11}$$

is L_2-stable if $0 < \lambda \leqslant (d\rho(A_j))^{-1}$. It was observed by Lax (1961) that this criterion is of limited value in applications because it cannot in general be combined with accuracy higher than first order.

The necessary and sufficient conditions for L_2-stability of the Kreiss stability lemma are of the nature that the necessary von Neumann condition (3.5) has to be supplemented by conditions assuring that the eigenvalues of $E_k(\xi)$ sufficiently well describe the growth behavior of $E_k(\xi)^n$. We now quote some such criteria which utilize relations between the behavior of the eigenvalues of $E_k(\xi)$ for small ξ and the accuracy of E_k. In Kreiss (1964) E_k is defined to be dissipative of order v, v even, if $\rho(\tilde{E}_k(h^{-1}\xi)) \leqslant 1 - \delta|\xi|^v$ for $\xi \in Q$ with $\delta > 0$, and it is shown that if E_k is consistent with the strongly hyperbolic system (3.10), accurate of order $v - 1$, and dissipative of order v, then it is L_2-stable. Further, it was shown by Parlett (1966) that if the system is symmetric hyperbolic, and E_k is dissipative of order v it suffices that it is accurate of order $v - 2$, and by Yamaguti (1967) that if the system is strictly hyperbolic, then dissipativity of some order v is sufficient. For strictly hyperbolic systems Yamaguti also showed that if the eigenvalues of $E_k(\xi)$ are distinct for $Q \ni \xi \neq 0$, then von Neumann's condition is sufficient for stability in L_2.

For $d = 1$ (with $A = A_1$) the well-known Lax–Wendroff (1960) scheme

$$E_k v(x) = \tfrac{1}{2}(\lambda^2 A^2 + \lambda A)v(x+h) + (1 - \lambda^2 A^2)v(x) + \tfrac{1}{2}(\lambda^2 A^2 - \lambda A)v(x-h)$$

is L_2-stable for $\lambda\rho(A) \leqslant 1$ if the system approximated is strongly hyperbolic. L_2-stable finite difference schemes of arbitrarily high order for such systems were constructed in, e.g., Strang (1962).

It is also possible to analyze multistep methods, e.g., by rewriting them as single-step systems. A popular stable such method is the leapfrog method $\hat{\partial}_t U_j^n = \hat{\partial}_x U_j^n$ (or $U_j^{n+1} = U_j^{n-1} + \lambda\rho(U_{j+1}^n - U_{j-1}^n)$). The eigenvalues $\tau_1(\xi), \tau_2(\xi)$ appearing in the stability analysis then satisfy the characteristic equation $\tau - \tau^{-1} - 2\lambda\rho \sin\xi = 0$, and we find that the von Neumann condition, $|\tau_j(\xi)| \leqslant 1$ for $\xi \in \mathbb{R}$, $j = 1, 2$, is satisfied if $\lambda\rho \leqslant 1$, and using the Kreiss lemma one can show that L_2-stability holds for $\lambda\rho < 1$.

As an example for $d = 2$ we quote the operator introduced in Lax and Wendroff (1964) defined by

$$\begin{aligned} E_k(h^{-1}\xi) = I &+ \mathrm{i}\lambda(A_1 \sin\xi_1 + A_2 \sin\xi_2) \\ &- \lambda^2(A_1^2(1 - \cos\xi_1) + \tfrac{1}{2}(A_1A_2 + A_2A_1)\sin\xi_1 \sin\xi_2 + A_2^2(1 - \cos\xi_2)). \end{aligned}$$

Using the above stability criterion of Kato they proved stability for $\lambda|A_j| \leqslant 1/\sqrt{8}$ in the symmetric hyperbolic case.

Consider now a variable coefficient symmetric hyperbolic system

$$u_t = \sum_{j=1}^{d} A_j(x) u_{x_j}, \quad \text{where } A_j(x)^* = A_j(x). \tag{3.12}$$

Using an energy argument Friedrichs (1954) showed that the corresponding initial value problem is well-posed; the boundedness follows at once by noting that after multiplication by u, integration over $x \in \mathbb{R}^d$, and integration by parts,

$$\frac{\mathrm{d}}{\mathrm{d}t}\|u(t)\|^2 = -\int_{\mathbb{R}^d} \Big(\sum_j \partial A_j/\partial x_j u, u\Big)\, \mathrm{d}x \leqslant 2\kappa\|u(t)\|^2.$$

For finite difference approximations $E_k v(x) = \sum_j a_j(x) v(x - jh)$ which are consistent with (3.12) various sufficient conditions are available. Kreiss, et al. (1970), studied difference operators which are dissipative of order v, i.e., such that $\rho(E(x,\xi)) \leqslant 1 - c|\xi|^v$ for $x \in \mathbb{R}^d$, $\xi \in Q$, with $c > 0$, where $E(x,\xi) = \sum_j a_j(x)\mathrm{e}^{-\mathrm{i}j\xi}$. As in the constant coefficient case it was shown that E_k is stable in L_2 if the $a_j(x)$ are symmetric and if E_k is accurate of order $\nu - 1$ and dissipative of order ν; if (3.12) is strictly hyperbolic, accuracy of order $\nu - 2$ suffices. The proofs are based on the Kreiss stability lemma and a perturbation argument.

In an important paper it was proved by Lax and Nirenberg (1966) that if E_k is consistent with (3.12) and $|E(x,\xi)| \leqslant 1$ then E_k is strongly stable with respect to the L_2-norm, i.e., $\|E_k\| \leqslant 1 + Ck$. In particular, if the $a_j(x)$ are symmetric positive-definite stability follows by the result of Friedrichs quoted earlier for constant coefficients. Further, if E_k is consistent with (3.12) and $|E(x,\xi)v \cdot v| \leqslant 1$ for $|v| \leqslant 1$, then E_k is L_2-stable.

As mentioned above, higher-order equations in time may be expressed as systems of the form (3.1), and finite difference methods may be based on such formulations. In Section 1 we described an example of a difference approximation of a second-order wave equation in its original form.

We turn to estimates in L_p norms with $p \neq 2$. For the case when the equation in (3.10) is symmetric hyperbolic it was shown by Brenner (1966) that the initial value problem is well-posed in $L_p, p \neq 2$, if and only if the matrices A_j commute. This is equivalent to the simultaneous diagonalizability of these matrices, which in turn means that by introducing a new dependent variable it can be transformed to a system in which all the A_j are diagonal, so that the system consists of N uncoupled scalar equations.

Since stability of finite difference approximations can only occur for well-posed problems it is therefore natural to consider the scalar equations $u_t = \rho u_x$ with ρ real, and corresponding finite difference operators (3.3). With $a(\xi)$ as before we recall that such an operator is stable in L_2 if and only if $|a(\xi)| \leqslant 1$ for $\xi \in \mathbb{R}$. Necessary and sufficient conditions for stability in $L_p, p \neq 2$, as well as rates of growth of $\|E_k^n\|_{L_p}$ in the nonstable case have been given, e.g., in Brenner and Thomée (1970), cf. also references in Brenner, Thomée and Wahlbin [9]. If $|a(\xi)| < 1$ for $0 < |\xi| \leqslant \pi$ and $a(0) = 1$ the condition is that $a(\xi) = e^{i\alpha\xi - \beta\xi^\nu(1+o(1))}$ as $\xi \to 0$, with α real, $\operatorname{Re}\beta > 0$, ν even. Thus E_k is stable in L_p, $p \neq 2$, if and only if there is an even number ν such that E_k is dissipative of order ν and accurate of order $\nu - 1$. As an example of an operator which is stable in L_2 but not in L_p, $p \neq 2$, we may take the Lax–Wendroff operator introduced above (with A replaced by ρ). For $0 < \lambda|\rho| < 1$, this operator is stable in L_2, is dissipative of order 4, but accurate only of order 2. It may be proved that for this operator $cn^{1/8} \leqslant \|E_k^n\|_\infty \leqslant Cn^{1/8}$ with $c > 0$, which shows a weak instability in the maximum-norm.

An area where finite difference techniques continue to flourish and to form an active research field is for nonlinear fluid flow problems. Consider the scalar nonlinear conservation law

$$u_t + F(u)_x = 0 \quad \text{for } x \in \mathbb{R},\ t > 0, \text{ with } u(\cdot, 0) = v \text{ on } \mathbb{R}, \tag{3.13}$$

where $F(u)$ is a nonlinear function of u, often strictly convex, so that $F''(u) > 0$. The classical example is Burger's equation $u_t + uu_x = 0$, where $F(u) = u^2/2$. Discontinuous solutions may arise even when v is smooth, and one therefore needs to consider weak solutions. Such solutions are not necessarily unique, and to select the unique physically relevant solution one has to require so called entropy conditions. This solution may also be obtained as the limit as $\varepsilon \to 0$ of the diffusion equation with εu_{xx} replacing 0 on the right in (3.13).

One important class of methods for (3.13) are finite difference schemes in conservation form

$$U_j^{n+1} = U_j^n - \lambda(H_{j+1/2}^n - H_{j-1/2}^n) \quad \text{for } n \geqslant 0, \text{ with } U_j^0 = v(jh), \tag{3.14}$$

where $H^n_{j+1/2} = H(U^n_j, U^n_{j+1})$ is a numerical flux function which has to satisfy $H(V, V) = F(V)$ for consistency. Here stability for the linearized equation is neither necessary nor sufficient for the nonlinear equation in the presence of discontinuities and much effort has been devoted to the design of numerical flux functions with good properties. When the right-hand side of (3.14) is increasing in $U^n_{j+l}, l = -1, 0, 1$, the scheme is said to be monotone (corresponding to positive coefficients in the linear case) and such schemes converge for fixed λ to the entropy solution as $k \to 0$, but are at most first-order accurate. Examples are the Lax–Friedrichs scheme which generalizes (3.11), the scheme of Courant, Isaacson and Rees (1952), which is one sided (upwinding), and the Engquist–Osher scheme which pays special attention to changes in sign of the characteristic direction. Godunov's method replaces U^n by a piecewise constant function, solves the corresponding problem exactly from t_n to t_{n+1} and defines U^{n+1} by an averaging process.

Higher-order methods with good properties are also available, and are often constructed with an added artificial diffusion term which depends on the solution, or so-called total variation diminishing (TVD) schemes. Early work in the area was Lax and Wendroff (1960), and more recent contributions have been given by Engquist, Harten, Kuznetsov, MacCormack, Osher, Roe, Yee; for overviews and generalizations to systems and higher dimension, see Le veque [23] and Godlewski and Raviart [16].

4. Finite differences for mixed initial–boundary value problems

The pure initial value problem discussed in Section 3 is often not adequate to model a given physical situation and one needs to consider instead a problem whose solution is required to satisfy the differential equation in a bounded spatial domain $\Omega \subset \mathbb{R}^d$, as well as boundary conditions on $\partial\Omega$ for positive time, and to take on given initial values. For such problems the theory of finite difference methods is less complete and satisfactory. In the same way as for the stationary problem treated in Section 2 one reason for this is that for $d \geqslant 2$ only very special domains may be well represented by mesh-domains, and even when $d = 1$ the transition between the finite difference approximation in the interior and the boundary conditions may be complex both to define and to analyze. Again there are three standard approaches to the analysis, namely methods based on maximum principles, energy arguments, and spectral representation. We illustrate this first for parabolic and then for hyperbolic equations.

As a model problem for parabolic equations we shall consider the one-dimensional heat equation in $\Omega = (0, 1)$,

$$u_t = u_{xx} \quad \text{in } \Omega, \quad u = 0 \quad \text{on } \partial\Omega = \{0, 1\}, \quad \text{for } t \geqslant 0, \;\; u(\cdot, 0) = v \text{ in } \Omega. \tag{4.1}$$

For the approximate solution we introduce mesh-points $x_j = jh$, where $h = 1/M$, and time levels $t_n = nk$, where k is the time step, and denote the approximate solution at (x_j, t_n) by U^n_j. As for the pure initial value problem we may then approximate (4.1) by means of the explicit forward Euler difference scheme

$$\partial_t U^n_j = \partial_x \bar{\partial}_x U^n_j \quad \text{in } \Omega, \text{ with } U^{n+1}_j = 0 \text{ on } \partial\Omega, \quad \text{for } n \geqslant 0$$

with $U^0_j = V_j = v(x_j)$ in Ω. For U^{n-1} given this defines U^n through

$$U^n_j = \lambda(U^{n-1}_{j-1} + U^{n-1}_{j+1}) + (1 - 2\lambda)U^{n-1}_j, \quad 0 < j < M, \;\; U^n_j = 0, \; j = 0, M.$$

For $\lambda = k/h^2 \leqslant 1/2$ the coefficients are nonnegative and their sum is 1 so that we conclude that $|U^{n+1}| \leqslant |U^n|$ where $|U| = \max_{x_j \in \Omega} |U_j|$. It follows that $|U^n| \leqslant |V|$, and the scheme is thus stable in maximum norm. Under these assumptions one shows as for the pure initial value problem that $|U^n - u(t_n)| \leqslant C(u)(h^2 + k) \leqslant C(u)h^2$. It is easy to see that $k \leqslant h^2/2$ is also a necessary condition for stability.

To avoid to have to impose the quite restrictive condition $\lambda \leqslant \frac{1}{2}$, Laasonen (1949) proposed the implicit backward Euler scheme

$$\bar{\partial}_t U_j^n = \partial_x \bar{\partial}_x U_j^n \quad \text{in } \Omega, \text{ with } U_j^n = 0 \text{ on } \partial\Omega, \quad \text{for } n \geqslant 1$$

with $U_j^0 = v(x_j)$ as above. For U^{n-1} given one now needs to solve the linear system

$$(1 + 2\lambda)U_j^n - \lambda(U_{j-1}^n + U_{j+1}^n) = U_j^{n-1}, \ 0 < j < M, \quad U_j^n = 0, \ j = 0, M.$$

This method is stable in maximum norm without any restrictions on k and h. In fact, we find at once, for suitable k,

$$|U^n| = |U_k^n| \leqslant \frac{2\lambda}{1 + 2\lambda} |U^n| + \frac{1}{1 + 2\lambda} |U^{n-1}|$$

and hence $|U^n| \leqslant |U^{n-1}| \leqslant |V|$. Here, the error is of order $O(h^2 + k)$.

Although the backward Euler method is unconditionally stable, it is only first-order accurate in time. A second-order accurate method was proposed by Crank and Nicolson (1947) which uses the equation

$$\bar{\partial}_t U_j^n = \partial_x \bar{\partial}_x U_j^{n-1/2}, \quad 0 < j \leqslant M, \text{ where } U_j^{n-1/2} = \tfrac{1}{2}(U_j^n + U_j^{n-1}). \tag{4.2}$$

The above approach will now show $|U^n| \leqslant |U^{n-1}|$ only if $\lambda \leqslant 1$.

For this problem, however, the energy method may be used to show unconditional stability in l_2-type norms: With $(V, W) = h \sum_{j=1}^{M-1} V_j W_j$ and $\|V\| = (V, V)^{1/2}$ one finds, upon multiplication of (4.2) by $2U_j^{n-1/2}$, summation, and summation by parts,

$$\bar{\partial}_t \|U^n\|^2 = 2(\bar{\partial}_t U^n, U^{n-1/2}) = -2h^{-2} \sum_{j=1}^{M} (U_j^{n-1/2} - U_{j-1}^{n-1/2})^2 \leqslant 0,$$

which shows $\|U^n\| \leqslant \|U^{n-1}\|$, i.e., stability in l_2 holds for any $\lambda > 0$. In the standard way this yields the error estimate $\|U^n - u(t_n)\| \leqslant C(u)(h^2 + k^2)$; a corresponding estimate in a discrete H^1-norm may be obtained similarly, and this also yields a maximum-norm estimate of the same order by using a discrete Sobolev inequality. The energy approach was developed by, e.g., Kreiss (1959a), Lees (1960) and Samarskii (1961).

Stability in l_2 may also be deduced by spectral analysis, as observed in O'Brien, Hyman and Kaplan (1951). We illustrate this for the Crank–Nicolson method: Representing the mesh-functions vanishing at $x = 0$ and 1 in terms of the eigenfunctions φ_p of $\partial_x \bar{\partial}_x$, the vectors with components $\varphi_{pj} = \sqrt{2} \sin \pi pjh$, $0 \leqslant j \leqslant M$, and eigenvalues $\mu_p = -2h^{-2}(1 - \cos \pi ph)$, one finds

$$U^n = \sum_{p=1}^{M-1} (V, \varphi_p) E(\pi ph)^n \varphi_p \quad \text{where } E(\xi) = \frac{1 - \lambda(1 - \cos \xi)}{1 + \lambda(1 - \cos \xi)}$$

and $\|U^n\| \leqslant \|V\|$ follows by Parseval's relation since $|E(\xi)| \leqslant 1$ for $\xi \in \mathbb{R}$, $\lambda > 0$. This is analogous to the Fourier analysis of Section 3 for pure initial value problems.

We remark that although the maximum-principle-type argument for stability for the Crank–Nicolson method requires $\lambda \leqslant 1$, it was shown by Serdjukova (1964) using Fourier analysis that the maximum-norm bound $|U^n| \leqslant 23|V|$ holds. Precise convergence analyses in maximum-norm for initial data with low regularity were carried out in, e.g., Juncosa and Young (1957).

For the pure initial value problem, difference schemes or arbitrarily high order of accuracy can be constructed by including the appropriate number of terms in a difference operator of the form $U_j^{n+1} = \sum_{l=-q}^{s} a_l U_{j-l}^n$. For application to the mixed initial–boundary value problem (4.1) such a formula would require additional equations for U_j^n near $x = 0$ or $x = 1$ when $s \geqslant 2$ or $q \geqslant 2$. For the semiinfinite interval $(0, \infty)$, Strang (1964) showed that with $s = 1$ any order of accuracy may be achieved together with stability by choosing an "unbalanced" operator with $q \geqslant 2$. The stability of schemes with additional boundary conditions has been analyzed in the parabolic case by Varah (1970) and Osher (1972) using the GKS-technique which we briefly describe for hyperbolic equations below.

We note that the above methods may be written as $U^{n+1} = E_k U^n$, where E_k acts in different spaces $\mathcal{N}_k$ of vectors $V = (V_0, \ldots, V_M)^{\mathrm{T}}$ with $V_0 = V_M = 0$, where M depends on k. In order to deal with the stability problem in such situations, Godunov and Ryabenkii (1963) introduced a concept of spectrum of a family of operators $\{E_k\}$, with E_k defined in a normed space $\mathcal{N}_k$ with norm $\|\cdot\|_k$, k small. The spectrum $\sigma(\{E_k\})$ is defined as the complex numbers z such that for any $\varepsilon > 0$ and sufficiently small k there is a $U_k \in \mathcal{N}_k$, $U_k \neq 0$, such that $\|E_k U_k - z U_k\|_k \leqslant \varepsilon \|U_k\|_k$, and the following variant of von Neumann's criterion holds: If $\{E_k\}$ is stable in the sense that $\|E_k^n V\|_k \leqslant C\|V\|_k$ for $t_n \leqslant T$, with C independent of k, then $\sigma(\{E_k\}) \subset \{z; |z| \leqslant 1\}$. It was demonstrated that the spectrum of a family such as one of the above is the union of three sets, one corresponding to the pure initial value problem and one to each of the one-sided boundary value problems for the differential equation in $\{x \geqslant 0, t \geqslant 0\}$ and in $\{x \leqslant 1, t \geqslant 0\}$, with boundary conditions given at $x = 0$ and 1, respectively. For instance, in the example of the explicit method, $\sigma(\{E_k\})$ equals the set of eigenvalues of the operator $\tilde{E}_k$ associated with the pure initial value problem, which is easily shown to be equal to the interval $[1 - 4\lambda, 1]$, and hence $\lambda \leqslant \frac{1}{2}$ is a necessary condition for stability. The proof of the equality between these sets is nontrivial as the eigenfunctions of $\tilde{E}_k$ do not satisfy the boundary conditions. Using instead a boundary condition of the form $u_0 - \gamma u_1 = 0$ at $x = 0$, will result in instability for certain choices of γ.

We now turn to the two-dimensional model problem in the square $\Omega = (0,1)^2$,

$$u_t = \Delta u \quad \text{in } \Omega, \qquad u = 0 \text{ on } \partial\Omega, \qquad \text{for } t > 0, \text{ with } u(\cdot, 0) = v \text{ in } \Omega. \tag{4.3}$$

Again with $h = 1/M$ we use mesh-points $x_j = jh$, now with $j \in \mathbb{Z}^2$. We consider the above methods collectively as the ϑ-method, with $0 \leqslant \vartheta \leqslant 1$,

$$\bar{\partial}_t U_j^n = \Delta_h(\vartheta U_j^n + (1 - \vartheta) U_j^{n-1}) \quad \text{in } \Omega, \quad U_j^n = 0 \text{ on } \partial\Omega,$$

where $\vartheta = 0, 1$, and $\frac{1}{2}$ for the forward and backward Euler methods and the Crank–Nicolson method, respectively. The above stability and error analysis carries over to this case; the ϑ-method is unconditionally stable in l_2 for $\frac{1}{2} \leqslant \vartheta \leqslant 1$ whereas for $0 \leqslant \vartheta < \frac{1}{2}$ the mesh-ratio condition $\lambda(1 - 2\vartheta) \leqslant \frac{1}{4}$ has to be imposed.

For the model problem (4.3) we consider also the alternating direction implicit (ADI) scheme of Peaceman and Rachford (1955). Noting that the Crank–Nicolson scheme requires the solution at time t_n of the two-dimensional elliptic problem $(I - \frac{1}{2}k\Delta_h)U^n = (I + \frac{1}{2}k\Delta_h)U^{n-1}$, the purpose of the

ADI method is to reduce the computational work by solving only one-dimensional problems. This is done by introducing the intermediate value $U^{n-1/2}$ for the solution at $t_{n-1/2}=t_n-k/2$ by the equations

$$\frac{U^{n-1/2}-U^{n-1}}{k/2}=\partial_1\bar{\partial}_1U^{n-1/2}+\partial_2\bar{\partial}_2U^{n-1},$$

$$\frac{U^{n}-U^{n-1/2}}{k/2}=\partial_1\bar{\partial}_1U^{n-1/2}+\partial_2\bar{\partial}_2U^{n}.$$

Elimination of $U^{n-1/2}$ gives, since the various operators commute,

$$U^n=E_kU^{n-1}=(I-\tfrac{k}{2}\partial_1\bar{\partial}_1)^{-1}(I+\tfrac{k}{2}\partial_1\bar{\partial}_1)(I-\tfrac{k}{2}\partial_2\bar{\partial}_2)^{-1}(I+\tfrac{k}{2}\partial_2\bar{\partial}_2)U^{n-1}.$$

By either energy arguments or by the spectral method one sees easily that this method is stable in the discrete l_2-norm and since it is second-order accurate in both space and time one finds $\|U^n-u(t_n)\|\leqslant C(u)(h^2+k^2)$. We may also express the definition of E_k (using a different $U^{n-1/2}$) by

$$U^{n-j/2}=(I-\tfrac{k}{2}\partial_j\bar{\partial}_j)^{-1}(I+\tfrac{k}{2}\partial_j\bar{\partial}_j)U^{n-(j-1)/2}\quad\text{for } j=0,1.$$

In this form, which generalizes in an obvious way to more dimensions, it is referred to as a fractional step method and depends on the splitting of the operator Δ_h into $\partial_1\bar{\partial}_1$ and $\partial_2\bar{\partial}_2$. This has been a very active area of research during the 1960s, with contributions by, e.g., Douglas, Kellogg, Temam, Wachspress, Dyakonov, Samarskii, Marchuk, and Yanenko, see the survey article by Marchuk [25].

In the same way as for the elliptic problems studied in Section 2, complications arise when the boundary mesh-points do not fall exactly on $\partial\Omega$, which is the case in the presence of a curved boundary. Using, e.g., linear interpolation or Shortley–Weller-type approximations of the Laplacian one may show $O(k+h^2)$ error estimates for the backward Euler method by means of a discrete maximum-principle, and one may also use energy arguments, with the crudest boundary approximation, to show $O(k+h^{1/2})$ error estimates, see Thomée [36,37].

We now turn to hyperbolic equations and consider first the spatially one-dimensional wave equation

$$u_{tt}=u_{xx}\quad\text{in } \Omega=(0,1),\text{ with } u(\cdot,0),\ u_t(\cdot,0)\text{ given.}$$

Here one may define the ϑ-method, $\vartheta\in[0,1]$, which is a special case of a family of schemes studied in Newmark (1959), by

$$\partial_t\bar{\partial}_tU_j^n=\partial_x\bar{\partial}_x((1-\vartheta)U_j^n+\tfrac{1}{2}\vartheta(U_j^{n+1}+U_j^{n-1}))\tag{4.4}$$

with $U_0^n=U_M^n=0$ for $n\geqslant 0$ and U_j^0 and U_j^1 given. This scheme is unconditionally stable for $\frac{1}{2}\leqslant\vartheta\leqslant 1$, and for $0\leqslant\vartheta<\frac{1}{2}$ it is stable if $\lambda^2=k^2/h^2<1/(1-2\vartheta)$; for $\vartheta=0$ we recognize the explicit scheme of Section 1, which can easily be shown to be unstable for $\lambda=1$, see Raviart and Thomas [27].

As a simple first-order hyperbolic model problem we consider, with $\rho>0$,

$$u_t=\rho u_x\quad\text{in } \Omega=(0,1),\quad u(1,t)=0,\quad\text{for } t>0,\text{ with } u(\cdot,0)=v\text{ in } \Omega.$$

Note that since the solution is constant along the characteristics $x+\rho t$=constant no boundary condition is needed at $x=0$.

Consider first the "upwind" scheme, see Courant, Isaacson and Rees (1952),

$$\partial_tU_j^n=\rho\partial_xU_j^n,\quad j=0,\dots,M-1,\qquad U_M^{n+1}=0,\quad\text{for } n\geqslant 0,\text{ with } U_j^0=v(x_j),$$

which may be written in explicit form as

$$U_j^{n+1} = (1 - \lambda\rho)U_j^n + \lambda\rho U_{j+1}^n, \quad 0 \leqslant j < M, \quad U_M^{n+1} = 0.$$

When $\lambda\rho \leqslant 1$ the method is stable in maximum norm; this condition may be expressed by saying that the characteristic traced back from (x_j, t_{n+1}) cuts $t = t_n$ in $[x_j, x_{j+1}]$, which we recognize as the CFL condition. By the lack of symmetry it is only first-order accurate. For $\rho\lambda > 1$ one can use instead the Carlson scheme (see [28])

$$\partial_t U_j^n = \rho\bar{\partial}_x U_j^{n+1}, \quad 1 \leqslant j \leqslant M, \qquad U_M^{n+1} = 0, \text{ for } n \geqslant 0, \text{ with } U_j^0 = v(x_j),$$

which determines U_j^{n+1} for decreasing j by $U_{j-1}^{n+1} = (1 - (\rho\lambda)^{-1})U_j^{n+1} + (\rho\lambda)^{-1}U_j^n$. Again this method is maximum-norm stable, but only first-order accurate.

A second-order method is the box scheme of Wendroff (1960),

$$\partial_t U_{j-1/2}^n = \rho\bar{\partial}_x U_j^{n+1/2}, \quad 1 \leqslant j \leqslant M, \qquad U_M^{n+1} = 0, \quad U_{j-1/2} = \tfrac{1}{2}(U_j + U_{j-1}).$$

With U^n and U_M^{n+1} given this determines U_{j-1}^{n+1} for decreasing j as a combination of U_j^{n+1}, U_{j-1}^n, and U_j^n. Stability in l_2 may be shown by an energy argument.

We end this section with two examples where the finite difference operators used in the interior of Ω require modification near the boundary or additional artificial boundary conditions. In our first example we describe a special case of an energy argument proposed by Kreiss and Scherer (1974), see also Gustafsson, Kreiss and Oliger [18]. We consider the initial–boundary value problem, with $\rho > 0$,

$$u_t - \rho u_x = f \quad \text{in } \Omega, \text{ with } u(1,t) = 0, \quad \text{for } t \geqslant 0, \ u(\cdot, 0) = v \text{ in } \Omega.$$

Assume that we want to apply the six-point Crank–Nicolson equation

$$\bar{\partial}_t U_j^n - \rho\hat{\partial}_x U_j^{n-1/2} = f_j^{n-1/2} \quad \text{for } 1 \leqslant j \leqslant M - 1. \tag{4.5}$$

At the right endpoint of Ω we set $U_M^n = 0$. For $x = 0$ the value of u is not given, and we therefore use the one-sided equation $\bar{\partial}_t U_0^n - \rho\partial_x U_0^{n-1/2} = f_0^{n-1/2}$. With the obvious definition of the composite difference operator D_x we may write $\bar{\partial}_t U_j^n - \rho D_x U_j^{n-1/2} = f_j^{n-1/2}$ for $0 \leqslant j < M$. Introducing temporarily the inner product $(U, V) = \frac{1}{2}hU_0V_0 + h\sum_{j=1}^{M-1} U_jV_j$ we have $(D_xU, U) = -\frac{1}{2}U_0^2$ if $U_M = 0$, which yields $(\bar{\partial}_t U^n, U^{n-1/2}) + \frac{1}{2}\rho(U_0^{n-1/2})^2 = (f^{n-1/2}, U^{n-1/2})$. Together with the inequality $hf_0U_0 \leqslant \rho U_0^2 + Ch^2 f_0^2$ this easily shows the stability estimate

$$\|U^n\|^2 \leqslant C\|U^0\|^2 + Ck\sum_{l=0}^{n-1} \|\tilde{f}^{l+1/2}\|^2, \quad \text{where } \tilde{f} = (hf_0, f_1, \ldots, f_{M-1}).$$

Note that the choice of the term with $j = 0$ in $(\cdot, \cdot)$ is essential for the argument.

Applying this to the error $U - u$, with the truncation error τ for f, and observing that $\tau_j = O(h^2)$ for $j \geqslant 1$, $\tau_0 = O(h)$, we find $\|U - u\| = O(h^2)$. This approach was also used in the reference quoted to construct higher-order schemes. We note that the modification of (4.5) for $j = 0$ may also be interpreted as using (4.5) for $j = 0$, and adding the boundary condition $h^2\partial_x\bar{\partial}_x U_0^n = U_1^n - 2U_0^n + U_{-1}^n = 0$.

We finally give an example of the stability analysis based on the use of discrete Laplace transforms developed by Kreiss (1968), Gustafsson, Kreiss and Sundström (1972), the so-called GKS-theory. Consider the initial–boundary value problem, again with $\rho > 0$,

$$u_t = \rho u_x \quad \text{for } x > 0, \ t > 0, \quad \text{with } u(x, 0) = v(x) \text{ for } x \geqslant 0.$$

Assume that we want to use the leapfrog scheme

$$\hat{\partial}_t U_j^n = \rho \hat{\partial}_x U_j^n \quad \text{for } j \geqslant 1, n \geqslant 1, \text{ with } U_j^0,\ U_j^1 \text{ given for } j \geqslant 0, \tag{4.6}$$

where we assume $\rho\lambda = \rho k/h < 1$ so that the corresponding scheme for the pure initial value problem is stable. Again an additional boundary condition is required for $j = 0$ in order to apply the equation at $j = 1$; following Strikwerda [35] we illustrate the theory by sketching the argument for stability in choosing the extrapolation $U_0^n = U_1^{n-1}$ for $n \geqslant 1$.

By subtracting a solution of the pure initial value problem one is reduced to assuming that $U_j^0 = U_j^1 = 0$ for $j \geqslant 0$, but then has to impose the inhomogeneous boundary condition $U_0^{n+1} = U_1^n + \beta^n$ for $n \geqslant 0$, and to show for this problem the stability estimate

$$\|U\|_\kappa \leqslant C_\kappa |\beta|_\kappa, \quad \text{with } \|U\|_\kappa^2 = k\sum_{n=0}^{\infty} \mathrm{e}^{-2\kappa t_n} h \sum_{j=0}^{\infty} |U_j^n|^2, \ \ |\beta|_\kappa^2 = k \sum_{n=0}^{\infty} \mathrm{e}^{-2\kappa t_n} |\beta^n|^2.$$

Note that κ is a parameter allowing for a certain exponential growth in time.

Applying discrete Laplace transforms in time to (4.6) one finds that the transformed solution $\tilde{U}_j(z)$ satisfies

$$(z - z^{-1})\tilde{U}_j = \lambda\rho(\tilde{U}_{j+1} - \tilde{U}_{j-1}), \quad \text{where } \tilde{U}(z) = k\sum_{n=0}^{\infty} z^{-n} U^n,$$

which is referred to as the resolvent equation. It is a second-order difference equation in j, and provided the two roots $\tau_1(z), \tau_2(z)$ of its characteristic equation $\lambda\rho(\tau - \tau^{-1}) = z - z^{-1}$ are distinct, the general solution is $\tilde{U}_j(z) = c_1(z)\tau_1(z)^j + c_2(z)\tau_2(z)^j$. It follows from the stability of the scheme for the initial value problem that for $|z| > 1$, with the proper ordering, $|\tau_1(z)| < 1$ and $|\tau_2(z)| > 1$. In fact, if this were not so and since $\tau_1(z)\tau_2(z) = 1$, we have $\tau_{1,2}(z) = \mathrm{e}^{\pm \mathrm{i}\xi}$ for some ξ, and z is therefore a solution of the characteristic equation $z - z^{-1} - 2\mathrm{i}\lambda\rho \sin\xi = 0$ of the leapfrog scheme for the pure initial value problem. By von Neumann's condition we therefore have $|z| \leqslant 1$ which contradicts our assumption. Since we want $\tilde{U}_j$ to be in $l_2(Z_+)$ we must have $c_2(z) = 0$, and taking the Laplace transform also at $j = 0$, we find $c_1(z)(z - \tau_1(z)) = z\tilde{\beta}(z)$, and thus $\tilde{U}_j(z) = z\tilde{\beta}(z)\tau_1(z)^j/(z - \tau_1(z))$. With $z = \mathrm{e}^{sk}$, $s = \kappa + \mathrm{i}\eta$ we obtain using Parseval's relation

$$\|U\|_\kappa^2 = kh\sum_j \int_{-\pi/k}^{\pi/k} |\tilde{U}_j(z)|^2 \,\mathrm{d}\eta = h\int_{-\pi/k}^{\pi/k} \frac{|z|^2 |\tilde{\beta}(z)|^2}{|z - \tau_1(z)|^2(1 - |\tau_1(z)|^2)} \,\mathrm{d}\eta.$$

By studying the behavior of $\tau_1(z)$ one may show that, $|z - \tau_1(z)| \geqslant c$ and $1 - |\tau_1(z)|^2 \geqslant 1 - |\tau_1(z)| \geqslant c(|z| - 1) = c(\mathrm{e}^{\kappa k} - 1) \geqslant c\kappa k$, with $c > 0$. Hence,

$$\|U\|_\kappa^2 \leqslant Ch(k\kappa)^{-1} \int_{-\pi/k}^{\pi/k} |\tilde{\beta}(z)|^2 \,\mathrm{d}\eta = C(\lambda\kappa)^{-1} |\beta|_\kappa^2,$$

which shows that the method is stable. Using similar arguments it is possible to show that the alternative extrapolation defined by $U_0^n = U_1^n$ is unstable.

5. Finite element methods for elliptic problems

In this section we summarize the basic definitions, properties, and successive development of the finite element method for elliptic problems. As a model problem we consider Dirichlet's problem for Poisson's equation in a domain $\Omega \subset \mathbb{R}^d$,

$$-\Delta u = f \quad \text{in } \Omega, \text{ with } u = 0 \text{ on } \partial\Omega. \tag{5.1}$$

The standard finite element method uses a variational formulation to define an approximate solution u_h of (5.1) in a finite-dimensional linear space S_h, normally consisting of continuous, piecewise polynomial functions on some partition of Ω: By Dirichlet's principle the solution u of (5.1) may be characterized as the function which minimizes $J(v) = \|\nabla v\|^2 - 2(f, v)$ over $H_0^1 = H_0^1(\Omega)$, where $(\cdot,\cdot)$ and $\|\cdot\|$ are the standard inner product and norm in $L_2 = L_2(\Omega)$. The Euler equation for this minimization problem is

$$(\nabla u, \nabla\varphi) = (f, \varphi), \quad \forall\varphi \in H_0^1; \tag{5.2}$$

this weak or variational form of (5.1) may also be derived by multiplying the elliptic equation in (5.1) by $\varphi \in H_0^1$, integrating over Ω, and applying Green's formula in the left-hand side. The standard finite element method assumes $S_h \subset H_0^1$ and defines the approximate solution u_h as the minimizer of $J(v)$ over S_h, or, equivalently,

$$(\nabla u_h, \nabla\chi) = (f, \chi), \quad \forall\chi \in S_h. \tag{5.3}$$

In terms of a basis $\{\Phi_j\}_{j=1}^{N_h}$ for S_h, our discrete problem (5.3) may be stated in matrix form as $A\alpha = \tilde{f}$, where A is the matrix with elements $a_{jk} = (\nabla\Phi_j, \nabla\Phi_k)$ (the stiffness matrix), $\tilde{f}$ the vector with entries $f_j = (f, \Phi_j)$, and α the vector of unknown coefficients α_j in $u_h = \sum_{j=1}^{N_h} \alpha_j\Phi_j$. Here A is a Gram matrix and thus, in particular, positive definite and invertible, so that (5.3) has a unique solution. From (5.2) and (5.3) follows that $(\nabla(u_h - u), \nabla\chi) = 0$ for $\chi \in S_h$, that is, u_h is the orthogonal projection of u onto S_h with respect to the Dirichlet inner product $(\nabla v, \nabla w)$.

We recall that defining u_h as the minimizer of $J(\chi)$ is referred to as the Ritz method, and using instead (5.3), which is suitable also for nonsymmetric differential equations, as Galerkin's method. Some further historical remarks are collected in the introduction to this paper.

For the purpose of error analysis we briefly consider the approximation in S_h of smooth functions in Ω which vanish on $\partial\Omega$. We first exemplify by the Courant elements in a convex plane domain Ω. For such a domain, let $\mathcal{T}_h$ denote a partition into disjoint triangles τ such that no vertex of any triangle lies on the interior of a side of another triangle and such that the union of the triangles determine a polygonal domain $\Omega_h \subset \Omega$ with boundary vertices on $\partial\Omega$. Let h denote the maximal length of the sides of the triangles of $\mathcal{T}_h$, and assume that the angles of the $\mathcal{T}_h$ are bounded below by a positive constant, independently of h. Let now S_h denote the continuous functions on the closure $\bar{\Omega}$ of Ω which are linear in each triangle of $\mathcal{T}_h$ and which vanish outside Ω_h. With $\{P_j\}_{j=1}^{N_h}$ the interior vertices of $\mathcal{T}_h$, a function in S_h is then uniquely determined by its values at the points P_j and thus $\dim(S_h) = N_h$. Let Φ_j be the "pyramid" function in S_h which takes the value 1 at P_j but vanishes at the other vertices; these functions form a basis for S_h. A given smooth function v on Ω which vanishes on $\partial\Omega$ may now be approximated by, e.g., its interpolant $I_h v = \sum_{j=1}^{N_h} v(P_j)\Phi_j \in S_h$, which agrees with v at the interior vertices, and one may show

$$\|I_h v - v\| \leqslant Ch^2\|v\|_2 \quad \text{and} \quad \|\nabla(I_h v - v)\| \leqslant Ch\|v\|_2, \quad \text{for } v \in H^2 \cap H_0^1, \tag{5.4}$$

where $\|\cdot\|_r$ denotes the norm in the Sobolev space $H^r = H^r(\Omega)$.

More generally we consider the case when $\Omega \subset \mathbb{R}^d$ and $\{S_h\}$ is a family of finite-dimensional subspaces of H_0^1 such that, for some integer $r \geqslant 2$,

$$\inf_{\chi \in S_h} \{\|v - \chi\| + h\|\nabla(v - \chi)\|\} \leqslant Ch^s\|v\|_s, \quad \text{for } 1 \leqslant s \leqslant r, \ v \in H^s \cap H_0^1. \tag{5.5}$$

The spaces S_h are thought of as consisting of piecewise polynomials of degree at most $r-1$ on a partition $\mathscr{T}_h$ of Ω, and bound (5.5) shown by exhibiting a $\chi = I_h u$ where $I_h : H^r \cap H_0^1 \to S_h$ is an interpolation type operator, see Zlámal (1968). The proof often involves the lemma of Bramble and Hilbert (1970):

Let $D \subset \mathbb{R}^d$ and assume that F is a bounded linear functional on $H^r(D)$ which vanishes for all polynomials of degree $< r$. Then $|F(u)| \leqslant C \sum_{|\alpha|=r} \|D^\alpha u\|_{L_2(D)}$.

To use this to show (5.4), e.g., one considers the difference $I_h u - u$ on an individual $\tau \in \mathscr{T}_h$, transforms this to a unit size reference triangle $\hat{\tau}$, invokes the Bramble–Hilbert lemma with $D = \hat{\tau}$, noting that $I_h u - u$ vanishes for linear functions, and transforms back to τ, using the fact that the bound for $|F(u)|$ in the lemma only contains the highest-order derivatives. In this example $\Omega_h \neq \Omega$ but the width of $\Omega \setminus \Omega_h$ is of order $O(h^2)$ and the contribution from this set is bounded appropriately. When $\partial\Omega$ is curved and $r > 2$, however, there are difficulties in the construction and analysis of such operators I_h near the boundary; we shall return to this problem below. When Ω is polygonal and $\Omega_h = \Omega$, the Bramble–Hilbert argument for (5.5) may be used also for $r > 2$, but in this case the solution of (5.1) will not normally have the regularity required. For comprehensive accounts of various choices of partitions $\mathscr{T}_h$ and finite element spaces S_h we refer to, e.g., Ciarlet [11], and Brenner and Scott [8].

We return to the finite element equation (5.3) using Courant elements. One way of triangulating $\Omega \subset \mathbb{R}^2$ is to start with the three families of straight lines $x_1 = lh$, $x_2 = lh$, $x_1 + x_2 = lh$, $l \in \mathbb{Z}$. The triangles thus formed may be used in the interior of Ω and then supplemented by other triangles near $\partial\Omega$ to form a triangulation $\mathscr{T}_h$ with the desired properties. With the notation (1.2) the equation corresponding to an interior vertex $x_j = jh$, $j \in \mathbb{Z}^2$ then takes the form

$$-\partial_1\bar{\partial}_1 U_j - \partial_2\bar{\partial}_2 U_j = h^{-2}(f, \Phi_j), \quad \text{where } U_j = u_h(x_j). \tag{5.6}$$

We recognize this as essentially the five-point finite difference equation (2.2), but with the right-hand side $f_j = f(x_j)$ replaced by an average of f over a neighborhood of x_j. Taking $f(x_j)$ may be considered as a quadrature rule for the right-hand side of (5.6). Recall that such averages were proposed also for finite difference methods.

Whereas a finite difference method may be obtained by replacing derivatives by finite differences, with some ad hoc modification near the boundary, the basic finite element method thus uses a variational formulation in a way that automatically accomodates the boundary conditions. We recall that the error analysis for the finite difference method uses a local estimate for the truncation error, together with some stability property, such as a discrete maximum principle. The finite element error analysis, as we shall now see, is based directly on the variational formulation and is global in nature. The difficulties in the construction of finite difference equations near the boundary are even greater for Neumann-type boundary conditions, whereas in the variational approach these are natural boundary conditions which do not have to be imposed on the approximating functions.

Under assumption (5.5) we now demonstrate the optimal order error estimate

$$\|u_h - u\| + h\|\nabla(u_h - u)\| \leqslant Ch^s\|u\|_s \quad \text{for } 1 \leqslant s \leqslant r. \tag{5.7}$$

Starting with the error in the gradient we note that since u_h is the orthogonal projection of u onto S_h with respect to $(\nabla v, \nabla w)$, we have, by (5.5),

$$\|\nabla(u_h - u)\| = \inf_{\chi\in S_h} \|\nabla(\chi - u)\| \leqslant Ch^{s-1}\|u\|_s \quad \text{for } 1\leqslant s\leqslant r \tag{5.8}$$

for linear finite elements this was observed in Oganesjan (1963). For the L_2-error we apply a duality argument by Aubin (1967) and Nitsche (1968): Let φ be arbitrary in L_2, take $\psi \in H^2 \cap H_0^1$ as the solution of

$$-\Delta\psi = \varphi \quad \text{in } \Omega, \quad \text{with } \psi = 0 \text{ on } \partial\Omega, \tag{5.9}$$

and recall the elliptic regularity inequality $\|\psi\|_2 \leqslant C\|\Delta\psi\| = C\|\varphi\|$. We then have for the error $e = u_h - u$, for any $\chi \in S_h$,

$$(e,\varphi) = -(e,\Delta\psi) = (\nabla e, \nabla\psi) = (\nabla e, \nabla(\psi - \chi)) \leqslant \|\nabla e\|\,\|\nabla(\psi - \chi)\| \tag{5.10}$$

and hence, using (5.8) and (5.5) with $s = 2$, the desired result follows from

$$(e,\varphi) \leqslant (Ch^{s-1}\|u\|_s)(Ch\|\psi\|_2) \leqslant Ch^s\|u\|_s\|\varphi\|.$$

In the case of a more general, not necessarily symmetric, elliptic equation, and an approximation by Galerkin's method, the estimate for the gradient may be obtained by application with $V = H_0^1$ of the lemma of Céa (1964):

Let V be a Hilbert space with norm $|\cdot|$ and let $A(u,v)$ be a continuous bilinear form on V such that $|A(u,v)| \leqslant M\,|u|\,|v|$ and $A(u,u) \geqslant \mu|u|^2$, $\mu > 0$. For F a continuous linear functional on V, consider the equation

$$A(u,\varphi) = F(\varphi), \quad \forall\varphi \in V. \tag{5.11}$$

Let $S_h \subset V$ and let *$u_h \in S_h$ be the solution of $A(u_h,\chi) = F(\chi)$ for $\chi \in S_h$. Then $|u_h - u| \leqslant M\mu^{-1}\inf_{\chi\in S_h}|\chi - u|$.*

Since $A(u_h - u, \chi) = 0$ for $\chi \in S_h$ this follows at once from

$$\mu|u_h - u|^2 \leqslant A(u_h - u, u_h - u) = A(u_h - u, \chi - u) \leqslant M|u_h - u|\,|\chi - u|.$$

Note that the problem (5.11) has a unique solution in V by the Lax–Milgram lemma.

We remark that the finite element error estimate for, e.g., the Courant elements, will require the solution to have two derivatives, whereas four derivatives were needed in the five-point finite difference method. This advantage of finite elements stems from the use of averages and disappears when a quadrature rule is used.

The error analysis given above assumed the approximation property (5.5) for some $r \geqslant 2$. The most natural example of such a family in a plane domain Ω would be to take for S_h the continuous piecewise polynomials of degree at most $r-1$ on a triangulation $\mathcal{T}_h$ of Ω of the type described above, which vanish on $\partial\Omega$. However, for $r > 2$ and in the case of a domain with curved boundary, it is then not possible, in general, to satisfy the homogeneous boundary conditions exactly, and the above analysis therefore does not apply. One method to deal with this difficulty is to consider elements near $\partial\Omega$ that are polynomial maps of a reference triangle $\hat{\tau}$, so called isoparametric elements, such that these elements define a domain Ω_h which well approximates Ω, and to use the corresponding maps of polynomials on $\hat{\tau}$ as approximating functions. Such finite element spaces were proposed by Argyris and by Fried, Ergatoudis, Irons, and Zienkiewicz, and Felipa and Clough, and analyzed in, e.g., Ciarlet and Raviart (1972), and other types of curved finite elements were considered by, e.g., Zlámal and Scott, see Ciarlet [11].

Another example of how to deal with the boundary condition is provided by the following method proposed by Nitsche (1971), again in a plane domain Ω. It uses a family $\mathcal{T}_h$ of triangulations which is quasi-uniform in the sense that area $(\tau) \geqslant ch^2$ for $\tau \in \mathcal{T}_h$, with $c > 0$ independent of h. In this case certain inverse inequalities hold, such as $\|\nabla\chi\| \leqslant Ch^{-1}\|\chi\|$ for $\chi \in S_h$; this follows at once from the corresponding result for each $\tau \in \mathcal{T}_h$, for which it is shown by transformation to a fixed reference triangle and using the fact that all norms on a finite-dimensional space are equivalent, see, e.g., [11] With $\langle\cdot,\cdot\rangle$ the L_2-inner product on $\partial\Omega$, the solution of (5.1) satisfies, for $\chi \in S_h$,

$$N_\gamma\ (u,\chi) := (\nabla u, \nabla\chi) - \left\langle \frac{\partial u}{\partial n}, \chi \right\rangle - \left\langle u, \frac{\partial \chi}{\partial n} \right\rangle + \gamma h^{-1}\langle u,\chi\rangle = -(\Delta u, \chi) = (f,\chi).$$

Using inverse and trace inequalities, the bilinear form $N_\gamma(\cdot,\cdot)$ is seen to be positive definite on S_h for γ fixed and sufficiently large, and we may therefore pose the discrete problem $N_\gamma(u_h,\chi) = (f,\chi)$ for $\chi \in S_h$. Nitsche showed

$$\|u_h - u\| + h\|\nabla(u_h - u)\| + h^{1/2}\|u_h\|_{L_2(\partial\Omega)} \leqslant Ch^r\|u\|_r.$$

The bound for the third term expresses that u_h almost vanishes on $\partial\Omega$.

Other examples of methods used to deal with curved boundaries for which $S_h \not\subset H_0^1$ include a method of Babuška (1973) with Lagrangian multipliers, the method of interpolated boundary conditions by Berger, Scott and Strang (1972), Scott (1975), and an approach by Bramble, Dupont and Thomée (1972) and Dupont (1974) where the finite element method is based on an approximating polygonal domain with a correction built into the boundary conditions.

In some situations one may want to use finite element spaces S_h defined by piecewise polynomial approximating functions on a partition $\mathcal{T}_h$ of Ω which are not continuous across interelement boundaries, so called nonconforming elements. Assuming Ω polygonal so that it is exactly a union of elements τ, one may introduce a discrete bilinear form by $D_h(\psi,\chi) = \sum_{\tau\in\mathcal{T}_h}(\nabla\psi,\nabla\chi)_\tau$. Provided S_h is such that $\|\chi\|_{1,h} = D_h(\chi,\chi)^{1/2}$ is a norm on S_h, a unique nonconforming finite element solution u_h of (5.1) is now defined by $D_h(u_h,\chi) = (f,\chi)$ for $\chi \in S_h$, and it was shown in Strang (1972) that

$$\|u_h - u\|_{1,h} \leqslant C \inf_{\chi\in S_h} \|u-\chi\|_{1,h} + C \sup_{\chi\in S_h} \frac{|D_h(u,\chi) - (f,\chi)|}{\|\chi\|_{1,h}}. \tag{5.12}$$

As an example, consider an axes parallel rectangular domain, partitioned into smaller such rectangles with longest edge $\leqslant h$, and let S_h be piecewise quadratics which are continuous at the corners of the partition. Then $\|\cdot\|_{1,h}$ is a norm on S_h. In Wilson's rectangle, the six parameters involved on each small rectangle are determined by the values at the corners plus the (constant) values of $\partial^2\chi/\partial x_l^2$, $l = 1,2$. The functions in S_h are not in $C(\bar{\Omega})$ but using (5.12) one may still show $\|u_h - u\|_{1,h} \leqslant C(u)h$.

The analysis above assumes that all inner products are calculated exactly. An analysis where quadrature errors are permitted was also worked out by Strang (1972). For instance, if (f,χ) is replaced by a quadrature formula $(f,\chi)_h$, a term of the form $C\sup_{\chi\in S_h}|(f,\chi) - (f,\chi)_h|/\|\nabla\chi\|$ has to be added to the bound for $\|\nabla(u_h - u)\|$. For example, if the quadrature formula is exact on each element for constants and if $f \in W_q^1(\Omega)$ with $q > 2$, then the O(h) error for $\|\nabla(u_h - u)\|$ is maintained. The situations when curved boundaries, nonconforming elements, or quadrature errors occur, so that the basic assumptions of the variational formulation are not satisfied, are referred to in Strang (1972), as variational crimes.

Because of the variational formulation of Galerkin's method, the natural error estimates are expressed in L_2-based norms. In the maximum-norm it was shown by Natterer (1975), Nitsche (1975), and Scott (1976), see Schatz and Wahlbin (1982), that, for piecewise linear approximating functions on a quasi-uniform family $\mathscr{T}_h$ in a plane domain Ω, we have

$$\|u_h - u\|_{L_\infty} \leqslant Ch^2 \log(1/h)\|u\|_{W^2_\infty}, \qquad \|\nabla(u_h - u)\|_{L_\infty} \leqslant Ch\|u\|_{W^2_\infty}.$$

For polygonal domains and with piecewise polynomials of degree $r - 1 > 1$,

$$\|u_h - u\|_{L_\infty} + h\|\nabla(u_h - u)\|_{L_\infty} \leqslant Ch^r\|u\|_{W^r_\infty},$$

but Haverkamp (1984) has proved that the above factor $\log(1/h)$ for piecewise linears may not be removed, even though it is not needed when estimating $I_h u - u$.

We shall now consider a finite element method for our model problem (5.1) which is based on a so called mixed formulation of this problem. Here the gradient of the solution u is introduced as a separate dependent variable whose approximation is sought in a different finite element space than the solution itself. This may be done in such a way that ∇u may be approximated to the same order of accuracy as u. With ∇u as a separate variable, (5.1) may thus be formulated

$$-\operatorname{div}\sigma = f \quad \text{in } \Omega, \quad \sigma = \nabla u \text{ in } \Omega, \text{ with } u = 0 \text{ on } \partial\Omega. \tag{5.13}$$

With $H = \{\omega = (\omega_1, \omega_2) \in L_2 \times L_2;\ \operatorname{div}\omega \in L_2\}$ we note that the solution $(u, \sigma) \in L_2 \times H$ also solves the variational problem

$$(\operatorname{div}\sigma, \varphi) + (f, \varphi) = 0, \quad \forall\varphi \in L_2, \quad (\sigma, \omega) + (u, \operatorname{div}\omega) = 0, \ \forall\omega \in H, \tag{5.14}$$

where the $(\cdot,\cdot)$ denote the appropriate L_2 inner products, and a smooth solution of (5.14) satisfies (5.13). Setting $L(v, \mu) = \frac{1}{2}\|\mu\|^2 + (\operatorname{div}\mu + f, v)$ the solution (u, σ) of (5.13) may also be characterized as the saddle-point satisfying

$$L(v, \sigma) \leqslant L(u, \sigma) \leqslant L(u, \mu), \quad \forall v \in L_2, \ \mu \in H \tag{5.15}$$

and the key to the existence of a solution is the inequality

$$\inf_{v\in L_2} \sup_{\mu\in H} \frac{(v, \operatorname{div}\mu)}{\|v\|\,\|\mu\|_H} \geqslant c > 0, \quad \text{where } \|\mu\|_H^2 = \|\mu\|^2 + \|\operatorname{div}\mu\|^2. \tag{5.16}$$

With S_h and H_h certain finite-dimensional subspaces of L_2 and H we shall consider the discrete analogue of (5.14) to find $(u_h, \sigma_h) \in S_h \times H_h$ such that

$$(\operatorname{div}\sigma_h, \chi) + (f, \chi) = 0, \quad \forall\chi \in S_h, \ (\sigma_h, \psi) + (u_h, \operatorname{div}\psi) = 0, \ \forall\psi \in H_h. \tag{5.17}$$

As in the continuous case this problem is equivalent to the discrete analogue of the saddle-point problem (5.15), and in order for this discrete problem to have a solution with the desired properties the choice of combinations $S_h \times H_h$ has to be such that the analogue of (5.16) holds, in this context referred to as the Babuška–Brezzi *inf–sup* condition (Babuška (1971); Brezzi (1974)).

One family of pairs of spaces which satisfy the *inf–sup* condition was introduced in Raviart and Thomas (1977); the first-order accurate pair of this family is as follows: With $\mathscr{T}_h$ a quasi-uniform family of triangulation of Ω, which we assume here to be polygonal, we set $S_h = \{\chi \in L_2; \chi|_\tau$ linear, $\forall\tau \in \mathscr{T}_h\}$, with no continuity required across inter-element boundaries. We then define

$H_h = \{\psi = (\psi_1, \psi_2) \in H; \psi|_\tau \in H(\tau), \quad \forall \tau \in \mathcal{T}_h\}$, where $H(\tau)$ denotes affine maps of quadratics on a reference triangle $\hat{\tau}$ of the form $(l_1(\xi) + \alpha\xi_1(\xi_1 + \xi_2), l_2(\xi) + \beta\xi_2(\xi_1 + \xi_2))$, with $l_1(\xi), l_2(\xi)$ linear, $\alpha, \beta \in R$. This space thus consists of piecewise quadratics on the triangulation $\mathcal{T}_h$ which are of the specific form implied by the definition of $H(\tau)$, and $\dim H(\tau) = 8$. As degrees of freedom for H_h one may use the values of $\psi \cdot n$ at two points on each side of τ (6 conditions) and in addition the mean values of ψ_1 and ψ_2 over τ (2 conditions). We note that the condition $\psi \in H$ in the definition of H_h requires that $\operatorname{div} \psi \in L_2$, which is equivalent to the continuity of $\chi \cdot n$ across inter-element boundaries. For the solutions of (5.17) and (5.13) holds

$$\|u_h - u\| \leqslant Ch^2 \|u\|_2 \quad \text{and} \quad \|\sigma_h - \sigma\| \leqslant Ch^s \|u\|_{s+1}, \quad s = 1, 2$$

and correspondingly higher-order estimates were derived for higher-order Raviart–Thomas elements.

We now turn to negative norm estimates and superconvergence. Recalling the error estimate (5.7) which holds for the model problem under the approximation assumption (5.5), we shall now see that for $r > 2$, the duality argument used to show the L_2-norm estimate yields an error estimate in a negative order norm. Introducing such negative norms by $\|v\|_{-s} = \sup_{\varphi \in H^s} (v, \varphi)/\|\varphi\|_s$ for $s \geqslant 0$, the error in u_h satisfies

$$\|u_h - u\|_{-s} \leqslant Ch^{q+s} \|u\|_q, \quad \text{for } 0 \leqslant s \leqslant r - 2, \ 1 \leqslant q \leqslant r. \tag{5.18}$$

In particular, $\|u_h - u\|_{-(r-2)} \leqslant Ch^{2r-2} \|u\|_r$. Since $2r - 2 > r$ for $r > 2$ the power of h in this estimate is higher than in the standard $O(h^r)$ error estimate in the L_2-norm. To show (5.18), we use the solution ψ of (5.9) and recall that $\|\psi\|_{s+2} \leqslant C\|\varphi\|_s$. This time (5.10) yields, for $0 \leqslant s \leqslant r - 2$,

$$|(e, \varphi)| \leqslant \|\nabla e\| \inf_{\chi \in S_h} \|\nabla(\psi - \chi)\| \leqslant \|\nabla e\| \, (Ch^{s+1} \|\psi\|_{s+2}) \leqslant Ch^{s+1} \|\nabla e\| \, \|\varphi\|_s.$$

By (5.8) this gives $|(e, \varphi)| \leqslant Ch^{q+s} \|u\|_q \|\varphi\|_s$ for $\varphi \in H^s$, which shows (5.18). As an application of (5.18), assume we want to evaluate the integral $F(u) = \int_\Omega u \psi \, dx = (u, \psi)$, where u is the solution of (5.1) and $\psi \in H^{r-2}$. Then for the obvious approximation $F(u_h) = (u_h, \psi)$ we find the superconvergent order error estimate

$$|F(u_h) - F(u)| = |(u_h - u, \psi)| \leqslant \|u_h - u\|_{-(r-2)} \|\psi\|_{r-2} \leqslant Ch^{2r-2} \|u\|_r \, \|\psi\|_{r-2}.$$

One more example of these ideas is provided by Douglas and Dupont (1974a), which concerns superconvergent nodal approximation in the two-point boundary value problem

$$Au = -\frac{d}{dx}\left(a \frac{du}{dx}\right) + a_0 u = f \quad \text{in } I = (0,1), \quad \text{with } u(0) = u(1) = 0. \tag{5.19}$$

Defining the partition $0 = x_0 < x_1 < \cdots < x_M = 1$, with $x_{i+1} - x_i \leqslant h$, we set

$$S_h = \{\chi \in C(\bar{I}); \ \chi|_{I_i} \in \Pi_{r-1}, \ 1 \leqslant i \leqslant M; \ \chi(0) = \chi(1) = 0\},$$

where $I_i = (x_{i-1}, x_i)$. Clearly this family satisfies our assumption (5.5). The finite element solution is now defined by $A(u_h, \chi) = (f, \chi)$ for $\chi \in S_h$, where $A(v, w) = (av_x, w_x) + (a_0 v, w)$, and the error estimate (5.18) holds.

Let $g = g^{\bar{x}}$ denote the Green's function of the two-point boundary value problem (5.19) with singularity at the partition point $\bar{x}$, which we now consider fixed, so that $w(\bar{x}) = A(w, g)$ for $w \in H_0^1 = H_0^1(I)$. Applied to the error $e = u_h - u$, and using the orthogonality of e to S_h with respect to $A(\cdot, \cdot)$, we find $e(\bar{x}) = A(e, g) = A(e, g - \chi)$ for $\chi \in S_h$, and hence that

$$|e(\bar{x})| \leqslant C\|e\|_1 \inf_{\chi \in S_h} \|g - \chi\|_1 \leqslant Ch^{r-1} \|u\|_r \inf_{\chi \in S_h} \|g - \chi\|_1.$$

Although $g^{\bar{x}}$ is not a smooth function at $\bar{x}$ it may still be approximated well by a function in S_h since it is smooth except at $\bar{x}$ and the discontinuity of the derivative at $\bar{x}$ can be accommodated in S_h. In particular, we have

$$\inf_{\chi \in S_h} \|g - \chi\|_1 \leqslant Ch^{r-1}(\|g\|_{H^r((0,\bar{x}))} + \|g\|_{H^r((\bar{x},1))}) \leqslant Ch^{r-1},$$

so that $|e(\bar{x})| \leqslant Ch^{2r-2}\|u\|_r$. Note that for $A = -\mathrm{d}^2/\mathrm{d}x^2$ the Green's function $g^{\bar{x}}$ is linear outside $\bar{x}$ and so $g^{\bar{x}} \in S_h$. We may then conclude that $e(\bar{x}) = 0$, which is a degenerate case.

We now touch on some superconvergent order estimates for the gradient in the two-dimensional model problem (5.1) using piecewise linear approximations for S_h in (5.3). It was shown in Oganesjan and Ruhovec (1969) that if the triangulations $\mathscr{T}_h$ are uniform then $\|\nabla(u_h - I_h u)\|_{L_2(\Omega_h)} \leqslant Ch^2\|u\|_{H^3}$, where as above I_h denotes the interpolant into S_h. This implies that at the midpoints of the edges of $\mathscr{T}_h$ the average of ∇u_h from the two adjacent triangles is a $\mathrm{O}(h^2)$ approximation to ∇u in a discrete l_2 sense. Such results have been improved to maximum-norm estimates and to triangulations that are perturbations in various ways of uniform triangulations by Chen, Lin, Xu, Zhou, Zhu, and others, and the approximation at other points than midpoints of edges has also been studied, see, e.g., references in Křížek and Neittaanmäki [22] or Wahlbin [41]. We remark that for uniform, axes parallel triangulations it follows from (5.6) that finite differences may be used as in (2.8) to approximate both the gradient and higher order derivatives to order $\mathrm{O}(h^2)$ in the interior of Ω.

All error estimates quoted above are a priori error estimates in that they depend on certain norms of the exact solution of the problem. In principle, these norms could be bounded in terms of norms of the data of the problem, but generally such bounds would be rather crude. During the last decades so-called a posteriori error estimates have been developed which depend directly on the computed solution, and on the data. Such estimates may be applied to include an adaptive aspect in the solution method, by detecting areas in a computational domain where the error is larger than elsewhere, and using this information to refine the mesh locally to reduce the error by an additional computation. Pioneering work is Babuška (1976) and Babuška and Rheinboldt (1978); for a recent survey, see Verfürth [40].

We illustrate this approach for the two-dimensional problem (5.1) in a polygonal domain Ω, using piecewise linear finite element approximations. With $\{\Phi_j\}$ the basis of pyramid functions we define Ω_j by $\bar{\Omega}_j = \operatorname{supp} \Phi_j$. Given the finite element solution $u_h \in S_h$ of (5.3), we now consider the local error equation

$$-\Delta w_j = f \quad \text{in } \Omega_j, \quad \text{with } w_j = u_h \text{ on } \partial\Omega_j.$$

It is then proved in Babuška and Babuška and Rheinboldt (1978), that, with c and C positive constants which depend on geometrical properties of the triangulations $\mathscr{T}_h$,

$$c\sum_j \eta_j^2 \leqslant \|\nabla(u_h - u)\|^2 \leqslant C\sum_j \eta_j^2, \quad \text{where } \eta_j = \|\nabla(w_j - u_h)\|.$$

The error in ∇u_h is thus bounded both above and below in terms of the local quantities η_j, which can be approximately determined. It is argued that a triangulation for which the quantities η_j are of essentially the same size gives a small error in ∇u_h, and this therefore suggests an adaptive strategy for the solution of (5.1).

Another approach was taken in Eriksson and Johnson (1991), showing an a posteriori error estimate of the form

$$\|\nabla(u_h - u)\| \leqslant C\left(\left(\sum_{\tau} h_\tau^2 \|f\|_{L_2(\tau)}^2\right)^{1/2} + \left(\sum_{\gamma} h_\gamma^2 \left|\left[\frac{\partial u_h}{\partial n}\right]_\gamma\right|\right)^{1/2}\right), \tag{5.20}$$

where the γ with length h_γ are the edges of $\mathscr{T}_h$ and $[\,\cdot\,]_\gamma$ denotes the jump across γ. Under a certain assumption on the local variation of h_τ, which is weaker than quasi-uniformity the a priori estimate

$$\|\nabla(u_h - u)\| \leqslant C\left(\sum_{\tau} h_\tau^2 \|u\|_{H^2(\tau)}^2\right)^{1/2} \tag{5.21}$$

is also derived. Together with (5.20) this may be used to justify an adaptive scheme with a given tolerance and an essentially minimal number of triangles. Analogous a posteriori and a priori bounds are demonstrated for $\|u_h - u\|$ and, in Eriksson (1994), also in maximum norm, where the analogue of (5.21) reads $\|\nabla(u_h - u)\|_{L_\infty} \leqslant C \max_\tau(h_\tau \|u\|_{W_\infty^2(\tau)})$.

Superconvergence of the error in the gradient has been used in, e.g., Zienkiewicz and Zhu (1992) to derive a posteriori error bounds for adaptive purposes.

We finally mention the p- and h–p-versions of the finite element method: So far it has been assumed that the approximating subspaces S_h are piecewise polynomial spaces of a fixed degree based on partitions $\mathscr{T}_h$ with $\max_{\tau\in\mathscr{T}_h} \operatorname{diam}(\tau) \leqslant h$, and higher accuracy is achieved by refining the partition. An alternative approach proposed in Babuška, Szabó and Katz (1981) is to fix the mesh and then let the degree of the polynomials grow. The two approaches are referred to as the h-version and the p-version of the finite element method, respectively. A combination of the two methods, the h–p-version has been studied in Babuška and Dorr (1981). For more material about the p- and h–p-methods, see Babuška and Suri (1990).

6. Finite element methods for evolution equations

This section is concerned with the application of the finite element method to time dependent problems. We begin with the model heat equation and discuss then the wave equation and finally some simple first-order hyperbolic model problems.

We consider thus first the approximate solution of the parabolic problem

$$u_t - \Delta u = f(t) \quad \text{in } \Omega, \text{ with } u = 0 \text{ on } \partial\Omega,\ t > 0,\ u(\cdot, 0) = v \text{ in } \Omega, \tag{6.1}$$

in a finite-dimensional space S_h belonging to a family satisfying (5.5). As a first step we discretize this problem in the spatial variable by writing it in variational form and defining $u_h = u_h(\cdot, t) \in S_h$ for $t \geqslant 0$ by

$$(u_{h,t}, \chi) + (\nabla u_h, \nabla \chi) = (f(t), \chi), \quad \forall \chi \in S_h,\ t > 0,\ u_h(0) = v_h \approx v. \tag{6.2}$$

With respect to a basis $\{\Phi_j\}_{j=1}^{N_h}$ of S_h this may be written as a system of ordinary differential equations $B\alpha' + A\alpha = \tilde{f}$ where A is the stiffness matrix introduced in Section 5 and where $B = (b_{jk})$, $b_{jk} = (\Phi_j, \Phi_k)$, is referred to as the mass matrix. A fully discrete time-stepping scheme may then be obtained by discretization of this system in time, using, e.g., the single-step ϑ-method:

With k the time step, $t_n = nk$, $\bar{\partial}_t U^n = (U^n - U^{n-1})/k$, and with $\vartheta \in [0,1]$, the approximation $U^n \in S_h$ of $u(t_n)$ for $n \geqslant 1$ is then defined by

$$(\bar{\partial}_t U^n, \chi) + (\nabla(\vartheta U^n + (1-\vartheta)U^{n-1}), \nabla\chi) = (f^{n-1+\vartheta}, \chi), \quad \forall \chi \in S_h,\ n \geqslant 1 \tag{6.3}$$

with $f^s = f(sk)$ and U^0 given. For $\vartheta = 0$ and 1 these are the forward and backward Euler methods and for $\vartheta = \frac{1}{2}$ the Crank–Nicolson method. Note that the forward Euler method is not explicit because the matrix B is nondiagonal.

For the semidiscrete problem (6.2) Douglas and Dupont (1970) showed that

$$\|u_h(t) - u(t)\| + \left(\int_0^t \|\nabla(u_h - u)\|^2 \, ds\right)^{1/2} \leqslant \|v_h - v\| + C(u)h^{r-1}. \tag{6.4}$$

For a proof we note that the error $e = u_h - u$ satisfies $(e_t, \chi) + (\nabla e, \nabla\chi) = 0$ for $\chi \in S_h, t > 0$, and hence $(e_t, e) + (\nabla e, \nabla e) = (e_t, \chi - u) + (\nabla e, \nabla(\chi - u))$, from which the result follows after integration and with the appropriate choice of χ. Because of the contribution from $\|\nabla(\chi - u)\|$ on the right, (6.4) is of suboptimal order in L_2-norm. In this regard the estimate was improved by Wheeler (1973) to

$$\|u_h(t) - u(t)\| \leqslant \|v_h - v\| + Ch^r \left(\|v\|_r + \int_0^t \|u_t\|_r \, ds\right) \quad \text{for } t > 0. \tag{6.5}$$

This was done by introducing the elliptic or Ritz projection, the orthogonal projection $R_h : H_0^1 \to S_h$ with respect to the Dirichlet inner product, thus defined by $(\nabla(R_h u - u), \nabla\chi) = 0$ for $\chi \in S_h$, and writing $e = (u_h - R_h u) + (R_h u - u) = \theta + \rho$. Here, by the error estimate (5.7) for the elliptic problem, $\|\rho(t)\| \leqslant Ch^r \|u(t)\|_r$, which is bounded as desired, and one also finds $(\theta_t, \chi) + (\nabla\theta, \nabla\chi) = -(\rho_t, \chi)$ for $\chi \in S_h$, $t > 0$. Choosing $\chi = \theta$ and integrating this yields $\|\theta(t)\| \leqslant \|\theta(0)\| + \int_0^t \|\rho_t\| \, ds$ which is easily bounded as desired. In particular, for $v_h \in S_h$ suitably chosen, this shows an optimal order error estimate in L_2.

Defining the discrete Laplacian $\Delta_h : S_h \to S_h$ by $-(\Delta_h \psi, \chi) = (\nabla\psi, \nabla\chi)\ \forall \psi, \chi \in S_h$ and using the L_2-projection P_h onto S_h the above equation for θ may be written as $\theta_t - \Delta_h \theta = -P_h \rho_t$, and, with $E_h(t) = e^{\Delta_h t}$ the solution operator of (6.2) with $f = 0$, we find by Duhamel's principle that $\theta(t) = E_h(t)\theta(0) + \int_0^t E_h(t-s)P_h \rho_t(s) \, ds$. An obvious energy argument shows the stability property $\|E_h(t)v_h\| \leqslant \|v_h\|$, which again gives the above bound for θ. The error estimate for the semidiscrete problem thus follows from the stability of $E_h(t)$ together with error estimates for the elliptic problem; for finite difference methods stability was similarly combined with a bound for the truncation error.

The use of the elliptic projection also yields an estimate of superconvergent order for $\nabla\theta$. In fact, by choosing this time $\chi = \theta_t$ in the variational equation for θ, we find after integration and simple estimates that $\|\nabla\theta(t)\| \leqslant C(u)h^r$ if $v_h = R_h v$. For piecewise linears ($r = 2$) this may be combined with the superconvergent second-order estimate for $\nabla(R_h u - I_h u)$ quoted in Section 5 to bound $\nabla(u_h - I_h u)$, with similar consequences as in the elliptic case, see Thomée, Xu and Zhang (1989).

Estimates for the fully discrete ϑ-method (6.3) were also shown in Douglas and Dupont (1970) and Wheeler (1973). The contribution from the time discretization that has to be added to (6.5) at $t = t_n$ is then $Ck \int_0^{t_n} \|u_{tt}\| \, ds$, with a stability condition $k \leqslant \gamma_\vartheta h^2$ for $0 \leqslant \vartheta < \frac{1}{2}$, and $Ck^2 \int_0^{t_n} (\|u_{ttt}\| + \|\Delta u_{tt}\|) \, ds$ for $\vartheta = \frac{1}{2}$.

The ϑ-method for the homogeneous equation may be defined by $U^n = E_{kh}^n v_h$ where $E_{kh} = r(-k\Delta_h)$, $r(\lambda) = (1 + (1 - \vartheta)\lambda)/(1 + \vartheta\lambda)$. Two-level schemes using more general rational functions $r(\lambda)$ of arbitrary order of accuracy were constructed in Baker, Bramble and Thomée (1977), under the stability assumption $|r(\lambda)| \leqslant 1$ for $\lambda \in \sigma(-k\Delta_h)$. Stable two-level time-stepping methods for the inhomogeneous equation of arbitrary order of accuracy may be constructed in the form $U^{n+1} = r(-k\Delta_h)U^n + k\sum_{j=1}^m q_j(-k\Delta_h)f(t_n + \tau_j k)$ where $r(\lambda)$ and the $q_j(\lambda)$ are rational functions, with, e.g., the backward Euler method included for $m=1$, $\tau_1 = 0$, $r(\lambda) = q_1(\lambda) = 1/(1+\lambda)$, see Brenner, Crouzeix and Thomée (1982). Stable multistep time discretization schemes of accuracy of order $q \leqslant 6$ have also been derived by Le Roux (1979) and others, see Thomée [38].

The regularity requirements needed for optimal order convergence in some of the above error estimates make it natural to enquire about error estimates under weaker regularity assumptions on data or on the solution. To illustrate this we now consider the solution of the homogeneous equation, i.e., (6.1) with $f=0$, and recall that the solution of this problem is smooth for $t > 0$ even if v is only in L_2, say, and satisfies $\|u(t)\|_s \leqslant Ct^{-s/2}\|v\|$. Similarly, for the semidiscrete solution, $\|\Delta_h^{s/2}u_h(t)\| \leqslant Ct^{-s/2}\|v_h\|$, and using this one may show the nonsmooth data error estimate

$$\|u_h(t) - u(t)\| \leqslant Ch^r t^{-r/2}\|v\| \quad \text{for } t > 0 \quad \text{if } v_h = P_h v,$$

so that optimal order $O(h^r)$ convergence holds for $t > 0$, without any regularity assumptions on v. The corresponding result for the backward Euler method reads

$$\|U^n - u(t_n)\| \leqslant C(h^r t_n^{-r/2} + kt_n^{-1})\|v\|, \quad \text{for } n \geqslant 1 \text{ if } v_h = P_h v.$$

Results of this type were shown by spectral representation in, e.g., Blair (1970), Helfrich (1974), and later by energy methods, permitting also time-dependent coefficients, in Huang and Thomée (1981), Luskin and Rannacher (1982), Sammon (1983a,b), see [38]. For stable fully discrete approximations of the form $U^n = E_{kh}^n v_h$ with $E_{kh} = r(-k\Delta_h)$ one then has to require $|r(\infty)| < 1$, see Baker, Bramble and Thomée (1977). The Crank–Nicolson method lacks this smoothing property, but Rannacher (1984) showed that using this method with two initial steps of the backward Euler method, one has $\|U^n - u(t_n)\| \leqslant C(h^r t_n^{-r/2} + k^2 t_n^{-2})\|v\|$ for $n \geqslant 1$.

The methods quoted in Section 5 for handling the difficulty of incorporating homogeneous Dirichlet boundary conditions in the approximating spaces S_h have been carried over from the elliptic to the semidiscrete parabolic case in Bramble, Schatz, Thomée and Wahlbin (1977). This may be accomplished by replacing the gradient term in (6.2) and (6.3) by more general bilinear forms such as $N_\gamma(\cdot,\cdot)$ described there, or by using other approximations of the Laplacian than the above Δ_h. Within this framework negative norm estimates and superconvergence results were derived in Thomée (1980). We also quote Johnson and Thomée (1981) where the mixed method discussed in Section 5 is applied to (6.1).

In the fully discrete schemes discussed above, Galerkin's method was applied in space but a finite-difference-type method was used in time. We shall now describe an approach which uses a Galerkin-type method also in time, the discontinuous Galerkin time-stepping method. This method was introduced and analyzed in Lesaint and Raviart (1974) and Jamet (1978), and generalized in the case of ordinary differential equations in Delfour, Hager and Trochu (1981). In the present context it was studied in Eriksson, Johnson and Thomée (1985). With a not necessarily uniform partition of $[0,\infty)$ into intervals $J_n = [t_{n-1}, t_n)$, $n \geqslant 1$, let $\mathscr{S}_h = \{X = X(x,t); X|_{J_n} = \sum_{j=0}^{q-1} \chi_j t^j, \chi_j \in S_h\}$, where S_h are finite element spaces satisfying (5.5). Since the elements $X \in \mathscr{S}_h$ are not required to be continuous at the t_j we set $U^n_{\pm} = U(t_n \pm 0)$ and $[U^{n-1}] = U^{n-1}_+ - U^{n-1}_-$. The discontinuous Galerkin method may then be stated: With $U^0_- = v_h$ given, find $U \in \mathcal{S}_h$ such that, for $n \geqslant 1$,

$$\int_{J_n} [(U_t, X) + (\nabla U, \nabla X)]\,ds + ([U^{n-1}], X^{n-1}_+) = \int_{J_n} (f, X)\,ds, \quad \forall X \in \mathcal{S}_h. \tag{6.6}$$

In the piecewise constant case, $q = 0$, this may be written, with $k_n = t_n - t_{n-1}$,

$$(\bar{\partial}_{t_n} U^n, \chi) + (\nabla U^n, \nabla \chi) = \left(k_n^{-1} \int_{J_n} f\,ds, \chi\right), \quad \bar{\partial}_{t_n} U^n = (U^n - U^{n-1})/k_n, \tag{6.7}$$

this reduces to the standard backward Euler method when the average of f is replaced by $f(t_n)$. It was shown in Eriksson and Johnson (1991) that for the error in the time discretization in (6.7)

$$\|U^n - u_h(t_n)\| \leqslant C\ell_n \max_{j \leqslant n}(k_j \|u_{h,t}\|_{J_j}), \quad \text{where } \ell_n = (1 + \log(t_n/k_n))^{1/2},$$

with u_h the solution of the semidiscrete problem (6.2) and $\|\varphi\|_{J_j} = \sup_{t \in J_j} \|\varphi(t)\|$.

For $q = 1$ the method requires the determination on J_n of $U(t)$ of the form $U(t) = U^{n-1}_+ + (t - t_{n-1})/k_n V_n$ with $U^{n-1}_+, V_n \in S_h$, and such that (6.6) holds, which gives a 2×2 system for these elements in S_h. In this case we have

$$\|U - u_h\|_{J_n} \leqslant C\ell_n \max_{j \leqslant n}(k_j^2 \|u_{h,tt}\|_{J_j}), \quad \|U^n_- - u_h(t_n)\| \leqslant C\ell_n \max_{j \leqslant n}(k_j^3 \|\Delta_h u_{h,tt}\|_{J_j}),$$

thus with third-order superconvergence at the nodal points. For the total error in the fully discrete scheme, with, e.g., piecewise linear elements in space, one has

$$\|U^n_- - u(t_n)\| \leqslant C\ell_n \max_{j \leqslant n}(k_j^3 \|\Delta u_{tt}\|_{J_j} + h^2 \|u\|_{2,J_j}), \quad \|u\|_{2,J_j} = \sup_{t \in J_j} \|u(t)\|_2.$$

All our error estimates so far have been a priori error estimates, expressed in terms of the unknown exact solution of our parabolic problem. We close by mentioning briefly some a posteriori estimates by Eriksson and Johnson (1991), based on an idea of Lippold (1991), where the error bounds are expressed in terms of the data and the computed solution. For $q = 0$, $r = 2$ such an estimate is

$$\|U^n_- - u(t_n)\| \leqslant C\ell_n \max_{j \leqslant n}((h^2 + k_j)\|f\|_{J_j} + k_j \|\bar{\partial}_t U^j\| + h^2 \|U^j\|_{2,h}), \tag{6.8}$$

where $\|\cdot\|_{2,h}$ is a discrete H^2-norm defined by $\|U\|_{2,h} = (\sum_\gamma |[\partial U/\partial n]_\gamma|^2)^{1/2}$, with γ denoting the edges of $\mathcal{T}_h$ and $[\,\cdot\,]_\gamma$ the jumps across γ. Error bounds are also available for $q = 1$, and the estimates generalize to variable h.

Estimates such as (6.8) may be used to design adaptive schemes in which the time step is successively chosen so that the error is bounded by a given tolerance. The earlier a priori estimates are then needed to show that such a procedure will end in a finite number of steps, cf. Eriksson and Johnson (1992). This approach was further developed in a sequence of paper by Eriksson and Johnson, see the survey paper Eriksson, Estep, Hansbo and Johnson [13]. A Petrov–Galerkin method with continuous in time approximations was studied by Aziz and Monk (1989).

We now briefly consider the question of maximum-norm stability for the finite element scheme. For the solution operator $E(t)$ of the homogeneous equation in (6.1) ($f = 0$) the maximum-principle shows at once that $\|E(t)v\|_{L_\infty} \leqslant \|v\|_{L_\infty}$, and the smoothing estimate $\|\Delta^{s/2} E(t)v\|_{L_\infty} \leqslant Ct^{-s/2}\|v\|_{L_\infty}$ also holds for $s > 0$. However, considering the case $r = d = 2$ one can easily see (cf. [38]) that the maximum principle does not apply for the semidiscrete finite element analogue. This is in contrast to the corresponding finite difference method and is related to the fact that the mass matrix B is nondiagonal. In this regard, it was shown by Fujii (1973) that if B is replaced by a diagonal matrix whose diagonal elements are the row sums of B, and if all angles of the triangulation are nonobtuse, then the maximum principle holds and hence $\|u_h(t)\|_{L_\infty} \leqslant \|v_h\|_{L_\infty}$ for $t > 0$. This method is called the lumped mass method and can also be defined by

$$(u_{h,t}, \chi)_h + (\nabla u_h, \nabla \chi) = 0, \quad \forall \chi \in S_h, \ t > 0,$$

where the first term has been obtained by replacing the first term in (6.2) by using the simple quadrature expression $Q_{h\tau}(u_{h,t}\chi)$ on each τ, where $Q_{h\tau}(\varphi) = \frac{1}{3}\,\mathrm{area}(\tau)\sum_{j=1}^{3}\varphi(P_j)$ with P_j the vertices of τ.

Even though the maximum principle does not hold for (6.2), it was shown in Schatz, Thomée and Wahlbin (1980) that, for $d = r = 2$ and quasi-uniform $\mathscr{T}_h$,

$$\|E_h(t)v_h\|_{L_\infty} + t\|E_h'(t)v_h\|_{L_\infty} \leqslant C\ell_h\|v_h\|_{L_\infty}, \quad \text{for } t > 0, \quad \ell_h = \log(1/h).$$

The proof uses a weighted norm technique to estimate a discrete fundamental solution. For $d = 1, 2, 3$ and $r \geqslant 4$, Nitsche and Wheeler (1981–82) subsequently proved stability without the factor ℓ_h, and this and the corresponding smoothing result were shown for $d = 1$ and $r \geqslant 2$ in Crouzeix, Larsson and Thomée (1994). Recently, logarithm-free stability and smoothness bounds have been shown for general d and r, first for Neumann boundary conditions in Schatz, Thomée and Wahlbin (1998), and then for Dirichlet boundary conditions in Thomée and Wahlbin (1998). We note that the combination of stability and smoothing shows that the semigroup $E_h(t)$ is analytic, and via a resolvent estimate for its infinitesimal generator Δ_h this may be used to derive stability estimates also for fully discrete approximations of the form $U^n = r(-k\Delta_h)^n v_h$ where $r(z)$ is a rational function with the appropriate stability and consistency properties, see Palencia (1992) and Crouzeix, Larsson, Piskarev and Thomée (1993). Other maximum-norm error bounds have been given in the literature by, e.g., Dobrowolski (1978), Nitsche (1979), and Rannacher (1991).

We now turn to hyperbolic equations and begin with a brief discussion of semidiscrete and fully discrete finite element schemes for the initial–boundary value problem for the wave equation

$$u_{tt} - \Delta u = f \quad \text{in } \Omega \subset \mathbb{R}^d, \ \text{with } u = 0 \text{ on } \partial\Omega, \ \text{for } t > 0$$

$$u(\cdot, 0) = v, \quad u_t(\cdot, 0) = w \quad \text{in } \Omega.$$

Assuming as usual that $S_h \subset H_0^1$ satisfies (5.5), the semidiscrete analogue of our problem is to find $u_h(t) \in S_h$ for $t \geqslant 0$ from

$$(u_{h,tt}, \chi) + (\nabla u_h, \nabla \chi) = (f, \chi) \quad \forall \chi \in S_h, \ t > 0, \quad \text{with } u_h(0) = v_h, \ u_{h,t}(0) = w_h.$$

Similarly to the parabolic case this problem may be written in matrix form, this time as $B\alpha'' + A\alpha = \tilde{f}$ for $t > 0$, with $\alpha(0)$ and $\alpha'(0)$ given, where B and A are the mass and stiffness matrices.

Analogously to the analysis in the parabolic case it was shown in Dupont (1973a), with a certain improvement in Baker (1976), that under natural regularity assumptions and with appropriate choices of v_h and w_h,

$$\|u_h(t) - u(t)\| + h\|\nabla(u_h(t) - u(t))\| \leqslant C(u)h^r.$$

One possible fully discrete method for the wave equation is, cf. the case $\vartheta = \frac{1}{2}$ of the Newmark-type method (4.4),

$$(\partial_t\bar{\partial}_t U^n, \chi) + (\nabla(\tfrac{1}{4}U^{n+1} + \tfrac{1}{2}U^n + \tfrac{1}{4}U^{n-1}), \nabla\chi) = (f(t_n), \chi), \quad \forall \chi \in S_h, \ n \geqslant 1,$$

where U^0 and U^1 are given approximations of $u(0)=v$ and $u(k)$, respectively. Setting $U^{n+1/2}=(U^n+U^{n+1})/2$ one shows for the homogeneous equation ($f=0$) that the energy $\|\partial_t U^n\|^2+\|\nabla U^{n+1/2}\|^2$ is conserved for $n\geqslant 0$, and, also in the general case, that $\|U^{n+1/2}-u(t_n+\frac{1}{2}k)\|=O(h^r+k^2)$ for appropriate initial values U^0 and U^1, u sufficiently regular, and t_n bounded. Although the error is then estimated at the points $t_n+\frac{1}{2}k$ it is easy to derive approximations also at the points t_n. For the homogeneous equation more general time-stepping schemes based on rational functions of the discrete Laplacian Δ_h were studied in Baker and Bramble (1979), where the second-order wave equation was written as a first-order system.

We proceed with some results for first-order hyperbolic equations, and begin with the periodic model problem

$$u_t+u_x=f \quad \text{for } x\in\mathbb{R},\ t>0,\ \text{ with } u(\cdot,0)=v \text{ on } \mathbb{R}, \tag{6.9}$$

where f, v, and the solution sought are 1-periodic in x; the L_2 inner products and norms used below are based on intervals of length 1.

To define an approximate solution, let $S_h\subset C^k(\mathbb{R})$, with $k\geqslant 0$, denote 1-periodic splines of order r (i.e., piecewise polynomials of degree $r-1$) based on a partition with maximal interval length h. The standard Galerkin method for (6.9) is then to find $u_h(t)\in S_h$ for $t\geqslant 0$ such that

$$(u_{h,t},\chi)+(u_{h,x},\chi)=(f,\chi),\quad \forall\chi\in S_h,\ t>0,\ \text{ with } u_h(0)=v_h.$$

The equation may again be written in the form $B\alpha'+A\alpha=\tilde{f}$ where as usual B is the mass matrix but where the matrix A now has elements $a_{jk}=(\Phi_j',\Phi_k)$ and is skew-symmetric.

We first establish the simple error estimate, cf. Swartz and Wendroff (1969),

$$\|u_h(t)-u(t)\|\leqslant\|v_h-v\|+C(u)h^{r-1},\quad \text{for } t\geqslant 0. \tag{6.10}$$

For this, we use an interpolation operator Q_h into S_h which commutes with time differentiation and is such that $\|Q_hv-v\|+h\|(Q_hv-v)_x\|\leqslant Ch^r\|v\|_r$. It remains to bound $\theta=u_h-Q_hu$, which satisfies $(\theta_t,\chi)+(\theta_x,\chi)=-(\rho_t+\rho_x,\chi)$ for $\chi\in S_h$. Setting $\chi=\theta$ and observing that $(\theta_x,\theta)=0$ by periodicity, we conclude

$$\frac{1}{2}\frac{\mathrm{d}}{\mathrm{d}t}\|\theta\|^2=\|\theta\|\frac{\mathrm{d}}{\mathrm{d}t}\|\theta\|\leqslant(\|\rho_t\|+\|\rho_x\|)\|\theta\|\leqslant C(u)(h^r+h^{r-1})\|\theta\|, \tag{6.11}$$

which shows $\|\theta(t)\|\leqslant\|\theta(0)\|+C(u)h^{r-1}\leqslant\|v_h-v\|+C(u)h^{r-1}$, and yields (6.10).

We observe that estimate (6.10) is of nonoptimal order $O(h^{r-1})$, because the first derivative of the error in Q_hu occurs on the right side of (6.11). For special cases more accurate results are known. For example, for the homogeneous equation, with S_h consisting of smooth splines ($k=r-2$) on a uniform partition, the last term in (6.10) may be replaced by the optimal order term $C(u)h^r$. In this case superconvergence takes place at the nodes in the sense that $\|u_h(t)-I_hu(t)\|\leqslant C(u)h^{2r}$ if $v_h=I_hv$, where I_hv denotes the interpolant of v in S_h. This follows from Fourier arguments, see Thomée (1973), after observing that the Galerkin method may be interpreted as a finite difference method for the coefficients with respect to a basis for S_h. This was generalized to variable coefficients in Thomée and Wendroff (1974). It was shown by Dupont (1973b), however, that the improvement to optimal order is not always possible. In fact, if S_h is defined by a uniform partition with $r=4$, $k=1$ (Hermite cubics), and if v is a nonconstant analytic 1-periodic function and $v_h\in S_h$ is arbitrary, then $\sup_{t\in(0,t^*)}\|u_h(t)-u(t)\|\geqslant ch^3$, $c>0$, for any $t^*>0$.

Leaving the standard Galerkin method, it was shown in Wahlbin (1974) that with S_h defined on a uniform partition and $0 \leqslant k \leqslant r-2$, optimal order convergence holds for the Petrov–Galerkin method

$$(u_{h,t} + u_{h,x}, \chi + h\chi_x) = (f, \chi + h\chi_x), \quad \forall \chi \in S_h,\ t > 0, \text{ with } u_h(0) = v_h.$$

In Dendy (1974) and Baker (1975) nonstandard variational schemes with optimal order convergence without requiring uniform meshes were exhibited for the initial–boundary value problem

$$u_t + u_x = f \quad \text{for } x \in I = (0,1), \quad u(0,t) = 0, \text{ for } t \geqslant 0, \text{ with } u(\cdot, 0) = v.$$

We now quote some space–time methods for this initial–boundary value problem for $(x,t) \in \Omega = I \times J\ (J = (0,T))$, and introduce the characteristic directional derivative $Du = u_t + u_x$. Let $\mathscr{T}_h = \{\tau\}$ be a quasi-uniform triangulation of Ω with max diam$(\tau) = h$ and let $\mathscr{S}_h = \{\chi \in C(\bar{\Omega}); \chi|_\tau \in \Pi_{r-1} : \chi = 0 \text{ on } \partial\Omega^-\}$, where $\partial\Omega^-$ is the inflow boundary $(\{0\} \times J) \cup (I \times \{0\})$. With $((v,w)) = \int_0^T \int_0^1 v\,w\,\mathrm{d}x\,\mathrm{d}t$, the standard Galerkin method in this context is then to find $u_h \in \mathscr{S}_h$ such that

$$((Du_h, \chi)) = ((f, \chi)), \quad \forall \chi \in \mathscr{S}_h. \tag{6.12}$$

Standard arguments show as above the nonoptimal order error estimate

$$\|u_h - u\| + \|u_h - u\|_{L_2(\partial\Omega^+)} \leqslant Ch^{r-1}\|u\|_r, \tag{6.13}$$

where $\partial\Omega^+ = \partial\Omega \backslash \partial\Omega^-$ is the outflow boundary. This method does not work well in the case of discontinuous solutions. To stabilize the scheme one could consider an artificial dissipation in the form of an additional term $h((\nabla u_h, \nabla \chi))$ on the left in (6.12), but such a method would be at most first-order accurate.

The so-called streamline diffusion method was introduced by Hughes and Brooks (1979) and analyzed in Johnson and Pitkäranta (1986). It consists in substituting $\chi + hD\chi$ for the test function χ in (6.12), and (6.13) then holds with h^{r-1} replaced by $h^{r-1/2}$. In the discontinuous Galerkin method studied by Lesaint and Raviart (1974), and Johnson and Pitkäranta (1986), cf. the corresponding method for parabolic equations introduced above, one determines $u_h \in \mathscr{S}_h = \{\chi \in L_2(\Omega); \chi|_\tau \in \Pi_{r-1}; \chi = 0 \text{ on } \partial\Omega^-\}$, thus without requiring $u_h \in C(\bar{\Omega})$, from

$$((Du_h, \chi))_\tau - \int_{\partial\tau^-} [u_h]_{\partial\tau}(n_t + n_x)\,\mathrm{d}s = ((f, \chi))_\tau, \quad \forall \chi \in \Pi_{r-1},\ \tau \in \mathscr{T}_h,$$

where $((\cdot,\cdot))_\tau$ denotes restriction to τ. This method also satisfies (6.12), with $h^{r-1/2}$ instead of h^{r-1}, and here the error in Du_h is of optimal order $\mathrm{O}(h^{r-1})$.

Winther (1981) investigated a Petrov–Galerkin method defining u_h in continuous, piecewise Π_{r-1} spaces $\mathscr{S}_h$ based on rectangles $\tau = I_j \times J_l$, where I_j and J_l are partitions of I and J, by the equation $((Du_h, \chi))_\tau = ((f, \chi))_\tau$ for all $\chi \in \Pi_{r-2}$ and all τ, and proved optimal order $\mathrm{O}(h^r)$ convergence. This method coincides with the cell vertex finite-volume method, and is associated with the Wendroff box scheme, see Morton [26].

These types of approaches have been developed also for advection dominated diffusion problems, see, e.g., Johnson, Nävert and Pitkäranta (1984), Hughes, Franca and Mallet (1987) and to non-linear conservation laws, Johnson and Szepessy (1987), Szepessy (1989). For such time-dependent problems Pirroneau (1982), Douglas and Russel (1982), and others have also analyzed the so-called characteristic Galerkin method in which one adopts a Lagrangian point of view in the time stepping, following an approximate characteristic defined by the advection term, in combination with a finite element approximation in the diffusive term.

7. Some other classes of approximation methods

Methods other than finite difference and finite element methods, but often closely related to these, have also been developed, and in this section we sketch briefly four such classes of methods, namely collocation methods, spectral methods, finite volume methods, and boundary element methods.

In a *collocation method* one seeks an approximate solution of a differential equation in a finite-dimensional space of sufficiently regular functions by requiring that the equation is satisfied exactly at a finite number of points. Such a procedure for parabolic equations in one space variable was analyzed by Douglas (1972), Douglas and Dupont (1974b); we describe it for the model problem

$$u_t = u_{xx} \quad \text{in } I=(0,1), \qquad u(0,t)=u(1,t)=0 \ \text{ for } t>0, \ \text{with } u(\cdot,0)=v \text{ in } I.$$

Setting $h=1/M$, $x_j=jh$, $j=0,\dots,M$, and $I_j=(x_{j-1},x_j)$, we introduce the piecewise polynomial space $S_h=\{\chi\in C^1(\bar I); v|_{I_j}\in \Pi_{r-1},\ v(0)=v(1)=0\}$, with $r\geqslant 4$. Letting ξ_i, $i=1,\dots,r-2$, be the Gaussian points in $(0,1)$, the zeros of the Lagrange polynomial P_{r-2}, we define the collocation points $\xi_{ji}=x_{j-1}+h\xi_i$ in I_j, and pose the spatially semidiscrete problem to find $u_h\in S_h$ such that

$$u_{h,t}(\xi_{ji},t)=u_{h,xx}(\xi_{ji},t), \quad \text{for } j=1,\dots,M,\ i=1,\dots,r-2,\ t>0,$$

with $u_h(\cdot,0)=v_h$ an approximation of v. This method may be considered as a Galerkin method using a discrete inner product based on a Gauss quadrature rule. For v_h appropriately chosen one may then show the global error estimate

$$\|u_h(t)-u(t)\|_{L_\infty}\leqslant Ch^r\left(\max_{s\leqslant t}\|u(s)\|_{r+2}+\left(\int_0^t\|u_t\|_{r+2}^2\,\mathrm{d}s\right)^{1/2}\right).$$

Further, for $r>4$, and with a more refined choice of initial approximation v_h, superconvergence takes place at the nodes,

$$|u_h(x_j,t)-u(x_j,t)|\leqslant C_T h^{2r-4}\sup_{s\leqslant t}\sum_{p+2q\leqslant 2r-1}\|u^{(q)}(s)\|_p \quad \text{for } t\leqslant T.$$

We note the more stringent regularity requirements than for the Galerkin methods discussed in Section 6. These results carry over to fully discrete methods using both finite difference approximations and collocation in time.

For a two-point boundary value problem, results of a similar nature were derived by de Boor and Swartz (1973).

Spectral methods are in many ways similar to Galerkin/collocation methods. The main difference is in the choice of finite-dimensional approximating spaces. We begin by considering an evolution equation in a Hilbert space framework.

Let thus H be a Hilbert space with inner product $(\cdot,\cdot)$ and norm $\|\cdot\|$, and assume L is a nonnegative operator defined in $D(L)\subset H$, so that $(Lu,u)\geqslant 0$. Consider the initial value problem

$$u_t+Lu=f \quad \text{for } t>0, \ \text{with } u(0)=v. \tag{7.1}$$

Let now $\{\varphi_j\}_{j=1}^{\infty}\subset H$ be a sequence of linearly independent functions in $D(L)$ which span H and set $S_N=\operatorname{span}\{\varphi_j\}_{j=1}^N$. We define a "spatially" semidiscrete approximation $u_N=u_N(t)\in S_N$ of (7.1) by

$$(u_{N,t},\chi)+(Lu_N,\chi)=(f,\chi) \quad \forall\chi\in S_N,\ t\geqslant 0, \quad \text{with } u_N(0)=v_N. \tag{7.2}$$

Introducing the orthogonal projection $P_N : H \to S_N$ we may write (7.2) as

$$u_{N,t} + L_N u_N = f_N := P_N f, \quad \text{for } t \geqslant 0, \text{ where } L_N = P_N L P_N.$$

Clearly $(L_N\chi, \chi) = (LP_N\chi, P_N\chi) \geqslant 0$. With $u_N(t) = \sum_{j=1}^{N} \alpha_j(t)\varphi_j$, this equation may be written $B\alpha' + A\alpha = \tilde{f}$ for $t \geqslant 0$, where the elements of the matrices A and B are $(L\varphi_i, \varphi_j)$ and (φ_i, φ_j), respectively. Clearly B is a Gram matrix and so positive definite.

As a simple example, let $L = -(\mathrm{d}/\mathrm{d}x)^2$ on $I = (0, 1)$ and $H = L_2(I)$, $D(L) = H^2 \cap H_0^1$, and let the $\varphi_j(x) = c_j \sin \pi jx$ be the normalized eigenfunctions of L. Then $B = I, A$ is positive definite and P_N is simply the truncation of the Fourier series, $P_N v = \sum_{j=1}^{N}(v, \varphi_j)\varphi_j$, with $L_N = \sum_{j=1}^{N}(\pi j)^2(v, \varphi_j)\varphi_j = P_N L v$.

We note that the error $e_N = u_N - u$ satisfies

$$e_{N,t} + L_N e_N = f_N - f + (L_N - L)u \quad \text{for } t > 0, \quad e_N(0) = v_N - v$$

and hence, since $E_N(t) = \mathrm{e}^{-L_N t}$ is bounded,

$$\|e_N\| \leqslant \|v_N - v\| + \int_0^t (\|(P_N - I)f\| + \|(L_N - L)u\|)\,\mathrm{d}s. \tag{7.3}$$

It follows that the error is small with $v_N - v$, $(P_N - I)f$, and $(L_N - L)u$.

In our above example we see that if $v_N = P_N v$, and if the Fourier series for v, f, and Lu converge, then the error is small. In particular, the convergence is of order $\mathrm{O}(N^{-r})$ for any r provided the solution is sufficiently regular.

Another way to define a semidiscrete numerical method employing the space S_N of our example is to make S_N a Hilbert space with the inner product $(v, w)_N = h\sum_{j=0}^{N-1} v(x_j)w(x_j)$ where $x_j = j/(N-1)$. This gives rise to a projection P_N defined by $P_N u(x_j) = u(x_j), j = 0, \ldots, N-1$, and the semidiscrete equation (7.2) now becomes the collocation equation

$$u_{N,t}(x_j, t) + Lu_N(x_j, t) = f(x_j, t) \quad \text{for } j = 0, \ldots, N-1, \; t \geqslant 0.$$

This is also referred to as a pseudospectral method and the error estimate (7.3) will be valid in the discrete norm corresponding to $(\cdot, \cdot)_N$.

Spectral and pseudospectral methods using the above sinusoidal basis functions are particularly useful for periodic problems. For initial–boundary value problems for hyperbolic equations basis functions related to Chebyshev and Lagrange polynomials are sometimes useful. Such methods are successfully applied in fluid dynamics calculations. Spectral methods have been studied since the 1970s, see Gottlieb and Orszag [17], Canuto, Hussaini, Quarteroni and Zhang [10], Boyd [6], and references therein.

We now turn to the *finite volume method* which we exemplify for the model problem

$$-\Delta u = f \text{ in } \Omega, \text{ with } u = 0 \text{ on } \partial\Omega, \tag{7.4}$$

where Ω is a convex polygonal domain in $\mathbb{R}^2$. The basis for this approach is the observation that for any $V \subset \Omega$ we have by Green's formula that

$$\int_{\partial V} \frac{\partial u}{\partial n}\,\mathrm{d}s = \int_V f\,\mathrm{d}x. \tag{7.5}$$

Let now $\mathscr{T}_h = \{\tau_j\}_{j=1}^{N_h}$ be a triangulation of Ω and consider (7.5) with $V = \tau_j, j = 1, \ldots, N_h$. Let Q_j be the center of the circumscribed circle of τ_j. If τ_i has an edge γ_{ji} in common with τ_j, then $Q_i - Q_j$ is orthogonal to γ_{ji}, and $\partial u/\partial n$ in (7.5) may be approximated by the difference quotient

$(u(Q_i)-u(Q_j))/|Q_i-Q_j|$. This produces a finite difference scheme on the nonuniform mesh $\{Q_j\}$; for the boundary triangles one may use the boundary values in (7.4). Writing the discrete problem as $AU=F$ the matrix A is symmetric positive definite, and the solution satisifies a discrete maximum principle. When the $\mathscr{T}_h$ is quasi-uniform (and such that the Q_j are on $\partial\Omega$) one has $\|U-u\|_{1,h}\leqslant Ch^{s-1}\|u\|_s$ for $s=2$ in a certain discrete H^1-norm, and, under an additional symmetry assumption on $\mathscr{T}_h$, also for $s=3$. This method may be described as cell centered and goes back to Tikhonov and Samarskii (1961) in the case of rectangular meshes; for further developments, see Samarskii, Lazarov and Maharov [30]. For such meshes it was used in Varga [39] to construct finite difference schemes.

An associated method is the following vertex centered method, also referred to as the finite volume element method: Let $S_h\subset H_0^1$ be the piecewise linear finite element space defined by $\mathscr{T}_h$. The straight lines connecting a vertex of $\tau\in\mathscr{T}_h$ with the midpoint of the opposite edge intersect at the barycenter of τ and divide τ into six triangles. Let $B_{j,\tau}$ be the union of the two of these which have P_j as a vertex. For each interior vertex P_j we let B_j be the union of the corresponding $B_{j,\tau}$, and let $\bar{S}_h$ denote the associated piecewise constant functions. Motivated by (7.5) we then pose the Petrov-Galerkin method to find $u_h\in S_h$ such that

$$\bar{A}(u_h,\psi):=\sum_j\psi_j\int_{\partial B_j}\frac{\partial u_h}{\partial n}\,\mathrm{d}s=(f,\psi)\quad\forall\psi\in\bar{S}_h, \tag{7.6}$$

this may also be thought of as a finite difference scheme on the irregular mesh $\{P_j\}$. The B_j are referred to as control volumes; they were called mesh regions in Mac Neal (1953). Associating with $\chi\in S_h$ the function $\bar{\chi}\in\bar{S}_h$ which agrees with χ at the vertices of $\mathscr{T}_h$ one finds that $\bar{A}(\psi,\bar{\chi})=A(\psi,\chi)$ so that (7.6) may be written $A(u_h,\chi)=(f,\bar{\chi})$ for $\chi\in S_h$. In particular, (7.6) has a unique solution, and using the Babuška–Brezzi *inf–sup* condition it was shown in Bank and Rose (1987) that the standard error estimate $\|u_h-u\|_1\leqslant Ch\|u\|_2$ holds for this method.

Finite volume methods are useful for operators in divergence form and have also been applied to time-dependent conservation laws, see Heinrich [20], Morton [26]. For the model heat equation the vertex centered method is similar to the lumped mass finite element method.

In a *boundary integral method* a boundary value problem for a homogeneous partial differential equation in a domain Ω is reformulated as an integral equation over the boundary $\partial\Omega$. This equation may then be used as a basis for numerical approximation. We shall illustrate this approach for the model problem

$$\Delta u=0\quad\text{in }\Omega\subset\mathbb{R}^d,\text{ with }u=g\text{ on }\partial\Omega, \tag{7.7}$$

$\partial\Omega$ smooth. To pose the boundary integral equation, let $\Gamma(x)=-(2\pi)^{-1}\log|x|$ for $d=2$ and $\Gamma(x)=c_d|x|^{-d+2}$ for $d>2$ be the fundamental solution of the Laplacian in $\mathbb{R}^d$. For any u with $\Delta u=0$ on $\partial\Omega$ we have by Green's formula

$$u(x)=\int_{\partial\Omega}\Gamma(x-y)\frac{\partial u}{\partial n_y}\,\mathrm{d}s_y-\int_{\partial\Omega}\frac{\partial\Gamma}{\partial n_y}(x-y)u(y)\,\mathrm{d}s_y,\quad x\in\Omega. \tag{7.8}$$

With x on $\partial\Omega$ the integrals on the right define the single- and double-layer potentials $V\partial u/\partial n$ and Wu (note that $K(x,y)=(\partial\Gamma/\partial n_y)(x-y)=\mathrm{O}(|x-y|^{-(d-2)})$ for $x,y\in\partial\Omega$). For $x\in\Omega$ approaching $\partial\Omega$ the two integrals tend to $V\partial u/\partial n$ and $\frac{1}{2}u+Wu$, respectively, so that (7.8) yields $\frac{1}{2}u=V\partial u/\partial n+Wu$. With $u=g$ on $\partial\Omega$ this is a Fredholm integral equation of the first kind to determine $\partial u/\partial n$ on $\partial\Omega$, which inserted into (7.8) together with $u=g$ on $\partial\Omega$ gives the solution of (7.7).

Instead of this direct method one may use the indirect method of assuming that the solution of (7.8) may be represented as a potential of a function on $\partial\Omega$, so that

$$u(x) = \int_{\partial\Omega} \Gamma(x-y)v(y)\,\mathrm{d}s_y \quad \text{or } u(x) = \int_{\partial\Omega} \frac{\partial \Gamma}{\partial n_y}(x-y)w(y)\,\mathrm{d}s_y, \quad x \in \Omega.$$

With V and W as above, if such functions v and w exist, they satisfy the first and second kind Fredholm integral equations

$$Vv = g \quad \text{and} \quad \frac{1}{2}w + Ww = g. \tag{7.9}$$

Writing $H^s = H^s(\partial\Omega)$, V and W are pseudodifferential operators of order -1, bounded operators $H^s \to H^{s+1}$, in particular compact on H^s; for $d=2$ the kernel of W is actually smooth. The first kind equation is uniquely solvable provided a certain measure, the transfinite diameter $\delta_{\partial\Omega}$ of $\partial\Omega$, is such that $\delta_{\partial\Omega} \neq 1$, and the second kind equation in (7.8) always has a unique solution. Similar reformulations may be used also for Neumann boundary conditions, for a large number of other problems involving elliptic-type equations, and for exterior problems; in fact, this approach to the numerical solution is particularly useful in the latter cases.

The use of boundary integral equations, particularly of the second kind, to study boundary value problems for elliptic equations has a long history, and includes work of Neumann and Fredholm. We shall not dwell on this here but refer to, e.g., Atkinson [2].

In the boundary element method (BEM) one determines the approximate solution in a piecewise polynomial finite element-type space of a boundary integral formulation such as the above using the Galerkin or the collocation method.

The numerical solution of second kind equations by projection methods, which include both Galerkin and collocation methods, were studied in an abstract Banach space setting in the important paper by Kantorovich (1948), and their convergence was established under the appropriate assumptions on the projection operator involved. Consider, e.g., the second kind equation in (7.9) (with $d=2$) in $C(\partial\Omega)$ with the maximum norm $|\cdot|_\infty$, and let $S_h \subset C(\partial\Omega)$ be finite dimensional. With $P_h : C(\partial\Omega) \to S_h$ a projection operator, the corresponding discrete problem is $\frac{1}{2}w_h + P_h W w_h = P_h g$, and if $|P_h W - W|_\infty \to 0$ in operator norm one may show that $|w_h - w|_\infty \leqslant C|P_h w - w|_\infty$, so that the discrete solution converges as fast as the projection as $h \to 0$.

The collocation method may also be combined with quadrature, as suggested in Nyström (1930): as an example we may use, e.g., the composite trapezoidal rule on a uniform partition, and then apply collocation at the nodal points so that, with $\partial\Omega = \{x(s);\ 0 \leqslant s \leqslant l\}$ and $K(s,t)$ the kernel of W, the discrete solution is

$$w_h(x(s)) = 2g(x(s)) - 2h\sum_{j=1}^{N_h} K(s,s_j)w_j \quad \text{for } 0 \leqslant s \leqslant l,$$

where the w_j are determined by setting $w_h(x(s_i)) = w_i$ for $i = 1, \ldots, N_h$. It is not difficult to see that since the trapezoidal rule is infinitely accurate for smooth periodic functions we have $|w_h - w|_\infty = O(h^r)$ for any $r > 0$.

For the second kind equation in (7.9), using Galerkin's method and a finite dimensional subspace S_h of $L_2(\partial\Omega)$, the discrete approximation $w_h \in S_h$ to w is determined from

$$\tfrac{1}{2}\langle w_h, \chi\rangle + \langle W w_h, \chi\rangle = \langle g, \chi\rangle, \quad \forall \chi \in S_h, \quad \text{where } \langle\cdot,\cdot\rangle = (\cdot,\cdot)_{L_2(\partial\Omega)}.$$

Writing $|\cdot|_s$ for the norm in H^s, one has $|w_h - w|_0 \leqslant C_r(u)h^r$ if S_h is accurate of order $O(h^r)$, and by a duality argument one may show the superconvergent order negative norm estimate $|w_h - w|_{-r} \leqslant C_r(u)h^{2r}$, see Sloan and Thomée (1985); using an iteration argument by Sloan (1976) this may be used, in principle, to define an approximate solution $\tilde{w}_h$ with $|\tilde{w}_h - w|_0 = O(h^{2r})$.

After early work of Wendland (1968) and Nedelec and Planchard (1973) the study of the Galerkin finite element approach using first kind equations was pursued from the mid 1970s by Hsiao, Le Roux, Nedelec, Stephan, Wendland, and others, see the surveys in Atkinson [2], Sloan [31], and Wendland [42]. Within this framework we consider the numerical solution of the first kind equation in (7.9) with $d = 2$ in the finite-dimensional space S_h of periodic smoothest splines of order r, i.e., $S_h \subset C^{r-2}$ consists of piecewise polynomials in Π_{r-1}. Our discrete problem is then to find $v_h \in S_h$ such that

$$\langle Vv_h, \chi \rangle = \langle g, \chi \rangle, \quad \forall \chi \in S_h.$$

It was shown in the work quoted above that the bilinear form $\langle Vv, w \rangle$ associated with $V : H^{-1/2} \to H^{1/2}$ is symmetric, bounded, and coercive in $H^{-1/2}$, i.e.,

$$\langle Vv, w \rangle = \langle v, Vw \rangle \leqslant C|v|_{-1/2}|w|_{-1/2} \quad \text{and} \quad \langle Vv, v \rangle \geqslant c|v|_{-1/2}^2, \text{ with } c > 0.$$

An application of Céa's lemma and approximation properties of S_h then show

$$|v_h - v|_{-1/2} \leqslant C \inf_{\chi \in S_h} |\chi - v|_{-1/2} \leqslant Ch^{r+1/2}|v|_r$$

and an Aubin–Nitsche-type duality argument first used by Hsiao and Wendland (1981) implies $|v_h - v|_{-r-1} \leqslant Ch^{2r+1}|v|_r$. For x an interior point of Ω we therefore find for $u_h = Vv_h$ that $|u_h(x) - u(x)| \leqslant C_x|v_h - v|_{-r-1} \leqslant Ch^{2r+1}$, since $\Gamma(x - y)$ is smooth when $y \neq x$.

Expressed in terms of a basis $\{\Phi_j\}$ of S_h this problem may be written in matrix form as $A\alpha = \tilde{g}$ where A is symmetric positive definite. However, although the dimension of A has been reduced by the reduction of the original two-dimensional problem to a one-dimensional one, in contrast to the finite element method for a differential equation problem, the matrix A is now not sparse. We also note that the elements $\langle V\Phi_i, \Phi_j \rangle$ require two integrations, one in forming $V\Phi_i$ and one in forming the inner product.

In order to reduce this work the collocation method has again been considered by Arnold and Wendland (1983); here v_h is determined from $Vv_h(x(s_j)) = g(x(s_j))$ at N_h quadrature points s_j in $[0, l]$, where $N_h = \dim S_h$. Applied to our above model problem this method, using smoothest splines of even order r, has a lower order of maximal convergence rate, $O(h^{r+1})$ rather than $O(h^{2r+1})$; if r is odd and the mesh uniform Saranen (1988) has shown $O(h^{r+1})$ in $\|\cdot\|_{-2}$. A further step in the development is the qualocation method proposed by Sloan (1988), which is a Petrov–Galerkin method, thus with different trial and test spaces. For S_h the smoothest splines of order r on a uniform mesh (so that Fourier analysis may be applied) and with the quadrature rule suitably chosen, negative norm estimates of order $O(h^{r+3})$ for even r and $O(h^{r+4})$ for odd r may be shown.

In the vast literature on the numerical boundary integral methods much attention has been paid to the complications arising when our above regularity assumptions fail to be satisfied, such as for domains with corners in which case V and W are not compact.

8. Numerical linear algebra for elliptic problems

Both finite difference and finite element methods for elliptic problems such as (2.1) lead to linear algebraic systems

$$AU = F, \tag{8.1}$$

where A is a nonsingular matrix. When Ω is a d-dimensional domain, using either finite differences or finite elements based on quasi-uniform triangulations, the dimension N of the corresponding finite-dimensional problem is of order $O(h^{-d})$, where h is the mesh-width, and for $d > 1$ direct solution by Gauss elimination is normally not feasible as this method requires $O(N^3) = O(h^{-3d})$ algebraic operations. Except in special cases one therefore turns to iterative methods. In this section we summarize the historical development of such methods.

As a basic iterative method we consider the Picard method

$$U^{n+1} = U^n - \tau(AU^n - F) \quad \text{for } n \geqslant 0, \text{ with } U^0 \text{ given}, \tag{8.2}$$

where τ is a positive parameter. With U the exact solution of (8.1) we have

$$U^n - U = R(U^{n-1} - U) = \cdots = R^n(U^0 - U), \quad \text{where } R = I - \tau A$$

and hence the rate of convergence of the method depends on $\|R^n\|$ where $\|\cdot\|$ is the matrix norm subordinate to the Euclidean norm in R^N. When A is symmetric positive definite (SPD) we have $\|R^n\| = \rho^n$ where $\rho = \rho(R) = \max_i|1 - \tau\lambda_i|$ denotes the spectral radius of R, and (8.2) converges if $\rho < 1$. The optimal choice is $\tau = 2/(\lambda_1 + \lambda_N)$, which gives $\rho = (\kappa - 1)/(\kappa + 1)$, where $\kappa = \kappa(A) = \lambda_N/\lambda_1$ is the condition number or A; note, however, that this choice of τ requires knowledge of λ_1 and λ_N which is not normally at hand. In applications to second-order elliptic problems one often has $\kappa = O(h^{-2})$ so that $\rho \leqslant 1 - ch^2$ with $c > 0$. Hence with the optimal choice of τ the number of iterations required to reduce the error to a small $\varepsilon > 0$ is of order $O(h^{-2}|\log \varepsilon|)$. Since each iteration uses $O(h^{-d})$ operations in the application of $I - \tau A$ this shows that the total number of operations needed to reduce the error to a given tolerance is of order $O(h^{-d-2})$, which is smaller than for the direct solution when $d \geqslant 2$.

The early more refined methods were designed for finite difference methods of positive type for second-order elliptic equations, particularly the five-point operator (2.2). The corresponding matrix may then be written $A = D - E - F$ where D is diagonal and E and F are (elementwise) nonnegative and strictly lower and upper triangular. The analysis was often based on the Perron–Frobenius theory of positive matrices. A commonly used property is diagonal dominance: $A = (a_{ij})$ is diagonally dominant if $\sum_{j \neq i} |a_{ij}| \leqslant |a_{ii}|, = 1 \ldots, N$, irreducibly diagonally dominant if it is also irreducible, so that (8.1) cannot be written as two lower-order systems, and strictly diagonally dominant if there is strict inequality for at least one i. Examples are the Jacobi (after Jacobi (1845)) and Gauss–Seidel (Gauss (1823); Seidel (1874) or Liebmann (1918)) methods which are defined by

$$U^{n+1} = U^n - B(AU^n - F) = RU^n + BF, \quad \text{with } R = I - BA, \tag{8.3}$$

in which $B = B_J = D^{-1}$ or $B = B_{GS} = (D - E)^{-1}$ with $R_J = D^{-1}(E + F)$ and $R_{GS} = (D - E)^{-1}F$, respectively. In the application to the model problem (2.1) in the unit square, using the five-point operator, the equations may be normalized so that $D = 4I$ and the application of R_J simply means that the new value at any interior mesh-point x_j is obtained by replacing it by the average of the old values at the four neighboring points $x_{j\pm e_l}$. The Gauss–Seidel method also takes averages, but with the mesh-points taken in a given order, and successively uses the values already obtained in forming the averages. The methods were referred to in Geiringer (1949) as the methods of simultaneous and successive displacements, respectively. For the model problem one may easily determine the eigenvalues and eigenvectors of A and show that with $h = 1/M$ one has $\rho(R_{\mathrm{J}}) = \cos \pi h = 1 - \frac{1}{2}\pi^2 h^2 + \mathrm{O}(h^4)$ and $\rho(R_{\mathrm{GS}}) = \rho(R_{\mathrm{J}})^2 = 1 - \pi^2 h^2 + \mathrm{O}(h^4)$ so that the number of iterates needed to reduce the error to ε is of the orders $2h^{-2}\pi^2|\log \varepsilon|$ and $h^{-2}\pi^2|\log \varepsilon|$. The Gauss–Seidel method thus requires about half as many iterations as the Jacobi method.

If A is irreducibly diagonally dominant then $\rho(R_{\mathrm{J}}) < 1$ and $\rho(R_{\mathrm{GS}}) < 1$ so that both methods converge, see Geiringer (1949); for A strictly diagonally dominant this was shown in Collatz (1942). Further, Stein and Rosenberg (1948) showed that if D is positive and E and F nonnegative and $\rho(B_{\mathrm{J}}) < 1$ then $\rho(B_{\mathrm{GS}}) < \rho(B_{\mathrm{J}})$, i.e., the Gauss–Seidel method converges faster than the Jacobi method.

Forming the averages in the Jacobi and Gauss–Seidel methods may be thought as relaxation; in the early work by Gauss and Seidel this was not done in a cyclic order as described above, and which is convenient on computers, but according to the size of the residual or other criteria, see Southwell [33] and Fox (1948). It turns out that one may obtain better results than those described above by overrelaxation, i.e., choosing $B_\omega = (D - \omega E)^{-1}$ and $R_\omega = (D - \omega E)^{-1}((1-\omega)E + F)$ with $\omega > 1$. These methods were first studied by Frankel (1950) in the case of the model problem, and Young (1950, 1954) in more general cases of matrices satisfying his property $\mathscr{A}$, which holds for a large class of difference approximations of elliptic problems in general domains. Frankel proved that for the model problem the optimal choice of the parameter is $\omega_{\mathrm{opt}} = 2/(1 + \sqrt{1 - \rho^2})$ where $\rho = \rho(B_{\mathrm{J}}) = \cos \pi h$, i.e., $\omega_{\mathrm{opt}} = 2/(1 + \sin \pi h) = 2 - 2\pi h + \mathrm{O}(h^2)$, and that correspondingly $\rho(R_{\omega_{\mathrm{opt}}}) = \omega_{\mathrm{opt}} - 1 = 1 - 2\pi h + \mathrm{O}(h^2)$. The number of iterations required is thus then of order $\mathrm{O}(h^{-1})$, which is significally smaller than for the above methods. It was shown by Kahan (1958), also for nonsymmetric A, that $\rho(R_\omega) \geqslant |\omega - 1|$ so that convergence can only occur for $0 < \omega < 2$. On the other hand, Ostrowski (1954) showed that if A is SPD, then $\rho(R_\omega) < 1$ if and only if $0 < \omega < 2$.

We consider again an iterative method of the form (8.3) with $\rho(R) < 1$, and introduce now the new sequence $V^n = \sum_{j=0}^n \beta_{nj} U^j$ where the β_{nj} are real. Setting $p_n(\lambda) = \sum_{j=0}^n \beta_{nj}\lambda^j$, and assuming $p_n(1) = \sum_{j=0}^n \beta_{nj} = 1$ for $n \geqslant 0$, we obtain easily $V^n - U = p_n(R)(U^0 - U)$. For V^n to converge fast to U one wants to choose the β_{nj} in such a way that $\rho(p_n(R))$ becomes small with n. By the Cayley–Hamilton theorem $p_n(R) = 0$ if p_n is the characteristic polynomial of R, and hence $V^n = U$ if $n \geqslant N$, but this is a prohibitively large number of iterations. For $n < N$ we have by the spectral mapping theorem that $\rho(p_n(R)) = \max_{\lambda \in \sigma(R)} |p_n(\lambda)|$. In particular, if R is symmetric and $\rho = \rho(R)$, a simple calculation shows that, taking the maximum instead over $[-\rho, \rho] \supset \sigma(R)$, the optimal polynomial is $p_n(\lambda) = T_n(\lambda/\rho)/T_n(1/\rho)$ where T_n is the nth Chebyshev polynomial, and the corresponding value of $\rho(p_n(R))$ is bounded by

$$T_n(1/\rho)^{-1} = 2\left(\left(\frac{1+\sqrt{1-\rho^2}}{\rho}\right)^n + \left(\frac{1+\sqrt{1-\rho^2}}{\rho}\right)^{-n}\right)^{-1} \leqslant 2\left(\frac{\rho}{1+\sqrt{1-\rho^2}}\right)^n.$$

For the model problem using the Gauss–Seidel basic iteration we have as above $\rho = 1 - \pi^2 h^2 + \mathrm{O}(h^4)$ and we find that the average error reduction factor per iteration step in our present method is bounded by $1 - \sqrt{2}\pi h + \mathrm{O}(h^2)$, which is of the same order of magnitude as for SOR. The use of the sequence

V^n instead of the U^n was called linear acceleration by Forsythe (1953) and is sometimes attributed to Richardson (1910); in [39] it is referred to as a semiiterative method.

We now describe the Peaceman–Rachford alternating direction implicit iterative method for the model problem (2.1) on the unit square, using the five-point discrete elliptic equation with $h=1/M$. In this case we may write $A=H+V$ where H and V correspond to the horizontal and vertical difference operators $-h^2\partial_1\bar{\partial}_1$ and $-h^2\partial_2\bar{\partial}_2$. Note that H and V are positive definite and commute. Introducing an acceleration parameter τ and an intermediate value $U^{n+1/2}$ we may consider the scheme defining U^{n+1} from U^n by

$$(\tau+H)U^{n+1/2}=(\tau-V)U^n+F, \quad (\tau+V)U^{n+1}=(\tau-H)U^{n+1/2}+F,$$

or after elimination, with G_τ appropriate and using that H and V commute,

$$U^{n+1}=R_\tau U^n+G_\tau, \quad \text{where } R_\tau=(\tau I-H)\,(\tau I+H)^{-1}(\tau I-V)\,(\tau I+V)^{-1}.$$

The error satisfies $U^n-U=R_\tau^n(U^0-U)$, and with μ_i the (common) eigenvalues of H and V, $\|R_\tau\|\leqslant \max_i|(\tau-\mu_i)/(\tau+\mu_i)|^2<1$, and it is easy to see that the maximum occurs for $i=1$ or M. With $\mu_1=4\sin^2(\frac{1}{2}\pi h)$, $\mu_M=4\cos^2(\frac{1}{2}\pi h)$ the optimal τ is $\tau_{\text{opt}}=(\mu_1\mu_M)^{1/2}$ with the maximum for $i=1$, so that, with $\kappa=\kappa(H)=\kappa(V)=\mu_M/\mu_1$,

$$\|R\|_{\tau_{\text{opt}}}\leqslant\left(\frac{(\mu_1\mu_M)^{1/2}-\mu_1}{(\mu_1\mu_M)^{1/2}+\mu_1}\right)^{1/2}=\frac{\kappa^{1/2}-1}{\kappa^{1/2}+1}=1-\pi h+O(h^2).$$

This again shows the same order of convergence rate as for SOR.

A more efficient procedure is obtained by using varying acceleration parameters τ_j, $j=1,2,\ldots$, corresponding to the n step error reduction matrix $\tilde{R}_n=\prod_{j=1}^n R_{\tau_j}$. It can be shown that the τ_j can be chosen cyclically with period m in such a way that $m\approx c\log\kappa\approx c\log(1/h)$ and

$$\|\tilde{R}_m\|^{1/m}=\max_{1\leqslant i\leqslant M}\left(\prod_{j=0}^{m-1}\left|\frac{\tau_j-\mu_i}{\tau_j+\mu_i}\right|\right)^{2/m}\leqslant 1-c(\log(1/h))^{-1}, \quad c>0.$$

The analysis indicated depends strongly on the fact that H and V commute, which only happens for rectangles and constant coefficients, but the method may be defined and shown convergent for more general cases, see Birkhoff and Varga (1959). We remark that these iterative schemes may often be associated with time-stepping methods for parabolic problems and that our discussion in Section 4 of fractional step and splitting methods are relevant also in the present context. For a comprehensive account of the above methods for solving systems associated with finite difference methods, including historical remarks, see Varga [39].

We now turn to the development of iterative methods for systems mainly associated with the emergence of the finite element method. We begin by describing the conjugate gradient method by Hestenes and Stiefel (1952), and assume that A is SPD. Considering the iterative method

$$U^{n+1}=(I-\tau_n A)U^n+\tau_n F \quad \text{for } n\geqslant 0, \text{ with } U^0=0,$$

we find at once that, for any choice of the parameters τ_j, U^n belongs to the Krylov space $K_n(A;F)=\text{span}\{F,AF,\dots,A^{n-1}F\}$. The conjugate gradient method defines these parameters so that U^n is the best approximation of U in $K_n(A;F)$ with respect to the norm defined by $|U|=(AU,U)^{1/2}$, i.e., as the orthogonal projection of U onto $K_n(A;F)$ with respect to the inner product (AV,W). By our above discussion it follows that, with $\kappa=\kappa(A)$ the condition number of A,

$$|U^n - U| \leqslant (T_n(1/\rho))^{-1}|U| \leqslant 2\left(\frac{\kappa^{1/2}-1}{\kappa^{1/2}+1}\right)^n |U|. \tag{8.4}$$

The computation of U^n can be done by a two-term recurrence relation, for instance in the following form using the residuals $r^n = F - AU^n$ and the auxiliary vectors $q^n \in K_{n+1}(A;F)$, orthogonal to $K_n(A;F)$,

$$U^{n+1} = U^n + \frac{(r^n,q^n)}{(Aq^n,q^n)}q^n, \quad q^{n+1} = r^{n+1} - \frac{(Ar^{n+1},q^n)}{(Aq^n,q^n)}q^n, \quad U^0 = 0,\ q^0 = F.$$

In the preconditioned conjugate gradient (PCG) method the conjugate gradient method is applied to Eq. (8.1) after multiplication by some easy to determine SPD approximation B of A^{-1} and using the inner product $(B^{-1}V,W)$; we note that BA is SPD with respect to this inner product. The error estimate (8.4) is now valid with $\kappa=\kappa(BA)$; B would be chosen so that this condition number is smaller than $\kappa(A)$. For the recursion formulas the only difference is that now $r^n = B(F-AU^n)$ and $q^0 = BF$. An early application of PCG to partial differential equations is Wachspress (1963) and it is systematically presented in Marchuk [24] and Axelsson and Barker [3], where reference to other work can be found.

One way of defining a preconditioner is by means of the multigrid method. This method is based on the observation that large components of the errors are associated with low frequencies in a spectral representation. The basic idea is then to work in a systematic way with a sequence of triangulations and reduce the low-frequency errors on coarse triangulations, which corresponds to small size problems, and higher frequency residual errors on finer triangulations by a smoothing operator, such as a step of the Jacobi method, which is relatively inexpensive.

One common situation is as follows: Assuming Ω is a plane polygonal domain we first perform a coarse triangulation of Ω. Each of the triangles is then divided into four similar triangles, and this process is repeated, which after a finite number M of steps leads to a fine triangulation with each of the original triangles devided into 4^M small triangles. Going from one level of fineness to the next the procedure may be described in three steps: (1) presmoothing on the finer triangulation, (2) correction on the coarser triangulation by solving a residual equation, (3) postsmoothing on the finer triangulation. This procedure is then used recursively between the levels of the refinement leading to, e.g., the V- or W-cycle algorithms. It turns out that under some assumptions the error reduction matrix R corresponding to one sweep of the algorithm satisfies $\|R\| \leqslant \rho < 1$, with ρ independent of M, i.e., of h, and that the number of operations is of order $O(N)$ where $N=O(h^{-2})$ is the dimension of the matrix associated with the finest triangulation.

The multigrid method was first introduced for finite difference methods in the 1960s by Fedorenko (1964) and Bahvalov (1966) and further developed and advocated by Brandt in the 1970s, see, e.g., Brandt (1977). For finite elements it has been intensely pursued by, e.g., Braess and Hackbusch, Bramble and Pasciak, Mandel, McCormick and Bank; for overviews with further references, see Hackbusch [19], Bramble [7].

A class of iterative methods that have attracted a lot of attention recently is the so-called domain decomposition methods. These assume that the domain Ω in which we want to solve our elliptic problem may be decomposed into subdomains Ω_j, $j=1,\ldots,M$, which could overlap. The idea is to reduce the boundary value problem on Ω into problems on each of the Ω_j, which are then coupled by their values on the intersections. The problems on the Ω_j could be solved independently on parallel processors. This is particularly efficient when the individual problems may be solved very fast, e.g., by fast transform methods. Such a case is provided by the model problem (2.1) on the unit square which may be solved directly by using the discrete Fourier transform, defined by $\hat{F}_m=\sum_j F^j e^{-2\pi i\, mjh}$. In fact, we then have $(-\Delta_h U)\hat{}_m = 2\pi^2|m|^2\hat{U}_m$ and hence $\hat{U}_m=(2\pi^2|m|^2)^{-1}\hat{F}_m$ so that by the inverse discrete Fourier transform $U^j=\sum_m(2\pi^2|m|^2)^{-1}\hat{F}_m e^{2\pi i\, mjh}$. Using the fast Fourier transform both $\hat{F}_m$ and U^j may be calculated in $O(N\log N)$ operations.

The domain decomposition methods go back to the alternating procedure by Schwarz (1869), in which $\Omega=\Omega_1\cup\Omega_2$. Considering the Dirichlet problem (2.1) on Ω one defines a sequence $\{u^k\}$ starting with a given u^0 vanishing on $\partial\Omega$, by

$$-\Delta u^{2k+1}=f \quad \text{in } \Omega_1, \text{ with } u^{2k+1}=u^{2k} \text{ on } \partial\Omega_1\cap\Omega_2,\ u^{2k+1}=0 \text{ on } \partial\Omega_1\cap\partial\Omega,$$

$$-\Delta u^{2k+2}=f \quad \text{in } \Omega_2, \text{ with } u^{2k+2}=u^{2k+1} \text{ on } \partial\Omega_2\cap\Omega_1,\ u^{2k+2}=0 \text{ on } \partial\Omega_2\cap\partial\Omega$$

and this procedure can be combined with numerical solution by, e.g., finite elements. A major step in the analysis of this so-called multiplicative form of the Schwarz alternating procedure was taken by Lions (1988). A modification referred to as the additive form was first studied by Matsokin and Nepomnyashchikh (1985) and Dryja and Widlund (1987).

The following alternative approach may be pursued when Ω_1 and Ω_2 are disjoint but with a common interface $\partial\Omega_1\cap\partial\Omega_2$: If u_j denotes the solution in $\Omega_j, j=1,2$, transmission conditions $u_1=u_2$, $\partial u_1/\partial n=\partial u_2/\partial n$ have to be satisfied on the interface. One method is then to reduce the problem to an integral-type equation on the interface and use this as a basis of an iterative method. For a survey of domain decomposition techniques, see Smith, Bjørstad and Gropp [32].

References

Arnold, D.N., Wendland, W.L., 1983. On the asymptotic convergence of collocation methods. Math. Comp. 41, 349–381.

Aronson, D.G., 1963. The stability of finite difference approximations to second order linear parabolic differential equations. Duke Math. J. 30, 117–128.

Aubin, J.P., 1967. Behavior of the error of the approximate solution of boundary value problems for linear elliptic operators by Galerkin's and finite difference methods. Ann. Scuola Norm. Sup. Pisa Cl. Sci. 21, 599–637.

Aziz, A.K., Monk, P., 1989. Continuous finite elements in space and time for the heat equation. Math. Comp. 52, 255–274.

Babuška, I., 1971. Error bounds for finite element method. Numer. Math. 16, 322–333.

Babuška, I., 1973. The finite element method with Lagrangian multipliers. Numer. Math. 20, 179–192.

Babuška, I., 1976. The selfadaptive approach in the finite element method. In: Whiteman, J.R. (Ed.), The Mathematics of Finite Elements and Applications, II. Academic Press, New York, pp. 125–142.

Babuška, I., Dorr, M.R., 1981. Error estimates for the combined h and p-version of the finite element method. Numer. Math. 37, 257–277.

Babuška, I., Rheinboldt, W.C., 1978. Error estimates for adaptive finite element computations. SIAM J. Numer. Anal. 15, 736–754.

Babuška, I., Suri, M., 1990. *The p- and h-p versions of the finite element method, an overview* In: Canuto, C., Quarteroni, A., (Eds.), Spectral and High Order Methods for Partial Differential Equations. North-Holland, Amsterdam, pp. 5–26.

Babuška, I., Szabó, B.A., Katz, I.N., 1981. The p-version of the finite element method. SIAM J. Numer. Anal. 18, 512–545.

Bahvalov, N.S., 1959. Numerical solution of the Dirichlet problem for Laplace's equation. Vestnik Moskov. Univ. Ser. Math. Meh. Astronom. Fiz. Him. 3, 171–195.

Bahvalov, N.S., 1966. On the convergence of a relaxation method with natural constraints on the elliptic operator. USSR Comput. Math. and Math. Phys. 6, 101–135.

Baker, G.A., 1975. A finite element method for first order hyperbolic equations. Math. Comp. 20, 995–1006.

Baker, G.A., 1976. Error estimates for finite element methods for second order hyperbolic equations. SIAM J. Numer. Anal. 13, 564–576.

Baker, G.A., Bramble, J.H., 1979. Semidiscrete and single step fully discrete approximations for second order hyperbolic equations. RAIRO Anal. Numér. 13, 75–100.

Baker, G.A., Bramble, J.H., Thomée, V., 1977. Single step Galerkin approximations for parabolic problems. Math. Comp. 31, 818–847.

Bank, R.E., Rose, D.J., 1987. Some error estimates for the box method. SIAM J. Numer. Anal. 24, 777–787.

Berger, A.E., Scott, L.R., Strang, G., 1972. *Approximate boundary conditions in the finite element method.* Symposia Mathematica, Vol. 10. Academic Press, New York, 295–313.

Birkhoff, G., Schultz, M.H., Varga, R.S., 1968. *Piecewise Hermite interpolation in one and two variables with applications to partial differential equations.* Numer. Math., 232–256.

Blair, J., 1970. Approximate solution of elliptic and parabolic boundary value problems. Ph.D. Thesis. University of California, Berkeley.

Bramble, J.H., Hilbert, S.R., 1970. Estimation of linear functionals on Sobolev spaces with application to Fourier transforms and spline interpolation. SIAM J. Numer. Anal. 7, 113–124.

Bramble, J.H., Hubbard, B.E., 1962. On the formulation of finite difference analogues of the Dirichlet problem for Poisson's equation. Numer. Math. 4, 313–327.

Bramble, J. H., Hubbard, B.E., 1963. A theorem on error estimation for finite difference analogues of the Dirichlet problem for elliptic equations. Contrib. Differential Equations 2, 319–340.

Bramble, J.H., Dupont, T., Thomée, V., 1972. Projection methods for Dirichlet's problem in approximating polygonal domains with boundary value corrections. Math. Comp. 26, 869–879.

Bramble, J.H., Hubbard, B.E., Thomée, V., 1969. Convergence estimates for essentially positive type discrete Dirichlet problems. Math. Comp. 23, 695–710.

Bramble, J.H., Kellogg, R.B., Thomée, V., 1968. On the rate of convergence of some difference schemes for second order elliptic equations. Nordisk Tidskr. Inform. Behandling (BIT) 8, 154–173.

Bramble, J.H., Schatz, A.H., Thomée, V., Wahlbin, L.B., 1977. Some convergence estimates for semidiscrete Galerkin type approximations for parabolic equations. SIAM J. Numer. Anal. 14, 218–241.

Brandt, A., 1977. Multi-level adaptive solution to boundary-value problems. Math. Comp. 31, 333–390.

Brenner, P., 1966. The Cauchy problem for symmetric hyperbolic systems in L_p. Math. Scand. 19, 27–37.

Brenner, P., Thomée, V., 1970. Stability and convergence rates in L_p for certain difference schemes. Math. Scand. 27, 5–23.

Brenner, P., Crouzeix, M., Thomée, V., 1982. Single step methods for inhomogeneous linear differential equations in Banach space. RAIRO Anal. Numér. 16, 5–26.

Brezzi, F., 1974. On the existence, uniqueness and approximation of saddle-point problems arising from Lagrangian mulitpliers. RAIRO Anal. Numér. 8, 129–151.

Céa, J., 1964. Approximation variationelle des problèmes aux limites. Ann. Inst. Fourier. (Grenoble) 14, 345–444.

Clough, R.W., 1960. *The finite element method in plane stress analysis.* Proceedings of Second ASCE Conference on Electronic Computation, Vol. 8, Pittsburg, Pennsylvania pp. 345–378.

Collatz, L., 1933. Bemerkungen zur Fehlerabschätzung für das Differenzenverfahren bei partiellen Differentialgleichungen. Z. Angew. Math. Mech. 13, 56–57.

Collatz, L., 1942. Fehlerabschätzung für das Iterationsverfahren zur Auflösung linearer Gleichungssysteme. Z. Angew. Math. Mech. 22, 357–361.

Courant, R., 1943. Variational methods for the solution of problems of equilibrium and vibrations. Bull. Amer. Math. Soc. 49, 1–23.

Courant, R., Friedrichs, K.O., Lewy, H., 1928. *Über die partiellen Differenzengleichungen der Mathematischen Physik.* Math. Ann. 100, 32–74 (English translation, with commentaries by Lax, P.B., Widlund, O.B., Parter, S.V., in IBM J. Res. Develop. 11 (1967)).
Courant, R., Isaacson, E., Rees, M., 1952. On the solution of nonlinear hyperbolic differential equations by finite differences. Comm. Pure Appl. Math. 5, 243–255.
Crank, J., Nicolson, P., 1947. A practical method for numerical integration of solution of partial differential equations of heat-conduction type. Proc. Cambridge Philos. Soc. 43, 50–67.
Crouzeix, M., Larsson, S., Thomée, V., 1994. Resolvent estimates for elliptic finite element operators in one dimension. Math. Comp. 63, 121–140.
Crouzeix, M., Larsson, S., Piskarev, S., Thomée, V., 1993. The stability of rational approximations of analytic semigroups. BIT 33, 74–84.
Dahlquist, G., 1954. Convergence and stability for a hyperbolic difference equation with analytic initial-values. Math. Scand. 2, 91–102.
de Boor, D., Swartz, B., 1973. Collocation at Gaussian points. SIAM J. Numer. Anal. 10, 582–606.
Delfour, M., Hager, W., Trochu, F., 1981. Discontinuous Galerkin method for ordinary differential equations. Math. Comp. 36, 453–473.
Demjanovič, Ju.K., 1964. The net method for some problems in mathematical physics. Dokl. Akad. Nauk SSSR 159, 250–253.
Dendy, J.E., 1974. Two methods of Galerkin type achieving optimum L^2 rates of convergence for first order hyperbolics. SIAM J. Numer. Anal. 11, 637–653.
Dobrowolski, M., 1978. L^∞-convergence of linear finite element approximations to quasi-linear initial boundary value problems. RAIRO Anal. Numér. 12, 247–266.
Douglas Jr. J., 1956. On the relation between stability and convergence in the numerical solution of linear parabolic and hyperbolic differential equations. J. Soc. Indust. Appl. Math. 4, 20–37.
Douglas Jr. J., 1972. A superconvergence result for the approximate solution of the heat equation by a collocation method. In: Aziz, A.K. (Ed.), The Mathematical Foundations of the Finite Element Method with Applications to Partial Differential Equations. Academic Press, New York, pp. 475–490.
Douglas Jr. J., Dupont, T., 1970. Galerkin methods for parabolic equations. SIAM J. Numer. Anal. 7, 575–626.
Douglas Jr. J., Dupont, T., 1974a. Galerkin approximations for the two point boundary problem using continuous, piecewise polynomial spaces. Numer. Math. 22, 99–109.
Douglas Jr. J., Dupont, T., 1974b. Collocation Methods for Parabolic Equations in a Single Space Variable. Lecture Notes in Mathematics, Vol. 385. Springer, Berlin.
Douglas Jr. J., Russel, T.F., 1982. Numerical methods for convection-dominated diffusion problems based on combining the method of characteristics with finite element or finite difference procedures. SIAM J. Numer. Anal. 19, 871–885.
Dryja, M., Widlund, O.B., 1987. An additive variant of the Schwarz alternating method for the case of many subregions. Technical Report, Department of Computer Science, Courant Institute of Mathematical Sciences New York University.
Dupont, T., 1973a. L^2-estimates for Galerkin methods for second order hyperbolic equations. SIAM J. Numer. Anal. 10, 880–889.
Dupont, T., 1973b. Galerkin methods for first order hyperbolics: an example. SIAM J. Numer. Anal. 10, 890–899.
Dupont, T., 1974. L_2 error estimates for projection methods for parabolic equations in approximating domains. In: C. de Boor, Mathematical Aspects of Finite Elements in Partial Differential Equations. Academic Press, New York, pp. 313-352.
Eriksson, K., 1994. An adaptive finite element method with efficient maximum norm error control for elliptic problems. Math. Models Methods Appl. Sci. 4, 313–329.
Eriksson, K., Johnson, C., 1991. Adaptive finite element methods for parabolic problems. I: A linear model problem. SIAM J. Numer. Anal. 28, 43–77.
Eriksson, K., Johnson, C., Thomée, V., 1985. A discontinuous in time Galerkin method for parabolic type problems. RAIRO Math.Anal. 19, 611–643.
Faedo, S., 1949. Un nuovo metodo de l'analisi esistenziale e quantitative dei problemi di propagazione. Ann. Scuola Norm. Sup Pisa Cl. Sci. 1 (3), 1–40.
Fedorenko, R., 1964. The speed of convergence of one iterative process. USSR Comput. Math. Math. Phys. 4, 227–235.
Feng, K., 1965. Finite difference schemes based on variational principles. Appl. Math. Comput. Math. 2, 238–262.

Fox, L., 1948. A short account of relaxation methods. Quart. J. Mech. Appl. Math. 1, 253–280.

Frankel, S.P., 1950. Convergence rates of iterative treatments of partial differential equations. Math. Tables Aids Comput. 4, 65–75.

Friedrichs, K.O., 1954. Symmetric hyperbolic linear differential equations. Comm. Pure Appl. Math 7, 345–392.

Friedrichs, K.O., 1962. A finite difference scheme for the Neumann and Dirichlet problem. Report No NYO-9760, AEC Comp. and Appl. Math. Center, Courant Inst., Math. Sci., New York Univ.

Friedrichs, K.O., Keller, H.B., 1966. A finite difference scheme for generalized Neumann problems. Numerical Solution of Partial Differential Equations. Academic Press, New York. pp. 1-19.

Friedrichs, K.O., Lewy, H., 1926. Über die Eindeutigkeit und das Abhängigkeitsgebiet der Lösungen beim Anfangswertproblem linearer hyperbolischer Differentialgleichungen. Math. Ann. 98, 192–204.

Fujii, H., 1973. Some remarks on finite element analysis of time-dependent field problems. Theory and Practice in Finite Element Structural Analysis. University of Tokyo, pp. 91–106.

Galerkin, B.G., 1915. Rods and plates. Series occurring in various questions concerning the elastic equilibrium of rods and plates. Vestnik Insz. 19, 897–908.

Gauss, C.F., 1823. Brief an Gerling. Werke 9, 278–281.

Geiringer, H., 1949. On the solution of systems of linear equations by certain iterative methods. Reissner Anniversary Volume. J.W. Edwards, Ann Arbor, Mi., pp. 365–393.

Gerschgorin, S., 1930. Fehlerabschätzung für das Differenzenverfahren zur Lösung partieller Differentialgleichungen. Z. Angew. Math. Mech. 10, 373–382.

Godunov, S.K., Ryabenkii, V.S., 1963. Special criteria of stability of boundary-value problems for non-selfadjoint difference equations. Uspekhi Mat. Nauk 18, 3–14.

Gustafsson, B., Kreiss, H.O., Sundström, A., 1972. Stability theory for difference approximations of mixed initial boundary value problems. Math. Comp. 26, 649–686.

Haverkamp, R., 1984. Eine Aussage zur L_∞-Stabilität und zur genauen Konvergenzordnung der H_0^1-Projektionen. Numer. Math. 44, 393–405.

Hedstrom, G.W., 1968. The rate of convergence of some difference schemes. SIAM J. Numer. Anal. 5, 363–406.

Helfrich, H.-P., 1974. Fehlerabschätzungen für das Galerkinverfahren zur Lösung von Evolutions-gleichungen. Manuscripta Math. 13, 219–235.

Hestenes, M., Stiefel, E., 1952. Methods of conjugate gradients for solving linear systems. J. Res. Nat. Bur. Standards 49, 409–436.

Hsiao, G., Wendland, W., 1981. The Aubin-Nitsche lemma for integral equations. J. Integral Equations 3, 299–315.

Hughes, T.J.R., Brooks, A., 1979. A multidimensional upwind scheme with no crosswind diffusion. In: Hughes, T.J.R. (Ed.), Finite Element Methods for Convection Dominated Flows, Vol. 34. American Society of Mechanical Engineers, New York, pp. 19–35.

Hughes, T.J.R., Franca, L.P., Mallet, M., 1987. A new finite element formulation for computational fluid dynamics: VI. Convergence analysis of the generalized SUPG formulation for linear time dependent multi-dimensional advective-diffusive systems. Comput. Meth. Appl. Mech. Eng. 63, 97–112.

Jamet, P., 1978. Galerkin-type approximations which are discontinuous in time for parabolic equations in a variable domain. SIAM J. Numer. Anal. 15, 912–928.

John, F., 1952. On integration of parabolic equations by difference methods. Comm. Pure Appl. Math. 5, 155–211.

Johnson, C., Pitkäranta, J., 1986. An analysis of the discontinuous Galerkin method for a scalar hyperbolic equation. Math. Comp. 46, 1–26.

Johnson, C., Szepessy, A., 1987. On the convergence of a finite element method for a nonlinear hyperbolic conservation law. Math. Comp. 49, 427–444.

Johnson, C., Thomée, V., 1981. Error estimates for some mixed finite element methods for parabolic type problems. RAIRO Modél. Math. Anal. Numér. 15, 41–78.

Johnson, C., Nävert, U., Pitkäranta, J., 1984. Finite element methods for linear hyperbolic problems. Comput. Methods Appl. Mech. Eng. 45, 285–312.

Juncosa, M.L., Young, D.M., 1957. On the Crank-Nicolson procedure for solving parabolic partial differential equations. Proc. Cambridge Philos. Soc. 53, 448–461.

Kantorovich, L., 1948. Functional analysis and applied mathematics. Uspehi Mat. Nauk 3, 89–185.

Kato, T., 1960. Estimation of iterated matrices, with application to the von Neumann condition. Numer. Math. 2, 22–29.

Kreiss, H.O., 1959a. Über die Lösung des Cauchyproblems für lineare partielle Differenzialgleichungen mit Hilfe von Differenzengleichungen. Acta Math. 101, 179–199.

Kreiss, H.O., 1959b. Über Matrizen die beschränkte Halbgruppen erzeugen, Math. Scand. 7, 72–80.

Kreiss, H.O., 1962. Über die Stabilitätsdefinition für Differenzengleichungen die partielle Differentialgleichungen approximieren. Nordisk Tidskr. Inform.-Behandling 2, 153–181.

Kreiss, H.O., 1964. On difference approximations of dissipative type for hyperbolic differential equations. Comm. Pure Appl. Math. 17, 335–353.

Kreiss, H.O., 1968. Stability theory for difference approximations of mixed initial boundary value problems. Math. Comp. 22, 703–714.

Kreiss, H.O., Scherer, G., 1974. Finite Element and Finite Difference Methods for Hyperbolic Partial Differential Equations. Academic Press, New York, 195–212.

Kreiss, H.O., Thomée, V., Widlund, O.B., 1970. Smoothing of initial data and rates of convergence for parabolic difference equations. Comm. Pure Appl. Math. 23, 241–259.

Laasonen, P., 1949. Über eine Methode zur Lösung der Wärmeleitungsgleichung. Acta Math. 81, 309–317.

Laasonen, P., 1958. On the solution of Poisson's difference equation. J. Assoc. Comput. Mach. 5, 370–382.

Lax, P.D., 1961. On the stability of difference approximations to solutions of hyperbolic differential equations. Comm. Pure Appl. Math. 14, 497–520.

Lax, P.D., Nirenberg, L., 1966. On stability for difference schemes: a sharp form of Gårding's inequality. Comm. Pure Appl. Math. 19, 473–492.

Lax, P.D., Richtmyer, R.D., 1959. Survey of the stability of linear finite difference equations. Comm. Pure Appl. Math. 9, 473–492.

Lax, P.D., Wendroff, B., 1960. Systems of conservation laws. Comm. Pure Appl. Math. 13, 217–237.

Lax, P.D., Wendroff, B., 1964. Difference schemes for hyperbolic equations with high order of accuracy. Comm. Pure Appl. Math. 17, 381–398.

Lees, M., 1960. A priori estimates for the solution of difference approximations to parabolic differential equations. Duke Math. J. 27, 297–311.

Le Roux, M.N., 1979. Semidiscretization in time for parabolic problems. Math. Comp. 33, 919–931.

Lesaint, P., Raviart, P.-A., 1974. On the finite element method for solving the neutron transport equation. In: de Boor, C. (Ed.), Mathematical Aspects of Finite Elements in Partial Differential Equations. Academic Press, New York, pp. 89–123.

Liebmann, H., 1918. Die angenäherte Ermittlung harmonischer Funktionen und konformer Abbildung (nach Ideen von Boltzmann und Jacobi). Sitzungsberichte der Beyer. Akad. Wiss., Math.-Phys. Kl. 47, 385–416.

Lions, P.-L., 1988. On the Schwarz alternating method I. In: Glowinski, R., Golub, G.H., Meurant, G.A., Périaux, J. (Eds.), First International Symposium on Domain Decomosition Methods. SIAM, Philadelphia, pp. 1–42.

Lippold, G., 1991. Error estimates and step-size control for the approximate solution of a first order evolution equation. RAIRO Modél. Math. Anal. Numér. 25, 111–128.

Löfström, J., 1970. Besov spaces in the theory of approximation. Ann. Mat. Pura Appl. 85, 93–184.

Luskin, M., Rannacher, R., 1982. On the smoothing property of the Crank-Nicolson scheme. Appl. Anal. 14, 117–135.

Mac Neal, R.H., 1953. An asymmetric finite difference network. Quart. Appl. Math. 1, 295–310.

Matsokin, A.M., Nepomnyashchikh, S.V., 1985. A Schwarz alternating method in a subspace. Izv. VUZ Mat. 29 (10), 61–66.

Motzkin, T.S., Wasow, W., 1953. On the approximation of linear elliptic differential equations by difference equations with positive coefficients. J. Math. Phys. 31, 253–259.

Natterer, F., 1975. Über die punktweise Konvergenz finiter Elemente. Numer. Math. 25, 67–77.

Nedelec, J.C., Planchard, J., 1973. Une méthode variationelle d'élements finis pour la résolution numérique d'un problème extérieur dans R^3. RAIRO, Anal. Numér. 7, 105–129.

Newmark, N.M., 1959. A method of computation for structural dynamics. Proc. ASCE, J. Eng. Mech. (EM3) 85, 67–94.

Nitsche, J., 1968. Ein Kriterium für die Quasioptimalität des Ritzschen Verfahrens. Numer. Math. 11, 346–348.

Nitsche, J., 1971. Über ein Variationsprinzip zur Lösung von Dirichlet-problemen bei Verwendung von Teilräumen, die keinen Randbedingungen unterworfen sind. Abh. Math. Semin. Univ. Hamb. 36, 9–15.

Nitsche, J.A., 1975. L_∞-convergence of finite element approximation. Second Conference on Finite Elements. Rennes, France.

Nitsche, J.A., 1979. L_∞-convergence of finite element Galerkin approximations for parabolic problems. RAIRO Anal. Numér. 13, 31–54.

Nitsche, J.A., Wheeler, M.F., 1981–82. L_∞-boundedness of the finite element Galerkin operator for parabolic problems. Numer. Funct. Anal. Optim. 4, 325–353.

Nyström, E., 1930. Über die praktische Auflösung von Integralgleichungen mit Anwendungen auf Randwertaufgaben. Acta Math. 54, 185–204.

O'Brien, G.G., Hyman, M.A., Kaplan, S., 1951. A study of the numerical solution of partial differential equations. J. Math. Phys. 29, 223–251.

Oganesjan, L.A., 1963. Numerical Solution of Plates. Lensovnarchoz. CBTI, Sbornik.

Oganesjan, L.A., 1966. Convergence of difference schemes in case of improved approximation of the boundary. Ž. Vyčisl. Mat. Mat. Fiz. 6, 1029–1042.

Oganesjan, L.A., Ruhovec, L.A., 1969. An investigation of the rate of convergence of variational-difference schemes for second order elliptic equations in a two-dimensional region with smooth boundary. Zh. Vychisl. Mat. Mat. Fiz. 9, 1102–1120.

Osher, S., 1972. Stability of parabolic difference approximations to certain mixed initial boundary value problems. Math. Comp. 26, 13–39.

Ostrowski, A.M., 1954. On the linear iteration procedures for symmetric matrices. Rend. Mat. Appl. 14, 140–163.

Palencia, C., 1992. A stability result for sectorial operators in Banach spaces. SIAM J. Numer. Anal. 30, 1373–1384.

Parlett, B., 1966. Accuracy and dissipation in difference schemes. Comm. Pure Appl. Math. 19, 111–123.

Peaceman, D.W., Rachford Jr. H.H., 1955. The numerical solution of parabolic and elliptic differential equations. J. Soc. Indust. Appl. Math. 3, 28–41.

Peetre, J., Thomée, V., 1967. On the rate of convergence for discrete initial-value problems. Math. Scand. 21, 159–176.

Pirroneau, O., 1982. On the transport-diffusion algorithm and its applications to the Navier-Stokes equations. Numer. Math. 38, 309–332.

Rannacher, R., 1984. Finite element solution of diffusion problems with irregular data. Numer. Math. 43, 309–327.

Rannacher, R., 1991. L^∞-stability and asymptotic error expansion for parabolic finite element equations. Bonner Math. Schriften 228.

Raviart, P., Thomas, J., 1977. A mixed finite element method for 2nd order elliptic problems. Proceedings of the Symposium on the Mathematical Aspects of the Finite Element Method, Vol. 606, Rome, December, 1975. Springer Lecture Notes in Mathematics, Springer, Berlin, pp. 292–315.

Lord Rayleigh, Theory of Sound (1894, 1896), Vols. I, II., Macmillan, London.

Ritz, W., 1908. Über eine neue Methode zur Lösing gewisser Variationsprobleme der mathematischen Physik. J. Reine Angew. Math. 135, 1–61.

Saranen, J., 1988. The convergence of even degree spline collocation solution for potential problems in smooth domains of the plane. Numer. Math. 53, 299–314.

Samarskii, A.A., 1961. A priori estimates for the solution of the difference analogue of a parabolic differential equation. Zh. Vychisl. Mat. Mat. Fiz, 1, 441–460 (in Russian) (English translation in U.S.S.R. Comput. Math. Math. Phys. 1 487–512.)

Sammon, P., 1983a. Convergence estimates for semidiscrete parabolic equation approximations. SIAM J. Numer. Anal. 19, 68–92.

Sammon, P., 1983b. Fully discrete approximation methods for parabolic problems with nonsmooth initial data. SIAM J. Numer. Anal. 20, 437–470.

Saulev, V.K., 1957. On a class of elliptic equations solvable by the method of finite-differences. Vycisl. Mat. 1, 81–86.

Schatz, A.H., Wahlbin, L.B., 1982. On the quasi-optimality in L_∞ of the $H^\circ 1$-projection into finite element spaces. Math. Comp. 38, 1–22.

Schatz, A.H., Thomée, V., Wahlbin, L.B., 1980. Maximum-norm stability and error estimates in parabolic finite element equations. Comm. Pure Appl. Math. 33, 265–304.

Schatz, A.H., Thomée, V., Wahlbin, L.B., 1998. Stability, analyticity, and almost best approximation in maximum-norm for parabolic finite element equations. Comm. Pure Appl. Math. 51, 1349–1385.

Schwarz, H.A., 1869. Über einige Abbildungsaufgaben. J. Reine Angew. Math 70, 105–120.

Scott, L.R., 1975. Interpolated boundary conditions in the finite element method. SIAM J. Numer. Anal. 12, 404–427.

Scott, L.R., 1976. Optimal L^∞ estimates for the finite element method. Math. Comp. 30, 681–697.

Seidel, L., 1874. Über ein Verfahren, die Gleichungen, auf welche die Methode der kleinsten Quadrate führt, sowie lineäre Gleichungen überhaupt, durch successive Annäherung aufzulösen. Abh. Math.-Phys. Kl., Bayerische Akad. Wiss. München 11, 81–108.

Serdjukova, S.J., 1964. *The uniform stability with respect to the initial data of a sixpoint symmetrical scheme for the heat conduction equation.* Numerical Methods for the Solution of Differential Equations and Quadrature Formulae. Nauka, Moscow, pp. 212–216.

Shortley, G.H., Weller, R., 1938. The numerical solution of Laplace's equation. J. Appl. Phys. 9, 334–344.

Sloan, I.H., 1976. Improvement by iteration for compact operator equations. Math. Comp. 30, 758–764.

Sloan, I.H., 1988. A quadrature-based approach to improving the collocation method. Numer. Math. 54, 41–56.

Sloan, I.H., Thomée, V., 1985. Superconvergence of the Galerkin iterates for integral equations of the second kind. J. Integral Equations 9, 1–23.

Stein, P., Rosenberg, R.L., 1948. On the solution of linear simultaneous equations by iteration. J. London Math. Soc. 23, 111–118.

Strang, G., 1962. Trigonometric polynomials and difference methods of maximum accuracy. J. Math. Phys. 41, 147–154.

Strang, G., 1964. Unbalanced polynomials and difference methods for mixed problems. SIAM J. Numer. Anal. 2, 46–51.

Strang, G., 1965. Necessary and insufficient conditions for well-posed Cauchy problems. J. Differential Equations 2, 107–114.

Strang, G., 1967. On strong hyperbolicity. J. Math. Kyoto Univ. 6, 397–417.

Strang, G., 1972. Variational crimes in the finite element method. In: Aziz, A.K. (Ed.), The Mathematical Foundations of the Finite Element Method with Applications to Partial Differential Equations. Academic Press, New York, pp. 689–710.

Swartz, B., Wendroff, B., 1969. Generalized finite-difference schemes. Math. Comp. 23, 37–49.

Szepessy, A., 1989. Convergence of a shock-capturing streamline diffusion finite element method for a scalar conservation law in two space dimensions. Math. Comp. 53, 527–545.

Tikhonov, A.N., Samarskii, A.A., 1961. Homogeneous difference schemes. Zh. Vychisl. Mat. Mat. Fiz. 1, 5–63, (U.S.S.R. Comput. Math. Math. Phys. 1, 5–67).

Thomée, V., 1964. Elliptic difference equations and Dirichlet's problem. Contrib. Differential Equations 3, 301–324.

Thomée, V., 1973. Convergence estimates for semi-discrete Galerkin methods for initial-value problems. Numerische, insbesondere approximationstheoretische Behandlung von Funktionalgleichungen, Vol. 333, Springer Lecture Notes in Mathematics. Springer, Berlin, pp. 243–262.

Thomée, V., 1980. Negative norm estimates and superconvergence in Galerkin methods for parabolic problems. Math. Comp. 34, 93–113.

Thomée, V., Wahlbin, L.B., 1974. Convergence rates of parabolic difference schemes for non-smooth data. Math. Comp. 28, 1–13.

Thomée, V., Wahlbin, L.B., 1998. Stability and analyticity in maximum-norm for simplicial Lagrange finite element semidiscretizations of parabolic equations with Dirichlet boundary conditions. Preprint no. 1998-44, Department of Mathematics, Göteborg.

Thomée, V., Wendroff, B., 1974. Convergence estimates for Galerkin methods for variable coefficient initial value problems. SIAM J. Numer. Anal. 1, 1059–1068.

Thomée, V., Westergren, B., 1968. Elliptic difference equations and interior regularity. Numer. Math. 11, 196–210.

Thomée, V., Xu, J.C., Zhang, N.Y., 1989. Superconvergence of the gradient in piecewise linear finite-element approximation to a parabolic problem. SIAM J. Numer. Anal. 26, 553–573.

Turner, M.J., Clough, R.W., Martin, H.C., Topp, L., 1956. Stiffness and deflection analysis of complex structures. J. Aero. Sci. 23, 805–823.

Varah, J.M., 1970. Maximum norm stability of difference approximations to the mixed initial boundary value problem for the heat equation. Math. Comp. 24, 31–44.

Volkov, E.A., 1966. *Obtaining an error estimate for a numerical solution of the Dirichlet problem in terms of known quantities.* Z. Vyčisl. Mat. Mat. Fiz. 6 (4)(suppl)., 5–17.

Wahlbin, L.B., 1974. A dissipative Galerkin method applied to some quasilinear hyperbolic equations. RAIRO Anal. Numér. 8, 109–117.

Wasow, W., 1952. On the truncation error in the solution of Laplace's equation by finite differences. J. Res. Nat. Bur. Standards 48, 345–348.

Wachspress, E., 1963. Extended application of alternating direction implicit iteration model problem theory. SIAM J. Appl. Math. 11, 994–1016.

Wendland, W., 1968. Die Behandlung von Randwertaufgaben im R_3 mit Hilfe von Einfach- und Doppelschichtpotentialen. Numer. Math. 11, 380–404.

Wendroff, B., 1960. On central difference equations for hyperbolic systems. J. Soc. Indust. Appl. Math. 8, 549–555.

Wendroff, B., 1968. Well-posed problems and stable difference operators. SIAM J. Numer. Anal. 5, 71–82.

Wheeler, M.F., 1973. A priori L_2 error estimates for Galerkin approximations to parabolic partial differential equations. SIAM J. Numer. Anal. 10, 723–759.

Widlund, O.B., 1966. Stability of parabolic difference schemes in the maximum-norm. Numer. Math. 8, 186–202.

Winther, R., 1981. A stable finite element method for initial-boundary value problems for first order hyperbolic equations. Math. Comp. 36, 65–86.

Yamaguti, M., 1967. Some remarks on the Lax-Wendroff finite-difference scheme for nonsymmetric hyperbolic systems. Math. Comp. 21, 611–619.

Young, Jr. D.M., Iterative methods for solving partial differential equations of elliptic type Thesis, Harvard University, 1950, reprinted in `http://www-sccm.stanford.edu/pub/sccm/`.

Young, D.M., 1954. Iterative methods for soliving partial difference equations of ellipic type. Trans. Amer. Math. Soc. 76, 92–111.

Zienkiewicz, O.C., Zhu, J.Z., 1992. The superconvergent patch recovery and a posteriori error estimates. Internat. J. Numer. Methods Eng. 33, 1331–1382.

Zlámal, M., 1968. On the finite element method. Numer. Math. 12, 394–409.

Survey articles and books

[1] J.H. Argyris, Energy Theorems and Structural Analysis. A Generalised Discourse with Applications on Energy Principles of Structural Analysis Including the Effects of Temperature and Non-linear Stress-Strain Relations, Butterworths, London, 1960.

[2] K.E. Atkinson, The Numerical Solution of Integral Equations of the Second Kind, Cambridge University Press, Cambridge, UK, 1997.

[3] O. Axelsson, V.A. Barker, Finite Element Solution of Boundary Value Problems, Academic Press, 1984.

[4] I. Babuška, Courant element: before and after, in: M. Křížek, P. Neittaanmäki, R. Stenberg (Eds.), Finite Element Methods: Fifty Years of the Courant Element, Decker, New York, 1994, pp. 37–51.

[5] I. Babuška, A.K. Aziz, Survey lectures on the mathematical foundation of the finite element method, in: A.K. Aziz (Ed.), The Mathematical Foundations of the Finite Element Method with Applications to Partial Differential Equations, Academic Press, New York, 1972, pp. 5–359.

[6] J.P. Boyd, Chebyshev & Fourier Spectral Methods, Lecture Notes in Engineering, Vol. 49, Springer, Berlin, 1989.

[7] J.H. Bramble, Multigrid Methods, Pitman Research Notes in Mathematics, Longman, London, 1993.

[8] S.C. Brenner, L.R. Scott, The Mathematical Theory of Finite Element Methods, Springer, New York, 1994.

[9] P. Brenner, V. Thomée, L.B. Wahlbin, in: Besov Spaces and Applications to Difference Methods for Initial Value Problems, Springer Lecture Notes in Mathematics, Vol. 434, Springer, Berlin, 1975.

[10] C. Canuto, Y.M. Hussaini, A. Quarteroni, T.A. Zang, Spectral Methods in Fluid Dynamics, Springer, Berlin, 1988.

[11] P.G. Ciarlet, The Finite Element Method for Elliptic Problems, North-Holland, Amsterdam, 1978.

[12] L. Collatz, Numerische Behandlung von Differentialgbeichungen, Springer, Berlin, 1955.

[13] K. Eriksson, D. Estep, P. Hansbo, C. Johnson, Introduction to adaptive methods for differential equations, Acta Numer. 4 (1995) 105–158.

[14] G.E. Forsythe, W.R. Wasow, Finite Difference Methods for Partial Differential Equations, Wiley, New York, 1960.

[15] A. Friedman, Partial Differential Equations of Parabolic Type, Prentice-Hall, Englewood Cliffs, NJ, 1964.

[16] E. Godlewski, P.-A. Raviart, Hyperbolic Systems of Conservation Laws, Ellipse, Paris, 1991.

[17] D. Gottlieb, S.A. Orszag, Numerical Analysis of Spectral Methods: Theory and Applications, SIAM, Philadelphia, 1977.

[18] B. Gustafsson, H.-O. Kreiss, J. Oliger, Time Dependent Problems and Difference Methods, Wiley, New York, 1995.

[19] W. Hackbusch, in: Multi-Grid Methods and Applications, Springer Series in Computational Mathematics, Vol. 4, Springer, New York, 1985.
[20] B. Heinrich, Finite Difference Methods on Irregular Networks, Akademie-Verlag, Berlin, 1987.
[21] L.V. Kantorovich, V.I. Krylov, Approximate Methods in Higher Analysis, Interscience, New York, 1958.
[22] M. Križek, P. Neittaanmäki, On superconvergence techniques, Acta Appl. Math. 9 (1987) 175–198.
[23] R.J. Le Veque, Numerical Methods for Conservation Laws, Birkhäuser, Basel, 1990.
[24] G.I. Marchuk, Methods of Numerical Mathematics, Springer, New York, 1975.
[25] G.I. Marchuk, Splitting and alternating direction methods, in: P.G. Ciarlet, J.L. Lions (Eds.), Handbook of Numerical Analysis, Vol. I. Finite Difference Methods (Part 1), North-Holland, Amsterdam, 1990, pp. 197–462.
[26] K.W. Morton, Numerical Solution of Convection-Diffusion Problems, Chapman & Hall, London, 1996.
[27] P.A. Raviart, J.M. Thomas, Introduction à l'analyse numérique des équations aux dérivées partielles, Masson, Paris, 1983.
[28] R.D. Richtmyer, K.W. Morton, Difference Methods for Initial-Value Problems, Interscience, New York, 1967.
[29] V.S. Ryabenkii, A.F. Filippov, Über die Stabilität von Differenzengleichungen, Deutscher Verlag der Wissenschaften, Berlin, 1960.
[30] A.A. Samarskii, R.D. Lazarov, V.L. Makarov, Difference Schemes for Differential Equations with Generalized Solutions, Vysshaya Shkola, Moscow, 1987 (in Russian).
[31] I.H. Sloan, Error analysis of boundary integral methods, Acta Numer. 1 (1992) 287–339.
[32] B.F. Smith, P.E. Bjørstad, W.D. Gropp, Domain Decomposition: Parallel Multilevel Methods for Elliptic Partial Differential Equations, Cambridge University Press, Cambridge, 1996,.
[33] R.V. Southwell, Relaxation Methods in Theoretical Physics, I, II, Clarendon Press, Oxford, 1946, 1956.
[34] G. Strang, G.J. Fix, An Analysis of the Finite Element Method, Prentice-Hall, Englewood Cliffs, NJ, 1973.
[35] J.C. Strikwerda, Finite Difference Schemes and Partial Differential Equations, Chapman & Hall, New York, 1990.
[36] V. Thomée, Stability theory for partial difference operators, SIAM Rev. 11 (1969) 152–195.
[37] V. Thomée, Finite difference methods for linear parabolic equations, in: P.G. Ciarlet, J.L. Lions (Eds.), Handbook of Numerical Analysis, Vol. I. Finite Difference Methods (Part 1), North-Holland, Amsterdam, 1990, pp. 5–196.
[38] V. Thomée, Galerkin Finite Element Methods for Parabolic Problems, Springer Series in Computational Mathematics, Vol. 25, Springer, Berlin, 1997.
[39] R.S. Varga, Matrix Iterative Analysis, Prentice-Hall, Englewood Cliffs, NJ, 1962.
[40] R. Verfürth, A Review of A Posteriori Error Estimation and Adaptive Mesh-Refinement Techniques, Wiley, New York, Teubner, Stuttgart, 1996.
[41] L.B. Wahlbin, Superconvergence in Galerkin Finite Element Methods, Lecture Notes in Mathematics, Vol. 1605, Springer, Berlin, 1995.
[42] W.L. Wendland, Boundary element methods for elliptic problems, in: A.H. Schatz, V. Thomée, W.L. Wendland (Eds.), Mathematical Theory of Finite and Boundary Element Methods, Birkhäuser, Basel, 1990.
[43] O.C. Zienkiewicz, The Finite Element Method in Engineering Science, McGraw-Hill, London, 1977.

A perspective on the numerical treatment of Volterra equations

Christopher T.H. Baker [1]

Department of Mathematics, The University of Manchester, Oxford Road, Manchester M13 9PL, UK

Abstract

We discuss the properties and numerical treatment of various types of Volterra and Abel–Volterra integral and integro-differential equations.

Keywords: Volterra and Abel equations; Quadrature, Runge–Kutta and collocation methods; Convergence; Superconvergence; Stability

1. Introduction

Volterra's book *Leçons sur les équations intégrales et intégro-différentielles* appeared in 1913. Since then, a considerable literature on the theory and on applications (which include elasticity, semi-conductors, scattering theory, seismology, heat conduction, metallurgy, fluid flow, chemical reactions, population dynamics, etc.) — e.g., [52,55,57,58,76,77,95,108,111,117,118] — and on the numerics — e.g., [1,25,30,41,49,92,96,109] — has appeared.

The obligation to provide a perspective on the subject in the year 2000 has settled on this author. A paper such as this, in which we seek to convey basic theoretical and computational features, could have been written from many different viewpoints, including that of *mathematical modelling*, that of *robust numerical algorithms*, or that of the *mathematical analysis* of numerical methods. Each standpoint has its own relevance to the numerical simulation of real phenomena, and each relies on rather different foundations and methodologies. In particular, the mathematical theory of numerical formulae relies on detailed and sometimes intricate analytical arguments to generate a formal theory (cf. [49] and its extensive references and bibliographical notes) that has mathematical significance

[1] Professor of Mathematics, & Director, Manchester Center for Computational Mathematics (MCCM).

E-mail address: cthbaker@ma.man.ac.uk (C.T.H. Baker).

in its own right but, at least in part, attempts to generate insight into the performance of real-life algorithms.

Within the space available, we give an introduction to the numerical treatment of classical Volterra equations. It is hoped that the author, while avoiding pedantry, is sufficiently careful to avoid mathematical pitfalls or ambiguities occasionally found in the literature that a reader may use the work as a basis for further critical reading and research. The mathematical style may be considered to be intermediate between that of the excellent introduction of Linz [96] (Math. Rev. 86m:65163) and the majestic detail provided by Brunner and van der Houwen [49] (Math. Rev. 88g:65136), which is currently out of print.

2. Basic theory

The classical forms of *Volterra integral equation of the first and second kind* and of *Volterra integro-differential equations* are, respectively,

$$\int_{t_0}^{t} K(t,s,y(s))\,\mathrm{d}s = g(t) \quad (t \in [t_0,T]), \tag{2.1a}$$

$$y(t) = g(t) + \int_{t_0}^{t} K(t,s,y(s))\,\mathrm{d}s \quad (t \in [t_0,T]), \tag{2.1b}$$

$$y'(t) = g(t) + \int_{t_0}^{t} K(t,s,y(s))\,\mathrm{d}s \quad (t \in [t_0,T)) \text{ with } y(t_0) = y_0 \tag{2.1c}$$

(given $g(\cdot)$ and $K(\cdot,\cdot,\cdot)$, find $y(\cdot)$) where the interval $[t_0,T]$ (interpreted as $[t_0,\infty)$ if T is unbounded) is prescribed, and where $g(\cdot) \in C[t_0,T]$ and

Hypothesis 1. (a) $K(t,s,v)$ is continuous for $t_0 \leqslant s \leqslant t \leqslant T$, $g(\cdot) \in C[t_0,T]$;
(b) $K(t,s,v)$ satisfies a uniform Lipschitz condition in v for $t_0 \leqslant s \leqslant t \leqslant T$.

As an alternative to (2.1c) we may encounter

$$y'(t) = F\left(t, y(t), \int_{t_0}^{t} K(t,s,y(s))\,\mathrm{d}s\right) \quad (t \in [t_0,T]) \quad \text{with } y(t_0) = y_0 \tag{2.1d}$$

(given $F(\cdot,\cdot,\cdot)$ and $K(\cdot,\cdot,\cdot)$, find $y(\cdot)$) where, in addition to Hypothesis 1 itself, $F(\cdot,\cdot,\cdot)$ satisfies appropriate Lipschitz conditions. The functions involved here can be real or complex valued. Higher-order integro-differential equations (say generalizing (2.1d)) also arise. Alternative assumptions to those in Hypothesis 1 allow us to consider equations of Abel type. Each of the Volterra integral equations (2.1a)–(2.1d), gives a corresponding *Abel* (or *Abel–Volterra*) equation when

Hypothesis 2. (a) $K(t,s,v) = (t-s)^{-\nu}H(t,s,v)$ where $0 < \nu < 1$ and $H(t,s,v)$ is continuous for $t_0 \leqslant s \leqslant t \leqslant T$, $g(\cdot) \in C[t_0,T]$; (b) $H(t,s,v)$ satisfies a uniform Lipschitz condition in v for $t_0 \leqslant s \leqslant t \leqslant T$.

Thus, the Abel equation of the second kind has the form

$$y(t) = g(t) + \int_{t_0}^{t} (t-s)^{-\nu} H(t,s,y(s))\,\mathrm{d}s \quad (t \in [t_0, T)). \tag{2.1e}$$

The condition $0 < \nu < 1$ is customary when referring to Abel equations.

Remark 2.1. (a) In the theory, the uniformity of the Lipschitz conditions may be *relaxed.* However, to obtain the benefits of the numerical processes described later, one often requires continuity properties *stronger* than those required in Hypotheses 1 and 2 (e.g., higher-order differentiability). Moreover, one sometimes (e.g., in certain Runge–Kutta methods discussed later) requires a suitable smooth *extension* of $K(t,s,v)$ (or of $H(t,s,v)$) defined on an extended domain $t_0 \leqslant s \leqslant t + \Delta \leqslant T + \Delta$ with $\Delta > 0$:

$$K_{\text{ext}}(t,s,v) = \begin{cases} K(t,s,v) & \text{if } s \leqslant t, \\ K_{\text{new}}(t,s,v) & \text{if } s > t \end{cases} \tag{2.2}$$

or

$$H_{\text{ext}}(t,s,v) = \begin{cases} H(t,s,v) & \text{if } s \leqslant t, \\ H_{\text{new}}(t,s,v) & \text{if } s > t. \end{cases} \tag{2.3}$$

(b) If $K(t,s,v) = \mathsf{k}(t,s)v$, Eqs. (2.1a)–(2.1c) are *linear equations.*

(c) An integral of the form $\int_{t_0}^{t} k(t-s)\phi(s)\,\mathrm{d}s$ is called a convolution of $k(\cdot)$ with $\phi(\cdot)$ — taking $k(t) \equiv 1$ is a special case. (Corresponding sums $\sum_{j=0}^{n} \omega_{n-j}\phi_j$, and the matrix–vector counterparts $\sum_{j=0}^{n} \boldsymbol{\Omega}_{n-j}\boldsymbol{\phi}_j$, are *discrete convolutions.*) If $K(t,s,v) = k(t-s)v$ or $K(t,s,v) = k(t-s)\varphi(v)$ for some continuous function $k(\cdot)$ and (usually Lipschitz–continuous) $\varphi(\cdot)$ the Volterra equations are (linear or nonlinear) *convolution equations.* Under additional conditions, the Laplace transform provides a tool for analysing linear convolution equations.

The ordinary differential equation (ODE) $y'(t) = f(t, y(t))$ $(t \geqslant t_0)$ can be re-written as a Volterra integral equation of the second kind:

$$y(t) = \int_{t_0}^{t} f(s, y(s))\,\mathrm{d}s + y(t_0) \quad (t \geqslant t_0). \tag{2.4}$$

Furthermore, Eq. (2.1d) can be written as a system of equations, one of which is an ODE and the other a Volterra integral equation:

$$y'(t) = F(t, y(t), z(t)), \tag{2.5a}$$

$$z(t) = \int_{t_0}^{t} K(t,s,y(s))\,\mathrm{d}s. \tag{2.5b}$$

Alternatively, by integrating (2.5a) and expressing it as

$$y(t) = y(t_0) + \int_{t_0}^{t} F(s, y(s), z(s))\,\mathrm{d}s, \tag{2.5c}$$

we obtain the system of Volterra integral equations (2.5b), (2.5c). The intimate relation between systems of Volterra equations of the second kind and Volterra integro-differential equations encourages us (though sacrificing some insight and detail) to concentrate upon the former.

A starting point for any mathematical discussion is to ask what is known about the well-posedness of the problems under consideration. (Rigorous concepts of well-posedness depend upon choices of underlying function spaces. Sometimes, a solution is required to be *positive*.) If the Volterra equation of the first kind is to be interpreted as implying equality everywhere, we require $g(t_0)=0$ and, clearly, a solution does not exist for all $g(\cdot)$. If $g(t_0)=0$ and we have sufficient differentiability, we can deduce

$$K(t,t,y(t)) + \int_{t_0}^{t} K_t(t,s,y(s))\,\mathrm{d}s = g'(t).$$

In particular, if $K(t,s,v)=\mathsf{k}(t,s)v$ and $\mathsf{k}(t,t)$ is nonvanishing, we have

$$y(t) + \int_{t_0}^{t} \frac{\mathsf{k}_t(t,s)}{\mathsf{k}(t,t)} y(s)\,\mathrm{d}s = \frac{g'(t)}{\mathsf{k}(t,t)} \quad (\text{for } t \geqslant t_0), \tag{2.6}$$

a linear Volterra equation of the second kind. On the other hand, if $\mathsf{k}(t,t)$ vanishes, but $g''(t)$ and $\mathsf{k}_{tt}(t,s)$ exist and $\mathsf{k}_t(t,t)$ is nonvanishing, we can obtain an equation of the second kind by a further differentiation. Our remarks serve to demonstrate that the first-kind equation can suffer various degrees of ill-posedness in the sense that if nonsmooth perturbations are made to $g(\cdot)$ a solution (other than in a generalized sense) may fail to exist. Inexact data therefore raises problems. Theoretical results on numerical procedures for first-kind equations usually assume a condition such as $\inf|\mathsf{k}(t,t)| > 0$, or an analogous condition in the nonlinear case. Some numerical methods require one to supply the value $y(t_0)$ (when (2.6) is valid, $y(t_0)=g'(t_0)/\mathsf{k}(t_0,t_0)$).

Example 2.2. When $\int_{t_0}^{t} y(s)\,\mathrm{d}s = g(t)$, the solution, if it exists, is $y(t)=g'(t)$. For Abel equations of the first kind, similar remarks are possible; e.g., if

$$\frac{1}{\Gamma(1-\nu)} \int_{t_0}^{t} (t-s)^{-\nu} y(s)\,\mathrm{d}s = g(t) \quad \text{for } \nu \in (0,1) \tag{2.7a}$$

($\nu = \frac{1}{2}$ has special interest) then the solution $y(\cdot)$, if it exists, is expressible

$$y(t) = \frac{1}{\Gamma(\nu)} \int_{t_0}^{t} (t-s)^{\nu-1} g'(s)\,\mathrm{d}s \tag{2.7b}$$

and $y(\cdot)$ is now a *fractional derivative* (or a *fractional integral*) of $g(\cdot)$ (cf. [49, p.8; 76, p.3; 96, p.73]). One can write (2.7) symbolically as $J^{(1-\nu)}y = g$, and $y = J^{\nu}Dg$ (where J is an operator of indefinite integration and D is the operator of differentiation and DJ is the identity).

Remark 2.3. On the face of it, there could be a link between the solution $y(\cdot)$ of the equation of the first kind (2.1a) and the solution $y_\varepsilon(\cdot)$ of the singularly perturbed equation $\varepsilon y_\varepsilon(t) = \int_{t_0}^{t} K(t,s,y_\varepsilon(s))\,\mathrm{d}s - g(t)$ of the second kind (cf. [91]), either as $\varepsilon \nearrow 0$ or $\varepsilon \searrow 0$. However, such links can only be established under special hypotheses. This said, one can develop discretization formulae that are *candidates* for application to equations of the first kind by constructing them for the singularly perturbed equation of the second kind above and, formally, setting $\varepsilon = 0$, provided one appreciates that whether or not these formulae are useful has to be investigated independently.

The theory for linear equations of the second kind is simpler than that of equations of the first kind. In particular, for the Volterra equation

$$y(t) = g(t) + \lambda \int_{t_0}^{t} \mathsf{k}(t,s)y(s)\,\mathrm{d}s \quad (\text{for } t \geqslant t_0), \tag{2.8a}$$

we have

$$y(t) = g(t) + \lambda \int_{t_0}^{t} \mathsf{r}^{\{\lambda\}}(t,s)g(s)\,\mathrm{d}s \quad (\text{for } t \geqslant t_0), \tag{2.8b}$$

where $\mathsf{r}^{\{\lambda\}}(\cdot,\cdot)$ is the *resolvent kernel* for (2.8a). The Neumann series gives the expression $\mathsf{r}^{\{\lambda\}}(t,s) = \sum_{r=1}^{\infty} \lambda^{r-1}\mathsf{k}^r(t,s)$ where $\mathsf{k}^r(t,s) = \int_{t_0}^{t} \mathsf{k}(t,\sigma)\mathsf{k}^{r-1}(\sigma,s)\,\mathrm{d}\sigma$ $(r = 2,3,\ldots)$, and $\mathsf{k}^1(t,s) = \mathsf{k}(t,s)$. One has

$$\mathsf{r}^{\{\lambda\}}(t,\sigma) = \lambda \int_{t_0}^{t} \mathsf{k}(t,s)\mathsf{r}^{\{\lambda\}}(s,\sigma)\,\mathrm{d}s + \mathsf{k}(t,\sigma) \quad (\text{for } t_0 \leqslant \sigma \leqslant t). \tag{2.9}$$

Since $\mathsf{r}^{\{\lambda\}}(s,\sigma) = 0$ if $\sigma > s$, the lower limit of integration in (2.9) can be replaced by σ. Smoothness properties of $\mathsf{r}^{\{\lambda\}}(t,s)$ follow from those of $\mathsf{k}(t,s)$, and the degree of smoothness of $y(\cdot)$ (which has an impact on the suitability of various numerical techniques) follows from the properties of $g(\cdot)$ and the resolvent. If $\mathsf{k}(t,s)$ is of Abel type, $\mathsf{k}(t,s) = (t-s)^{-\nu}\mathsf{h}(t,s)$, for $\nu \in (0,1)$, the resolvent is expressible [49, p.16] as $\mathsf{r}^{\{\lambda\}}(t,s) = (t-s)^{-\nu}\mathsf{p}^{\{\lambda,\nu\}}(t,s)$ with continuous $\mathsf{p}^{\{\lambda,\nu\}}(\cdot,\cdot)$. One deduces that the solution of an Abel equation of the second kind has unbounded derivatives at t_0 when $g(t)$ is smooth (asymptotic expansions as $t \to t_0$ can be found).

The Neumann series is related to an iteration that, for the more general nonlinear equation (2.1b), reads

$$y_{k+1}(t) = g(t) + \int_{t_0}^{t} K(t,s,y_k(s))\,\mathrm{d}s \quad (k = 0,1,2,\ldots), \tag{2.10}$$

often with $y_0(t) = g(t)$. (The iteration may also be used to "refine" an approximation $\tilde{y}(t)$ by setting $y_0(\cdot) = \tilde{y}(\cdot)$ and computing the corresponding iterate $y_1(\cdot)$.) Iteration (2.10) collapses to the Picard iteration in the case of an integrated form of an ODE (2.4). It extends in an obvious manner to systems of Volterra integral equations, and, thence, through its application to (2.5b), (2.5c) to the integro-differential equation (2.1d).

Remark 2.4. For the linear equation $y'(t) = a(t)y(t) + \int_{t_0}^{t} \mathsf{k}(t,s)y(s)\,\mathrm{d}s + g(t)$ $(t \geqslant t_0)$ subject to the initial condition $y(t_0) = y_0$, where $a(\cdot)$ is continuous, we can establish the existence of a *resolvent* $\mathsf{u}(t,s)$ such that $y(t) = \mathsf{u}(t,t_0)y_0 + \int_{t_0}^{t} \mathsf{u}(t,s)g(s)\,\mathrm{d}s$; $(\partial/\partial s)\mathsf{u}(t,s) + a(t)\mathsf{u}(t,s) + \int_{s}^{t} \mathsf{k}(t,\sigma)\mathsf{u}(\sigma,s)\,\mathrm{d}\sigma = 0$, $\mathsf{u}(t,t) = 1$.

We have introduced the resolvent kernels $\mathsf{r}^{\{\lambda\}}(t,s)$, and $\mathsf{u}(t,s)$, into our discussion because they have a rôle in a discussion of *stability of solutions* (see [49, p.493–494; 118]), and in an *error analysis of certain approximate solutions* — both studies involving aspects of perturbation theory. Clearly, if $g(\cdot)$ in (2.8a) is perturbed to $g(\cdot) + \delta g(\cdot)$ then, by (2.8b), $y(\cdot)$ suffers a consequent change

$$\delta y(t) = \lambda \int_{t_0}^{t} \mathsf{r}^{\{\lambda\}}(t,s)\delta g(s)\,\mathrm{d}s + \delta g(t). \tag{2.11}$$

With appropriate conditions, we have a similar result [2] for the nonlinear equation (2.1b). Suppose that $y(\cdot)$ is the unique solution of (2.1b) and we have

$$u(t) = g(t) + \delta(t) + \int_{t_0}^{t} K(t,s;u(s))\,\mathrm{d}s, \tag{2.12a}$$

$$u(t) \equiv y(t) + \delta y(t) \tag{2.12b}$$

(where we assume that $\delta(t) \equiv \delta(u(\cdot);t)$). Then there exists a corresponding $R(t,s;u(s))$ such that

$$\delta y(t) = \delta(t) + \int_{t_0}^{t} R(t,s;u(s))\delta(s)\,\mathrm{d}s. \tag{2.13}$$

Let us obtain one such result, under assumptions that are obvious. Subtract from (2.12a) the unperturbed equation (2.1b) and we obtain

$$\delta y(t) = \delta(t) + \int_{t_0}^{t} \{K(t,s;u(s)) - K(t,s;y(s))\}\,\mathrm{d}s. \tag{2.14}$$

Hence, with $K_\natural(t,s;u(s)) = K_3(t,s;(1-w(s))y(s) + w(s)u(s))$ for some appropriate $w(s) \in [0,1]$, where $K_3(t,s;v) = (\partial/\partial v)K_3(t,s;v)$, we have

$$\delta y(t) = \delta(t) + \int_{t_0}^{t} K_\natural(t,s;u(s))\delta y(s)\,\mathrm{d}s, \tag{2.15}$$

which provides a result of the form (2.13) with $R(t,s;u(s))$ as the resolvent kernel satisfying $R(t,s;u(s)) = K_\natural(t,s;u(s)) + \int_{t_0}^{t} R(t,\sigma;u(\sigma))K_\natural(\sigma,s;u(s))\,\mathrm{d}\sigma$.

3. Some numerical methods

We review, selectively, numerical methods for Volterra integral and integro-differential equations, concentrating on the classical forms. Discrete methods for the solution of Volterra integral and integro-differential equations are based upon the development of a grid or mesh:

$$\mathscr{T} := \{t_0 < t_1 < t_2 < \cdots < t_{n-1} < t_n < \cdots\}, \quad h_n := t_{n+1} - t_n. \tag{3.1}$$

Remark 3.1. (a) The *width* (or *diameter*) of a grid $\mathscr{T}$ is $h(\mathscr{T}) := \sup\{h_n : t_n \in \mathscr{T}\}$. $\mathscr{T}$ is called *uniform* if $h_n \equiv h$ for all n, and *quasi-uniform* if there exists a finite κ such that $\sup h_n \leqslant \kappa \inf h_n$. If $t_N := \max_n t_n = T < \infty$ then $\mathscr{T}$ is a finite grid that provides a partition of the bounded interval $[t_0, T]$. Such a partition is called (*i*) *graded* with *grading exponent* α if $t_n - t_0 = (n/N)^\alpha$ [49, p.349 et seq.], or (*ii*) *geometric* [43] if $t_n - t_0 = \beta^{N-n}\{T - t_0\}$ for some $\beta \in (0,1)$. (b) Later, in connection with Runge–Kutta (RK) or collocation processes, we shall introduce a set of indexed abscissae $\{\vartheta_i\}_{i=1}^{m}$ and the points $t_{n,i} = t_n + \vartheta_i h_n$. In the case $0 \leqslant \vartheta_1 < \vartheta_2 < \cdots < \vartheta_m \leqslant 1$ these points define an

[2] Various similar formulae appear in the literature [11,24,94], sometimes under the heading of variation of constants formulae for integro-differential equations obtained by differentiating the integral equation and with restrictions on the form of perturbation. In this paper, we are trying to capture the spirit of the approach and we shall not need detailed information about $R(t,s;u(s))$ in (2.13), other than knowledge of its continuity properties, which are obvious in the linear case.

ordered set of abscissae $\mathscr{T}^{\#}(\vartheta):=\{t_0 \leqslant t_0+\vartheta_1 h_0 < t_0+\vartheta_2 h_0 < \cdots < t_0+\vartheta_m h_0 \leqslant t_1 \leqslant t_1+\vartheta_1 h_1 < t_1+\vartheta_2 h_1 < \cdots < t_1+\vartheta_m h_1 \leqslant t_2 < \cdots\}$. We shall usually need to note whether $\vartheta_1=0$, $\vartheta_m=1$.

We now consider some *primitive quadrature methods.* A simple family of methods for classical Volterra equations can be built from the quadrature rule

$$\int_{t_j}^{t_{j+1}} \phi(s)\,\mathrm{d}s \approx h_j\{(1-\theta)\phi(t_j)+\theta\phi(t_{j+1})\} \tag{3.2a}$$

(where $\theta \in [0,1]$), which yield, in particular, the Euler rule ($\theta=0$), the backward Euler rule ($\theta=1$), and the trapezium rule ($\theta=\frac{1}{2}$). From the primitive rule (3.2a) one obtains the basic components for use with Volterra integral and integro-differential equations. In particular, one obtains by repeated application of the basic rules the approximations

$$\int_{t_0}^{t_n} \phi(s)\,\mathrm{d}s \approx \sum_{j=0}^{n-1} h_j\{(1-\theta)\phi(t_j)+\theta\phi(t_{j+1})\}. \tag{3.2b}$$

If we proceed formally, we can discretize (2.1b) using (for $n=0,1,2,\ldots$)

$$\tilde{y}(t_{n+1})=g(t_{n+1})+\sum_{j=0}^{n} h_j\{(1-\theta)K(t_{n+1},t_j,\tilde{y}(t_j))+\theta K(t_{n+1},t_{j+1},\tilde{y}(t_{j+1}))\}. \tag{3.3}$$

We need to establish that Eqs. (3.3) do have a solution and that it provides a good approximation to $y(\cdot)$. However, the basic idea underlying a wide class of numerical methods is already present since Eqs. (3.3) have the form

$$\tilde{y}(t_{n+1})=g(t_{n+1})+\sum_{j=0}^{n+1} \Omega_{n+1,j}K(t_{n+1},t_j,\tilde{y}(t_j)), \tag{3.4}$$

which are *discrete Volterra equations* that arise from a family of quadrature rules using weights $\{\Omega_{n,j}\}_{j=0}^{n}$ ($n=1,2,3,\ldots$) and abscissae $\{t_j\}\subset\mathscr{T}$.

Example 3.2. For the *test equation* $y(t)-\lambda\int_{t_0}^{t} y(s)\,\mathrm{d}s=g(t)$, the θ-rule with uniform step $h_r=h$ yields a recurrence that simplifies to

$$\frac{\tilde{y}(t_{n+1})-\tilde{y}(t_n)}{h}-\lambda\{\theta\tilde{y}(t_{n+1})+(1-\theta)\tilde{y}(t_n)\}=\frac{g(t_{n+1})-g(t_n)}{h} \tag{3.5}$$

(which is solvable if $\lambda\theta h \neq 1$) that simulates the corresponding analytical result $y'(t)-\lambda y(t)=g'(t)$. Those familiar with the numerics of ODEs can, in this example, readily infer properties from (3.5).

The θ-rules have relatively low order, and one may turn to alternatives. The set of weights that arise on taking $h_0=h_1=h'$, $h_2=h_3=h''$, $h_4=h_5=h'''$,... and combining Simpson's rule (repeated as necessary) with the trapezium rule is indicated in the following tableau:

Tableau of weights $\Omega_{n,j}$

$j =$	0	1	2	3	4	5	6	...
$n = 1$	$\frac{h'}{2}$	$\frac{h'}{2}$						
$n = 2$	$\frac{h'}{3}$	$\frac{4h'}{3}$	$\frac{h'}{3}$					
$n = 3$	$\frac{h'}{3}$	$\frac{4h'}{3}$	$\frac{h'}{3}+\frac{h''}{2}$	$\frac{h''}{2}$				
$n = 4$	$\frac{h'}{3}$	$\frac{4h'}{3}$	$\frac{h'}{3}+\frac{h''}{3}$	$\frac{4h''}{3}$	$\frac{h''}{3}$			
$n = 5$	$\frac{h'}{3}$	$\frac{4h'}{3}$	$\frac{h'}{3}+\frac{h''}{3}$	$\frac{4h''}{3}$	$\frac{h''}{3}+\frac{h'''}{2}$	$\frac{h'''}{2}$		
		Simpson's rule		Simpson's rule	trapezium rule			
$n = 6$	$\frac{h'}{3}$	$\frac{4h'}{3}$	$\frac{h'}{3}+\frac{h''}{3}$	$\frac{4h''}{3}$	$\frac{h''}{3}+\frac{h'''}{3}$	$\frac{4h'''}{3}$	$\frac{h'''}{3}$	
$\vdots$		Simpson's rule ...		Simpson's rule ...		Simpson's rule ...		

(3.6)

Consider the rather special case of a *uniform mesh* ($h_r \equiv h$): the corresponding quadrature weights $\{\Omega_{n,j}\}$ for $j \leqslant n$ is said to have a *repetition factor* r if there is a least integer r such that, for $n \geqslant n_0$, $\Omega_{n+r,j} = \Omega_{n,j}$ for $r_0 \leqslant j \leqslant n - r_1$ where n_0, r_0, r_1 are fixed. Where there is no exact repetition, the weights are said to have *asymptotic repetition factor* r' provided there exists a least integer r' such that, where r_0, r_1 are fixed, $\lim_{n\to\infty} \sup_{r_0 \leqslant j \leqslant n-r_1} |\Omega_{n+r',j} - \Omega_{n,j}| = 0$. With $h' = h'' = h''' = \cdots = h$, the weights in (3.6) have repetition factor unity.

The discretization in (3.6) may be thought of as based on a repeated application of Simpson's rule to approximate the "history" and an application of the trapezium rule or another Simpson's rule for the "local" or "incremental" term. This viewpoint is lost if, instead, we generate the weights $\Omega_{n,j}$ for odd values of n by applying the trapezium rule to the first sub-interval (where needed) rather than the last. Retain a uniform step h and we now have weights

$n = 3$	$\frac{h}{2}$	$\frac{h}{2}+\frac{h}{3}$	$\frac{4h}{3}$	$\frac{h}{3}$		
$n = 5$	$\frac{h}{2}$	$\frac{h}{2}+\frac{h}{3}$	$\frac{4h}{3}$	$\frac{h}{3}+\frac{h}{3}$	$\frac{4h}{3}$	$\frac{h}{3}$,
	trapezium rule		Simpson's rule		Simpson's rule	

(3.7)

with a repetition factor 2. *Although one can establish convergence results of the same order*, for both cases, the stability properties of weights (3.7) have led to them being rejected (see [110] for more insight) as unsuitable for numerical computation in the solution of arbitrary Volterra integral equations of the second kind; the scheme suffers the same defect as Simpson's rule for solving ODEs (under certain assumptions, it is weakly unstable).

As the last remark indicates, certain 'plausible' weights (that yield convergence) may be rejected for practical computation in the numerical solution of Volterra equations of the second kind. For equations of the first kind (2.1a), the situation is more delicate: Whilst the formal analogue of (3.4) reads

$$\sum_{r=0}^{n+1} \Omega_{n+1,j} K(t_{n+1}, t_j, \tilde{y}(t_j)) = g(t_{n+1}) \quad (n = 0, 1, 2, \ldots) \tag{3.8}$$

(cf. Remark 2.3) one has to be more circumspect in the choice of quadrature rules to obtain a satisfactory (or even convergent) scheme for (2.1a).

Example 3.3. Consider, for the case $\int_{t_0}^{t} y(s)\,\mathrm{d}s = g(t)$ (where $g(t_0)=0$ and $g'(t)$ exists for $t \in [t_0, \infty)$), the application of a θ-rule with a uniform step $h_r \equiv h$:

$$(1-\theta)h\tilde{y}(t_0) + h\tilde{y}(t_1) + \cdots + h\tilde{y}(t_{n-1}) + \theta h\tilde{y}(t_n) = g(t_n) \quad (n = 1, 2, 3, \ldots).$$

For $\theta \in (0, 1]$, we deduce that (for $n = 0, 1, 2, \ldots$)

$$\tilde{y}(t_{n+1}) = \mu\tilde{y}(t_n) + \frac{g(t_{n+1}) - g(t_n)}{\theta h} \quad \text{where } \mu := \frac{\theta - 1}{\theta}, \tag{3.9}$$

which is an unstable recurrence if $|\mu| > 1$, namely if $\theta \in (0, \frac{1}{2})$. Unless $\theta = 1$ the value $\tilde{y}(t_0) \approx g'(t_0)$ in these equations has to be supplied independently. Clearly, if $\theta = 1$ (the backward or implicit Euler rule) then we deduce $\tilde{y}(t_{n+1}) = \{g(t_{n+1}) - g(t_n)\}/h$ which is a local approximation to $g'(t_{n+1})$ that clearly converges to the correct value as $h \to 0$. For $\theta \in (0, 1)$,

$$\tilde{y}(t_{n+1}) = \sum_{r=0}^{n} \mu^{n-r} \frac{g(t_{r+1}) - g(t_r)}{\theta h} + \mu^{n+1}\tilde{y}(t_0). \tag{3.10}$$

A difficulty with (3.10) is that it seeks to approximate $g'(t_{n+1})$ in terms of $g(t_0), g(t_1), g(t_2), \ldots, g(t_{n+1})$, and if $|\mu| > 1$ it gives increasing weight to $g(t_0)$ as n increases. In contrast, the value $g'(t_{n+1}) = \lim_{\delta \to 0}\{g(t_{n+1} + \delta) - g(t_{n+1})\}/\delta$ depends only upon limiting behaviour of $g(\cdot)$ in the neighbourhood of t_{n+1}. Given $g(t)$, one can (*without changing the derivative* $g'(t_{n+1})$) add an arbitrary multiple of any smooth function $\gamma(t)$ that has support (t_0, t_{n+1}), thereby *changing the computed approximation* (3.10) *by an arbitrary amount.* If $\theta = 0$, the coefficient matrix in our equations is formally singular, and to be useful the equations have to be re-interpreted:
$h\tilde{y}(t_1) = g(t_2) - h\tilde{y}(t_0)$; $h\tilde{y}(t_1) + h\tilde{y}(t_2) = g(t_3) - h\tilde{y}(t_0)$; $h\tilde{y}(t_1) + h\tilde{y}(t_2) + h\tilde{y}(t_3) = g(t_4) - h\tilde{y}(t_0), \ldots$ (where $\tilde{y}(t_0)$ is obtained independently).
We now find $\tilde{y}(t_n) = \{g(t_{n+1}) - g(t_n)\}/h$ which converges to $g'(t_\star)$ as $h \to 0$ and $n \to \infty$ with $t_n = t_\star \in (t_0, \infty)$ fixed.

One can generalize the composite Euler and trapezium rules in order to treat Abel equations; corresponding to (3.2b) one has, in particular, formulae

$$\int_{t_0}^{t_n} (t_n - s)^{-\nu}\phi(s)\,\mathrm{d}s \approx \sum_{j=0}^{n-1} \left\{ \int_{t_j}^{t_{j+1}} (t_n - s)^{-\nu}\,\mathrm{d}s \right\} \phi(t_j), \tag{3.11a}$$

$$\int_{t_0}^{t_n} (t_n - s)^{-\nu}\phi(s)\,\mathrm{d}s \approx \sum_{j=0}^{n-1} \left\{ \int_{t_j}^{t_{j+1}} (t_n - s)^{-\nu}\,\mathrm{d}s \right\} \phi(t_{j+1}), \tag{3.11b}$$

$$\int_{t_0}^{t_n} (t_n - s)^{-\nu} \phi(s)\, ds$$
$$\approx \sum_{j=0}^{n-1} \left\{ \int_{t_j}^{t_{j+1}} (t_n - s)^{-\nu} \frac{t_{j+1} - s}{h_j}\, ds\; \phi(t_j) + \int_{t_j}^{t_{j+1}} (t_n - s)^{-\nu} \frac{s - t_j}{h_j}\, ds\; \phi(t_{j+1}) \right\} \tag{3.11c}$$

and the integrals that occur here as weights multiplying values $\phi(t_j)$ and $\phi(t_{j+1})$ can be written down very simply, given $\mathscr{T}$. Bounds on the error in such approximations are readily obtained in terms of the error in piecewise-constant or piecewise-linear interpolation to $\phi(\cdot)$. With a uniform grid $\mathscr{T}$, the above approximations yield (on summation) convolution sums approximating the convolution integrals $\int_{t_0}^{t_n} (t_n - s)^{-\nu} \phi(s)\, ds$.

Remark 3.4. When solving Volterra integro-differential equations we can combine methods for ODEs and integral equations. Thus, if $\theta', \theta'' \in [0,1]$ we might write, discretizing (2.5a), (2.5b),

$$\tilde{y}(t_{n+1}) = \tilde{y}(t_n) + h_n\{(1-\theta')F(t_n, \tilde{y}(t_n), \tilde{z}(t_n)) + \theta' F(t_{n+1}, \tilde{y}(t_{n+1}), \tilde{z}(t_{n+1}))\},$$
$$\tilde{z}(t_k) := \sum_{r=0}^{k-1} h_r\{(1-\theta'')K(t_k, t_r, \tilde{y}(t_r)) + \theta'' K(t_k, t_{r+1}, \tilde{y}(t_{r+1}))\}, \quad k \in \{1,2,3,\ldots,(n+1)\}. \tag{3.12}$$

(We intend to indicate some flexibility of approach which might not be transparent if one viewed the integro-differential equation as a system of two integral equations: one may reasonably ask whether there is a purpose to selecting $\theta' \neq \theta''$.) If the integral term is, instead, of Abel type, there is a rôle for approximations (3.11) in the term for $\tilde{z}(t_r)$.

4. Relationship to ODE methods

The integrated form of the initial-value problem for the equation $y'(t) = f(t, y(t))$ (for $t \geqslant t_0$) is (2.4), and it is not surprising to find that (although Volterra equations have more complex character) there is a close connection between methods for the ODE and for classical Volterra integral equations of the second kind. We enumerate some features:

(1) For certain types of kernel $K(t,s,v)$ (in particular, $\sum_{j=1}^{N} T_j(t)S_j(s,v)$, or $(t-s)^n \exp\{-v(t-s)\}v$) we find that (2.1b) and (2.1c), (2.1d) can be reduced to a system of ODEs (see [20] for an exploitation of such a result);

(2) An *embedding* relationship has a rôle in the theory of numerical methods. Consider $G(t,s)$ such that (one may compare with Eqs. (8.7)):

$$\frac{\partial}{\partial t} G(t,s) = K(t,s,G(t,t)) \quad (t_0 \leqslant s \leqslant t), \tag{4.1a}$$

$$G(t_0, s) = g(s) \quad (s \in [t_0, T]). \tag{4.1b}$$

Then the solution of (2.1b) satisfies $y(t) = G(t,t)$. We refer to [114,123].

(3) In principle, every ODE method (multistep, cyclic multistep, RK method, hybrid method, general linear method) generates a corresponding integral equation method for (2.1b), or (2.1d). By (2.4), every integral equation method for (2.1b) generates a (possibly novel) ODE method.

Remark 4.1. Jones and McKee [88] used a family of predictor–corrector methods of *variable order* and *variable step-size* to devise a method for nonlinear second kind Volterra integral equations with smooth kernels. Their strategy for changing the step size followed closely one employed for ODEs. (b) Bownds and Applebaum [21] have contributed a code based on separable kernels. (c) Shampine [116] addressed problems associated with the special nature of the ODE approach to Volterra equations.

Variable step ODE methods provide approximations to integrals over intervals $[t_0, t_n]$ in the form of sums, similar to (3.2b) (which originates with the θ-method). In particular, with a *uniform mesh*, the linear multistep method with first and second characteristic polynomials $\rho(\zeta){:=}\alpha_0\zeta^k + \alpha_1\zeta^{k-1} + \cdots + \alpha_k$, $\sigma(\zeta){:=}\beta_0\zeta^k + \beta_1\zeta^{k-1} + \cdots + \beta_k$ $(\alpha_0 \neq 0)$ generates, with appropriate starting values, an approximation (termed a $\{\rho, \sigma\}$-reducible quadrature) of the form

$$\int_{t_0}^{t_n} \phi(s)\,\mathrm{d}s \approx \underbrace{h\sum_{j=0}^{n_0} w_{n,j}\phi(t_j)}_{\text{starting terms}} + \underbrace{h\sum_{j=n_0+1}^{n} \omega_{n-j}\phi(t_j)}_{\text{convolution sum}} \; (t_n \equiv t_0 + nh) \tag{4.2}$$

for $n \geqslant n_0$. An example arises if we take a uniform mesh in (3.2b). Formally,

$$\sum_{j=0}^{\infty} \omega_j \zeta^j = \frac{\zeta^k \sigma(\zeta^{-1})}{\zeta^k \rho(\zeta^{-1})} = \frac{\beta_0 + \beta_1\zeta + \cdots + \beta_k\zeta^k}{\alpha_0 + \alpha_1\zeta + \cdots + \alpha_k\zeta^k} \quad (\alpha_0 \neq 0) \tag{4.3}$$

is the *generating function* (related to the *Z-transform*) for the sequence of weights $\{\omega_\ell\}_{\ell=0}^{\infty}$. To ensure uniform bounds on the weights we require $\rho(\cdot)$ to be simple von Neumann (i.e., $\rho(\cdot)$ must have all its zeros on the closed unit disk centred on the origin, any on its boundary being simple). Adapting (4.2) to the discretization of nonlinear Volterra convolution integrals we obtain

$$\underbrace{\int_{t_0}^{t_n} k(nh - s)\varphi(y(s))\,\mathrm{d}s}_{\text{continuous convolution}} \approx h\sum_{j=0}^{n_0} \mathfrak{s}_{n,j}\varphi(\tilde{y}(jh)) + \underbrace{h\sum_{j=n_0+1}^{n} \mathfrak{w}_{n-j}\varphi(\tilde{y}(jh))}_{\text{discrete convolution}}, \tag{4.4}$$

where $\mathfrak{s}_{n,j} := k((n-j)h)w_{n,j}$, $\mathfrak{w}_{n-j} := k((n-j)h)\omega_{n-j}$. The *discrete convolution property* in (4.4) facilitates the use of FFT techniques (cf. [3,4,81,97,101]) in solving Volterra convolution integral and integro-differential equations. Gregory rules of a fixed order correspond to Adams-Moulton methods. The BDF formulae rules generate rules (4.2) with asymptotic repetition factor unity.

Remark 4.2. Suitable linear multistep formulae generate fractional integration rules for use with a uniform mesh in discretizing Abel equations:

$$\frac{1}{\Gamma(1-\nu)}\int_{t_0}^{t_n} (t_n - s)^{-\nu}\phi(s)\,\mathrm{d}s \approx h^{1-\nu}\sum_{j=0}^{n_0} w_{n,j}^{\{\nu\}}\phi(t_j) + h^{1-\nu}\sum_{j=n_0+1}^{n} \omega_{n-j}^{\{\nu\}}\phi(t_j) \; (t_n \equiv t_0 + nh). \tag{4.5}$$

We refer to [3,4,81,97,101]. It is noteworthy that the starting weights can be chosen so that (4.5) is exact for selected nonpolynomial terms (e.g., $\phi(s) = (s - t_0)^{\mu_i}$, with $\mu_i \in \mathbb{R}$ and where $y(t) \sim a_0 + a_1(t-t_0)^{\mu_1} + a_2(t-t_0)^{\mu_2} + a_3(t-t_0)^{\mu_3} + \cdots$, as $t \to t_0$, with $\mu_1 < \mu_2 < \mu_3 < \cdots < \mu_m$).

5. RK processes

We now consider *RK methods* with nonuniform meshes for Volterra equations of the second kind. At the risk of offending the purist, we develop these from a heuristic viewpoint. The early work was due to Pouzet [114] and exploited RK parameters arising for ODEs; Bel'tyukov [14] introduced schemes with additional parameters. For the rigorous foundations of RK-type methods in terms of order conditions and from a modern perspective, see [12,42]; for extensions to Abel equations see [100]; for integro-differential equations see [99].

The literature contains examples of *RK formulae for ODEs*, generated by the *RK triple* $(\boldsymbol{\vartheta}, \boldsymbol{A}, \boldsymbol{b})$ where $\boldsymbol{\vartheta} = [\vartheta_1, \vartheta_2, \ldots, \vartheta_m]^{\mathrm{T}} \in \mathbb{R}^m$, $\boldsymbol{A} = [a_{i,j}] \in \mathbb{R}^{m \times m}$, $\boldsymbol{b} = [b_1, b_2, \ldots, b_m]^{\mathrm{T}}$, commonly represented as an *RK* (or *Butcher*) *tableau* $\begin{array}{c|c} \boldsymbol{\vartheta} & \boldsymbol{A} \\ \hline & \boldsymbol{b}^{\mathrm{T}} \end{array}$. Examples of RK tableaux are

$$\begin{array}{c|ccc} 0 & 0 & & \\ \frac{1}{2} & \frac{1}{4} & \frac{1}{4} & \\ 1 & \frac{1}{6} & \frac{2}{3} & \frac{1}{6} \\ \hline (1) & \frac{1}{6} & \frac{2}{3} & \frac{1}{6} \end{array}, \quad \begin{array}{c|cc} \frac{1}{2} - \frac{\sqrt{3}}{6} & \frac{1}{4} & \frac{1}{4} - \frac{\sqrt{3}}{6} \\ \frac{1}{2} + \frac{\sqrt{3}}{6} & \frac{1}{4} + \frac{\sqrt{3}}{6} & \frac{1}{4} \\ \hline (1) & \frac{1}{2} & \frac{1}{2} \end{array}, \quad \begin{array}{c|cc} \frac{1}{3} & \frac{5}{12} & -\frac{1}{12} \\ 1 & \frac{3}{4} & \frac{1}{4} \\ \hline (1) & \frac{3}{4} & \frac{1}{4} \end{array}, \quad \begin{array}{c|ccc} 0 & 0 & & \\ \frac{1}{2} & \frac{5}{24} & \frac{1}{3} & -\frac{1}{24} \\ 1 & \frac{1}{6} & \frac{2}{3} & \frac{1}{6} \\ \hline (1) & \frac{1}{6} & \frac{2}{3} & \frac{1}{6} \end{array} \tag{5.1}$$

(a) Trapezium & Simpson; (b) Gauss family (collocation RK); (c) Radau IIA family (collocation RK); (d) Lobatto IIA family (collocation RK)

(the parenthetical "(1)" denoting $\vartheta_{m+1} = 1$, that has been inserted here, is usually omitted).

Remark 5.1. (i) The choice of $\boldsymbol{\vartheta}$ can be indicated by a subscript, as in $\boldsymbol{\vartheta}_{[\mathrm{NC}_m]}$, $\boldsymbol{\vartheta}_{(\mathrm{Gauss}_m)}$, $\boldsymbol{\vartheta}_{(\mathrm{Radau}_m]}$, $\boldsymbol{\vartheta}_{[\mathrm{Lobatto}_m]}$ (the m quadrature points for closed Newton–Cotes formulae, open Gauss–Legendre formulae, Radau right-hand formulae, and closed Lobatto formulae, such as appear in (5.1)(a)–(d), respectively) to denote the abscissae. Note that (a) and (d) in (5.1) have the same $\boldsymbol{\vartheta}$. (ii) The RK parameters are *explicit* (or, more accurately, "*formally explicit*", since some reordering is possible) if $a_{i,j}=0$ for $j \geqslant i$, and *semi-implicit* if $a_{i,i} \neq 0$ for some i and $a_{i,j}=0$ for $j > i$. Other features to note are whether $\vartheta_i \in [0,1]$ for all i; whether $\{\vartheta_i\}_1^m$ are distinct; whether $0 \leqslant \vartheta_1 < \vartheta_2 < \cdots < \vartheta_m < \vartheta_m \leqslant 1$; whether $\boldsymbol{b}(\vartheta_i) = [a_{i,1}, a_{i,2}, \ldots, a_{i,m}]$; whether, if $\vartheta_m = 1$, $\boldsymbol{b} = [a_{m,1}, a_{m,2}, \ldots, a_{m,m}]$. (iii) The 'stability' and 'convergence' properties of the RK method when applied to an ODE have a bearing on our methods.

The abscissae ϑ_i and the points $t_n \in \mathcal{T}$ together define

$$\mathcal{T}^{\#}(\boldsymbol{\vartheta}) := \{t_{n,i} = t_n + \vartheta_i h_n\}, \tag{5.2}$$

the set of points anticipated in Remark 3.1. We shall wish to discretize the integrals $\int_{t_0}^{t_{n,i}} K(t_{n,i}, s, y(s))\,\mathrm{d}s$, and more generally $\int_{t_0}^{t_n+\sigma h_n} K(t_n + \sigma h_n, s, y(s))\,\mathrm{d}s$, in (2.1). Now any RK triple defines a family of quadrature rules

$$\int_0^{\vartheta_i} \phi(s)\,\mathrm{d}s \approx \sum_{j=1}^{m} a_{i,j} \phi(\vartheta_j). \tag{5.3a}$$

This family, if we set $\vartheta_{m+1} = 1, a_{m+1,j} = b_j$, includes

$$\int_0^1 \phi(s)\,\mathrm{d}s \approx \sum_{j=1}^{m} b_j \phi(\vartheta_j). \tag{5.3b}$$

In addition, it is possible to provide a vector $\boldsymbol{b}(\sigma) = [b_1(\sigma), b_2(\sigma), \ldots, b_m(\sigma)]^{\mathrm{T}}$, defined for $\sigma \in [0,1]$, that yields quadrature

$$\int_0^{\sigma} \phi(s)\,\mathrm{d}s \approx \sum_{j=1}^{m} b_j(\sigma)\phi(\vartheta_j) \quad \text{for all } \sigma \in [0,1], \tag{5.3c}$$

where $\boldsymbol{b}(1) \equiv \boldsymbol{b}$. The parameters $(\boldsymbol{\vartheta}, \boldsymbol{A}, \boldsymbol{b}(\sigma))$ define a *continuous RK process.* For those tableaux labelled "collocation-RK" in (5.1) (which are linked to piecewise-polynomial collocation processes) the polynomials $b_j(\cdot)$ can be obtained by indefinite integration of the Lagrangean form of the polynomial interpolating $\phi(\cdot)$ at $\vartheta_1, \ldots, \vartheta_m$ and we find that $a_{i,j} = b_j(\theta_i)$ for $i, j = 1, 2, \ldots, m$.

Remark 5.2. By an affine transformation, Eq. (5.3b) gives, as an alternative to (5.3a),

$$\int_0^{\vartheta_i} \phi(s)\,\mathrm{d}s \approx \sum_{j=1}^{m} \vartheta_i b_j \phi(\vartheta_i \vartheta_j) \tag{5.4}$$

that is useful when $\phi(\cdot)$ is smooth on $[0, \vartheta_i]$ but not on all of $[0,1]$.

We observe that

$$\int_{t_0}^{t_{n,i}} \phi(s)\,\mathrm{d}s = \sum_{j=0}^{n-1} \int_{t_j}^{t_{j+1}} \phi(s)\,\mathrm{d}s + \int_{t_n}^{t_{n,i}} \phi(s)\,\mathrm{d}s \tag{5.5a}$$

for continuous $\phi(\cdot)$, and the extended use of RK parameters yields

$$\int_{t_0}^{t_{n,i}} \phi(s)\,\mathrm{d}s \approx \sum_{j=0}^{n-1} h_j \sum_{k=1}^{m} b_k \phi(t_{j,k}) + h_n \sum_{k=1}^{m} a_{i,k} \phi(t_{n,k}) \tag{5.5b}$$

and, for $\sigma \in [0,1]$,

$$\int_{t_0}^{t_n + \sigma h_n} \phi(s)\,\mathrm{d}s \approx \sum_{j=0}^{n-1} h_j \sum_{k=1}^{m} b_k \phi(t_{j,k}) + h_n \sum_{k=1}^{m} b_k(\sigma) \phi(t_{n,k}). \tag{5.5c}$$

If $a_{i,j} \neq b_j(\vartheta_i)$, (5.5c) does *not* reduce to (5.5b) on setting σ to ϑ_i. Formally, the RK integration formulae (5.5b) allow one to discretize a Volterra integral or integro-differential equation with an "*extended RK method*". Thus, we may seek approximations $\tilde{y}_{n,i} \approx y(t_{n,i})$ to the solution $y(\cdot)$ of (2.1b), on $\mathscr{T}^{\#}(\boldsymbol{\vartheta})$, via the equations

$$\tilde{y}_{n,i} = g(t_{n,i}) + \underbrace{\sum_{j=0}^{n-1} h_j \sum_{k=1}^{m} b_k K(t_{n,i}, t_{j,k}, \tilde{y}_{j,k})}_{\text{history term ('lag' or 'tail' term)}} + \underbrace{h_n \sum_{k=1}^{m} a_{i,j} K(t_{n,i}, t_{n,k}, \tilde{y}_{n,k})}_{\text{'incremental term'}} \tag{5.6a}$$

($i = 1, 2, \ldots, m, m+1$) in which the equation with $i = m+1$ provides

$$\tilde{y}(t_{n+1}) = g(t_{n+1}) + \sum_{j=0}^{n} h_j \sum_{k=1}^{m} b_k K(t_{n+1}, t_{j,k}, \tilde{y}_{j,k}). \tag{5.6b}$$

Example 5.3. Compare the use of the left-most RK tableau in (5.1) with quadrature based on the weights in (3.6), taking steps $h_1 = 2h'$, $h_2 = 2h''$, $h_3 = 2h'''$, ...

The formulae $\tilde{y}_{n,i} \approx y(t_{n,i})$ have to be interpreted appropriately if one has an RK triple with $\vartheta_i = \vartheta_k$ for $i \neq k$ but (5.6) are essentially the *extended RK formulae* obtained in [114] using an RK triple ($\boldsymbol{\vartheta}$, $\boldsymbol{A}$, $\boldsymbol{b}$). Where the formulae involve $K(t_{n,i}, t_{n,j}, \tilde{y}_{n,j})$ with $t_{n,j} > t_{n,i}$, we are required to define a *smooth* extension (2.2) of $K(t,s,v)$ for values of $s > t$. Though this is not transparent from the above derivation, the order of convergence on $\mathcal{T}$ (defined in Section 7) is inherited from the ODE RK process *provided* $K(t,s,y(s))$ (or $K_{\mathrm{ext}}(t,s,y(s))$ in (2.2), if it is required) is sufficiently smooth. Unless the RK formula is explicit (Remark 5.1), we have to solve for a block of unknown values $\{\tilde{y}_{n,i} : i = 1, 2, \ldots, m\}$, simultaneously. There is a unique solution of the equations given an appropriate Lipschitz condition (Remark 2.1) and a sufficiently small width of $\mathcal{T}$. An appropriate choice of nonlinear equation solver depends on various factors including the step size h_n; some questions remain open.

We can modify the above approach slightly and obtain an approximation $\tilde{y}(\cdot)$ defined on $[t_0, T]$, if we use continuous RK parameters ($\boldsymbol{\theta}$, $\boldsymbol{A}, \boldsymbol{b}(\sigma)$) and write

$$\tilde{y}(t_n + \sigma h_n) = g(t_n + \sigma h_n) + \sum_{j=0}^{n-1} h_j \sum_{k=1}^{m} b_k K(t_n + \sigma h_n, t_{j,k}, \tilde{y}_{j,k})$$
$$+ h_n \sum_{k=1}^{m} b_k(\sigma) K(t_n + \sigma h_n, t_{n,k}, \tilde{y}_{n,k}), \tag{5.7}$$

which, having obtained $\{\tilde{y}_{n,i}\}$, yields a densely defined function $\tilde{y}(\cdot)$.

Evaluating the term $\sum_{j=0}^{n-1} h_j \sum_{k=1}^{m} b_k K(t_n + \sigma h_n, t_{j,k} \tilde{y}_{j,k})$ in (5.6) or (5.7) involves a considerable computational expense, which prompts a modification (a *mixed quadrature-RK method*) based upon the use of quadrature formulae of the type arising in (3.4). The discretized equations now read

$$\tilde{y}_{n,i} = \underbrace{g(t_{n,i}) + \sum_{j=0}^{n} \Omega_{n,j} K(t_{n,i}, t_j, \tilde{y}(t_j))}_{\text{quadrature history term}} + \underbrace{h_n \sum_{k=1}^{m} a_{i,k} K(t_{n,i}, t_{n,k}, \tilde{y}_{n,k})}_{\text{'incremental term'}}, \tag{5.8a}$$

from which we obtain

$$\tilde{y}(t_{n+1}) = g(t_{n+1}) + \sum_{j=0}^{n} \Omega_{n,j} K(t_{n+1}, t_j, \tilde{y}(t_j)) + h_n \sum_{k=1}^{m} b_k K(t_{n+1}, t_{n,k}, \tilde{y}_{n,k}) \tag{5.8b}$$

with a corresponding densely defined extension

$$\tilde{y}(t_{n+\sigma}) = g(t_{n+\sigma}) + \sum_{j=0}^{n} \Omega_{n,j} K(t_{n+\sigma}, t_j, \tilde{y}(t_j)) + h_n \sum_{k=1}^{m} b_k(\sigma) K(t_{n+\sigma}, t_{n,k}, \tilde{y}_{n,k})$$

for $\sigma \in (0, 1]$. The incremental terms together define a one-step method with an increment function similar to that for an ODE. It will be clear that the history term and the increment function both contribute to the overall discretization error on $\mathscr{T}$. For variations on these methods, see [86]. Wolkenfelt [121] also discusses, *inter alia*, multilag variations of the mixed RK methods.

Remark 5.4. (a) Note that if one has constructed a dense-output approximation $\tilde{y}(t)$ for $t \in [t_0, t_n]$ then it is possible to compute the history terms $g(t)+\int_{t_0}^{t_n} K(t,s,\tilde{y}(s))\,ds$ by adaptive quadrature. (b) In terms of controlling accuracy, a major problem to be faced is that the accuracy in approximating $y(\cdot)$ by $\tilde{y}(\cdot)$ 'in the past' may prove inadequate to allow a good approximation to $g(t)+\int_{t_0}^{t} K(t,s,y(s))\,ds$ 'in the future'. (c) The structure of the extended and mixed (Pouzet) RK formulae discussed above makes them suitable for the application of FFT techniques in the case of convolution integral or integro-differential equations and a uniform mesh. (d) One may use embedded RK formulae and quadrature correction terms to try to control the discretization error. (e) The RK discretizations discussed here do not apply to Abel type equations, see [80] for a discussion of RK-type parameters for Abel equations. (f) We restricted attention to the classical (Pouzet) RK parameters. Some mathematical ingenuity has been invested into the derivation of Bel'tyukov RK-type formulae (these are generalizations [14,49, p.175] of the RK triples that were adapted by Pouzet from the ODE context).

For Volterra integro-differential equations, the extended use of $(\boldsymbol{\vartheta}, \boldsymbol{A}, \boldsymbol{b})$ for (2.5a),(2.5b), gives an approximating set of equations

$$\tilde{y}(t_{n,i}) = \tilde{y}(t_n) + h_n \sum_{j=1}^{m} a_{i,j} F(t_{n,j}, \tilde{y}(t_{n,j}),\ \tilde{z}(t_{n,j})), \tag{5.9a}$$

$$\tilde{z}(t_{n,r}) = \sum_{j=0}^{n-1} h_j \sum_{k=1}^{m} b_k K(t_{n,r}, t_{j,k}, \tilde{y}(t_{j,k})) + h_n \sum_{k=1}^{m} a_{i,j} K(t_{n,r}, t_{n,k}, \tilde{y}(t_{n,k})), \quad i,r \in \{1,2,\ldots,m\} \tag{5.9b}$$

with $\tilde{y}(t_{n+1}) = \tilde{y}(t_{n,m+1})$ there is a related mixed quadrature-RK formula.

6. Collocation and related methods

Collocation and Galerkin methods in the literature are, frequently though not exclusively, based upon polynomial spline or piecewise-polynomial densely defined approximations $\tilde{y}(t)$. In electronic databases for tracing the literature, perhaps a quarter of the citations on numerical methods for Volterra equations refer to collocation methods. This statistic may, however, be misleading, because the collocation methods described are often discretized to produce block-by-block methods that are intimately related to RK methods.

Given $\mathscr{T}$ in (3.1) we assume that the restriction of $\tilde{y}(\cdot)$ on $[t_0, t_1]$ and subsequently, for $n=1,2,\ldots$, each subinterval $(t_n, t_{n+1}]$ is a polynomial of degree q_n $(n = 0, 1, 2, \ldots)$. For simplicity we suppose $q_n = q$ for all n. Clearly, the restriction of $\tilde{y}(\cdot)$ on $(t_n, t_{n+1}]$ has associated limits $\tilde{y}(t_n+)$ and derivatives $\tilde{y}'(t_n+)$, $\tilde{y}''(t_n+), \ldots$, at each point t_n; the continuity class of $\tilde{y}(\cdot)$ for $t \geqslant t_0$ is determined by the requirement that left- and right-hand limits agree. To simplify, it is assumed that the order of the derivatives that exist (and are continuous) at t_n is independent of n, say $\kappa_n = \kappa$ ($\kappa = -1$ implies lack of continuity at points t_n), and

$$\tilde{y}(t_n+) = \tilde{y}(t_n-), \quad \tilde{y}'(t_n+) = \tilde{y}'(t_n-), \ldots, \tilde{y}^{(\kappa)}(t_n+) = \tilde{y}^{(\kappa)}(t_n-). \tag{6.1}$$

The approximation[3] is then termed a *polynomial spline of degree m* and of *continuity class* κ with *knots* $\mathcal{T}_N \equiv \{t_n\}_1^N \subset \mathcal{T}$. The linear space of such functions, denoted $S_m^\kappa(\mathcal{T}_N)$ has dimension $d = N(m-\kappa) + \kappa + 1$.

Any approximation to the solution of (2.1b), $\hat{y}(t) \approx y(t)$, gives rise to a corresponding *defect* $\delta(\hat{y}(\cdot); t)$ defined as

$$\delta(\hat{y}(\cdot); t) := \hat{y}(t) - \{g(t) + \int_{t_0}^{t} K(t, s, \hat{y}(s))\, ds\} \quad (t \in [t_0, T)). \tag{6.2}$$

Clearly, if Eq. (2.1b) has a unique solution, then $\hat{y}(\cdot) = y(\cdot)$ if and only if $\delta(\hat{y}(\cdot); t) \equiv 0$, and this suggests that one should seek an approximation that gives rise to a small defect. Analogous comments apply to (2.1d) and, provided that one interprets "smallness" of the defect in an appropriate sense, to some first-kind equations (2.1a). In collocation methods one seeks approximations of a particular type such that the defect vanishes at a set of points known as the *collocation points*. For Galerkin methods, one instead asks that the *moments of the defect*, taken with a prescribed set of functions, should vanish. "Refinement" (of the collocation or the Galerkin approximation) by iteration based on (2.10) can be useful.

We now return to the determination, through collocation-type techniques, of an approximation $\tilde{y}(\cdot) \in S_m^\kappa(\mathcal{T}_N)$. An arbitrary element of $S_m^\kappa(\mathcal{T}_N)$ is determined uniquely by d parameters where d is the dimension of the space $S_m^\kappa(\mathcal{T}_N)$. Given $\mathcal{T}^{\#}(\boldsymbol{\vartheta})$, we may therefore expect to be able to chose $\tilde{y}(\cdot)$ in order to satisfy the $\kappa+1$ continuity conditions (6.1) at each point t_n ($n = 1, 2, \ldots, N-1$) and an additional Nm collocation conditions of the type

$$\delta(\tilde{y}(\cdot); t_{n,r}) = 0, \qquad n = 0, 1, \ldots, N-1; \quad r \in \{1, 2, \ldots, m\}, \tag{6.3}$$

in which we ensure the defect vanishes at selected collocation points $t_{n,r} = t_n + \vartheta_r h_n$, here assumed to be distinct. From (6.2) we have, for $t \in [t_n, t_{n+1}]$,

$$\delta(\tilde{y}(\cdot); t) = \tilde{y}(t) - \left\{ g(t) + \sum_{j=0}^{n-1} \int_{t_j}^{t_{j+1}} K(t, s, \tilde{y}(s))\, ds + \int_{t_n}^{t} K(t, s, \tilde{y}(s))\, ds \right\}, \tag{6.4}$$

where, in each interval $(t_j, t_{j+1}]$, $\tilde{y}(\cdot)$ is a polynomial (of degree q, say).

We consider polynomial spline collocation with $\tilde{y}(\cdot) \in S_{m-1}^{-1}(\mathcal{T}_N)$, following the technique of Blom and Brunner [19]. Given $0 \leqslant \vartheta_1 < \vartheta_2 < \cdots < \vartheta_m \leqslant 1$, and the grid $\mathcal{T}$, $\tilde{y}(\cdot)$ is constructed to be a polynomial of degree $q = m-1$ in every interval $[t_n, t_{n+1}]$. Then $\tilde{y}(\cdot)$ is such that (for $\sigma \in (0,1)$)

$$\tilde{y}(t_n + \sigma h_n) = \sum_{i=1}^{m} \prod_{\substack{\ell \neq i \\ \ell = 1}}^{m} \frac{(\sigma - \vartheta_\ell)}{(\vartheta_i - \vartheta_\ell)} \tilde{y}(t_{n,i}) \quad \text{where } t_{n,i} = t_n + \vartheta_i h_n. \tag{6.5}$$

In general, $\tilde{y}(\cdot) \notin C[t_0, T]$ except when $\vartheta_1 = 0$ and $\vartheta_m = 1$.

---

[3] We have piecewise constant approximations if $m=1, \kappa=-1$; piecewise linear continuous approximations if $m=1, \kappa=0$; classical cubic splines if $m=3, \kappa=1$.

Taking the set $\{t_{n,r}\}$ as collocation points we have the *exact collocation equations* of the form

$$\tilde{y}(t_{n,r}) = g(t_{n,r}) + \sum_{j=0}^{n-1} h_j \int_0^1 K\left(t_{n,r}, t_j + \sigma h_j, \sum_i \prod_{\ell \neq i} \frac{(\sigma - \vartheta_\ell)}{(\vartheta_i - \vartheta_\ell)} \tilde{y}(t_{j,i})\right) \mathrm{d}\sigma$$

$$+ h_n \int_0^{\vartheta_r} K\left(t_{n,r}, t_n + \sigma h_n, \sum_i \prod_{\ell \neq i} \frac{(\sigma - \vartheta_\ell)}{(\vartheta_i - \vartheta_\ell)} \tilde{y}(t_{n,i})\right) \mathrm{d}\sigma. \tag{6.6}$$

For $n = 0, 1, 2, \ldots$, in turn, we fix n and run through the values of r in (6.6), to obtain sufficient equations to determine (if Hypothesis 1 holds and h_n is sufficiently small) the values $\{\tilde{y}(t_{n,r})\}$.

An unfortunate obstacle to implementation of the exact collocation scheme is the need to compute the integrals occurring in (6.6). We may therefore consider *discretized collocation equations*, in which the integrals are replaced by interpolatory quadrature and the formulae become more tractable. The presentation in [19] is based on the use of M-point quadrature, in the form

$$\int_0^1 \phi(s)\,\mathrm{d}s \approx \sum_{k=1}^{M} b_k \phi(\vartheta_k) \quad \text{and} \quad \int_0^{\vartheta_r} \phi(s)\,\mathrm{d}s \approx \sum_{k=1}^{M} \vartheta_r b_k \phi(\vartheta_r \vartheta_k) \tag{6.7a}$$

with $M = m - 1$ (if $\vartheta_m < 1$) or with $M = m$ (if $\vartheta_m = 1$), where

$$b_k = \int_0^1 \prod_{\substack{\ell \neq k \\ \ell = 1}}^{M} \frac{(s - \vartheta_\ell)}{(\vartheta_k - \vartheta_\ell)} \mathrm{d}s \quad (k = 1, 2, \ldots, M) \tag{6.7b}$$

(cf. Remark 5.2). Thus our discretized equations read

$$\tilde{y}(t_{n,r}) = g(t_{n,r}) + \sum_{j=0}^{n-1} h_j \sum_{k=1}^{M} b_k K(t_{n,r}, t_{j,k}, \tilde{y}(t_{j,k}))$$

$$+ h_n \vartheta_r \sum_{k=1}^{M} b_k K\left(t_{n,r}, t_n + \vartheta_k \vartheta_r h_n, \sum_{i=1}^{m} \prod_{\ell \neq i} \frac{(\vartheta_k \vartheta_r - \vartheta_\ell)}{(\vartheta_i - \vartheta_\ell)} \tilde{y}(t_{n,i})\right), \tag{6.8}$$

$$\tilde{y}(t_n + s h_n) = \sum_{i=1}^{m} \prod_{\substack{\ell \neq i \\ \ell = 1}}^{m} \frac{(s - \vartheta_\ell)}{(\vartheta_i - \vartheta_\ell)} \tilde{y}(t_{n,i}). \tag{6.9}$$

One may further refine the approximations $\tilde{y}(\cdot)$ by computing a discretized *iterated collocation approximation*, which can possess improved continuity and convergence properties. The abscissae $\boldsymbol{\vartheta}_{(\mathrm{Gauss}_m)}$, $\boldsymbol{\vartheta}_{(\mathrm{Radau}_m]}$, $\boldsymbol{\vartheta}_{[\mathrm{Lobatto}_m]}$ referred to earlier are candidates for use. Defining a new set of abscissae by supplementation of $\boldsymbol{\vartheta}_{(\mathrm{Gauss}_m)}$ with the points $\{0, 1\}$, or of $\boldsymbol{\vartheta}_{(\mathrm{Radau}_m]}$ with $\{0\}$, has also been suggested. The formulae above may be regarded as a generalization of the RK methods discussed previously.

Remark 6.1. (a) Hermann Brunner has authored or co-authored numerous papers in the extensive literature on collocation, including the recent report [43]. (b) For collocation methods for a wide class of integro-differential equations, see, for example, [36]. (c) Brunner [27] investigated the convergence rate of the numerical solution, by polynomial spline collocation, of the Abel integral equation $y(t)=g(t)+\int_0^t(t-s)^{-\nu}\mathsf{h}(t,s)y(s)\,ds$, $t\in I:=[0,T]$, $T<\infty$, $0<\nu<1$, assuming g and h to be smooth. Brunner showed that, independent of the degree of the polynomials, the convergence rate with respect to a *quasi-uniform mesh* is only $N^{\nu-1}$, where N denotes the number of subintervals on I. The optimal rate of convergence N^{-m}, where $m-1$ is the degree of the polynomials, is attainable using the special *graded mesh* $t_n=(n/N)^rT$, $r=m/(1-\nu)$, $n=0,1,\ldots,N$ (and similar results are valid for the approximate solution of the nonlinear equation $y(t)=g(t)+\int_0^t(t-s)^{-\nu}H(t,s,y(s))\,ds$). (d) For a recent paper on collocation for Abel equations (which opens with a succinct review of previous results and of some practical difficulties), see [87]. (e) Brunner [26] introduced nonpolynomial spline functions in a discretized collocation method to obtain high-order approximations to the solutions of integral and integro-differential equations having Abel-type kernels. (f) For a paper that closes a number of long-standing gaps in our understanding of collocation for equations of the first kind, see [92].

7. Numerical analysis

Mathematical numerical analysis encompasses, *inter alia*, a study of convergence, order of convergence, and superconvergence of approximations, and their stability properties. There is a considerable literature in this area. A rigorous study must commence with the issue of existence and uniqueness of the true solution and of the approximate solution (or, in the case of nonuniqueness, clarification of which solution or approximation is being discussed).

The term *convergence* has to be interpreted in context. It can apply, with given assumptions (which may include specification of ϑ and restrictions on $\mathscr{T}$ — requiring it to be *uniform* or *quasi-uniform*, or a *graded partition* of $[t_0,T]$ – see Remark 3.1) to, e.g., convergence on an interval or on mesh points:

$$\sup_{t\in[t_0,T]}|y(t)-\tilde{y}(t)|\to 0 \quad \text{as } h(\mathscr{T})\to 0 \quad \text{with } \mathscr{T}\in\mathfrak{T}, \tag{7.1a}$$

$$\sup_{t\in\mathscr{T}}|y(t)-\tilde{y}(t)|\to 0 \quad \text{as } h(\mathscr{T})\to 0 \quad \text{with } \mathscr{T}\in\mathfrak{T}. \tag{7.1b}$$

Here $\mathscr{T}\in\mathfrak{T}$ denotes that relevant restrictions on $\mathscr{T}$ are satisfied, and $h(\mathscr{T}):=\sup_{t_r\in[t_0,T]}\{t_r-t_{r-1}\}$ is the width (or diameter) of $\mathscr{T}$. As for *order of convergence*, we may have, first (order of convergence of the densely defined approximation on the whole interval $[t_0,T]$)

$$\sup_{t\in[t_0,T]}|y(t)-\tilde{y}(t)|=\mathcal{O}(\{h(\mathscr{T})\}^{\rho}) \quad \text{with } \mathscr{T}\in\mathfrak{T}, \tag{7.1c}$$

second (order of convergence on the points $\{t_n\}$ and the points $\{t_{n,r}\}$)

$$\sup_{t\in\mathscr{T}^{\#}(\vartheta')}|y(t)-\tilde{y}(t)|=\mathcal{O}(\{h(\mathscr{T})\}^{\varrho}) \quad \text{with } \mathscr{T}\in\mathfrak{T}, \tag{7.1d}$$

third (order of convergence on the points $\{t_n\}$ of $\mathcal{T}$)

$$\sup_{t\in\mathcal{T}}|y(t)-\tilde{y}(t)| = \mathcal{O}(\{h(\mathcal{T})\}^{\varrho'}) \quad \text{with } \mathcal{T}\in\mathfrak{T}. \tag{7.1e}$$

If the value of ρ in (7.1c) is optimal, and $\varrho > \rho$, or if $\varrho' > \varrho$ where ϱ is optimal or $\varrho' > \rho$, we speak of *superconvergence* on $\mathcal{T}^{\#}(\vartheta')$ or on $\mathcal{T}$, respectively. Superconvergence results arise, in particular, in the context of RK methods (they should be familiar for RK methods for ODEs) and of collocation.

Remark 7.1. Statement (7.1c) is simply short hand for the claim that there exist $h^\star > 0$ and a constant M_ρ (that is finite, and independent of $h(\mathcal{T})$ and of $\mathcal{T}\in\mathfrak{T}$) such that

$$\sup_{t\in[t_0,T]}|y(t)-\tilde{y}(t)|\leqslant M_\rho\{h(\mathcal{T})\}^{\rho} \quad \text{when} \quad h(\mathcal{T})\leqslant h^\star. \tag{7.2}$$

It does not follow from (7.2) that a reduction in the size of $h(\mathcal{T})$ will reduce $\sup_{t\in[t_0,T]}|y(t)-\tilde{y}(t)|$; nor can we infer that $\sup_{t\in[t_0,T]}|y(t)-\tilde{y}(t)|$ is proportional to $h(\mathcal{T})$. However, statements of orders of convergence can often be strengthened — so that, with a uniform mesh of width h, one may have, e.g.,

$$y(t)-\tilde{y}_h(t) = h^\rho e_\rho(h;t) + \mathcal{O}(h^{\rho^\star}), \quad \tilde{y}_h(t)\equiv\tilde{y}(t) \tag{7.3}$$

with $|e_\rho(h;t)|$ uniformly bounded for $0<h\leqslant h^\star$, $t\in[t_0,T]$, where $\rho^\star>\rho$. The vanishing of the nonnull function $e_\rho(h;t)$ on a set of points $\mathcal{S}_h\subset[t_0,T]$ $(0<h\leqslant h^\star)$ implies superconvergence on $\{\mathcal{S}_h\}$.

Occasionally, one sees estimates of the convergence rate at the point t of the form

$$\rho_{\text{computed}}^{h',\kappa h',\kappa_\star h'}(t) := \log_\kappa\left|\frac{y_{\kappa_\star h'}(t)-y_{\kappa h'}(t)}{y_{\kappa_\star h'}(t)-y_{h'}(t)}\right| \quad \text{where } \kappa_\star\ll\kappa<1.$$

Such estimates may not have been reconciled with the theory. However, if, in (7.3), $e_\rho(h;t)\neq 0$ is independent of h ($e_\rho(h;t)\equiv e_\rho(t)$, say), then $y(t)-\tilde{y}_h(t)$ is, *for sufficiently small h*, monotonically decreasing as $h\to 0$ and *asymptotically proportional* to h^ρ (here, $|y(t)-\tilde{y}_h(t)|/h^\rho \to |e_\rho(t)|$ for some bounded nonnull $e_\rho(t)$, where $t\in[t_0,T]$, as $h\to 0$).

Whilst controlling the accuracy of numerical computation is, in fact, a complicated matter, the mathematical community finds delight in the precision of mathematical theorems and "concise" results (e.g., 'superconvergence occurs at the collocation points' or 'algebraic stability implies C-stability' — see below) have a particular appeal. Such theories are both elegant and rewarding and they can give confidence in the underlying algorithms, but real-life computational conditions and practical adaptive algorithms do not always conform to the theoretical conditions assumed and it is worthwhile reviewing, from time to time, the exact practical significance of the theory. As an example, in his paper on collocation for integro-differential equations with nonsmooth solutions, Brunner [29] remarked that even though the use of suitably graded meshes yields optimal convergence rates, such meshes have their limitations (e.g., they may demand a very small initial step size, and carry a risk of contamination by rounding error).

8. Some approaches to convergence analysis

We shall indicate some of the tools used in establishing convergence and orders of convergence. We commence with a *Gronwall-type inequality*, that follows from a result quoted in [49, p.41]:

Lemma 8.1. *For* $\Lambda > 0$ *and* $z_\ell > 0$ $(\ell = 0, 1, 2, \ldots)$, *suppose that*

$$|\xi_n| \leqslant \Lambda \sum_{\ell=0}^{n-1} z_\ell |\xi_\ell| + |\eta_n| \quad (\mathit{for}\ n = 0, 1, \ldots), \tag{8.1}$$

then $|\xi_n| \leqslant |\eta_n| + \Lambda \sum_{r=0}^{n-1} \{z_r \prod_{q=r+1}^{n-1} (1 + \Lambda z_q)\} |\eta_r|$ *for* $n = 0, 1, 2, \ldots$.

An example will indicate an application of (8.1). Assume that $\mathscr{T}$ is given (with a sequence of positive steps $\{h_n\}$ and $h(\mathscr{T}) = \sup h_r$), and consider a set of approximating formulae (3.4), which we assume have a unique solution $\{\tilde{y}(t_0),\ \tilde{y}(t_1),\ \tilde{y}(t_2), \ldots\}$. We find that the true solution of (2.1b) gives

$$y(t_{n+1}) = g(t_{n+1}) + \sum_{j=0}^{n+1} \Omega_{n+1,j} K(t_{n+1}, t_j, y(t_j)) + \eta_n, \tag{8.2}$$

where $\{\eta_n\}$ are the *local truncation errors*. If we subtract the corresponding equations (3.4) and write $e_n := \tilde{y}(t_n) - y(t_n)$ we have

$$e_{n+1} = \sum_{j=0}^{n+1} \Omega_{n+1,j} \{K(t_{n+1}, t_j, \tilde{y}(t_j)) - K(t_{n+1}, t_j, y(t_j))\} - \eta_n \tag{8.3}$$

and hence, by Hypothesis 1,

$$|e_{n+1}| \leqslant \sum_{j=0}^{n+1} \Omega_{n+1,j} L |e_j| + |\eta_n|. \tag{8.4}$$

Now an examination of the weights in (3.6) reveals a property

$$|\Omega_{n,j}| \leqslant W\{h_{j-1} + h_j\} \quad \text{for some finite absolute constant } W, \tag{8.5}$$

when $j = 0, 1, 2, \ldots, n$ that holds for more general weights, and we therefore take (8.5) as an assumption in the case of quadrature methods for Volterra equations. From Lemma 8.1 above, applying (8.1) with $z_\ell = h_\ell + h_{\ell+1}$, assuming $Wh(\mathscr{T}) \leqslant \frac{1}{4}$, we obtain without effort, on setting $\Lambda = 2LW$,

$$|e_n| \leqslant \sup_{r \in \{0,1,\ldots,n\}} |\eta_r| \{1 + 2\Lambda(T - t_0) \exp\{2\Lambda(T - t_0)\}\} \quad \text{for } t_n \in [t_0, T]. \tag{8.6}$$

This establishes that the order of convergence of quadrature approximations $\tilde{y}(t)$ to $y(t)$, for $t \in \mathscr{T}$, is determined by the local truncation errors. (Actually, this result can be refined to take account of starting procedures.)

Remark 8.2. For quadrature generated by cyclic multistep formulae and for RK formulae, *convergence* can be established by similar techniques though results on the order of accuracy tend to be pessimistic and a more refined analysis is required. For Abel equations of the second kind (2.1e) the weights analogous to $\Omega_{n,j}$ behave like $(n-j)^{-\nu}h^{1-\nu}$, and McKee [107] developed an appropriate Gronwall-type lemma for this case (see also [49, pp. 42–44]).

8.1. One-step methods

The discussion of one-step methods on a uniform mesh $\mathscr{T}$ by Hairer et al. [80] applies to various methods of RK type, and merits attention. To indicate something of their viewpoint, we re-write (2.1b) in the form

$$y(t) = G_n(t) + \int_{t_n}^{t} K(t,s,y(s))\,\mathrm{d}s \tag{8.7a}$$

for $t \geqslant t_n$, where $G_n(t) = g(t) + \int_{t_0}^{t_n} K(t,s,y(s))\,\mathrm{d}s$, so that

$$G_k(t) = G_{k-1}(t) + \int_{t_{k-1}}^{t_k} K(t,s,y(s))\,\mathrm{d}s \quad (k = 1,2,3,\ldots) \tag{8.7b}$$

with $G_0(t) = g(t)$. A discrete analogue of (8.7a) is provided by

$$\tilde{y}(t_{n+1}) = \tilde{G}_n(t_{n+1}) + h_n\Phi(t_{n+1}, \tilde{G}(\cdot); h_n). \tag{8.8a}$$

Here (cf. the theory of one-step methods for ODEs) $\Phi(\cdot,\cdot;\cdot)$ is an *increment function* (whose evaluation involves internal stage values) and $\{\tilde{G}_k(\cdot)\}$ are approximations to $\{G_k(\cdot)\}$. One assumes that the increment functions inherit, from the properties of $K(\cdot,\cdot,\cdot)$, a uniform Lipschitz condition with respect to its second argument (uniform with regard to the index and the step size).

For the development of this approach to a wide class of RK methods and various approximations to the history term, consult [80]. One should not forget that some[4] RK methods require a *smooth* extension such as (2.2) to achieve their potential in terms of accuracy.

For extended (Pouzet-)RK methods, $\tilde{G}_{n+1}(\cdot)$ is itself computed via the incremental function:

$$\tilde{G}_k(t) = \tilde{G}_{k-1}(t) + h_k\Phi(t, \tilde{G}_{k-1}(\cdot); h_k) \tag{8.8b}$$

in analogy with (8.7a), and moreover $\tilde{y}(t_{n+1}) = \tilde{G}_{n+1}(t_{n+1})$. In this case, the link with an embedding approach (cf. (4.1a)) is apparent.

8.2. Resolvents and variation of parameters in the analysis of collocation

There have been numerous results [49] on convergence and superconvergence properties of collocation-type methods. Some results [24] were achieved by relying on a (nonlinear) variation of constants formula (2.13) due to Brauer (the reader should also consult [11] which corrects errors in [15] where, in turn, a slip in [22] was noted) to relate the true and approximate solution. Our purpose here is to draw attention to the rôle of a suitable formula of type (2.13), without entering into the detail which has to be investigated to place the theory on a rigorous foundation.

---

[4] An extension of the values $K(t,s,v)$ for $s > t$ is *not required* in the Pouzet-type RK formulae if $a_{i,j} = 0$ when $\vartheta_j > \vartheta_i$ and $\vartheta_\ell \in [0,1]$ for all ℓ.

We appeal to (2.13) (with minimal assumptions concerning the nature of $R(t,s,u(s))$) to compare the solution $\tilde{y}(\cdot)$ of the collocation equations with the solution $y(\cdot)$ of (2.1b). We let $\delta(\tilde{y}(\cdot);t)$ in (6.4) denote the defect, which vanishes at the points $\{t_{n,i}\}_{i=1}^{m}$ for $n=0,1,2,\ldots,N-1$, and we have

$$\tilde{y}(t) - y(t) = \delta(\tilde{y}(\cdot);t) + \int_{t_0}^{t} R(t,s;\tilde{y}(s))\delta(\tilde{y}(\cdot);s)\,\mathrm{d}s \tag{8.9}$$

and, in particular,

$$\tilde{y}(t_n) - y(t_n) = \delta(\tilde{y}(\cdot);t_n) + \sum_{j=0}^{n-1} \int_{t_j}^{t_{j+1}} R(t_n,s,\tilde{y}(s))\delta(\tilde{y}(\cdot);s)\,\mathrm{d}s. \tag{8.10}$$

Using interpolatory quadrature with abscissae $t_{j,k}$ (the collocation points in the jth subinterval, or even a subset of these abscissae) and weights $h_j w_k$ we deduce that

$$\int_{t_j}^{t_{j+1}} R(t_n,s,\tilde{y}(s))\delta(\tilde{y}(\cdot);s) = h_j \sum_k w_k R(t_n,t_{j,k},\tilde{y}(t_{j,k}))\delta(\tilde{y}(\cdot);t_{j,k}) + E_{n,j}, \tag{8.11}$$

where $E_{n,j}$ denotes the error in the quadrature, and the sum in (8.11) vanishes because the defect is zero at the collocation points. Thus,

$$\tilde{y}(t_n) - y(t_n) = \delta(\tilde{y}(\cdot);t_n) + \sum_{j=0}^{n-1} E_{n,j}.$$

To obtain high order of convergence on $\mathscr{T}$ under suitable smoothness conditions (the integrand in (8.11) should be shown to be smooth on $[t_j,t_{j+1}]$) one should therefore ensure (i) that $\delta(\tilde{y}(\cdot);t)$ vanishes on $\mathscr{T}$ (the points $\{t_n\}$ should be included in the collocation points), and (ii) that the points $t_{j,k}$ (or a subset thereof) give interpolatory quadrature with an error of high order (Gaussian-type quadrature, e.g., taking ϑ as $[\vartheta_{(\mathrm{Gauss}_m)},1]$, $\vartheta_{(\mathrm{Radau}_m]}$, or $\vartheta_{[\mathrm{Lobatto}_m]}$). For a complete analysis for linear equations see [49], cf. [23,51].

9. Stability

The stability theory for numerical methods for Volterra equations is still incomplete though considerable advances have been made. Stability is concerned with the effect on a solution (here defined over $[t_0,\infty)$) of perturbations in the problem, and differing definitions of stability arise when one considers different types of *admissible perturbations*. Let us therefore consider

$$y(t) = g(t) + \int_{t_0}^{t} K(t,s,y(s))\,\mathrm{d}s \quad (t \in [t_0,\infty)). \tag{9.1a}$$

With a choice of a class $\mathfrak{G}$ of admissible perturbations defined on $[t_0,\infty)$, consider the introduction of a perturbation $\delta g(\cdot) \in \mathfrak{G}$ to give

$$y(t) + \delta y(t) = g(t) + \delta g(t) + \int_{t_0}^{t} K(t,s,y(s)+\delta y(s))\,\mathrm{d}s \quad (t \in [t_0,\infty)). \tag{9.1b}$$

Possible choices of $\mathfrak{G}$ are

$$\mathfrak{G} := \{\delta g(t) \in BC[t_0,\infty)\}, \text{ i.e., } \delta g(t) \in C[t_0,\infty),\ \sup_{t\geqslant t_0} |\delta g(t)| < \infty, \tag{9.2a}$$

$$\mathfrak{G}:=\{\delta g(t) \in BC_0[t_0,\infty)\}, \text{ i.e., } \delta g(t) \in BC[t_0,\infty), \lim_{t\to\infty} \delta g(t) = 0, \tag{9.2b}$$

$$\mathfrak{G}:=\{\delta g(t) \in \mathscr{L}_1[t_0,\infty)\}, \text{ i.e., } \int_{t_0}^{\infty} |\delta g(s)|\,\mathrm{d}s < \infty. \tag{9.2c}$$

The notion of the stability of the solution $y(\cdot)$ of (9.1a) is related to the effect $\delta y(\cdot)$ that results from making the perturbation $\delta g(\cdot)$, e.g. *The solution $y(t)$ is globally stable with respect to perturbations $\delta g(\cdot) \in \mathfrak{G}$ in the case that $\sup_{t\in[t_0,\infty)}|\delta y(t)|$ is bounded whenever $\delta g(\cdot) \in \mathfrak{G}$; it is globally asymptotically stable if it is stable and in addition $\delta y(t) \to 0$ as $t \to \infty$ whenever $\delta g(\cdot) \in \mathfrak{G}$.*

Concepts of the stability of the *numerical solution* $\tilde{y}(t)$ follows analogous lines. These reflect whether $\tilde{y}(\cdot)$ is densely defined on $[t_0,\infty)$ or is regarded as a mesh function defined on $\mathscr{T}$ (or on $\mathscr{T}^{\#}(\boldsymbol{\vartheta})$ — for definiteness assume the former) and whether we take $\mathfrak{G}$ as in (9.2) or instead choose a corresponding $\tilde{\mathfrak{G}}$ comprising mesh functions, such as

$$\tilde{\mathfrak{G}}:=\{\{\delta g(t_0),\delta g(t_1),\delta g(t_2),\ldots\} \in \ell_\infty\}, \text{ i.e., } \sup_{t_n\in\mathscr{T}}|\delta g(t)| < \infty, \tag{9.3a}$$

$$\tilde{\mathfrak{G}} := \{\{\delta g(t_0),\delta g(t_1),\delta g(t_2),\cdots\} \in \ell^0_\infty\},$$
$$\text{i.e., } \sup_{t_n\in\mathscr{T}} |\delta g(t_n)| < \infty, \quad \text{and} \quad \lim_{t_n\to\infty} \delta g(t_n) = 0, \tag{9.3b}$$

$$\tilde{\mathfrak{G}}:=\{\{\delta g(t_0),\delta g(t_1),\delta g(t_2),\ldots\} \in \ell_1\}, \text{ i.e., } \sum_{n=0}^{\infty}|\delta g(t_n)| < \infty. \tag{9.3c}$$

We shall denote the choice by suffices ($\mathfrak{G}_{(9.2a)}$ through $\tilde{\mathfrak{G}}_{(9.3c)}$). Note the duplication in the sense that, e.g., $\delta g \in \mathfrak{G}_{(9.2a)}$ implies that $\delta g \in \tilde{\mathfrak{G}}_{(9.3a)}$ (but not vice versa). As an example of definitions of numerical stability we have

Suppose $\tilde{y}(\cdot)$ satisfies (3.4), *and*

$$\tilde{y}(t_{n+1}) + \delta\tilde{y}(t_{n+1}) = \sum_{r=0}^{n+1} \Omega_{n+1,j} K(t_{n+1}, t_j, \tilde{y}(t_j) + \delta\tilde{y}(t_j)) + g(t_{n+1}) + \delta g(t_n). \tag{9.4}$$

The solution $\tilde{y}(t)$ ($t \in \mathscr{T}$) of (3.8) *is globally stable with respect to perturbations $\delta g(\cdot) \in \tilde{\mathfrak{G}}$ in the case that $\sup_{t_n\in\mathscr{T}}|\delta\tilde{y}(t_n)|$ is bounded whenever $\delta g(\cdot) \in \tilde{\mathfrak{G}}$; it is globally asymptotically stable if it is stable and in addition $\delta\tilde{y}(t_n) \to 0$ as $t_n \to \infty$ with $t_n \in \mathscr{T}$ whenever $\delta g(\cdot) \in \tilde{\mathfrak{G}}$; it is* exponentially stable *if it is asymptotically stable and there exist $M, z > 0$ (corresponding to $\delta g(\cdot) \in \tilde{\mathfrak{G}}$) such that $\delta\tilde{y}(t_n) \leqslant M \exp\{-z(t_n - t_0)\}$ for $t_n \in \mathscr{T}$.*

With favourable assumptions (though *not* invariably), investigation of stability of a solution of a nonlinear equation can be reduced to that of the solutions of a corresponding linear equation (stability in the first approximation). Alternative approaches invoke Lyapunov theory or ad hoc qualitative arguments. The stability of any solution $y(t)$ of

$$y(t) = \lambda \int_{t_0}^{t} \mathsf{k}(t,s)y(s)\,\mathrm{d}s + g(t) \tag{9.5}$$

can be settled by investigation of

$$\delta y(t) = \lambda \int_{t_0}^{t} \mathsf{r}^{\{\lambda\}}(t,s)\delta g(s)\,\mathrm{d}s + \delta g(t) \quad (\text{for } t \geqslant t_0), \tag{9.6}$$

where $\mathsf{r}^{\{\lambda\}}(\cdot,\cdot)$ is the resolvent kernel. One has only to determine the boundedness and asymptotic behaviour of $\delta y(\cdot)$ given the condition $\delta g(\cdot) \in \mathfrak{G}$ and conditions on $\mathsf{r}^{\{\lambda\}}(t,s)$ that can be deduced from assumed conditions on $\mathsf{k}(t,s)$. E.g., boundedness of $\delta y(t)$ for all $t \geqslant t_0$ follows from the property $\sup_{t\geqslant t_0}|\delta g(t)| < \infty$ if the resolvent condition $\sup_{t\geqslant t_0} \int_{t_0}^t |\mathsf{r}^{\{\lambda\}}(t,s)|\,\mathrm{d}s < \infty$ is satisfied, and one has stability under this condition if $\delta g(\cdot) \in \mathfrak{G}_{(9.2a)}$.

Remark 9.1. In the integro-differential equation in Remark 2.4, the effect $\delta y(t)$ $(t > t_0)$ of perturbing the initial value $y(t_0)$ to $y(t_0)+\delta y(t_0)$ and of perturbing $g(\cdot)$ by a uniformly bounded function $\delta g(\cdot)$ to give $g(\cdot) + \delta g(\cdot)$ can be investigated in terms of the function $\mathsf{u}(\cdot,\cdot)$ discussed in that remark.

For discretized equations one has a similar result to (9.6). Consider, for example, *discrete Volterra equations* of the form

$$\tilde{y}_{n+1} = \lambda \sum_{i=0}^{n+1} \mathsf{a}_{n+1,j}\tilde{y}_j + g_{n+1} \; (n = 0, 1, \ldots) \tag{9.7}$$

with $\tilde{y}_0 = g_0$, that correspond when $\mathsf{a}_{n,j} = \Omega_{n,j}\mathsf{k}(t_n, t_j)$, to the application of the scheme in (3.4) to (9.5). Then (9.7) provides an infinite system of equations with a lower triangular coefficient matrix for which the solution (a) exists if $\lambda\mathsf{a}_{n,n} - 1 \neq 0$ for all n and (b) can then be expressed, if required, as

$$\tilde{y}_{n+1} = g_{n+1} + \lambda \sum_{j=0}^{n+1} \mathsf{b}^{\{\lambda\}}_{n+1,j} g_j \quad (n = 0, 1, \ldots). \tag{9.8}$$

Thus, $\delta\tilde{y}_{n+1} = \delta g_{n+1} + \lambda \sum_{j=0}^{n+1} \mathsf{b}^{\{\lambda\}}_{n+1,j}\delta g_j$ and conditions for the stability of (9.7) with $\delta g \in \tilde{\mathfrak{G}}_{(9.3a)}$ can be expressed in terms of the condition $\sup_n \sum_{j=0}^{n+1} |\mathsf{b}^{\{\lambda\}}_{n+1,j}| < \infty$. (For RK and block-by-block formulae one may use vectors and submatrices in an analogous discussion.) In previous literature, this underlying approach is frequently concealed from view, because special structure of (9.7) is exploited to reduce Eqs. (9.7) to a finite-term recurrence relation or to one in which other structures can be exploited.

For equations of the second kind, numerical stability analysis focussed initially on methods with a uniform grid $\mathcal{T}$ applied to the equation

$$y(t) - \lambda \int_{t_0}^t y(s)\,\mathrm{d}s = g(t), \tag{9.9}$$

which (Example 3.2) reduces to an ODE if $g'(\cdot)$ exists. The corresponding stability results were therefore related to those for (possibly novel) ODE methods for $y'(t) = \lambda y(t) + g'(t)$, but of course a perturbation $\delta g(\cdot)$ in the integral equation corresponds to a change in the ODE of $\delta g'(\cdot)$ in $g'(\cdot)$ and of $\delta g(t_0)$ in the initial value. However, without assuming differentiability we can still introduce a continuous perturbation $\delta g(\cdot)$ in (9.9) and use the resolvent to write

$$\delta y(t) = \delta g(t) + \lambda \int_{t_0}^t \exp\{\lambda(t-s)\}\delta g(s)\,\mathrm{d}s. \tag{9.10}$$

Note that (i) when $\Re(\lambda) \leqslant 0$ then $\delta y(\cdot)$ is bounded when $\delta g(\cdot) \in \mathfrak{G}_{(9.2a)}$ (giving stability) and (ii) when $\Re(\lambda) < 0$ then $\delta y(t) \to 0$ as $t \to \infty$ when $\delta g(\cdot) \in \mathfrak{G}_{(9.2b)}$ (giving asymptotic stability). For the corresponding discrete equations based on an application of a $\{\rho,\sigma\}$-reducible quadrature (4.2) to (9.9) we find that perturbations satisfy

$$\delta\tilde{y}_n - \lambda h \sum_{j=n_0+1}^{n} \omega_{n-j}\delta\tilde{y}_j = \Delta_n \quad (n = n_0+1, n_0+2, \ldots), \tag{9.11}$$

where $\Delta_n := \delta g(nh) + h\sum_{j=0}^{n_0} w_{n,j}\delta\tilde{y}_j$ and where the boundedness of $\{w_{n,j}\}$ implies the boundedness of $\{\Delta_n\}$ when $\delta g(\cdot) \in \tilde{\mathfrak{G}}_{(9.3a)}$. Then, if $\lambda h\omega_0 \neq 1$,

$$\delta\tilde{y}_n = \Delta_n + \lambda h \sum_{j=n_0+1}^{n} \varpi_{n-j}^{\{\lambda h\}}\Delta_j \quad (n = n_0+1, n_0+2, \ldots), \tag{9.12}$$

where $(1 + \lambda h\{\varpi_0^{\{\lambda h\}} + \varpi_1^{\{\lambda h\}}\zeta + \varpi_2^{\{\lambda h\}}\zeta^2 + \cdots\})(1 - \lambda h\{\omega_0 + \omega_1\zeta + \omega_2\zeta^2 + \cdots\}) = 1$, and hence, by virtue of (4.3),

$$1 + \lambda h\{\varpi_0^{\{\lambda h\}} + \varpi_1^{\{\lambda h\}}\zeta + \varpi_2^{\{\lambda h\}}\zeta^2 + \cdots\} = \frac{\zeta^k\rho(\zeta^{-1})}{\zeta^k\rho(\zeta^{-1}) - \lambda h\zeta^k\sigma(\zeta^{-1})}. \tag{9.13}$$

Using partial fractions and expanding, it follows that if the polynomial $\rho(\zeta) - \lambda\sigma(\zeta)$ is *Schur* [5] then the coefficients $\{\varpi_n^{\{\lambda\}}\}$ satisfy $\sum_{n=0}^{\infty}|\varpi_n^{\{\lambda\}}| < \infty$, and we deduce *stability* for perturbations $\delta g(\cdot) \in \tilde{\mathfrak{G}}_{(9.3a)}$. If $\{\rho,\sigma\}$ is A-stable, this property holds whenever $\Re(\lambda) < 0$ for all h. Note, however, that when $\delta g \in \tilde{\mathfrak{G}}_{(9.3b)}$, we find $|\Delta_n| \nrightarrow 0$ and we do not deduce *asymptotic stability* without further restriction of the admissible δg; this minor difficulty disappears later when we introduce a suitable $\mathscr{L}_1$-kernel $k(\cdot)$ into the equation.

Remark 9.2. (a) Some of the techniques used for (9.9) can be modified for separable kernels, polynomial convolution kernels (e.g., $k(t) = \xi + \eta t$, for which (9.14) below reduces to $y''(t) = \xi y'(t) + \eta y(t) + g''(t)$), etc. (b) For integro-differential equations, $y'(t) = \xi y(t) + \eta\int_{t_0}^{t} y(s)\,\mathrm{d}s + g(t)$ plays a similar rôle to (9.9). (c) Stability and stability regions for test equations are discussed in, for example, [8,7,10,50].

A natural question to ask is whether results for (9.9) provide any insight into results for more general equations; we indicate one approach to answering this question by considering the discretization of the linear convolution equation

$$y(t) - \lambda\int_{t_0}^{t} k(t-s)y(s)\,\mathrm{d}s = g(t) \quad \text{with } \lambda \in \mathbb{C}, \tag{9.14}$$

Let us keep in mind the substitutions $\mathfrak{a}_n = hk(nh)\omega_n$, $\tilde{y}_j = \tilde{y}(jh)$, $g_n = g(nh)$, and consider the discrete convolution equations

$$\tilde{y}_{n+1} = \lambda\sum_{j=0}^{n+1} \mathfrak{a}_{n+1-j}\tilde{y}_j + g_{n+1} \quad (n = 0, 1, \ldots). \tag{9.15}$$

[5] All its zeros lie inside the unit circle centered on the origin, or, equivalently, $z^k\rho(z^{-1}) - \lambda hz^k\sigma(z^{-1}) = 0$ implies $|z| \geqslant 1$.

If $\mathfrak{a}(\zeta)$ denotes the formal power series $\sum_{j=0}^{\infty} \mathfrak{a}_j \zeta^j$, and if $\mathfrak{b}^{\{\lambda\}}(\zeta)$ denotes the formal power series $\sum_{j=0}^{\infty} \mathfrak{b}_j^{\{\lambda\}} \zeta^j$ where $(1 - \lambda \mathfrak{a}(\zeta))(1 + \lambda \mathfrak{b}^{\{\lambda\}}(\zeta)) = 1$ then

$$\tilde{y}_{n+1} = g_{n+1} + \lambda \sum_{j=0}^{n+1} \mathfrak{b}_{n+1-j}^{\{\lambda\}} g_{n+1} \quad (n = 0, 1, \ldots). \tag{9.16}$$

If, further,

$$\sum_{j=0}^{\infty} |\mathfrak{a}_j| < \infty \quad \text{and} \quad 1 \neq \lambda \sum_{j=0}^{\infty} \mathfrak{a}_j \zeta^j \text{ whenever } |\zeta| \leqslant 1, \tag{9.17}$$

then $\sum_{j=0}^{\infty} |\mathfrak{b}_j^{\{\lambda\}}| < \infty$ and hence the solution of (9.15) is stable for perturbations δg_n in $\tilde{\mathfrak{G}}_{(9.3a)}$ or in $\tilde{\mathfrak{G}}_{(9.3b)}$ and is asymptotically stable for perturbations δg_n in $\tilde{\mathfrak{G}}_{(9.3b)}$ or, indeed, in $\tilde{\mathfrak{G}}_{(9.3c)}$. Under assumptions on the 'starting weights' $\mathfrak{s}_{n+1,j}$ the analysis can be extended to discuss equations, cf. (4.2), of the form

$$\tilde{y}_{n+1} = \lambda \left\{ \sum_{j=0}^{n_0} \mathfrak{s}_{n+1,j} \tilde{y}_j + \sum_{j=n_0+1}^{n+1} \mathfrak{a}_{n+1-j} \tilde{y}_j \right\} + g_{n+1} \; (n \geqslant n_0). \tag{9.18}$$

Normally $\mathfrak{s}_{n,j} = w_{n,j} k((n-j)h)$, and $\mathfrak{a}_{n,j} = \omega_{n-j} k((n-j)h)$; hence if $k(\cdot) \in C[0,\infty) \cap L_1[0,\infty)$ (so that $k(t) \to 0$ as $t \to \infty$) and if the weights ω_ℓ are uniformly bounded and

$$\sum_{\ell=0}^{\infty} |k(\ell h)| < \infty, \tag{9.19}$$

we find $\mathfrak{s}_{n,j} = 0$; $\sum_{\ell=0}^{\infty} |\mathfrak{a}_\ell| < \infty$. Condition (9.19) is a minor annoyance but should not be overlooked; it holds under our previous assumptions when $k(\cdot)$ is monotonic decreasing or is monotonic decreasing for sufficiently large arguments or when h is sufficiently small (say $h \leqslant h^\star(k(\cdot))$).

In our discussion of (9.9) we relied upon the simple nature of the kernel. Whilst (9.17) and (9.19) allow one to compute approximate stability regions for special $k(\cdot)$, what we seek now is a class of integral equations for which stability is readily analysed and classes of methods that simulate the stability properties of the true solution. A theory can be constructed if we consider *positive quadrature* and *completely monotone* or *positive definite functions*.

Remark 9.3. (a) A convolution quadrature is called *positive* (i) if and only if $\sum\sum \omega_{i-j} z_i \bar{z}_j \geqslant 0$ for all finite sequences of complex-valued z_ℓ. (ii) A quadrature is positive if it is a reducible quadrature generated by an A-stable LMF (ρ, σ). (b) A function $\psi : [0,\infty) \to \mathbb{C}$ is *positive definite* if (i) on defining $\phi(t) = \psi(t)$, $t \geqslant 0$, $\phi(t) = -\overline{\psi(-t)}$, $t \leqslant 0$, we have $\sum_i \sum_j \phi(t_i - t_j) z_i \overline{z_j} \geqslant 0$ for all t_ℓ and all finite sequences of complex valued z_ℓ, or (*ii*) equivalently, by a result of Bochner, if (a Laplace transform condition) $\Re\{\int_0^\infty \psi(s)\exp(-zs)\mathrm{d}s\} \geqslant 0$ if $\Re(z) > 0$. (c) A function ψ that is *completely monotone* (that is, $\psi(\cdot) \in C^\infty[0,\infty)$ and $(-1)^j k^{(j)}(t) > 0$ for $j = 0, 1, 2, \ldots$) is positive definite. (d) When the convolution quadrature with weights ω_ℓ is positive and $k(\cdot)$ is a positive-definite function, the sequence $\{\mathfrak{a}_\ell\}$ (with $\mathfrak{a}_\ell = hk(\ell h)\omega_\ell$) is a positive definite sequence and the analogue of Bochner's Laplace transform result (in effect, a Z-transform condition) holds: $\Re\{\sum_{n=0}^{\infty} \mathfrak{a}_n \zeta^n\} \geqslant 0$ for $|\zeta| \leqslant 1$.

Consider the linear equation (9.14) for which a resolvent convolution kernel $r^{\{\lambda\}}(\cdot) \in C[0,\infty)$ provides the solution $y(t) = g(t) + \lambda \int_{t_0}^{t} r^{\{\lambda\}}(t-s)g(s)\,\mathrm{d}s$. If

Hypothesis 3. $k(\cdot) \in C[0,\infty) \cap \mathscr{L}_1[0,\infty)$, and $\Re(\lambda) < 0$, and either
(i) $k(\cdot)$ is *completely monotone*, or, more generally,
(ii) $k(\cdot)$ is *positive definite*,

then we have $\int_0^{\infty} |r^{\{\lambda\}}(s)\,\mathrm{d}s| < \infty$ and hence *all solutions of* (9.14) *are stable for perturbations* $\delta g(\cdot)$ in $\mathfrak{G}_{(9.2a)}$ (or a fortiori for perturbations $\delta g(\cdot)$ in $\mathfrak{G}_{(9.2b)}$), and asymptotically stable for perturbations $\delta g(\cdot)$ in $\mathfrak{G}_{(9.2b)}$ (or a fortiori for perturbations $\delta g(\cdot)$ in $\mathfrak{G}_{(9.2c)}$). The preceding remarks inspire the definition of a *C-stable* RK or quadrature method as one that, when applied to (9.14) with $\Re(\lambda) < 0$ and a bounded continuous *positive-definite* kernel $k(\cdot)$ yields *for all fixed h* approximations $\{\tilde{y}_n\}$ such that $\tilde{y}_n \to 0$ whenever $g(t_n) \to 0$ as $n \to \infty$. In fact, when using the concept of C-stability, we have to impose condition (9.19) or a similar condition, thereby restricting slightly the class of functions $k(\cdot)$. Expressed in our terms, the method is C-stable if, given Hypothesis 3 along with (9.19) (or along with an analogous assumption in the case of RK methods) the mesh-functions $\tilde{y}(t)$ $(t \in \mathscr{T})$ are asymptotically stable for all h with respect to perturbations in $\tilde{\mathfrak{G}}_{(9.3b)}$. It was Lubich who showed (*inter alia*) that A-stable $\{\rho,\sigma\}$-reducible quadrature formulae applied to (9.14) are C-stable, using assumption (9.19) (see [102]), in particular, therefore, under the assumption that $k(\cdot)$ is completely monotone. Given (9.19), $\{\mathfrak{a}_\ell\}_0^{\infty}$ is a positive definite ℓ_1 sequence, which (with our previous remarks) is sufficient to establish the claim.

Remark 9.4. (a) The work of Nevanlinna (cf. [112,113]) stimulated a strand of research in the stability analysis for numerical methods for Volterra equations. (b) Hairer and Lubich [78] showed that, provided one employs the correct extension of the positive-definite function $k(\cdot)$ to negative arguments, *algebraically stable* extended Pouzet–R–K methods are *C-stable under an assumption analogous to* (9.19). An order barrier was given. (c) The stability of collocation methods has been discussed in [17,18,65] etc. (d) Nonlinear equations were discussed in [72], where an approach of Corduneanu in [56] is exploited and analogous results are obtained. The discussion in [73] deals with a class of nonlinear convolution equations and refers to use that can be made of *strong* and *strict positivity*.

10. Concluding remarks

The term "Volterra equations" encompasses a variety of "nonanticipative" functional equations that generalize the classical equations we have studied here. It has been necessary to limit our discussion and to omit a number of growth areas, and we have concentrated on classical Volterra equations. This leaves some noteworthy omissions, including integro-differential equations with discretely distributed delays, such as

$$y'(t) = \lambda y(t)\left\{A - \int_{t_0}^{t} K(t,s,y(s))\,\mathrm{d}s - \sum_{i=1}^{N} B_i y(t-\tau_i(t))\right\},$$

where $\tau_i(t) \geqslant 0$ $(i = 1, 2, \ldots, N)$ and partial integro-differential equations (see e.g., [106,117] and their references).

Equations with infinite memory that have the form

$$y(t) = \gamma(t) + \int_{-\infty}^{t} K(t, s, y(s))\,\mathrm{d}s \quad (t \in [t_0, T]), \tag{10.1a}$$

$$y(t) = \psi(t) \quad (-\infty < t \leqslant t_0), \tag{10.1b}$$

reduce, formally, to (2.1b) if, in the classical case, we set

$$g(t) := \gamma(t) + \int_{-\infty}^{t_0} K(t, s, \psi(s))\,\mathrm{d}s. \tag{10.2}$$

This begs the question of the accurate evaluation of (10.2) which can be computationally expensive. One can also consider equations with a lower limit of integration $\alpha(t)$ and an upper limit of integration $\beta(t)$ (where $\alpha(t) \leqslant \beta(t) < t$),

$$y(t) = \gamma(t) + \int_{\alpha(t)}^{\beta(t)} K(t, s, y(s))\,\mathrm{d}s \quad (t \in [t_0, T]), \tag{10.3}$$

$$y(t) = \psi(t) \quad \left(\inf_{t' \geqslant t_0} \alpha(t') \leqslant t \leqslant t_0\right), \tag{10.4}$$

if we define $K(t, s, u) := 0$ for $s \notin [\alpha(t), \beta(t)]$. However, in order to treat "nonclassical" limits of integration, we must admit that $K(t, s, v)$ may suffer a jump discontinuity when $s = \alpha(t)$ or $\beta(t)$. It is clear that the form of $\alpha(t)$ has an impact upon the type of initial condition required to determine a unique solution. Whereas the analytical theories can to some extent unified, numerical practice should reflect the presence of discontinuities and it seems better to treat the original nonclassical equations directly. A number of papers in the literature do attempt that. Further variants include equations

$$y(t) = g(t) + \int_{\alpha(t)}^{\beta(t)} \mathscr{K}(t, s, y(t), y(s), y(a(t)), y(b(s)))\,\mathrm{d}s \quad (t \geqslant t_0) \tag{10.5}$$

with $\alpha(t) \leqslant \beta(t) \leqslant t$ and with $a(s) \leqslant t$, and $b(s) \leqslant t$ for $s \in [\alpha(t), \beta(t)]$.

As regards future work, there remain interesting theoretical questions about the numerical simulation of qualitative properties (e.g., the effect of different types of memory, the onset of periodicity through bifurcation, and [115] blow-up in solutions); there exist opportunities for the application of mathematical arguments to the design and validation of robust and efficient *adaptive codes* that are suited to numerical simulation of real-life models (particularly Volterra equations arising in the biosciences).

We conclude with some references to the literature. The material herein should provide an adequate base for the interested reader to pursue the above topics. The book of Brunner and van der Houwen [49] has an excellent bibliography to the date of its completion in 1985. Since the output referred to there is repeated here when we wish to draw particular attention to it, our choice of Refs. [1–123] may be regarded as reflecting idiosyncrasies of the author.

Acknowledgements

The author thanks Prof. Pieter van der Houwen (CWI., Amsterdam) and a number of his own students (in particular Yihong Song and Hongjiong Tian) and colleagues in MCCM for reading and commenting on drafts of this paper.

References

[1] C.T.H. Baker, Numerical analysis of Volterra functional and integral equations, in: I.S. Duff, G.A. Watson (Eds.), The State of the Art in Numerical Analysis, Oxford University Press, New York, 1997, pp. 193–222.
[2] C.T.H. Baker, M.S. Derakhshan, FFT techniques in the numerical solution of convolution equations, J. Comput. Appl. Math. 20 (1987) 5–24.
[3] C.T.H. Baker, M.S. Derakhshan, Computed approximations to some power series, Internat. Schriftenreihe Numer. Math. 81 (1986) 11–20.
[4] C.T.H. Baker, M.S. Derakhshan, Stability barriers to the construction of $\{\rho, \sigma\}$-reducible and fractional quadrature rules, Internat. Schriftenreihe Numer. Math. 85 (1988) 1–15.
[5] C.T.H. Baker, A. Tang, Stability analysis of continuous implicit Runge–Kutta methods for Volterra integro-differential systems with unbounded delay, Appl. Numer. Math. 24 (1997) 153–173.
[6] C.T.H. Baker, G.F. Miller (Eds.), Treatment of Integral Equations by Numerical Methods, Academic Press, London, 1982.
[7] C.T.H. Baker, M.S. Keech, Stability regions in the numerical treatment of Volterra integral equations, SIAM J. Numer. Anal. 15 (1978) 394–417.
[8] C.T.H. Baker, A. Makroglou, E. Short, Regions of stability in the numerical treatment of Volterra integro-differential equations, SIAM J. Numer. Anal. 16 (1979) 890–910.
[9] C.T.H. Baker, A. Tang, Numerical solution of Volterra integro-differential equations with an unbounded delay, Hellenic European Research on Mathematics and Informatics '94, Vols. 1, 2, Hellenic Mathematical Society, Athens, 1994, pp. 129–136.
[10] C.T.H. Baker, J.C. Wilkinson, Stability analysis of Runge–Kutta methods applied to a basic Volterra integral equation, J. Austral. Math. Soc. Ser. B 22 (1980/81) 515–538.
[11] P.R. Beesack, On some variation of parameter methods for integro-differential, integral, and quasilinear partial integro-differential equations, Appl. Math. Comput. 22 (1987) 189–215.
[12] A. Bellen, Z. Jackiewicz, R. Vermiglio, M. Zennaro, Natural continuous extensions of Runge–Kutta methods for Volterra integral equations of the second kind and their applications, Math. Comp. 52 (1989) 49–63.
[13] A. Bellen, Z. Jackiewicz, R. Vermiglio, M. Zennaro, A stability analysis of the trapezoidal method for Volterra integral-equations with completely positive kernels, J. Math. Anal. Appl. 152 (1990) 324–342.
[14] B.A. Bel'tyukov, Analogue of the Runge–Kutta method for the solution of nonlinear integral equations of Volterra type, Differential Equations 1 (1965) 417–433.
[15] S.H. Bernfeld, M.E. Lord, A nonlinear variation of constants method for integro differential and integral equations, Appl. Math. Comput. 4 (1978) 1–14.
[16] J.G. Blom, H. Brunner, The numerical solution of nonlinear Volterra integral equations of the second kind by collocation and iterated collocation methods, SIAM J. Sci. Statist. Comput. 8 (1987) 806–830.
[17] L. Blank, Stability of collocation for weakly singular Volterra equations, IMA J. Numer. Anal. 15 (1995) 357–375.
[18] L. Blank, Stability results for collocation methods for Volterra integral equations, Appl. Math. Comput. 79 (1996) 267–288.
[19] J.G. Blom, H. Brunner, Algorithm 689: collocation for nonlinear Volterra integral equations of the second kind, ACM Trans. Math. Software 17 (1991) 167–177.
[20] J.M. Bownds, Theory and performance of a subroutine for solving Volterra integral equations, Computing 28 (1982) 317–332.
[21] J.M. Bownds, L. Applebaum, Algorithm 627: A FORTRAN subroutine for solving Volterra integral equations, ACM Trans. Math. Software 11 (1985) 58–65.

[22] F. Brauer, A nonlinear variation of constants formula for Volterra equations, Math. Systems Theory 6 (1972) 226–234.

[23] H. Brunner, Superconvergence in collocation and implicit Runge–Kutta methods for Volterra-type integral equations of the second kind, Internat. Schriftenreihe Numer. Math. 53 (1980) 54–72.

[24] H. Brunner, The variation of constants formulas in the numerical analysis of integral and integro-differential equations, Utilitas Math. 19 (1980) 255–290.

[25] H. Brunner, A survey of recent advances in the numerical treatment of Volterra integral and integro-differential equations, J. Comput. Appl. Math. 8 (1982) 213–229.

[26] H. Brunner, Nonpolynomial spline collocation for Volterra equations with weakly singular kernels, SIAM J. Numer. Anal. 20 (1983) 1106–1119.

[27] H. Brunner, The numerical solution of weakly singular Volterra integral equations by collocation on graded meshes, Math. Comp. 45 (1985) 417–437.

[28] H. Brunner, The approximate solution of Volterra equations with nonsmooth solutions, Utilitas Math. 27 (1985) 57–95.

[29] H. Brunner, On the numerical solution by collocation of Volterra integro-differential equations with nonsmooth solutions, Internat. Schriftenreihe Numer. Math. 73 (1985) 74–92.

[30] H. Brunner, On the history of numerical methods for Volterra integral equations, CWI Newslett. 11 (1986) 3–20.

[31] H. Brunner, High-order methods for the numerical solution of Volterra integro-differential equations, J. Comput. Appl. Math. 15 (1986) 301–309.

[32] H. Brunner, Polynomial spline collocation methods for Volterra integro-differential equations with weakly singular kernels, IMA J. Numer. Anal. 6 (1986) 221–239.

[33] H. Brunner, Implicit Runge–Kutta–Nyström methods for general second-order Volterra integro-differential equations, Comput. Math. Appl. 14 (1987) 549–559.

[34] H. Brunner, The approximate solution of initial-value problems for general Volterra integro-differential equations, Computing 40 (1988) 125–137.

[35] H. Brunner, The numerical treatment of Volterra integro-differential equations with unbounded delay, J. Comput. Appl. Math. 28 (1989) 5–23.

[36] H. Brunner, Collocation methods for nonlinear Volterra integro-differential equations with infinite delay, Math. Comp. 53 (1989) 571–587.

[37] H. Brunner, Direct quadrature methods for nonlinear Volterra integro-differential equations with infinite delay, Utilitas Math. 40 (1991) 237–250.

[38] H. Brunner, Implicitly linear collocation methods for nonlinear Volterra equations, Appl. Numer. Math. 9 (1992) 235–247.

[39] H. Brunner, On discrete superconvergence properties of spline collocation methods for nonlinear Volterra integral equations, J. Comput. Math. 10 (1992) 348–357.

[40] H. Brunner, Open problems in the discretization of Volterra integral equations, Numer. Funct. Anal. Optim. 17 (1996) 717–736.

[41] H. Brunner, 1896–1996: one hundred years of Volterra integral equations of the first kind, Appl. Numer. Math. 24 (1997) 83–93.

[42] H. Brunner, E. Hairer, S.P. Nørsett, Runge–Kutta theory for Volterra integral equations of the second kind, Math. Comp. 39 (1982) 147–163.

[43] H. Brunner, Q. Hu, Q. Lin, Geometric meshes in collocation methods for Volterra integral equations with proportional time delays, Report 99-25 Seminar für Angewandte Mathematik, ETH, Zurich, 1999.

[44] H. Brunner, Y. Lin, S. Zhang, Higher accuracy methods for second-kind Volterra integral equations based on asymptotic expansions of iterated Galerkin methods, J. Integral Equations Appl. 10 (1998) 375–396.

[45] H. Brunner, A. Makroglou, R.K. Miller, On mixed collocation methods for Volterra integral equations with periodic solution, Appl. Numer. Math. 24 (1997) 115–130.

[46] H. Brunner, A. Pedas, G. Vainikko, The piecewise polynomial collocation method for nonlinear weakly singular Volterra equations, Math. Comp. 68 (1999) 1079–1095.

[47] H. Brunner, L. Qun, Y. Ningning, The iterative correction method for Volterra integral equations, BIT 36 (1996) 221–228.

[48] H. Brunner, Y. Ningning, On global superconvergence of iterated collocation solutions to linear second-kind Volterra integral equations, J. Comput. Appl. Math. 67 (1996) 185–189.
[49] H. Brunner, P.J. van der Houwen, The Numerical Solution of Volterra Equations, CWI Monographs, North-Holland, Amsterdam, 1986.
[50] H. Brunner, J.D. Lambert, Stability of numerical methods for Volterra integro-differential equations, Computing 12 (1974) 75–89.
[51] H. Brunner, S.P. Nørsett, Superconvergence of collocation methods for Volterra and Abel integral equations of the second kind, Numer. Math. 36 (1980/81) 347–358.
[52] T.A. Burton, Volterra Integral and Differential Equations, Academic Press, New York, 1983.
[53] B. Cahlon, Numerical solution of nonlinear Volterra integral equations, J. Comput. Appl. Math. 7 (1981) 121–128.
[54] B. Cahlon, On the stability of Volterra integral-equations with a lagging argument, BIT 35 (1995) 19–29.
[55] C. Corduneanu, Principles of Differential and Integral Operators, Chelsea, New York, 1988.
[56] C. Corduneanu, Integral Equations and Stability of Feedback Systems, Academic Press, New York, 1973.
[57] C. Corduneanu, Functional Equations with Causal Operators, Gordon and Breach, in preparation.
[58] C. Corduneanu, V. Lakshmikantham, Equations with unbounded delay: a survey, Nonlinear Anal. 4 (1980) 831–877.
[59] C. Corduneanu, I.W. Sandberg (Eds.), Volterra Equations and Applications, Gordon and Breach, Amsterdam, 2000.
[60] M.R. Crisci, V.B. Kolmanovskii, E. Russo, A. Vecchio, Stability of difference Volterra equations: direct Liapunov method and numerical procedure, Comput. Math. Appl. 36 (1998) 77–97.
[61] M.R. Crisci, V.B. Kolmanovskii, E. Russo, A. Vecchio, Boundedness of discrete Volterra equations, J. Math. Anal. Appl. 211 (1997) 106–130.
[62] M.R. Crisci, V.B. Kolmanovskii, E. Russo, A. Vecchio, Stability of continuous and discrete Volterra integro-differential equations by Liapunov approach, J. Integral Equations Appl. 7 (1995) 393–411.
[63] M.R. Crisci, E. Russo, A. Vecchio, Stability of collocation methods for Volterra integro-differential equations, J. Integral Equations Appl. 4 (1992) 491–507.
[64] M.R. Crisci, E. Russo, Z. Jackiewicz, A. Vecchio, Global stability of exact collocation methods for Volterra integro-differential equations, Atti Sem. Mat. Fis. Univ. Modena 39 (1991) 527–536.
[65] M.R. Crisci, E. Russo, A. Vecchio, Stability results for one-step discretized collocation methods in the numerical treatment of Volterra integral equations, Math. Comp. 58 (1992) 119–134.
[66] J. Dixon, S. McKee, R. Jeltsch, Convergence analysis of discretization methods for nonlinear first kind Volterra integral equations, Numer. Math. 49 (1986) 67–80.
[67] J. Dixon, S. McKee, A unified approach to convergence analysis of discretization methods for Volterra-type equations, IMA J. Numer. Anal. 5 (1985) 41–57.
[68] I.S. Duff, G.A. Watson (Eds.), The State of the Art in Numerical Analysis, Oxford University Press, New York, 1997.
[69] P.P.B. Eggermont, Improving the accuracy of collocation solutions of Volterra integral equations of the first kind by local interpolation, Numer. Math. 48 (1986) 263–279.
[70] P.P.B. Eggermont, On monotone Abel–Volterra integral equations on the half line, Numer. Math. 52 (1988) 65–79.
[71] P.P.B. Eggermont, Uniform error estimates of Galerkin methods for monotone Abel–Volterra integral equations on the half-line, Math. Comp. 53 (1989) 157–189.
[72] N.J. Ford, C.T.H. Baker, Qualitative behaviour and stability of solutions of discretised nonlinear Volterra integral equations of convolution type, J. Comput. Appl. Math. 66 (1996) 213–225.
[73] N.J. Ford, C.T.H. Baker, Preserving transient behaviour in numerical solutions of Volterra integral equations of convolution type, in: R. Agarwal, D. O'Regan (Eds.), Integral and Integro-Differential Equations, Gordon and Breach, Lausanne, 1999, pp. 77–89.
[74] N.J. Ford, C.T.H. Baker, J.A. Roberts, A. Nonlinear Volterra integro-differential equations — stability and numerical stability of θ-methods, J. Integral Equations Appl. 10 (1998) 397–416.
[75] R. Gorenflo, F. Mainardi, Fractional calculus: integral and differential equations of fractional order, in: A. Carpinteri, F. Mainardi (Eds.), Fractals and Fractional Calculus in Continuum Mechanics, CISM Courses and Lectures, Vol. 378, Springer, Vienna, 1997, pp. 223–276.
[76] R. Gorenflo, S. Vessela, Abel Integral Equations, Springer, Berlin, 1991.
[77] G. Gripenberg, S.-O. Londen, O. Staffans, Volterra Integral and Functional Equations, Cambridge University Press, Cambridge, 1990.

[78] E. Hairer, C. Lubich, On the stability of Volterra–Runge–Kutta methods, SIAM J. Numer. Anal. 21 (1984) 123–135.
[79] E. Hairer, C. Lubich, An extension of ODE-methods to Volterra integral equations, Teubner-Texte Math. 82 (1986) 55–63.
[80] E. Hairer, C. Lubich, S.P. Nørsett, Order of convergence of one-step methods for Volterra integral equations of the second kind, SIAM J. Numer. Anal. 20 (1983) 569–579.
[81] E. Hairer, C. Lubich, M. Schlichte, Fast numerical solution of weakly singular Volterra integral equations, J. Comput. Appl. Math. 23 (1988) 87–98.
[82] E. Hairer, C. Lubich, M. Schlichte, Fast numerical solution of nonlinear Volterra convolution equations, SIAM J. Sci. Statist. Comput. 6 (1985) 532–541.
[83] P.J. van der Houwen, Convergence and stability results in Runge–Kutta type methods for Volterra integral equations of the second kind, BIT 20 (1980) 375–377.
[84] P.J. van der Houwen, H.J.J. te Riele, Backward differentiation type formulas for Volterra integral equations of the second kind, Numer. Math. 37 (1981) 205–217.
[85] P.J. van der Houwen, P.H.M. Wolkenfelt, On the stability of multistep formulas for Volterra integral equations of the second kind, Computing 24 (1980) 341–347.
[86] P.J. van der Houwen, P.H.M. Wolkenfelt, C.T.H. Baker, Convergence and stability analysis for modified Runge–Kutta methods in the numerical treatment of second-kind Volterra integral equations, IMA J. Numer. Anal. 1 (1981) 303–328.
[87] Q. Hu, Superconvergence of numerical solutions to Volterra integral equations with singularities, SIAM J. Numer. Anal. 34 (1997) 1698–1707.
[88] H.M. Jones, S. McKee, Variable step size predictor–corrector schemes for second kind Volterra integral equations, Math. Comp. 44 (1985) 391–404.
[89] J.-P. Kauthen, Implicit Runge–Kutta methods for some integrodifferential-algebraic equations, Appl. Numer. Math. 13 (1993) 125–134.
[90] J.-P. Kauthen, Implicit Runge–Kutta methods for singularly perturbed integro-differential systems, Appl. Numer. Math. 18 (1995) 201–210.
[91] J.-P. Kauthen, A survey of singularly perturbed Volterra equations, Appl. Numer. Math. 24 (1997) 95–114.
[92] J.-P. Kauthen, H. Brunner, Continuous collocation approximations to solutions of first kind Volterra equations, Math. Comp. 66 (1997) 1441–1459.
[93] D. Kershaw, Some results for Abel–Volterra integral equations of the second kind, in: C.T.H. Baker, G.F. Miller (Eds.), Treatment of Integral equations by Numerical Methods, Academic Press, London, 1982, pp. 273–282.
[94] V. Lakshmikantham, S.G. Deo, Method of Variation of Parameters for Dynamic Systems, Gordon and Breach, Lausanne, 1998.
[95] V. Lakshmikantham, M. Rama Mohana Rao, Theory of Integro-differential Equations, Gordon and Breach, Lausanne, 1995.
[96] P. Linz, Analytical and Numerical Methods for Volterra Equations, SIAM, Philadelphia, PA, 1985.
[97] C. Lubich, Fractional linear multistep methods for Abel–Volterra integral equations of the first kind, IMA J. Numer. Anal. 7 (1987) 97–106.
[98] C. Lubich, A stability analysis of convolution quadratures for Abel–Volterra integral equations, IMA J. Numer. Anal. 6 (1986) 87–101.
[99] C. Lubich, Runge–Kutta theory for Volterra integro-differential equations, Numer. Math. 40 (1982) 119–135.
[100] C. Lubich, Runge–Kutta theory for Volterra and Abel integral equations of the second kind, Math. Comp. 41 (1983) 87–102.
[101] C. Lubich, Fractional linear multistep methods for Abel–Volterra integral equations of the second kind, Math. Comp. 45 (1985) 463–469.
[102] C. Lubich, On the stability of linear multistep methods for Volterra convolution equations, IMA J. Numer. Anal. 3 (1983) 439–465.
[103] C. Lubich, Runge–Kutta theory for Volterra and Abel integral equations of the second kind, Math. Comp. 41 (1983) 87–102.
[104] C. Lubich, Runge–Kutta theory for Volterra integro-differential equations, Numer. Math. 40 (1982) 119–135.
[105] C. Lubich, On the stability of linear multistep methods for Volterra integral equations of the second kind, in: C.T.H. Baker, G.F. Miller (Eds.), Treatment of Integral equations by Numerical Methods, Academic Press, London, 1982, pp. 233–238.

[106] C. Lubich, I.H. Sloan, V. Thomée, Nonsmooth data error estimates for approximations of an evolution equation with a positive-type memory term, Math. Comp. 65 (1996) 1–17.

[107] S. McKee, Generalized discrete Gronwall lemmas, Z. Angew. Math. Mech. 62 (1982) 429–434.

[108] S. McKee, Volterra integral and integro-differential equations arising from problems in engineering and science, Bull. Inst. Math. Appl. 24 (1988) 135–138.

[109] S. McKee, A review of linear multistep methods and product integration methods and their convergence analysis for first kind Volterra integral equations, in: C.T.H. Baker, G.F. Miller (Eds.), Treatment of Integral equations by Numerical Methods, Academic Press, London, 1982, pp. 153–161.

[110] S. McKee, H. Brunner, The repetition factor and numerical stability of Volterra integral equations, Comput. Math. Appl. 6 (1980) 339–347.

[111] R.K. Miller, Nonlinear Volterra Integral Equations, W.A. Benjamin, Menlo Park, CA, 1972.

[112] O. Nevanlinna, On the numerical solutions of some Volterra equations on infinite intervals, Rev. Anal. Numér. Theéorie Approx. 5 (1976) 31–57.

[113] O. Nevanlinna, On the stability of discrete Volterra equations, in: C.T.H. Baker, G.F. Miller (Eds.), Treatment of Integral equations by Numerical Methods, Academic Press, London, 1982, pp. 139–147.

[114] P. Pouzet, Etude, en vue de leur approximation numérique, des solutions d'équations intégrales et intégro-différentielles de type Volterra pour des problèmes de conditions initiales, Thése, Université de Strasbourg, 1962.

[115] A.A. Roberts, Analysis of explosion for nonlinear Volterra equations, J. Comput. Appl. Math. 97 (1998) 153–166.

[116] L.F. Shampine, Solving Volterra integral equations with ODE codes, IMA J. Numer. Anal. 8 (1988) 37–41.

[117] S. Shaw, J.R. Whiteman, Applications and numerical analysis of partial differential Volterra equations: a brief survey, Comput. Methods Appl. Mech. Eng. 150 (1997) 397–409.

[118] Z.B. Tsalyuk, Volterra integral equations, J. Soviet Math. 12 (1979) 715–758.

[119] A. Vecchio, Stability results on some direct quadrature methods for Volterra integro-differential equations, Dynamics Systems Appl. 7 (1998) 501–518.

[120] V. Volterra, Leçons sur les équations intégrales et intégro-différentielles, Gauthier-Villars, Paris, 1913.

[121] P.H.M. Wolkenfelt, The numerical analysis of reducible quadrature methods for Volterra integral and integro-differential equations — Academisch Proefschrift, Mathematisch Centrum, Amsterdam, 1981.

[122] P.H.M. Wolkenfelt, The construction of reducible quadrature-rules for Volterra integral and integro-differential equations, IMA J. Numer. Anal. 2 (1982) 131–152.

[123] P.H.M. Wolkenfelt, P.J. van der Houwen, C.T.H. Baker, Analysis of numerical methods for second kind Volterra equations by imbedding techniques, J. Integral Equations 3 (1981) 61–82.

Numerical methods for ordinary differential equations in the 20th century

J.C. Butcher *

The University of Auckland, Department of Mathematics, Private Bag 92019, Auckland, New Zealand

Abstract

Numerical methods for the solution of initial value problems in ordinary differential equations made enormous progress during the 20th century for several reasons. The first reasons lie in the impetus that was given to the subject in the concluding years of the previous century by the seminal papers of Bashforth and Adams for linear multistep methods and Runge for Runge–Kutta methods. Other reasons, which of course apply to numerical analysis in general, are in the invention of electronic computers half way through the century and the needs in mathematical modelling of efficient numerical algorithms as an alternative to classical methods of applied mathematics. This survey paper follows many of the main strands in the developments of these methods, both for general problems, stiff systems, and for many of the special problem types that have been gaining in significance as the century draws to an end.

Keywords: Initial value problems; Adams–Bashforth method; Adams–Moulton method; Runge–Kutta method; Consistency; Stability and convergence; Order of methods; Stiff problems; Differential equation software

1. Introduction

It is not possible to assess the history of this subject in the 20th century without first recognizing the legacy of the previous century on which it has been built. Notable are the 1883 paper of Bashforth and Adams [5] and the 1895 paper of Runge [57]. Not only did the former present the famous Adams–Bashforth method, which plays an essential part in much modern software, but it also looked ahead to the Adams–Moulton method and to the practical use of Taylor series methods. The paper by Runge is now recognized as the starting point for modern one-step methods. These early contributions, together with a brief introduction to the fundamental work of Euler, will form the subject matter of Section 2.

* Tel.: +64-9-373-7599; fax: +64-9-373-7457.

E-mail address: butcher@scitec.auckland.ac.nz (J.C. Butcher).

These early papers each formulates the general initial value problem in much the same form. That is, given a function $f(x, y)$ and an "initial value" y_0, corresponding to a solution value at x_0, we seek to evaluate numerically the function y satisfying

$$y'(x) = f(x, y(x)), \quad y(x_0) = y_0. \tag{1}$$

The basic approach is to extend the set of x values for which an approximation to $y(x)$ is known, in a step-by-step fashion.

In the early writing on this problem, y is regarded as a scalar value function but the generalization to more general problems is suggested by a consideration of a pair of simultaneous equations

$$y'(x) = f(x, y(x), z(x)), \quad y(x_0) = y_0,$$

$$z'(x) = g(x, y(x), z(x)), \quad z(x_0) = z_0.$$

Today it is more natural to use formulation (1) but to interpret y as a vector-valued function. In this case, it is even possible to consider an autonomous system of differential equations

$$y'(x) = f(y(x)), \tag{2}$$

because, if necessary, x can be appended to $y(x)$ as an additional component satisfying the trivial differential equation $\mathrm{d}x/\mathrm{d}x = 1$.

After the section dealing with 19th century contributions, this review paper is divided into a number of further sections dealing either with specific periods of time or with contributions with a unifying theme. The development of algorithms based on linear multistep methods continued with the paper of Moulton [49] and to the predictor–corrector formulation together with local error estimation using Milne's device. This will be discussed in Section 2.6.

Sections follow on Runge–Kutta methods and on Taylor series methods. Special methods are needed for stiff problems, and we review some of the stability and other issues involved with the phenomenon of stiffness in Section 6. The development of software to solve initial value problems is discussed in Section 7. Finally, we discuss in Section 8 a number of identifiable problem classes that call for special techniques and special methods.

2. Early work on numerical ordinary differential equations

2.1. *The Adams–Bashforth paper*

The famous booklet by Bashforth and Adams [5] has a very long title but, when this is broken into two halves, as it appears on the title page, Fig. 1, the authorship of the two distinct aspects of the work is clearly ascribed to the separate authors. Thus, we may assume that the numerical component of this work is due to Mr Adams.

The numerical discussion begins by pointing out that given, for example, a second-order differential equation

$$\frac{\mathrm{d}^2 y}{\mathrm{d}t^2} = f\left(\frac{\mathrm{d}y}{\mathrm{d}t}, y, t\right),$$

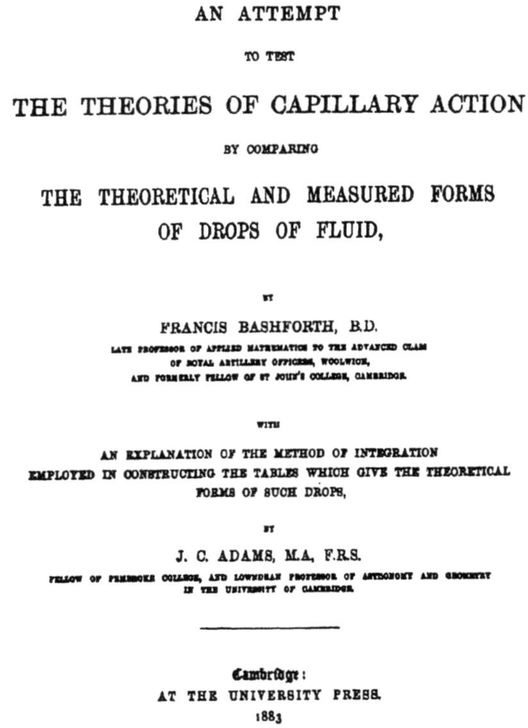
AN ATTEMPT

TO TEST

THE THEORIES OF CAPILLARY ACTION

BY COMPARING

THE THEORETICAL AND MEASURED FORMS
OF DROPS OF FLUID,

BY

FRANCIS BASHFORTH, B.D.

LATE PROFESSOR OF APPLIED MATHEMATICS TO THE ADVANCED CLASS
OF ROYAL ARTILLERY OFFICERS, WOOLWICH,
AND FORMERLY FELLOW OF ST JOHN'S COLLEGE, CAMBRIDGE.

WITH

AN EXPLANATION OF THE METHOD OF INTEGRATION
EMPLOYED IN CONSTRUCTING THE TABLES WHICH GIVE THE THEORETICAL
FORMS OF SUCH DROPS,

BY

J. C. ADAMS, M.A., F.R.S.

FELLOW OF PEMBROKE COLLEGE, AND LOWNDEAN PROFESSOR OF ASTRONOMY AND GEOMETRY
IN THE UNIVERSITY OF CAMBRIDGE.

Cambridge:
AT THE UNIVERSITY PRESS.
1883

Fig. 1. The title page of the Adams–Bashforth paper.

it is possible to find, by repeated differentiation and substitution of $\mathrm{d}^2y/\mathrm{d}t^2$ into the result, formulas for

$$\frac{\mathrm{d}^3y}{\mathrm{d}t^3}, \frac{\mathrm{d}^4y}{\mathrm{d}t^4}, \ldots .$$

From these data evaluated at the initial value, the solution may then be advanced using the Taylor series. Hence, after a small time-step, values of y and of $\mathrm{d}y/\mathrm{d}t$ can be found. Further steps can then be taken in the same manner until a desired value of t is reached.

After these remarks, Adams goes on to derive the Adams–Bashforth method, as we know it today, in the form

$$y_1 - y_0 = \omega\left(q_0 + \tfrac{1}{2}\Delta q_0 + \tfrac{5}{12}\Delta^2 q_0 + \cdots\right), \tag{3}$$

where ω is the stepsize and $q_0, q_{-1}, \ldots$ denote the derivatives computed at the points $t_0, t_{-1}, \ldots$ where the solution values are $y_0, y_{-1}, \ldots$. In the Adams notation, Δ denotes the backward difference $\Delta q_0 = q_0 - q_{-1}$, in contrast to the modern terminology of reserving Δ for the forward difference and using ∇ for the backward difference.

Adams goes on to discuss the relative merits of using, instead of (3), the formula

$$y_0 - y_{-1} = \omega\left(q_0 - \tfrac{1}{2}\Delta q_0 - \tfrac{1}{12}\Delta^2 q_0 + \cdots\right). \tag{4}$$

He correctly observes the advantages of (4) in terms of magnitudes of the error constants. The use of this implicit form of the Adams method was revisited and developed many years later by Moulton [49].

Eine ähnliche Ueberlegung führt nun auch für die Differentialgleichungen zu einer wesentlichen Verbesserung des Euler'schen Verfahrens. Ich will mich zunächst auf Differentialgleichungen erster Ordnung beschränken.

Statt

(1) $$\Delta y = f(x_0 y_0)\,\Delta x \quad \text{u. s. w.}$$

ist es schon viel besser wenn man

(2) $$\Delta y = f\left(x_0 + \tfrac{1}{2}\Delta x,\ y_0 + \tfrac{1}{2} f(x_0 y_0)\,\Delta x\right)\Delta x$$

u. s. w.

setzt. Diese Art der Berechnung entspricht dem aus der Summe der Tangententrapeze gebildeten Näherungswerthe eines Integrals und deckt sich völlig damit, wenn $f(xy)$ von y unabhängig vorausgesetzt wird.

Oder man kann der Summe der Sehnentrapeze entsprechend setzen:

(3) $$\Delta y = \frac{f(x_0 y_0) + f\left(x_0 + \Delta x,\ y_0 + f(x_0 y_0)\,\Delta x\right)}{2}\,\Delta x$$

u. s. w.

Fig. 2. An extract from the Runge paper.

2.2. The Runge paper

The second great legacy of the 19th century to numerical methods for ordinary differential equations was the work of Runge [57]. Whereas the Adams method was based on the approximation of the solution value for given x, in terms of a number of previously computed points, the approach of Runge was to restrict the algorithm to being "one step", in the sense that each approximation was based only on the most recent point already computed in a previous step. To achieve the required accuracy, approximations are found at a number of internal points within each step and the final result is computed in terms of these various stage values. The short extract from Runge's paper given in Fig. 2, includes the formulations of methods with two derivative calculations per step, based on the mid-point and trapezoidal quadrature rules, respectively.

2.3. The contributions of Heun and Kutta

Following the important and prophetic work of Adams and of Runge, the new century began with further contributions to what is now known as the Runge–Kutta method, by Heun [40] and Kutta [45]. In particular, the famous method in Kutta's paper is often known as *the* Runge–Kutta method. Heun's contribution was to raise the order of the method from two and three, as in Runge's paper, to four. This is an especially significant contribution because, for the first time, numerical methods for differential equations went beyond the use of what are essentially quadrature formulas. Even though second-order Runge methods can be looked at in this light, because the derivatives of the solution are computed from accurate enough approximations so as not to disturb the second-order behaviour, this is no longer true for orders greater than this. Write a three stage method in the form

$$Y_1 = y_0, \quad F_1 = f(x_0, Y_1),$$

$$Y_2 = y_0 + ha_{21}F_1, \quad F_2 = f(x_0 + hc_2, Y_2),$$

$$Y_3 = y_0 + h(a_{31}F_1 + a_{32}F_2), \quad F_3 = f(x_0 + hc_3, Y_3),$$

$$y_1 = y_0 + h(b_1F_1 + b_2F_2 + b_3F_3),$$

where a_{21}, a_{31}, a_{32}, b_1, b_2, b_3, c_2, c_3 are constants that characterize a particular method in this family. We can view computation of the stage values Y_1, identical to the initial value for the step, Y_2, which approximates the solution at x_0+hc_2 and Y_3, which approximates the solution at x_0+hc_3 as temporary steps, whose only purpose is to permit the evaluation of F_1, F_2 and F_3 as approximations to $y'(x_0)$, $y'(x_0+hc_2)$ and $y'(x_0+hc_3)$, respectively. From these derivative approximations, the result at the end of the step is found from the quadrature approximation

$$y(x_0+h) \approx y(x_0) + h(b_1 y'(x_0) + b_2 y'(x_0+hc_2) + b_3 y'(x_0+hc_3)).$$

It is essential that this quadrature formula be sufficiently accurate to integrate polynomials of degree up to 2 exactly. This gives the conditions

$$b_1 + b_2 + b_3 = 1,$$

$$b_2c_2 + b_3c_3 = \tfrac{1}{2},$$

$$b_2c_2^2 + b_3c_3^3 = \tfrac{1}{3}.$$

However, because of possible inaccuracies in the computation of Y_2 and Y_3 as approximations to $y(x_0+hc_2)$ and $y(x_0+hc_3)$, respectively, the quadrature conditions are not enough and it is also necessary that

$$b_3a_{32}c_2 = \tfrac{1}{6},$$

to obtain third-order behaviour.

An example of a method due to Heun which satisfies the four conditions for this order uses the coefficients

$$c_2 = \tfrac{1}{3}, \quad c_3 = 1, \quad a_{21} = \tfrac{1}{3}, \quad a_{31} = 0, \quad a_{32} = \tfrac{2}{3}, \quad b_1 = \tfrac{1}{4}, \quad b_2 = 0, \quad b_3 = \tfrac{3}{4}.$$

Kutta took this investigation further and found a complete classification of the solutions to the eight conditions for four-stage methods with order 4. He also derived the 16 conditions for order 5.

The extract of Kutta's paper given in Fig. 3, includes the formulation of the method, together with the order conditions and the first line of the solution in the case that 0, c_2, c_3 and c_4 are all distinct numbers. In his notation we see that $\kappa = c_2$, $\lambda = c_3$ and $\mu = c_4$. It is an interesting consequence of these order conditions, that μ is necessarily equal to 1.

Of the various four stages, fourth-order methods derived by Kutta, the most famous, and also the most widely used, is

$$Y_1 = y_0, \quad F_1 = f(x_0, Y_1),$$

$$Y_2 = y_0 + \tfrac{1}{2}hF_1, \quad F_2 = f(x_0 + \tfrac{h}{2}, Y_2),$$

$$Y_3 = y_0 + \tfrac{1}{2}hF_2, \quad F_3 = f(x_0 + \tfrac{h}{2}, Y_3),$$

$$Y_4 = y_0 + hF_3, \quad F_4 = f(x_0 + h, Y_4),$$

$$y_1 = y_0 + h(\tfrac{1}{6}F_1 + \tfrac{1}{3}F_2 + \tfrac{1}{3}F_3 + \tfrac{1}{6}F_4).$$

The set of conditions for fifth-order methods is actually a little more complicated than Kutta realised, because there are actually 17 conditions. The reason for the discrepancy is that he was dealing with scalar differential equations, rather than vector-valued differential equations, and for orders five

Zu den Näherungen vierter Ordnung übergehend finden wir die Berechnung von vier Funktionswerten nötig, und die Vergleichung des Taylorschen Satzes ergiebt die folgenden acht Bedingungsgleichungen für die Koeffizienten:

$$a + b + c + d = 1,$$
$$b\varkappa + c\lambda + d\mu = \tfrac{1}{2},$$
$$b\varkappa^2 + c\lambda^2 + d\mu^3 = \tfrac{1}{3},$$
$$c\varrho\varkappa + d(\sigma\lambda + \tau\varkappa) = \tfrac{1}{6},$$
$$b\varkappa^3 + c\lambda^3 + d\mu^3 = \tfrac{1}{4},$$
$$c\varrho\varkappa\lambda + d(\sigma\lambda + \tau\varkappa)\mu = \tfrac{1}{8},$$
$$c\varrho\varkappa^2 + d(\sigma\lambda^2 + \tau\varkappa^2) = \tfrac{1}{12},$$
$$d\varrho\sigma\varkappa = \tfrac{1}{24},$$

wenn als gewünschte Näherung angesetzt ist:

$$\Delta y = a\Delta' + b\Delta'' + c\Delta''' + d\Delta'''',$$
$$\Delta' = f(x, y)\Delta x,$$
$$\Delta'' = f(x + \varkappa\Delta x,\ y + \varkappa\Delta')\Delta x,$$
$$\Delta''' = f(x + \lambda\Delta x,\ y + \varrho\Delta'' + (\lambda - \varrho)\Delta')\Delta x,$$
$$\Delta'''' = f(x + \mu\Delta x,\ y + \sigma\Delta''' + \tau\Delta'' + (\mu - \sigma - \tau)\Delta')\Delta x.$$

Hier läſst sich das Lösungssystem auch noch niederschreiben. Wenn man $\varkappa$ und λ willkürlich läſst, erhält man nach einiger Rechnung:

$$c = \frac{1 - 2\varkappa}{12\lambda(\lambda - \varkappa)(1 - \lambda)},\ b = \frac{1 - 2\lambda}{12\varkappa(\varkappa - \lambda)(1 - \varkappa)},\ d = \frac{6\varkappa\lambda - 4(\varkappa + \lambda) + 3}{12(1 - \lambda)(1 - \varkappa)};$$

Fig. 3. An extract from the Kutta paper.

or greater the conditions become different. Another difficulty is in actually finding solutions to the algebraic conditions and Kutta presented methods that are slightly incorrect. It is interesting that once the correction is made, the additional condition, to make the method applicable to high-dimensional problems, happens to be satisfied.

2.4. The contributions of E.J. Nyström

The early history of Runge–Kutta methods culminated in the work of Nyström [53] in 1925. He was able to correct some of the fifth-order methods of Kutta and he also showed how to apply the Runge–Kutta method to second-order differential equation systems.

At first sight this is quite straightforward, because every second-order system can be re-formulated as a first-order system with additional dependent variables. However, solving such a problem directly may be much more efficient and the great prevalence of second-order problems in physical modelling makes this sort of gain in efficiency of considerable practical significance.

2.5. Moulton's paper and predictor–corrector methods

Implicit versions of Adams methods were first suggested in the Adams–Bashforth paper, but not studied in their own right until the paper of Moulton [49]. These so-called Adams–Moulton methods have two great advantages over the original explicit methods. The first is that they do not need to use so many past values to obtain the same order and they have smaller error constants. To use them in practice, however, one first has to overcome the difficulty associated with their implicit nature. This difficulty hinges on the fact that y_n is not given in terms of rational operations on known data, but as the solution to an algebraic equation. For example, consider the third-order Adams–Bashforth and Adams–Moulton methods given by

$$y_n = y_{n-1} + h(\tfrac{23}{12}f(x_{n-1}, y_{n-1}) - \tfrac{4}{3}f(x_{n-2}, y_{n-2}) + \tfrac{5}{12}f(x_{n-3}, y_{n-3})), \qquad (5)$$

$$y_n = y_{n-1} + h(\tfrac{5}{12}f(x_n, y_n) + \tfrac{2}{3}f(x_{n-1}, y_{n-1}) - \tfrac{1}{12}f(x_{n-2}, y_{n-2})). \qquad (6)$$

It is known that the error introduced into the result in a single step is $-\frac{3}{8}y^{(4)}h^4 + O(h^5)$ for the Adams–Bashforth method and $\frac{1}{24}y^{(4)}h^4 + O(h^5)$ for the Adams–Moulton method. The way that advantage is gained from the desirable properties of each of the methods is to use them in "predictor–corrector mode". This means that a predicted value of y_n is first found using the explicit form of the method. The implicit or Moulton form of the method is then used with the term $f(x_n, y_n)$ replaced by the value calculated using the *predicted* value of y_n. There are many variants of this method in common use, but the most popular is the so-called PECE mode. In this mode, $f(x_n, y_n)$ is re-evaluated for use in later steps using y_n found from the Adams–Moulton method. Thus each step requires two evaluations of the function f and is thus twice as expensive as the simple use of the Adams–Bashforth formula alone. However, the advantages in terms of stability and accuracy resulting from the use of this PECE predictor–corrector mode are usually regarded as well worth the additional computing cost.

2.6. The Milne device

Although Milne preferred methods based on Newton–Cotes quadrature formulas, methods which are largely abandoned today in favour of Adams methods, a proposal he made [47] has been adapted to other situations and widely used. In the context of the predictor pair (5) and (6), implemented, for example in PECE mode, there are two approximations to $y(x_n)$ computed in each step. Since the local truncation errors of the two approximations are in the ratio -9 to 1, it is proposed that the *difference* of the two approximations divided by 10 should be used as an estimate of the error in the corrected formula.

Milne, of course, intended this device to be used to check the accuracy of hand-computed results, but today it is used in automatic solvers, not just to verify the accuracy of any completed step, but also to adjust the size of a subsequent step in the interests both of efficiency and robustness.

Many modern computer codes implement predictor–corrector methods in a different manner than we have described. Specifically, the step number k is chosen to be the same for both the predictor and corrector formulas. This means that the order of the predictor will be k and the order of the corrector, which becomes the overall order of the combined method, will be $p = k + 1$. Even though the difference between the predicted and corrected solutions is no longer asymptotically equal to a multiple of the local truncation error, this difference is still used as the basis for stepsize control.

3. The modern theory of linear multistep methods

The modern analysis of linear multistep methods is intimately bound up with the work of Dahlquist [21,22]. This large body of work is in several parts, of which the first deals with the concepts of consistency, stability and convergence, expressed in terms of generating functions for the coefficients of the method. The key result in this phase of the work, is that consistency and stability are together equivalent to convergence. The second principle phase relates order of accuracy to stability and culminates in the famous "Dahlquist barrier" result, which limits the order of a convergent linear k-step method to $k+1$ (if k is odd) and to $k+2$ (if k is even). The remaining phase of Dahlquist's work is more appropriately discussed in Section 6.

3.1. Generating functions

Consider a linear multistep method of the form

$$\begin{aligned}&\alpha_k y_n + \alpha_{k-1} y_{n-1} + \alpha_{k-2} y_{n-2} + \cdots + \alpha_0 y_{n-k}\\ &\quad = h(\beta_k f(x_n, y_n) + \beta_{k-1} f(x_{n-1}, y_{n-1}) + \beta_{k-2} f(x_{n-2}, y_{n-2}) + \cdots + \beta_0 f(x_{n-k}, y_{n-k})),\end{aligned}$$

assuming that $\alpha_k \neq 0$ and that α_0 and β_0 are not both zero (otherwise the value of k could be reduced). Such a method is known as a "linear k-step method" because the solution at step number n depends on exactly k previous step values. Dahlquist introduced polynomials ρ and σ to characterize the method as follows:

$$\begin{aligned}\rho(z) &= \alpha_k z^k + \alpha_{k-1} z^{k-1} + \alpha_{k-2} z^{k-2} + \cdots + \alpha_0,\\ \sigma(z) &= \beta_k z^k + \beta_{k-1} z^{k-1} + \beta_{k-2} z^{k-2} + \cdots + \beta_0.\end{aligned}$$

Although Dahlquist allowed for the generality of allowing the coefficient of z^k to take on any non-zero value, in an actual computation with the method, the value of α_k has to be cancelled out from both polynomials.

It is clear that given any linear multistep method, the corresponding pair of polynomials (ρ, σ) can be written down automatically and, given the polynomials, the method is completely specified. Hence, it has become customary to identify the methods with the pair of polynomials and we can speak of "the method (ρ, σ)". It is convenient to assume that ρ and σ have no common polynomial factor, since it would be possible to describe most aspects of the computational behaviour of the method in terms of simpler polynomials. Following Dahlquist, we will make this assumption.

3.2. Consistency, stability and convergence

There are some natural assumptions that can be made about linear multistep methods to guarantee that they can at least solve certain specific problems. We will consider these one, by one.

The first problem is $y'(x) = 0$, with initial value $y(0) = 0$. Since we are given only this single initial value we will need an algorithm to generate $y_0, y_1, y_2, \ldots, y_{k-1}$ which is, in the limit as $h \to 0$, consistent with the given initial data. Choose some $x > 0$, for example $x = 1$, as the point where the numerical result approximating the solution is supposed to be found. We would like our method to be able to compute $y(1)$ exactly in the limiting case as $x \to 0$.

This requirement is equivalent to the "stability condition": A linear multistep method (ρ, σ) is stable if all zeros of ρ lie in the unit disc and all zeros on the boundary are simple.

The second initial value problem is also based on the equation $y'(x) = 0$ but with $y(0) = 1$. To compute the correct result $y(1) = 1$, in the limit, it is necessary that $\rho(1) = 0$. We will refer to this as the "pre-consistency condition".

Finally, consider the initial value problem $y'(x)=1$, $y(0)=0$. If a method is stable and pre-consistent, then its ability to solve this problem in the limit hinges on the requirement that $\rho'(1) = \sigma(1)$. This condition, when combined with the pre-consistency condition, is known as the "consistency condition".

The definition of convergence is rather technical but deals with the ability of the linear multistep method to solve *any* differential equation system on condition only that f is continuous in its first variable and satisfies a Lipschitz condition in its second variable. The k initial approximations required to start the numerical process must converge to the given initial value as the stepsize tends to zero. This class of problems might seem restrictive but it is easy to extend it to many situations where the Lipschitz condition is replaced by a *local* Lipschitz condition.

The basic theorem connecting these concepts is that a method is convergent if and only if it is both stable and consistent. Of course convergence is not enough to ensure that the method is computationally efficient. In the next section we look at the criteria for the method to have some specific order of accuracy and we review a famous result of Dahlquist which imposes a barrier on what order is really achievable.

3.3. The order of linear multistep methods

Given a linear multistep method characterized by the polynomials ρ and σ, define the operator L on the continuously differentiable functions $I \to \mathbb{R}^N$ by the formula

$$L(y)(x) = \sum_{i=0}^{k} \alpha_{k-i} y(x - ih) - h \sum_{i=0}^{k} \beta_{k-i} y'(x - ih). \tag{7}$$

A method is said to be of order p if $L(P) = 0$ for P any polynomial of degree not exceeding p.

To understand the significance of this definition, assume that y is continuously differentiable at least $p+1$ times and expand the right-hand side of (7) in a Taylor series about x_n. We have

$$L(y)(x_n) = \sum_{i=0}^{p+1} C_i h^i y^{(i)}(x_n) + O(h^{p+2}),$$

where

$$C_0 = \sum_{i=0}^{k} \alpha_i,$$

$$C_1 = -\sum_{i=1}^{k} i\alpha_{k-i} - \sum_{i=0}^{k} \beta_{k-i},$$

$$C_j = \frac{(-1)^j}{j!} \left(\sum_{i=1}^{k} i^j \alpha_{k-i} + j \sum_{i=1}^{k} i^{j-1} \beta_{k-i} \right), \quad j = 2, 3, \ldots, p+1.$$

If y is replaced by a polynomial of degree p, then

$$L(P)(x_n) = \sum_{i=0}^{p} C_i h^i P^{(i)}(x_n)$$

and, for this to vanish for all such polynomials, it is necessary and sufficient that

$$C_0 = C_1 = C_2 = \cdots = C_p = 0.$$

We derive the two methods (5) and (6) using these expressions. The Adams–Bashforth method of order 3 requires $k = 3$ and assumes that $\alpha_3 = 1$, $\alpha_2 = -1$, $\alpha_1 = \alpha_0 = \beta_3 = 0$. We have

$$C_0 = \alpha_3 + \alpha_2 + \alpha_1 + \alpha_0 \quad = 0,$$

$$C_1 = -\alpha_2 - 2\alpha_1 - 3\alpha_0 - \beta_2 - \beta_1 - \beta_0 \quad = 1 - \beta_2 - \beta_1 - \beta_0,$$

$$C_2 = \tfrac{1}{2}(\alpha_2 + 4\alpha_1 + 9\alpha_0 + 2(\beta_2 + 2\beta_1 + 3\beta_0)) \quad = \beta_2 + 2\beta_1 + 3\beta_0 - \tfrac{1}{2},$$

$$C_3 = -\tfrac{1}{6}(\alpha_2 + 8\alpha_1 + 27\alpha_0 - 3(\beta_2 + 4\beta_1 + 9\beta_0)) \quad = \tfrac{1}{6} - \tfrac{1}{2}(\beta_2 + 4\beta_1 + 9\beta_0).$$

The solution of $C_1 = C_2 = C_3 = 0$ is $\beta_2 = \frac{23}{12}$, $\beta_1 = -\frac{4}{3}$, $\beta_0 = \frac{5}{12}$, with the first nonzero coefficient in the Taylor expansion of $L(y)(x_n)$ given by

$$C_4 = -\tfrac{1}{24}(1 - 4(\beta_2 + 8\beta_1 + 27\beta_0)) = \tfrac{3}{8}.$$

The value of this quantity is closely related to the "error constant" for the method which is actually given by $C_4/\rho'(1)$. Note that, in this case, and also for the Adams–Moulton method which we will discuss next, $\rho'(1) = 1$.

For the Adams–Moulton method of order 3, an additional nonzero parameter β_k is available and $k = 2$ is sufficient for this order. We find $C_0 = 0$, $C_1 = 1 - \beta_2 - \beta_1 - \beta_0$, $C_2 = \beta_1 + 2\beta_0 - \frac{1}{2}$ and $C_3 = \frac{1}{6} - \frac{1}{2}(\beta_1 + 4\beta_0)$ and $C_1 = C_2 = C_3 = 0$ implies $\beta_2 = \frac{5}{12}$, $\beta_1 = \frac{2}{3}$, $\beta_0 = -\frac{1}{12}$, with

$$C_4 = -\tfrac{1}{24}(1 - 4(\beta_1 + 8\beta_0)) = -\tfrac{1}{24}.$$

To investigate the order conditions further, it is convenient to consider the expression $(\rho(\exp(z)) - z\sigma(\exp(z)))\exp(-kz)$ which can be expanded by Taylor series

$$\begin{aligned}(\rho(\mathrm{e}^z) - z\sigma(\mathrm{e}^z))\,\mathrm{e}^{-kz} &= \alpha_k + \alpha_{k-1}\mathrm{e}^{-z} + \alpha_{k-2}\mathrm{e}^{-2z} + \cdots \\ &\quad + z(\beta_k + \beta_{k-1}\mathrm{e}^{-z} + \beta_{k-2}\mathrm{e}^{-2z} - \cdots) \\ &= C_0 + C_1 z + C_2 z^2 + \cdots \\ &= C_{p+1} z^{p+1} + \mathrm{O}(z^{p+2}),\end{aligned}$$

if the order is p. The number C_{p+1} does not vanish unless the order is actually higher than p. Hence

$$\rho(\mathrm{e}^z) - z\sigma(\mathrm{e}^z) = \mathrm{O}(z^{p+1}).$$

Because $\rho(1) = 0$ for a consistent method, we can divide by z and we find

$$\frac{\rho(\mathrm{e}^z)}{z} - \sigma(\mathrm{e}^z) = \mathrm{O}(z^p)$$

and substituting $\exp(z)$ by $1 + z$

$$\frac{\rho(1+z)/z}{\ln(1+z)/z} - \sigma(1+z) = \mathrm{O}(z^p), \tag{8}$$

where $\ln(1+z)/z$ is defined in a neighbourhood of 0 by the series

$$\frac{\ln(1+z)}{z} = 1 - \frac{z}{2} + \frac{z^2}{3} - \cdots$$

so that

$$\left(\frac{\ln(1+z)}{z}\right)^{-1} = 1 + \frac{z}{2} - \frac{z^2}{12} + \frac{z^3}{24} - \frac{19z^4}{720} + \frac{3z^5}{160} + O(z^6).$$

Using this expression, (8) can be used to derive methods with specific choices of ρ. Rewriting in the form

$$\rho(1+z) = \sigma(1+z)\left(z - \frac{z^2}{2} + \frac{z^3}{3} - \cdots\right)$$

enables coefficients to be found for the backward difference and similar methods in which the form of σ is prescribed.

3.4. The Dahlquist barrier

Even though it is possible, in principle, for linear multistep methods to have order as high as $2k$, this does not yield stable methods if $k > 2$. This is a consequence of the so-called "Dahlquist barrier" [21], which states that

Theorem 1. *The order of a stable linear k-step method is bounded by*

$$p \leqslant \begin{cases} k+2, & k \text{ even}, \\ k+1, & k \text{ odd}. \end{cases}$$

Proof. We will give here a vastly abbreviated proof, along the same lines as originally given by Dahlquist. Let

$$r(z) = \rho\left(\frac{1+z}{1-z}\right)\left(\frac{1-z}{2}\right)^k,$$

$$s(z) = \sigma\left(\frac{1+z}{1-z}\right)\left(\frac{1-z}{2}\right)^k,$$

where we note that the order conditions can be rewritten in the form

$$\frac{r(z)/z}{\ln((1+z)/(1-z))/z} - s(z) = O(z^p). \tag{9}$$

Let $r(z) = a_0 + a_1 z + a_2 z^2 + \cdots + a_k z^k$ and $s(z) = b_0 + b_1 z + b_2 z^2 + \cdots + b_k z^k$, where $a_0 = 0$ by the consistency condition. By the stability condition, $a_1 \neq 0$ and no two of the coefficients in r can have opposite signs. If

$$\frac{1}{\ln((1+z)/(1-z))/z} = c_0 + c_2 z^2 + c_4 z^4 + \cdots,$$

it can be shown that $c_2, c_4, \ldots$ are all negative [21,39]. If (9) is to hold for $p > k+1$, then the coefficient of z^{p+1} in

$$(c_0 + c_2z^2 + c_4z^4 + \cdots)(a_1 + a_2z + \cdots + a_kz^{k-1}), \tag{10}$$

must vanish. If the order is $p > k+2$, then the coefficient of z^{k+2} in (10) must also vanish. The two coefficients are respectively

$$a_kc_2 + a_{k-2}c_4 + \cdots, \tag{11}$$

$$a_{k-1}c_4 + a_{k-3}c_6 + \cdots. \tag{12}$$

If k is odd, (11) cannot vanish because this would imply that

$$a_k = a_{k-2} = \cdots = a_1 = 0.$$

On the other hand, if k is even, then (12) cannot vanish because we would then have

$$a_{k-1} = a_{k-3} = \cdots = a_1 = 0.$$

4. The modern theory of Runge–Kutta methods

The meaning of order looks quite different and is relatively complicated for one-step methods, for the very good reason that the result computed in a step is built up from the derivatives evaluated sequentially from the stages values and, at least for the early stages, these have low accuracy. In contrast, the result computed in linear multistep methods makes use of derivatives evaluated from a number of step values, which themselves have been evaluated in previous steps and all share the same order.

The basic approach to the analysis of Runge–Kutta methods is to obtain the Taylor expansions for the exact and computed solutions at the end of a single step and to compare these series term by term. This idea dates back to Runge, Heun, Kutta and Nyström and we will give as an example the derivation of the conditions for order 3.

For the scalar differential equation

$$y'(x) = f(x, y(x)), \tag{13}$$

we calculate in turn

$$y'' = \frac{\partial f}{\partial x} + \frac{\partial f}{\partial y} f, \tag{14}$$

$$y''' = \frac{\partial^2 f}{\partial x^2} + 2\frac{\partial^2 f}{\partial x \partial y} f + \frac{\partial^2 f}{\partial y^2} f^2 + \frac{\partial f}{\partial x}\frac{\partial f}{\partial y} + \left(\frac{\partial f}{\partial y}\right)^2 f, \tag{15}$$

where we have substituted $y' = f$ in the formula $y'' = (\partial f/\partial x) + (\partial f/\partial y)y'$ to obtain (14) and made similar substitutions in the derivation of (15). From these expressions we can write down the first few terms of the Taylor expansion $y(x_0 + h) = y(x_0) + hy'(x_0) + \frac{1}{2}h^2y''(x_0) + \frac{1}{6}h^3y'''(x_0) + O(h^4)$.

Complicated though these expressions are, they are simple in comparison with the corresponding formulas for the fourth and higher derivatives. To obtain conditions for order 3 we also need the

Table 1
Details of Taylor expansions up to order 3

	Y_1	hF_1	Y_2	hF_2	Y_3	hF_3	y_1	$y(x_0+h)$
y	1	0	1	0	1	1	1	1
hf	0	1	a_{21}	1	$a_{31}+a_{32}$	1	$b_1+b_2+b_3$	1
$h^2\dfrac{\partial f}{\partial x}$	0	0	0	c_2	$a_{32}c_2$	c_3	$b_2c_2+b_3c_3$	$\frac{1}{2}$
$h^2\dfrac{\partial f}{\partial y}f$	0	0	0	a_{21}	$a_{32}a_{21}$	$a_{31}+a_{32}$	$b_2a_{21}+b_3(a_{31}+a_{32})$	$\frac{1}{2}$
$h^3\dfrac{\partial^2 f}{\partial x^2}$	0	0	0	$\frac{1}{2}c_2^2$	$\frac{1}{2}a_{32}c_2^2$	$\frac{1}{2}c_3^2$	$\frac{1}{2}(b_2c_2^2+b_3c_3^2)$	$\frac{1}{6}$
$h^3\dfrac{\partial^2 f}{\partial x\partial y}f$	0	0	0	c_2a_{21}	$a_{32}c_2a_{21}$	$c_3(a_{31}+a_{32})$	$b_2c_2a_{21}+b_3c_3(a_{31}+a_{32})$	$\frac{1}{3}$
$h^3\dfrac{\partial^2 f}{\partial y^2}f^2$	0	0	0	$\frac{1}{2}a_{21}^2$	$\frac{1}{2}a_{32}a_{21}^2$	$\frac{1}{2}(a_{31}+a_{32})^2$	$\frac{1}{2}(b_2a_{21}^2+b_3(a_{31}+a_{32})^2)$	$\frac{1}{6}$
$h^3\dfrac{\partial f}{\partial x}\dfrac{\partial f}{\partial y}$	0	0	0	0	0	$a_{32}c_2$	$b_3a_{32}c_2$	$\frac{1}{6}$
$h^3\left(\dfrac{\partial f}{\partial y}\right)^2 f$	0	0	0	0	0	$a_{32}a_{21}$	$b_3a_{32}a_{21}$	$\frac{1}{6}$

formulas for the first, second and third derivatives of the approximation computed by a Runge–Kutta method, which we will assume is explicit and has exactly 3 stages.

To simplify notation we will denote x, y, f and the various partial derivatives, as being evaluated at the initial point (x_0, y_0) in a step and we will then find Taylor expansions in turn for $Y_1, hF_1, \dots, Y_2$, hF_2, Y_3, hF_3 and finally y_1. We will express the sequence of calculations in tabular form in Table 1, where the coefficients of y, hf, etc. are shown. In addition to the coefficients in the expansion of y_1, we append the corresponding coefficients for the exact solution at x_0+h.

By equating the last two columns of this table, we obtain conditions for order 3. These imply that

$$a_{21} = c_2, \tag{16}$$

$$a_{31} + a_{32} = c_3 \tag{17}$$

and that

$$b_1 + b_2 + b_3 = 1,$$

$$b_2c_2 + b_3c_3 = \tfrac{1}{2},$$

$$b_2c_2^2 + b_3c_3^3 = \tfrac{1}{3},$$

$$b_3a_{32}c_2 = \tfrac{1}{6}.$$

If $s = p$, which turns out to be possible for orders up to 4, conditions such as (16) and (17) always hold. Even for higher orders, where the argument is a little more complicated, there is never any reason for not assuming that

$$\sum_{j=1}^{s} a_{ij} = c_i, \quad i = 1, 2, \ldots, s, \tag{18}$$

where we adopt a convention that $a_{ij} = 0$ for $j \geqslant i$ in explicit methods. For the more general implicit methods, we will continue to assume (18).

There are three reasons for abandoning (1) as the standard problem and replacing it instead by (2), where the values of $y(x)$ are now in a finite-dimensional vector space rather than scalars. The first reason for the change to a high-dimensional *autonomous* problem is that there is no need to retain x as an argument of f in the vector case, because nonautonomous problems can always be transformed into equivalent autonomous problems by adding an additional component which always has a value exactly the same as x. A consideration of this formal re-formulation can be used to justify the assumption (18). The second reason is that the analysis is actually more straightforward in the autonomous vector case. Finally, it is found that the conditions for order as derived using the scalar first-order problem (13) are inadequate for specifying the order requirements for the general vector case. The two theories do not diverge until the fifth-order case is reached but after that the families of order conditions for the scalar and vector cases become increasingly different.

4.1. The order of Runge–Kutta methods

The analysis of order for the vector case that we present here is due to the present author [9] and is related to earlier work by Gill [33] and Merson [46]. Since it relates the various terms in the Taylor series expansion of both the exact solution and the approximation computed by a Runge–Kutta method, to the graphs known as "rooted trees" or arborescences, we briefly review rooted trees.

A rooted tree is simply a connected directed graph for which each vertex, except the root, has a single predecessor (or parent). The root has no predecessor. The order of a rooted tree t is the number of vertices. Denote this by $r(t)$. Clearly, the number of arcs in the rooted tree is $r(t) - 1$. Let a_n denote the number of distinct rooted trees with order n. Table 2 gives the first few values of a_n together with the sums $\sum_{i=1}^{n} a_i$.

The eight rooted trees for which $r(t) \leqslant 4$ are shown in Table 3, together with the values of $\sigma(t)$, the "symmetry" of t and $\gamma(t)$ the "density" of t. The quantity $\sigma(t)$ is the order of the group of permutations of the vertices which leave the structure unchanged, while $\gamma(t)$ is the product over all vertices of t of the total number of descendents (including the vertex itself) of this vertex. Also

Table 2
Numbers of trees and accumulated sums up to order 8

n	1	2	3	4	5	6	7	8
a_n	1	1	2	4	9	20	48	115
$\sum_{i=1}^{n} a_i$	1	2	4	8	17	37	85	200

Table 3
Various functions on trees

t	$\cdot$							
$r(t)$	1	2	3	3	4	4	4	4
$\sigma(t)$	1	1	2	1	6	1	2	1
$\gamma(t)$	1	2	3	6	4	8	12	24
$\alpha(t)$	1	1	1	1	1	3	1	1
$\beta(t)$	1	2	3	6	4	24	12	24

shown are the values of $\alpha(t)=r(t)!/\gamma(t)\sigma(t)$ and $\beta(t)=r(t)!/\sigma(t)$. The values of α and β have simple interpretations in terms of possible labellings of the vertices of a tree under various restrictions.

It can be shown that the Taylor expansion of the exact solution has the form

$$y(x_0+h)=y(x_0)+\sum_{t\in T}\frac{\alpha(t)h^{r(t)}}{r(t)!}F(t)(y_0) \tag{19}$$

and that the corresponding expansion for the solution computed using a Runge–Kutta method is

$$y(x_0)+\sum_{t\in T}\frac{\beta(t)\Phi(t)h^{r(t)}}{r(t)!}F(t)(y_0). \tag{20}$$

In each of these formulas, $F(t)$ is the "elementary differential" which we will define below and $\Phi(t)$ is the "elementary weight". The formula for $F(t)(y)$ is defined in terms of the differential equation and $\Phi(t)$ in terms of the Runge–Kutta method being used. Each of these quantities can be defined recursively but, for our present purposes, it will be enough to present one example, using a tree of order 7 and to list these quantities for all trees of order up to 4. In the special example, the tree t is shown with labels i, j, k, l, m, n, o attached to the vertices. The formula for $F(t)$, is given in terms of an expression for component number i, written as a superscript. The summation convention is assumed and f^i_{jkl} denotes the third partial derivative, $\partial^3 f^i/\partial y^j\partial y^k\partial y^l$, of f^i, evaluated at y, with similar meanings for f^j_m, f^k_{no}. The summations in the formula for $\Phi(t)$ are over all subscripts running from 1 to s. Note that the formula is simplified from $\Phi(t)=\sum b_i a_{il}a_{ij}a_{jm}a_{ik}a_{kn}a_{ko}$ by summing over l, m, n and o.

$t=$ 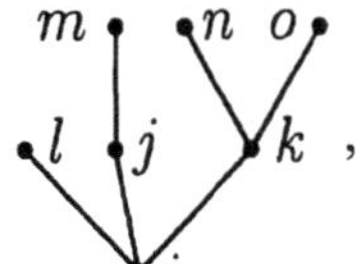

,

$$F^i(t)=f^i_{jkl}f^l f^j_m f^m f^k_{no} f^n f^o,$$

$$\Phi(t)=\sum b_i c_i a_{ij} c_j a_{ik} c_k^2.$$

Because the elementary differentials are independent, in the sense that, given any set of n rooted trees, $t_1, t_2, \ldots, t_n$ and any sequence of real numbers $q_1, q_2, \ldots, q_n$, it is possible to find a function f such that for some specific value of y and some specific coordinate direction, say e_1^{T}, all the equations

$$e_1^{\mathrm{T}}F(t_i)(y)=q_i,\quad i=1,2,\ldots,n,$$

Table 4
Elementary differential and weights up to order 4

t								
$F(t)$	f^i	$f^i_j f^j$	$f^i_{jk} f^j f^k$	$f^i_j f^j_k f^k$	f^i_{jkl}	$f^i_{jk} f^j f^k_l f^l$	$f^i_j f^j_{kl} f^k f^l$	$f^i_j f^j_k f^k_l f^l$
$\Phi(t)$	$\sum b_i$	$\sum b_i c_i$	$b_i c_i^2$	$\sum b_i a_{ij} c_k$	$\sum b_i c_i^3$	$\sum b_i c_i a_{ij} c_j$	$\sum b_i a_{ij} c_j^2$	$\sum b_i a_{ij} a_{jk} c_k$

can be satisfied simultaneously, it is only possible that (19) and (20) agree to within $O(h^{p+1})$ if

$$\alpha(t) = \beta(t)\Phi(t)$$

for every tree t with no more than p vertices.

Inserting the formulas for α and β, we find that

$$\Phi(t) = \frac{1}{\gamma(t)} \tag{21}$$

as the condition corresponding to this tree (Table 4).

It is interesting that, for the single first-order scalar differential equation (13), the independence of the elementary differentials breaks down and it turns out to be possible to obtain methods that have some specified order in this case, but a lower order for the more general system of equations given by (2). This effect occurs for order 5 and higher orders.

Other interpretations of order are of course possible. An alternative derivation of the order conditions, due to Albrecht [1], is based on expressions arising from the Taylor series for

$$y(x_0) + h\sum_{j=1}^{s} a_{ij} y'(x_0 + hc_j) - y(x_0 + hc_i) = \varepsilon^{(2)} h^2 + \varepsilon^{(3)} h^3 + \cdots,$$

where

$$\varepsilon_i^{(2)} = \sum_j a_{ij} c_j - \frac{1}{2} c_j^2,$$

$$\varepsilon_i^{(3)} = \sum_j a_{ij} c_j^2 - \frac{1}{3} c_j^3,$$

$$\vdots \quad \vdots$$

For order 4 for example, it is found to be necessary that

$$\sum_i b_i \varepsilon_i^{(2)} = 0, \quad \sum_i b_i c_i \varepsilon_i^{(2)} = 0, \quad \sum_i b_i a_{ij} \varepsilon_j^{(2)} = 0, \quad \sum_i b_i \varepsilon_i^{(3)} = 0,$$

which, together with the quadrature conditions

$$\sum_i b_i c_i^{k-1} = \frac{1}{k}, \quad k = 1, 2, 3, 4,$$

are equivalent to (21), up to order four. A third approach, due to Hairer and Wanner [36], is based on the use of B-series. This theory, used to study compositions of Runge–Kutta methods, is related to [13], and has applications also to more general problems and methods.

Table 5
Minimum s to obtain order p

p	1	2	3	4	5	6	7	8
s	1	2	3	4	6	7	9	11

4.2. Attainable order of Runge–Kutta methods

For explicit Runge–Kutta methods with s stages, there are $s(s+1)/2$ free parameters to choose. It is easy to show that an order p is possible only if $s \geqslant p$. Up to order 4, $s = p$ is actually possible. However, for $p > 4$, the relationship between the minimum s to obtain order p is very complicated but is partly given in Table 5. The results given for $p > 4$ were proved in [11,15].

For implicit Runge–Kutta methods, which we will discuss below, the relationship is much simpler. In fact, order p can be obtained with s stages if and only if $p \leqslant 2s$.

4.3. Implicit Runge–Kutta methods

One of the earliest references to implicitness, as applied to Runge–Kutta methods, was in the book by Kunz [44] where the method of Clippinger and Dimsdale was quoted. This method with tableau

$$\begin{array}{c|ccc} 0 & 0 & 0 & 0 \\ \frac{1}{2} & \frac{5}{24} & \frac{1}{3} & -\frac{1}{24} \\ 1 & \frac{1}{6} & \frac{2}{3} & \frac{1}{6} \\ \hline & \frac{1}{6} & \frac{2}{3} & \frac{1}{6} \end{array}$$

is the forerunner both of Lobatto methods and of block methods [59].

The method of Hammer and Hollingsworth [38] will be explored in some detail. It is the forerunner of Gauss and other important classes of methods. The coefficients for the method are

$$\begin{array}{c|cc} \frac{1}{2}-\frac{\sqrt{3}}{6} & \frac{1}{4} & \frac{1}{4}-\frac{\sqrt{3}}{6} \\ \frac{1}{2}+\frac{\sqrt{3}}{6} & \frac{1}{4}+\frac{\sqrt{3}}{6} & \frac{1}{4} \\ \hline & \frac{1}{2} & \frac{1}{2} \end{array}$$

This method has order 4. This is a little surprising because the eight conditions for this order have been seemingly satisfied using only the six free parameters in A and b^{T}. Although the order conditions are trivial to check, we will verify them below using an argument that illustrates what happens much more generally.

If the coefficient matrix A is allowed to be fully implicit, that is any element on or above the diagonal may have a non-zero value, then there are clearly more free parameters available to satisfy the order conditions. The advantages, in terms of order, are even greater than might be expected from a mere comparison of the number of conditions with the number of free parameters, because various simplifying assumptions are easier to satisfy. These simplifying assumptions bring about a drastic lowering of the number of further conditions required for order; furthermore they interact and reinforce each other.

The simplifying assumptions we will use, are denoted by $C(\eta)$, $D(\xi)$ and $E(\xi,\eta)$, where we have used the notation of [10]. In each case ξ and η are positive integers and the assumptions refer to some equalities involving the coefficients of a specific method. The specific meanings are

$$C(\eta):\ \sum_{j=1}^{s} a_{ij}c_j^{l-1} = \frac{1}{k}c_i^l, \quad i=1,2,\ldots,s,\ l=1,2,\ldots,\eta,$$

$$D(\xi):\ \sum_{i=1}^{s} b_i c_i^{k-1} a_{ij} = \frac{1}{k}b_j(1-c_j^k), \quad j=1,2,\ldots,s,\ k=1,2,\ldots,\xi,$$

$$E(\xi,\eta):\ \sum_{i=1}^{s} b_i c_i^{k-1} a_{ij} c_j^{l-1} = \frac{1}{l(k+1)}, \quad k=1,2,\ldots,\xi,\ l=1,2,\ldots,\eta.$$

Let us consider the relationship between these assumptions in the case that $\eta=\xi=s$ and the further assumption that c and b^{T} are chosen, as in the two-stage method we are considering, so that the c_i are the zeros of the degree s Legendre polynomial, shifted to the interval $[0,1]$, and b_i are the corresponding Gaussian weights. These assumptions on c and b^{T} will guarantee that $\sum_i b_i c_i^{k-1} = 1/k$ for $k=1,2,\ldots,2s$. Under this condition, $E(s,s)$ follows from $D(s)$ and because the linear combinations used to verify this have coefficients in a nonsingular (Vandermonde) matrix, the argument can be reversed. Similarly, $C(s)$ is also equivalent to $E(s,s)$.

In what has come to be referred to as a Gauss method, the b^{T} and c vectors are chosen to satisfy the requirements of Gaussian quadrature and the elements in each row of A are chosen so that $C(s)$ is satisfied. It then follows that $D(s)$ also holds. The method formed in this way always has order $2s$ and we will verify this for $s=2$ in Table 6. Where no reason is given, the result is because of Gaussian quadrature. In other cases the manipulations are based on $C(2)$ or $D(2)$ and make use of order conditions already verified earlier in Table 6. Gauss methods for arbitrary $s \geqslant 1$ were introduced in [10,17].

Methods also exist with order $2s-1$ based on Radau quadrature of type I ($c_1=0$) or type II ($c_s=1$). The most important of these are the Radau IIA methods. Some variants of Lobatto methods ($c_1=0$ and $c_s=1$) with order $2s-2$, were once considered attractive for practical computation but have been superseded by other implicit methods.

It is now generally believed that the proper role of implicit Runge–Kutta methods is in the solution of stiff problems (see Section 6). There is a conflict between the three aims of high accuracy, good stability, and low implementation cost. Gauss methods seem to be ideal from the stability and accuracy points of views but they are very expensive, because of the fully implicit structure of the coefficient matrix. The accuracy is not as good as might be expected from order considerations alone because of an "order reduction" phenomenon [55,29], but the cost alone is enough to make alternative methods more attractive.

4.4. DIRK and SIRK methods

An obvious alternative to fully implicit methods, is to insist that the coefficient matrix have a lower triangular structure, because in this case the stages can be evaluated sequentially and the cost of each is relatively low. It turns out to be an advantage to have the diagonal elements equal and this

Table 6
Verification of order conditions for 2 stage Gauss method

	$b_1 + b_2 = 1$
	$b_1c_1 + b_2c_2 = \frac{1}{2}$
	$b_1c_1^2 + b_2c_2^2 = \frac{1}{3}$
	$b_1(a_{11}c_1 + a_{12}c_2) + b_2c_2(a_{11}c_1 + a_{12}c_2)$ $= \frac{1}{2}(b_1c_1^2 + b_2c_2^2) = \frac{1}{6}$
	$b_1c_1^3 + b_2c_2^3 = \frac{1}{4}$
	$b_1c_1(a_{11}c_1 + a_{12}c_2) + b_2c_2(a_{11}c_1 + a_{12}c_2)$ $= \frac{1}{2}(b_1c_1^3 + b_2c_2^3) = \frac{1}{8}$
	$(b_1a_{11} + b_2a_{21})c_1^2 + (b_1a_{12} + b_2a_{22})c_2^2$ $= b_1(1 - c_1)c_1^2 + b_2(1 - c_2)c_2^2 = \frac{1}{12}$
	$\sum b_ia_{i1}(a_{11}c_1 + a_{12}c_2) + \sum b_ia_{i2}(a_{21}c_1 + a_{22}c_2)$ $= \frac{1}{2}(\sum b_ia_{i1}c_1^2 + \sum b_ia_{i2}c_2^2) = \frac{1}{24}$

additional requirement has little impact on the availability of methods of a required order with good stability. Methods of this type have been variously named "semi-implicit" [10], "semi-explicit" [51] and "diagonally implicit" or "DIRK" [2]. Although equal diagonals were originally built into the DIRK formulation, common usage today favours using this name more widely and using "SDIRK" (or "singly diagonally implicit") in the more restricted sense. Other key references concerning these methods are [3,18].

Singly implicit methods, *without* necessarily possessing the DIRK structure are those for which A has only a single eigenvalue $\sigma(A) = \{\gamma\}$ [7]. If the stage order is s, it turns out that the abscissae for the method satisfy $c_i = \gamma\xi_i$, where $\xi_1, \xi_2, \ldots, \xi_s$ are the zeros of the Laguerre polynomial L_s. The advantage of these methods is that for many large problems, the component of the computer cost devoted to linear algebra is little more than for a DIRK method. Various improvements to the design of SIRK methods have been proposed.

5. Nontraditional methods

While the traditional methods, linear multistep and Runge–Kutta, are widely used and are generally regarded as satisfactory for solving a wide variety of problems, many attempts have been made to extend the range of available methods. Some of these will be discussed in this section.

5.1. *Taylor series methods*

Because the Euler method is based on approximations to the Taylor expansion

$$y(x_0+h)\approx y(x_0)+hy'(x_0),$$

it is natural to ask if it is possible to take this expansion further by evaluating $y''(x_0)$, and possibly higher derivatives, by algebraic means. Algorithms for carrying out this process can be constructed using a recursive evaluation scheme. We mention two important early papers which exploit this idea, using a classical view of computer arithmetic but combined with this non-classical method of solution [4,32].

A second interesting and important contribution to Taylor series that has a further distinctive feature is the work of Moore [48]. The distinctive feature is that the work is carried out in the context of interval arithmetic. This means that it becomes possible, not only to advance the solution step-by-step in a relatively efficient manner, but it also becomes possible, owing to the standard bounds on the truncation error of a Taylor expansion, to obtain *rigorous* error bounds. Thus, in principle, it became possible to obtain intervals in which each component of the solution is *certain* to lie for any particular value of x. The difficulty is, of course, that the lengths of these intervals can grow rapidly as x increases.

5.2. *Hybrid methods*

These methods are similar to linear multistep methods in predictor–corrector mode, but with one essential modification: an additional predictor is introduced at an *offstep* point. This means that the final (corrector) stage has an additional derivative approximation to work from. This greater generality allows the consequences of the Dahlquist barrier to be avoided and it is actually possible to obtain convergent k-step methods with order $2k+1$ up to $k=7$. Even higher orders are available if two or more offstep points are used. The three independent discoveries of this approach were reported in [34,30,12]. Although a flurry of activity by other authors followed, these methods have never been developed to the extent that they have been implemented in general purpose software.

5.3. *Cyclic composite methods*

It is remarkable that even though a number of individual linear multistep methods may be unstable, it is possible to use them cyclically to obtain a method which, overall, is stable. An example of a fifth-order method given in the key paper on this subject [26] is as follows:

$$\begin{aligned} y_n = {} & -\tfrac{8}{11}y_{n-1}+\tfrac{19}{11}y_{n-2} \\ & +h(\tfrac{10}{33}f(x_n,y_n)+\tfrac{19}{11}f(x_{n-1},y_{n-1})+\tfrac{8}{11}f(x_{n-2},y_{n-2})-\tfrac{1}{33}f(x_{n-3},y_{n-3})), \end{aligned} \tag{22}$$

$$\begin{aligned} y_n = {} & \tfrac{449}{240}y_{n-1}+\tfrac{19}{30}y_{n-2}-\tfrac{361}{240}y_{n-3}+h(\tfrac{251}{720}f(x_n,y_n)+\tfrac{19}{30}f(x_{n-1},y_{n-1})-\tfrac{449}{240}f(x_{n-2},y_{n-2}) \\ & -\tfrac{35}{72}f(x_{n-3},y_{n-3})), \end{aligned} \tag{23}$$

$$\begin{aligned} y_n = {} & -\tfrac{8}{11}y_{n-1}+\tfrac{19}{11}y_{n-2} \\ & +h(\tfrac{10}{33}f(x_n,y_n)+\tfrac{19}{11}f(x_{n-1},y_{n-1})+\tfrac{8}{11}f(x_{n-2},y_{n-2})-\tfrac{1}{33}f(x_{n-3},y_{n-3})). \end{aligned} \tag{24}$$

The method is used cyclically in the sense that in each of three steps, the first uses (22), the second uses (23) and the third uses (24) (which in this particular example happens to be the same as the first member of the cycle). Methods of this family have been studied by a number of authors and have also been discovered for use with stiff problems.

5.4. *Rosenbrock methods*

Rosenbrock in his 1963 paper [56], discusses the problem of evaluating the stages of a diagonally implicit Runge–Kutta methods. Normally this is carried out by an iteration process based on Newton's method. For each stage and each iteration, an evaluation of f is carried out together with a solution of a linear equation system with matrix of coefficients of the form $I - h\lambda J$, where J denotes an approximation to the Jacobian matrix. The question was then asked if improved performance can be obtained by an alternative procedure in which exactly the same amount of work is performed but *only once* per stage, with the proviso that J is exactly the Jacobian evaluated at y_{n-1}. Amongst the examples of this type of "Rosenbrock method" given in the original paper, the following is identified as having order 2 and possessing L-stability:

$$\begin{aligned}
F_1 &= (I - h(1 - \tfrac{\sqrt{2}}{2})J)^{-1} f(y_{n-1}),\\
F_2 &= (I - h(1 - \tfrac{\sqrt{2}}{2})J)^{-1} f(y_{n-1} + h(\tfrac{1}{2}\sqrt{2} - \tfrac{1}{2})F_1),\\
y_n &= y_{n-1} + hF_2.
\end{aligned}$$

Amongst the many further contributions to the study of Rosenbrock methods, and their generalizations, we refer to [16,52,43].

6. Methods for stiff problems

The paper by Curtiss and Hirschfelder [20] is usually acknowledged as introducing numerical analysts to the phenomenon of stiffness. Much has been written about what "stiffness" really means but the property is generally understood in terms of what goes wrong when numerical methods not designed for such problems are used to try to solve them. For example, classical explicit Runge–Kutta methods were not intended to solve stiff problems but, when one attempts to use them, there is a severe restriction on stepsize that must be imposed, apparently because of stability rather than accuracy requirements. It is easy to see how this can come about for linear problems of the form

$$y'(x) = My(x),$$

if the matrix M happens to have all its eigenvalues close to zero or else in the left half complex plane and with a large magnitude. Assuming for simplicity that M can be diagonalized and that the problem is solved in its transformed form, the accuracy is determined by the ability of the numerical method to solve problems of the form

$$y'(x) = \mu y(x),$$

where $|\mu|$ is small. However, the stability of the numerical approximation is limited by the fact that we are simultaneously trying to solve a problem of the form

$$y'(x) = \lambda y(x),$$

where $|\lambda|$ is large. In the exact solution, terms of the second type correspond to rapidly decaying transients, whereas in the computed solution they represent unstable parasitic solutions, unless h is so small that $h\lambda$ lies in what is known as the "stability region".

To find the stability region for a numerical method it is necessary to consider the behaviour of the numerical method with a problem of just this type. For classical methods the behaviour depends on the product of h and λ which we will write as z. For the classical fourth-order Runge–Kutta method, the numerical solution for this problem satisfies

$$y_n = R(z)y_{n-1}, \tag{25}$$

where

$$R(z) = 1 + z + \frac{z^2}{2} + \frac{z^3}{6} + \frac{z^4}{24}$$

and the stability region is the set of points in the complex plane for which $|R(z)| \leqslant 1$.

If $z = h\lambda$ is outside this set, as it might well be, then selecting h to make $h\mu$ of a reasonable size will not be satisfactory because the unstable behaviour of the component of the solution associated with λ will swamp the solution and destroy its accuracy.

To analyse this type of possible instability, Dahlquist [23] introduced the concept of A-stability. A numerical method is said to be "A-stable" if its stability region includes all of the left half-plane. Even though the definition was first framed in the context of linear multistep methods, it was soon applied to Runge–Kutta methods, for which it takes a particularly simple form. Write Y for the vector of stage values, then this vector and the output approximation are related by

$$Y = y_{n-1}\mathrm{e} + zAY, \quad y_n = y_{n-1} + zb^{\mathrm{T}}Y,$$

leading to (25) with the stability function given by

$$R(z) = 1 + zb^{\mathrm{T}}(I - zA)^{-1}\mathrm{e}.$$

For explicit methods the stability region is always a bounded set and these methods cannot be A-stable. On the other hand for an implicit method, $R(z)$ has the form $N(z)/D(z)$ where the polynomials N and D can have degrees as high as s. Methods of arbitrarily high orders can be A-stable.

For a general linear multistep method, defined by polynomials ρ and σ, the stability region is the set of points z for which the polynomial $\rho(w) - z\sigma(w)$, of degree k in w satisfies the root condition. It was shown by Dahlquist [23] that for these methods, A-stability cannot exist for orders greater than 2.

6.1. Order stars

Runge–Kutta methods of Gauss type have stability functions of the form

$$R(z) = \frac{N(z)}{N(-z)},$$

where the polynomial N has degree s. Furthermore, $R(z)=\exp(z)+O(z^{2s+1})$. This is an example of a "Padé approximation" to the exponential function, in the sense that the order of approximation is exactly the sum of the numerator and denominator degrees. Not only are the diagonal members of the Padé table significant, but the same can be said of the first subdiagonal (with degree $s-1$ for the numerator and s for the denominator, because these correspond to the stability functions for the Radau IA methods, and for the practically important Radau IIA methods. The second subdiagonals are also important because they are the stability functions for the Lobatto IIIC methods. It is known that the Padé approximations to the exponential function, in each of these three diagonals, correspond to A-stable methods. It is also clear that the approximations above the main diagonal cannot share this property but what can be said about approximations *below* the second subdiagonal? Considerable evidence existed for the "Ehle conjecture" [27] which claimed that *none* of these stability functions can correspond to an A-stable method or, in Ehle's terminology, that they are not A-acceptable.

In 1978 a new method was discovered for settling this, and many related questions. This approach introduced "order stars" [60], based on relative stability regions. Rather than study the regions of the complex plane for which $|R(z)| < 1$, the regions are studied for which $|\exp(-z)R(z)| < 1$. Since A-stable methods are those for which the stability function has no poles in the left half-plane and has its magnitude bounded by 1 on the imaginary axis, changing from the stability function $R(z)$ to the relative stability function $\exp(-z)R(z)$ leaves this criterion unchanged, but introduces much more structure, because $\exp(-z)R(z) = 1 + Cz^{p+1} + O(z^{p+2})$, when z is small.

Consider Fig. 4 taken from [60]. Shown in this figure are the order stars of four Padé approximation, with degrees j (denominator) and k (numerator). The shaded parts of the figures, known as the fingers and characterized by $|\exp(-z)R(z)| > 1$ and the unshaded parts, the dual fingers characterized by $|\exp(-z)R(z)| < 1$ meet at 0 in a pattern determined by the sign of the real part of Cz^{p+1}, for $|z|$ small. This means that there will be exactly $p+1$ fingers and the same number of dual fingers meeting at zero. Furthermore, the angles subtended by each finger is the same $\pi/(p+1)$. It can be shown that all the bounded fingers contain poles and the bounded dual fingers contain zeros. The two upper figures are for A-stable methods, in which all the poles are in the right half-plane and no finger crosses the imaginary axis. The two lower figures, for which $j-k > 2$, cannot meet these requirements, because there are too many bounded fingers for it to be possible for them all to leave zero in the right-hand direction. Some of these fingers must leave zero in the left-hand direction and either reach poles in the left half-plane or cross the imaginary axis to reach poles in the right-hand half-plane. A rigorous form of this argument is used to prove the Ehle conjecture and to prove a number of other results concerning both one step and multistep methods.

A recent study of order stars, which reviews most of the work up to the present time, is available in the book by Iserles and Nørsett [42].

6.2. Nonlinear stability

To obtain a deeper understanding of the behaviour of stiff problems, and of the numerical methods used to solve them, Dahlquist in 1975 [24], studied nonlinear problems of the form

$$y'(x) = f(x, y(x)), \tag{26}$$

where f satisfies the dissipativity condition

$$\langle f(x,u) - f(x,v), u - v\rangle \leqslant 0 \tag{27}$$

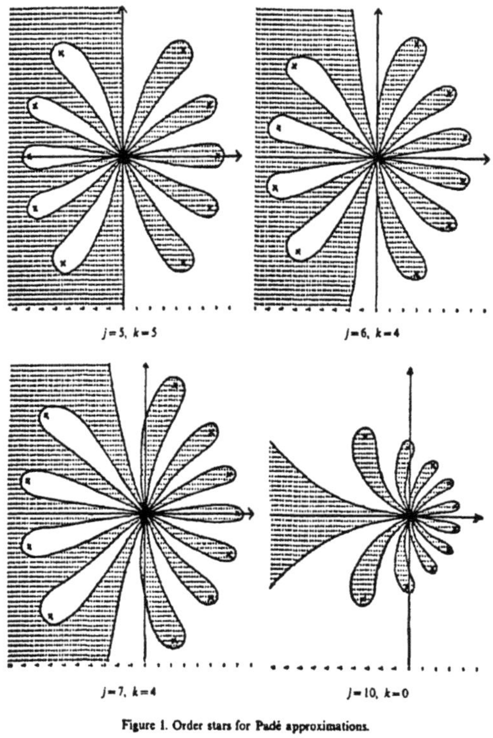

Fig. 4. An extract from the order star paper.

and $\langle\cdot\rangle$ denotes an inner product, with $\|\cdot\|$ the corresponding norm. It is easy to see that two exact solutions to (26) possess the property that

$$\|y(x)-z(x)\| \leqslant \|y(x_0)-z(x_0)\|, \quad \text{for } x \geqslant x_0. \tag{28}$$

The aim is now to find conditions on a method such that a discrete analogue of (28) holds. It turns out to be more convenient to consider instead of a linear multistep method

$$\begin{aligned}&\alpha_k y_n + \alpha_{k-1} y_{n-1} + \alpha_{k-2} y_{n-2} + \cdots + \alpha_0 y_{n-k}\\ &\quad = h(\beta_k f(x_n, y_n) + \beta_{k-1} f(x_{n-1}, y_{n-1}) + \cdots + \beta_0 f(x_{n-k}, y_{n-k})),\end{aligned}$$

the corresponding "one-leg method"

$$\begin{aligned}&\alpha_k y_n + \alpha_{k-1} y_{n-1} + \alpha_{k-2} y_{n-2} + \cdots + \alpha_0 y_{n-k}\\ &\quad = h\left(\sum_{i=0}^{k}\beta_i\right) f\left(\frac{\beta_k}{\sum_{i=0}^{k}\beta_i} y_n + \frac{\beta_{k-1}}{\sum_{i=0}^{k}\beta_i} y_{n-1} + \cdots + \frac{\beta_0}{\sum_{i=0}^{k}\beta_i} y_{n-k}\right).\end{aligned}$$

For this type of method, Dahlquist considered contractivity in the sense that

$$\|Y_n\| \leqslant \|Y_{n-1}\|,$$

where

$$Y_n = \begin{bmatrix} y_n \\ y_{n-1} \\ \vdots \\ y_{n-k+1} \end{bmatrix}$$

and

$$\left\| \begin{bmatrix} \eta_1 \\ \eta_2 \\ \vdots \\ \eta_k \end{bmatrix} \right\| = \sum_{i,j=1}^{k} g_{ij} \langle \eta_i, \eta_j \rangle .$$

It is assumed that

$$G = \begin{bmatrix} g_{11} & g_{12} & \cdots & g_{1k} \\ g_{21} & g_{22} & \cdots & g_{2k} \\ \vdots & \vdots & & \vdots \\ g_{k1} & g_{k2} & \cdots & g_{kk} \end{bmatrix}$$

is a positive-definite matrix.

It is explained in Dahlquist's paper how results for one-leg methods can be interpreted as having a significance also for the corresponding linear multistep methods. He also found necessary and sufficient conditions for this property to hold. In a later paper [25], he showed that for one-leg methods, A-stability and G-stability are essentially equivalent properties.

The corresponding theory for Runge–Kutta methods [14,19,8], leads to a consideration of a matrix M with (i,j) element equal to $b_i a_{ij} + b_j a_{ji} - b_i b_j$. Assuming that this matrix is positive semi-definite, and the same is true for $\operatorname{diag}(b_1, b_2, \ldots, b_s)$, then a Runge–Kutta method applied to two distinct solutions of (26), satisfying (27), satisfies the contractivity property

$$\| y_n - z_n \| \leqslant \| y_{n-1} - z_{n-1} \| .$$

It is interesting that M has a more modern role in connection with symplectic methods for Hamiltonian problems.

A further development, initiated in the paper [29], is connected with the behaviour of truncation error for Runge–Kutta methods applied to stiff problems.

7. The beginnings of differential equation software

Programs to solve differential equations are as old as modern computers themselves. Today, a central aim in the design of differential equation software is the building of general purpose codes, specific only as regards stiffness versus nonstiffness, which adapt their behaviour to that of the computed solution dynamically. Variable stepsize is a characteristic feature of this software and usually variable order is used as well.

The most famous of the early codes in this tradition is the FORTRAN subroutine named by its designer, Gear, as "DIFSUB" [31]. Actually, this name was used generally by Gear for a range of possible subroutines using a range of different methods. We will concentrate for the moment on the linear multistep version of DIFSUB.

As with all linear multistep implementations, the characteristic problems of starting values, local error estimation, change of stepsize and output interpolation have to be solved. A basic principle used in DIFSUB are the use of the Nordsieck representation of the data passed between steps, and this plays a crucial role in the solution of all these implementation questions, as well as the further problem of variable order.

The single paper of Nordsieck [50] explains how it is possible to rewrite a k-step Adams methods so that information on the values of $hf(x_{n-i}, y_{n-i})$ for $i=1,2,\dots,k$ is organized as approximations to $hy'(x_{n-1}), \frac{1}{2!}h^2y''(x_{n-1}),\dots,\frac{1}{k!}h^ky^{(k)}(x_{n-1})$. The rules for integration to the next step, are particularly simple in the case of the Adams–Bashforth method. The solution is first extrapolated using the approximation

$$\begin{bmatrix} y(x_n) \\ hy'(x_n) \\ \frac{1}{2!}h^2y''(x_n) \\ \vdots \\ \frac{1}{k!}h^ky^{(k)}(x_n) \end{bmatrix} \approx \begin{bmatrix} 1 & 1 & 1 & 1 & \cdots & 1 \\ 0 & 1 & 2 & 3 & \cdots & k \\ 0 & 0 & 1 & 3 & \cdots & \binom{k}{2} \\ \vdots & \vdots & \vdots & \vdots & & \vdots \\ 0 & 0 & 0 & 0 & \cdots & 1 \end{bmatrix} \begin{bmatrix} y(x_{n-1}) \\ hy'(x_{n-1}) \\ \frac{1}{2!}h^2y''(x_{n-1}) \\ \vdots \\ \frac{1}{k!}h^ky^{(k)}(x_{n-1}) \end{bmatrix}$$

and a correction is then made to each component using a multiple of $hf(x_n, y_n) - hy'(x_n)$, so as to ensure that the method is equivalent to the Adams–Bashforth method. Adding an Adams–Moulton corrector to the scheme, is equivalent to adding further corrections.

Using the Nordsieck representation, it is possible to change stepsize cheaply, by simply rescaling the vector of derivative approximations. It is possible to estimate local truncation error using the appropriately transformed variant of the Milne device. It is also possible to measure the accuracy of lower and one higher-order alternative methods so that the appropriateness of order-changing can be assessed. Thus the ingredients are present to build a completely adaptive nonstiff solver. By adapting the backward difference methods to a similar form, it is possible to allow also for stiffness.

The DIFSUB program of Gear uses these techniques to obtain an efficient solver and many later programs are based on similar ideas. The first general purpose solver for differential–algebraic equations, the DASSL subroutine of Petzold [54], is also closely related to DIFSUB.

Early success was also achieved in the algorithm of Bulirsch and Stoer [6]. This used extrapolation in a similar manner to the quadrature algorithm of Romberg. The main difference between differential equations and quadrature is that significant efficiency gains are made by reusing some of the abscissae in a quadrature formula; this happens in the traditional version of the Romberg method because the mesh size is halved in each iteration. For differential equations there is no advantage in this, because reuse is not possible. Hence, in later developments of extrapolation methods, for both nonstiff and stiff problems, various sequences of stepsizes have been considered, where the aim is to balance computational cost against the quality of the convergence.

As many programs became available, using a variety of methods and variety of implementations of the same basic method, it became appropriate to consider what is really expected of these automatic solvers. Each person who develops this software needs to apply quality tests and to compare any new implementation against existing codes. In the interests of providing objective standards, a number of test sets have been developed. The earliest of these that has become widely adopted, and which in fact serves as a de facto standard, is the DETEST problem set [41]. While this is appropriate for testing and comparing nonstiff solvers, a stiff counterpart, known as STIFF DETEST, [28] became available a little later.

8. Special problems

While the general aim of providing accurate and efficient general purpose numerical methods and algorithms has been a central activity in the 20th century, there has always been a realization that some problem types have such distinctive features that they will need their own special theory and techniques. Stiff problems were recognized approximately half way through the century as such a problem type and these have received considerable attention, especially in the last 30 years.

Another of the special problem types that has a separate claim for its own special methods, has been second-order differential equations and systems. These have a natural importance as arising in classical mechanical modelling and they were treated as a particular case by Nyström and others. While any problem of this type can be rewritten as a first-order system, it is found that treating them directly can lead to substantial gains in efficiency, especially if the second-order system takes the special form

$$y''(x) = f(x, y(x)),$$

where we note that $y'(x)$ does not occur as an argument of f. The Runge–Kutta approach to this type of problem was studied by Nyström [53] and has been of interest ever since. A modern theory of these methods is given in [37]. Linear multistep methods for this problem were studied as part of an investigation of a more general situation

$$y^{(n)}(x) = f(x, y(x)),$$

by Dahlquist [22].

It is interesting that one of the most active areas of modern research is closely related to this long-standing problem. Mechanical problems that can be expressed in a Hamiltonian formulation, rather than as a second-order system, can be studied in terms of the preservation of qualitative properties. It is found that the symplectic property can be preserved by the use of specially designed Runge–Kutta methods. The burgeoning subject of geometric integration, started from the study of Hamiltonian systems by Feng Kang, J.M. Sanz-Serna and others, and is now a central activity as the century closes. Although it is too early to view geometric integration from a historical perspective, it is at least possible to refer to a recent review of this subject [58].

There are several other evolutionary problems that can be solved by methods closely related to ordinary differential equation methods. Delay differential equations, and other types of functional differential equations can be solved using a combination of a differential equation solver, an interpolator and a means of handling discontinuities.

We have already noted that algebraic differential equations, especially those of low index, can be effectively solved using linear multistep methods. Implicit Runge–Kutta methods also have an important role in the numerical treatment of differential–algebraic equations. The theory of order of these methods can be extended to allow for the inclusion of algebraic constraints in the formulation, using generalizations of rooted trees [35].

References

[1] P. Albrecht, A new theoretical approach to Runge–Kutta methods, SIAM J. Numer. Anal. 24 (1987) 391–406.
[2] R. Alexander, Diagonally implicit Runge–Kutta methods for stiff ODEs, SIAM J. Numer. Anal. 14 (1977) 1006–1021.
[3] R. Alt, Deux théorèms sur la A-stabilité des schémas de Runge–Kutta simplement implicites, Rev. Francais d'Automat. Recherche Opérationelle Sér. R-3 6 (1972) 99–104.
[4] D. Barton, I.M. Willers, R.V.M. Zahar, The automatic solution of ordinary differential equations by the method of Taylor series, Comput. J. 14 (1971) 243–248.
[5] F. Bashforth, J.C. Adams, An attempt to test the theories of capillary action by comparing the theoretical and measured forms of drops of fluid, with an explanation of the method of integration employed in constructing the tables which give the theoretical forms of such drops, Cambridge University Press, Cambridge, 1883.
[6] R. Bulirsch, J. Stoer, Numerical treatment of ordinary differential equations by extrapolation methods, Numer. Math. 8 (1966) 1–13.
[7] K. Burrage, A special family of Runge–Kutta methods for solving stiff differential equations, BIT 18 (1978) 22–41.
[8] K. Burrage, J.C. Butcher, Stability criteria for implicit Runge–Kutta methods, SIAM J. Numer. Anal. 16 (1979) 46–57.
[9] J.C. Butcher, Coefficients for the study of Runge–Kutta integration processes, J. Austral. Math. Soc. 3 (1963) 185–201.
[10] J.C. Butcher, Implicit Runge–Kutta processes, Math. Comp. 18 (1964) 50–64.
[11] J.C. Butcher, On the attainable order of Runge–Kutta methods, Math. Comp. 19 (1965) 408–417.
[12] J.C. Butcher, A modified multistep method for the numerical integration of ordinary differential equations, J. Assoc. Comput. Mach. 12 (1965) 124–135.
[13] J.C. Butcher, An algebraic theory of integration methods, Math. Comp. 26 (1972) 79–106.
[14] J.C. Butcher, The non-existence of ten stage eighth order explicit Runge–Kutta methods, BIT 25 (1985) 521–540.
[15] J.C. Butcher, A stability property of implicit Runge–Kutta methods, BIT 15 (1975) 358–361.
[16] D.A. Calahan, A stable, accurate method of numerical integration for nonlinear systems, Proc. IEEE 56 (1968) 744.
[17] F. Ceschino, J. Kuntzmann, Problèmes Différentiels de Conditions Initiales, Dunod, Paris, 1963.
[18] M. Crouzeix, Sur les méthodes de Runge–Kutta pour l'approximation des problèmes d'évolution, Lecture Notes in Economics and Mathematical Systems, Vol. 134, Springer, Berlin, 1974, pp. 206–223.
[19] M. Crouzeix, Sur la B-stabilité des méthodes de Runge–Kutta, Numer. Math. 32 (1979) 75–82.
[20] C.F. Curtiss, J.O. Hirschfelder, Integration of stiff equations, Proc. Nat. Acad. Sci. 38 (1952) 235–243.
[21] G. Dahlquist, Convergence and stability in the numerical integration of ordinary differential equations, Math. Scand. 4 (1956) 33–53.
[22] G. Dahlquist, Stability and error bounds in the numerical integration of ordinary differential equations, Kungl. Tekn. Högsk. Handl. Stockholm 130 (1959) 1–87.
[23] G. Dahlquist, A special stability property for linear multistep methods, BIT 3 (1963) 27–43.
[24] G. Dahlquist, Error Analysis for a Class of Methods for Stiff Non-Linear Initial Value Problems, Lecture Notes in Mathematics, Vol. 506, Springer, Berlin, 1976, pp. 60–72.
[25] G. Dahlquist, G-stability is equivalent to A-stability, BIT 18 (1978) 384–401.
[26] J. Donelson, E. Hansen, Cyclic composite multistep predictor-corrector methods, SIAM J. Numer. Anal. 8 (1971) 137–157.
[27] B.L. Ehle, A-stable methods and Padé approximations to the exponential, SIAM J. Numer. Anal. 4 (1973) 671–680.
[28] W.H. Enright, T.E. Hull, B. Lindberg, Comparing numerical methods for stiff systems of ODEs, BIT 15 (1975) 10–48.
[29] R. Frank, J. Schneid, C.W. Ueberhuber, The concept of B-convergence, SIAM J. Numer. Anal. 18 (1981) 753–780.
[30] C.W. Gear, Hybrid methods for initial value problems in ordinary differential equations, SIAM J. Numer. Anal. 2 (1965) 69–86.
[31] C.W. Gear, Algorithm 407, DIFSUB for solution of ordinary differential equations, Comm. ACM 14 (1971) 447–451.
[32] A. Gibbons, A program for the automatic integration of differential equations using the method of Taylor series, Comput. J. 3 (1960) 108–111.

[33] S. Gill, A process for the step-by-step integration of differential equations in an automatic digital computing machine, Proc. Cambridge Philos. Soc. 47 (1951) 96–108.
[34] W.W. Gragg, H.J. Stetter, Generalized multistep predictor-corrector methods, J. Assoc. Comput. Mach. 11 (1964) 188–209.
[35] E. Hairer, C. Lubich, M. Roche, in: The Numerical Solution of Differential–Algebraic Systems by Runge–Kutta Methods, Lecture Notes in Mathematics, Vol. 1409, Springer, Berlin, 1989.
[36] E. Hairer, G. Wanner, On the Butcher group and general multi-value methods, Computing 13 (1974) 1–15.
[37] E. Hairer, G. Wanner, A theory for Nyström methods, Numer. Math. 25 (1976) 383–400.
[38] P.C. Hammer, J.W. Hollingsworth, Trapezoidal methods of approximating solutions of differential equations, MTAC 9 (1955) 269–272.
[39] P. Henrici, Discrete Variable Methods in Ordinary Differential Equations, Wiley, New York, 1962.
[40] K. Heun, Neue Methoden zur approximativen Integration der Differentialgleichungen einer unabhängigen Veränderlichen, Z. Math. Phys. 45 (1900) 23–38.
[41] T.E. Hull, W.H. Enright, B.M. Fellen, A.E. Sedgwick, Comparing numerical methods for ordinary differential equations, SIAM J. Numer. Anal. 9 (1972) 603–637.
[42] A. Iserles, S.P. Nørsett, Order Stars, Chapman & Hall, London, 1991.
[43] P. Kaps, G. Wanner, A study of Rosenbrock type methods of high order, Numer. Math. 38 (1981) 279–298.
[44] K.S. Kunz, Numerical solution of ordinary differential equations: methods of starting the solution, in: Numerical Analysis, McGraw-Hill, New York, 1957.
[45] W. Kutta, Beitrag zur näherungsweisen Integration totaler Differentialgleichungen, Z. Math. Phys. 46 (1901) 435–453.
[46] R.H. Merson, An operational method for the study of integration processes, Proceedings of the Symposium on Data Processing, Weapons Research Establishment, Salisbury, South Australia, 1957.
[47] W.E. Milne, A note on the numerical integration of differential equations, J. Res. Nat. Bur. Standards 43 (1949) 537–542.
[48] R.E. Moore, The automatic anlyasis and control of error in digital computation based on the use of interval numbers, Error in Digital Computation, Vol. 1, Wiley, New York, 1964, pp. 61–130.
[49] F.R. Moulton, New Methods in Exterior Balistics, University of Chicago, Chicago, 1926.
[50] A. Nordsieck, On numerical integration of ordinary differential equations, Math. Comp. 16 (1962) 22–49.
[51] S.P. Nørsett, One-step methods of Hermite type for numerical integration of stiff systems, BIT 14 (1974) 63–77.
[52] S.P. Nørsett, A. Wolfbrandt, Order conditions for Rosenbrock type methods, Numer. Math. 32 (1979) 1–15.
[53] E.J. Nyström, Über die numerische Integration von Differentialgleichungen, Acta Soc. Sci. Fennicae 50 (13) (1925).
[54] L.R. Petzold, A description of DASSL: a differential/algebraic system solver, in: Scientific Computing, Montreal, Quebec, North-Holland, Amsterdam, 1983, 65–68.
[55] A. Prothero, A. Robinson, On the stability and accuracy of one-step methods for solving stiff systems of ordinary differential equations, Math Comp. 28 (1974) 145–162.
[56] H.H. Rosenbrock, Some general implicit processes for the numerical solution of differential equations, Comput. J. 5 (1963) 329–330.
[57] C. Runge, Über die numerische Auflösung von Differentialgleichungen, Math. Ann. 46 (1895) 167–178.
[58] J.M. Sanz-Serna, Geometric integration, The State of the Art in Numerical Analysis, Oxford University Press, Oxford, 1997, pp. 121–143.
[59] L.F. Shampine, H.A. Watts, Block implicit one-step methods, Math. Comp. 23 (1969) 731–740.
[60] G. Wanner, E. Hairer, S.P. Nørsett, Order stars and stability theorems, BIT 18 (1978) 475–489.

Retarded differential equations

Christopher T.H. Baker [1]

Department of Mathematics, The University of Manchester, Oxford Road, Manchester M13 9PL, UK

Abstract

Retarded differential equations (RDEs) are differential equations having retarded arguments. They arise in many realistic models of problems in science, engineering, and medicine, where there is a time lag or after-effect. Numerical techniques for such problems may be regarded as extensions of dense-output methods for ordinary differential equations (ODEs), but scalar RDEs are inherently infinite dimensional with a richer structure than their ODE counterparts. We give background material, develop a theoretical foundation for the basic numerics, and give some results not previously published.

Keywords: Retarded differential equations; Delay and neutral delay differential equations; Continuity and stability; Numerics; Mesh and densely defined approximations; Convergence; Order of convergence; Numerical stability

1. Introduction

Many real-life problems that have, in the past, sometimes been modelled by initial-value problems for differential equations actually involve a significant memory effect that can be represented in a more refined model, using a differential equation incorporating retarded or delayed arguments (arguments that 'lag behind' the current value). The last few decades have seen an expanding interest in problems variously classified as delay differential equations (DDEs), retarded differential equations (RDEs), retarded functional differential equations (RFDEs), or neutral delay differential equations (NDDEs). (Stochastic DDEs, whose basic numerics are addressed in [6], also arise.)

Amongst the application areas are the biosciences, economics, materials science, medicine, public health, and robotics; in a number of these there is an underlying problem in control theory. Regarding the independent variable (t, say) as representing "time" in an evolutionary problem, the significance

E-mail address: cthbaker@maths.man.ac.uk (C.T.H. Baker).

[1] Professor of Mathematics & Director, Manchester Centre for Computational Mathematics (MCCM).

of any "time-lag" is in a sense determined by its size relative to the underlying "time-scales". Parameter estimation [3,4,9,11] in prospective models can assist in assessing the need to incorporate time-lags. To the extent that numerical analysts should be influenced by applications (see, for example, [5]) and by the theoretical background, the starting point for a study of numerical methods lies with mathematical modelling and in dynamical systems defined by Volterra (that is, nonanticipative) equations. The advent of robust general numerical routines for DDEs and NDDEs then changes the opportunities for mathematical modelling (through numerical simulation) using equations for which closed-form solutions do not exist.

Our main task is to convey the essence of the subject to the reader by means of a rigorous presentation that foregoes any attempt at a complete discussion (this would require a book rather than a paper), and the structure of this paper is as follows. Following this introduction, we provide some background theory (i) concerning the functional equations and (ii) concerning the background numerical analysis. We then discuss (emphasizing relatively simple formulae to advance the development, but indicating extensions) the major theoretical issues: *existence of an approximate solution*, *convergence to the true solution*, and *numerical stability*; the aim, always, is to adopt a mathematically sound viewpoint whilst noting the limitations of the discussion. We conclude with a mention of some further important issues. Though the material is to a large extent available in the literature, our personal viewpoint pervades the presentation — and we were unable to resist the temptation to include new material.

To illustrate problems of the type that interest us, we may consider

$$y'(t) = f(t, y(t), y(\alpha(t))), \quad t \geqslant t_0 \quad \text{where } \alpha(t) \leqslant t, \tag{1.1a}$$

$$y(t) = \psi(t), \quad t \in [t_{\min}, t_0] \quad \text{where } t_{\min} := \inf_{t \geqslant t_0} \alpha(t) \tag{1.1b}$$

($t_{\min}$ denotes a generic value, dependent on the problem) with one retarded or 'lagging' argument $\alpha(t)$. In general, the equations of interest present an *initial function* problem rather than an initial-value problem familiar in ordinary differential equations (ODEs): the solution $y(t)$ of (1.1) is defined by $\psi(t)$ on an initial interval depending on the initial point t_0. Thus, $y(t) = y(\psi, t_0; t)$.

Eq. (1.1a) is an example of a DDE. By way of illustration, one may take $\alpha(t) = t - \tau_\star$ where the *lag* $\tau_\star > 0$ is fixed and $t_{\min} = t_0 - \tau_\star$. That this form of lag is common in the modelling literature may owe more to the difficulty of treating more general equations analytically than to the realism of the model. In any event, (1.1a) may itself be generalized. Systems with multiple delays e.g., $y'(t) = G(t, y(t), y(\alpha_1(t)), y(\alpha_2(t)), \dots, y(\alpha_m(t)))$, $t \geqslant t_0$, $y(t) \in \mathbb{R}^N$, $(\alpha_i(t) \leqslant t;\ i = 1, 2, \dots, m)$ also occur. We may also consider a system of *neutral* differential equations, or NDDEs, say (y, F vector-valued)

$$y'(t) = F(t, y(t), y(\alpha(t)), y'(\beta(t))), \quad t \geqslant t_0 \tag{1.2a}$$

with $\alpha(t), \beta(t) \leqslant t$; if $t^0_{\min} = \inf_{t \geqslant t_0} \alpha(t)$ and $t^1_{\min} = \inf_{t \geqslant t_0} \beta(t)$ then the initial conditions are that

$$y(t) = \psi_0(t), \quad t \in [t^0_{\min}, t_0] \quad \text{and} \quad y'(t) = \psi_1(t), \ t \in [t^1_{\min}, t_0]. \tag{1.2b}$$

We may set $t_{\min} := \min\{t^0_{\min}, t^1_{\min}\}$. We term $\alpha(t)$, $\beta(t)$ the *delayed arguments*, and $\tau(t) := t - \alpha(t) \geqslant 0$, $\zeta(t) := t - \beta(t) \geqslant 0$ the corresponding *lags*. An NDDE (1.2a) is characterized by the dependence of $y'(t)$ on $y'(\beta(t))$, as an argument of F. Frequently, but not always, $\psi_1(t) = \psi_0'(t)$. We concentrate for simplicity on DDEs and NDDEs. The theory of NDDEs is rather less straightforward than that of DDEs, and it is commonplace to impose sufficient, rather than necessary, conditions on F for existence and uniqueness. We observe, in passing, that an alternative standard type of NDDE is represented by

$$\{y(t) - \gamma(t, y(\beta(t)))\}' = \Gamma(t, y(t), y(\alpha(t))), \tag{1.3}$$

which with suitable assumptions is in "Hale's form". The theory of (1.3) seems to be better developed than that for (1.2a), and its numerical treatment can merit separate investigation.

A classification of the types of delayed argument is important in the modelling, the analysis, and the numerics. We refer to $\alpha(t) = t - \tau(t)$ in (1.2a), and by analogy to $\beta(t) = t - \zeta(t)$, to distinguish (i) *fading* or *persistent* memory, respectively, $\alpha(t) \to \infty$ as $t \to \infty$ or $\alpha(t) \not\to \infty$ as $t \to \infty$; (ii) *bounded lag* if $\sup \tau(t) < \infty$; (iii) *constant* and *state-dependent* lag if, respectively, $\alpha(t) = t - \tau_\star$, with $\tau_\star$ fixed, or $\alpha(t) \equiv \alpha(t, y(t)) := t - \tau(t, y(t))$. Finally, we have (iv) a *vanishing lag* if $\alpha(t_\star) = t_\star$. (Analogously, (i) $\beta(t) \to \infty$ or $\beta(t) \not\to \infty$ as $t \to \infty$, (ii) $\sup \zeta(t) < \infty$, (iii) $\zeta(t) \equiv \zeta_\star$, (iv) $\beta(t_\star) = t_\star$.)

Remark 1.1. Where left- and right-hand derivatives of the solution $y(\cdot)$ do not agree, in an RDE such as that in (1.1) or (1.2), $y'(t)$ is interpreted as the right-hand derivative. The right- and left-hand derivatives of a function $\phi(\cdot)$, are (provided the required limit exists), respectively

$$\phi'_+(t) = \lim_{\delta \to 0} \frac{\phi(t + |\delta|) - \phi(t)}{|\delta|} \quad \text{and} \quad \phi'_-(t) = \lim_{\delta \to 0} \frac{\phi(t) - \phi(t - |\delta|)}{|\delta|}. \tag{1.4}$$

2. Background theory

We touch on some theoretical issues, asking the reader to consult the literature[2] (see, e.g., citations [16, 78, 83, 102, 104, 125, 151, 153, 200] in [30] and [59]) for more detail. We give some very simple DDEs to illustrate interesting features. Our first example is the scalar, linear, DDE

$$y'(t) = \lambda_\star y(t) + \mu_\star y(t - \tau_\star), \quad t \geqslant t_0 \quad \text{with} \quad y(t) = \psi_\star(t),\ t \in [t_0 - \tau_\star, t_0]. \tag{2.1a}$$

By a simple change of variable, one can (if appropriate) normalize the lag $\tau_\star$ to unity and obtain the equation

$$u'(t) = \lambda u(t) + \mu u(t - 1), \quad t \geqslant t_0, \quad \text{with} \quad u(t) = \psi(t), \quad t \in [t_0 - 1, t_0], \tag{2.1b}$$

where $u(t) := y(\tau_\star t)$, $\psi(t) = \psi_\star(\tau_\star t)$, $\lambda = \lambda_\star \tau_\star$, $\mu = \mu_\star \tau_\star$. On the other hand, the further substitution $v(t) = \exp(-\lambda t) u(t)$ then gives, with $\hat{\lambda} := \mu \exp(-\lambda)$, the equation $v'(t) = \hat{\lambda} v(t - 1)$ $(t \geqslant t_0)$, a "*pure delay equation*". This can be solved, in principle, by repeated integration: we have $v(t) = \hat{\lambda} \int_{t_0+(n-1)}^{t-1} v(s)\, ds + v(t_0 + n)$, with $t \in [t_0 + n, t_0 + (n + 1)]$, for $n = 0, 1, 2, \ldots$.

Our second example is the popular *delayed logistic equation*

$$y'(t) = \Lambda_\star y(t)\{1 - y(t - \tau_\star)\}, \quad \tau_\star > 0 \tag{2.2}$$

[2] We are constrained in this presentation by pressures on space, and this limits the comprehensiveness of our references (the references [1–77] that are listed here carry additional citations), and limits our discussion (our main aim is to convey, without over-simplification, a rigorous underlying approach to theory and practice). For future reading, we refer to a forthcoming book in [26].

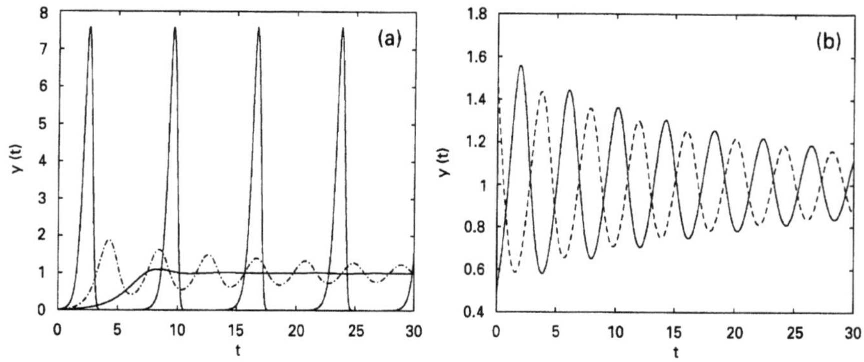

Fig. 1. Some solutions of $u'(t) = \Lambda u(t)\{1 - u(t-1)\}$ in (2.3): (a) for $\Lambda = 0.75, 1.5$ and 3.0, and a fixed $\psi(\cdot)$; (b) for $\Lambda = 1.5$; $\psi(t) = 0.5$ and $\psi(t) = 1.5$ ($t \in [-1, 0]$).

with an initial function defined on $[t_0 - \tau_\star, t_0]$, or (normalizing $\tau_\star$ to unity):

$$u'(t) = \Lambda u(t)\{1 - u(t-1)\} \quad (t \geqslant t_0), \tag{2.3a}$$

$$u(t) = \psi(t), \quad t \in [t_0 - 1, t_0], \quad \Lambda = \Lambda_\star \tau_\star \in \mathbb{R}. \tag{2.3b}$$

For $\Lambda > 0$, a positive initial function $\psi(t) > 0$ gives a positive solution $u(t)$ ($t \geqslant t_0$) and with $w(t) := \ln u(t)$ one has the DDE $w'(t) = \Lambda\{1 - \exp(w(t-1))\}$, another "pure delay equation". Numerical solutions of (2.3) are plotted in Fig. 1. The realism of the model is open to question but it has some typical features: (a) The equation has a positive equilibrium solution $u(t) = 1$ for all Λ, but the qualitative behaviour of solutions of the scalar equation depends upon Λ and chaotic behaviour can arise. For large Λ, a solution can remain small for a substantial interval in t before increasing to a large value and then decaying again (cf. the case $\Lambda = 3$ which has high peaks, in Fig. 1(a)). The work [37] yields asymptotic expressions (valid as Λ becomes large) for maxima and minima and ultimate periodicity. (b) Solutions corresponding to different *initial functions* can assume the same value at a point $t = t_\star$ though they are not identical; see Fig. 1(b). For ODEs where the solution is uniquely determined by initial data at an arbitrary initial point this cannot happen. A further example is the *pantograph equation* $y'(t) = y(\gamma t)$ ($t \geqslant t_0 \geqslant 0$); $y(t) = \psi(t)$, $t \in [\gamma t_0, t_0]$, $\gamma \in (0, 1)$. If $t_0 = 0$ this is an *initial-value* problem (cf. [29] for an extension). In certain limiting cases (when $\tau_\star = 0$ or $\gamma = 0$), equations mentioned above reduce to ODEs, e.g., $y'(t) = \Lambda_\star y(t)\{1 - y(t)\}$, the logistic ODE. If we set $\gamma = 0$ in the pantograph equation we obtain $y'(t) = y(0)$ which displays a persistent memory ($y(0)$ is never "forgotten").

Remark 2.1. Eq. (2.3a) arises from $x'(t) = \rho x(t)\{1 - K^{-1}x(t - \tau_\star)\}$, on a change of variables. Sir Robert May, FRS (at the time of writing, the Chief Scientific Adviser to HM Government) proposed, in 1973, study of a related *system* (with $\kappa, \xi, \rho, \omega, \tau_\star > 0$, $K > 0$) $y_1'(t) = \rho y_1(t)\{1 - K^{-1}y_1(t - \tau_\star)\} - \kappa y_1(t)y_2(t)$; $y_2'(t) = -\omega y_2(t) + \xi y_1(t)y_2(t)$. Other similar systems arise.

Theorem 2.2 (Existence; cf. [57]). *With* $\delta > 0$, $t_0 \in [t_{\min}, t^{\max}]$, *there is a unique solution of* (1.2a) *on* $[t_{\min}, t^{\max}]$ *if, whenever* $u, U \in C^1[t_{\min}, t^{\max}]$ *and* $v, V \in C[t_{\min}, t^{\max}]$, (i) $F(t, u(t), u(\alpha(t)), u'(\beta(t)))$ *is* *continuous on* $[t_{\min}, t^{\max}]$ *and, further,* (ii) $\|F(t,u(t),u(\alpha(t)),v(\beta(t)))-F(t,U(t),U(\alpha(t)),V(\beta(t)))\| \leqslant L_1(\sup_{s\in[t_{\min},t]}|u(s)-U(s)| + \sup_{s\in[t_{\min},t-\delta]}|v(s)-V(s)|) + L_2\sup_{s\in[t-\delta,t]}|v(s)-V(s)|$ *holds for* $t\in[t_0,t^{\max}]$ *with* $L_{1,2}\geqslant 0$ *and* $L_2<1$.

One of the important characteristics of an RDE is the *sensitivity of a particular solution* to changes in the problem. This may be sensitivity to changes in the parameters [11] in the equation ("structural stability"), or sensitivity to changes in the initial conditions, or to persistently acting disturbances, and each gives rise to a definition of *stability*. To give definitions we shall refer to a solution of the NDDE problem (1.2) (then the DDE problem (1.1) is a special case), and we consider the perturbed problem

$$y'(t)+\delta y'(t) = F(t, y(t)+\delta y(t), y(\alpha(t))+\delta y(\alpha(t)), y'(\beta(t))+\delta y'(\beta(t))) + \delta F(t) \quad (t\geqslant t_0), \tag{2.4a}$$

$$y(t)+\delta y(t) = \psi_0(t)+\delta\psi_0(t),\quad y'(t)+\delta y'(t) = \psi_1(t)+\delta\psi_1(t) \quad (t\leqslant t_0), \tag{2.4b}$$

conditions (2.4b) being valid on the appropriate initial intervals. We can measure the overall size of the input perturbations by, for example,

$$\Delta = \sup_{t\geqslant t_0}\|\delta F(t)\| + \max\left\{\sup_{t\leqslant t_0}\|\delta\psi_0(t)\|, \sup_{t\leqslant t_0}\|\delta\psi_1(t)\|\right\}, \tag{2.5}$$

where $\|\cdot\|$ is a vector norm. It may be appropriate to restrict the classes of perturbations (the *admissible perturbations*); for example, we may require $\delta F(t)\to 0$ as $t\to\infty$ or $\delta F(\cdot)\in\mathscr{L}[t_0,\infty)$, or (if $t_{\min}$ is $-\infty$) $\delta\psi(t)\to 0$ as $t\to -\infty$. For nonneutral equations, $\psi_1(\cdot)$ is absent and $\delta\psi_1(\cdot)$ vanishes. Stability concerns the boundedness of $\delta y(\cdot)$ on $[0,\infty)$, and the limiting behaviour of $\delta y(t)$ as $t\to\infty$, when the problem is perturbed; note that a solution of a *nonlinear* problem may be *unbounded but stable*. In consequence, to attempt (without further comment) to define stability of a solution of a *linear homogeneous* equation in terms of boundedness of its solutions $y(\cdot)$ can lead to misunderstanding of the nonlinear case. The unqualified term *stability* often refers to *stability with respect to perturbed initial conditions* in which we require that $\delta F(\cdot)$ vanishes, so the only perturbations are in $\psi_0(\cdot)$, $\psi_1(\cdot)$. We now give formal definitions of selected concepts.

Definition 2.3 (Stability). (a) Given t_0 and a class of admissible perturbations satisfying (2.5), a solution $y(\cdot)$ of the neutral equation (1.2) is (i) stable, if for each $\varepsilon>0$ there exists $\Delta^*=\Delta^*(\varepsilon,t_0)>0$ such that $\|\delta y(t)\|<\varepsilon$, when $t\geqslant t_0$ whenever $\Delta<\Delta^*$; (ii) asymptotically stable, if it is stable and there exists $\Delta^\dagger=\Delta^\dagger(t_0)$ such that $\delta y(t)\to 0$ as $t\to\infty$ whenever $\Delta<\Delta^\dagger$.

There are many complementary theoretical tools for analysing stability; stability theory for bounded lags may not extend to unbounded lags and theory for DDEs does not always extend to NDDEs.

Remark 2.4. (a) For linear equations (2.1), some results are well known. Thus, the use of Laplace transforms and an investigation of the zeros of the quasi-polynomial that serves as a stability function allows one to establish that *all solutions of the DDE* (2.1a) *are stable* with respect to perturbed initial conditions *when the point* $(\lambda,\mu):=(\lambda_\star\tau_\star,\mu_\star\tau_\star)$ *lies in the stability region* Σ which is the

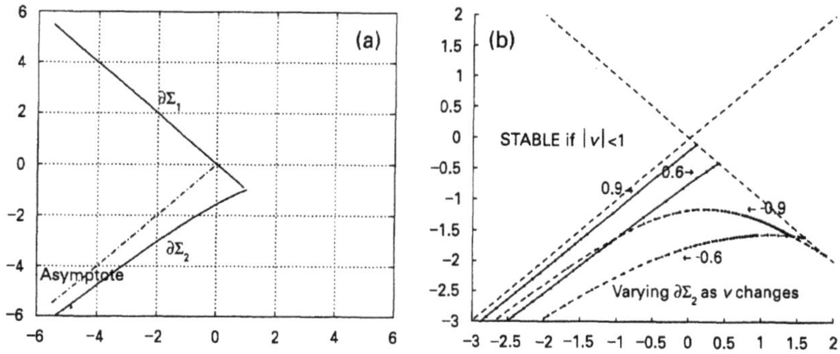

Fig. 2. Representation of the boundary $\partial\Sigma_1 \cup \partial\Sigma_2$ of the unbounded exact stability region Σ in the (λ, μ)-plane: (a) as given in (2.6), showing that Σ contains the wedge $\mathscr{W}$ of points (λ, μ) such that $|\mu| < -\lambda$ and $\tau_\star > 0$; $\lambda = \mu$ is an asymptote to $\partial\Sigma_2$; (b) showing for comparison the corresponding boundary (2.10) for the NDDE (2.9) in the case $v = -0.9, -0.6, 0.6, 0.9$; the stability regions again include $\mathscr{W}$ (for $|v| < 1$).

region of the (λ, μ)-plane that includes the half-line $\lambda < 0$, $\mu = 0$ and whose boundary

$$\partial\Sigma = \partial\Sigma_1 \cup \partial\Sigma_2 \text{ is formed by the loci} \tag{2.6a}$$

$$\partial\Sigma_1 := \{\mu = -\lambda\}, \tag{2.6b}$$

$$\partial\Sigma_2 := \{(\lambda = \omega \cot \omega,\ \mu = -\omega \operatorname{cosec} \omega);\ 0 < \omega < \pi\}. \tag{2.6c}$$

In particular, therefore, we obtain a stability condition

$$|\mu_\star| \leqslant -\lambda_\star \tag{2.7}$$

that is independent of $\tau_\star > 0$ from the observation $\Sigma \supset \mathscr{W} := \{(\lambda, \mu) \text{ such that } |\mu| < -\lambda\}$. For *complex-valued* $\lambda_\star, \mu_\star$, an analogue of (2.7) is that solutions are stable if $|\mu_\star| < -\Re\, \lambda_\star$; Guglielmi [41, p. 409] indicates the region $\Sigma^\dagger$ of complex parameters $(\lambda_\star\tau_\star, \mu_\star\tau_\star)$ for which all solutions of (2.1a) are stable.

(b) The above results extend after some precautions to certain NDDEs such as

$$y'(t) = \lambda_\star y(t) + \mu_\star y(t - \tau_\star) + v_\star y'(t - \tau_\star) \quad (\lambda_\star, \mu_\star, v_\star \in \mathbb{C}) \text{ with } |v_\star| < 1. \tag{2.8}$$

For the linear neutral delay differential equation, (2.8) a sufficient condition for stability is $|\lambda_\star \bar{v}_\star - \bar{\mu}_\star| + |\lambda_\star v_\star - \mu_\star| < -2\Re\, \lambda_\star$; if $\lambda_\star, \mu_\star, v_\star \in \mathbb{R}$, it suffices that $|\mu_\star| < -\lambda_\star$ and $|v_\star| < 1$. The lag can be normalized to unity if we consider

$$y'(t) = \lambda y(t) + \mu y(t-1) + v y'(t-1) \quad (\lambda := \lambda_\star\tau_\star, \mu := \mu_\star\tau_\star, v = v_\star) \tag{2.9}$$

Boundaries of the stability region are presented graphically (after some re-labelling) in [60, p. 117]. We here note that the locus corresponding to $\partial\Sigma_2$ in (2.6c), for v not necessarily 0, is the set of parameterized points

$$\{(\lambda, \mu):\ \lambda = \omega \cot \omega - v\omega \operatorname{cosec}(\omega), \quad \mu = -\omega \operatorname{cosec}(\omega) + v\omega \cot(\omega)\}, \tag{2.10}$$

and sample loci are drawn in Fig. 2(b) for $v = -0.9, -0.6, 0.6$, and 0.9.

(c) For systems of ODEs, the test equation $y'(t) = \Lambda y(t)$ with $\Lambda \in \mathbb{C}$ (rather than with $\Lambda \in \mathbb{R}$) can be motivated by consideration of the equations $y_i'(t) = \sum a_{i,j} y_j(t)$, or, in compact notation,

$y'(t) = \boldsymbol{A}y(t)$, where the real matrix $\boldsymbol{A} = [a_{i,j}]$ is supposed to be similar to the diagonal matrix $\boldsymbol{diag}(\Lambda_1, \Lambda_2, \ldots, \Lambda_m)$ with complex $\{\Lambda_\ell\}$. The corresponding test equation (2.8) with complex coefficients should probably be considered less significant than in the complex-coefficient ODE case, because (i) it will be a rare occasion for three matrices $\boldsymbol{L} = [\lambda_{i,j}]$, $\boldsymbol{M} = [\mu_{i,j}]$, $\boldsymbol{N} = [\nu_{i,j}]$ to be simultaneously diagonalizable — i.e., under the same similarity transform — and (ii) the system of NDDEs that one would prefer to consider has the form

$$y_i'(t) = \sum_j \lambda_{i,j} y_j(t) + \sum_j \mu_{i,j} y_j(t - \tau_{i,j}) + \sum_j \nu_{i,j} y_j'(t - \zeta_{i,j}) \tag{2.11}$$

where the positive values $\{\tau_{i,j}, \zeta_{i,j}\}$ are generally all *different* (and may be t– or state–dependent).

If we turn to nonlinear equations, the solution $y(t) \equiv 1$ of (2.3) is readily shown to have different stability properties from those of the solution $y(t) \equiv 0$ and this illustrates the general observation that stability properties attach to a particular solution $y(t) \equiv y(\psi, t_0; t)$ of a nonlinear equation. Let us observe also that there are conditions in the literature under which stability with respect to certain types of persistent perturbations may be deduced from asymptotic stability with respect to initial perturbations.

Emphasis on test equations, such as (2.1) has limited interest unless one can relate the results to more general equations. In this respect, Baker and Tang [12] established results on stability with respect to initial perturbations, that include the following as a special case, and they investigated analogues for some numerical approximations.

Theorem 2.5. *Define* $\|u(\cdot)\|^{[t-k\tau_\star, t]} := \sup_{\sigma \in [t-k\tau_\star, t]} |u(\sigma)|$. *Suppose* $\alpha(t) \equiv t - \tau(t, u(t)) \leqslant t$ $(t \geqslant t_0)$ *and* $|\tau(t, v) - \tau_\star| \to 0$ *uniformly as* $t \to \infty$, $v \to 0$, *and, for some finite* $k \geqslant 1$, $|f(t, u(t), u(\alpha(t, u(t)))) - \{\lambda_\star u(t) + \nu_\star u(t - \tau_\star)\}| = o(\|u(\cdot)\|^{[t-k\tau_\star, t]})$ *uniformly as* $\|u(\cdot)\|^{[t-k\tau_\star, t]} \to 0, t \to \infty$. *Then, the zero solution of* (1.1) *is asymptotically stable if the zero solution of* (2.1*a*) *is asymptotically stable* (*and unstable if the zero solution of* (1.1) *is unstable*).

We turn to some further key features that are of interest in the context of numerical analysis for RDEs. These include (i) propagated discontinuities in the solution or its derivatives, or *large* derivative values; (ii) delay terms that can *stabilize* or can *destabilize* solutions of the problem; (iii) onset of periodicity or chaotic behaviour as a parameter in the defining equations is varied. We can illustrate these by reference to scalar equations, such as (2.3).

Firstly, consider the question of smoothness. Suppose that the initial function $\psi(t) \in C^1[t_0 - \tau_\star, t_0]$ in (2.2) is such that its (left-hand) derivative $\psi_\star'(t_0)$ is not equal to $\Lambda_\star \psi_\star(t_0)\{1 - \psi_\star(t_0 - \tau_\star)\}$. Then (since $y(\cdot)$ and $\psi_\star(\cdot)$ coincide on $[t_0 - \tau_\star, t_0]$), it follows that the right-hand derivative $y_+'(t_0)$ does not agree with the left-hand derivative $y_-'(t_0)$. In this case, we say that the derivative of y has a jump at t_0. We deduce, from (2.2), that $y''(t)$ has a similar jump at $t_0 + \tau_\star$, $y'''(t)$ has a jump at $t_0 + 2\tau_\star$, and so on. Here, the solution becomes smoother as t increases. There exists a well-developed theory of tracking of the points where derivatives of the solution have jump discontinuities, see [69,75] and other citations in [10]. The reader may consider a modification of (2.2),

$$y'(t) = \Lambda_\star y(t)\{1 - y(t) - \tau_\star y'(t - \tau_\star)\} \qquad (t \geqslant 0) \tag{2.12}$$

which is an NDDE, as an example where smoothing does not take place. Note that the type of approximation that purports to relate (2.2)–(2.12) is most suspect!

Secondly, we address stability. Consider the equilibrium $y(t) \equiv 1$ for the ODE $y'(t) = \Lambda^\star y(t)\{1 - y(t)\}$ which, by 'stability in the first approximation', is stable with respect to initial perturbations if $\Lambda^\star > 0$. However, $y(t) \equiv 1$ is a stable solution of $y'(t) = \Lambda_\star y(t)\{1 - y(t - \tau_\star)\}$ under the introduction of (small) perturbations $\varepsilon(t)$ for $t \in [-\tau_\star, 0]$ if (by a refined version of 'stability in the first approximation') corresponding solutions of $\varepsilon'(t) = -\Lambda_\star \varepsilon(t - \tau_\star)\{1 + \varepsilon(t)\}$ are bounded. A *stability condition* now is that $\Lambda_\star \tau_\star \in (0, \frac{1}{2}\pi)$ ($y(t) \equiv 1$ is unstable for $\Lambda_\star \tau_\star > \frac{1}{2}\pi$). Note the continuous dependence on $\tau_\star$ as $\tau_\star \searrow 0$.

Finally, consider, in association with bifurcation theory (see [30, p. 301]), the qualitative behaviour of the solutions of the delayed logistic equation as $\Lambda_\star \tau_\star$ varies. The solutions of (2.2) with $\Lambda_\star \tau_\star > 0$ and a positive $\psi(t)$ ($t \in [-\tau_\star, 0]$) are positive, converge monotonically to 1 if $0 \leqslant \Lambda_\star \tau_\star \leqslant e^{-1}$, converge to 1 in an oscillatory fashion if $e^{-1} \leqslant \Lambda_\star \tau_\star < \pi/2$, and display oscillatory behaviour for $\Lambda_\star \tau_\star \geqslant \pi/2$ (a Hopf bifurcation point — it is not a coincidence that the locus $\partial \Sigma_2$ in Fig. 2(a) crosses $\lambda = 0$ where $\mu = -\pi/2$). The numerical analysis of Hopf bifurcation is addressed in the literature (cf. [32,38], where differing perspectives are presented, and [34,35,64,65,76] along with the citations given therein). This area merits the reader's further consideration: Bifurcation, the onset of periodicity, and the behaviour of periodic solutions, are of practical interest.

3. Background numerical analysis

Numerical computation is designed to give quantitative insight (and, thereby, qualitative insight) into the solution of various mathematical models. As in the numerical solution of ODEs, there are two inter-related types of question that concern us: the one relates to the design of algorithms, and the other relates to what can be established rigorously about their properties and performance. This requirement of rigour imposes assumptions that in general reduce the degree of realism but nevertheless (one hopes) allows some practical insight to be gained. Additional valuable citations will be found in [3,10,26,43,77].

The choice of numerical techniques for the treatment of RDEs (DDEs and NDDEs) relies heavily on the construction of densely-defined *continuous extensions*. (Similar constructions produce dense-output in the numerics of ODEs.) The method of defining the continuous extension can, in addition to the properties of the other components of our methods, *affect both the accuracy and the stability of the numerical method* [10,26,77]. We concentrate upon the scalar version of (1.2a), for definiteness, so that

$$y'(t) = F(t, y(t), y(\alpha(t)), y'(\beta(t))) \quad \text{for } t \geqslant t_0, \tag{3.1}$$

(with $y(t) = \psi_0(t)$, $y'(t) = \psi_1(t)$, $t \leqslant t_0$). The restriction to DDEs, and the extension to systems, will be obvious.

Using the initial conditions $\tilde{y}(t) = \psi_0(t)$, $\widetilde{y'}(t) = \psi_1(t)$ for $t \leqslant t_0$, our numerical solution proceeds, usually with adaptive step sizes $\{h_n\}$, with a mesh

$$\mathcal{T} := \{t_0 < t_1 < t_2 < \cdots < t_{n-1} < t_n < \cdots\}, \quad h_n := t_{n+1} - t_n, \tag{3.2}$$

by obtaining $\tilde{y}(t), \widetilde{y'}(t)$ for $t \geqslant t_0$. The *width* of $\mathcal{T}$ is $\max_n h_n$. One advances, from t_n to t_{n+1} at the nth stage, in 'evolutionary' mode. The choice of $\mathcal{T}$ is clearly important and should take account of the nature of the solution. (Some of the published analysis treats the uniform step-size case, $h_n \equiv h$.) Where one seeks the solution on $[t_0, t^{\max}]$, we suppose that $t^{\max}$ is the only possible point of accumulation of $\{t_\ell\}$. In practice, the steps h_n are chosen on the basis of one or more of the following: (a) knowledge (to within a given tolerance) of the points where the solution or a "significant" derivative has a jump discontinuity (or is relatively large), (b) estimates of the *local truncation error*, (c) estimates of the *defect* (see [36,45]; and citation [46] of [10]). Determination of the points referred to in (a) can be attempted using *tracking* theory or by using indicators from estimates obtained under (b) or (c). For some problems, the solution becomes smoother as t increases, and the problem of discontinuities is transient; for certain NDDEs, and certain DDEs with unbounded lag, the problem of lack of smoothness persists. Where discontinuities cause a problem, the author's view is that modest-order one-step methods, such as those based on Runge–Kutta (RK) processes with RK abscissae in $[0, 1]$, and *local* approximation techniques for providing a densely defined approximation, together with control of the step sizes, have great appeal (they are self-starting and can more readily avoid derivative discontinuities). Some RK methods are equivalent to a form of collocation. RK methods with abscissae outside $[0, 1]$, linear multistep methods, and nonlocal extensions, can have rôles in the particular case of problems with smooth (or ultimately smooth) solutions.

3.1. The construction of some approximating formulae

For state-dependent $\alpha(t) \equiv \alpha(t, y(t)), \beta(t) \equiv \beta(t, y(t))$ we write

$$\tilde{\alpha}(t) = \hat{\alpha}(t, \tilde{y}(t)) \equiv \min(t, \alpha(t, \tilde{y}(t))), \tag{3.3a}$$

$$\tilde{\beta}(t) = \hat{\beta}(t, \tilde{y}(t)) \equiv \min(t, \beta(t, \tilde{y}(t))). \tag{3.3b}$$

Write $t_n + sh_n$ as t_{n+s}; as the calculation proceeds, one retains information to generate approximations $\tilde{y}(t_{n+s})$ (and $\widetilde{y'}(t_{n+s})$), $s \in (0, 1)$, when required.

We orientate the reader by considering *the θ-method* applied, with fixed $\theta \in [0, 1]$, to (3.1); later, we consider RK methods. With $\tilde{y}_k := \tilde{y}(t_k) \approx y(t_k)$, $\tilde{F}_k := F(t_k, \tilde{y}(t_k), \tilde{y}(\tilde{\alpha}(t_k)), \widetilde{y'}(\tilde{\beta}(t_k)))$, suppose we have computed and stored

$$\tilde{y}_0, \tilde{F}_0; \quad \tilde{y}_1, \tilde{F}_1; \quad \tilde{y}_2, \tilde{F}_2; \quad \cdots \quad \tilde{y}_n, \tilde{F}_n \tag{3.4}$$

(see Remark 3.2 for a generalization) and require to advance from t_n to t_{n+1}. The θ-formula for $\tilde{y}_{n+1}$ reads

$$\tilde{y}_{n+1} := \tilde{y}_n + (1-\theta)h_n\tilde{F}_n + \theta h_n F(t_{n+1}, \tilde{y}_{n+1}, \tilde{y}(\tilde{\alpha}(t_{n+1})), \widetilde{y'}(\tilde{\beta}(t_{n+1}))). \tag{3.5}$$

Of course, $\widetilde{y'_k} = \widetilde{y'}(t_k) := \tilde{F}_k$ approximates $y'(t_k)$. The θ-formula is implicit if $\theta \neq 0$ in which case one needs the delayed function value $\tilde{y}(\tilde{\alpha}(t_{n+1}))$ or $\widetilde{y'}(\tilde{\beta}(t_{n+1}))$. Taking $\theta = \frac{1}{2}$ defines *the trapezium rule*

$$\tilde{y}_{n+1} := \tilde{y}_n + \tfrac{1}{2}h_n\tilde{F}_n + \tfrac{1}{2}h_n F(t_{n+1}, \tilde{y}_{n+1}, \tilde{y}(\tilde{\alpha}(t_{n+1})), \widetilde{y'}(\tilde{\beta}(t_{n+1}))). \tag{3.6}$$

In the *special case* where $\alpha(t) = t - \tau_\star$, $\beta(t) = t - \zeta_\star$ and h can be (and is) fixed so that $\tau_\star/h = M \in \mathbb{N}$ and $\zeta_\star/h = M' \in \mathbb{N}$, the natural complete definition is

$$\tilde{y}_{n+1} := \tilde{y}_n + \tfrac{1}{2}h_n\tilde{F}_n + \tfrac{1}{2}h_nF(t_{n+1}, \tilde{y}_{n+1}, \tilde{y}_{n+1-M}, \tilde{F}_{n+1-M'}), \tag{3.7a}$$

$$\widetilde{y'}_{n+1} \equiv \tilde{F}_{n+1} := F(t_{n+1}, \tilde{y}_{n+1}, \tilde{y}_{n+1-M}, \tilde{F}_{n+1-M'}). \tag{3.7b}$$

In general, however, $\tilde{\alpha}(t_{n+1})$ or $\tilde{\beta}(t_{n+1}) \notin \mathcal{T}$. One then needs approximation formulae that extend the solution defined on the mesh $\mathcal{T}$, to compute $\tilde{y}(\tilde{\alpha}(t))$ at $\tilde{\alpha}(t) = t_\kappa + sh_\kappa$ for $s = s(\mathcal{T}; t)$, and to compute $\widetilde{y'}(\tilde{\beta}(t))$ with $\tilde{\beta}(t) = t_{\kappa'} + qh_{\kappa'}$ for $q = q(\mathcal{T}; t)$ $(s, q \in (0, 1))$. Generic relationships of the type

$$\varphi(t_{k+s}) = A(s)\varphi(t_k) + B(s)\varphi(t_{k+1}) + h_k\{C(s)\varphi'(t_k) + D(s)\varphi'(t_{k+1})\} + \mathcal{O}(h_k^{\varrho_k}),$$

$$\varphi'(t_{k+s}) = \frac{1}{h_k}\{A'(s)\varphi(t_k) + B'(s)\varphi(t_{k+1})\} + C'(s)\varphi'(t_k) + D'(s)\varphi'(t_{k+1}) + \mathcal{O}(h_k^{\varrho'_k})$$

(where ϱ_k, ϱ'_k depend on the smoothness of φ on $[t_k, t_{k+1}]$) generate such approximations on omitting the Landau order terms. We obtain piecewise-constant, piecewise-linear, piecewise-quadratic and piecewise-cubic approximations interpolating $\varphi(t)$ at $\{t_k\}$ if, respectively, $A(s) = 1, B(s) = C(s) = D(s) = 0$; $A(s) = 1 - s, B(s) = s$, $C(s) = D(s) = 0$;

$$A(s) = 1, \quad B(s) = 0, \quad C(s) = (s - \tfrac{1}{2}s^2), \quad D(s) = \tfrac{1}{2}s^2 \tag{3.8}$$

and (the piecewise cubic case)

$$\begin{aligned} A(s) &= \{1 + 2s\}(1 - s)^2, \qquad B(s) = 1 - A(s), \\ C(s) &= s(1 - s)^2, \qquad D(s) = s - C(s). \end{aligned} \tag{3.9}$$

As a simple low-order method, we might employ the Euler formula ($\theta = 0$ in the θ-method) with piecewise-linear interpolation for both $\tilde{y}(t_{k+s})$ and $\widetilde{y'}(t_{k+s})$. In this manner, we obtain

$$\tilde{y}(t_{n+1}) := \tilde{y}_n + h_n\tilde{F}_n, \tag{3.10a}$$

$$\begin{aligned} \tilde{y}(t_{k+s}) &:= \tilde{y}_k + h_n\{(1 - s)\tilde{F}_k + (1 - s)\tilde{F}_{k+1}\} \\ &= (1 - s)\tilde{y}_k + s\tilde{y}_{k+1} \quad (k \leqslant n,\ s \in [0, 1]), \end{aligned} \tag{3.10b}$$

$$\widetilde{y'}(t_{k+s}) := (1 - s)\tilde{F}_k + s\tilde{F}_{k+1} \quad (k \leqslant n,\ s \in [0, 1]). \tag{3.10c}$$

An alternative to the expression in (3.10c) is, of course, the derivative obtained from (3.10b), namely

$$\{\tilde{y}_{k+1} - \tilde{y}_k\}/h, \quad s \in [0, 1). \tag{3.10d}$$

A method based on (3.10d) will, in general, have different properties from one based on (3.10c).

For each choice of $\{A(s), B(s), C(s), D(s)\}$, we obtain extensions of the form $\tilde{y}(t_{k+s}) = A(s)\tilde{y}_k + B(s)\tilde{y}_{k+1} + h_k\{C(s)\widetilde{y'_k} + D(s)\widetilde{y'_{k+1}}\}$ (and similarly for $\widetilde{y'}(t_{k+s})$) if we set $\varphi(t_\ell)$ to $\tilde{y}_\ell$, and $\varphi'(t_\ell)$ to $\tilde{F}_\ell$. By (3.5), $\tilde{y}_{k+1}$ can be eliminated and (since $A(s) + B(s) = 1$) we deduce, given θ, approximations of the type

$$\tilde{y}(t_k + sh_k) \equiv \tilde{y}(t_{k+s}) = \tilde{y}_k + h_k\{\mathfrak{b}_1(s)\tilde{F}_k + \mathfrak{b}_2(s)\tilde{F}_{k+1}\}, \tag{3.11a}$$

$$\widetilde{y'}(t_k + sh_k) \equiv \widetilde{y'}(t_{k+s}) = \mathfrak{c}_1(s)\tilde{F}_k + \mathfrak{c}_2(s)\tilde{F}_{k+1}. \tag{3.11b}$$

All choices above give *local* approximations on $[t_k, t_{k+1}]$ using computed values of $\tilde{y}(\cdot)$ and $\widetilde{y'}(\cdot)$ at arguments in $[t_k, t_{k+1}]$. For convenience, we suppose we take one type of approximation consistently for all k.

If we combine (3.9) and (3.6), we find $\mathfrak{b}_1(s)=s-\frac{1}{2}s^2$, $\mathfrak{b}_2(s)=\frac{1}{2}s^2$. The same result arises on using (3.8). If we differentiate (3.9) and use (3.6), we find $\mathfrak{c}_1(s)=1-s$, $\mathfrak{c}_2(s)=s$. Here, $\mathfrak{c}_\ell(s)=\mathfrak{b}_\ell'(s)$, but this is not (cf. (3.10b)–(3.10c)) essential in general. The trapezium rule gives, with the chosen extensions,

$$\tilde{y}(t_{n+s}) := \tilde{y}_n + h_n\{(s-\tfrac{1}{2}s^2)\tilde{F}_n + \tfrac{1}{2}s^2 F(t_{n+1}, \tilde{y}_{n+1}, \tilde{y}(\tilde{\alpha}(t_{n+1})), \widetilde{y'}(\tilde{\beta}(t_{n+1})))\}, \tag{3.12a}$$

$$\widetilde{y'}(t_{k+s}) := (1-s)\tilde{F}_k + s\tilde{F}_{k+1} \quad (k=0,1,\ldots,n). \tag{3.12b}$$

Note that we have $\tilde{y}(t_\ell)=\tilde{y}_\ell$, $\widetilde{y'}(t_\ell)=\tilde{F}_\ell$, for $\ell\in\{k,k+1\}$. Approximation theory yields (*inter alia*) the following results related to (3.12).

Lemma 3.1. *If, for* $r\in\{0,1,2\}$, $y(\cdot)\in C^{2-r}[t_k,t_{k+1}]$ *is Lipschitz continuous* (*in particular if* $y(\cdot)\in C^{3-r}[t_k,t_{k+1}]$) *and if*

$$y(t_{k+s}) = y(t_k) + h_k\{(s-\tfrac{1}{2}s^2)y'(t_k) + \tfrac{1}{2}s^2 y'(t_{k+1})\} + \eta_{h_k}(y;s), \tag{3.13a}$$

$$y'(t_{k+s}) = (1-s)y'(t_k) + sy'(t_{k+1}) + \eta^\natural_{h_k}(y;s), \tag{3.13b}$$

then $\sup_{s\in[0,1]}|\eta_{h_k}(y;s)| = \mathcal{O}(h_k^{3-r})$, $\sup_{s\in[0,1]}|\eta^\natural_{h_k}(y;s)| = \mathcal{O}(h_k^{2-r})$.

Our discussion illustrates the general approach to adapting an ODE solver to treat a DDE or NDDE using auxiliary approximations to compute solution or derivative values at retarded arguments and there exists a wide choice for the latter. We progress from the θ-methods to consider *RK methods*. The ODE literature contains examples of *continuous RK formulae* that incorporate an inbuilt method for generating dense output. Such a formula is generated by the *continuous RK triple* $(\boldsymbol{\vartheta}, \boldsymbol{A}, \boldsymbol{b}(s))$ featured, with an example, in the tableau in (3.14):

$$\begin{array}{c|c} \boldsymbol{\vartheta} & \boldsymbol{A} \\ \hline s & \boldsymbol{b}^{\mathrm{T}}(s), \end{array} \quad \text{for example,} \quad \begin{array}{c|cc} 0 & 0 & \\ 1 & \frac{1}{2} & \frac{1}{2} \\ \hline s & s-\frac{1}{2}s^2 & \frac{1}{2}s^2 \end{array}, \quad s\geqslant 0. \tag{3.14}$$

We have $\boldsymbol{A}=[a_{i,j}]\in\mathbb{R}^{m\times m}$, $\boldsymbol{b}(s)=[b_1(s),b_2(s),\ldots,b_m(s)]^{\mathrm{T}}$ and $\boldsymbol{\vartheta}=[\vartheta_1,\vartheta_2,\ldots,\vartheta_m]^{\mathrm{T}}\in\mathbb{R}^m$. The RK parameters are "*formally explicit*" if $a_{i,j}=0$ for $j\geqslant i$ and will be called "*local*" if $\vartheta_i\in[0,1]$ for all i. An RK triple allows one to obtain a formula for the numerical solution of a DDE.

For an NDDE one requires a further vector of weights $\boldsymbol{c}^{\mathrm{T}}(s)$, conveniently but not necessarily taken as the derivative of $\boldsymbol{b}^{\mathrm{T}}(s)$ (cf. (3.11), but note (3.10b), (3.10c)). Such parameters define a *RK-quadruple* $(\boldsymbol{\vartheta}, \boldsymbol{A}, \boldsymbol{b}(s), \boldsymbol{c}(s))$ corresponding to an augmented tableau:

$$\begin{array}{c|c} \boldsymbol{\vartheta} & \boldsymbol{A} \\ \hline s & \boldsymbol{b}^{\mathrm{T}}(s) \\ \hline & \boldsymbol{c}^{\mathrm{T}}(s), \end{array} \quad \text{for example,} \quad \begin{array}{c|cc} 0 & 0 & \\ 1 & \frac{1}{2} & \frac{1}{2} \\ \hline s & s-\frac{1}{2}s^2 & \frac{1}{2}s^2 \\ \hline & 1-s & s \end{array}, \quad s\geqslant 0. \tag{3.15}$$

We write $t_{n,i} := t_n + \vartheta_i h_n$, $\tilde{y}_{n,i} := \tilde{y}(t_{n,i})$ (also, as above, $t_{n+s} = t_n + sh_n$ for $s \in [0,1]$), and the continuous RK discretization of (3.1) is given by

$$\tilde{y}(t_{n+s}) := \tilde{y}_n + h_n \sum_{j=1}^{m} b_j(s)\tilde{F}_{n,j}, \tag{3.16a}$$

$$\widetilde{y'}(t_{n+s}) := \sum_{j=1}^{m} c_j(s)\tilde{F}_{n,j}, \tag{3.16b}$$

$$\tilde{y}_{n,i} = \tilde{y}_n + h_n \sum_{j=1}^{m} a_{i,j}\tilde{F}_{n,j}, \qquad \tilde{F}_{n,j} = F(t_{n,j}, \tilde{y}_{n,j}, \tilde{y}(\tilde{\alpha}(t_{n,j})), \widetilde{y'}(\tilde{\beta}(t_{n,j}))). \tag{3.16c}$$

Eqs. (3.16c) together constitute equations to be solved for $\{\tilde{F}_{n,i}\}_{i=1}^{m}$ for substitution in (3.16a), (3.16b). For compatibility, we evaluate past values using

$$\tilde{y}(\tilde{\alpha}(t_{n,j})) = \tilde{y}_{k_{n,j}} + h_{k_{n,j}} \sum_{\ell=1}^{m} b_\ell(s_{n,j})\tilde{F}_{k_{n,j},\ell} \quad (\tilde{\alpha}(t_{n,j}) = t_{k_{n,j}} + s_{n,j}h_{k_{n,j}}), \tag{3.16d}$$

$$\widetilde{y'}(\tilde{\beta}(t_{n,j})) = \sum_{\ell=1}^{m} c_\ell(q_{n,j})\tilde{F}_{k'_{n,j},\ell} \quad (\tilde{\beta}(t_{n,j}) = t_{k'_{n,j}} + q_{n,j}h_{k'_{n,j}}). \tag{3.16e}$$

Alongside the general form in (3.15), we gave an illustration (the trapezium rule). That tableau produces a continuous RK method equivalent to Eqs. (3.12a)–(3.12b). A Dormand & Prince RK tableau and an extension due to Shampine are employed in the code `Archi` [71].

There may be complications in the use of (3.16). Eqs. (3.16a), (3.16b) may become implicit if $\tau(t) < h_n$ for some $t \in [t_n, t_{n+1}]$ (*e.g.*, for vanishing lag), even if the RK parameters are formally explicit. If α (or β) is state-dependent ($\tilde{\alpha}(t_{n,j})$ signifies $\hat{\alpha}(t_{n,j}, \tilde{y}(t_{n,j}))$), (3.16) are implicit and must be solved iteratively.

Remark 3.2. For RK methods, one may store $\tilde{y}_0, \{\tilde{F}_{0,j}\}_{j=1}^{m}$; $\tilde{y}_1, \{\tilde{F}_{1,j}\}_{j=1}^{m}$; $\tilde{y}_2, \{\tilde{F}_{2,j}\}_{j=1}^{m}$; $\cdots$; $\tilde{y}_n$, $\{\tilde{F}_{n,j}\}_{j=1}^{m}$; .... If the equations are solved iteratively, one records $\tilde{y}_n^{[r]}$ and $\tilde{F}_{n,j}^{[r]}$ where $\tilde{F}_{n,j}^{[r]} := F(t_{n,j}, \tilde{y}_{n,j}^{[r]}, \tilde{y}(\tilde{\alpha}^{[r]}(t_{n,j})), \widetilde{y'}(\tilde{\beta}^{[r]}(t_{n,j})))$ (or alternatively $\tilde{y}_n^{[r+1]}$ and $\tilde{F}_{n,j}^{[r]}$), with $\tilde{\alpha}^{[r]}(t_{n,k}) = \hat{\alpha}(t, \tilde{y}^{[r]}(t_{n,k}))$, and $\tilde{\beta}^{[r]}(t_{n,k}) = \hat{\beta}(t, \tilde{y}^{[r]}(t_{n,k}))$ $(r \equiv r_n)$.

4. Approximate solutions — existence, uniqueness, convergence

Consider a sequence $\mathfrak{T} = \{\mathscr{T}^{[0]}, \mathscr{T}^{[1]}, \mathscr{T}^{[2]}, \ldots\}$ of meshes (3.2) whose widths $\{h_{\max}^{[0]} \geqslant h_{\max}^{[1]} \geqslant h_{\max}^{[2]}, \ldots\}$ tend to zero. The approximation $\tilde{y}(t)$, if it exists, denotes $\tilde{y}(\mathscr{T}^{[m]}; t)$ for some $\mathscr{T}^{[m]} \in \mathfrak{T}$ with width $h_{\max}^{[m]} \equiv h_{\max}(\mathscr{T}^{[m]})$.

Definition 4.1. Given $\tilde{y}(t) \equiv \tilde{y}(\mathscr{T}^{[m]}; t)$, the error on the grid points and the global error, are, respectively,

$$e^{[m]}([t_0, t^{\max}]) := \sup\{|y(t_n) - \tilde{y}(t_n)| \text{ for } t_n \in \mathscr{T}^{[m]} \cap [t_0, t^{\max}]\},$$

$$\mathrm{e}^{[m]}([t_0, t^{\max}]) := \sup\{|y(t) - \tilde{y}(t)| \text{ for } t \in [t_0, t^{\max}]\}.$$

For fixed $t^{\max} < \infty$, the approximations $\tilde{y}(\cdot)$ are (i) *convergent on grid-points in* $[t_0, t^{\max}]$ if $\lim_{m\to\infty} e^{[m]}([t_0, t^{\max}]) = 0$, (ii) *convergent of order* ρ *on grid-points in* $[t_0, t^{\max}]$ if $e^{[m]}([t_0, t^{\max}]) = \mathcal{O}((h^{[m]}_{\max})^{\rho})$ as $m \to \infty$, and (iii) *convergent of order* ϱ *on the interval* $[t_0, t^{\max}]$ if $\mathrm{e}^{[m]}([t_0, t^{\max}]) = \mathcal{O}((h^{[m]}_{\max})^{\varrho})$ as $m \to \infty$.

Clearly, $\varrho \leqslant \rho$; there are cases where $\varrho < \rho$. Further, there are cases where the order of convergence depends upon the sequence of meshes $\{\mathscr{T}^{[m]}\}$. (Consider the method defined by (3.6) together with the low-order extension $\tilde{y}(t_{n+s}) := \tilde{y}(t_n)$ for $s \in (0,1)$. Apply it to $y'(t) = y(t-1)$ with $t_i = t_0 + ih$ and uniform step h: compare the case $\mathscr{T}^{[m]} \in \mathfrak{T}$ if $h = 1/m$ $(m \in \mathbb{N})$, where no extension is needed, and the case $\mathscr{T}^{[m]} \in \mathfrak{T}$ if $h = 1/(\sqrt{2}m)$, $m \in \mathbb{N}$.) One *caveat* is in order: theories that apply for small discretization parameters ("as $h \to 0$") may not give the desired insight in real life (e.g., if h is large relative to the time-scale of the problem). The concept of stiff order in the study of ODEs (where practical step sizes do not produce asymptotically correct convergence behaviour) reflects this observation.

We illustrate some more general results by reference to method (3.12) for (3.1). Recall that $t_{n+s} = t_n + sh_n$, for $s \in (0,1]$. We use (if the solution $y(\cdot)$ and the approximation $\tilde{y}(\cdot)$ exist and are unambiguously defined) the notation

$$F_{n+s} = F(t_{n+s}, y(t_{n+s}), y(\alpha(t_{n+s})), y'(\beta(t_{n+s}))), \tag{4.1a}$$

$$\tilde{F}_{n+s} = F(t_{n+s}, \tilde{y}(t_{n+s}), \tilde{y}(\tilde{\alpha}(t_{n+s}), \widetilde{y'}(\tilde{\beta}(t_{n+s})))). \tag{4.1b}$$

Henceforth, we presume the existence of a unique true solution $y(t) \equiv y(\psi, t_0; t)$ but discuss existence of $\tilde{y}(\cdot)$. We detect a difficulty on considering (2.12) with an initial function $\psi_\star(t)=1$, $\psi'_\star(t)=0$ and solution $y(t)=1$. Taking a uniform mesh with $h_n = h$ and setting $\Lambda := \Lambda_\star h$, (3.12) gives, with $\tilde{y}(0) = 1$, a quadratic equation for $\tilde{y}(h)$ of which one solution is $\tilde{y}_1 = 1$ corresponding to which $\tilde{y}(t) = 1$, $\widetilde{y'}(t) = 0$ for $t \leqslant h$. By consistently taking the "appropriate" root, we have as one solution $\tilde{y}(t)=1$, $\widetilde{y'}(t)=0$ for $t \leqslant nh$. At the nth stage the equation for $\tilde{y}_{n+1}$ is then $\frac{1}{2}\Lambda\tilde{y}^2_{n+1} + (1-\Lambda)\tilde{y}_{n+1} - 1 = 0$. (The path from $\tilde{y}_n$ to $\tilde{y}_{n+1}$ is in general multi-valued though the existence of a real sequence $\{\tilde{y}_\ell\}_{\ell\geqslant 0}^{\infty}$ on every path is not guaranteed.) The preceding quadratic equation has as its solutions $\tilde{y}^{\pm}_{n+1} = \{\Lambda - 1 \pm |1+\Lambda|\}/\{2\Lambda\}$, namely the value 1 (the true solution) and the value $-1/\Lambda$ (which becomes infinite as $h \to 0$). This example suggests that *an implicit formula, such as* (3.12), *if applied to a nonlinear NDDE, may give rise to a multi-valued approximation* $\tilde{y}(t)$ *of which only some branches are defined for all* $t \geqslant t_0$ *and also satisfy* $\lim_{h\to 0} \sup_{t\in[t_0, t^{\max}]} |\tilde{y}(t) - y(t)| = 0$.

Suppose $\tilde{y}(t)$ exists for $t \in [t_{\min}, t_n]$. Now, $\tilde{y}_{n+1} = \tilde{y}(t_{n+1})$, if it exists, satisfies $\tilde{y}_{n+1} = \varphi_n(\tilde{y}_{n+1})$ where (with notation (3.3))

$$\varphi_n(x) := \tilde{y}_n + \tfrac{1}{2}h_n\tilde{F}_n + \tfrac{1}{2}h_n F(t_{n+1}, x, \tilde{y}(\hat{\alpha}(t_{n+1}, x)), \widetilde{y'}(\hat{\beta}(t_{n+1}, x))). \tag{4.2}$$

The function $\varphi_n(\cdot)$ satisfies a global Lipschitz condition with constant $\frac{1}{2}h_n L$ if, with $\Phi(x) := F(t_{n+1}, x, \tilde{y}(\hat{\alpha}(t_{n+1}, x)), \widetilde{y'}(\hat{\beta}(t_{n+1}, x)))$, we have $|\Phi(x') - \Phi(x'')| \leqslant L|x' - x''|$ uniformly for all x', x''.

A line of enquiry is to discover reasonable conditions that determine a suitable L; we can then use the following lemma which may be found in a student text.

Lemma 4.2 (Fixed-point iteration). *Consider the general fixed-point iteration* $x_{k+1}=\phi(x_k)$. *Suppose (given* $\varepsilon > 0$*) that* $\phi(\cdot)$ *continuous on* $[x_\star-\varepsilon, x_\star+\varepsilon]$ *and* $|\phi(x')-\phi(x'')| \leqslant \Lambda|x'-x''|$ *for all* $x', x'' \in [x_\star-\varepsilon, x_\star+\varepsilon]$ *where* $0 \leqslant \Lambda < 1$ *and let* $x_0 \in [x_\star-\varepsilon, x_\star+\varepsilon]$ *be such that* $|x_0-\phi(x_0)| \leqslant (1-\Lambda)\varepsilon$. *Then* $\lim_{r\to\infty} x_r$ *exists and is the only fixed point of* $\phi(\cdot)$ *lying in* $[x_\star-\varepsilon, x_\star+\varepsilon]$.

If the *global* Lipschitz condition holds, we can apply this result to the iteration $x_{r+1}=\varphi_n(x_r)$, with arbitrary x_0 and ε as large as necessary, by ensuring (say) that $h_n L < 1$, so there exists a unique value $\tilde{y}(t_{n+1})$ for all h_n sufficiently small, and a unique $\tilde{y}(t)$ exists for $t \in [t_{\min}, t_{n+1}]$. We have (when $h_{\max}^{[m]}$ is sufficiently small) the basis of a proof, by induction, of the existence of a unique $\tilde{y}(t)$ for all $t \in [t_{\min}, t^{\max}]$. To bound the error, compare (3.12) with

$$y(t_{n+s}) = y(t_n) + h_n\{(s-\tfrac{1}{2}s^2)F_n + \tfrac{1}{2}s^2F_{n+1}\} + \eta_{h_n}(y;s), \tag{4.3a}$$

$$y'(t_{k+s}) = (1-s)F_k + sF_{k+1} + \eta^\natural_{h_k}(y;s) \quad (k=0,1,\ldots,n), \tag{4.3b}$$

use (cf. Eq. (3.13)) theoretical bounds on $\{\eta_{h_n}(y;s), \eta^\natural_{h_k}(y;s)\}$, apply Lipschitz conditions and use inequalities familiar in the numerics of ODEs.

4.1. *Theory based on local Lipschitz conditions*

Global conditions are not always realistic and we will request local Lipschitz conditions (this material appears to be a novel extension of previously published results). We sacrifice generality by imposing conditions on the neutral term: in particular, we suppose β to be state-independent and consider

$$y'(t) = F(t, y(t), y(\alpha(t, y(t))), y'(\beta(t))) \quad \text{for } t \geqslant t_0. \tag{4.4}$$

Thus, *scalar state-dependent DDEs* and *state-independent NDDEs* are covered by our discussion. (The rather intricate details that we supply allow the reader to construct the extension to state-dependent $\beta(t, y(t))$ if Lipschitz conditions apply globally.)

Our main task is to use conditions on the discretization error to establish, for $h_{\max}^{[m]}$ sufficiently small, the existence of $\tilde{y}(t)$, $\widetilde{y'}(t)$ in a region in which local Lipschitz conditions apply. We focus on formulae (3.12), and use the notation $\eta_{h_k}(y;s)$, $\eta^\natural_{h_k}(y;s)$ in Lemma 3.1 and

$$\eta_n(y) := \sup_{\ell\in\{0,1,\ldots,(n-1)\}} |\eta_{h_\ell}(y)|, \qquad \eta_{h_\ell}(y) = \sup_{s\in[0,1]} |\eta_{h_\ell}(y;s)|, \tag{4.5a}$$

$$\eta^\natural_n(y) := \sup_{\ell\in\{0,1,\ldots,(n-1)\}} |\eta^\natural_{h_\ell}(y)|, \qquad \eta^\natural_{h_\ell}(y) = \sup_{s\in[0,1]} |\eta^\natural_{h_\ell}(y;s)| \tag{4.5b}$$

and, for a given $\omega \in (0,\infty)$,

$$\eta^\omega_{h_\ell}(y) := |\eta_{h_\ell}(y)| + h_\ell\omega|\eta^\natural_\ell(y)|. \tag{4.5c}$$

To impose *local* conditions for u, w we define the neighbourhoods

$$\mathscr{D}(\varepsilon, \varepsilon_1; s) := \{(s, u, w);\ |u-y(s)| \leqslant \varepsilon(s), |w-y'(s)| \leqslant \varepsilon_1(s)\} \tag{4.6}$$

for $s \in [t_0, t^{\max}]$ ($\varepsilon(\cdot)$, $\varepsilon_1(\cdot)$ continuous and strictly positive on $[t_0, t^{\max}]$). We ask (i) that $\sup_{t\in[t_{\min},t^{\max}]} |y(t)| \leqslant M$, $\sup_{t\in[t_{\min},t^{\max}]} |y'(t)| \leqslant M$, (ii) that $y'(\cdot)$ should be Lipschitz-continuous on $[t_{\min}, t^{\max}]$ with Lipschitz constant M; (iii) (uniform boundedness of F) that

$$|F(t,u(t),u(\alpha(t,u(t))),w(\beta(t)))| \leqslant M \tag{4.7}$$

for $t \in [t_0, t^{\max}]$ when $|u(t) - y(t)| \leqslant \varepsilon(t)$ and $|w(\sigma) - y'(\sigma)| \leqslant \varepsilon_1(\sigma)$ for $\sigma \in [t_0, t]$, $t_0 \leqslant t \leqslant t^{\max}$. We require (iv) Lipschitz conditions $|\alpha(\sigma,u) - \alpha(\sigma,U)| \leqslant \mu_\alpha |u - U|$, $|F(\sigma,U,v,w) - F(\sigma,u,v,w)| \leqslant \mu_2 |U - u|$, $|F(\sigma,u,V,w) - F(\sigma,u,v,w)| \leqslant \mu_3 |V - v|$, $|F(\sigma,u,v,W) - F(\sigma,u,v,w)| \leqslant \mu_4 |W - w|$. *We ask that* $\mu_4 < 1$ (*cf. Theorem* 2.2, *with the condition* $L_2 < 1$). The conditions hold for $u, U \in [y(\sigma) - \varepsilon(\sigma), y(\sigma) + \varepsilon(\sigma)]$, $v, V \in [y(\sigma_\alpha) - \varepsilon_\alpha, y(\sigma_\alpha) + \varepsilon_\alpha]$ ($\varepsilon_\alpha = \varepsilon(s_\alpha)$), and $w, W \in [y'(\sigma_\beta) - \varepsilon_\beta, y'(\sigma_\beta) + \varepsilon_\beta]$, with $\varepsilon_\beta = \varepsilon_1(\sigma_\beta)$, whenever $\sigma_\alpha, \sigma_\beta \leqslant \sigma$, and $\sigma \in [t_0, t^{\max}]$. We ask (if necessary by a redefinition) that when $t \in [t_n, t_{n+1}]$,

$$\varepsilon_1(t) \geqslant \mu \sup_{\sigma\in[t_0,t_{n+1}]} \varepsilon(\sigma) + \varepsilon_\star, \quad \text{where } \varepsilon_\star > 0, \tag{4.8}$$

where $\mu = \omega\mu_\star$, with $\mu_\star = \mu_2 + \mu_3 + M\mu_3\mu_\alpha$, $\omega = (1 - \mu_4)^{-1}$. In relation to $\mathscr{T}$, we also ask for continuity of $y''(t)$ on each interval $[t_k, t_{k+1}]$ and that $h^{[m]}_{\max}\mu < 1$, $h^{[m]}_{\max}L < 1$, where $L = \mu_2 + \mu_3\mu_\alpha M$. Setting $\hat{\mu} := \mu_3 + \mu(1 + \mu_4)$ and $\delta_n = \{h_n\eta^\natural_{h_n}(y) + \eta_{h_n}(y;1)\}$ we require that, for $n = 0, 1, \ldots$,

$$2\exp\{2\mu(t_n - t_0)\} \sum_{\ell=0}^{n} \eta^\omega_{h_\ell} \leqslant \varepsilon(t) \quad \text{when } t \in [t_n, t_{n+1}], \tag{4.9a}$$

$$\left(1 - \frac{1}{2}h_nL\right)\left\{(1 + \hat{\mu}h_n)\exp\{2\mu(t_n - t_0)\} \sum_{k=0}^{n-1} \eta_{h_k}(y) + |\delta_n|\right\} \leqslant \varepsilon(t_{n+1}) \tag{4.9b}$$

and

$$\omega \sup_{t_\ell \leqslant t^{\max}} |\eta^\natural_{h_\ell}(y)| \leqslant \varepsilon_\star. \tag{4.9c}$$

Conditions (4.9) are satisfied on taking $h^{[m]}_{\max}$ sufficiently small, since $\eta^\omega_{h_\ell}(y) = \mathcal{O}(h_\ell h_{\max})$, and $\sum_k \eta^\omega_{h_k} = \{\sum_k h_k\}\,\mathcal{O}(h^{[m]}_{\max})$. To simplify, we assume $h^{[m]}_{\max}$ is smaller than a "minimum possible lag":

$$0 < \max_\ell h_\ell < \inf_{t\in[t_0,t^{\max}]} \min\left(\inf_{|u-y(t)|\leqslant\varepsilon(t)} (t - \alpha(t,u)), t - \beta(t)\right). \tag{4.10}$$

Then, $\tilde{\alpha}(t_{\ell+1}) \leqslant t_\ell$, $\tilde{\beta}(t_{\ell+1}) \leqslant t_\ell$ for $\ell \in \mathbb{N}$. (If $\alpha(t) = t - \tau(t)$, $\beta(t) = t - \zeta(t)$ with $\xi := \min\{\inf_t \tau(t), \inf_t \zeta(t)\} > 0$ it suffices to take $h^{[m]}_{\max} < \xi$.)

Theorem 4.3. *If all the preceding assumptions hold, there is a unique approximate solution* $\tilde{y}(t)$ *defined on* $[t_0, t^{\max}]$ *and satisfying* $|y(t) - \tilde{y}(t)| \leqslant \varepsilon(t)$. *Furthermore,* $\sup_{[t_0,t^{\max}]} |y(t) - \tilde{y}(t)| \to 0$ *as* $h^{[m]}_{\max} \to 0$ (*that is, as* $m \to \infty$).

4.1.1. The proof of Theorem 4.3

We shall indicate a proof of the above, via some lemmas. When brevity requires, we write

$$e(t) = y(t) - \tilde{y}(t) \quad \text{and} \quad e_1(t) = y'(t) - \widetilde{y'}(t), \tag{4.11a}$$

if defined. For $r \in \mathbb{N}$, $\mathfrak{I}_r$ will denote the set

$$\mathfrak{I}_r := [t_{\min}, t_r] \cup \{t_{r+1}\}. \tag{4.11b}$$

Lemma 4.4. *Suppose, for some $r \in \mathbb{N}$, that there exists a unique $\tilde{y}(t)$ that satisfies $|y(t)-\tilde{y}(t)| \leqslant \varepsilon(t)$ for $t \in \mathfrak{I}_r$, and $|y'(t) - \widetilde{y'}(t)| \leqslant \varepsilon_1(t)$ for $t \in [t_{\min}, t_r]$. Then* (a) *$\tilde{y}(t)$, $\widetilde{y'}(t)$ exist on $[t_{\min}, t_{r+1}]$, and $|\widetilde{y'}(t)| \leqslant M$ for $t \in [t_{\min}, t_{r+1}]$.* (b) *With $\mu = \omega\mu_\star$ defined above, and $\ell = 0, 1, \ldots, r$,*

$$\sup_{t\in[t_0,t_{\ell+1}]} |y'(t) - \widetilde{y'}(t)| \leqslant \mu \sup_{t\in[t_0,t_{\ell+1}]} |y(t) - \tilde{y}(t)| + \omega|\eta^{\natural}_{\ell+1}(y)| \tag{4.12}$$

and, uniformly for $\ell \in \{0, 1, \ldots, r\}$,

$$\sup_{t\in[t_0,t_{\ell+1}]} |y(t) - \tilde{y}(t)| \leqslant 2\exp\{2\mu(t_\ell - t_0)\} \sum_{k=0}^{\ell} \eta^{\omega}_{h_\ell}(y) = \mathcal{O}(h^{[m]}_{\max}). \tag{4.13}$$

Finally, $|y(t) - \tilde{y}(t)| \leqslant \varepsilon(t)$, and $|y'(t) - \widetilde{y'}(t)| \leqslant \varepsilon_1(t)$ for $t \in [t_0, t_{r+1}]$.

Proof. We outline a proof of the above, but first note that a similar result (the value of μ is changed) holds even if β is state-dependent. By assumption, the trapezium-rule approximation $\tilde{y}(t)$ is unambiguously defined for $t \in [t_0, t_r]$ and at t_{r+1}, so $\tilde{F}_\ell$ is defined for $\ell = 0, 1, \ldots, (r+1)$ and hence there exists a unique twice-differentiable approximation $\tilde{y}(t)$ on $[t_{\min}, t_{r+1}]$ satisfying $|y(t)-\tilde{y}(t)| \leqslant \varepsilon(t)$ on $[t_0, t_r] \cup \{t_{r+1}\}$, and $\widetilde{y'}(t_\ell) = \tilde{F}_\ell \equiv F(t_\ell, \tilde{y}(t_\ell), \tilde{y}(\hat{\alpha}(t_\ell, \tilde{y}(t_\ell))), \widetilde{y'}(\beta(t_\ell)))$ if $\ell \leqslant r+1$ (cf. (3.3)). By (3.12b) with (4.7), $|\widetilde{y'}(t)| \leqslant M$ $(t \in [t_0, t_{r+1}])$ and (a) follows.

Now $y'(t_\ell) = F_\ell \equiv F(t_\ell, y(t_\ell), y(\alpha(t_\ell, y(t_\ell))), y'(\beta(t_\ell)))$, and $\widetilde{y'}(t_\ell) = \tilde{F}_\ell$ as above, and $|e_1(t_\ell)| = |F_\ell - \tilde{F}_\ell|$. If we use $|y(\alpha(t_\ell, y(t_\ell))) - \tilde{y}(\hat{\alpha}(t_\ell, \tilde{y}(t_\ell)))| \leqslant |y(\alpha(t_\ell, y(t_\ell))) - y(\hat{\alpha}(t_\ell, \tilde{y}(t_\ell)))| + |y(\hat{\alpha}(t_\ell, \tilde{y}(t_\ell))) - \tilde{y}(\hat{\alpha}(t_\ell, \tilde{y}(t_\ell)))| \leqslant M|\alpha(t_\ell, y(t_\ell)) - \alpha(t_\ell, \tilde{y}(t_\ell))| + |e(\hat{\alpha}(t_\ell, \tilde{y}(t_\ell)))|$ we find

$$|e_1(t_\ell)| \leqslant \mu_2|e(t_\ell)| + \mu_3 \left\{ \sup_{t\in[t_0,t_{\ell-1}]} |e(t)| + \mu_\alpha M|e(t_\ell)| \right\} + \mu_4 \sup_{t\in[t_0,t_{\ell-1}]} |e_1(t)|. \tag{4.14}$$

If $s \in [0,1]$, then $|e_1(t_{\ell+s})| \leqslant (1-s)|e_1(t_\ell)| + s|e_1(t_{\ell+1})| + |\eta^{\natural}_\ell(y;s)| \leqslant \max_{j\in\{\ell,\ell+1\}} |e_1(t_j)| + |\eta^{\natural}_{h_\ell}(y;s)|$. Thus, with $\ell \in \{0, 1, \ldots, r\}$, $\mathfrak{I}_\ell \equiv [t_0, t_\ell] \cup \{t_{\ell+1}\}$, $\mu_\star = \mu_2 + \mu_3 + \mu_3\mu_\alpha M$, $0 < \mu_4 < 1$,

$$|e_1(t_{\ell+s})| \leqslant \mu_\star \sup_{t\in\mathfrak{I}_\ell} |e(t)| + \mu_4 \sup_{t\in[t_0,t_\ell]} |e_1(t)| + |\eta^{\natural}_{h_\ell}(y;s)|. \tag{4.15}$$

From (4.15) we deduce $\sup_{t\in[t_0,t_{\ell+1}]} |e_1(t)| \leqslant \mu_4 \sup_{t\in[t_0,t_{\ell+1}]} |e_1(t)| + \mu_\star \sup_{t\in\mathfrak{I}_\ell} |e(t)| + |\eta^{\natural}_{\ell+1}(y)| \leqslant \mu_4 \sup_{t\in[t_0,t_{\ell+1}]} |e_1(t)| + \mu_\star \sup_{t\in[t_0,t_{\ell+1}]} |e(t)| + |\eta^{\natural}_{\ell+1}(y)|$, and so (4.12) follows.

For $\ell = 0, 1, \ldots, r$, $|y(t_{\ell+s}) - \tilde{y}(t_{\ell+s})| \leqslant |y(t_\ell) - \tilde{y}(t_\ell)| + h_\ell\{(s - \frac{1}{2}s^2)|e_1(t_\ell)| + \frac{1}{2}s^2|e_1(t_{\ell+1})|\} + |\eta_{h_\ell}(y)| \leqslant |y(t_\ell) - \tilde{y}(t_\ell)| + \mu h_\ell\{(s - \frac{1}{2}s^2)\sup_{t\in[t_0,t_\ell]} |e(t)| + \frac{1}{2}s^2 \sup_{t\in[t_0,t_{\ell+1}]} |e(t)|\} + |\eta^{\omega}_{h_\ell}(y)|$. We deduce that $(1 - \frac{1}{2}\mu h_\ell)\sup_{t\in[t_\ell,t_{\ell+1}]} |e(t)| \leqslant (1 + \frac{1}{2}\mu h_\ell)\sup_{t\in[t_0,t_\ell]} |e(t)| + \eta^{\omega}_{h_\ell}(y)$, and, if $\mu h_\ell < 1$,

$$\sup_{t\in[t_0,t_{\ell+1}]} |e(t)| \leqslant (1 + 2\mu h_\ell) \sup_{t\in[t_0,t_\ell]} |e(t)| + 2\eta^{\omega}_{h_\ell}(y), \tag{4.16}$$

where, by definition, $\eta^{\omega}_{h_\ell}(y) = |\eta_{h_\ell}(y)| + \omega h_\ell|\eta^{\natural}_\ell(y)| = \mathcal{O}(h_\ell h^{[m]}_{\max})$. By induction, since $1 + 2\mu_\star h_\ell \leqslant \exp(2\mu_\star h_\ell)$, result (4.13) follows. The remainder of the lemma follows by reference to (4.9a), (4.9c). □

Consider, with notation (3.3) the Lipschitz-continuity of the functions

$$\varphi_n(x) := \tilde{y}_n + \tfrac{1}{2}h_n\tilde{F}_n + \tfrac{1}{2}h_nF(t_{n+1}, x, \tilde{y}(\hat{\alpha}(t_{n+1}, x)), \widetilde{y'}(\hat{\beta}(t_{n+1}))). \tag{4.17}$$

Lemma 4.5. *Suppose* $|y(t)-\tilde{y}(t)|\leqslant\varepsilon(t)$ *for* $t\in\mathfrak{I}_{n-1}$, *and* $|y'(t)-\widetilde{y'}(t)|\leqslant\varepsilon_1(t)$ *for* $t\in[t_{\min},t_{n-1}]$. *Then there exists a finite L such that* $|\varphi_n(x')-\varphi_n(x'')|\leqslant\frac{1}{2}h_nL|x'-x''|$, *whenever* $x',x''\in[y(t_{n+1})-\varepsilon(t_{n+1}),y(t_{n+1})+\varepsilon(t_{n+1})]$, $t_{n+1}\in[t_1,t^{\max}]$.

Proof. We indicate a proof, taking advantage of the state-independence of β, and invoking (4.10). By Lemma 4.4 $|e(t)|\leqslant\varepsilon(t)$, $|e_1(t)|\leqslant\varepsilon_1(t)$, and $|\widetilde{y'}(t)|\leqslant M$, $t\in[t_{\min},t_n]$. Now, $|\tilde{y}(\hat{\alpha}(t_{n+1},x'))-\tilde{y}(\hat{\alpha}(t_{n+1},x''))|\leqslant|x'-x''|\mu_\alpha\sup_{\sigma\leqslant t_n}|\widetilde{y'}(\sigma)|$ giving the bound $\mu_\alpha M|x'-x''|$ for this quantity. By the repeated use of the triangle inequality, we have the stated result for $\varphi_n(\cdot)$ in (4.17), with $L=\mu_2+\mu_3\mu_\alpha M$.

We outline a proof of Theorem 4.3. We consider the stage at which we compute $\tilde{y}(t)$ on $[t_n,t_{n+1}]$ and wish to validate the conditions of Lemma 4.4 with $r=n$, assuming that they (and hence the conclusions of the Lemma) hold with $r=n-1$. To achieve our objective we have to establish the existence of $\tilde{y}(t_{n+1})$ such that $|y(t)-\tilde{y}(t)|\leqslant\varepsilon(t)$ for $t\in\mathfrak{I}_n$, and we apply Lemma 4.2 to $x_{k+1}=\varphi_n(x_k)$. For theoretical purposes, take $x_\star=x_0=y(t_{n+1})$ (which exists by assumption). Now $|e(t)|\leqslant\varepsilon(t)$, $|e_1(t)|\leqslant\varepsilon_1(t)$ $(t\leqslant t_n)$, $\alpha(t_{n+1},y(t_{n+1}))$, $\beta(t_{n+1})\leqslant t_n$; if we set $E_n:=F(t_{n+1},y(t_{n+1}),\tilde{y}(\alpha(t_{n+1},y(t_{n+1}))),\widetilde{y'}(\beta(t_{n+1}))))-F(t_{n+1},y(t_{n+1}),\ y(\alpha(t_{n+1},y(t_{n+1}))),y'(\beta(t_{n+1})))$ then

$$|E_n|\leqslant\left\{\mu_3\sup_{t\in[t_0,t_n]}|y(t)-\tilde{y}(t)|+\mu_4\sup_{t\in[t_0,t_n]}|y'(t)-\widetilde{y'}(t)|\right\},\tag{4.18}$$

Lemma 4.4 allows us to bound (4.18) in terms of $\sup_{t\in[t_0,t_n]}|y(t)-\tilde{y}(t)|$; we have $|E_n|\leqslant\{\mu_3+\mu\mu_4\}\sup_{t\in[t_0,t_n]}|e(t)|+\omega\mu_4|\eta_n^\natural(y)|$. The value $\varphi_n(y(t_{n+1}))$ is

$$\begin{aligned}
&\tilde{y}_n+\tfrac{1}{2}h_n\tilde{F}_n+\tfrac{1}{2}h_nF(t_{n+1},y(t_{n+1}),\tilde{y}(\hat{\alpha}(t_{n+1},y(t_{n+1}))),\widetilde{y'}(\hat{\beta}(t_{n+1},y(t_{n+1}))))\\
&\quad=y(t_n)+\tfrac{1}{2}h_n\{F_n+F_{n+1}\}-e(t_n)+\tfrac{1}{2}h_n\{\tilde{F}_n-F_n+E_n\} &(4.19)\\
&\quad=y(t_n)+\int_{t_n}^{t_{n+1}}y'(s)\,\mathrm{d}s+\eta_{h_n}(y;1)-e(t_n)+\tfrac{1}{2}h_n\{\tilde{F}_n-F_n+E_n\}.\\
&\quad=y(t_{n+1})+\mathscr{E}_{n+1} &(4.20)
\end{aligned}$$

with $\mathscr{E}_{n+1}=\eta_{h_n}(y;1)-e(t_n)+\frac{1}{2}h_n\{\tilde{F}_n-F_n+E_n\}$. We have, with $\delta_n=\{h_n\eta_{h_n}^\natural(y)+\eta_{h_n}(y;1)\}$ and $\hat{\mu}:=\mu_3+\mu(1+\mu_4)$,

$$\begin{aligned}
|\mathscr{E}_{n+1}|&\leqslant|e(t_n)|+\tfrac{1}{2}h_n\{\mu_3+\mu(1+\mu_4)\}\sup_{t\in[t_0,t_n]}|e(t)|+\{\tfrac{1}{2}h_n(1+\mu_4)|\eta_n^\natural(y)|+|\eta_{h_n}(y;1)|\}, &(4.21)\\
&\leqslant(1+\tfrac{1}{2}\hat{\mu}h_n)\sup_{t\in[t_0,t_n]}|e(t)|+\{\tfrac{1}{2}h_n(1+\mu_4)\eta_{h_n}^\natural(y)+|\eta_{h_n}(y;1)|\}\\
&\leqslant(1+\hat{\mu}h_n)\exp\{2\mu(t_{n-1}-t_0)\}\sum_{k=0}^{n-1}\eta_{h_k}^\omega(y)+|\delta_n|, &(4.22)\\
&=\mathcal{O}(h_{\max}^{[m]}). &(4.23)
\end{aligned}$$

Now $|y(t_{n+1})-\varphi_n(y(t_{n+1}))|=|\mathscr{E}_{n+1}|$. In Lemma 4.2, concerning fixed-point iterations, set $x_\star=x_0=y(t_{n+1})$ and $\Lambda=\frac{1}{2}h_nL$ (where $h_n<1/L$ and where L is given by Lemma 4.5), and we determine that there is a unique value $\tilde{y}_{n+1}$ with $\tilde{y}_{n+1}=\varphi_n(\tilde{y}_{n+1})$ and satisfying

$$|\tilde{y}_{n+1}-y(t_{n+1})|\leqslant\varepsilon(t_{n+1})\quad\text{in the case }|\mathscr{E}_{n+1}|\leqslant\{1-\tfrac{1}{2}h_nL\}^{-1}\varepsilon(t_{n+1}).\tag{4.24}$$

We then have $|\tilde{y}(t_{n+s}) - y(t_{n+s})| \leqslant \varepsilon(t_{n+s})$ if, in particular, Eq. (4.9b) holds and, by (4.8), (4.9c) and (4.12), $|\widetilde{y'}(t_{n+s}) - y(t_{n+s})| \leqslant \varepsilon_1(t_{n+s})$, $s \in [0,1]$. □

Theorem 4.3 now follows by induction, since the assumptions in Lemma 4.4(c) follow from our discussion. We note that our results provide a rate of convergence for the error on the mesh-points (Definition 4.1), $\sup_n |y(t_{n+1}) - \tilde{y}(t_{n+1})| = \mathcal{O}(h_{\max}^{[m]2})$, in terms of the global error $\sup_t |y(t) - \tilde{y}(t)| = \mathcal{O}(h_{\max}^{[m]})$. Note that it is useful to have asymptotically correct expansions of the error, in addition to orders of convergence, but convergence can be provable with relaxed assumptions, using Lemma 3.1, with $r \in \{1,2\}$, or a suitable extension.

5. Numerical stability

Stability (with or without qualifying adjectives or acronyms — e.g., "absolute stability", "relative stability", "stiff stability", "A-stability"), and its opposite instability, are amongst the most over-used (perhaps misused) terms in the literature on evolutionary problems. Numerical stability, applied to a class of discretization methods, can refer to *zero stability* as in "stability plus consistency implies convergence". In *strong stability*, dominant terms in the error in $\tilde{y}(t)$ satisfy an equation naturally associated with perturbations in the original problem. In contrast, Simpson's rule for $y'(t) = \lambda y(t)$ with constant step h is "unstable" as it has a spurious "growth parameter" that gives rise to a contribution of the form $-(1 - \frac{1}{3}\lambda h + \mathcal{O}(h^2))^n$ in $\tilde{y}(t_n)$; this clearly has no association with the original problem (and can grow in magnitude for negative λ). For selected λ, μ, it can readily be shown that this analysis extends to Simpson's rule for $y'(t) = \lambda y(t) + \mu y(t - \tau_\star)$ where $h = 1/N$, $N \in \mathbb{N}$; this does not concern us further.

The preponderance of work on numerical stability has related to numerical solutions of DDEs rather than NDDEs. We will provide a definition of stability that parallels Definition 2.3. Concerning initial perturbations one addresses the case where the initial functions are perturbed to $\psi_0(t) + \delta\psi_0(t)$ and $\psi_1(t) + \delta\psi_1(t)$. To simplify, we consider the trapezium rule formulated as in (3.12a)–(3.12b) and leave the reader to generalize. For persistent (or steady acting) perturbations $\Delta_{0,1}(t)$ we use the notation

$$\tilde{\mathfrak{F}}_k := F(t_k, \tilde{y}_k + \delta\tilde{y}_k, \tilde{y}(\tilde{\alpha}(t_k)) + \delta\tilde{y}(\tilde{\alpha}(t_k)), \widetilde{y'}(\tilde{\beta}(t_k)) + \delta\widetilde{y'}(\tilde{\beta}(t_k)))$$

and consider

$$\tilde{y}(t_{n+s}) + \delta\tilde{y}(t_{n+s}) = \tilde{y}_n + \delta\tilde{y}_n + h_n\{(s - \tfrac{1}{2}s^2)\tilde{\mathfrak{F}}_n + \tfrac{1}{2}s^2\tilde{\mathfrak{F}}_{n+1}\} + \Delta_0(t_{n+s}), \tag{5.1a}$$

$$\widetilde{y'}(t_k + sh_k) = (1 - s)\tilde{\mathfrak{F}}_k + s\tilde{\mathfrak{F}}_{k+1} + \Delta_1(t_{n+s}). \tag{5.1b}$$

Given a vector norm $||\cdot||$, we can measure the input perturbations by

$$\tilde{\Delta} = \sup_{t \geqslant t_0} \max\{||\Delta_0(t)||, ||\Delta_1(t)||\} + \max\left\{\sup_{t \leqslant t_0} ||\delta\psi_0(t)||, \sup_{t \leqslant t_0} ||\delta\psi_1(t)||\right\}. \tag{5.2}$$

For nonneutral DDEs, Δ_1 is absent (in general, Δ_1 could be eliminated by incorporation into a revised definition of Δ_0). If one wishes to cover numerical perturbations, one does not require the derivative of Δ_0 to equal Δ_1 (an unsuitable requirement if $\widetilde{y}'(\cdot) \neq \widetilde{y'}(\cdot)$ in the defining extensions), nor even continuity of $\Delta_0(\cdot), \Delta_1(\cdot)$. However, one can by suitable choices attempt to simulate the

Table 1
Some of the stability terms in previous literature

Test equation (subject to conditions:)	Conditions ($h_n \equiv h$ fixed)	Type of stability	Reduces to the ODE concept
$y'(t) = \lambda_\star y(t) + \mu_\star y(t-\tau_\star)$ $\lvert\mu_\star\rvert < -\Re(\lambda_\star)$ cf. (2.7)	$\tau_\star/h \in \mathbb{N}$ arbitrary h	P-stability GP-stability	A-stability
$y'(t) = \lambda_\star(t)y(t) + \mu_\star(t)y(t-\tau_\star)$ $\lvert\mu_\star(t)\rvert < -\Re(\lambda_\star(t))$	$\tau_\star/h \in \mathbb{N}$ arbitrary h	PN-stability GPN-stability	AN-stability
A system $y'(t) = \mathfrak{f}(t, y(t), y(t-\tau_\star))$ subject to (5.3)	$\tau_\star/h \in \mathbb{N}$ arbitrary h	RN-stability GRN-stability See (5.4).	BN-stability

perturbations $\delta F(t)$ in Definition 2.3 The accompanying Table 1 presents some of the terms that appear in the literature. The natural definitions of numerical stability are analogues of Definition 2.3, and concern boundedness of $\delta\tilde{y}(\cdot)$ on $[0,\infty)$, and behaviour of $\delta\tilde{y}(t)$ as $t \to \infty$.

Definition 5.1 (Numerical stability). (a) For given admissible perturbations, a solution $\tilde{y}(\cdot)$ of (3.12) is termed (i) *stable*, if for each $\varepsilon > 0$ there exists $\tilde{\Delta}^* = \tilde{\Delta}^*(\varepsilon, t_0) > 0$ such that $\|\delta y(t)\| < \varepsilon$ when $t \geqslant t_0$ for any $\tilde{\Delta} < \tilde{\Delta}^*$; (ii) *asymptotically stable*, if it is stable and, given t_0, there exists $\tilde{\Delta}^\dagger = \tilde{\Delta}^\dagger(t_0)$ such that $\delta\tilde{y}(t) \to 0$ as $t \to \infty$ when $\tilde{\Delta} < \tilde{\Delta}^\dagger$; (iii) *uniformly asymptotically stable*, if the number $\tilde{\Delta}^\dagger$ in definition (ii) is independent of t_0; (iv) *ξ-exponentially stable*, if it is *asymptotically stable* and, given t_0, there exist finite $K = K(t_0)$ and $\tilde{\Delta}^\ddagger = \tilde{\Delta}^\ddagger(t_0)$ such that $\delta\tilde{y}(t) \leqslant K\exp\{-\xi(t-t_0)\}$ (with $\xi > 0$) for $t \geqslant t_0$ when $\tilde{\Delta} < \tilde{\Delta}^\ddagger$. (b) The stability is "*stability with respect to perturbed initial conditions*" if we require that $\delta F(\cdot)$ vanishes, so that the only perturbations are in $\psi_0(\cdot)$, $\psi_1(\cdot)$.

The condition $|\mu_\star(t)| < -\Re(\lambda_\star(t))$ is sufficient to ensure *asymptotic stability* (with respect to perturbed initial perturbations) of every solution of $y'(t) = \lambda_\star(t)y(t) + \mu_\star(t)y(t-\tau_\star)$ for *every* $\tau_\star > 0$. Recalling the concept of contractivity when studying ODEs, we note that nonuniqueness of solutions passing through a point (cf. Fig. 1(b)) makes the definition of contractivity for ODEs inappropriate for DDEs. The condition imposed on $\mathfrak{f}$ in $y'(t) = \mathfrak{f}(t, y(t), y(t-\tau_\star))$ for RN- and GRN-stability is (compare with $|\mu_\star(t)| < -\Re(\lambda_\star(t))$, used above)

$$\sup_{y,x_1 \neq x_2} \frac{\|\mathfrak{f}(t,y,x_1) - \mathfrak{f}(t,y,x_2)\|}{\|x_1 - x_2\|} \leqslant - \sup_{x,y_1 \neq y_2} \frac{\Re\,\langle(\mathfrak{f}(t,y_1,x) - \mathfrak{f}(t,y_2,x), y_1 - y_2)\rangle}{\|y_1 - y_2\|^2}, \tag{5.3}$$

where $\langle\cdot,\cdot\rangle$ is a vector inner product and $\|u\| := \langle u,u\rangle^{1/2}$. This ensures that

$$\|y(\psi;t) - y(\psi + \delta\psi;t)\| \leqslant \sup_{t \leqslant t_0} \|\delta\psi(t)\| \tag{5.4}$$

and RN- and GRN-stable methods preserve property (5.4) under discretization, with the respective constraints on h; clearly, $\tilde{y}(\cdot)$ is then *stable with respect to initial perturbations*.

Some theoretical tools for analysing stability are Z-transform theory (in the case of constant-coefficient constant-lag DDEs), boundary-locus techniques, linearized stability, the fundamental matrix, comparison theorems, Lyapunov functions and functionals (the work [62] echos the analytical theory in an attractive manner), and Halanay inequalities. It cannot be overemphasized that *such tools complement each other* in providing insight. The stability properties attach to a particular solution $\tilde{y}(t) \equiv \tilde{y}(\mathcal{T}, \psi, t_0; t)$, and as in the analytic case a solution of a nonlinear problem may be unbounded but stable. It is often assumed, without loss of generality, that the null function is a solution so that stability can be defined in terms of stability of the null solution (one considers the effect of "small nonnull initial functions").

For *linear* equations all solutions corresponding to a given $\mathcal{T}$ simultaneously have the same stability properties. We indicate some features by considering the case of the test equation

$$y'(t) = \lambda_\star y(t) + \mu_\star y(t - \tau_\star) + \nu_\star y'(t - \zeta_\star) \quad \text{with } \tau_\star = Nh,\ \zeta_\star = N'h \tag{5.5}$$

(where $N, N' \in \mathbb{N}$). The general θ-method yields

$$\begin{aligned}(1 - \theta\lambda_\star h)\tilde{y}_{n+1} = {} & (1 + (1-\theta)\lambda_\star h)\tilde{y}_n + \theta\mu_\star h\tilde{y}_{n+1-N} \\ & + (1-\theta)\mu_\star h\tilde{y}_{n-N} + \theta\nu_\star h\widetilde{y'}_{n+1-N'} + (1-\theta)\nu_\star h\widetilde{y'}_{n-N'},\end{aligned} \tag{5.6a}$$

$$\widetilde{y'}_{n+1} = \lambda_\star \tilde{y}_{n+1} + \mu_\star \tilde{y}_{n+1-N} + \nu_\star \widetilde{y'}_{n+1-N'} \tag{5.6b}$$

and in the present case we avoid the need for interpolation for past values and past derivative values. With $\theta = \frac{1}{2}$ we obtain an example of (3.7).

We can now derive (if $\theta\lambda_\star h \neq 1$) a vector recurrence between the vectors $\boldsymbol{\phi}_k = [\tilde{y}_k, \widetilde{y'_k}]^{\mathrm{T}}$ $(k \in \mathbb{N})$ of the form

$$\boldsymbol{\phi}_{n+1} + \boldsymbol{A}_1\boldsymbol{\phi}_n + \boldsymbol{A}_N\boldsymbol{\phi}_{n-N} + \boldsymbol{A}_{N'}\boldsymbol{\phi}_{n-N'} = 0. \tag{5.7}$$

Here, $\boldsymbol{A}_\ell \equiv \boldsymbol{A}_\ell(\lambda_\star, \mu_\star, \nu_\star; h)$, for $\ell = 1, 2, \ldots, \max(N, N')$. A slightly different recurrence is obtained if the derivative values $\widetilde{y'_\ell}$ are obtained using numerical differentiation

$$\begin{aligned}(1 - \theta\lambda h)\tilde{y}_{n+1} = {} & (1 + (1-\theta)\lambda h)\tilde{y}_n + \theta\mu h\tilde{y}_{n+1-N} + (1-\theta)\mu h\tilde{y}_{n-N} \\ & + \theta\nu h\widetilde{y'}_{n+1-N'} + (1-\theta)\nu h\widetilde{y'}_{n-N'},\end{aligned} \tag{5.8a}$$

$$\widetilde{y'}_{n+1-N} = \{\tilde{y}_{n+1-N} - \tilde{y}_{n-N}\}/h. \tag{5.8b}$$

If desired, a vector recurrence such as (5.7) can be re-expressed, using a block companion matrix, as a two-term autonomous vector recurrence of the form $\boldsymbol{\varphi}_{n+1} = \boldsymbol{M}\boldsymbol{\varphi}_n$ with an amplification matrix $\boldsymbol{M} \equiv \boldsymbol{M}(\lambda_\star, \mu_\star, \nu_\star; h)$ of order $2 \times \max\{N, N'\}$ — e.g., if $N' > N$:

$$\underbrace{\begin{pmatrix} \boldsymbol{\phi}_{n+1} \\ \boldsymbol{\phi}_n \\ \vdots \\ \boldsymbol{\phi}_{n-N'+1} \end{pmatrix}}_{\boldsymbol{\varphi}_{n+1}} = \underbrace{\begin{pmatrix} -\boldsymbol{A}_1 & -\boldsymbol{A}_2 & \cdots & -\boldsymbol{A}_N & \cdots & -\boldsymbol{A}_{N'} \\ \boldsymbol{0} & \boldsymbol{I} & \cdots & \boldsymbol{0} & \cdots & \boldsymbol{0} \\ \vdots & & & & & \\ \boldsymbol{0} & \boldsymbol{0} & \cdots & \cdots & \boldsymbol{I} & \boldsymbol{0} \end{pmatrix}}_{\text{amplification matrix } \boldsymbol{M}} \times \underbrace{\begin{pmatrix} \boldsymbol{\phi}_n \\ \boldsymbol{\phi}_{n-1} \\ \vdots \\ \boldsymbol{\phi}_{n-N'} \end{pmatrix}}_{\boldsymbol{\varphi}_n}. \tag{5.9}$$

Solutions of the latter recurrence are (uniformly) *stable with respect to initial perturbations* if $\rho(\boldsymbol{M}) \leqslant 1$ and any eigenvalues of modulus unity are semi-simple; uniformly *asymptotically stable with respect to initial perturbations* if $\rho(\boldsymbol{M}) < 1$ (and *ξ-exponentially stable* if $\rho(\boldsymbol{M}) \leqslant 1 - \xi h$, $\xi > 0$). *Stability with respect to uniformly bounded persistent perturbations* is assured when

$\rho(M) < 1$. *Note that since the dimensionality of* $\boldsymbol{M}$ *varies with* h, *stability for* $h = h'$ *and for* h'' *need not be quantitatively comparable properties.* The eigenvalues of $\boldsymbol{M}$ are, of course, the zeros of the characteristic polynomial

$$\begin{aligned}\mathscr{P}(z) &:= \det\{z^{N^*+1}\boldsymbol{I} + z^{N^*}\boldsymbol{A}_1 + z^{N^*-N}\boldsymbol{A}_N + z^{N^*-N'}\boldsymbol{A}_{N'}\},\\ \mathscr{P}(z) &\equiv \mathscr{P}(\lambda,\mu,v,h;z), \quad N^* = \max\{N,N'\}.\end{aligned} \tag{5.10}$$

Regions of stability can be computed using the boundary-locus technique. To illustrate, such a region can be associated with the parameters of the test equation (2.1a) if we fix $v_\star = 0$ in the above, and the region of stability $\mathscr{S}(h)$ can then be defined[3] as

$$\mathscr{S}(h) := (\lambda = \lambda_\star\tau_\star, \mu = \mu_\star\tau_\star) \in \mathbb{R}\times\mathbb{R} \tag{5.11a}$$

such that

$$\mathscr{P}(z) = 0 \text{ implies } \quad |z| < 1 \quad \text{or} \quad |z| = 1 \text{ and } z \text{ is semi-simple,} \tag{5.11b}$$

with a corresponding definition of $\mathscr{S}_\star(h)$ when λ, $\mu \in \mathbb{C}$. Guglielmi [41] offered the following definitions in terms of the exact stability regions Σ and $\Sigma^\dagger$ referred to in Remark 2.4:

Definition 5.2. A numerical method is called (i) $\tau(0)$-*stable* if, for $v = 0$,

$$S(\tau_\star/N) \supseteq \Sigma \text{ whenever } 1 \leqslant N \in \mathbb{N}; \tag{5.12}$$

(ii) τ-*stable* if, for $v = 0$, $\mathscr{S}(\tau_\star/N) \supseteq \Sigma^\dagger$ whenever $1 \leqslant N \in \mathbb{N}$.

Guglielmi [41] showed that the trapezium rule is $\tau(0)$-stable but not τ-stable. We may contrast the rather strict requirements of τ- or $\tau(0)$-stability with the observation that, even for the Euler rule, given arbitrary fixed $(\lambda,\mu) \in \Sigma$ there exists a corresponding $M = M(\lambda,\mu) \in \mathbb{N}$ such that $(\lambda,\mu) \in \mathscr{S}(\tau_\star/N)$ for all $N \geqslant M$ (here, $v = 0$).

Remark 5.3. (a) Regions of stability can be computed using the boundary-locus technique, in which one seeks the loci on which $\mathscr{P}(\cdot)$ has a zero of modulus unity. Such a region can be obtained in (λ,μ)-plane for each parameter v of the test equation (2.1a) or of (2.8) or (5.5). Further, one can use this approach for the cases where $\tau_\star/h$ or $\zeta_\star/h$ are noninteger, so that interpolation is required for the lagging value, and its effect on stability observed. The bounded stability regions in Fig. 3 correspond to the use of the Euler method. From Fig. 3(a) it may be observed that for $h = 1/(m+\varsigma)$ with modest $m \in \mathbb{N}$, $\varsigma \in (0,1)$, the stability region for the test equation

$$y'(t) = \lambda y(t) + \mu y(t-1) \tag{5.13}$$

depends in a pronounced manner on ς. The effect is much less pronounced for implicit formulae such as the trapezium rule.

(b) In Fig. 3(b) the stability regions are those for the Euler method applied with $h = \frac{1}{25}$ to the test equation

$$y'(t) = \lambda y(t) + \mu y(t-1) + v y'(t-1). \tag{5.14}$$

[3] A zero z of $\mathscr{P}(\cdot)$ defined by (5.10) is semi-simple if $\mathscr{P}'(z) \neq 0$ (i.e., if it is simple) or if its geometric and algebraic multiplicities, as an eigenvalue of $\boldsymbol{M}$ are the same.

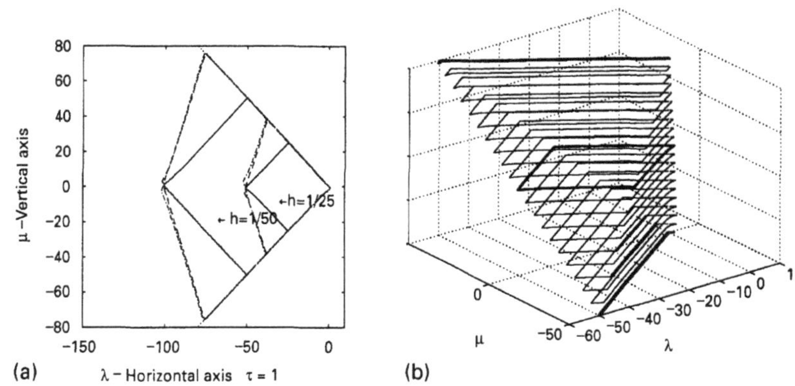

Fig. 3. Boundaries of stability regions for the explicit Euler method: (a) applied to the DDE (2.1a) with step size h ($h=\frac{1}{25}$ and $h=\frac{1}{50}$ — unbroken lines, $h=\frac{3}{72}$, $\frac{3}{69}$ and $\frac{3}{153}$, $\frac{3}{156}$ — broken lines), in the (λ,μ)-plane; (b) applied to the NDDE (2.8) with step $h=\frac{1}{25}$, in (λ,μ,ν)-space with $\lambda\in(-60,0)$, $\mu\in(-60,60)$ and (vertical axis) $\nu\in(-1,1)$. See Remark 5.3.

The "lozenge" shown emphasized for $\nu=0$ in Fig. 3(b) corresponds to the case shown in Fig. 3(a) with $h=\frac{1}{25}$. As ν varies from -1 to $+1$ in Fig. 3(b) the stability region in the (λ,μ)-plane first expands (until $\nu=0$) and then contracts, and as it does so it moves from proximity to $\lambda=-\mu$ towards proximity to $\lambda=\mu$. The exact stability regions were indicated in Fig. 2(b), and it seems that it has to be concluded that *the Euler formulae given above seem unsuitable for NDDEs.* This and similar observations are being pursued elsewhere.

In an elegant paper, Guglielmi and Hairer [42] develop the relationship between stability analysis and order stars.

Clearly, some interesting results can be obtained with the approaches indicated above but it is already clear that difficulties will be encountered if (i) step-sizes h_n are non-constant, and (ii) the lags $\tau_\star,\zeta_\star$ are replaced by time- or state-dependent values. In general, the two-term recurrence $\boldsymbol{\varphi}_{n+1}=\boldsymbol{M}\boldsymbol{\varphi}_{n+1}$ introduced above has to be replaced by a local recurrence $\boldsymbol{\psi}_{n,n+1}=\boldsymbol{M}_n\boldsymbol{\psi}_{n,n+1}$ where the dimension of the vectors $\boldsymbol{\psi}_{n,n+1},\boldsymbol{\psi}_{n,n}$ and the order of $\boldsymbol{M}_n$ changes with n, and the stability analysis is less straightforward.

A rather different approach relies upon a generalization of an inequality of Halanay (see [44, pp. 377–378 et seq.]) in the study of stability of DDEs, to which we restrict ourselves, though extensions are possible. The basic result of Halanay states that if $p(t)$ is a positive scalar-valued function satisfying $p'(t)\leqslant -Ap(t)+B\sup_{s\in -\tau_\star\leqslant s\leqslant t}p(t)$ for $t\geqslant t_0$, where $A>B>0$ then there exist positive k,ξ such that $p(t)\leqslant k\exp\{-\xi(t-t_0)\}$. In applications, this result may be used to establish exponential stability. A useful extension is to the case where $\alpha(t)\to\infty$ as $t\to\infty$ and (with $A>B>0$)

$$p'(t)\leqslant -Ap(t)+B\sup_{s\in[\alpha(t),t]}p(s)\quad \text{for } t\geqslant t_0 \tag{5.15}$$

whereupon we have (a result helpful in establishing asymptotic stability, but not exponential stability) $p(t)\to 0$ as $t\to\infty$. We note a further extension [13] to a *nonlinear inequality of Halanay type.* For the analysis of numerical stability, one can seek *discrete analogues* of the Halanay inequalities indicated above, in which a right-hand derivative is replaced by a difference $\Delta p(t)$. The approach is suggested by comparison of $y'(t) = \lambda y(t) + \mu y(t - \tau(t))$ with the trapezium rule applied with arbitrary step h, which yields expressions of the form $\Delta \tilde{y}_n = \frac{1}{2}\lambda h \tilde{y}_{n+1} + \frac{1}{2}\lambda h \tilde{y}_n + \frac{1}{2}\mu h\{\gamma_{n+1,\ell_n} \tilde{y}_{\ell_{n+1,n}} + \gamma^*_{n+1,\ell_n} \tilde{y}_{\ell_{n+1,n}+1}\} + \frac{1}{2}\mu h\{\gamma_{n,\ell_n} \tilde{y}_{\ell_{n,n}} + \gamma^*_{n,\ell_n} \tilde{y}^*_{\ell_{n,n}+1}\}$ where $\Delta \tilde{y}_n = \tilde{y}_{n+1} - \tilde{y}_n$ and where $\ell_{n+1,n}$ and $\ell_{n,n} \to \infty$ as $n \to \infty$.

6. Further issues and concluding remarks

Here, we shall address some problems that are in our view ongoing and therefore merit further attention, and greater space than we can afford here.

We turn first to stiffness in retarded equations, on which further study is required. The concept of stiffness in numerical analysis is, whilst the terminology is widely used in ODEs, open to varying (frequently controversial) mathematical interpretations [1]. For systems of ODEs, some would define stiffness in terms of the occurrence of different time-scales. Since a scalar DDE is infinite dimensional and in some sense incorporates a countable set of time-scales (the solution of many delayed equations can be expressed in the form of an infinite sum of exponential terms with differing time-scales) generalization of this idea seems unfruitful. Stiffness is sometimes related to the situation where explicit methods do not work, or where stability rather than accuracy constrains the step size. A *particular solution* $y(t)$ of a given DDE or NDDE will be regarded by this author as *stiff in the neighbourhood of a point* $t_\star$ when (i) $y(t)$ is smooth in the neighbourhood of $t_\star$, but (ii) accurate numerical simulation of the behaviour of $y(\cdot)$ can only be achieved by constraining the step size or the choice of method so that local errors are not amplified. In colloquial terms, (ii) relates to stability, but stability is defined formally in terms of behaviour of perturbations as $t \to \infty$ rather than in the shorter term as implied here. One naturally turns to experience with ODEs on detection of stiffness and for treatment using numerical methods based upon highly stable implicit formulae. However, some care has to be exercised here, because including delay terms modifies the behaviour of the solution and the numerical solution. For example, a solution of the equation $y'(t) = \lambda y(t)$ with $\lambda \ll 0$ is generally regarded as stiff, but solutions of $y'(t) = \lambda y(t) - 2\lambda y(t-1)$ (with $\lambda \ll 0$) are not even stable. The design of methods for detecting stiffness, and methods that switch in or out of stiff mode, therefore merit further examination.

We introduce a class of problem not discussed above, *constrained DDEs*, or *delay differential algebraic equations* (DDAEs) of the form (say)

$$\begin{aligned} u'(t) &= f(t, u(t), v(t), u(\alpha_1(t)), v(\alpha_1(t)), \ldots, u(\alpha_q(t)), v(\alpha_q(t))), \\ 0 &= g(t, u(t), v(t), u(\alpha_1(t)), v(\alpha_1(t)), \ldots, u(\alpha_q(t)), v(\alpha_q(t))), \end{aligned} \tag{6.1}$$

that is, a system of DDEs coupled with constraints. Such systems are modifications, incorporating delayed arguments, of differential algebraic equations — hence, the description DDAEs. The constraints need not be algebraic constraints, and problems with inequality rather than equality constraints also arise. For the numerics of DDAEs, see [2] and its citations, and [46,63]; for neutral DDAEs see [58]. One difficulty is that DDAEs can be equivalent to neutral differential equations with deviating arguments in which some arguments are advanced ("$\alpha(t) > t$, $\beta(t) > t$") rather than delayed. If one excludes this possibility, it appears that concerns should focus on the numerics of the related NDDEs, the problem of overcoming apparent "stiffness" (depending on the local index of the solution) and the question of error control in the presence of poorly behaved derivatives.

Closely allied to DDAEs are *singular perturbation problems*, such as

$$\begin{aligned} u_\varepsilon'(t) &= f(t, u_\varepsilon(t), v_\varepsilon(t)), \\ \varepsilon v_\varepsilon'(t) &= g(t, u_\varepsilon(t), v_\varepsilon(t), u_\varepsilon(\alpha_1(t)), v_\varepsilon(\alpha_1(t)), \ldots, u_\varepsilon(\alpha_q(t)), v_\varepsilon(\alpha_q(t))) \end{aligned} \tag{6.2}$$

or, a simpler example, $\varepsilon u_\varepsilon'(t) = f(t, u_\varepsilon(t), u_\varepsilon(t-\tau_\star))$. For a preview of some results on the numerics see [3]. Finally, we concentrated on DDEs and NDDEs, but one can discuss Volterra integro-differential equations with delays.

We conclude with a reference to a different viewpoint, recently introduced in [23] in their analysis of a linear system of DDEs with a fixed lag $\tau_\star$. The basis of the work [23] is the observation that the problem

$$y'(t) = f(t, y(t), y(t - \tau_\star)), \quad t \in [-\tau, \infty), \tag{6.3a}$$

$$y(t) = \psi(t), \quad t \in [t_0 - \tau_\star, t_0] \tag{6.3b}$$

may be solved — in particular if

$$\psi'(t_0) = f(t_0, \psi(t_0), \psi(t_0 - \tau_\star)), \tag{6.3c}$$

by constructing the solution of the PDE

$$\frac{\partial}{\partial t}u(t,s) = \frac{\partial}{\partial s}u(t,s), \quad t \geqslant t_0, \; s \in [t_0 - \tau_\star, t_0], \tag{6.4a}$$

$$\frac{\partial}{\partial s}u(t,t_0) = f(t, u(t,t_0), u(T, t_0 - \tau_\star)), \tag{6.4b}$$

$$u(t_0, s) = \psi(s), \quad s \in [t_0 - \tau_\star, t_0], \tag{6.4c}$$

on setting $y(t+s) = u(t,s)$. Bellen and Maset relate this to an abstract Cauchy problem that can be treated numerically, and consider stability and convergence for linear systems ($y'(t) = My(t) + Ny(t - \tau_\star)$); the approach has further potential.

Acknowledgements

The author acknowledges funding from EPSRC, the London Mathematical Society, NAG Ltd. and the University of Manchester, which supported fruitful collaboration with co-workers and former and current PhD students, a number of whom kindly commented on a draft of this work. He also gives thanks to F. Rihan for assisting in preparation of some of the figures, and to an anonymous referee for helpful remarks.

References

[1] R.C. Aitken (Ed.), Stiff Computation, Oxford University Press, Oxford, 1985.

[2] U.M. Ascher, L.R. Petzold, The numerical solution of delay–differential–algebraic equations of retarded and neutral type, SIAM J. Numer. Anal. 32 (1995) 1635–1657.
[3] C.T.H. Baker, G.A. Bocharov, A. Filiz, N.J. Ford, C.A.H. Paul, F.A. Rihan, A. Tang, R.M. Thomas, H. Tian, D.R. Willé, Numerical modelling by delay and Volterra functional differential equations, in: E.A. Lipitakis (Ed.), Topics in Computer Mathematics and its Applications, LEA Athens, Hellas, 1999, pp. 113–137. ISBN 960-85176-6-4.
[4] C.T.H. Baker, G.A. Bocharov, C.A.H. Paul, F.A. Rihan, Modelling and analysis of time-lags in cell proliferation, J. Math. Biol. 37 (1998) 341–371.
[5] C.T.H. Baker, G.A. Bocharov, F.A. Rihan, A report on the use of delay differential equations in numerical modelling in the biosciences, MCCM Technical Report 343, 1999. ISSN 1360–1725.
[6] C.T.H. Baker, E. Buckwar, Introduction to the numerical analysis of stochastic delay differential equations. MCCM Technical Report 345, 1999. ISSN 1360–1725.
[7] C.T.H. Baker, C.A.H. Paul, Computing stability regions — Runge–Kutta methods for delay differential equations, IMA J. Numer. Anal. 14 (1994) 347–362.
[8] C.T.H. Baker, C.A.H. Paul, A global convergence theorem for a class of parallel continuous explicit Runge–Kutta methods and vanishing lag delay differential equations, SIAM J. Numer. Anal. 33 (1996) 1559–1576.
[9] C.T.H. Baker, C.A.H. Paul, Pitfalls in parameter estimation for delay differential equations, SIAM J. Sci. Comput. 18 (1997) 305–314.
[10] C.T.H. Baker, C.A.H. Paul, D.R. Willé, Issues in the numerical solution of evolutionary delay differential equations, Adv. Comput. Math. 3 (1995) 171–196.
[11] C.T.H. Baker, F.A. Rihan, Sensitivity analysis of parameters in modelling with delay-differential equations. MCCM Technical Report 349, 1999 ISSN 1360–1725.
[12] C.T.H. Baker, A. Tang, Stability of non-linear delay integro-differential equations via approximating equations, MCCM Technical Report 327, 1998. ISSN 1360–1725.
[13] C.T.H. Baker, A. Tang, Generalized Halanay inequalities for Volterra functional differential equations and discretized versions, in: C. Corduneanu, I.W. Sandberg, (Eds.), Volterra Equations and Applications, Gordon and Breach, Amsterdam, 1999, pp. 39–55. ISBN 1023-6155.
[14] V.L. Bakke, Z. Jackiewicz, Stability analysis of linear multistep methods for delay differential equations, Internat. J. Math. Math. Sci. 9 (1986) 447–458.
[15] V.K. Barwell, Special stability problems for functional differential equations, BIT 15 (1975) 130–135.
[16] A. Bellen, One-step collocation for delay differential equations, J. Comput. Appl. Math. 10 (1984) 275–283.
[17] A. Bellen, Constrained mesh methods for functional differential equations, Internat. Schriftenreihe Numer. Math. 74 (1985) 52–70.
[18] A. Bellen, N. Guglielmi, A. Ruehli, On a class of stable methods for linear systems of delay differential equations of neutral type, IEEE Trans. Circuits Systems 46 (1999) 212–216.
[19] A. Bellen, N. Guglielmi, L. Torelli, Asymptotic stability properties of Theta-methods for the pantograph equation, Appl. Numer. Math. 24 (1997) 279–293.
[20] A. Bellen, N. Guglielmi, M. Zennaro, On the contractivity and asymptotic stability of systems of delay differential equations of neutral type, BIT 39 (1999) 1–24.
[21] A. Bellen, N. Guglielmi, M. Zennaro, Numerical stability of nonlinear delay differential equations of neutral type, this issue, J. Comput. Appl. Math. 125 (2000) 251–263.
[22] A. Bellen, Z. Jackiewicz, M. Zennaro, Stability analysis of one-step methods for neutral delay-differential equations, Numer. Math. 52 (1988) 605–619.
[23] A. Bellen, S. Maset, Numerical solution of constant coefficient linear delay differential equations as abstract Cauchy problems Numer. Math. 84 (2000) 351–374.
[24] A. Bellen, M. Zennaro, Numerical solution of delay differential equations by uniform corrections to an implicit Runge–Kutta method, Numer. Math. 47 (1985) 301–316.
[25] A. Bellen, M. Zennaro, Strong contractivity properties of numerical methods for ordinary and delay differential equations, Appl. Numer. Math. 9 (1992) 321–346.
[26] A. Bellen, M. Zennaro, Numerical Methods for Delay Differential Equations, Oxford University Press, Oxford, in preparation.
[27] H. Brunner, Collocation Methods for Volterra Integral and Related Functional Differential Equations, book in preparation.

[28] M.D. Buhmann, A. Iserles, Numerical analysis of functional equations with a variable delay, Pitman Research Notes Mathematics Series, vol. 260, (1991), pp. 17–33, Longman Sci. Tech., Harlow, 1992.
[29] M.D. Buhmann, A. Iserles, S.P. Nørsett, Runge–Kutta methods for neutral differential equations, WSSIAA 2 (1993) 85–98.
[30] O. Diekmann, S.A. van Gils, S.M. Verduyn Lunel, H-O. Walther, Delay Equations, Springer, New York, 1994. ISBN 0-387-94416-8.
[31] L.E. El'sgolts, S.B. Norkin, Introduction to the Theory and Application of Differential Equations with Deviating Arguments, Academic Press, New York, 1973. LIBR. OF CONGRESS CAT. NO. 73-811.
[32] K. Engelborghs, T. Luzyanina, D. Roose, Numerical bifurcation analysis of delay differential equations, this issue, J. Comput. Appl. Math. 125 (2000) 265–275.
[33] K. Engelborghs, D. Roose, Smoothness loss of periodic solutions of a neutral functional-differential equation: on a bifurcation of the essential spectrum, Dynamics Stability Systems 14 (1999) 255–273.
[34] K. Engelborghs, D. Roose, Numerical computation of stability and detection of Hopf bifurcations of steady state solutions of delay differential equations, Adv. Comput. Math. 10 (1999) 271–289.
[35] K. Engelborghs, D. Roose, T. Luzyanina, Bifurcation analysis of periodic solutions of neutral functional-differential equations: a case study, Internat. J. Bifur. Chaos Appl. Sci. Eng. 8 (1998) 1889–1905.
[36] W.H. Enright, H. Hayashi, Convergence analysis of the solution of retarded and neutral delay differential equations by continuous numerical methods, SIAM J. Numer. Anal. 35 572–585.
[37] A.C. Fowler, An asymptotic analysis of the delayed logistic equation when the delay is large, IMA J. Appl. Math. 28 (1982) 41–49.
[38] N.J. Ford, V. Wulf, How do numerical methods perform for delay differential equations undergoing a Hopf bifurcation? this issue, J. Comput. Appl. Math. 125 (2000) 277–285.
[39] K. Gopalsamy, Stability and Oscillations in Delay Differential Equations of Population Dynamics, Kluwer Academic, London, 1992. ISBN 0-7923-1594-4.
[40] N. Guglielmi, On the asymptotic stability properties of Runge–Kutta methods for delay differential equations, Numer. Math. 77 (1997) 467–485.
[41] N. Guglielmi, Delay dependent stability regions of theta-methods for delay differential equations, IMA J. Numer. Anal. 18 (1998) 399–418.
[42] N. Guglielmi, E. Hairer, Order stars and stability for delay differential equations, Numer. Math. 83 (1999) 371–383.
[43] E. Hairer, S.P. Nørsett, G. Wanner, Solving Ordinary Differential Equations – I, Springer, Berlin, 1987. ISBN 3-540-17145-2.
[44] A. Halanay, Differential Equations — Stability, Oscillations, Time Lags Academic Press, New York, 1966. LIBR. OF CONGRESS CAT. NO. 65-25005.
[45] H. Hayashi, Numerical solution of delay differential equations, Ph.D. Thesis, University of Toronto, 1995.
[46] R. Hauber, Numerische Behandlung von retardierten differential-algebraischen Gleichungen, Doctoral Thesis, Universität München, 1994.
[47] K.J. in 't Hout, The stability of a class of Runge–Kutta methods for delay differential equations, Appl. Numer. Math. 9 (1992) 347–355.
[48] K.J. in 't Hout, A new interpolation procedure for adapting Runge–Kutta methods to delay differential equations, BIT 32 (1992) 634–649.
[49] K.J. in 't Hout, Runge–Kutta methods in the numerical solution of delay differential equations, Ph.D. Thesis, The University of Leiden, 1992.
[50] K.J. in 't Hout, C. Lubich, Periodic solutions of delay differential equations under discretization, BIT 38 (1998) 72–91.
[51] K.J. in 't Hout, M.N. Spijker, Stability analysis of numerical methods for delay differential equations, Numer. Math. 59 (1991) 807–814.
[52] C. Huang, Dissipativity of Runge–Kutta methods for dynamical systems with delay, IMA J. Numer. Anal. 20 (2000) 153–166.
[53] C. Huang, H. Fu, S. Li, G. Chen, Stability analysis of Runge–Kutta methods for non-linear delay differential equations, BIT 39 (1999) 270–280.
[54] C. Huang, S. Li, H. Fu, G. Chen, Stability and error analysis of one-leg methods for nonlinear delay differential equations, J. Comput. Appl. Math. 103 (1999) 263–279.

[55] A. Iserles, Numerical analysis of delay differential equations with variable delays, Ann. Numer. Math. 1 (1994) 133–152.
[56] A. Iserles, Exact and discretized stability of the pantograph equation, Appl. Numer. Math. 24 (1997) 295–308.
[57] Z. Jackiewicz, One-step methods of any order for neutral functional differential equations, SIAM J. Numer. Anal. 21 (1984) 486–511.
[58] T. Jankowski, M. Kwapisz, Convergence of numerical methods for systems of neutral functional-differential-algebraic equations, Appl. Math. 40 (1995) 457–472.
[59] V.B. Kolmanovskii, A. Myskhis, Applied Theory of Functional Differential Equations, Kluwer Academic, Dordrecht, 1992. ISBN 0-7923-2013-1.
[60] V.B. Kolmanovskii, V.R. Nosov, Stability of Functional Differential Equations, Academic, London, 1986. ISBN 0-12-417941-X.
[61] T. Koto, A stability property of A-stable collocation-based Runge–Kutta methods for neutral delay differential equations, BIT 36 (4) (1996) 855–859.
[62] T. Koto, The stability of natural Runge–Kutta methods for nonlinear delay differential equations, Japan J. Ind. Appl. Math. 14 (1997) 111–123.
[63] Y. Liu, Numerical solution of implicit neutral functional-differential equations, SIAM J. Numer. Anal. 36 (1999) 516–528.
[64] T. Luzyanina, K. Engelborghs, K. Lust, D. Roose, Computation, continuation and bifurcation analysis of periodic solutions of delay differential equations, Internat. J. Bifur. Chaos Appl. Sci. Eng. 7 (1997) 2547–2560.
[65] T. Luzyanina, D. Roose, Numerical stability analysis and computation of Hopf bifurcation points for delay differential equations, J. Comput. Appl. Math. 72 (1996) 379–392. 6
[66] G. Meinardus, G. Nürnberger, approximation theory and numerical methods for delay differential equations, Internat. Schriftenreihe Numer. Math. 74 (1985) 13–40.
[67] K.W. Neves, Automatic integration of functional differential equations: an approach, ACM Trans. Math. Software 1 (1975) 357–368.
[68] K.W. Neves, Algorithm 497 — automatic integration of functional differential equations, ACM Trans. Math. Software 1 (1975) 369–371.
[69] K.W. Neves, A. Feldstein, Characterization of jump discontinuities for state dependent delay differential equations, J. Math. Anal. Appl. 56 (1976) 689–707.
[70] K.W. Neves, S. Thompson, Software for the numerical solution of systems of functional differential equations with state dependent delays, Appl. Numer. Math. 9 (1992) 385–401.
[71] C.A.H. Paul, A User Guide to ARCHI, MCCM Technical Report 283, 1995. ISSN 1360–1725.
[72] L. Qiu, B. Yang, J. Kuang, The NGP-stability of Runge–Kutta methods for systems of neutral delay differential equations, Numer. Math. 81 (1999) 451–459.
[73] L. Torelli, Stability of numerical methods for delay differential equations, J. Comput. Appl. Math. 25 (1989) 15–26.
[74] L. Torelli, A sufficient condition for GPN-stability for delay differential equations, Numer. Math. 59 (1991) 311–320.
[75] D.R. Willé, C.T.H. Baker, The tracking of derivative discontinuities in systems of delay-differential equations, Appl. Num. Math. 9 (1992) 209–222.
[76] V. Wulf, Numerical analysis of delay differential equations undergoing a Hopf bifurcation Ph.D. Thesis, University of Liverpool, 1999.
[77] M. Zennaro, Delay differential equations: theory and numerics, in: M. Ainsworth, L. Levesley, W.A. Light, M. Marletta (Eds.), Theory and Numerics of Ordinary and Partial Differential Equations, Oxford University Press, Oxford, 1995. ISBN 0-19-851193-0.

www.ingramcontent.com/pod-product-compliance
Ingram Content Group UK Ltd.
Pitfield, Milton Keynes, MK11 3LW, UK
UKHW050418051026
15470UKWH00022B/1724

9 780444 506177